Fundamental Processes
of Atomic Dynamics

NATO ASI Series

Advanced Science Institutes Series

A series presenting the results of activities sponsored by the NATO Science Committee, which aims at the dissemination of advanced scientific and technological knowledge, with a view to strengthening links between scientific communities.

The series is published by an international board of publishers in conjunction with the NATO Scientific Affairs Division

A	**Life Sciences**	Plenum Publishing Corporation
B	**Physics**	New York and London
C	**Mathematical**	Kluwer Academic Publishers
	and Physical Sciences	Dordrecht, Boston, and London
D	**Behavioral and Social Sciences**	
E	**Applied Sciences**	
F	**Computer and Systems Sciences**	Springer-Verlag
G	**Ecological Sciences**	Berlin, Heidelberg, New York, London,
H	**Cell Biology**	Paris, and Tokyo

Recent Volumes in this Series

Series B: Physics

Fundamental Processes of Atomic Dynamics

Edited by

J. S. Briggs

University of Freiburg
Freiburg, Federal Republic of Germany

H. Kleinpoppen

University of Stirling
Stirling, Scotland

and

H. O. Lutz

University of Bielefeld
Bielefeld, Federal Republic of Germany

Plenum Press
New York and London
Published in cooperation with NATO Scientific Affairs Division

Proceedings of a NATO Advanced Study Institute on
Fundamental Processes of Atomic Dynamics,
held September 21–October 2, 1987,
in Maratea, Italy

ISBN-13: 978-1-4684-5546-5   e-ISBN-13: 978-1-4684-5544-1
DOI: 10.1007/978-1-4684-5544-1

Library of Congress Cataloging in Publication Data

NATO Advanced Study Institute on Fundamental Processes of Atomic Dynamics
(1987: Maratea, Italy)
    Fundamental processes of atomic dynamics / edited by J. S. Briggs, H. Klein-
poppen, and H. O. Lutz.
        p.   cm.—(NATO ASI series. Series B, Physics; v. 181)
    "Proceedings of a NATO Advanced Study Institute on Fundamental Processes
of Atomic Dynamics, held September 21–October 2, 1987, in Maratea, Italy"—
Verso t.p.
    "Published in cooperation with NATO Scientific Affairs Division."
    Includes bibliographical references and index.

    1. Collisions (Nuclear physics)—Congresses. 2. Electron-atom collisions—
Congresses. 3. Photonuclear reactions—Congresses. 4. Heavy-ion colli-
sions—Congresses. I. Briggs, J. S. II. Kleinpoppen, H. (Hans) III. Lutz, H. O. IV.
North Atlantic Treaty Organization. Scientific Affairs Division. V. Title. VI. Series.
QC794.6.C6N42   1987   88-21804
539.7′54—dc19   CIP

© 1988 Plenum Press, New York
Softcover reprint of the hardcover 1st edition 1988
A Division of Plenum Publishing Corporation
233 Spring Street, New York, N.Y. 10013

Ugo Fano, Maratea, Italy, September 1987

PREFACE

This volume contains the lectures presented at the NATO Advanced study Institute "Fundamental Processes of Atomic Dynamics" held in Maratea, Italy from September 20th to October 2nd 1987. The institute and this volume were conceived as a natural complement to previous institutes held in Maratea (1982) and in Santa Flavia (1984) whose proceedings are to be found in NATO ASI Series B vol. 103 and 134 respectively. The subject matter of these institutes was the study of the fundamental processes occurring in the interactions of atoms with photons, electrons and heavy-ions. The aim has been to unify these processes in a coherent experimental and theoretical approach. The present volume brings this approach up to date and contains in addition, for contrast and variety, a description of similar dynamical processes in the study of clusters and surfaces.

The institute was opened with a lecture by Joe Macek in which he summarised the current status of atomic collision research. propounded the philosophy of a unified approach to structure, fragmentation and collision and posed the outstanding questions in the field. This lecture forms the introduction to this volume.

The subject matter was divided into experiment and theory with the lectures inter-linked so that the one could re-inforce the other. The whole of the theoretical part of the institute was organised by Ugo Fano as an on-going symposium. He, more than any other individual, has strived to develop a consistent and unified theory of the fundamental processes of atomic dynamics. His lectures on introducing the theory symposium were the highlight of the institute and form a fitting introduction to the theory part of this volume. To emphasise this consistency and unity the theory lectures were given mostly by former students or co-workers of Ugo Fano and the occasion of the institute was used unashamedly to celebrate his seventy-fifth birthday year. As a tribute to Ugo Fano, the whole of this theory symposium forms the first part of this volume. To allow the younger participants to fully appreciate the impact made by Fano, not only in atomic physics but in other disciplines. Mitio Inokuti was invited by the directors to deliver an informal evening lecture on Fano's scientific accomplishments. This memorable address is printed here in its entirety together with, for the first time, a complete list of Ugo Fano's publications.

The chapters on experimental atomic physics are presented by a distinguished list of researchers all of whom have made significant advances in their respective fields. The reader can thus obtain a detailed impression of the state-of-the-art in atomic and molecular spectroscopy, laser-atom interactions. the collision of electrons and positrons with atoms and ion-atom collision processes, in addition to the afore-mentioned study of clusters and ion-surface interactions.

The directors are grateful to many people whose efforts made the institute a great success, not least of whom are Dr. Craig Sinclair and his staff of the NATO Scientific Affairs Division. Particular thanks go to John Broad, Uwe Thumm and Rüdiger Brenn of the University of Freiburg who organised all details of the meeting. This organisation was helped in no small measure by the friendly co-operation of Mr. A. Guzzardi and his staff of the Hotel Villa del Mare. Maratea. The directors acknowledge the advice of the International Advisory Committee and particularly the unselfish assistance of Eugen Merzbacher. The support of the National Science Foundation. Springer-Verlag, Lambda Physik GmbH, the Department of Tourism, Regione Basilicata and the Banca di Lucania were much appreciated.

John Briggs
Hans Kleinpoppen
Hans Lutz

Freiburg. December 1987

CONTENTS

EXPERIMENT

UGO FANO'S SCIENTIFIC ACCOMPLISHMENTS[*]

Mitio Inokuti

Argonne National Labortory
Argonne, IL 60439, U.S.A.

Prologue

Doing science is great fun because it is a delight to discover
something new and true. It is equally a delight to become acquainted
with other good scientists, to learn from their work, and to share ideas
and thoughts with them. It is even ecstatic to interact with an
exceptionally strong scientist such as Fano.

In the spring of 1962 shortly after my arrival in the U.S., I
attended the Annual Meeting of the Radiation Research Society held at
Colorado Springs. At the reception of the meeting, I wandered alone
through a big crowd to see many scientists whom I had known only through
their papers. Thus, I met Fano for the first time, recognizing him only
by a little name tag on the lapel of his jacket. I introduced myself and
began a scientific conversation. I hardly remember what we talked
about. Yet I do remember the enthusiasm with which he talked about
science, and the warmth with which he treated me, then a novice from
Japan. He kindly introduced me to many leaders of radiation research.
This was a crucial experience in my career. Since then I have maintained
most valuable associations with him, for which I am ever thankful.

What follows is my personal view of his scientific accomplishments. My view is undoubtedly biased and limited by my perspective.
Merely for the sake of presentation I classify Fano's accomplishments
into several broad areas. Actually his accomplishments and ideas are
almost always closely related to one another, and cannot be simply
separated. Indeed, they are united toward an overriding theme, i.e., the
persistent urge to find out how atoms, molecules, and their aggregates
behave when they acquire some energy, and thus transform from one form to
another. In other words, he has pursued the physical basis of chemistry
and biology in a broad sense.

---

[*]Work supported in part by the U.S. Department of Energy, Office of
Health and Environmental Research, under Contract W-31-109-Eng-38.
Accordingly, the U.S. Government retains a nonexclusive, royalty-
free license to publish or reproduce the published form of this
contribution, or allow others to do so, for U.S. Government purposes.

1. Atomic Dynamics and Spectroscopy: Earlier Years

Fano began his career under the influence of Racah (a cousin of his on his mother's side) and Persico first, and of Fermi later. The best known of his earlier accomplishments is the interpretation [3]* of the asymmetric profile of spectral lines of rare gases in the autoionization region, seen in measurements by Beutler.[1] Fano showed that the peculiar profile arises from an interference between a quasi-bound state and a background continuum, and gave the now well-known expression $(\varepsilon + q)^2/(1 + \varepsilon^2)$, where ε is the energy measured in an appropriate scale and q is an index representing the degree of the interference. This recognition turned out to have far-reaching consequences. A follow-up of the same idea in a generalized and modernized formalism was reported in his 1961 paper [84]. This paper was prompted by a question by Platzman, who noted Lassettre's observation[2] of doubly excited states of helium by high-resolution electron energy-loss spectroscopy, a technique new at that time. Fortunately, systematic studies of the absorption spectroscopy in far ultra-violet and soft x-ray domains began to be technically feasible at about the same time, with the use of synchrotron radiation. Fano was then at the National Bureau of Standards (NBS), and played a key role in urging Madden, Codling,[3] and others to develop spectroscopy with the use of synchrotron radiation. At the same time, Simpson, Kuyatt, and others had developed high-resolution electron-impact spectroscopy. Stimulated by Fano [91], they also studied resonances and doubly excited states of various atoms.[4,5] Fano [169] gives an account of the exciting period in the early 1960's, with several interesting observations.

The three-way collaboration at the NBS (theory by Fano and his co-worker, Cooper, the synchrotron-radiation work, and the electron-impact work) led to spectacular successes. One of them was the discovery of the new quantum number, + or -, which classifies doubly excited states of helium [90]. This was the beginning of the vast knowledge we now have about doubly excited states of atoms and molecules in general, as seen in the proceedings of the present NATO Institute and also in the recent treatise by Fano and Rau [194].

Another success of the early 1960's was the work with Cooper on the general characteristics of the optical oscillator-strength spectra of atoms in general, as seen in a thorough analysis of the autoionization spectra [98] and in the classic review article [111,117]. One of the notable findings was the frequent occurrence of the Cooper zero of the dipole matrix element.[6] Certainly examples of the zeros had been reported by Bates[7] and by Seaton,[8] as fully documented by Kim et al.[9] in their footnote 3. But Fano and Cooper pursued this matter much deeper. Thus, they are stimulating even now extensive studies by many workers.[9-14] An example of Fano's persistent quests into a subject is seen in the prediction of the spin orientation of photoelectrons ejected by unpolarized light [113,116], a phenomenon that occurs at photon energies near a Cooper minimum. This arises from the spin-orbit coupling, which causes a difference in the dipole matrix element at the same photon energy depending upon the photoelectron spin orientation. This phenomenon is often referred to as a <u>Fano effect</u>, and is sometimes used as a method for the production of spin-polarized electrons.[15]

---

*Throughout the present article, Fano's papers are cited with numbers in square brackets; for full citations, see the current list of his publications following the present article. Other references are indicated with superscripts.

Another important accomplishment at NBS was the discovery of what is now known as the Fano-Lichten electron-promotion effect in ion-atom collisions [101,114]. Up to the early 1960's it was the commonly accepted view that collisions between atomic particles carrying electrons were largely elastic if the collision speed is less than the orbital speed of the electrons. This view stems from the concept of adiabaticity of the nuclear motion in the well-known Born-Oppenheimer theory of molecular structure. The same concept was also used extensively in the theory of ion penetration through matter as seen in Bohr.[16]

Fano noted the experiments by Everhart and co-workers,[17-19] as well as by Fedorenko and co-workers[20] on ion-atom collisions, e.g., Ar^+ - Ar collisions. Those experiments showed abundant production of electrons, i.e., an obvious signature of inelasticity. Moreover, those electrons had kinetic energies that were not continuously distributed but were grouped into a few distinct ranges. Fano recognized the promotion of the $4f\sigma$ orbital as a crucial mechanism, with the help of Lichten, who drew the relevant molecular-orbital diagram. The Fano-Lichten work laid a foundation for many current studies on ion-atom collision.[21]

I cannot help adding a footnote to the early history of ion-atom collision physics. In 1950, Fano gave a lecture [47] on energy losses from particles penetrating in matter. In the discussion that followed, Platzman[22] said, "To say that a heavy particle of velocity v_0 (v_0 = c/137 = velocity of the electron in the lowest orbit of the hydrogen atom), which has a most appreciable energy (25 keV for a proton, 99 keV for an alpha particle) does not ionize with good efficiency is not justified by any established knowledge and is probably also incorrect."

2. Atomic Dynamics and Spectroscopy: Work in Chicago

Fano's work went deeper and deeper after his move from NBS to the University of Chicago in 1966. Together with graduate students, Rau, Starace, Lu, Dehmer, Dill, Li, Lin, Greene, and others, he developed the theory of photoionization, electron collisions and other related topics. See the Appendix for a full list of the co-workers at Chicago.

One focus of attention was effects of a potential barrier on the motion of electrons leaving an atom or molecule and their "diverse manifestations" [148] in photoabsorption spectra and other contexts. For atoms, potential barriers occur in d and f states as a result of an interplay between the electrostatic potential and the centrifugal potential, as noted by Fermi[23] and by Goeppert-Mayer.[24] After a survey of atomic potentials [108,152], Fano and associates found evidence for the frequent occurrence of barrier effects not only in atoms, but also in molecules and solids where molecular effects were often found to be decisive. A conspicuous consequence of a barrier is a shape resonance, i.e., a temporary trapping of an electron [128,132,133]. Studies on this topic turned out to be extremely fruitful, as seen in the lecture by Dehmer[25] in the present Institute.

Another direction of work concerned the basics of photoabsorption by crystals [140,153]. The standard band theory uses Bloch waves that are fully delocalized over the whole crystal, while the photoabsorption in an inner shell of an atom in the crystal leads to an electron state referred to the particular atomic site. Considerations of this problem led to a theory of multipole waves to represent electron states in a crystal. This topic has been fully pursued and developed by Strinati.[26-29]

An overriding theme of Fano's research in this period is the quantum-defect theory, which is a treatment of a single electron interacting with an ion core. Interestingly, his first paper [1] dealt with quantum-defect calculations of atoms with the use of the Thomas-Fermi potential. Elements of the quantum-defect theory, which was modernized by Ham[30] in the context of solid-state electronic structure, and in electron-ion interactions by Seaton[31] and co-workers, are now well known.[32] Together with some of the graduate students, Fano extended the quantum-defect theory in several directions [147]. Perhaps the most notable accomplishment is the application to many-channel cases especially of molecules. The paper [124,160] on the spectrum of molecular hydrogen measured by Herzberg[33,34] opened up a new perspective [189]. An extension of the quantum-defect approach to still other cases, such as the dipole potential (relevant to electron collisions with a polar molecule) is also important [166,173,178,182]. Eventually, the work led to the frame transformation [125,130], i.e., a new key idea for accurate treatments of collision problems in general. The idea of the frame transformation is a natural extension of the R-matrix theory [142], which stems from nuclear physics and has proved to be highly effective in atomic collision theory[35] as well as in chemical-reaction theory.

In the study of the quantum-defect theory we see an example of Fano's persistent quest into a broader perspective. He started with the traditional atomic spectroscopy in which the basic issue is the interactions of a single electron with an atomic ion core. Then he considered a more general problem in which the electron interacts with a molecular-ion core, which has vibrational and rotational degrees of freedom. So far the electron interactions with the ion core are simply Coulombic at large distances. In the latest work [166,178], Fano treated those cases in which the interactions at large distances are non-Coulombic, e.g., the electric dipole type.

Fano generalized his quest in another direction, i.e., the treatment of interactions of two or more electrons with an ionic core. The work is a natural extension of his work at NBS on doubly excited states of He, which I discussed earlier. Perhaps it was also prompted by Rau,[36] who wrote a thesis on the behavior of two energetic electrons in the field of an ion. Work on this problem had been pioneered by Wannier.[37]

Wannier pointed out several key aspects of the correlated motion of two electrons in a Coulomb field having near-zero kinetic energy (i.e., much higher than the ground-state energy but low enough to be greatly influenced by the Coulomb field), and derived several important consequences including the threshold law for electron-impact ionization. Since the Wannier idea was foreign to the standard collision theory as described in the treatise by Mott and Massey,[38] its importance was not generally appreciated until the 1970's, when Cvejanović and Read[39] carried out crucial experiments. During this period Fano, Rau, and others were the first to recognize the importance of incorporating elements of the Wannier idea into accurate calculations on inelastic collisions of low-energy electrons with atoms, double ionization of atoms by photons, and indeed all processes in which two or more electrons emerge under the influence of a central Coulomb field. Efforts in this direction are still continuing [157,171,179,193,194].

A notable element in this context is the appropriateness of the hyperspherical coordinates for describing electron correlation. Although these coordinates had been used earlier by Fock[40] in a treatment of the ground state of the helium atom, Fano was the first to pursue their systematic use in treatments of doubly excited states and continuum states [139,144,149,156,158,159,165,179]. He urged Macek[41,42] to develop an initial formulation of such a treatment. Later developments amply show how far-sighted Fano was indeed. In particular, the treatment led to the recognition that the hyperspherical radius is an adiabatic variable, i.e., a quantity that varies in general much slower than individual coordinates of each electron. It is also important to recognize, as he first did, that the hyperspherical radius is the only variable that gives rise to the continuous spectrum, all other variables of the hyperspherical coordinates giving rise to discrete spectra only. The situation closely parallels with the well-known Born-Oppenheimer theory of molecular structure, which identifies the nuclear coordinates as adiabatic variables. Later developments by Fano himself as well as Macek, Lin, Klar, Greene, Watanabe, and others amply show the fruitfulness of the method of the hyperspherical coordinates, as seen in Chapters 10 and 11 of Fano and Rau [194].

3. Radiation Biology and Radiological Physics

Perhaps, the audience in the present NATO Institute may be unaware of Fano's contributions to an important branch of biology and medicine, i.e., radiological physics and dosimetry. Recent events exemplified by the Chernobyl accident shows the great importance of the full understanding of radiation effects on materials in general and on the human body in particular.

Fano got his first job in the U.S. at a genetics laboratory at Cold Spring Harbor to work with Demerec on radiation biology and radiological physics. Indeed, his papers of the period [12-26] represent pioneering studies in this general area of research. Demerec, Fano, and co-workers studied the differences in the biological effects of x-rays and neutrons. They also found that the same amount of absorbed energy (i.e., dose) led to different effects depending upon the time-variation of irradiation (which are now referred to as the dose-rate effect and the dose-fractionation effect). Fano also came up with theoretical interpretation of those effects [19,29].

This line of work eventually led to a quest for the material basis of genetics in bacteria [14,17,21,22,25,41]. We now know that the material basis is the organized structure of the DNA, as shown by Delbruck and others.

Upon the staff appointment at NBS in 1946 Fano began to concentrate on the physics of radiation actions on matter. The earliest work concerned the yield of ionization in gases subjected to radiation, i.e., a fundamental quantity in radiation dosimetry. Fano [27] gave a pioneering theory based on full consideration of electronic structure of atoms and molecules. He [28] was also the first to evaluate the statistical fluctuations of the ionization yield. The magnitude of these fluctuations is given by a quantity now called the Fano factor, and determines the ultimate precision of the energy determination through ionization measurements.

Fano made other lasting contributions to theoretical dosimetry, i.e., the evaluation of the spatial distribution of the radiation energy absorbed in matter [39,40,56]. An example is the moment method for treating transport equations, which was first conceived by Spencer and was fully formulated by Fano. The 1959 article in Handbuch der Physik with Spencer and Berger [69] is a classic on the theory of penetration and diffusion of x-rays. There one sees already most of the current physico-mathematical techniques discussed at great depth, including analytic methods and Monte Carlo methods. Another notable contribution is the Fano theorem in dosimetry [59]. Roughly stated, it says that, in a medium exposed to a uniform flux of primary radiation, the flux of secondary radiation is also uniform and is independent of the density of the medium or its spatial variation. The theorem is a generalization of the Bragg-Gray principle, and constitutes a logical justification of radiation dose measurements under a wide range of conditions.

The introduction [49,57] of the slowing-down spectrum (or degradation spectrum) is a milestone in the rigorous theory of radiation effects. For the description of slowing-down processes of a fast ion in matter, it is appropriate to use the notion of the pathlength, because the ion is only slightly deflected and the energy loss in a collision is usually a small fraction of the kinetic energy. Then, an ion of energy T traverses a distance $[S(T)]^{-1}dT$ while it loses energy of an amount dT. Here S(T) is the stopping power as given by

$$S(T) = N \sum_n E_n \sigma_n(T),$$

where $\sigma_n(T)$ is the cross section for an inelastic collision in which the energy loss is E_n, and N is the number of targets per unit volume. It is then straightforward to calculate the yield of a particular product s resulting from a collision with cross section $\sigma_s(T)$. The yield is given by $N \int [\sigma_s(T)/S(T)]dT$, according to the definition of the cross section. This is called the continuous-slowing-down approximation (CSDA).

For ions of lower energies, the CSDA is less justifiable, because some of the energy losses are appreciable compared to the kinetic energy. For electrons (which are abundantly generated as a secondary radiation from any primary radiation), the CSDA is inappropriate in general, because the electrons generate further electrons through ionizing collisions. A rigorous description, beyond the limitation of the CSDA, is possible with the use of the slowing-down spectrum. For electrons in particular, Spencer and Fano [57] defined the spectrum y(T) as follows: y(T)dT is the total pathlength of all the electrons that are present in a medium under stationary irradiation and that have kinetic energies ranging from T to T + dT. They also showed that y(T) satisfies a linear integral equation, and gave solutions to simpler cases. The crux of the idea is that it is appropriate to consider all the electrons of different generations together and not separately. Here we see an example of Fano's clear thought and his ability to recognize the essence of physics. The idea laid a foundation to important numerical work by Spencer, Berger, McGinnies, and others, and then led to many fruitful developments, as summarized elsewhere.[43]

Theory of electron slowing-down phenomena was thus given a general framework. However, to carry out numerical solution, one must have knowledge of cross-sections for all major collision processes involved. Fano's work in the late 1950's and early 1960's at NBS was devoted to the understanding of the collision processes of fast electrons and ions. A notable accomplishment in this respect is the re-examination of the Bethe theory,[44] and its generalizations [58,61,62,74,75,76,77,79]. For instance, he introduced [58] what is known as the Fano plot[45] for analyzing the ionization and other cross sections. In addition, he showed how the theory could be extended to treat charged-particle interactions with condensed matter. Among other things he obtained a microscopic definition of the dielectric susceptibility [62], and also gave a general criterion for the occurrence of collective excitation in condensed matter [77]. His 1963 article on stopping power [89] gives a summary of his effort in this direction, and remains now as a standard reference on the subject matter. Further developments in this topic are summarized elsewhere.[45,46]

As Fermi[47] first pointed out and Bethe[44] formulated wave-mechanically, there is a close connection between inelastic collisions of fast charged particles and photoabsorption. Therefore, in radiation physics[48,49] it is of fundamental importance to know the photoabsorption cross section, viz., the oscillator-strength distribution of relevant materials over the whole range of excitation energies, especially in the far ultraviolet and soft x-ray domains, where the absorption by most material is strongest. Fano was among the few to recognize this, and thus urged measurements of the photoabsorption cross sections in those domains [104,111]. The campaign led by Fano and Platzman [146] in the 1960's for this purpose was an impetus toward the use of synchrotron radiation, whose importance is now well known.

His interest in some of the basic issues of radiation physics [119,145] is maintained up to now. One area of his recent work concerns the behavior of slow electrons in condensed matter, which are not energetic enough to cause electronic excitation but are nevertheless responsible for certain chemical effects such as negative-ion formation. Those electrons lose energies through interactions with molecular vibration and rotation or with phonons in solids. In part stimulated by recent experiments (as exemplified by Refs. 50-52) by Sanche and others, Fano continues to work in this area [196,197,202,204].

Work on radiation physics necessarily leads one to consider the fate of energy absorbed in matter. To be more specific, one needs to learn how the electronic excitation in a molecule or in condensed matter is dissipated, i.e., is transformed into other degrees of freedom, e.g., vibration and rotation in a molecule or phonons in a solid. This is an intrinsically complicated subject, because one must deal with a many-body problem in the true sense of the term. Here again, Fano was a pioneer, as I sketch in the next section.

4. Statistical Physics and Description of Relaxation Phenomena

His work in this broad area began in 1951 [46], continued with the 1954 paper [55], and attained a full formulation in the 1957 paper [63] entitled "Description of states in quantum mechanics by density matrix and operator techniques." The theme of the work is to describe the state of matter, i.e., a many-body system, when it is not in a pure eigenstate. When we know the complete set of quantum numbers in the Dirac sense, it is straightforward in principle to calculate any property and its time development. However, when we have only "less-than-maximum information" [63], it is highly nontrivial to predict how the system will develop in the future, or even to know how to talk about it at all. There are numerous examples in nuclear, atomic, molecular, and condensed-matter physics, in which one faces problems of this kind. The examples include the relaxation of electronic excitation, the relaxation of electron or nuclear spin orientation in electron or nuclear spin resonance, and dissipation phenomena.

A key idea in the theory of dissipation phenomena, or irreversible processes, is to distinguish two sets of variables. One set of variables that are amenable to measurements should be treated explicitly. Another set of variables should be eliminated from explicit treatment. To give a simple example, let us consider the motion of a pendulum in air. Obviously we treat the position and momentum of the pendulum explicitly. But we do not want to treat the positions and momenta of all molecules in air. First, we may set up a set of equations of motion involving the pendulum and all the air molecules. Then, we eliminate variables concerning air molecules, and thus arrive at a microscopic definition of the coefficient of air friction. Fano [60,88,92] was one of the founders of what is now known as the statistical-mechanical theory of irreversible processes, together with Kubo,[53-55] and Zwanzig,[56-58] and others. Perhaps the most important among the notions Fano and others introduced is the general theorem that relates the susceptibility (i.e., an index of changes of a variable A as a result of external perturbation) with the time-correlation function of A (with itself or with another variable, depending on the nature of the problem).

5. Mathematical Physics of Symmetry under Rotations and Other Operations

Another field of Fano's work closely related to both atomic dynamics and density matrices is the mathematical physics of symmetry under rotation and other geometrical operations. Perhaps his interest in this field began with the treatment of the angular correlation of successive gamma rays emitted from excited nuclei [48,65], and with the quantum-mechanical formulation of the Stokes parameters for the description of the photon polarization in general [55]. His key contribution is to recognize the tensorial aspects of the ket-bra product $|a><b|$, and to relate it to the multipole moment. Exploitation of the tensorial aspects enables one to carry out various calculations transparently and efficiently. Work until the late 1950's is summarized in the well-known treatise [67] by Fano and Racah. Later on, Fano came back from time to time to this field as opportunities arose [78,86,100,].

To name a few more examples, there are papers with Dill [131,135], with Chang [151], and most recently with C. W. Lee [195,200,201]. Fano also stimulated others to work in this area; a notable example is Briggs' paper[59] on the graphical treatment of angular-momentum algebra, and another is the Fano-Macek article [138], which is now a standard reference on the alignment and spin orientation resulting from atomic collisions.

Epilogue

How has Fano accomplished so much? To help answer this question I offer my observations of his characteristics as follows.

First of all, he maintains a total devotion to science as well as a strong curiosity about scientific matters. His zeal for learning something true has been intense, and shows no sign of decline even now. This aspect of his personality is best exemplified by his question, "What's new in physics?", as he addresses his students virtually every day. I got this question many times, too; and if I managed to point out an item that deserved his interest, he would say, "Okay, that's just fine. What's next?"

Second, Fano always tries to find out essentials involved in something seemingly complicated, which may be an observed phenomenon or even someone's theory. Sometimes when I tried to explain to him something in science, I received the sharp question, "Is that idea good or bad? Should I learn it?" Perhaps the most important thing I learned from him is the unpretentious attitude toward scientific matters, yet accompanied with a sharp value judgement.

Third, once he captures the essentials, he tries to express them in the clearest possible terms. This may be illustrated with his frequent demand, "Say it in your own words."

Fourth, he always strives for rigor in physics, rather than in mathematics or other technical aspects. He uses all the intuitive and intellectual power to perceive something right in physics first, and then tries to give a mathematical formulation afterwards (rather than the other way around). This style of approach was also Fermi's, as pointed out by Segre.[60] Fano's words about this approach are the following, "Let's see if the equations say what we want." If mathematics leads him to something unexpected, he asks, "What does this mean in physics?" This attitude may be contrasted to that of many computationists who say, "This is what the computer gives me, and it must be right." They do not know nor even attempt to learn why and how that is so.

Finally, I wish to stress the great influence he has rightly exerted on his many students, associates, (see Appendix) as well as colleagues, and friends in conversations, correspondence, and many other ways. I am sure that all those who have had direct contact with him agree with me that we had much to learn from him. Hopefully, his influence will extend further for many years to come through his papers, and his monographs [67,68,137,194], which amply convey his insight, his originality, and above all, his enthusiasm at teaching the reader what is essential in physics.

Appendix

Ph.D.'s at University of Chicago

Name (year of Ph.D.)	Present affiliation
H. A. Stewart (1970)	Tennessee Valley Authority
A.R.P. Rau (1970)	Louisiana State University
K. T. Lu (1971)	Graduate School of Science, Beijing
A. F. Starace (1971)	University of Nebraska
J. L. Dehmer (1971)	Argonne National Laboratory
Dan Dill (1972)	Boston University
C. M. Lee [Li, Jia-ming] (1974)	Institute of Physics, Beijing
C. D. Lin (1974)	Kansas State University
C. E. Theodosiou (1977)	University of Toledo
G. Strinati (1977)	Scuola Normale Superiore, Pisa
C. W. Clark (1978)	National Bureau of Standards
C. H. Greene (1980)	Louisiana State University
S. Watanabe (1981)	Observatoire de Paris, Meudon
D. A. Harmin (1981)	University of Kentucky
P. F. O'Mahony (1984)	Royal Holloway College
M. Cavagnero (1986)	University of Nebraska
C. W. Lee (1986)	Max-Planck-Institut, Göttingen

Postdocs and Visitors

Name (years in Chicago)	Present affiliation
P. L. Altick (1967-8)	University of Nevada
J. S. Briggs (1969)	Universität Freiburg
E. S. Chang (1969-70)	University of Massachusetts
A.R.P. Rau (1970)	Louisiana State University
T. Åberg (1971)	Technical University of Helsinki
Dan Dill (1972-3)	Boston University
T. N. Chang (1974-6)	University of Southern California
C. M. Lee [Li, Jia-ming](1974)	Institute of Physics, Beijing
Y.-K. Kim (1974)	National Bureau of Standards
H. Klar (1975-6)	Universität Freiburg
G. Strinati (1977-8)	Scuola Normale Superiore, Pisa
C. W. Clark (1979)	National Bureau of Standards
R. Colle (1980)	Scuola Normale Superiore, Pisa

A. Giusti-Suzor (1981-2)   Université Paris-Sud, Orsay
D. A. Harmin (1982)   University of Kentucky
R. E. Johnson (1982)   University of Virginia
Z. Deng (1983)   Graduate School of Science,
                                   Beijing

J. A. Stephens (1985-6)   California Institute of
                                   Technology

M. Zarcone (1986)   Università di Palermo

References

1. H. Beutler, Z. Phys. 93:177 (1935).
2. S. Silverman and E. N. Lassettre, J. Chem. Phys. 40:1265 (1964).
3. R. P. Madden and K. Codling, Astrophys. J. 141:364 (1965).
4. C. E. Kuyatt, J. A. Simpson, and S. R. Mielczarek, Phys. Rev. 138:A385 (1965).
5. J. A. Simpson, G. E. Chamberlain, and S. R. Mielczarek, Phys. Rev. 139:A1039 (1965).
6. J. W. Cooper, Phys. Rev. 128:681 (1962).
7. D. R. Bates, Proc. Roy. Soc. A 188:350 (1947).
8. M. J. Seaton, Proc. Roy. Soc. A 208:418 (1951).
9. Y. S. Kim, A. Ron, R. H. Pratt, B. R. Tambe, and S. T. Manson, Phys. Rev. Lett. 46:1326 (1981).
10. S. T. Manson, Phys. Rev. A 31:3698 (1985).
11. S. D. Oh and R. H. Pratt, Phys. Rev. A 34:2486 (1986).
12. R. Y. Yin and R. H. Pratt, Phys. Rev. A 35:1149 (1987).
13. R. Y. Yin and R. H. Pratt, Phys. Rev. A 35:1154 (1987).
14. R. H. Pratt, R. Y. Yin, and X. Liang, Phys. Rev. 35:1450 (1987).
15. J. Kessler, Polarized Electrons, 2nd Edition, (Springer-Verlag, Berlin, 1985).
16. N. Bohr, Kgl. Dansk. Videnskab. Selskab, Mat.-Fys. Medd. 18:No. 8 (1948).
17. G. H. Morgan and E. Everhart, Phys. Rev. 128: 667 (1962).
18. E. Everhart and Q. C. Kessel, Phys. Rev. Lett. 14:247 (1965).
19. Q. C. Kessel, A. Russek, and E. Everhart, Phys. Rev. Lett. 14:484 (1965).
20. V. V. Afrosimov, Yu. S. Gordeev, M. N. Panov, and N. V. Fedorenko, Zh. Tekhn. Fiz. 34:1613; 1624; 1637 (1964) [Sov. Phys. - Tech. Phys. 9:1248; 1256; 1265 (1965)].
21. J. S. Briggs, Rep. Prog. Phys. 39:217 (1976).
22. R. L. Platzman, in Symposium on Radiobiology, edited by J. J. Nickson (John Wiley and Sons, Inc., New York, 1952) p. 20.
23. E. Fermi, in Quantentheorie und Chemie, Leipziger Vorträge, edited by H. Falkenhagen (S. Hirzel Verlag, Leipzig, 1928), p. 95.
24. M. Goeppert-Mayer, Phys. Rev. 60:184 (1941).
25. J. L. Dehmer, the present Proceedings.
26. G. Strinati, Phys. Rev. B 18:4096 (1978).
27. G. Strinati, Phys. Rev. B 18:4104 (1978).
28. G. Strinati, J. Math. Phys. 20:188 (1979).
29. G. Strinati, Phys. Rev. B 29:5718 (1984).
30. F. S. Hamm, in Solid State Physics. Vol. 1, edited by F. Seitz and D. Turnbull, (Academic Press, New York, 1969), p. 127.
31. M. J. Seaton, Mon. Not. R. Astron. Soc. 118:508 (1958).
32. M. J. Seaton, Rep. Prog. Phys. 46:167 (1983).
33. G. Herzberg, Phys. Rev. Lett. 23:1081 (1969).
34. G. Herzberg and Ch. Jungen, J. Mol. Spectrosc. 41:425 (1972).
35. P. G. Burke, in Physics of Electronic and Atomic Collisions, edited by S. Datz, (North-Holland, Amsterdam, 1982), p. 447.

36. A.R.P. Rau, Phys. Rev. A 4:207 (1971).
37. G. H. Wannier, Phys. Rev. 90:817 (1953).
38. N. F. Mott and H.S.W. Massey, The Theory of Atomic Collisions, Third Edition, (Oxford University Press, London, 1965).
39. S. Cvejanović and F. H. Read, J. Phys. B 7:1841 (1974).
40. V. A. Fock, Izv. Akad. Nauk S.S.S.R. Ser. Fiz. 18:161 (1954) [Kgl. Norske Vidensk Selskabs. Forh. 31:138 (1958)].
41. J. H. Macek, Phys. Rev. 160:170 (1967).
42. J. H. Macek, J. Phys. B 1:831 (1968).
43. M. Inokuti, M. A. Dillon, M. Kimura, Int. J. Quantum Chem. Symp. Ser., in press.
44. H. Bethe, Ann. Phys. (Leipzig) 5:325 (1930).
45. M. Inokuti, Rev. Mod. Phys. 43:297 (1971).
46. M. Inokuti, Y. Itikawa, and J. E. Turner, Rev. Mod. Phys. 50:23 (1978).
47. E. Fermi, Z. Physik 29:315 (1924).
48. R. L. Platzman, in Radiation Research 1966. Proceedings of the Third International Congress of Radiation Research, Cortina d'Ampezzo, Italy, June 1966, edited by G. Silini (North-Holland Publishing Company, Amsterdam, 1967) p. 20.
49. M. Inokuti, in Applied Atomic Collision Physics, Vol. 4. Condensed Matter, edited by S. Datz, (Academic Press, New York, 1983) p. 179.
50. M. Michaud and L. Sanche, J. Chem. Phys. 30:6067 (1984).
51. L. Sanche and M. Michaud, J. Chem. Phys. 30:6078 (1984).
52. L. G. Caron, G. Perluzzo, G. Bader, and L. Sanche, Phys. Rev. B 33:3027 (1986).
53. R. Kubo, J. Phys. Soc. Japan 12:570 (1957).
54. R. Kubo, in 1965 Tokyo Summer Lectures in Theoretical Physics, Part 1, edited by R. Kubo, (Shokabo, Tokyo and Benjamin, New York, 1966) p. 1.
55. R. Kubo, Rep. Prog. Phys. 29/1:255 (1966).
56. R. Zwanzig, J. Chem. Phys. 33:1338 (1960).
57. R. Zwanzig, in Lectures in Theoretical Physics, Vol. 3, edited by W. E. Brittin, (Wiley Interscience, New York, 1961), p. 106.
58. R. Zwanzig, Phys. Rev. 124:983 (1961).
59. J. S. Briggs, Rev. Mod. Phys. 43:189 (1971).
60. E. Segrè, Enrico Fermi. Physicist (The University of Chicago Press, Chicago, Illinois, 1970) p. 59.

1. Sul calcolo dei termini ottici, e in particolare dei potenziali di ionizzazione dei metalli bivalenti, per mezzo del potenziale statistico di Fermi [On the calculation of optical terms, and in particular of the ionization potentials of bivalent metals, by Fermi's method of the statistical potential] Atti Accad. Naz. Lincei, Cl. Sci. Fis. Mat. Nat., Rend. 20, 35 (1934).

2. Lo stato attuale del problema del calcolo dei termini spettrali [The present status of the problem of computation of spectral terms], Nuovo Cimento 11, 550 (1934).

3. Sullo spettro di assorbimento dei gas nobili presso il limite dello spettro d'arco [On the absorption spectrum of noble gas near the limit of the discrete spectrum], Nuovo Cimento 12, 154 (1935).

4. Some theoretical considerations on anomalous diffraction gratings, Phys. Rev. 50, 573 (1936).

5. On the anomalous diffraction gratings. II, Phys. Rev. 51, 288 (1937).

6. Zur Deutung der elektrischen Quadrupolomente der Atomkerne [Interpretation of the quadrupole moments of atomic nuclei], Naturwissenschaften 25, 602 (1937).

7. Zur Theorie der Intensitätsanomalien der Beugung [Theory of anomalies in the intensity of diffraction], Ann. Physik 32, 393 (1938).

8. L'introduzione di concetti termodinamici nella fisica nucleare [Introduction of thermodynamical concepts in nuclear physics], Nuovo Cimento 15, 343 (1938).

9. Sur la possibilité de décomposition de noyaux tres lourds en deux noyaux de poids moyen [On the possibility of the decomposition of very heavy nuclei into two lighter nuclei], J. Phys. Radium 10, 229 (1939).

10. A theory on cathode luminescence, Phys. Rev. 58, 544 (1940).

11. The theory of anomalous diffraction gratings and of quasi-stationary waves on metallic surfaces (Sommerfeld's waves), J. Opt. Soc. Am. 31, 213 (1941).

12. Mechanism of the origin of x-ray induced Notch deficiencies in Drosophila melanogaster, Proc. Nat. Acad. Sci. U.S. 27, 24 (1941), with M. Demerec.

13. On the analysis and interpretation of chromosomal changes in Drosophila, in Cold Spring Harbor Symposia on Quantitative Biology, Vol. 9. Genes and Chromosomes, Structure and Organization (The Biological Laboratory, Cold Spring Harbor, Long Island, New York, 1941), p. 113.

---

[*]Prepared in September 1987 by M. Inokuti from materials given by U. Fano. Short abstracts of conference papers are omitted.

14. The gene, in <u>Carnegie Institution of Washington Year Book</u> No. 40, p. 225 (1941), with M. Demerec, B. P. Kaufmann and E. Sutton.

15. An autosomal recessive factor inducing semi-sterility in Drosophila malanogaster females, Proc. Nat. Acad. Sci. U.S. <u>28</u>, 119 (1942).

16. On the interpretation of radiation experiments in genetics, Quart. Rev. Biol. <u>17</u>, 244 (1942).

17. The gene, in <u>Carnegie Institution of Washington Year Book</u> No. 41, p. 190 (1942), with M. Demerec, B. P. Kaufmann, E. Sutton and E. R. Sansome.

18. Mechanism of induction of gross chromosomal rearrangements in Drosophila sperms. Proc. Nat. Acad. Sci. U.S. <u>29</u>, 12 (1943).

19. Note on the time-intensity factor in radiobiology, Proc. Nat. Acad. Sci. U.S. <u>29</u>, 59 (1943), with L. D. Marinelli.

20. Production of ion clusters by X-rays, Nature <u>151</u>, 698 (1943).

21. The gene, in <u>Carnegie Institution of Washington Year Book</u> No. 42, p. 139 (1943), with M. Demerec, B. P. Kaufmann, E. R. Sansome, and H. Gay.

22. Genetics: Physical aspects, in <u>Medical Physics</u>, edited by Otto Glasser, (The Year Book Publishers, Inc., Chicago, Ill., 1944), p. 495, with M. Demerec.

23. Frequency of dominant lethals induced by radiation in sperms of Drosophila melanogaster. Genetics <u>29</u>, 348 (1944), with M. Demerec.

24. Experiments on mutations induced by neutrons in Drosophila melanogaster sperms. Genetics <u>29</u>, 361 (1944).

25. The gene, in <u>Carnegie Institution of Washington Year Book</u> No. 43, p. 103 (1944), with M. Demerec and E. R. Sansome.

26. Bacteriophage-resistant mutants in Escherichia coli, Genetics <u>30</u>, 119 (1945), with M. Demerec.

27. On the theory of the ionization yield of radiations in different substances, Phys. Rev. <u>70</u>, 44 (1946).

28. Ionization yield of radiations. II. The fluctuations of the number of ions, Phys. Rev. <u>72</u>, 26 (1947).

29. Note on the theory of radiation-induced lethals in Drosophila, Science <u>106</u>, 87 (1947).

30. A possible contributing mechanism of catalysis, J. Chem. Phys. <u>15</u>, 845 (1947).

31. Electric quadrupole coupling of the nuclear spin with the rotation of a polar diatomic molecule in an external electric field, J. Res. Nat. Bur. Std. <u>40</u>, 215 (1948).

32. On the theory of imperfect gratings, J. Opt. Soc. Am. <u>38</u>, 921 (1948).

33. Meson mass and range of nuclear forces, Am. J. Phys. <u>17</u>, 318 (1949).

34. Interpretation of the Poisson brackets, Am. J. Phys. 17, 449 (1949).

35. Penetration and diffusion of hard X-rays through thick barriers. I. The approach to spectral equilibrium, Phys. Rev. 76, 538 (1949), with H. A. Bethe and P. R. Karr.

36. Penetration and diffusion of X-rays through thick barriers. II. The asymptotic behavior when pair production is important, Phys. Rev. 76, 739 (1949).

37. Remarks on the classical and quantum-mechanical treatment of partial polarization, J. Opt. Soc. Am. 39, 859 (1949).

38. Penetration and diffusion of X-rays. V. Effect of small deflections upon the asymptotic behavior, Phys. Rev. 77, 425 (1950), with H. Hurwitz, Jr. and L. V. Spencer.

39. Atomic definition of primary standards, Nature 166, 167 (1950), with R. D. Huntoon.

40. Dosage units for high-energy radiation, Radiology 55, 743 (1950), with L. S. Taylor.

41. Genetics, in Medical Physics, Vol. 2, edited by Otto Glasser, (The Year Book Publishers, Inc., Chicago, Illinois, 1950), p. 365, with E. Caspari and M. Demerec.

42. Note on quantum effects in optics, J. Opt. Soc. Am. 41, 58 (1951).

43. Directional distribution of 1040-keV radiation from a high voltage X-ray tube, J. Appl. Phys. 22, 417 (1951), with N. Goldstein and H. O. Wyckoff.

44. Penetration and diffusion of x-rays. Calculation of spatial distributions by polynomial expansion, Phys. Rev. 81, 464 (1951), with L. V. Spencer.

45. Penetration and diffusion of x-rays. Calculation of spatial distributions by polynomial expansion, J. Res. Nat. Bur. Std. 46, 446 (1951), with L. V. Spencer.

46. Statistical matrix techniques and their application to the directional correlations of radiations, NBS Report 1214 (1951).

47. Secondary electrons: Average energy loss per ionization, in Symposium on Radiobiology, edited by J. J. Nickson, (John Wiley and Sons, Inc., New York 1952), p. 13.

48. Geometrical characterization of nuclear states and the theory of angular correlations, Phys. Rev. 90, 577 (1953).

49. Degradation and range straggling of high energy radiation, Phys. Rev. 92, 328 (1953).

50. Penetration of x- and gamma rays to extremely great depths, J. Res. Nat. Bur. Std. 51, 95 (1953).

51. Gamma-ray attenuation. Part I. Basic processes, Nucleonics 11, No. 8, p. 8 (1953).

52. Gamma-ray attenuation. Part II. Analysis of penetration, Nucleonics 11, No. 9, p. 55 (1953).

53. Principles of radiological physics, in Radiation Biology, edited by A. Hollaender, (McGraw-Hill Book Company, Inc., New York, 1954), p. 1.

54. Inelastic collisions and the Molière theory of multiple scattering, Phys. Rev. 93, 117 (1954).

55. A Stokes-parameter technique for the treatment of polarization in quantum mechanics, Phys. Rev. 93, 121 (1954).

56. Introductory remarks on the dosimetry of ionizing radiations. Radiat. Res. 1, 3 (1954).

57. Energy spectrum resulting from electron slowing down, Phys. Rev. 93, 1172 (1954), with L. V. Spencer.

58. Ionizing collisions of very fast particles and the dipole strength of optical transitions, Phys. Rev. 95, 1198 (1954).

59. Note on the Bragg-Gray cavity principle for measuring energy dissipation, Radiat. Res. 1, 237 (1954).

60. Note on the quantum theory of irreversible processes, Phys. Rev. 96, 869 (1954).

61. Differential inelastic scattering of relativistic charged particles, Phys. Rev. 102, 385 (1956).

62. Atomic theory of electromagnetic interactions in dense materials, Phys. Rev. 103, 1202 (1956).

63. Description of states in quantum mechanics by density matrix and operator techniques, Rev. Mod. Phys. 29, 74 (1957).

64. An approximate expression for gamma-ray degradation spectra, J. Res. Nat. Bur. Std. 59, 207 (1957), with A. T. Nelms.

65. Angular correlation of radiations with parallel angular momenta, Nuovo Cimento 5, 1358 (1957).

66. Evaluation of bremsstrahlung cross sections at the high-frequency limit, Phys. Rev. 112, 1679 (1958), with H. W. Koch and J. W. Motz.

67. Irreducible Tensorial Sets, (Academic Press, New York, 1959), with G. Racah, 171 pages.

68. Basic Physics of Atoms and Molecules, (Wiley, New York, 1959), with L. Fano, 414 pages.

69. Penetration and diffusion of X-rays, in Handbuch der Physik, edited by S. Flügge, 38/2, 660 (1959), with L. V. Spencer and M. J. Berger.

70. Sauter theory of the photoelectric effect, Phys. Rev. 116, 1147 (1959), with K. W. McVoy and J. R. Albers.

71. High-frequency limit of bremsstrahlung in the Sauter approximation, Phys. Rev. 116, 1156 (1959).

72. Interference of orbital and spin currents in bremsstrahlung and photoelectric effect, Phys. Rev. 116, 1159 (1959), with K. W. McVoy and J. R. Albers.

73. Bremsstrahlung and photoelectric effect as inverse processes, Phys. Rev. 116, 1168 (1959), with K. W. McVoy.

74. Theoretical reviews: A summary, in Penetration of Charged Particles in Matter. Proceedings of an Informal Conference, Gatlinburg, Tennessee, 1958, edited by E. A. Uehling, National Academy of Science – National Research Council Publication 752 (1960), p. 144.

75. Compilation of dielectric constant formulas presented by Nozieres and Pines, in Penetration of Charged Particles in Matter. Proceedings of an Informal Conference, Gatlinburg, Tennessee, 1958, edited by E. A. Uehling, National Academy of Science – National Research Council Publication 752 (1960), p. 158.

76. Comments on "collective" effects in atoms and in extended media, in Penetration of Charged Particles in Matter. Proceedings of an Informal Conference, Gatlinburg, Tennessee, 1958, edited by E. A. Uehling, National Academy of Science – National Research Council Publication 752 (1960), p. 163.

77. Normal modes of a lattice of oscillators with many resonances and dipolar coupling, Phys. Rev. 118, 451 (1960).

78. Real representations of coordinate rotations, J. Math. Phys. 1, 417 (1960).

79. Collective effects in absorption of energy from ionizing radiation, in Comparative Effects of Radiation, edited by M. Burton, J. S. Kirby-Smith, and J. L. Magee, (Wiley and Sons, New York, 1960) p. 14.

80. On the scattering of γ rays by nuclei, NBS Technical Note No. 83, (1960).

81. Deep penetration of radiation, in Proceedings of Symposia in Applied Mathematics. Vol. II. Nuclear Reactor Theory, edited by G. Birkhoff and E. P. Wigner, (American Mathematical Society, Providence, Rhode Island, 1961) p. 43, with M. J. Berger.

82. Quantum theory of interference effects in the mixing of light from phase-independent sources, Am. J. Phys. 29, 539 (1961).

83. Introductory remarks, in Report of Conference on Coherence Properties of Electromagnetic Radiation, 27-29 June 1960, Technical Note No. 5 published by the University of Rochester, April 1961, AFOSR-583, p. 31.

84. Effects of configuration interactions on intensities and phase shifts, Phys. Rev. 124, 1866 (1961).

85. On a qualitative feature of photoionization spectra, in Proceedings of the Second International Conference on Electronic Atomic Collisions, Boulder, Colorado (W. A. Benjamin, Inc., New York, 1961) p. 10.

86. Interaction matrix element in a shell model, Phys. Rev. 129, 2643 (1963), with F. Prats and Z. Goldschmidt.

87. On the connection between the theories of collisions and of atomic spectra, J. Nat. Acad. Sci. (India), $\underline{33}$, Pt. IV, 553 (1963), with F. Prats.

88. Pressure broadening as a prototype of relaxation, Phys. Rev. $\underline{131}$, 259 (1963).

89. Penetration of protons, α particles, and mesons, Ann. Rev. Nucl. Sci. $\underline{13}$, 1 (1963).

90. Classification of two-electron excitation levels of helium, Phys. Rev. Lett. $\underline{10}$, 518 (1963), with J. W. Cooper and F. Prats.

91. Classification of resonances in the electron scattering cross sections of Ne and He, Phys. Rev. Lett. $\underline{11}$, 158 (1963), with J. A. Simpson.

92. Liouville representation of quantum mechanics with application to relaxation processes, in Lectures on the Many-Body Problem, Vol. 2, edited by E. R. Caianiello (Academic Press, New York, 1964) p. 217.

93. Metastable levels in the continuum and the independent particle model, in Proceedings of the Third International Conference on the Physics of Electronic and Atomic Collisions, University College, London, 22-26 July 1963, edited by M.R.C. McDowell, (North-Holland Publishing Company, Amsterdam, 1964) p. 600, with F. Prats.

94. Precession equation of a spinning particle in non-uniform fields, Phys. Rev. $\underline{133}$, B828 (1964).

95. Contributions to the theory of shell corrections, in Studies in Penetration of Charged Particles in Matter, National Academy of Sciences-National Research Council Publication 1133, Report 39, (1964), p. 49, with J. E. Turner.

96. A list of currently unsolved problems, in Studies in Penetration of Charged Particles in Matter, National Academy of Sciences-National Research Council Publication 1133, Report 39, (1964), p. 281.

97. Exclusion of parity unfavored transitions in forward scattering, Phys. Rev. $\underline{135}$, B863 (1964).

98. Line profiles in the far-uv absorption spectra of the rare gases, Phys. Rev. $\underline{137}$, A1364 (1965), with J. W. Cooper.

99. Identification of energy levels of negative ions, Phys. Rev. $\underline{138}$, A400 (1965), with J. W. Cooper.

100. Interaction between configurations with several open shells, Phys. Rev. $\underline{140}$, A67 (1965).

101. Interpretation of Ar^+ - Ar collisions at 50 keV, Phys. Rev. Lett. $\underline{14}$, 627 (1965), with W. Lichten.

102. Connection between different approximations to the excited states of helium, Nat. Bur. Stds. Report No. 8993 (1965), with J. H. Macek.

103. Interference in the photo-ionization of molecules, Phys. Rev. $\underline{150}$, 30 (1966), with H. D. Cohen.

104. Radiation absorption between the ultraviolet and x-ray bands, Science <u>153</u>, 522 (1966).

105. Introductory remarks (at the symposium on energy deposition at the atomic level, in <u>Radiation Research 1966, Proceedings of the Third International Congress of Radiation Research, Cortina d'Ampezzo, Italy, June 1966</u>, edited by G. Silini, (North-Holland Publishing Company, Amsterdam, 1967) p. 13.

106. Transition matrix elements for large momentum or energy transfer, Phys. Rev. <u>162</u>, 68 (1967), with A. R. P. Rau.

107. An analysis of data on double transitions in He. Atomic Energy Commission Report No. COO-1674-2 (1967), with D. P. Chock.

108. Atomic potential wells and the periodic table, Phys. Rev. <u>167</u>, 7 (1968), with A. R. P. Rau.

109. Field configurations and parameters that identify states with j = 1, in <u>Spectroscopic and Group Theoretical Methods in Physics. Racah Memorial Volume</u>, edited by F. Bloch, S. G. Cohen, A. de-Shalit, S. Sambursky, and I. Talmi (North-Holland Publishing Company, Amsterdam, 1968) p. 153.

110. Ionization of the inner shells of atoms, in <u>The Physics of Electronic and Atomic Collisions; Invited Papers from the Fifth International Conference, Leningrad, July 1967</u>, edited by L. M. Branscomb, (Joint Institute for Laboratory Astrophysics, Boulder, Colorado, 1968) p. 150.

111. Spectral distribution of atomic oscillator strengths, Rev. Mod. Phys. <u>40</u>, 441 (1968), with J. W. Cooper.

112. Doubly excited states of atoms, in <u>Atomic Physics. Proceedings of the First International Conference on Atomic Physics, New York, 1968</u>, edited by B. Bederson, V. W. Cohen, and F. M. J. Pichanick, (Plenum Press, N.Y., 1969), p. 209.

113. Spin orientation of photoelectrons ejected by circularly polarized light, Phys. Rev. <u>178</u>, 131 (1969).

114. Inelastic collisions between slow atoms or ions, Comments At. Mol. Phys. <u>1</u>, 1 (1969).

115. Compound states in electron-atom processes, Comments At. Mol. Phys. <u>1</u>, 45 (1969).

116. Spin orientation of photoelectrons: Erratum and Addendum, Phys. Rev. <u>184</u>, 250 (1969).

117. Addendum to: Spectral distribution of atomic oscillator strengths [Rev. Mod. Phys. <u>40</u>, 441 (1968)], Rev. Mod. Phys. <u>41</u>, 724 (1969).

118. Effect of the centrifugal force on electron-molecule interactions, Comments At. Mol. Phys. <u>1</u>, 140 (1969).

119. The formulation of track structure theory, in <u>Charged Particle Tracks in Solids and Liquids. Proceedings of the Second L. H. Gray Conference, Cambridge, England, April 1969</u>, edited by G. E. Adams and D. K. Bewley, (Institute of Physics, London, 1970) p. 1.

120. Remarks on the theory of inelastic electron-atom collisions, Comments At. Mol. Phys. $\underline{1}$, 159 (1970).

121. Graphic analysis of perturbed Rydberg series, Phys. Rev. A $\underline{2}$, 81 (1970), with K. T. Lu.

122. Spin-orbit coupling: a weak force with conspicuous effects, Comments At. Mol. Phys. $\underline{2}$, 30 (1970).

123. Phase-amplitude method in atomic physics. I. Basic formulas for an electron in an ionic potential, Phys. Rev. A $\underline{2}$, 304 (1970), with J. L. Dehmer.

124. Quantum defect theory of ℓ uncoupling in H_2 as an example of channel interaction treatment, Phys. Rev. A $\underline{2}$, 353 (1970).

125. Frame transformations in electron-molecule collisions, Comments At. Mol. Phys. $\underline{2}$, 47 (1970).

126. Z-Dependence of spin-orbit coupling, in Topics in Modern Physics. A Tribute to Edward U. Condon, edited by W. E. Brittin and H. Odabasi, (Colorado Associated University Press, Boulder, Colorado, 1971) p. 147, with W. C. Martin.

127. Branching ratios: Statistical and otherwise, Comments At. Mol. Phys. $\underline{2}$, 171 (1971).

128. Raising of discrete levels into the far continuum, Phys. Rev. Lett. $\underline{26}$, 1521 (1971), with J. L. Dehmer, A. F. Starace, J. Sugar, and J. W. Cooper.

129. Theory of photodetachment near fine-structure thresholds, Phys. Rev. A $\underline{4}$, 1751 (1971), with A. R. P. Rau.

130. Theory of electron-molecule collisions by frame transformations, Phys. Rev. A $\underline{6}$, 173 (1972), with E. S. Chang.

131. Angular momentum transfer in the theory of angular distributions, Phys. Rev. A $\underline{6}$, 185 (1972), with Dan Dill.

132. Virtual electronic levels in molecules and crystals, Comments At. Mol. Phys. $\underline{3}$, 75 (1972).

133. Barrier to electron passage through electronegative atoms in BF_3, Chem. Phys. Lett. $\underline{17}$, 15 (1972), with B. Cadioli, U. Pincelli, E. Tosatti, and J. L. Dehmer.

134. Tensor polarizability and related parameters, Comments At. Mol. Phys. $\underline{3}$, 187 (1972), with R. D. Deslattes.

135. Parity unfavoredness and the distribution of photofragments, Phys. Rev. Lett. $\underline{29}$, 1203 (1972), with Dan Dill.

136. Quantomeccanica [Quantum mechanics], in Enciclopedia della Chimica, Vol. IX (Uses Edizioni Scientifiche, Firenze, Italy, 1972), p. 272.

137. Physics of Atoms and Molecules. An Introduction to the Structure of Matter (University of Chicago Press, Chicago, Illinois, 1972), with L. Fano, 592 pages.

138. Impact excitation and polarization of the emitted light, Rev. Mod. Phys. $\underline{45}$, 553 (1973), with J. H. Macek.

139. Correlations in He**, e + He and in related systems, in <u>The Physics of Electronic and Atomic Collisions. Invited Lectures and Progress Reports of the VIIIth International Conference on the Physics of Electronic and Atomic Collisions, Belgrade, Yugoslavia, 1973</u>, edited by B. C. Cobić and M. V. Kurepa (Institute of Physics, Beograd, 1973), p. 229, with C. D. Lin.

140. Multipole waves in periodic lattices, Phys. Rev. Lett. $\underline{31}$, 234 (1973).

141. Quasi-resonant coupling of atomic shells. Amusia's Report at the VIII I.C.P.E.A.C., Comments At. Mol. Phys. $\underline{4}$, 119 (1973).

142. Variational calculation of R matrices. Application to Ar photo-absorption. Phys. Rev. Lett. $\underline{31}$, 1573 (1973), with C. M. Lee.

143. Excitation of atoms to states of high orbital momentum, J. Phys. B $\underline{7}$, L401 (1974).

144. Correlations of excited electrons, in <u>Atomic Physics 4. Proceedings of the Fourth International Conference on Atomic Physics, Heidelberg, Germany, July 22-26, 1974</u>, edited by G. zu Putlitz, E. W. Weber, and A. Winnacker, (Plenum Press, New York, 1974) p. 47, with C. D. Lin.

145. Quasi-scaling of electron degradation spectra, Int. J. Radiat. Phys. Chem. $\underline{7}$, 63 (1975), with L. V. Spencer.

146. Platzman's analysis of the delivery of radiation energy to molecules, Radiat. Res. $\underline{64}$, 217 (1975).

147. Unified treatment of perturbed series, continuous spectra and collisions, J. Opt. Soc. Am. $\underline{65}$, 979 (1975).

148. Diverse manifestations of barriers to atomic electrons, Phys. Rev. A $\underline{12}$, 2638 (1975).

149. Analysis of electron correlations, in <u>The Physics of Electronic and Atomic Collisions. Invited Lectures, Review Papers, and Progress Reports of the IX International Conference on the Physics of Electronic and Atomic Collisions</u>, edited by J. S. Risley and R. Geballe (University of Washington, Seattle, 1975), p. 27.

150. Many-body theory of atomic transitions, Phys. Rev. A $\underline{13}$, 263 (1976), with T. N. Chang.

151. Transition matrices for the theory of spectra. Techniques for their construction and calculation, Phys. Rev. A $\underline{13}$, 282 (1976), with T. N. Chang.

152. Electron-optical properties of atomic fields, Rev. Mod. Phys. $\underline{48}$, 49 (1976), with C. E. Theodosiou and J. L. Dehmer.

153. Multipole expansion of the density of states about a crystal cell, J. Math. Phys. $\underline{17}$, 434 (1976), with G. Strinati.

154. Spectral and electron collision properties of atomic ions, in <u>Beam-Foil Spectroscopy Vol. 2. Proceedings of the Fourth International</u>

Conference on Beam-Foil Spectroscopy at Gatlinburg, Tennessee, September 1975, edited by I. A. Sellin and D. J. Pegg, (Plenum Publishing Corporation, New York, 1976), p. 637, with K. D. Chao, J. L. Dehmer, M. Inokuti, S. T. Manson, A. Msezane, R. F. Reilman, and C. E. Theodosiou.

155. Introduction to the school program, in Photoionization and Other Probes of Many-Electron Interactions, ed. F. J. Wuilleumier (Plenum Press, New York, NATO-ASI Series, Vol. 18 Series B, 1976) p. 1.

156. Correlations of excited electrons, in Photoionization and Other Probes of Many-Electron Interactions, ed. F. J. Wuilleumier (Plenum Press, New York, NATO-ASI Series, Vol. 18 Series B, 1976) p. 11.

157. On the theory of ionization by electron collisions, Argonne National Laboratory Report ANL-76-80 (1976), with M. Inokuti.

158. Dynamics of electron excitation, Phys. Today, 29, No. 9, p. 32, (1976).

159. Post-adiabatic analysis of atomic collisions, Phys. Rev. Lett. 37, 1132 (1976), with H. Klar.

160. Erratum: Quantum defect theory of ℓ uncoupling in H_2 as an example of channel-interaction treatment [Phys. Rev. A 2, 353 (1970)], Phys. Rev. A 15, 817 (1977).

161. The manifold aspects of Rydberg states, invited paper at the Colloque International n°273, "Atomic and Molecular States Coupled to a Continuum Highly Excited Atoms and Molecules" June 13-17, 1977, Aussois, France (Centre National de la Recherche Scientifique, Paris, France, 1977), p. 127.

162. Remarks on predissociation and autoionization, presented at the Colloque International n°273, "Atomic and Molecular States Coupled to a Continuum Highly Excited Atoms and Molecules" June 13-17, 1977, Aussois, France (Centre National de la Recherche Scientifique, Paris, France, 1977), p. 139.

163. Connection between configuration mixing and quantum defect treatments, Phys. Rev. A 17, 93 (1978).

164. Perspectives and prospectives, (on the R-matrix method), in Electronic and Atomic Collisions. Proceedings of the 10th International Conference on the Physics of Electronic and Atomic Collisions, Paris, 21-27 July 1977, edited by G. Watel (North-Holland Publishing Co., Amsterdam, 1978), p. 271.

165. Adiabatic analysis of collisions. III. Remarks on the spin model, Phys. Rev. A 19, 410 (1979).

166. General form of the quantum defect theory, Phys. Rev. A 19, 1485 (1979), with C. Greene and G. Strinati.

167. Identification of states excited by collision: Current state of theory, in Proceedings of the International Workshop on Coherence and Correlation in Atomic Collisions, University College London, 18-20 September 1978, ed. H. Kleinpoppen and J. F. Williams (Plenum Press, New York, 1979) p. 223.

168. Motional Stark effect on Li vapor photoabsorption in high magnetic fields, Phys. Rev. Lett. 42, 963 (1979), with H. Crosswhite, K. T. Lu, and A. R. P. Rau.

169. More on atomic resonances (Letter comment on an article by M. Biondi, et al.), Phys. Today 33, No. 4, p. 15 (1980).

170. Unfavored angular distributions, Phys. Rev. A 22, 1760 (1980), with C. H. Greene.

171. Wave propagation and diffraction on a potential ridge, Phys. Rev. A 22, 2660 (1980).

172. Formation of Landau standing waves in Rydberg spectra, J. Phys. B 13, L519 (1980).

173. Evolution of quantum-defect methods, Comments At. Mol. Phys. 10, 223 (1981).

174. Correlation of excited electrons: Progress in the alkaline earth and other spectra, Phys. Scr. 24, 656 (1981).

175. Stark effect of nonhydrogenic Rydberg spectra, Phys. Rev. A 24, 619 (1981).

176. Orientation by collision, J. Phys. B 14, L447. (1981), with M. Kohmoto.

177. Unified treatment of collisions, Phys. Rev. A 24, 2402 (1981).

178. General form of the quantum-defect theory. II, Phys. Rev. A 26, 2441 (1982), with C. H. Greene and A. R. P. Rau.

179. Correlations of two excited electrons, Rep. Prog. Phys. 46, 97 (1983).

180. Dynamics of resonant states, in Atomic Physics 8. Proceedings of the Eighth International Conference on Atomic Physics, August 2-6, 1982, Göteborg, Sweden, edited by Ingvar Lindgren, Arne Rosen, and Sune Svanberg (Plenum Press, New York, 1983), p. 5.

181. Erratum: Unified treatment of collisions [Phys. Rev. A 24, 2402 (1981)], Phys. Rev. A 27, 1208 (1983).

182. Evolution of quantum defect methods. II, Comments At. Mol. Phys. 13, 157 (1983).

183. Pairs of two-level systems, Rev. Mod. Phys. 55, 855 (1983).

184. Spin correlations in photodetachment, Phys. Rev. A 29, 177 (1984), with S. Watanabe and C. H. Greene.

185. Alternative parameters of channel interactions. I. Symmetry analysis of the two-channel coupling, J. Phys. B 17, 215 (1984), with A. Giusti-Suzor.

186. Alternative parameters of channel interactions. II. A Hamiltonian model, J. Phys. B 17, 4267 (1984), with A. Giusti-Suzor.

187. Alternative parameters of channel interactions. III. Note on a narrow band in the Ba J = 2 spectrum, J. Phys. B $\underline{17}$, 4277 (1984), with A. Giusti-Suzor.

188. Molecular complexes in collisions. Introductory remarks, in Electronic and Atomic Collisions, Invited Papers. Proceedings of the XIIIth International Conference on the Physics of Electronic and Atomic Collisions, Berlin, July 27-August 2, 1983, edited by J. Eichler, I.V. Hertel, and N. Stolterfoht (North-Holland, Amsterdam, 1984), p. 629.

189. Herzberg's impact on the physics of Rydberg states, Can. J. Phys. $\underline{62}$, 1264 (1984), with K. T. Lu.

190. Erratum: General form of the quantum-defect theory. II [Phys. Rev. A $\underline{26}$, 2441 (1982)], Phys. Rev. A $\underline{30}$, 3321 (1984), with C. H. Greene and A. R. P. Rau.

191. Propensity rules: An analytical approach, Phys. Rev. A $\underline{32}$, 617 (1985).

192. Resonances in electron collisions, Comments At. Mol. Phys. $\underline{16}$, 241 (1985), with A. R. P. Rau.

193. WKB Approach to nonseparable wave equations, in Semiclassical Descriptions of Atomic and Nuclear Collisions, edited by J. Bang and J. de Boer (Elsevier Science Publishers B. V., Amsterdam, 1985) p. 367.

194. Atomic Collisions and Spectra, (Academic Press, New York, 1986), with A. R. P. Rau, 409 pages.

195. New quantum numbers in collision theory, Phys. Rev. A $\underline{33}$, 921 (1986), with Chung-Woo Lee.

196. Slow electrons in condensed matter, Phys. Rev. B $\underline{34}$, 438 (1986), with J. A. Stephens.

197. Absence of resonances in the elastic scattering of electrons in molecular solids, J. Chem. Phys. $\underline{85}$, 6239 (1986), with J. A. Stephens and M. Inokuti.

198. Generalized WKB and Milne solutions to one-dimensional wave equations, Phys. Rev. A $\underline{35}$, 3619 (1987), with F. Robicheaux, M. Cavagnero, and D. A. Harmin.

199. Bypassing translation factors in molecular dissociation and reactions, Phys. Rev. A $\underline{35}$, 3940 (1987), with J. Macek, M. Cavagnero, and J. Jerjian.

200. New quantum numbers in collision theory. III. Symmetries of the scattering geometry, Phys. Rev. A $\underline{36}$, 66 (1987), with Chun-Woo Lee.

201. New quantum numbers in collision theory. IV. Separation of the first Born contributions, Phys. Rev. A $\underline{36}$, 74 (1987), with Chun-Woo Lee.

202. Short- and long-range interactions of slow electrons in condensed matter: Effects on reflection and transmission, Phys. Rev. A. $\underline{36}$, 1929 (1987).

203. Half-scattering, detailed balance and energy transfers, Comments At. Mol. Phys. <u>19</u>, 253 (1987).

204. Studies of slow electron action on condensed media, Radiat. Phys. Chem., in press.

205. Semi-analytical study of diamagnetism in a degenerate hydrogenic manifold, in press, with F. Robicheaux and A. R. P. Rau.

206. Quantum and classical treatments of Rydberg states in magnetic fields, in <u>Festschrift for Val Telegdi</u>, edited by K. Winter (North-Holland, Amsterdam), in press.

207. Implications of new evidence on quasi-Landau spectra for a unified treatment of collisions and spectra, in <u>Proceedings of the Conference on Atomic Spectra in External Fields. II</u>, edited by K. T. Taylor (Plenum, London), in press.

208. Introduction to the theory symposium, in the <u>Proceedings of the NATO Advanced Study Institute "Fundamental Processes of Atomic Dynamics," Maratea, Italy, 21 September - 2 October 1987</u>, edited by J. S. Briggs, H. Kleinpoppen, and H. Lutz (Plenum Press, New York), in press.

209. General features of collisions and spectra, in the <u>Proceedings of the NATO Advanced Study Institute "Fundamental Processes of Atomic Dynamics," Maratea, Italy, 21 September - 2 October 1987</u>, edited by J. S. Briggs, H. Kleinpoppen, and H. Lutz (Plenum Press, New York), in press.

FRONTIERS IN ATOMIC PHYSICS--INTRODUCTION AND OPENING REMARKS

Joseph Macek

Department of Physics and Astronomy
University of Nebraska
Lincoln, NE   68588-0111

ABSTRACT

Fragmentation, spectroscopy, and collisions represent three broad categories of atomic and molecular processes. We regard the formation of excited bound states in spectroscopic studies as fragmentations in which the fragments do not have enough energy to escape. Collisions are represented as the product of fragmentation reactions run in reverse in the initial state, and fragmentation reactions run in normal direction in the final state. The Jost functions form the mathematical framework for this picture. Issues that emerge naturally are the identification of appropriate reaction coordinates to describe the propagation of the system, the classification of the channels at large and small values of the reaction coordinate and the connection between the two regions.

INTRODUCTION

Much activity at the frontiers of atomic physics is characterized by an ever-expanding capability for researchers to control, in minute detail, initial and final states of atomic processes. With each advance in our ability to select initial states or measure final states has come a corresponding advance in our understanding of fundamental atomic processes. We will naturally be concerned with the latest advances in experimental technique since such advances always are the best indicators of future directions. Our broader objective is to fit the various fragments of the field into a single conceptual framework that will enable us to get some idea of a bigger picture underlying individual results. Many of the lecturers of this Advanced Study Institute are directed towards articulating this broader picture.

It is convenient to categorize experimental investigations into atomic processes into three broad categories, namely studies of fragmentation, spectroscopy and collisions. In studies of fragmentation we consider processes where energy, momentum or angular momentum is transferred, usually suddenly, to a bound atomic system. As a consequence of this input of energy the system expands and ultimately breaks up into one or more fragments that are detected by an instrument that analyzes the final states of the fragments and their relative motion. Immediately after the input of energy to the system, but before it has expanded, the state of the system is characterized by a set of channels α. These channels refer to a configuration where the system still has the spatial size of the original bound state and are therefore referred to as condensation channels. The condensation channels, of course, form a complete set.

Selection of an appropriate set is an important part of our study of atomic processes. Lectures by Dehmer on molecular photoionization, Connerade on resonances, and Greene on variational R-matrix theory deal with various aspects of these condensation channels. Electron correlations play a significant part here and will be discussed by Lin, Starace, Feagin, and Cavagnero. A particularly interesting example of the relevance of condensation channels will be discussed by Giusti-Suzor. Her lecture deals with the relatively new subject of multi-photon processes, where one not only transfers specific amounts of energy to systems, but specific, and controllable, amounts of angular momentum as well. The resulting condensation channels then pertain not only to states of atoms, but also to the photons absorbed.

At later times, after the bound system has absorbed energy it is in a non-stationary state which expands and separates into fragments. This expansion is represented by an outgoing wave in some appropriate coordinate r which I call a "reaction coordinate". This reaction coordinate is not specified; it could by any coordinate that we find convenient, indeed, selection of an appropriate reaction coordinate is an important part of our description of atomic processes. Usually, one chooses reaction coordinates which are identical to the separation distance between two fragments in processes leading to breakup into two fragments. For example, in the study of photoionization an electron separates from a positive ion and one often uses the radial coordinate of an electron as the reaction coordinate. This choice seems quite natural, indeed inevitable, but it does present certain conceptual problems for multi-electron systems. Because any one of the many electrons in a many-electron system could escape, it is necessary to treat all electrons on the same footing. Usually this is done by employing an explicitly antisymmetric wave function in accord with Pauli's principle. The need to treat all electrons equivalently while separating out an electron coordinate as a reaction coordinate leads to non-local potentials, i.e. exchange terms, in the equations of motion. Alternative choices of reaction coordinates which treat the electrons symmetrically do not give rise to non-local potentials. Quite generally, use of reaction coordinates adapted to a specific break-up channel, as in the case of single-particle electron coordinates for multi-electron systems, result in non-local interactions. Use of a reaction coordinate which is common to all channels avoids such non-local interactions. Two examples of these alternative reaction coordinates are discussed by Feagin for two-electrons atoms, and Fano and Cavagnero for multi-electron systems.

When the system has split into fragments separated by macroscopic distances it is analyzed by an analyzer. It is in this region, the asymptotic region, where the states of the fragment and the states of relative motion are measured that one obtains experimental information about the fragmentation process. Much of the frontier part of atomic physics concerns the detection and measurement of these atomic states. This is appropriate; after all, the final atomic states represent our only real window into the inner workings of atomic processes. Lectures by Kleinpoppen, Raith, Andersen, Campbell, Lutz, and Burgdörfer deal with various aspects of the analysis of atomic states. The lectures by Kleinpoppen and Raith discuss the characterization of atomic states in the context of electron-atom collisions. Other lectures deal primarily with ion-atom collisions and will be introduced in my second lecture. Underlying both sets of lectures is the general theory of angular correlations. Not to be overlooked is an equally important theme of these lectures, namely, the insights into atomic processes that have been extracted from the analysis of final states.

The second topic, spectroscopy, deals with excited bound states of atomic systems. For purposes of this introduction, we regard excited bound states as fragmentation states which do not have quite enough energy for the fragments to escape to macroscopic distances. Rather, the system expands but comes to a region of reaction coordinate r which is classically forbidden. That is, the system reaches a classical turning point in the reaction coordinate, where the outgoing wave is reflected. The superposition of outgoing and reflected wave sets up a standing wave, i.e. a bound state. Clearly these excited states exhibit many of the features of states where the fragments actually reach the asymptotic region. The relation between bound and fragmentation states represents the point of departure for multi-channel quantum defect theory (MQDT) and will be articulated in the lectures of Ravi Rau.

Our third subject, collisions, is viewed here as a combination of two fragmentation reactions. That is, the system is prepared in an initial state by a "state selector", allowed to propagate along a reaction coordinate r to a condensation configuration, and then separates to the asymptotic region where the fragments are again observed by an analyzer. The propagation inward to the condensation configuration is viewed as a fragmentation reaction run in the time reversed direction. Interpretation of collisions is thereby separated into interpretations of two fragmentation reactions. For this reason one often speaks of a fragmentation reaction as a "half collision". Alternatively, and this is point of view adapted here, a collision is regarded as the product of two fragmentation reactions. This enables us to employ a common language to describe fragmentation, spectroscopy, and collisions.

MATHEMATICAL FORMULATION

To put these ideas in quantitative form we follow the general approach of the book by Fano and Rau,[1] which emphasizes the special significance of Jost functions. These functions might better be described as Jost coefficients, since they represent the coefficients of incoming and outgoing waves in the asymptotic region. The designation "functions" is employed since the early application emphasized analytic properties as a function of wave number k. Here the analytic properties are not central, rather the emphasis is on the conceptual significance as coefficients of incoming and outgoing waves which describe how a system connects condensation channels α and asymptotic channels i. I will continue to employ the now standard designation Jost function and hope that will not lead to any confusion with special functions of mathematical physics.

Following ref. (1) we designated a reaction coordinate by r and the collection of all other coordinates of the system including direction coordinates of the direction of relative motion by the symbol ω. The asymptotic eigenstates of the system are represented by a $\phi_j(\omega)$. At any finite value of the reaction coordinate r the system which ultimately evolves into an eigenstate i is represented by superposition of eigenstates j. Propagation as an outgoing wave is represented by set of functions $f^+_{ji}(r)$ and propagation as an incoming wave is represented by $f^-_{ji}(r)$, where

$$f_{ji}^{\pm}(r) \sim \delta_{ji} e^{\pm i k_i r} \qquad \text{as } r \to \infty. \tag{1}$$

A wave function for the system which starts out in a specific condensation channel α near $r \approx 0$ is given by

$$\psi_\alpha(r,\omega) = \frac{1}{2i} \sum_j \phi_j(\omega) \sum_i \left[f_{ji}^+ (r) \left(\frac{2}{\pi k_i}\right)^{1/2} (-i)^{\ell_i} J_{i\alpha}^+ (k) \right.$$

$$\left. -f_{ji}^- (r) \left(\frac{2}{\pi k_i}\right)^{1/2} i^{\ell_i} J_{i\alpha}^- (k) \right] \tag{2}$$

Here it is convenient to suppose that the matrices $f^{\pm}_{ji}(r)$ are diagonal outside of a certain radius r=a.

$$f_{ji}^{\pm} (r) = \delta_{ji} f_j^{\pm}(r), \text{ for } r > a. \tag{3}$$

The coefficients $J^+_{j\alpha}(k)$ and $J^-_{i\alpha}(k)$ then denote amplitudes with which the condensation channel α populates the asymptotic states i and relative motion is described by outgoing and incoming waves, respectively.

Eq. (2) outlines a three-fold task for studies of atomic processes. Firstly, the condensation channels α must be classified, secondly, the reaction coordinate r and the Jost coefficients which connect condensation and asymptotic channels must be identified, and finally the asymptotic states must be characterized.

Bound states occur for energies such that all of the wave numbers k_i are purely imaginary $k_i \to i \kappa_i$, $\kappa_i, > 0$. The one has

$$f_i^{\pm} \to e^{\pm \kappa_i r}, \text{ as } r \to \infty. \tag{4}$$

In order to have acceptable physical solutions it is necessary that the coefficients of the exponentially increasing term vanishes for some linear combination of $\psi_\alpha(r,\omega)$. This is only possible if the determinant of the coefficients $J^-_{i\alpha}$, represented by $\underline{J}^-(\kappa_i)$ vanishes;

$$\text{Det}\left[\underline{J}^-(\kappa_i)\right] = 0. \tag{5}$$

We see that spectroscopic studies of bound states gives us information on the Jost functions via Eq. (5).

Scattering is described by wave functions ψ_i in which the amplitude of the incoming wave vanishes for all asymptotic channels j except one particular channel $j=i$. Such functions are constructed from the set by forming the linear combination $\sum_\alpha \psi_\alpha (\underline{J}^-)^{-1}_{\alpha i}$ where $\underline{J}^{-1}$ denotes the inverse of the $\underline{J}^-$ matrix;

$$\psi_i = \sum_\alpha \psi_\alpha \, (\underline{\underline{J}}^{-1})_{\alpha i} = \frac{1}{2i} \sum_j \phi_j (\omega) \left[f_j^+(r)(\frac{2}{\pi \kappa_j})^{1/2} (-1)^{\ell_j} \right.$$

$$\qquad \qquad (6)$$

$$x \sum_\alpha J_{j\alpha}^+ \, (\underline{J}^{-1})_{\alpha i} - f_j^-(r) \, (i)^{\ell_j} \Big]$$

The scattering matrix is defined by

$$S_{ji} = \sum_\alpha J_{j\alpha}^+ \, (\underline{J}^{-1})_{\alpha i}. \qquad \qquad (7)$$

Note that the "condensation channel" indices α are summed over in Eq. (7). This emphasizes the flexibility that one has in selecting such channels. Technically the only requirement on the set α is completeness, but physical content is brought out when a set can be found for which only a few terms give a good representation. Determining such sets continues to be a central task for atomic and molecular physics.

In the intermediate region, where the propagation of the system is described by the functions $f_{ij}^{\pm}(r)$, we have seen that selection of the appropriate reaction coordinate is the first task. A second task is to find representation in which $f_{ij}^{\pm}(r)$ is diagonal, i.e. we seek bases $\phi_j(\omega)$ in which $f_{ij}^{\pm}(r)$ takes the form

$$f_{ij}^{\pm}(r) = \delta_{ij} \, f_j(r) \qquad \qquad (8)$$

This frequently requires different bases in different regions of r and transformations between different bases at the boundaries separating the different regions. Such transformations are called "frame transformations" and have been extensively exploited in MQDT. The lectures of Jungen deal with such transformations in MQDT for molecules.

Frame transformations also enter into the description of atoms in strong external fields. Even "strong" external fields are weak compared to the internal fields of atoms; the external fields become effective

only at large distances. At small distances appropriate bases are constructed as for an atom in no external field, while at the largest distances the appropriate bases treat the external field as the determining
potential. The transformation between the two bases is one example of a
frame transformation. P. O'Mahoney will describe our current understanding of appropriate bases for atoms in strong fields.

The need for frame transformations implies that atomic and molecular
physics is rich in physical phenomena. Corresponding to each transformation there must exist phenomena which manifest the mathematical operation
of changing representation. As a particularly simple example consider
quantum beats. For r less than some radius R_1 it is usually appropriate
to ignore fine and hyperfine structure effects on atomic states. The
appropriate bases is then $|LM_LSM_SIM_I>$ where L, S, and I are orbital,
spin and nuclear spin quantum numbers respectively. For larger values of
r less than another radius R_2, the spin-orbit interaction becomes rele-
vant and the appropriate basis is $|LSJM_JIM_I>$, while for still larger
values of r greater than R_2 the hyperfine interaction becomes relevant so
that the appropriate basis is $|(LS)JIFM_F)>$. Thus we have for our
bases:

$$|LM_LSM_SIM_I> \quad , \quad r < R_1$$

$$|LSJM_JIM_I> \quad , \quad R_1 < r < R_2 \tag{9}$$

$$|(LS)JIFM_F> \quad , \quad R_2 < r$$

Let us label each state in the manifold of nearly degenerate states
by the index i or j. Then the scattered part of the coordinate wave
function becomes

$$\psi_i^{scat} \sim \sum_j \phi_j(\omega) \, S_{ji} \epsilon^{ik_jr} \left(\frac{2}{\pi\kappa j}\right)^{1/2} (-1)^{\ell_j} \tag{10}$$

Define a centroid of the fine structure levels as E and the corresponding
wave vector by $\bar{k}$, and measure the energy levels of the atom from the
centroid.

$$k_j = (\bar{k}^2 - 2mE_j)^{1/2}. \tag{11}$$

For collisions where the relative energy is much larger than any of
the fine or hyperfine structure splittings we may approximate k_j by

$$k_j \approx \bar{k} - \frac{ME_j}{\bar{k}} = \bar{k} - \frac{ME_j}{hp} = \bar{k} - \frac{Ej}{hv} ,$$

so that our scattered wave becomes

$$\psi^{scat} \sim e^{i\bar{k}r} \sum_j (-1)^{1j} \phi_j(\omega) (\frac{2}{\pi k j})^{1/2} S_{ji} e^{-iE_j r/hv} \qquad (12)$$

If we now introduce a variable t = r/v defined as one normally defines time in a time-of-flight experiment we see that our scattered wave function represents a system in a non-stationary state which exhibits quantum beats.

$$\psi^{scat} \sim e^{ikr} \sum_j (-1)^{1j} \phi_j(\omega) (\frac{2}{\pi k_j})^{1/2} S_{ji} e^{-iE_j t} \qquad (13)$$

Chris Greene will discuss the observational consequences of these beats. He will show, using transformations between different angular momentum representations, that anisotropic states exhibit oscillations characteristic of fine- and hyperfine structure splittings. These transformations maybe regarded as a particularly simple form of the general frame transformations discussed in a wider context by Rau and by Jungen.

ASYMPTOTIC STATES

Measurements are made in the asymptotic region, i.e. in the region where the reaction coordinate takes on macroscopic dimensions. In this region it is necessary to specify the states i and f in accord with experiment. Now generally, experimental arrangements do not select eigenstates specified by a wave function but mixed state specified by a density matrix ρ_{ij}. Usually the density matrix projects onto a set of energy eigenstates of the system. The density matrix selected by the "state preparation: device, i.e. the initial state, is denoted by ϱ^M and the state analyzed by the "state analyzer", i.e. the final state, is denoted by ϱ^A. Then the probability for the system to make a transition from the initial state to the final state is given by[2,3]

$$P = 2\pi \, Tr \, [\underline{T} \underline{\varrho}^M \, \underline{T}^+ \, \underline{\varrho}^A \,] \qquad (14)$$

where $\underline{T}$ is the transition matrix.

Since we are usually interested in reactions other than elastic scattering, it is permissible to replace the transition matrix $\underline{T}$ by the scattering matrix $\underline{S}$ in Eq. (14)

$$P = 2\pi \, Tr \, [\underline{S} \, \underline{\varrho}^M \, \underline{S}^+ \, \underline{\varrho}^{A+}] \qquad (15)$$

The initial and final density matrices represent the new aspect of Eq. (15). Determination of these quantities is a matter of experimental design. In this context we regard the density matrices ϱ^A and ϱ^M as descriptions of macroscopic pieces of equipment. Specific examples will be discussed by many lectures, in particular, Kleinpoppen and Raith in connection with electron scattering, and Andersen, Campbell, and Lutz in connection with ion-atom collisions. Some relevant theory will be discussed by Greene and by Burgdörfer.

It is appropriate to conclude our discussion by interpreting Eq. (15) in terms of the matrices $\underline{J}$ rather than the usual scattering matrices $\underline{S}$. Substituting Eq. (7) into Eq. (15) replaces $\underline{S}$ by $\underline{J}$;

$$ P = 2\pi \ \text{Tr} \ \left[\underline{J}^+ (\underline{J}^{-1}) \ \underline{\rho}^M \ (\underline{J}^{-1})^+ (\underline{J}^+)^+ \underline{\rho}^{A+} \right] \tag{16} $$

This equation may be rearranged to give;

$$ P = 2\pi \ \text{Tr} \ \left[(\underline{J}^{-1} \ \underline{\rho}^M \ \underline{J}^{-1})(\underline{J}^{++} \ \underline{\rho}^{A+} \ \underline{J}^+) \right] \tag{17} $$

We can now introduce the complete set of states at the boundary of the "inner" region and write

$$ P = \sum \ (\underline{J}^{-1} \ \underline{\rho}^M \ \underline{J}^{-1+})_{\alpha\beta} (\underline{J}^{++} \ \underline{\rho}^{A+} \ \underline{J}^+)_{\beta\alpha} \tag{18} $$

The first factor in Eq. (18) represents propagation from the "state selector", described by $\underline{\rho}^M$, inward to the reaction region where the fragmentation channels α and β are appropriate. The second factor represents propagation outward from the reaction region toward the "state analyzer", described by $\underline{\rho}^A$. Each individual terms represents a fragmentation reaction or a fragmentation reaction run in reverse. In this way we see that a collision is described by two fragmentations.

Eq. (18) summarizes our objectives; we seek appropriate "condensation" channels α, specified on some boundary close to the region where reactions take place. We also seek the Jost matrices $\underline{J}^\pm$ which describe the propagation of the system from the condensation limit to the asymptotic limit. To accomplish this task we employ experimental equipment which gives us control over $\underline{\rho}^A$ or $\underline{\rho}^M$. Measurements of P as a function of $\underline{\rho}^A$ or $\underline{\rho}^M$ provide us with information on the all important condensation channels.

ION-ATOM COLLISIONS

Ion-atoms collisions provide us with a protype of this program. Here it is often appropriate to select as a "reaction coordinate" the quantity z/v where z is the z-coordinate of the relative position of the projectile and target in a frame with the z-axis parallel to the velocity vector. In addition, we define time according to t = z/v. Then the time evolution of the system is given by the time evolution matrix $\underline{U}(t,t')$.

The S-matrix is then defined according to

$$\underline{S} = \underline{U}(+\infty, -\infty) = \underline{U}(+\infty, 0)\ \underline{U}(0, -\infty). \tag{19}$$

More explicitly we have

$$S_{ij} = \sum_{\alpha} U_{i\alpha}(+\infty, 0)\ U_{\alpha j}(0, -\infty). \tag{20}$$

The indices i and j in Eq. (20) refer to asymptotic states, however, the indices α represent any complete set of states. Note that the states of the complete set are those which best represent the system at t=0, that is at the distance of closest approach when z=0. Accordingly, these indices refer to the condensation states appropriate to the construction of the Jost matrices. Indeed, we see that Eq. (20) is analogous to Eq. (7) which has the same structure. For ion-atom studies at low and intermediate velocities these "condensation" channels are more commonly referred to as united atom channels. Much progress has been made in understanding the nature of these united atom channels, and Lin will discuss one method, the AO+ method, for incorporating appropriate united atom states in quantitative calculations.

Because one deals with the $\underline{U}$ matrix of ion-atom collisions, it is often possible to interpret the structure of the semiclassical $\underline{J}$ matrix, equal to $\underline{U}(\infty,0)$, in terms of simple models such as the Landau-Zener model. For that reason we have a better understanding of the structure of $\underline{J}$ for ion-atom collisions than for most other reactions. Even here many issues remain to be developed. The lectures in the second week emphasize ion-atom collisions and will consider some of their central issues.

ACKNOWLEDGMENT

Support by the National Science Foundation under grant PHY-8602988 is gratefully acknowledged.

REFERENCES

1. U. Fano and A.R.P. Rau, Atomic Collisions and Spectra, (Academic Press, New York 1986), pp. 84-110.

2. Joseph Macek in Fundamental Processes in Energetic Atomic Collisions, Ed. by H.O. Lutz, J.S. Briggs and H. Kleinpoppen, (Plenum Publishing Co., New York 1983), pp. 39-67.

3. Roger G. Newton, Scattering Theory of Particles and Waves, (McGraw-Hill, New York 1966), p. 230.

INTRODUCTION TO THE THEORY SYMPOSIUM

U. Fano

Dept. of Physics and James Franck Inst.
University of Chicago, Chicago, IL 60637

The concepts and methods to be reviewed in this Symposium were sparked by a burst of novel experimental evidence in the mid-'60s, on energy transfers in the range of $10-100eV$ among electrons, photons and atoms. A main feature of this evidence lay in the wealth of resonances discovered in this range, symptoms of a rich yet rather unexpected set of phenomena. The task of interpreting the manifold features of new observations and of the underlying structural changes of atoms or molecules led to a style of physico-chemical research aimed at identifying, evaluating and exploiting the parameters that control each process.

Mastery of such parameters served to correlate large amounts of data. It also served to condense the data into *few independent* entries and to evaluate additional observables. The program's motivation stems mainly from its chemical implications, namely, from the challenge of describing how the Coulomb forces among electrons and nuclei may combine to steer delicate chemical transformations. A curious aspect of the whole development lies in the crucial role of *larger* energy transfers in revealing mechanisms that are also central to thermal or epithermal reactions.

These developments contrast with the view of atomic theory that has prevailed among physicists since the '30s, which I summarize as follows: 1) Atomic theory has successfully completed its main task of accounting in considerable detail for the spectra and collisions of simple systems (e, H, H_2, He, alkalis) by quantum mechanics. (Further refinements by quantum electrodynamics or chromodynamics are always welcome). 2) More complex atoms and molecules are treated successfully by the independent-electron approximation complemented by perturbation methods and configuration mixing. 3) Still more complex systems should be treated by numerical engineering procedures beyond the scope of physics proper.

The phenomena discovered in the '60s involved atomic states with energy larger than previously considered. The representation of such states by configuration mixing would require the superposition of every large -- indeed infinite -- numbers of independent-electron configurations, thus introducing ill-surveyable, inconvenient features. The observed phenomena displayed nevertheless rather simple features, characterized quantitatively by appropriate parameters. The task of theory was accordingly to cast the quantum mechanical representation of each phenomenon in a form that would define each relevant parameter, often providing also a workable procedure to evaluate such parameters *ab initio*. It was a pleasant surprise that parametrization procedures -- typical of macroscopic physics -- proved applicable to the quantum treatment of rather simple atomic systems.

Qualitative analysis of the newly observed phenomena also led unexpectedly, time and again, to largely analytical formulations of theory requiring only modest complements of numerical work. This feature affords ready transferability of results among related substances. An additional feature of new theory was its dealing mainly with the quantum mechanics of observables rather than with the calculation of wave functions. On the other hand the introduction of novel concepts and procedures into chemical physics imposed a special burden to present them in readily accessible and sufficiently documented form. The organization of the present Symposium constitutes a step toward this goal. Earlier efforts in the same direction will be mentioned in the following sections and articles.

The several sections of this Introduction outline streams of research that are fairly distinct, even though interlinked, with references to their more detailed presentation in separate lectures.

GENERAL FEATURES OF PHOTOABSORPTION

The photoabsorption spectra of matter, and those of the associated radiation scattering, in the $10-1000eV$ range were virtually uncharted prior to 1960. Extrapolation of hydrogenic theory and of X-ray absorption data at energies $\gg 1keV$ suggested that absorption by each atomic shell would peak near its ionization threshold and decline thereafter smoothly with increasing photon energy. Two major systematic departures from this expectation were discovered by an initial calculation[1] followed by extensive experimental and theoretical studies.[2].

Absorption *minima*, noted occasionally long ago, were found to occur systematically in the spectra where a specific transition matrix element reverses its sign yielding a "Cooper zero". Such sign reversals are foreign to hydrogenic spectra where a sign difference occurs only between intra- and extra-shell transitions. They have been interpreted qualitatively in terms of mapping non-hydrogenic onto hydrogenic spectra, but a basic theory of this mapping has been approached only very recently[3] and is still evolving.

Photoabsorption is often strongly depressed near ionization thresholds, where it would otherwise peak, by centrifugal barriers that oppose the escape of low-energy photoelectrons through the outer layers of atoms or molecules.[2] Such "delayed onsets" of enhanced absorption depend sensitively on the balance between the centrifugal repulsion and the net Coulomb attraction on the escaping electron, a balance that tilts rapidly in favor of the attraction as one proceeds to positive ions along each isoelectronic sequence.

The influence of these major features (Cooper zeros and centrifugal barriers) can be assessed within the independent-electron approximation. The lesser effects of polarization of the ionic residue and of spin-orbit interaction on the photoelectron can also be treated by including appropriate terms in the effective field acting on a slow photoelectron.[4] Note that spin-orbit coupling acts mainly near the nucleus (specifically within the K-shell radius).[5] Its effect is, however, obscured -- indeed apparently suppressed -- by countervailing exchange interactions as a photoelectron escapes through an outer incomplete shell, reappearing eventually when that shell is left behind.[4]

On the other hand, important effects on photoabsorption spectra near threshold depend essentially on electron correlations. They will be outlined in Sections 3 and 4 and treated in relevant lectures.

QUANTUM DEFECT THEORIES

It is well known that the (singly excited) energy levels of the (monovalent) alkali atoms are represented by the hydrogen-like formula $-Ry/(n-\mu_{lj})^2$, whose parameter μ_{lj} is called a "Rydberg correction" or "quantum defect". The parameter μ_{lj} depends on the orbital quantum number l and on the spin-orbit quantum j but varies hardly at all from one level to another of the same series. Knowledge of μ_{lj} thus *summarizes* the results of measurements of all lines of a series; it also determines the phase shift $\delta_{lj}=\pi\mu_{lj}$ of an ionized electron's wave function.

It is somewhat less familiar that the *transition probabilities* to alternative levels of a series, in a photoabsorption or collision process, are given by the product of a single coefficient I_{lj} and of a standard function $\bar{F}(n,\mu_{lj})$. Parameters I_{lj}, weakly dependent on n or on energy, thus combine with the quantum defects μ_{lj} to parametrize the quantitative properties of all the states of a "channel", which consists of a Rydberg series of discrete levels and of its adjoining continuum, as detailed in Rau's lectures.

The analogous systematics of energy levels and transition probabilities, for atoms with more than one valence electron and for molecules, is less straightforward than for alkalis, mainly because the orbital and total angular momenta (l,j) of a photoexcited electron do not suffice to identify a channel. Excitation channels with the same conserved quantum numbers are generally "coupled" in the sense that excitation may pass from one channel to another. Such a group of channels should accordingly be treated as a unit. The relevant treatment has been developed *analytically* by Seaton in terms of a larger, but limited, set of parameters,[6] as described in Rau's lectures and outlined in later sections of this Introduction.

The conceptual approaches outlined in the present section have spawned a still broader and ever expanding body of theoretical developments, loosely labeled as quantum defect theories. The current diversity of subject and intent among these developments suggests referring to them as techniques, technologies or, simply, as "approaches".

SYMMETRY OF CORRELATED ELECTRONS

The correlations among electrons in closed shells are constrained by angular and spin symmetries to such an extent that their net effect is generally represented adequately by an adjusted effective potential acting on each electron. The correlation effects among electrons in one or more open shells are much more prominent and diverse. The correlations of angular and spin coordinates have long been represented in terms of angular momentum couplings of orbital and spin coordinates. Evidence of correlations among their *radial* coordinates failed instead to emerge clearly until the spectrum of an *electron pair* excited outside the ground state shell of He, at ~60eV, was first observed in 1963.[7]

This spectrum, obtained by absorption of synchrotron light, displayed a single major Rydberg series where three had been expected, corresponding to alternative distributions of angular momentum between the electrons. The concentration of spectral intensity into a single series implies the existence of a novel selection rule, which reflects the previously disregarded symmetry under the interchange of the radial coordinates of the electrons. Analogs of the correlated states thus observed by photoexcitation were soon identified in resonant states formed by electron collision with atoms and molecules, thus greatly increasing their interest.

Calculations of energy levels and intensities of two-electron transitions in He were carried out by existing channel-mixing procedures on the heels of the initial observations. They confirmed and extended the experimental evidence in remarkable detail without, however, *tracing the origin* of the newly discovered selection rules. This origin was traced later by a new approach to the solution of a two-electron, six coordinate, Schrödinger equation, which preserves manifestly its symmetry under permutation of the two excited electrons. This approach is developed in a group of lectures arranged by A. F. Starace; some of its broad implications will be anticipated in the next lecture.

THEORY OF EXCITATION INTO ALTERNATIVE CHANNELS

As an excited electron moves away from an ionic core, its interaction with the core often changes qualitatively with increasing radial distance. This phenomenon emerged sharply from Herzberg's study of photoabsorption by molecular hydrogen.[8] The excited electron rotates *with* the ionic core at short radial distances but *independently* of it when far away. The energy and the angular momentum of the excited molecule are thus allotted between electron and core as the electron travels away from the core in its radial motion, rather than in the initial process of photoabsorption. The existence of different regimes of interaction ("Hund's cases") between an electron and a molecular core had been known for a long time, but the transition between them had hardly been treated adequately prior to Herzberg's observations.

Two concurrent remarks provide now a quantitative treatment of the allotment of energy and angular momentum between electron and core in the excitation process: a) The wave functions appropriate to small and to large radial distances, respectively, are connected by a "frame transformation" which amounts mainly to a standard rearrangement in the combination of the angular momenta of different motions. b) Analytic interpolation between small and large radial distances is provided by Seaton's Multichannel Quantum Defect Theory (MQDT).[6] Both of these aspects are treated in lectures by Rau, Jungen and Greene.

The successful treatment of photoabsorption by H_2 [8] proved generally applicable not only to photoexcitation, photoionization and photodissociation but to collision processes as well. In each of these processes one deals with alternative sets of states appropriate to small and large radial distances, respectively. The total energy and angular momentum are variously allotted between two, or more, "fragments" that are well separated and interact weakly at large distances. At short ranges strong interactions prevail which weld the fragments into a single "complex"; energy and angular momentum are then variously distributed among internal degrees of freedom of the complex rather than among the fragments. These two sets of states are generally related by a frame transformation. Quantum defect procedures, or variants thereof, connect the two sets.

The essence of the frame transformation procedures lies in considering separately the internal mechanics of a complex, the fragmented states and their connections. Quantum numbers, quantum defects (or equivalent phase shifts) and frame transformation matrices serve to parametrize each process, together with transition parameters such as the I_{lj} introduced in Sec. 2.

CALCULATION OF MQDT AND ANALOGOUS PARAMETERS

Ability to treat the motion of an electron in the field of an ionic core *analytically* underlies the success of MQDT in correlating a large mass of spectral and collisional data. Ability to treat similarly the motion of any few separated fragments extends the success. Such treatments of well separated objects needs, however, to be complemented by the contrasting task of calculating the *parameters* of MQDT *ab initio,* unless one merely infers their values from experimental data.

Ab initio calculation of parameters deals generally with a *large number* of *narrowly spaced particles* forming a complex, i.e., confined within a *limited volume.* Dynamical parameters that depend on multiparticle interactions within a limited volume of space have been called "short range parameters".

An important subset of short range parameters consists of the element $K_{ij}^{(s)}$ of the "short range reaction matrix" (real and symmetric). Each of these elements determines the probability amplitude of fragment transitions from a channel j into a channel i under the influence of short-range interactions. Short range transition and scattering matrices, $T^{(s)}$ and $S^{(s)}$, are calculated from $K^{(s)}$. The corresponding matrices K,T and S, which include the effect of long range interactions, are obtained by combining $\{K^{(s)},T^{(s)},S^{(s)}\}$ with appropriate MQDT parameters.

Confinement within a *small* volume, comparable to that occupied in the complex' ground state, affords the opportunity of calculating parameters far more *economically* than would be feasible in a direct calculation of spectral levels or collision cross sections, a calculation that would extend over a far larger volume. Parameters calculated within a small volume, where interaction and kinetic energies are stronger, are correspondingly *insensitive* to variations of the total energy. These calculations amount, in the main, to solving a multiparticle Schrödinger equation within a *finite volume,* subject to boundary conditions on the volume's surface. Procedures for solving problems thus circumscribed are loosely called "R-matrix methods". These procedures are the subject of a lecture by C. H. Greene.

A particular factor is noted here that often limits the accuracy of short range parameters. Consider for example the probability amplitude for photo-excitation, -ionization, or -dissociation of an atom or molecule from its ground state, i.e., for processes whose initial step is confined within the volume occupied by the ground state. This amplitude is calculated usually from wave functions Ψ_g and Ψ_f of the ground and final states, obtained as *separate* solutions of a Schrödinger equation. Consistency and accuracy require that Ψ_g and Ψ_f embody *equivalent approximations,* e.g., consistent selection of configuration mixing and consistent allowance for electron exchange.[9] Procedures that enforce consistency have been developed, by which one calculates the "transition density" $\Psi_f^* \Psi_g$ in a single step rather than as a product of separately obtained wave functions. In spite of notable successes[10] such procedures have not yet been developed adequately[11] within the R-matrix framework.

GEOMETRICAL FEATURES OF ATOMIC PROCESSES

Geometrical aspects of atomic phenomena have provided both qualitative insight and quantitative information. They include the angular distribution of photoelectrons or other reaction fragments, the polarization of light emitted or scattered by atoms or molecules and the spin orientation and other multipole moments of reactants. Theoretical work has centered on sorting out the dependence of each measured variable on dynamical parameters from the influence of purely geometrical elements.

Theory is straightforward in simple cases. For example, the yield of photoelectrons ejected from atomic hydrogen (or from the valence shells of alkalis) is proportional to the squared product $(\vec{v} \cdot \vec{p})^2$ of the photoelectron's velocity and of the light polarization vector $\vec{p}$. On the other hand any anisotropy of the target and/or of the ionic residue must be averaged or else scored and analyzed. Analogous or greater complications arise in the presence of highly structured reactants.

The theoretical tools of the geometric analysis of nontrivial processes consist mainly of the Wigner-Racah algebra of angular momenta. Physical and geometric insight are nevertheless required for effective design and presentation of diverse applications. (The same tools serve to evaluate the angular integral in the matrix elements of particle interactions). Analysis of collision and of photoprocesses in terms of the angular momentum transferred between fragments as well as the classification of contributing mechanism as "parity favored" or "unfavored" have proved useful.[12]

Examples of the geometric analysis of molecular processes are described in a lecture by Greene.

RYDBERG STATES IN EXTERNAL FIELDS

The geometric aspects of atoms and molecules are enriched by the application of external (electric and/or magnetic) fields, which break the rotational invariance of isolated atomic systems. (Analogous, if more complex, breaches of symmetry occur for atomic processes embedded in a crystal[13] or in a cluster).

The MQDT approach proves appropriate to the study of atoms in external fields for the following reason: Fields of laboratory strength exert only a weak perturbation within the short ranges where the multi-particle aspects of atoms are relevant and atomic interactions are strong and complex. They are instead comparable to the weak fragment interactions at large ranges where the interactions are weaker and their geometrical aspects simpler. Rydberg states, where the latter circumstances prevail, have attracted much attention in recent years as described in O'Mahony's lecture. I anticipate here that the motion of a Rydberg electron is separable in an electric field though not so in a magnetic field.

The contrast of dynamical circumstances in the short range and in the long range ("Rydberg") regions is reflected in contrasting geometrical circumstances. On the one hand an external field barely perturbs the rotational invariance of an atomic system at short ranges. On the other hand the dynamical complexities prevailing at short ranges hardly perturb the combined action of the external field and of fragment interactions at long ranges. The interplay of influences prevailing in the two regions is treated properly by the relevant frame transformation.[14]

MULTIPHOTON PROCESSES

The action of electromagnetic radiations on matter was treated adequately by low-order perturbation theory prior to the advent of lasers. Laser action, however, requires at least high-order perturbation procedures, which are further complicated when the absorption of photons is accompanied by stimulated emission. Agostini's lectures in this volume describe the current status of multiphoton phenomenology within this framework.

A more basic approach views a laser beam as closely coupled to an atom or molecule, forming with it an *excited complex*. Each fragmentation channel of this complex is labelled by the number of photons in the beam (or by the number previously absorbed) in addition to the quantum numbers of its atomic component. MQDT procedures are appropriate to this approach because absorption and emission of a photon energy $\hbar\omega$ by any electron are confined to the region of space where the electron can exchange a sufficient momentum with the atomic field. The early development of this novel application of MQDT is reported in the lecture by A. Giusti-Suzor.

Note, however, that this new theoretical development does not bear directly on the phenomena described by Agostini, since it deals with phenomena in the "frequency domain" rather than in the "time domain". Specifically the present theory deals with the action of a single-mode cw laser on an atomic system. High power lasers are instead pulsed and multi-mode; their fields may be regarded in principle as partially coherent superpositions of single-mode cw fields but quantitative descriptions of such superpositions are not available in practice.

REFERENCES

1. J.W. Cooper, Phys. Rev. 128, 861 (1962).

2. (a) U. Fano and J.W. Cooper, Rev. Mod. Phys. 40, 441 (1968) and (b) 41, 724 (1969); c) A.F. Starace, in Encyclopedia of Physics, Vol. 31, W. Mehlhorn ed. (Springer, Berlin 1982) pp. 1-121.

3. S.D. Oh and R.H. Pratt, Phys. Rev. A, 34, 2486 (1986); R.Y. Yin and R.H. Pratt, Phys. Rev. A, 35, 1154 (1987); R.H. Pratt, R.Y. Yin and X. Liang, Phys. Rev. A, 35, 1450 (1987).

4. Section 6 of Ref. 2 (a).

5. U. Fano, C.E. Theodosion and J.L. Dehmer, Rev. Mod. Phys., 48, 49 (1976), Sec. 6A.

6. M.J. Seaton, Proc. Phys. Soc. (London) 88, 801 (1966); Rep. Prog. Phys., 46, 167 (1983).

7. R.P. Madden and K. Codling, Astrophys. J. 140, 364 (1965).

8. G. Herzberg and Ch. Jungen, J. Mol. Spectr. 41, 425 (1972); C.H. Greene and Ch. Jungen, Adv. in Atom. Mol. Phys. 21, 51 (1985).

9. A.F. Starace, Phys. Rev. A, 3, 1242 (1971).

10. M. Ya Amusia and N.A. Cherepkov, Case Studies in Atomic Physics, 5, 47-179 (1975).

11. T.N. Chang and U. Fano, Phys. Rev. A 13, 263 (1976).

12. C.W. Lee and U. Fano, Phys. Rev. A, 36, 66 (1987) and ref. therein.

13. U. Fano, Phys. Rev. Lett., 31, 234 (1973); G. Strinati, Phys. Rev. B, 18, 4096 and 4104 (1978).

14. D.A. Harmin, Phys. Rev. A, 26, 2656 (1982) and P. O'Mahony's lecture.

GENERAL FEATURES OF COLLISIONS AND SPECTRA

U. Fano

James Franck Institute and Department of Physics
University of Chicago, Chicago, IL 60637

FORMATION AND FRAGMENTATION OF COMPLEXES

This lecture will provide a further introduction to the Theory Symposium, describing features that are common to all collision and spectral phenomena of matter in gas phase. Many of these features are also relevant to clusters, viewed as large molecules, as well as to condensed phases of matter. Thus, e.g., ejection of a photoelectron from a crystal cell may be viewed as a photoionization of that cell modified by its crystalline environment.

Two reciprocal stages may be distinguished in a collision process, namely, the formation of a "complex" from initially separate reactants (one of which might be a single electron) and "fragmentation" of the complex into separate products. Both of these stages are also identified in a photoexcited state, with the variant that fragmentation lacks sufficient energy to reach completion, the fragments being forced back into the

complex again and again. An alternative view regards a photoprocess as the collision of a photon with matter; the complex resulting from photoabsorption may or may not reemit a photon and may or may not proceed to full fragmentation, depending on the available energy.

Even though this lecture centers on the formation and dynamics of complexes, it should stress that collisions do *not* lead to the formation of a complex when their impact parameter (or equivalent angular momentum) is sufficiently large. (The precise meaning of this condition requires analysis in each specific application, of course). The effects of collisions with large impact parameter result from long range fields acting between the reactants. They are accordingly analogous to the effects of photoabsorption and can often be treated similarly in the Born approximation. Collisions of fast charged particles afford a typical, familiar example, in which the weakness of perturbation rests on the short duration of the collision.

The process of formation of a complex from separate constituents involves generally a structural change, whose complication varies greatly depending on circumstances. When a single electron combines with an ion the structural change may reduce to a recoupling of orbital and spin angular momenta, together with a slight change of the ion's charge distribution to incorporate the electron's own charge with lower energy. Complex formation by two open-shell atoms with low kinetic energy leads readily to formation of a chemical bond, whereas a repulsive antibonding action necessarily results if neither atom has an open valence shell. Complex formation by two, or more, atoms with high kinetic energy leads instead to a deep rearrangement of their structures.

Analysis of the structural changes that may be involved in the formation or fragmentation of a complex is facilitated by the construction of "correlation diagrams." These diagrams connect alternative structures of the complex to those of fragments according to a scheme that allows the smoothest evolution of their respective orbital wavefunctions. Fig. 1 shows a very simple prototype diagram illustrating the correlation of two alternative states of a pair of He fragments that combine to form a complex which is isoelectronic (in its "united atom" limit) to a Be atom. Note particularly the crossing of the diagram lines which results from the following circumstance: The ground states of two He atoms have closed shells and hence combine -- if sufficient energy is available -- to form an *excited* state of the complex, with the united atom configuration $1s^2 2p^2$. (The excitation results from spin symmetry, reflecting the Pauli exclusion which requires some electrons to occupy antibonding orbitals. The configuration labels on successive steps of the diagram correspond to successive stages of the system's evolution from its united atom to its separate atom limit.) Conversely a pair of He atoms in excited metastable states can readily form two bonds evolving into the ground state configuration $1s^2 2s^2$ of the complex. Analogous circumstances occur most frequently, underlying the structure and energy evolutions in collision processes.

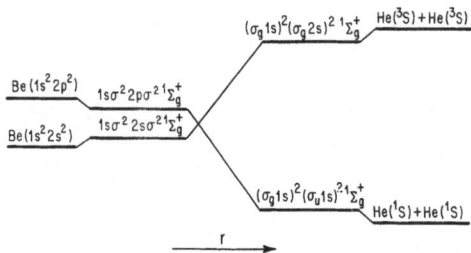

Fig.1   Correlation diagram of a 4-electron He_2 molecule (Adapted from G. Herzberg, Spectra of Diatomic Molecules, Van Nostrand, New York, 1950).

Developing a unified analytical representation of the manifold processes of formation and fragmentation of complexes is the task outlined in the following sections. Its initial step, rather familiar in special cases but less so in its full generality, will be introduced in the context of striking experimental evidence. The comprehensive view outlined in this lecture has not yet led to a fully developed theory. It should nevertheless provide a framework in which to place the subjects of following lectures.

Collisions will be treated here in the time-independent form that is equally appropriate to the theory of spectral levels. Time-dependent formulations are, however, frequently utilized in the treatment of collisions between atoms or molecules. These alternative formulations are basically equivalent but lend themselves to emphasizing different aspects of phenomena.

ADIABATIC AND NON-ADIABATIC EVOLUTION OF STRUCTURES. QUASI-DEGENERACIES AND AVOIDED CROSSINGS.

A striking example of the evolution of a complex' structure is afforded by $Ar^+ + Ar$ collisions with energies in the 10-100 keV range. Since these energies are sufficient to pulverize the electron structures of projectile and target, conventional wisdom might suggest that electrons would be stripped off massively in smoothly increasing average numbers as the impact parameter is reduced. Experiments revealed a quite different picture.

Measurements of the incident and final energies of the Ar^+ projectile and of its deflection θ, combined with conservation of momentum and energy in the collision, determine the energy Q transferred to the electronic structures in each collision. The mean value $\bar{Q}$ was measured as a function of θ, and θ was then converted to the distance r_0 of closest approach of the two Ar nuclei through a classical analysis of the nuclear trajectories. Remarkably the results for different incident energies were then found nearly to coincide in the plot of $\bar{Q}$ vs. r_0 shown in Fig. 2.

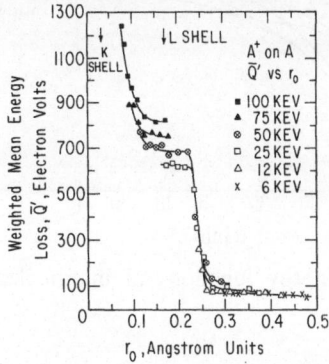

Fig.2   Energy transfer to electrons in $Ar^+ + Ar$ collisions as a function of closest internuclear distance (Ref. 1).

This most instructive plot[1] shows $\bar{Q}$ to remain nearly constant, as long as r_0 exceeds ~0.25 Å, at a level ≤100 Å sufficient to splash out a few valence electrons. On the other hand $\bar{Q}$ rises *suddenly* by a factor ~5, as r_0 drops below 0.25 by ≤10%. Ref. 1 attributed the discontinuity of $\bar{Q}$ correctly to onset of excitations from the L shell (the rise amounts to the binding of two L electrons) even though its sharpness contrasts with the smoothness of each shell's structure and the L shells already overlap substantially at $r_0 > 0.25$ Å. Fig. 2 thus documents a sharply defined response in the dynamics of seemingly smooth atomic structures.

The occurrence and significance of critical internuclear distances in the formation of complexes had been understood prior to Ref. 1, mainly in the context of valence shell phenomena. Fig. 3 shows a quantitative version of the correlation diagram of Fig. 1 pertaining to $He^+ + He$ collisions.[2] Each curve represents the energy level of a three-electron configuration of the He_2^+ molecular complex as a function of its internuclear distance; it also represents the potential energy that governs the internuclear motion of the He_2^+ molecule in the specified electronic structure. Analogous diagrams are provided by more laborious calculations of quantum chemistry for moderately complicated molecular complexes. Qualitative and semiquantitative rules developed by Mulliken long ago have also provided effective guidance in constructing such diagrams.

Two main lessons are drawn from a large body of collision experiments with reference to diagrams analogous to Fig. 3:

a)   The electronic structures of molecular complexes, identified by independent-electron and related quantum numbers, remain *largely invariant* under variations of internuclear distances, with energies accessible to rather easy estimation.

b)   Structural changes are then *largely confined* to *loci of degeneracy,* identified in Fig. 3 by *curve crossings.*

Degeneracy of two potential curves -- actually quasi-degeneracy in most cases -- permits structural changes to proceed by energy transfers *among electrons,* without requiring substantial energy exchanges with the motion of the (heavy and slow) nuclei. In accordance with b), the internuclear distance of the sharp increase of $\bar{Q}$ in Fig. 2 is seen to coincide with the crossing of the Ar+Ar energy levels $4f \sigma$ and $3s \sigma$ in Fig. 4; the σ label of these levels accounts for the fact that *just two* electrons are excited from the L to the M shell at this point of degeneracy. Structural changes also occur in violation of the rules a) and b), but those rules do provide a framework for the analysis of collision processes. Remarkably these rules prove also relevant to electron collisions as will emerge in Sec. 3.

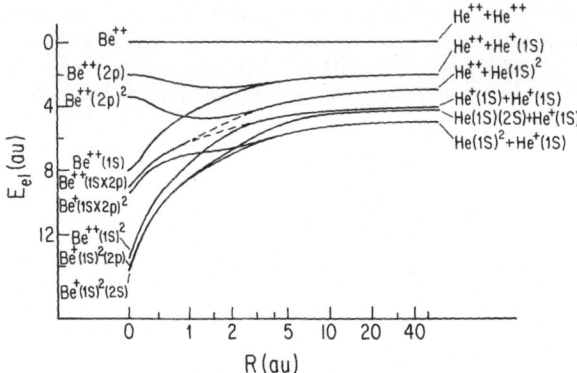

Fig. 3   Diabatic potential energy functions of internuclear distance for He_2^+ ions. (From Ref. 2).

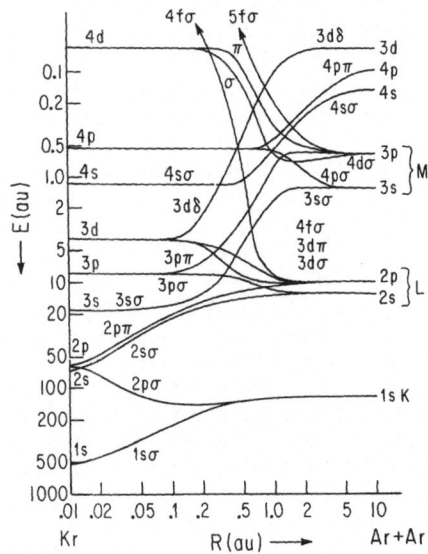

Fig.4   Diabatic potential energy functions for the Ar_2 complex. [Courtesy W. Lichten, from Phys. Rev. Letters <u>14</u>, 627 (1965)].

Level Crossing Effects

The potential curves of Figs. 3 and 4 have been drawn to represent the energy of an electronic structure of the complex that preserves its independent-electron quantum numbers. On the other hand, calculations of the electronic energy eigenvalues for each internuclear distance include electron interaction effects that cause a major departure from the independent-electron approximation near each point of degeneracy (unless the curves are labeled by different values of an *exact* constant of the motion). Plots of such "adiabatic" eigenvalues show *avoided crossings* of the type depicted schematically in Fig. 5.

The formation or fragmentation of a complex proceeds along the *adiabatic* path, accompanied by an interchange of structures, if the internuclear distance varies sufficiently *slowly*. It proceeds instead along the *diabatic* path which preserves the

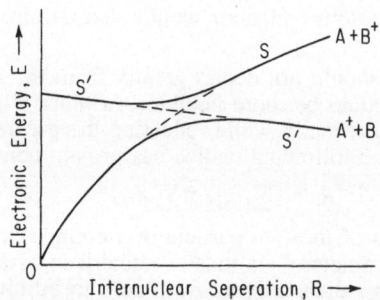

Fig.5   Crossing of potential curves.   Two potential curves for the states S′ and S
may cross in a certain approximation (such as in a single-configuration
molecular orbital theory).   In a higher approximation, the curves repel each
other.   If the atoms approach each other slowly in state S, an adiabatic transition from S to S′ will occur.   If they approach each other rapidly, a diabatic transition from S to S′ will occur.   (From Ref. 2).

electronic structure if that distance varies sufficiently *fast*. At intermediate speeds the
complex' wavefunction *splits* into two channels with complementary amplitudes. Such
splittings form the very core of inelastic and reactive collisions because *only elastic*
collisions result if the complex evolves only adiabatically or only diabatically.
Specific examples are treated in the lectures by Macek and Lin.

Splitting of the complex' wave function into two channels is qualitatively, and
also analytically, equivalent to the partial reflection and partial transmission of a wave
function by a potential barrier. The adiabatic approximation corresponds to full
reflection at low incident energy, the diabatic case to full transmission well above the
barrier top. In Fig. 5 the barrier crest would run between the two adiabatic curves.

Problems involving a symmetric potential barrier can be solved by constructing
symmetric and antisymmetric degenerate standing-wave solutions and then superposing
them to represent incidence from one side only, as detailed in the following lecture by
Rau. The example of a barrier with parabolic potential is solved analytically by Parabolic Cylinder functions which coincide with Coulomb field functions of the squared
abscissa. The best known Landau-Zener-Stückelberg[3] treatment of level crossing is
modeled to fit a parabolic barrier and leads to probability amplitudes expressed in
terms of the normalization coefficients of Coulomb functions. Extensions to asymmetric barriers can be treated in similar manner.[4]

The name "adiabatic procedure" refers strictly to the treatment of a process that
develops entirely along adiabatic potential curves. In practice, however, it has the
broader implication of treating the evolution of a process that follows largely adiabatic
potential curves with interruptions by localized non-adiabatic steps.

THE RADIAL PARAMETER

Sec. 2 has parametrized the evolution of a two-atom complex through the internuclear distance, familiar in molecular physics as the adiabatic parameter of the Born-
Oppenheimer approximation. This parametrization meets with three important limitations -- to be removed here through the introduction of a more general and flexible
parameter -- namely:

a)   The adiabatic Born-Oppenheimer approximation breaks down at very large internuclear distances where the electron motions are greatly influenced by the motion
of nuclei.

b)   Alternative fragmentations of a polyatomic complex should be parametrized by
different internuclear distances.

c) Ionization and electron-atom collisions would also require a different parametrization.

A desirable parameter should not depart greatly from the internuclear distance in the diatomic example but should be more flexible. In other words it should treat all particles of a complex symmetrically while reflecting the greater inertial role of nuclei in a molecular complex. The following choice has proven convenient,[5]

$$R = [\sum_i m_i r_i^2 / \sum_i m_i]^{1/2},$$ (1)

where m_i indicates the mass of the i-th particle of a complex and r_i its distance from the *center of mass*. The parameter R is thus characterized as an effective *radius of inertia* of the complex, with values ranging from zero to infinity. It may be viewed as an observable of a mechanical system and it can be defined without reference to the number or identity of constituent particles.

The name of the parameter R, "hyperspherical radius", originates from a formulation of the Schrödinger equation for He which views the pair of electron position vectors $(\vec{r}_1, \vec{r}_2)$ as a single 6-dimensional vector $\vec{R}$, with six independent hyper-polar coordinates $\{R, \alpha, \theta_1, \phi_1, \theta_2, \phi_2\}$. The novel element of this representation lies in the transformation of the separate radial coordinates (r_1, r_2) into the fully polar representation (R, α), with

$$r_1 = R\cos\alpha, \quad r_2 = R\sin\alpha, \quad \alpha = \arctan(r_2/r_1).$$ (2)

A wavefunction's dependence on the coordinate $0 \le \alpha \le \pi/2$ represents the *radial correlation* of the electron pair, as its dependence on the angle $\theta_{12} = \arccos(\hat{r}_1 \cdot \hat{r}_2)$ represents the pair's *angular correlation*. Ionization of the He atom corresponds to either limit $\alpha \to 0$ or $\alpha \to \pi/2$; identity of the electrons implies that the range of α may be restricted to $\alpha \le \pi/4$ with spin-dependent boundary conditions imposed at $\alpha = \pi/4$.

Extension of the hyperspherical treatment of He to a molecular complex consisting of N nuclei and electrons with different masses m_i involves the following:

1) Weighting the position vector $\vec{r}_i$ of each particle by the factor $\sqrt{m_i}$, a frequent practice in quantum chemistry.

2) Using the center of mass as the origin of the $3(N-1)$-dimensional vector $\vec{R}$.

3) Including, among the $3N-4$ angles that identify the direction of $\vec{R}$, $N-2$ angles $\alpha_j \le \pi/2$ which identify all ratios among the r_i, as well as the $2(N-1)$ independent angles that identify the N space directions $\hat{r}_i \equiv (\theta_i, \phi_i)$.

Key features of the resulting procedure are:

4) The angular coordinates $\{\alpha_j, \hat{r}_i\}$ are orthogonal to $\vec{R}$, being independent of the magnitude of R. This feature *removes the limitation* a) to the Born-Oppenheimer parameter whose variations influence instead each electron's distance from the nuclei.[6]

5) Transformations of the $\{\alpha_j, \hat{r}_i\}$, dubbed "kinematic rotations", are by the same token independent of R. These transformations, including changes of the coordinate system besides simple rotations, are represented by standard but elaborate analytical formulas.[7]

6) Different systems of coordinates $\{\alpha_j, \hat{r}_i\}$ characterize alternative fragmentations of a complex; one of the α_j vanishes in the fragmentation limit, $R \to \infty$.[7]

7) The treatment is equally applicable to electron and to atom collisions and to fragmentation into electrons, atoms or molecules.

The Hyperspherical Hamiltonian
_____

Treating the dynamics of a complex by means of one radial parameter and of $3N-4$ angular variables $\{\alpha_j, \hat{r}_i\}$ allows one to write its Schrödinger equation in a form parallel to that of the H atom, namely,

$$\left\{ \frac{\hbar^2}{2\sum_i m_i} \left[-\frac{\partial^2}{\partial R^2} + \frac{\vec{\Lambda}^2 + (3N-4)(3N-6)/4}{R^2} \right] + \frac{C(\alpha_j, \hat{r}_i)}{R} \right\} \Psi = E\Psi.$$ (3)

Note three main features of Eq. (3):

A) The kinetic energy operator separates into radial and centrifugal terms, the latter one $\propto R^{-2}$ and linear in a squared "grand angular momentum" $\vec{\Lambda}^2$, whose eigenvalues and eigenfunctions are *known analytically*[7] for any finite set of angles $\{\alpha_j, \hat{r}_i\}$. The centrifugal term added to $\vec{\Lambda}^2/R^2$ is *positive* for all $N > 2$ but it vanishes in the elementary case of two particles.

B) The potential energy of any molecular complex is $\propto R^{-1}$ because it consists entirely of Coulomb interactions among electrons and nuclei. It becomes negligible as compared to the centrifugal term as $R \to 0$. The key difference between Eq. (3) and the hydrogen equation lies in the noncommutability of $\vec{\Lambda}^2$ and $C(\alpha_j, \hat{r}_i)$ which controls the phenomenology of fragmentation as will be indicated in Sec. 5.

C) The spectrum of $\vec{\Lambda}^2$ is *discrete,* as it is for the squared orbital momentum of H, because the range of each of the angular coordinates $\{\alpha_j, \hat{r}_i\}$ is finite.

The separation in Eq. (3) of the kinetic energy operator into terms proportional to $\partial^2/\partial R^2$ and to $\vec{\Lambda}^2$, and the analogous separation of $\vec{\Lambda}^2$ into contributions from second order derivatives with respect to each of the variables $\{\alpha_j, \hat{r}_i\}$, require Ψ to be renormalized by the Jacobian factor $[\partial(\vec{r}_1, ..\vec{r}_i ..)/\partial(R, \alpha_j, \hat{r}_i)]^{1/2}$ analogous to the renormalization of the radial function of H by a factor r.

Note finally an additional implication of representing radial correlations in terms of the angular coordinates α_j in the hyperspherical formulation (3). As the α_j are treated much in the same way as the space coordinates $\hat{r}_i \equiv (\theta_i, \phi_i)$, the algebraic evaluation of the dependence of the Hamiltonian matrix on angular correlations, familiar in the theory of atomic spectra, extends now to its dependence on radial correlations. This feature, described in Cavagnero's lecture, contrasts with the earlier numerical evaluation of radial correlation integrals.

BOUND STATES, PHOTOFRAGMENTATION AND COLLISIONS

Wave functions that represent alternative processes involving a molecular complex and provide information on their respective probabilities are constructed from a *single complete set* of standing-wave solutions $\{\Psi_{E\alpha}\}$ of (3) for the relevant energies E. The index α represents a complete set of angular momentum and other quantum numbers sufficient to specify a particular solution of the indicial equation of (3) at $R \to 0$. Each Ψ_α resolves generally into a number of fragmentation channels at large R. The solution of (3) that represents a specific process is represented by a superposition $\sum_\alpha A_\alpha \Psi_{E\alpha}$ with coefficients A_α selected to satisfy the relevant boundary conditions at $R \to \infty$.

Bound state eigenfunctions must converge to zero in *each channel* as $R \to \infty$. This condition leads to a homogeneous system of algebraic equations in the A_α, which has solutions only for discrete eigenvalues of E, as usual.

Photofragmentation and collision processes involve wavefunctions for energies E in the continuous spectrum of Eq. (3), where fragmentation proceeds to $R \to \infty$ in one, or more, "open" channels. The wavefunction components in the remaining "closed" channels must still converge to zero at $R \to \infty$, as required for bound states, but this condition no longer suffices to identify the relevant state.

A collision process is generally specified by requiring two reactants to approach one another in a channel i with relative momentum $\vec{p}_i$ and with specified magnitudes and orientations of their respective angular momenta and of their internal energies. The wavefunction with these characteristics is then to be expanded into partial waves and to be represented in the form $\sum A_\alpha^{(i)} \Psi_\alpha$, with the additional condition that the amplitude of each *ingoing* component wave *vanish* in any channel *other than i*. The resulting amplitude of the *outgoing* component wave of any channel j determines the cross section for fragmentation into that channel.

In the case of photofragmentation, a particular wavefunction $\sum_\alpha A_\alpha^{(j)} \Psi_\alpha$ is determined by the requirement that the amplitude of the *outgoing* component waves vanish in *all but one* specified channel j.

Procedures to determine the coefficients A_α under such alternative circumstances are described in Rau's lectures. These procedures center on the *Jost matrices,* whose significant role is introduced in Macek's introductory lecture to this volume.

PHENOMENOLOGY OF ADIABATIC PROCEDURES

The adiabatic procedures originate from the Born-Oppenheimer approximation of molecular physics, usually justified by the small ratio of electron to nuclear masses. However, these procedures have also proved valuable in the context of multi-electron excitations where no such small ratio of masses occurs. (Indeed adiabatic procedures and hyperspherical coordinates were first combined in the treatment of two-electron excitations of He, as described in Starace's and later lectures.) Adiabatic treatment of Eq. (3) is now seen to rest on the slow shift in the balance of its terms proportional to R^{-1} and R^{-2}.

The adiabatic treatments developed thus far cast eigenfunctions Ψ_α of the *exact* Eq. (3) in the approximate separable form

$$\Psi_\alpha(R,\alpha_j,\hat{r}_i) = F_\alpha(R)\Phi_\alpha(R;\alpha_j,\hat{r}_i). \tag{4}$$

Here Φ_α depends on R only as a parameter, namely, as the *fixed* value of R in the equation obeyed by Φ_α, which is obtained from (3) by retaining only its dependence on $(\alpha_j,\vec{r}_i)$ at constant R, namely,

$$[\frac{\hbar^2\vec{\Lambda}^2}{2\sum_i m_i R^2} + \frac{C(\alpha_j,\hat{r}_i)}{R}]\, \Phi_\alpha(R;\alpha_j,\hat{r}_i) = U_\alpha(R)\Phi_\alpha(R;\alpha_j,\hat{r}_i). \tag{5}$$

Its eigenvalue $U_\alpha(R)$ serves then as the potential in the equation of motion along R with the eigenfunction $F_\alpha(R)$.

Construction of the eigenfunction (4) begins at $R\sim0$, where the potential C/R is negligible as compared to $1/R^2$ terms. In this limiting range $\Phi_\alpha(\alpha_j,\hat{r}_i)$ is an eigenfunction of $\vec{\Lambda}^2$ with quantum number λ_α, and eigenvalue $[\lambda_\alpha+(3N-5)/2]^2-1/4$. The indicial equation yields then $F_\alpha(R)\propto R^{\lambda_\alpha+(3N-4)/2}$. As R increases, the Coulomb potential $C(\alpha_j,\hat{r}_i)/R$ is taken into account by a degenerate perturbation treatment, that is, in the form C_α/R, where C_α is an eigenvalue of the matrix of $C(\alpha_j,\hat{r}_i)$ in the base of degenerate eigenfunctions of $\vec{\Lambda}^2$ with the same index λ_α. Remarkably, numerical solutions of (3) for doubly excited He have shown this perturbation treatment to remain valid up to moderately large values of R. Numerical solution becomes necessary as R increases further. Solution of (3) also requires an evaluation of non-adiabatic effects which cause each Ψ_α to resolve into different fragmentation channel as anticipated above. The component of Ψ_α in each fragmentation channel is generally confined within a "potential valley" where $-C(\alpha_j,\hat{r}_i)$ is large.

The effectiveness of this procedure has, however, been limited by inherent features, which make it now obsolescent regardless of its achievements described in following lectures, namely:

a) Immediately relevant is an accumulation of avoided crossings of potentials $U_\alpha(R)$. The resulting non-adiabatic processes are localized along "potential ridges" where the (positive) quantity $-C(\alpha_j,\hat{r}_i)$ is lowest. Their aggregate effect on the set of Ψ_α with the separable structure (4) has defied evaluation.

b) Evidence from observations of both collisions and spectra has shown resonant states to be also localized on potential ridges. Such a localization of wave functions in high potential regions of coordinates has also defied interpretation within the scope of the adiabatic approximation (4).

c) Last, but not least, the very structure of (4) appears clearly unrealistic because it implies uniformity of the wavelength of $F_\alpha(R)$ over each hypersphere $R = $ const, even though the potential $C(\alpha_j,\hat{r}_i)/R$ varies greatly over that surface.

A suitable point of attack, for correcting these and other minor deficiencies of the approximation (4), appeared to lie in adapting the coordinate system to the potential function $C(\alpha_j,\hat{r}_i)$, so as not to constrain constant-phase surfaces of the radial motion to a spherical shape. Establishing a correlation between the radial and angular coordinates must, however, involve the interplay of the terms of Eq. (3) with *different* dependence on R. Simultaneous resolution of all three difficulties -- a), b) and c) -- has now emerged from a perturbative, rather than adiabatic, approach in the context of the Schrödinger equation

$$[-\Delta - \frac{a^2}{r} + b^2 r^2 \sin^2\theta]\Psi(r,\theta) = E\Psi(r,\theta), \tag{6}$$

which shares with (3) the characteristics of nonseparability and of an infinite range of r. This equation, which pertains to a H atom in a diamagnetic field and is discussed further in O'Mahony's lecture, is known to have two sets of quasi-separable states localized in the potential valley at $\theta \sim 0^o$ or 180° and on a ridge at $\theta \sim 90^o$, respectively. The perturbation approach treats the coefficient b^2 of (6) as a small parameter and studies the effect of the nonhydrogenic potential $b^2 r^2 \sin^2\theta$ on manifolds of degenerate hydrogenic solutions $f_{nl}(r)P_l(\cos\theta)$ with fixed n and with $0 \le l < n$. This effect has been evaluated by constructing approximate eigenfunctions $\sum_l a_l f_{nl}(r)P_l(\cos\theta)$ and studying them semianalytically.[8] Two subsets of these eigenfunctions have been identified, which are localized in the potential valley and astride the ridge, respectively, and have constant phase surfaces departing appreciably from the circles $r = $ const. Weak dependence of these states on the parameter n indicates that analogous states could be constructed which propagate radially over large distances without experiencing substantial non-adiabatic effects, but this avenue remains unexplored.

More significant is the qualitative understanding of the localization process attained in Ref. 8, namely, that the set of eigenvectors of a matrix in a *finite base* includes subsets localized in regions of high and low potential respectively. This phenomenon appears to be an expression of the "spectral repulsion" widely observed in quantum phenomena.

REFERENCES

1. G.H. Morgan and E. Everhart, Phys. Rev. <u>128,</u> 667 (1962).

2. W. Lichten, Phys. Rev. <u>131,</u> 229 (1963).

3. L. Landau, Physik. Z. Sovietunion <u>2,</u> 46 (1932); C. Zener, Proc. Roy. Soc. (London) <u>A137,</u> 696 (1932); E.C.G. Stückelberg, Helv. Phys. Acta. <u>5,</u> 369 (1932).

4. F. Robicheaux *et al.,* Phys. Rev. A <u>35,</u> 3619 (1987).

5. U. Fano, Phys. Rev. A <u>24,</u> 2402 (1981).

6. J. Macek *et al.,* Phys. Rev. A <u>35,</u> 3940 (1987).

7. Yu. F. Smirnov and K.V. Shitikova, Fiz. Elem. Chastits At. Yadra <u>8,</u> 847 (1977) [Transl.: Sov. J. Part. Nucl. <u>8,</u> 344 (1977)].

8. U. Fano, F. Robicheaux and A.R.P. Rau, submitted to Phys. Rev. A.

A UNIFIED VIEW OF COLLISIONS AND SPECTRA

A. R. P. Rau

Department of Physics and Astronomy
Louisiana State University
Baton Rouge, LA   70803-4001

ABSTRACT

The interaction of electrons in an atomic system has historically and traditionally been viewed as falling into two main categories, spectroscopy and collisions. Pedagogically as well, most textbooks in quantum mechanics make this division, devoting their initial discussion to bound state problems and turning in later chapters to scattering. This dichotomy is artificial. Over the past two decades, research in atomic physics has in fact shown the usefulness, even as a point of view, of giving up the division into two categories and seeing both kinds of processes as essentially related. Thereby, links are established between elastic and inelastic electron atom scattering on the one hand and processes involving radiation such as photo absorption and ionization on the other.

These links are specifically codified in what has come to be called Quantum Defect Theory. The effects of interactions at short range are separated from those arising at larger values of the radial distance r of an electron from the rest of the atom. The latter is usually motion in a standard field such as the Coulomb field and can be handled analytically. It is also this latter motion which alone is sensitive to the asymptotic kinetic energy of the electron, so that discrete and continuum states are distinguished only in this large-r behavior. The short range behavior involves multi-particle aspects of the problem, nonlocal (exchange) forces and other such complications but, on the other hand, is insensitive to energy so that it can be parameterized by a limited set of quantities common to both bound and scattering states. Also, since this part of the study is restricted to a finite volume, the numerical calculations can be kept to a modest size.

This set of lectures will be an introduction to the unified view of collisions and spectroscopy. Starting with a simple illustration in potential scattering, we will move on to electron-atom scattering in both single and many channel contexts. Key elements to be presented are quantum defect theory and the basic parameters of that theory, and the so-called method of frame transformations for linking alternative descriptions that are suitable for describing the small- and large-r behavior. Negative ions and neutral atoms with closed shells are used for illustrative purposes.

I. INTRODUCTION

It is a truism that the bound states and scattering states of a potential constitute together the spectrum of the Hamiltonian, and as such one should expect intimate links between them. However, in most of our textbook learning and in research practice, we tend to make a sharp distinction between these two classes of states. Thus, most textbooks in quantum mechanics devote the initial discussion to the properties of bound states, taking up in a few final chapters quantum-mechanical scattering theory in a language that looks quite different from that used for bound states. Where books do point out the relationship between the two classes of states, it is usually in terms of an abstract discussion of the analytical structure of the scattering amplitude, poles of the S-matrix, etc. The practical benefits, therefore, of a unified view of the spectrum are scarcely noticed. Not surprisingly, research over the decades in atomic and nuclear physics has generally proceeded with little cross-fertilization between spectroscopy and collisions. Whereas at least in experiments, the techniques for these two areas tend to be rather different (although, even here, modern work has shown the usefulness of a collisional experiment for extracting spectroscopic information) there is really little justification for maintaining the dichotomy in theoretical research.

The central theme of these lectures is to emphasize a point of view that has developed especially in the last two decades in atomic physics which advocates an abandonment of sharp distinctions between the studies of collisions and of spectra. This point of view goes generally under the name of quantum defect theory (QDT). It actually arose in the work of Ham[1] in condensed matter theory but has been developed for atomic physics by M. J. Seaton and his school. This extensive and continuing work, whose beginnings lie about 25 years ago, has been recently reviewed by Seaton.[2] Over the last 15 years, U. Fano and his school have also made QDT a central element of their work with some specific emphasis and directions of their own.[3] It is this work--and, in particular, material in a recent book[4]--that I draw upon for these lectures. Among the points of emphasis of this approach to QDT are

1. The division of configuration space into long and short ranges with respect to the motion of one particle (from now on to be referred to as an electron) from the rest of the atomic system. The radial distance of this electron will be denoted by r.

2. The two regions thus specified have very different sensitivities to the energy. In particular, strong potential energies generally prevail at short range so that small (sometimes even substantial) changes in the asymptotic energy of the electron are irrelevant. Since different bound states (particularly at high quantum numbers) and the adjoining continuum differ only in these small variations in the asymptotic energy, all these states are described on a common footing at small r.

3. A distinction is, therefore, made between basis sets appropriate to the two regions. In particular, an energy-independent description through energy-independent boundary conditions at the origin defines solutions at small r which apply equally to bound and scattering states. The distinction between the states arises only at large-r and is introduced at the end through the imposition of the appropriate boundary conditions at infinity. This departs from the canonical practice of imposing the boundary conditions at infinity right from the start.

4. The connection between the small-r and large-r solutions involves geometrical and dynamical elements. The geometrical aspect arises from the orbital and spin motions reflecting that often different choices for these quantum numbers, different coupling schemes, etc., are appropriate to the two regions. Passage between them is accomplished by unitary transformations which can usually be worked out analytically and in a

very general way, independent of the specific atomic system under study.

5.   The dynamical aspect of the connection pertains to the radial wave function. Here the large-r study involves motion in a standard long range potential, whose solutions are again often known analytically. The small-r part of the problem is, therefore, the only one that requires numerical handling. But finiteness of the volume involved, and insensitivity to the energy, make for a considerable economy in this numerical work. Ab initio methods restricted to finite r, like the R-matrix method discussed in Greene's lecture, or semi-empirical fitting to data can provide the small-r parameters.

6.   The small-r and large-r wave functions are related through Jost functions or their matrix generalization in a multi-channel problem. (A channel refers to a whole set of states differing only in their energy but otherwise sharing all the discrete quantum numbers of angular and spin motion.) The Jost functions are, therefore, the key parameters for describing bound states, photoabsorption and electron scattering in atomic systems.

7.   Although QDT was originally developed for a system composed of electron + ion whose long range potential was Coulombic, the basic program sketched above applies more generally and its extensions[5] are of interest in atomic and molecular physics. Applications of QDT to molecular problems, involving both electronic and nuclear degrees of freedom, have been recently reviewed.[6]

These lectures, like most of the QDT literature, will be restricted to non-relativistic quantum mechanics although the formalism has been extended to the relativistic range.[7]

A simple illustration of some of the above points is already provided by the Rydberg formula for the energy levels of a one-electron atom like that of an alkali metal. The Rydberg progression of energy levels is described by

$$E_n = I - \frac{13 \cdot 6 eV}{(n - \mu_\ell)^2} , \tag{1}$$

with E_n the position of the nth level as measured from the ground state, I the ionization potential, and μ_ℓ the so-called quantum defect from which QDT takes its name. I and μ_ℓ are specific to each atom. μ_ℓ depends on the angular momentum of the outer electron but is weakly dependent on energy so that particularly at large n it attains a constant value. The presence of μ_ℓ , by which (1) differs from the Bohr formula for hydrogen, is an expression of the penetration of the outer electron into the inner core of the alkali ion where the prevailing Coulomb field is stronger than in H. μ_ℓ is a parameter that characterizes the core (small r) and is insensitive to energy. The major energy dependence in (1), the sharp variation from one level to the next, is given by the same Bohr $1/n^2$ dependence as in H and it depends on the long range - 1/r Coulomb field that is universal to all atoms. Also this dependence is analytically known from the properties of the Coulomb potential. On the other hand, μ_ℓ , the small-r quantum defect parameter, is generally determined for each atom either empirically or numerically.

Further aspects of QDT for a Coulomb field will be taken up in Sec. IV but, because of the complexities associated with Coulomb functions, we will first illustrate the basic elements of QDT in simpler contexts. Sec. II will consider a simple one-dimensional potential well and its three-dimensional analog. Multi-channel aspects will then be introduced in Sec. III through consideration of a negative ion in which case the long range field is exponentially vanishing. These two sections will serve to illustrate all the features of QDT mentioned above. The structure of the wave functions and QDT parameters for a Coulomb field, or a more general long range potential, will then be introduced in Sec. IV. Sec. V then puts all the elements together for a full multi-channel QDT analysis of bound states, photoabsorption and electron scattering in atoms. The final Sec. VI will give a specific two-channel illustration of how the ideas of Sec. V are implemented in practice, including also graphical presentations of spectroscopic data in a manner adapted to the QDT analysis.

II. A POTENTIAL WELL

Consider the one-dimensional square well, $V(x) = -V_0$, $|x| \leq a$; $= 0$, $|x| > a$. Stationary states, that is, real energy eigenstates of even (ψ_+) and odd (ψ_-) parity, can be defined quite generally by

$$
\begin{array}{llll}
\psi_+(x) & = N_E \cos(k_o x - \delta_+) & \psi_-(x) = -N_E \cos(k_o x - \delta_-) & x \leq -a \\
& = A \cos kx & = B \sin kx & |x| < a \\
& = N_E \cos(k_o x + \delta_+) & = N_E \cos(k_o x + \delta_-) & x \geq a \qquad (2)
\end{array}
$$

where N_E, A, B and $\delta_\pm$ are constants and we have defined

$$
E = \hbar^2 k_o^2/2\mu, \quad E+V_o = \hbar^2 k^2/2\mu. \tag{3}
$$

The same solutions are valid regardless of the sign of E. For E<0, k_o is replaced by $i\kappa_o$ and the trigonometric functions by rising and falling exponentials. The $\psi_\pm$ in (2) clearly satisfy evenness and oddness with respect to $x \to -x$. Continuity of the wave function at x=a expresses A and B in terms of the other unknowns N_E, δ_+ and δ_-. The "phase shifts" $\delta_\pm$ are determined by matching derivatives at x=a:

$$
k_o \tan(k_o a+\delta_+) = k \tan ka \tag{4a}
$$

$$
k_o \tan(k_o a+\delta_-) = -k \cot ka. \tag{4b}
$$

Finally, only one unknown parameter is left, the normalization constant N_E to be fixed by some appropriate convention. For continuum solutions, normalization to a Dirac delta function in energy is traditional and for this purpose the notation has employed the subscript E. Since normalization is irrelevant for the present discussion, we will not specify it any further here but return to it later in this section in the three dimensional context.

So far no reference has been made to boundary conditions at infinity. Since it is these that single out bound state vs scattering behavior, (2) and (4) apply equally to all states of the spectrum. Note in particular that the imposition of boundary conditions at $\pm\infty$ has no effect on the $|x|<a$ part of $\psi_\pm$ in (2). Together with the fact that V_o sets the scale of variation of the wave number k in this region so that when $|E| << V_o$, k is quite insensitive to E, bound and scattering states are essentially indistinguishable in this region. We now proceed to define these states by imposing the suitable boundary conditions at infinity.

Bound States

For the symmetric potential well under consideration, the bound states can be classified according to parity. Replacing k_o by $i\kappa_o$, the x≥a wave function takes the form:

54

$$\psi_\pm \to \tfrac{1}{2} N_E \left[e^{i\delta_\pm} e^{-\kappa_o x} + e^{-i\delta_\pm} e^{\kappa_o x} \right] \quad x \to \infty . \tag{5}$$

The boundary condition requires setting equal to zero the coefficient of the rising exponential:

$$e^{-i\delta_\pm} = 0 . \tag{6}$$

Folding this into Eqs. (4) gives the familiar form of the eigenvalue condition[8]

$$k_B \tan k_B a = \kappa_{oB} \qquad \text{even} \tag{7a}$$
$$k_B \cot k_B a = - \kappa_{oB} \qquad \text{odd} , \tag{7b}$$

where the subscript refers to insertion of the bound state energy E_B (<0) in (3).

Note the characteristic structure of the bound state condition in (6). The coefficients $e^{\pm i\delta}$ are replaced in more general contexts below by Jost functions $J^\pm$; bound states are quite generally given by the zeroes of J^- as we will see in later sections.

Continuum Problems

For $E>0$, the typical problem involves scattering from the potential well. Since the particle is incident either from $x=-\infty$ or $x=\infty$, these solutions break parity invariance and we have to consider general solutions that are not parity eigenstates. However, the ψ_+ solutions in (2) constitute a complete set at the energy E so that the general solution is of the form

$$\psi = c_+ \psi_+ + c_- \psi_- ; \tag{8}$$

here $c_\pm$ are complex constants satisfying

$$|c_+|^2 + |c_-|^2 = 1 , \tag{9}$$

so that ψ remains normalized.

In order to impose boundary conditions at infinity, we first write out explicitly from (8) and (2)

$$\psi \to \tfrac{1}{2} N_E \left[(c_+ e^{i\delta_+} + c_- e^{i\delta_-}) e^{ik_o x} + (c_+ e^{-i\delta_+} + c_- e^{-i\delta_-}) e^{-ik_o x} \right], \; x>a$$
$$\to \tfrac{1}{2} N_E \left[(c_+ e^{-i\delta_+} - c_- e^{-i\delta_-}) e^{ik_o x} + (c_+ e^{i\delta_+} - c_- e^{i\delta_-}) e^{-ik_o x} \right], \; x<-a .$$
$$\tag{10}$$

Scattering from the well: Consider now the situation when a particle is incident on the well from $x=-\infty$. This amounts to saying that the coefficient of $e^{-ik_o x}$ at $x>a$ is zero, that is,

$$c_+/c_- = -\exp i(\delta_+ - \delta_-) . \tag{11}$$

With this in (10), the reflection (R) and transmission (T) coefficients can be read off from the other terms that remain. Thus

$$R = |(c_+ e^{i\delta_+} - c_- e^{i\delta_-})/(c_+ e^{-i\delta_+} - c_- e^{-i\delta_-})|^2 = \cos^2(\delta_+ - \delta_-) , \qquad (12a)$$

and

$$T = \sin^2(\delta_+ - \delta_-) , \qquad (12b)$$

with $\delta_\pm$ of course given in (4).

The boundary condition (11) is an "outgoing wave boundary condition" in which outgoing waves exist in all directions (here two, $x=-\infty$ and $x=\infty$) but an ingoing wave only in one direction, namely at $x=-\infty$. This is the familiar scattering boundary condition when a particle is incident on a well. Escape from the well: Another class of continuum solutions, not typically encountered in discussions of the square well but as easily considered as the one above, is for a situation in which the particle leaves the well in a specific direction, say $x=-\infty$. In this case, it is the coefficient of $e^{ik_o x}$ at $x>a$ that vanishes:

$$c_+/c_- = - \exp i(\delta_- - \delta_+) . \qquad (13)$$

We now have an "ingoing wave boundary condition" in which ingoing waves are present in all directions (both $x=-\infty$ and $x=\infty$) but the outgoing wave is only in the one direction ($x=-\infty$) where the particle is indeed detected. Note that (13) is the complex conjugate of (11) and is a realization of the connection that exists through time reversal between the two kinds of problems. As we will see in later sections, it is precisely the ingoing wave condition that is relevant to problems such as photoionization in which an electron leaves from the atom as a result of the absorption of a photon. It should be clear, therefore, from the above that there is an intimate link between such a photoabsorption process and electron-atom scattering. In fact, the above discussion already suffices to show that the physics of escape, scattering and bound states are all contained in the same parameters, $\delta_\pm$, or in the Jost functions $e^{i\delta_\pm}$.

A three-dimensional spherical well

When the $\ell=0$ radial function in the spherically symmetric potential $V(r) = - V_o$ $r \leq r_o$, $V(r) = 0$ $r > r_o$, is multiplied by r, the problem becomes identical to the odd parity solutions considered above for the square well (but note r goes only from 0 to ∞):

$$F(r) = A \sin kr \qquad r \leq r_o$$

$$= N_E \sin(k_o r + \delta) \qquad r > r_o . \qquad (14)$$

Once again, matching wave functions at r_o expresses A in terms of N_E and δ whereas matching derivatives determines δ :

$$k_o \cot(k_o r_o + \delta) = k \cot k r_o . \qquad (15)$$

Bound states are again given by the condition $e^{-i\delta} = 0$ which leads precisely to (7b).

Consider scattering solutions with $E \gtrless 0$. With k_o small, it follows from (15) that

$$\delta \simeq - k_o a , \qquad (16)$$

where a, which has dimensions of length, is called the scattering length.[9]

56

The low energy cross-section for scattering, $\sigma = 4\pi \sin^2\delta/k_o^2$ reduces to $\pi(2a)^2$.

Looking back at (14) at asymptotically large distances, $F(r)$ is a superposition

$$F(r) \rightarrow \sin k_o r + \tan\delta \cos k_o r \, , \quad r > r_o \, , \quad (17)$$

of the two linearly independent solutions $(\sin k_o r, -\cos k_o r)$ of the radial Schrödinger equation at $r > r_o$: $(d^2/dr^2 + k_o^2) F(r) = 0$. However, when $k_o r \ll 1$, this equation can be simplified further to $d^2F/dr^2 = 0$ which has as a general solution a linear dependence for $F(r)$ in the vicinity of r_o. As an alternative to (17), therefore, when $k_o r \ll 1$ we can match the solution of $F(r)$ for $r < r_o$ to the form

$$F(r) \propto r + \tan\delta^{(o)} r_o \, . \quad (18)$$

In this alternative, the two independent solutions are r and a constant, chosen for dimensional reasons to be $-r_o$. Both are independent of the energy. Therefore, $\delta^{(o)}$ represents a phase shift defined in terms of such energy-independent basis solutions at small r and is likewise insensitive to the energy. Matching logarithmic derivatives of (18) and of the A sinkr within the well gives

$$k \cot kr_o = \left[r_o(1 + \tan\delta^{(o)}) \right]^{-1}. \quad (19)$$

Comparison with (15) and (16) gives

$$\tan\delta^{(o)} = -a/r_o \, , \quad (20)$$

which corresponds to the form of (18), $F(r) \sim r-a$.

With this perspective of an energy-independent phase shift $\delta^{(o)}$ defined at the boundary r_o through use of energy-independent basis functions, the result (16) for the full phase shift δ can be viewed as the product of two factors

$$\tan\delta \approx \delta \approx (k_o r_o)(-a/r_o) \approx (k_o r_o)\tan\delta^{(o)}. \quad (21)$$

This result exemplifies a feature central to QDT. Both δ and $\delta^{(o)}$ are phase shifts due to the short range potential, the difference lying in the base sets of solutions with respect to which they are defined, (17) and (18), respectively. δ is the true phase shift as measured at infinity and defined in terms of the asymptotic sine and cosine functions. However, sensitive energy dependences, like that of k_o in (16) merely reflect dependences contained in the base functions. At small r, the asymptotic energy is negligible when compared to V_o and it is possible to define base sets which are independent of the energy so that the corresponding "smooth" phase shift $\delta^{(o)}$ defined in terms of them near $r \approx r_o$ is also smooth in energy. The connection between $\delta^{(o)}$ so defined at r_o and δ as measured at infinity is provided by (21) and exhibits explicitly that the k_o dependence in δ merely reflects the subsequent passage from r_o to ∞. In fact, this $(k_o r_o)$ factor expresses the connection between the base sets $(\sin k_o r, -\cos k_o r)$ and $(r, -r_o)$. This connection has nothing to do with the specific short range potential under consideration and can be provided analytically by considering the solutions of the radial equation at long range, $r > r_o$. In this way, physical parameters can be decomposed into parts that are determined analytically, are often sharply varying and depend on the long range potentials (zero in the above example), and others (such as $\delta^{(o)}$ or a) which are truly representative of the short range forces. These have smooth dependences and can be readily extrapolated over an energy range. More complicated manifestations of (21) will be seen in later sections.

As a final remark for this section, the normalization for continuum functions will be throughout assumed to be on the unit energy scale, that is,

$$\int_0^\infty F(E,r) \, F(E',r) dr = \delta(E'-E) \ . \tag{22}$$

This requires[10]

$$N_E = (2\mu/\pi \hbar^2 k_o)^{\frac{1}{2}} \ . \tag{23}$$

For the rest of these lectures we will work in atomic units, $e = \hbar = \mu = 1$.

III. PHOTODETACHMENT OF A NEGATIVE ION

Definition of a Channel

The short range $\ell=0$ problem considered at the end of Sec. II has just one label, namely, energy (or k_o), that specifies the states in the spectrum. Considering this as a running index (over both discrete and continuum energies), Eqs. (14) and (15) describe all the states together. The entire spectrum is viewed thereby as one "channel" of states. More generally, we will refer to a channel of states as a set of states with unspecified energy but whose remaining discrete quantum numbers (such as the angular momentum ℓ in the above example) take specific values. A typical atomic system generally presents us with a multi-channel problem. As a simple illustration, we will now consider a negative ion such as F^- or I^- at low energies in which case the only additional aspect introduced beyond the $\ell=0$ spherical well example is that there are now two excitation channels. These negative ions have one stable bound state with quantum numbers $p^6 \, {}^1S_o$.

Consider a photodetachment process which leads to the ejection of low energy photoelectrons, leaving behind the neutral atom in either of its fine structure states $p^5 \, {}^2P_{1/2,3/2}$. Dipole selection rules restrict the orbital angular momentum of the photoelectron to $\ell=0$ or 2. Since the electron leaves behind a neutral atom, the long range interaction between them vanishes exponentially. (Because of the polarizability of the atom, there is a $1/r^4$ potential as well but this does not basically alter the arguments that follow.) In such a situation, only the $\ell=0$ component is appreciable for low photoelectron energies.[11] The process is described, therefore, by

$$F^- + h\nu \rightarrow F + e$$

$$p^6 \, {}^1S_o^e \rightarrow \left[p^5({}^2P_{1/2,3/2}) + e(\ell=0) \right] \, {}^1P_1^o \ . \tag{24}$$

There are now two channels. All their discrete quantum numbers are the same except for the two alternative j values of 1/2 and 3/2 for the core. Since these core levels have different energies, at a fixed photon energy the wave number of the photoelectron also takes alternative values k_j.

Fragmentation channels

At asymptotic separation of the core and electron, the channels can be described, therefore, by $|p^5({}^2P) jm_j; k_j, \ \ell=0, \ j_e=1/2, \ m_e; \ J=1,M\rangle$ with $j=(1/2, 3/2)$. Since the initial negative ion has all angular momentum quantum numbers equal to zero, the value of M in the final state will be fixed by the polarization of the photon. This description of the final state is appropriate to the fragmentation limit of large separation. As appropriate to this limit, the channels are specified by a jj-coupling scheme. The electric dipole matrix element for the process can be written as

$$\langle p^5({}^2P) jm_j; k_j s j_e m_e; J=1,M | \, P_M^{[1]} | \, p^6 \, {}^1S_o \rangle \ , \tag{25}$$

where $P_M^{[1]}$ is the photon operator. M=0 corresponds to linear polarization, M=±1 to circular.

Eigenchannels

The photon is actually absorbed at small r (separation of electron and core) because that is where the wave function of the negative ion is concentrated. This small-r region is one of strong coupling of all six electrons in (24). In fact, the appropriate description here is to couple all individual orbital angular momenta together and likewise all spins before finally combining them to J--this is a region of LS coupling. We use, therefore, an alternative set of channels (again two) described as $|p^5(^2P); \ell=0, s=1/2; ^{2S+1}P, J=1, M\rangle$ where S=0,1 are the two alternative values of total spin. Since this region has strong coupling forces, the asymptotic energy is ignorable and we, therefore, have no k_j label. In fact, the above small-r states describe equally the negative ion and the detached continuum states. These small-r channels will be termed "short range eigenchannels".

Description as "half-scattering"

Through the insertion of a complete set of small-r states, (25) can be rewritten as

$$\sum_S \langle p^5(^2P) jm_j, k_j s j_e m_e | p^5(^2P) s\ ^{2S+1}P \rangle \langle p^5(^2P) s\ ^{2S+1}P | P_M^{[1]} | p^6\ ^1S_o \rangle\ , \tag{26}$$

where we have dropped the common index JM on the states. This recasting of (25) into two factors in (26) describes the photodetachment process as proceeding through two steps. The factor on the right corresponds to the initial absorption of the photon to create eigenchannels of the core + electron complex. This part is restricted to small r, is independent of the energy and is a standard spectroscopic "strength".[12] In fact, in our example, the photon can only couple to S=0 thus involving a single matrix element d which we can call the spectroscopic "feed" into the small-r channels. Except for absolute measurements, we need not even evaluate d. In all measurements of relative cross-sections, d will be a common factor that cancels out.

The factor on the left in (26) contains, therefore, all the non-trivial dynamics in this process. It represents the fragmentation to large r of the complex initially created at small r and involves projecting eigenchannels onto fragmentation channels. This factor is like half of a scattering--in fact, precisely the second half of elastic scattering in the same electron + core system, were we to describe this scattering similarly as formation of a complex at small r from the incident channel, followed by fragmentation of the complex. Photodetachment (or photoionization) can, therefore, be regarded quite generally as half of a scattering and our analysis here and in later sections will bring out explicitly the implications and usefulness of such a picture.

Analysis of fragmentation

Frame transformations

The states in the left factor of (26) involve orbit, spin and radial dependences. Denoting the orbital and spin wavefunction of the fragmentation channels by ϕ_{JM}^j and that of the eigenchannels by χ_{JM}^S, passage between them is given by the LS $\rightarrow$ jj unitary transformation[13]

$$X^S_{JM} = \sum_j \phi^j_{JM} U^{(J)}_{jS} \quad. \tag{27}$$

In our 2 channel example, U is a 2x2 orthogonal matrix, a rotation between the two frames (LS and jj) through an angle θ . For the specific quantum numbers involved it takes the form

$$U = \begin{vmatrix} \cos\theta & \sin\theta \\ -\sin\theta & \cos\theta \end{vmatrix} \quad \begin{array}{c} = 1/2 \\ 3/2 \end{array} \quad \begin{vmatrix} \sqrt{2/3} & \sqrt{1/3} \\ -\sqrt{1/3} & \sqrt{2/3} \end{vmatrix} \quad, \tag{28}$$

(with column headings: j S 1 0)

with $\theta \approx 35^\circ$. Thus all the orbital and spin aspects of the projection of eigenchannels onto fragmentation channels are given by this one geometrical element, an angle θ , which depends only on the angular and spin quantum numbers and is thereby common to all atomic systems sharing those numbers. This simple U-matrix is an example of more general "frame transformations" which connect the alternative descriptions of fragmentation and eigenchannels. They incorporate the geometrical elements in the problem.

Radial functions

We have thus set the stage for a specific focus on the dynamics, namely on the radial behavior of the left factor in (26). For the small-r radial wave functions, we will use the scattering length description as in (18). Two scattering lengths a_S, for the two values S=0,1, are involved. The basis functions for the eigenchannels are, therefore, $X^S_{JM} (r - a_S)$. The full wave function will be a superposition with coefficients A_{Sj_f} chosen so as to satisfy the appropriate boundary conditions at infinity when the photoelectron is detected in the final channel j_f:

$$\psi = \sum_S X^S_{JM} (r - a_S) A_{Sj_f} \quad. \tag{29}$$

To continue this function to large r and recast it in the fragmentation channel description, we first use (27) and write

$$\psi = \sum_j \phi^j_{JM} \left[r \sum_S U_{jS} A_{Sj_f} - \sum_S U_{jS} a_S A_{Sj_f} \right] \quad. \tag{30}$$

The base set of radial functions at large r are the sines and cosines of (14) and (17). Specifically, according to the normalization in (23), they are

$$(2/\pi k_j)^{\frac{1}{2}} \sin k_j r \; , \; -(2/\pi k_j)^{\frac{1}{2}} \cos k_j r \quad. \tag{31}$$

Rewriting (30), therefore, in terms of these functions, we have

$$\psi = \sum_j \phi^j_{JM} k_j^{-\frac{1}{2}} \left[\sin k_j r \sum_S k_j^{-\frac{1}{2}} U_{jS} A_{Sj_f} - \cos k_j r \sum_S k_j^{\frac{1}{2}} U_{jS} a_S A_{Sj_f} \right] , \tag{32}$$

or

$$\psi = \sum_j \phi_{JM}^j (2/\pi k_j)^{\frac{1}{2}} (1/2i) \left[e^{ik_j r} k_j^{-\frac{1}{2}} \sum_S J_{jS}^+ A_{Sj_f} - e^{-ik_j r} k_j^{-\frac{1}{2}} \sum_S J_{jS}^- A_{Sj_f} \right] , \tag{33}$$

where

$$J_{jS}^+ = U_{jS} (1-ik_j a_S) (\pi/2)^{\frac{1}{2}} = (J_{jS}^-)^* \tag{34}$$

is the Jost matrix for this problem. Its first factor U_{jS} arises from projecting the eigenchannel frame onto the fragmentation frame, and its second factor comes from the connection between the radial functions in these two frames (note from (16) that $e^{i\delta} \simeq 1 - ika$).

Cross-sections

ψ in (33) is now in the form on which the boundary conditions at infinity can be imposed. For photodetachment into a specific channel j_f, that is, when the electron is recorded with the energy corresponding to j_f, the condition we require as in Sec. II is the "ingoing wave" one which says that there is an outgoing wave only in j_f. We have, therefore, to choose A_{Sj_f} in (33) so that

$$\sum_S J_{jS}^+ A_{Sj_f} = k_j^{\frac{1}{2}} \delta_{jj_f} . \tag{35}$$

The matrix A is thereby proportional to the inverse of J^+.

Returning to (26), the left factor is given by the solution of (35), A_{Sj_f}, except for a de-coupling of the core and electron total angular momenta for the purposes of the photoelectron detection. The cross-section for detachment to j_f is, therefore,

$$\sigma_{j_f} = |(j_f \, m_{j_f} \, \tfrac{1}{2} \, m_e | \, 1M) \, A_{0j_f} d|^2 . \tag{36}$$

Note that only the S=0 coefficient contributes because of the selection rule discussed earlier for the right factor in (26).

By constructing the J^+ matrix from (34) and (28) and inverting it to obtain A_{0j_f} in (35), we can write down explicitly the detachment cross-sections. Thus when the photon energy lies in a range that allows detachment to either channel, both $k_{1/2}$ and $k_{3/2}$ are real and the cross-sections are

$$\sigma_{3/2} = (2d^2/\pi) \, k_{3/2} \left[(2/3)(1+k_{1/2}^2 a_1^2)/\text{denom} \right] \tag{37a}$$

$$\sigma_{1/2} = (2d^2/\pi) \, k_{1/2} \left[(1/3)(1+k_{3/2}^2 a_1^2)/\text{denom} \right] , \tag{37b}$$

with

$$\text{denom} \equiv (1-k_{1/2} k_{3/2} a_0 a_1)^2 + (1/9)\left[k_{1/2}(2a_1+a_0) + k_{3/2}(2a_0+a_1) \right]^2 . \tag{37c}$$

When the photon energy lies in between the two thresholds for detachment, the lower lying channel is "open" and has $k_{3/2}$ real but the upper 1/2 channel is "closed" with $k_{1/2} = - i\kappa_{1/2}$. [Note that the wave function with ingoing wave boundary conditions has both $\exp(-ik_{3/2}r)$ and $\exp(-ik_{1/2}r)$. The latter must decay exponentially in the closed channel, requiring the choice $k_{1/2} = - i\kappa_{1/2}$. Contrast this with Sec. II where starting with an outgoing wave form, the passage to bound states or closed channels is rendered by $k = i\kappa$.] Between the thresholds then, we have

$$\sigma_{3/2} = (2d^2/\pi)\, k_{3/2}[(2/3)(1-\kappa_{1/2}\, a_1)^2/\{[1-(\kappa_{1/2}\,/3)(a_0+2a_1)]^2$$

$$+ k_{3/2}^2[(2a_0+a_1)/3 \,-\, \kappa_{1/2}\, a_0 a_1]^2\}] \ . \tag{37d}$$

Summary

Note the characteristic structure of the cross-section expressions in
(37); for specific examples, see refs. 14 and 15. The dominant energy
dependence is in the leading k_{j_f} behavior which is the Wigner threshold
law[11] for this process. This factor k can be traced back to the passage
between (30) and (32) and reflects the proportionality constants between the
large-r base pair in (31) and the small-r energy independent pair that enter
into (29). It appears just as it did in the single channel example in Sec.
II. The short range part of the problem only induces weaker energy
dependences in (37)--the short range parameters a_0 and a_1 always come
accompanied by a κ or k which are small in the range of energies considered.
[This aspect can be argued on dimensional grounds alone.]

The detailed analysis of a concrete problem presented in this section
exemplifies the various features of QDT listed in the introduction. The
division into small and large r isolates specific vs. general, and complicated
numerical vs. simple analytical handling. The scattering lengths a_0 and a_1
are specific to each system, and have to be calculated numerically or
extracted from other data such as electron-atom scattering. In this problem
the long range part corresponds to vanishing potential and is particularly
simple to handle analytically, dealing only with sines and cosines and with
their small argument limits. [For $\ell \neq 0$, the situation is not much more
complicated, with spherical Bessel functions replacing the sine and cosine.]
The small- and large-r regions have different quantum numbers appropriate to
their description, in this case S and j. Transformation between the regions
involves geometrical elements, here a LS $\rightarrow$ jj transformation, and leads to
characteristic numerical dependences such as the 2/3 and 1/3 factors in (37a)
and (37b). The connection of the radial functions involves Jost matrices
which become the key dynamical parameters of the theory. These functions and
their inverses describe photodetachment as well as electron scattering and the
bound states in the system. The same two scattering lengths are the crucial
dynamical parameters.

IV. COULOMB AND GENERAL LONG RANGE FIELDS

The examples considered in the previous sections with a vanishing long
range field have generally a limited number of bound states. Problems of
electron + positive ion, on the other hand, with a long range Coulomb field,
have a very rich structure of bound states in the spectrum. This fact has, of
course, been central to atomic spectroscopy. The Rydberg formula and the
quantum defect have already been introduced in Sec. I. Here we take up a more
complete study of the various base sets of solutions and the analytical
definition of QDT parameters for the Coulomb field, in a manner that admits
generalization to other long range potentials. Notable examples would be the
$1/r^2$ dipole potential for an electron + polar molecule, the $1/r^4$ potential due
to atomic polarizability, and the $1/R^6$ van der Waals potential between two
atoms.

Typical of atomic phenomena in which one electron reaches large values of
r relative to the rest (whether in a bound state of high excitation or in the

process of escape) is the potential

$$V(r) = V_s(r) - 1/r \quad r < r_o \quad , \quad V(r) = -1/r \quad r > r_o \ . \qquad (38)$$

The short range potential $V_s(r)$ is very complicated, often not even defined as a local single-particle potential but is an entanglement of many particle interactions (including non-local terms like exchange). What will be exploited, however, is that at large r the potential is Coulombic and that standard solutions are available.

An effective quantum number ν, defined through

$$\epsilon = -1/2\nu^2 \ , \qquad (39)$$

is introduced where ϵ is the energy measured from the ionization limit; and we use atomic units. This number takes the values

$$\nu = i/k \quad \epsilon > 0 \quad , \quad \nu = 1/\kappa \quad \epsilon < 0 \ . \qquad (40)$$

Note that ν is defined for all values of ϵ, positive or negative. Among the latter, ν coincides with the principal quantum number n at the bound state energy positions for the H atom.

A pair of independent solutions $\{f^+, f^-\}$ with large-r behavior

$$f^{\pm} \rightarrow e^{\mp r/\nu} e^{\pm\nu\ell n r} \qquad (41)$$

is defined in analogy to the $e^{\pm ikr}$ solutions of previous sections. The logarithmic dependence in (41) is a characteristic phase distortion in the presence of long range potentials that drop off at infinity like 1/r or slower. For $\epsilon < 0$, the pair (41) no longer oscillate but become rising and falling exponentials, and their definition requires care[4,5].

Regular solutions

Regular solutions near the origin are determined by the angular momentum potential so long as the external potential is not more singular than $1/r^2$ near the origin. The $1/r^2$ potential is of some interest and, for this and other singular potentials, a special handling is required,[5] but we will say no more about it here. For the Coulomb and most fields of interest then, neither the asymptotic energy nor the potential is important at the origin, the behavior there being determined entirely by the ℓ quantum number of the electron. A smooth, regular function satisfying energy-independent boundary conditions at the origin is defined through

$$f^o(\epsilon,\ell,r) \rightarrow r^{\ell+1} \qquad r \rightarrow 0 \ . \qquad (42)$$

The more familiar energy-normalized regular Coulomb solution $f(\epsilon,\ell,r)$ that is encountered in text-books[16] $[f(\epsilon,\ell,r) = (2/\pi k)^{1/2} \ F_\ell(-1/k,kr)$ of Ref. 16] is defined through its asymptotic behavior

$$f(\epsilon,\ell,r) \sim (2/\pi k)^{\frac{1}{2}} \sin[kr + (1/k)\ell n r + \eta(k,\ell)] \ , \qquad \epsilon > 0 \qquad (43)$$

where $\eta(k,\ell)$ is the Coulomb phase shift

$$\eta(k,\ell) = -\tfrac{1}{2} \ell\pi + (1/k)\ell n 2k + arg\Gamma(\ell+1-i/k). \qquad (44)$$

Since f and f^o are both regular solutions in the Coulomb field, they can differ only in a normalization factor

$$f^o(\epsilon,\ell,r) = B^{-\frac{1}{2}}(\epsilon,\ell)f(\epsilon,\ell,r); \qquad \epsilon > 0, \qquad (45)$$

$B^{-\frac{1}{2}}$ is one of the Coulomb QDT parameters. It is the analog of the $k^{-\frac{1}{2}}$ encountered for zero long range field in earlier sections. Its value for Coulomb and other fields has been tabulated.[4,5]

Jost functions and QDT parameters

The Jost functions are quite generally defined by expanding solutions f^o defined with energy-independent boundary conditions near the origin in terms of the energy normalized asymptotic solutions $f^{\pm}$:

$$f^o(\varepsilon,\ell,r) = (2/\pi k)^{\frac{1}{2}} (1/2i) [J^+(k,\ell)f^+(\varepsilon,\ell,r)(-i)^{\ell}$$

$$- J^-(k,\ell)f^-(\varepsilon,\ell,r)i^{-\ell}] , \quad \varepsilon > 0 . \tag{46}$$

Once again, $J^{\pm}$ have been tabulated for various fields.[5] (Warning: Definitions vary by multiplicative factors.) These Jost functions have a rich analytic structure[17]; in particular, $J^+(k) = [J^-(k)]^*$. By their very definition linking f^o to $f^{\pm}$, note that they are appropriate to a description of the "one-way traversal" from small to large r. It will, therefore, not be surprising that, in the examples already encountered and in the more general discussion of Sec. V, bound states and photo-absorption which deal with such a one-way traversal are naturally described in terms of Jost functions (zeroes of J^- and elements of $(J^+)^{-1}$ respectively for the two problems). On the other hand scattering, which involves a two-way traversal by the particle from large into small r and then back again to large r, will involve the factorization of the S-matrix, bilinear in $J^{\pm}$, $S = J^+(J^-)^{-1}$.

Comparison of (43) and (45) with (46) establishes the relationship between the Jost functions and the QDT parameters for $\varepsilon > 0$

$$J^{\pm}(k,\ell)(\mp i)^{\ell} = B^{-\frac{1}{2}} e^{\pm i\eta} . \tag{47}$$

Although $J^{\pm}$ carry all the information about the connection from small to large-r, the real functions B and η of QDT prove more convenient than the complex $J^{\pm}$. Note that these QDT parameters provide in essence the separation of the Jost functions into amplitude and phase.

Irregular solutions

Just as in earlier sections, the regular solutions f^o and f have to be complemented, to obtain a complete set, by irregular counterparts g^o and g, respectively. The definition of g is straightforward, it being constructed as the energy normalized form specified through an asymptotic behavior that lags (43) by 90° in phase (as in (31) for zero long range field):

$$g(\varepsilon,\ell,r) \sim -(2/\pi k)^{\frac{1}{2}} \cos[kr+(1/k)\ell n \, r + \eta(k,\ell)] , \quad \varepsilon > 0 . \tag{48}$$

The two linearly independent solutions (43) and (48) have the Wronskian

$$W(f,g) = 2/\pi . \tag{49}$$

The definition of g^o turns out to be a little more involved, is considered elsewhere,[4,5] and requires the introduction of a third QDT parameter $G(\varepsilon,\ell)$:

$$g^o(\varepsilon,\ell,r) = B^{\frac{1}{2}}(\varepsilon,\ell)g(\varepsilon,\ell,r) - G(\varepsilon,\ell)f^o(\varepsilon,\ell,r) , \quad \varepsilon > 0 . \tag{50}$$

Base sets at $\varepsilon < 0$.

For $\varepsilon < 0$, the $f^{\pm}$ in (41) are falling and rising exponentials at $r \rightarrow \infty$. f^o, being insensitive to energy, remains unchanged, defined again through (42). The expansion of f^o in terms of asymptotic solutions to obtain the Jost functions $J^{\pm}(\kappa,\ell)$ is now written as

$$f^0(\varepsilon,\ell,r) = (\pi\kappa)^{-\frac{1}{2}}\left[J^+(\kappa,\ell)f^+(\varepsilon,\ell,r) + J^-(\kappa,\ell)f^-(\varepsilon,\ell,r)\right], \quad \varepsilon<0 , \tag{51}$$

with $\nu = 1/\kappa$. These $J^{\pm}$ have been tabulated in ref. 5. The QDT parameters for $\varepsilon<0$ are defined through

$$f^0(\varepsilon,\ell,r) = A^{-\frac{1}{2}}(\varepsilon,\ell)f(\varepsilon,\ell,r) \tag{52a}$$

$$g^0(\varepsilon,\ell,r) = A^{\frac{1}{2}}(\varepsilon,\ell)g(\varepsilon,\ell,r) - G(\varepsilon,\ell)f^0(\varepsilon,\ell,r) , \quad \varepsilon<0 \tag{52b}$$

and

$$f(\varepsilon,\ell,r) = (\pi\kappa)^{-\frac{1}{2}}\left[\sin\beta(\kappa,\ell)D^{-1}(\kappa,\ell)f^-(\varepsilon,\ell,r)\right.$$
$$\left. - \cos\beta(\kappa,\ell)D(\kappa,\ell)f^+(\varepsilon,\ell,r)\right] \tag{53}$$

$$g(\varepsilon,\ell,r) = -(\pi\kappa)^{-\frac{1}{2}}\left[\cos\beta(\kappa,\ell)D^{-1}(\kappa,\ell)f^-(\varepsilon,\ell,r)\right.$$
$$\left. + \sin\beta(\kappa,\ell)D(\kappa,\ell)f^+(\varepsilon,\ell,r)\right] . \tag{54}$$

These connections are similar to the earlier ones for $\varepsilon>0$ although the separation into amplitude and phase proceeds slightly differently when $J^{\pm}(\kappa,\ell)$ are real. The parameter A is a normalization function analogous to B, and the phase parameter $\beta(\kappa,\ell)$ replaces the $\eta(k,\ell)$ above threshold. The parameter $D(\kappa,\ell)$ is new and specific to $\varepsilon<0$. It arises because $f^{\pm}$ for $\varepsilon<0$ have very different amplitudes unlike the corresponding pair for $\varepsilon>0$. D serves to "re-scale" the functions placing their amplitudes on a more nearly equal footing. The QDT parameters A, D and β for standard fields are given in refs. 4 and 5 along with a prescription for their general calculation. This prescription stems from a useful WKB interpretation as shown in Fig. 1. For a general potential of the form shown (note the small-r behavior depends on the angular momentum barrier), the various parameters correspond to the WKB phase integrals shown. For the important case of the Coulomb field, β is given by

$$\beta(\kappa,\ell) = \pi(-\ell+1/\kappa) . \tag{55}$$

Scattering phase shifts and bound states

For potentials of the form (38), where a short range interaction is superposed on the long range Coulomb potential, we can proceed as in previous sections but now in terms of the base sets $\{f^0,g^0\}$ and $\{f,g\}$ defined above. Thus a regular solution $F(\varepsilon,r)$ obtained for $r<r_0$ by some procedure (see Greene's lecture) can be expanded at r_0 in terms of either of these base pairs to get

$$F(\varepsilon,r) = f(\varepsilon,r) \cos\delta_\ell^{(s)} - g(\varepsilon,r) \sin\delta_\ell^{(s)} \tag{56a}$$

$$r>r_0$$

$$= N(\varepsilon)\left[f^0(\varepsilon,r)\cos\delta_\ell^{(so)} - g^0(\varepsilon,r) \sin\delta_\ell^{(so)}\right] . \tag{56b}$$

The full short-range phase shift $\delta_\ell^{(s)}$ then adds to the long range phase shift, to get the full phase shift at infinity when $\varepsilon>0$. The phase shift $\delta_\ell^{(so)}$, on the other hand, is smoother in energy and is the candidate for ready extrapolation across an energy range. Connections between $\delta_\ell^{(s)}$ and $\delta_\ell^{(so)}$ can be read off from (56), upon inserting the connection between the base pairs $\{f^0, g^0\}$ and $\{f, g\}$ given in (45) and (50) for $\epsilon > 0$ and in (52) for $\epsilon < 0$.

The bound state eigenvalue condition can be obtained from Eqs. (53)-(56). For a pure Coulomb field, the relevant solution is the regular $f(\epsilon, \ell, r)$. From (53), a bound state requires the vanishing of the coefficient of $f^-(\epsilon, \ell, r)$ so that

$$\sin\beta(\kappa, \ell) = 0. \tag{57a}$$

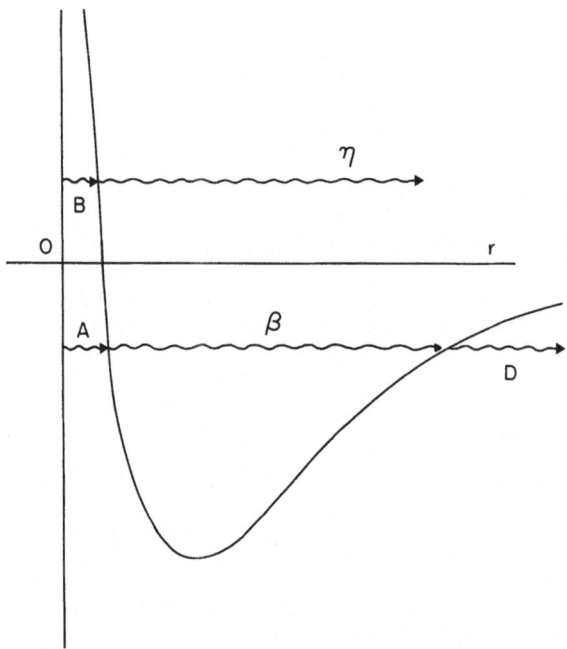

Figure 1. WKB interpretation of the QDT parameters for a general potential. The five parameters are given by phase integrals over the ranges shown.

Upon using (55), this gives

$$\beta = \pi(-\ell + 1/\kappa) = \pi(n_r + 1), \quad n_r = 0, 1, 2, \ldots \tag{57b}$$

and is equivalent to

$$\epsilon = -\tfrac{1}{2} \kappa^2 = -\tfrac{1}{2} (n_r + \ell + 1)^{-2}, \tag{57c}$$

the Bohr formula, with $n = n_r + \ell + 1$, n_r being the radial quantum number. On the other hand, when there is an additional short range potential, the regular solution is given by (56a), which upon combination with (53) and (54) gives for the coefficient of f^- to be equated to zero,

$$\sin\left[\beta(\kappa,\ell) + \delta_\ell^{(s)}\right] = 0 , \qquad (58a)$$

leading to

$$\varepsilon = - \tfrac{1}{2} (n-\delta_\ell^{(s)}/\pi)^{-2} . \qquad (58b)$$

This is the Rydberg formula (1) with the quantum defect $\mu_\ell = \delta_\ell^{(s)}/\pi$.

Extrapolations within the bound and continuous spectrum would, therefore, start for instance with semi-empirical values of $\pi\mu_\ell$ or scattering phase shifts $\delta_\ell^{(s)}$ at a given energy and convert them to the smooth phase shifts $\delta_\ell^{(so)}$. Alternatively, these smooth parameters may be extracted from an <u>ab initio</u> calculation of $F(\varepsilon,r)$ for $r<r_o$ with energy-independent boundary conditions at the origin and matched as in (56b). With the $\delta_\ell^{(so)}$ in hand, they may be extrapolated to a new energy range and then transformed again through (56) to the physical phase shifts $\delta_\ell^{(s)}$ or $\pi\mu_\ell$ at the new energy.

Normalization of discrete and continuum states

The QDT connection between discrete and continuum states applies not just to energies but also to wave functions and, therefore, to other physical properties. In fact, the very interpretation of the phase shift points to this. The presence of short range interactions pulls in or pushes out the wave functions in r, an effect common to all states. Thus, at large r, the radial function F is shifted in phase by $\delta_\ell^{(s)}$ or $\pi\mu_\ell$ relative to the purely long range solution f. It remains only to place discrete and continuum states on an equivalent footing in their normalization to have a complete link between their wave functions. This is accomplished[2,4] by multiplying discrete wave functions, normalized in a conventional way to unity upon volume integration, by $(d\varepsilon/dn)^{1/2}$ obtained from (58b), that is, by

$$N_n = (n-\delta_\ell^{(s)}/\pi)^{-3/2}\left[1+(n-\delta_\ell^{(s)}/\pi)^{-3}\pi^{-1}d\delta_\ell^{(s)}/d\varepsilon\right]^{-1/2} . \qquad (59)$$

Discrete state normalization is thereby placed on a unit energy scale, on a par with normalization in the continuum. The result (59) for Coulomb fields was derived by Seaton[2] and its counterpart for other long-range fields is given in refs. 4 and 5.

Matrix elements of any operator in the discrete and continuum parts of the spectrum can, therefore, be presented on a common footing by dividing the discrete ones by the corresponding factors (59). Such a "histogram plot" for oscillator strengths for photoabsorption from the ground state in hydrogen is shown in Fig. 2.[18] Note that the oscillator strengths f_n in the discrete, upon division by N_n^2, coincide in dimension with that of $df/d\varepsilon$ in the continuum. The heights of the histograms blocks are determined by such a division and their bases are given by N_n^2. The continuity of the plot across the ionization limit is evident. Placing matrix elements on a common footing in this way across the entire spectrum again permits extrapolation of data between different ranges of the spectrum.

V. MULTI-CHANNEL QUANTUM DEFECT THEORY

We now take up the study of an atomic problem with many channels in the presence of a non-trivial long range field like the Coulomb field[20]. All the elements are in place from the previous sections and need only to be put together in a very general context. In the two-channel problem considered in Sec. III, there was only one bound state in the negative ion system so that considerations of the discrete spectrum took a back seat. The spectrum is

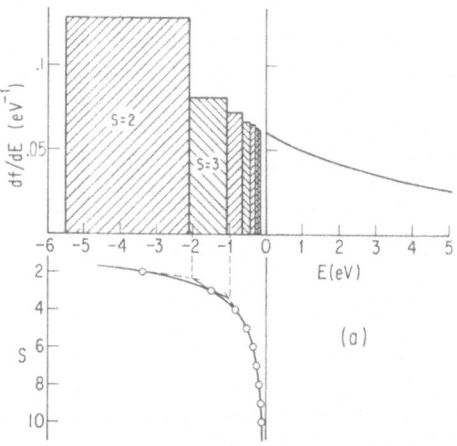

Figure 2. Discrete and continuum oscillator strengths for photoabsorption in hydrogen (from ref. 18, whose scale of ordinates is erroneous). The area of each histogram block represents the oscillator strength for that discrete transition. The base of each block is given by dE_s/ds and its geometrical construction through tangents to the E_s vs s curve is shown.

much richer for an $e + A^+$ system, with infinite numbers of levels converging to multiple ionization limits. From an actual spectroscopic observation of an energy level in a mixed spectrum it is not possible to assign a level generally to a specific series. A multichannel problem of this kind also introduces a new element, namely levels which lie in between thresholds. Those are, therefore, not strictly bound but mix with the continuum on the lower ionization limit--they are autoionization resonances. Multichannel QDT provides inter-connections between all these aspects: energy positions and oscillator strengths of the bound state levels; positions, intensities, widths and profiles in the autoionization region; and the distribution of photo-electrons between alternative channels in the full continuum.

Multi-channel spectrum of a rare gas

An example of an atomic spectrum with two ionization limits just as in Sec. III but with several channels and an asymptotic Coulomb field is provided by the spectrum of a rare gas. Thus, in place of F^- or I^- in Sec. III, if we consider photoionization of Ne or Xe, also $^1S_0$ ground states, the quantum numbers of the process remain the same. A major difference due to the Coulomb field is that the $\ell=2$ wave for the electron is now no longer suppressed even at low energies. In the presence of a longer range field like the Coulomb $-1/r$, the angular momentum barrier plays no role in the leading energy dependence of inelastic threshold cross-sections. We, therefore, have both outgoing s and d waves now.

In the enumeration of fragmentation channels we now have five

$$p^5(^2P_{3/2})s_{1/2}, \quad p^5(^2P_{1/2})s_{1/2}, \quad p^5(^2P_{3/2})d_{5/2}, \quad p^5(^2P_{3/2})d_{3/2}, \quad p^5(^2P_{1/2})d_{3/2},$$

$$(60)$$

all with J=1. The last three involving the d wave add to the earlier two considered in Sec. III. The jj coupling scheme can still be regarded as the appropriate one for the large-r limit. Likewise, at least in the lighter rare gases, LS coupling should provide the useful description of the eigenchannels at small r. Again there are five now

$$(p^5s)^3P, \quad (p^5s)^1P, \quad (p^5d)^3P, \quad (p^5d)^3D, \quad (p^5d)^1P, \quad (61)$$

the last three involving the d-wave. In a general case, we will refer to the fragmentation channels by the index i and the eigenchannels by the label α.

Steps of the analysis

The small-r α-channels have wave functions $F_\alpha(\omega,r)$, where ω stands for all the coordinates of the electrons in the complex except the radial coordinate r of the one electron which alone makes excursions to large r. Thus, the full N-electron wave function of the core as well as the orbital and spin functions of the outer electron are implicitly included through ω. The small-r solutions are expanded in terms of large-r energy-normalized solutions according to

$$F_\alpha(\omega,r) = \frac{1}{2i} \sum_i (2/\pi k_i)^{\frac{1}{2}} \, \Phi_i(\omega)[f^+(\epsilon_i,\ell_i,r)(-i)^{\ell_i} \, J^+_{i\alpha}(k_i)$$

$$- f^-(\epsilon_i,\ell_i,r)i^{\ell_i} J^-_{i\alpha}(k_i)] \, , \quad \epsilon > 0 \, . \tag{62a}$$

$$= \sum_i (1/\pi\kappa_i)^{\frac{1}{2}} \, \Phi_i(\omega)[f^+(\epsilon_i,\ell_i,r)J^+_{i\alpha}(\kappa_i) + f^-(\epsilon_i,\ell_i,r)J^-_{i\alpha}(\kappa_i)], \quad \epsilon_i < 0 \, .$$

$$(62b)$$

Contrast these with the single channel formulae in (46) and (53). In these expressions $\Phi_i(\omega)$ represents the fragmentation channel functions, the i label embracing all the discrete quantum numbers including those of orbit and spin for the outer electron. The energy label ε_i associated with the motion in r is the only running index along the channel. Each α channel is, of course, a superposition of all the i channels in the problem, and vice versa.

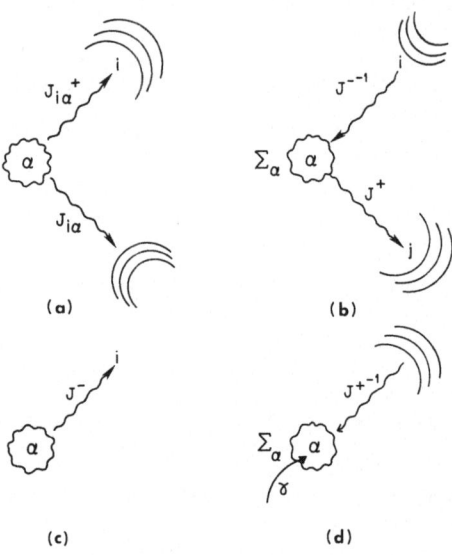

Fig. 3. Geometrical interpretation of the Jost matrices:

    a) $J_{i\alpha}^{\pm}$ as they connect short range α-channels to long range i-channels, the +(−) associated with outgoing(ingoing) waves, which are rendered pictorially by wavefronts pointing away from (into) the origin.

    b)   Scattering from i to j, mediated by the α channels.

    c)   Definition of a bound state. Following J^- from α to i connects to exponentially rising functions which have to be killed by selecting zeroes of J^-.

    d)   Photoionization into channel i.

Structure of the Jost matrices

For Coulomb and other long range potentials, the QDT parameters B and A do not have sensitive energy dependences near the threshold, $\varepsilon \approx 0$, unlike in the zero field problem of Sec. III. Therefore, it proves convenient in defining the solutions F_α at small r to work with the base set $\{f, g\}$ itself rather than $\{f^0, g^0\}$. This has been the common practice and avoids the carrying over of factors in B and A into the Jost matrices as in (47) although caution is necessary at the end in constructing cross-sections from the Jost matrices (see ref. 4). The Jost matrices take a transparently simple form, one that exhibits the generalization of the $e^{\pm i\delta}$ factors first encountered in Sec. II,

$$J_{i\alpha}^{\pm}(k_i) = (\pm i)^{\ell_i} e^{\pm i\eta_i(k_i)} U_{i\alpha} e^{\pm i\delta_\alpha^{(s)}} \qquad \varepsilon_i > 0 \qquad (63a)$$

$$J_{i\alpha}^{\pm}(\kappa_i) = \mp D^{\pm 1}(\kappa_i,\ell) U_{i\alpha} \frac{\cos}{\sin} [\beta(\kappa_i) + \delta_\alpha^{(s)}] \qquad \varepsilon_i < 0 . \qquad (63b)$$

Thus in (63a), the connection between α and i channels involves first the short range phase factor $e^{\pm i\delta_\alpha^{(s)}}$ which is insensitive to energy so that it carries no i dependence. Then a frame transformation $U_{i\alpha}$ achieves the geometrical projection onto the fragmentation channel space. Finally the phase shift accumulated in the long range propagation in the i-channel completes the definition. This η is a function of the energy in the i-channel and, therefore, carries a k_i dependence. Notice that the $(\pm i)^{\ell_i}$ factor serves to cancel the $-\ell_i \pi/2$ that is included in our definition of η in (44). Eq. (63b) for the negative energy values has a similar structure, $\beta(\kappa_i)$ now being the long range phase, trigonometric functions replacing the exponential oscillations, and finally a new element introduced by the D parameter which rescales the amplitudes of $f^{\pm}$ with an attendant rescaling in $J^{\pm}$. Fig. 3a gives a pictorial rendering of the Jost matrices in their role of connecting α to i channels.

The full wave function is a superposition over α-channels of the F_α in (62) with coefficients to be determined so that the appropriate boundary conditions at infinity are satisfied. The analysis of the three situations of interest, electron-ion scattering, bound states and photoionization will now be taken up one by one.

Scattering: The outgoing wave boundary condition required for scattering calls for a superposition of F_α with an ingoing wave $f^-(\varepsilon_i,\ell_i,r)$ *only* in the incident channel i. The coefficient of f^- in the superposition of (62a) must reduce, therefore, to δ_{ji} and this is accomplished by

$$\sum_\alpha F_\alpha(\omega,r) [(J^-)^{-1}]_{\alpha i} = \frac{1}{2i} \sum_j (2/\pi\kappa_j)^{\frac{1}{2}} \Phi_j(\omega) [-\delta_{ji} i^{\ell_i} f^-(\varepsilon_i,\ell_i,r)$$

$$+ (-i)^{\ell_j} f^+(\varepsilon_j,\ell_j,r) \sum_\alpha J_{j\alpha}^+ (J^{--1})_{\alpha i}] . \qquad (64)$$

There are outgoing waves in all the j channels and the coefficient of $f^+(\varepsilon_j,\ell_j,r)$ gives the S-matrix element for scattering from i into j

$$S_{ji} = \sum_\alpha J^+_{j\alpha} [(J^-)^{-1}]_{\alpha i} . \tag{65}$$

Fig. 3b gives an immediate geometrical visualization of this expression, the initial $(J^-)^{-1}$ passing <u>from</u> the i <u>to</u> the α channel (ingoing waves) and then J^+ (outgoing waves) taking the α into j; a summation over all α completes the description. The S matrix provides the full scattering information. The construction of cross-sections in terms of it proceeds in the usual way given in many a textbook.[4]

Bound states: The quantum mechanical analysis of bound states is always done by following an initial regular solution at the origin to asymptotic r and picking out those energy values for which it decays exponentially at infinity. This prescription now amounts to looking for the superposition $\Sigma_\alpha F_\alpha(\omega,r) A_\alpha$ such that in (62b), the coefficient of every $f^-(\varepsilon_i,\ell_i,r)$ vanishes; recall that, with $k_i = i\kappa_i$, f^- blows up at infinity. The condition $\Sigma_\alpha J^-_{i\alpha}(\kappa_i) A_\alpha = 0$ gives, therefore, the eigenvalue equation for bound states

$$\text{Det}|J^-_{i\alpha}(\kappa_i)| = 0 . \tag{66}$$

Zeroes of J^- determine, therefore, the bound states in the spectrum, again generalizing the condition (6) first encountered in Sec. II. Fig. 3c also makes plausible why it is J^- that determines bound states, given its role of going from small r to the exponentially rising functions at large-r.

In terms of the QDT parameters, the condition (66), upon folding in (63b), reduces to

$$\text{Det}|U_{i\alpha}\sin[\beta(\kappa_i) + \delta^{(s)}_\alpha]| = 0 . \tag{67}$$

For Coulomb problems $\beta(\kappa_i)$ may be replaced with $\pi\nu_i$ according to (55), so that (67) is the multi-channel generalization of (58a). Thus, the generalization of the single-channel Rydberg description of bound state energy levels assigns to each level a set of i-channel effective quantum numbers ν_i, defined as in (1) but with respect to the corresponding ionization limit of that channel. Thereby the different ν_i are implicitly related. Upon combining these relations with (67), which gives another relationship between the ν_i, the values that satisfy these equations simultaneously provide the bound levels. Sec. VI will give an illustration.

Photoionization: For the photoionization of an atom, the super-position of α channel functions $F_\alpha(\omega,r)$ must satisfy the ingoing wave boundary condition, wherein an outgoing wave f^+ exists only in the specific channel i in which the photoelectron is detected. We have, therefore,

$$\sum_\alpha F_\alpha(\omega,r) [(J^+)^{-1}]_{\alpha i} = \frac{1}{2i} \sum_j (2/\pi k_j)^{1/2} \, \Phi_j(\omega) [\delta_{ji}(-i)^{\ell_i} f^+(\varepsilon_i,\ell_i,r)$$

$$- i^{\ell_j} f^-(\varepsilon_j,\ell_j,r) \sum_\alpha J^-_{j\alpha}[(J^+)^{-1}]_{\alpha i}] . \tag{68}$$

Note the appearance of the inverse of the S-matrix as a coefficient of the ingoing wave piece. This is immediately plausible from the earlier discussion of a time reversal connection between outgoing and ingoing boundary conditions.

The half-scattering aspect of photoprocesses, also discussed earlier in Sec. III, can be seen from the left hand side of (68). J^+ describes the outgoing wave for the electron which is recorded in the i-th channel. Tracing it back to the small-r region requires the inverse of J^+. There, it is weighted by the dipole matrix elements which are evaluated between the initial state wave function and the $F_\alpha(\omega, r)$. Finally a summation over α gives the net contribution at infinity. Fig. 3d illustrates this point. The appearance of $(J^+)^{-1}$ in photoionization in contrast to $(J^-)^{-1}$ in scattering also has a ready physical picture. In the case of scattering, J^-, the coefficient of the ingoing f^-, is what one needs to create the α-channel complex in the first half of scattering. Since this is a traversal from i to α, it is $(J^-)^{-1}$ that enters. The complex then evolves outwards to j and this is governed by J^+. For photoionization, it is the inverse of this half of the scattering that one needs to relate the amplitude detected at infinity back to the α-channels where it is weighted by the spectroscopic strength of the channel. As a final remark, when considering photoionization with some thresholds energetically closed, the corresponding k_i are again replaced by $-i\kappa_i$ as in Sec. III. This ensures the exponential decay in those channels of the f^- piece in (68). More generally, when considering scattering with a certain number of channels closed, they may be eliminated by setting the coefficient of the corresponding f^- to zero, retaining only the exponentially decreasing f^+ term in those channels. The full S-matrix can thus be reduced to one with dimension equal to the number of open channels.[2,4] The closed channels may then manifest themselves as resonances in the scattering.

VI. AN ILLUSTRATION

In this final section, we illustrate through a simple example the general ideas developed so far, especially in Sec. V. We will also point here to a powerful graphical rendering of the analysis which has proved useful even as an organizing principle for experimental data. Our example is of a Coulomb spectrum with two channels and two limits. Our discussion will actually be for the Ne J=1 spectrum over energy ranges where the outgoing s and d waves mix slightly. Thus the full five-channel, two-ionization-limits problem of a rare gas in (60) and (61) can be simplified to one with two channels $p^5(^2P_{3/2,1/2})s$ J=1. In this way, we are back to the problem considered in Sec. III except that there is now a long range Coulomb field.[21]

Each bound state energy level in the spectrum is converted to two effective quantum numbers according to

$$E_n = I_1 - \frac{1}{2\nu_1^2} = I_2 - \frac{1}{2\nu_2^2} , \tag{69}$$

where I_1 and I_2 are the two ionization limits $I_{3/2}$ and $I_{1/2}$ and all energies are again measured from the ground state. The bound state condition (67) becomes

$$\text{Det}|U_{i\alpha}\sin\pi(\nu_i + \mu_\alpha)| = 0 , \tag{70}$$

with i=1,2 and α=I,II, and $\pi\mu_\alpha$ the short range eigen phase shifts $\delta_\alpha^{(s)}$. Explicitly, we have

$$\begin{vmatrix} \cos\theta \, \sin\pi(\nu_1+\mu_I) & \sin\theta \, \sin\pi(\nu_1+\mu_{II}) \\ -\sin\theta \, \sin\pi(\nu_2+\mu_I) & \cos\theta \, \sin\pi(\nu_2+\mu_{II}) \end{vmatrix} = 0 , \tag{71}$$

where θ is the rotation angle of the frame transformation as in (28).

Eqs. (69) and (71) provide two relations between ν_1 and ν_2. The procedure for reducing experimental spectral data is to reduce E_n's to ν_1 and ν_2 from (69) and plot them as in Fig. 4(a). The solid line that describes (71) is drawn through the points, adjusting θ and μ_α as necessary. In fact, Fig. 4 shows the plotting in two steps because (71) is periodic modulo 1 in both ν_1 and ν_2. Thus all the points and curves can be folded into the unit square plot shown in Fig. 4b. Such a plot is called a Lu-Fano

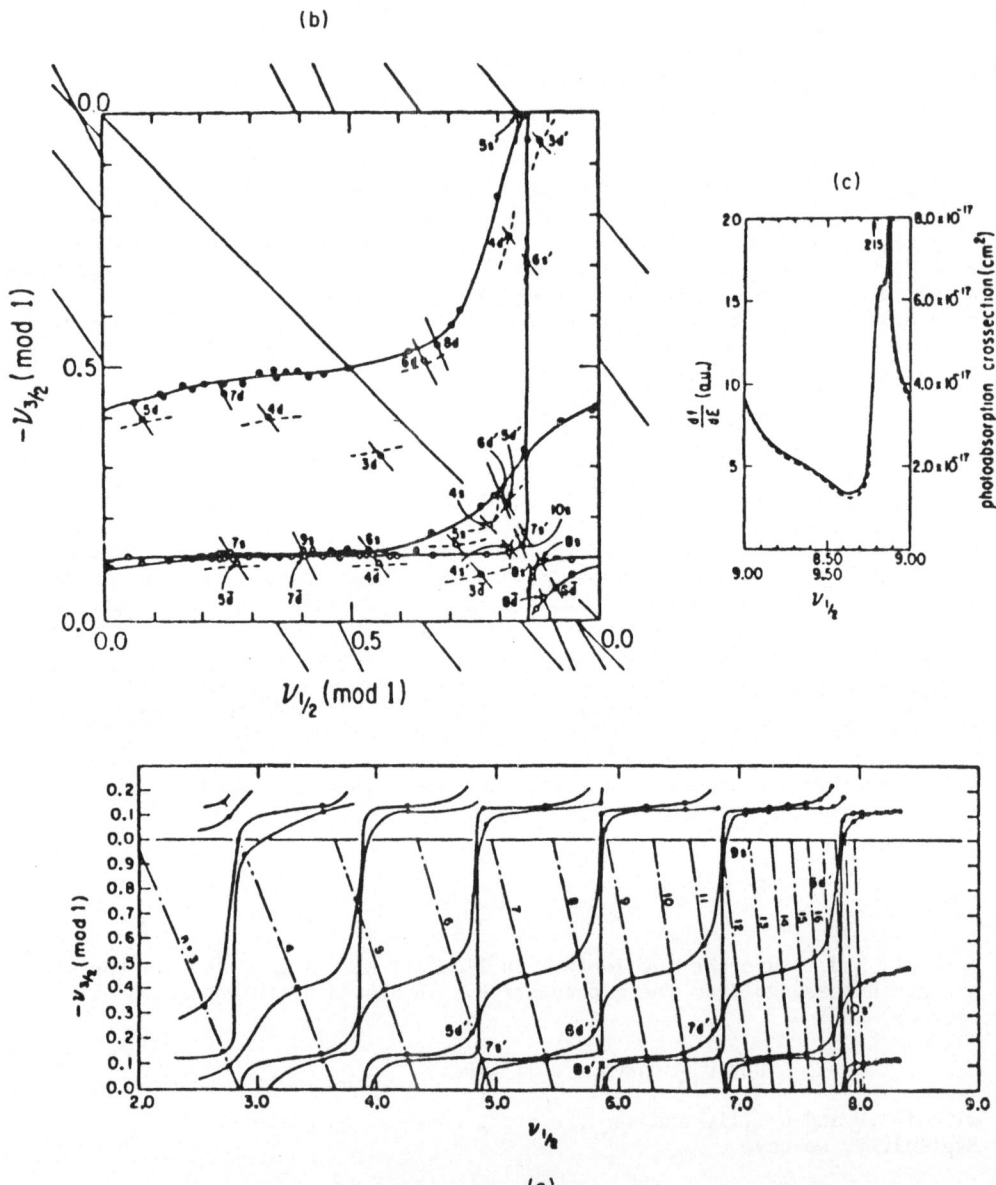

Fig. 4. Lu-Fano plots for photoabsorption in Ar.
  a) Extended plot of quantum defects. —— Eq. (70),
      —— - —— Eq. (69).
  b) Reduced (modulo 1) Lu-Fano plot.
  c) Autoionization profile. --- experiment, —— QDT fit. (from ref. 24).

74

plot.[22] Fig. 4 actually displays the d channels as well but the s channels of interest to our illustration are the two curves in Fig. 4b which run almost horizontally (at $-\nu_{3/2} \approx 0.1$) and vertically (at $\nu_{1/2} \approx 0.85$) with a sharp "avoided crossing" around the point $\nu_{1/2} \approx 0.85$, $-\nu_{3/2} \approx 0.1$. These two curves are actually a single curve because the periodic nature of the plot identifies points on opposite sides of the square. The dashed, slanted lines in Fig. 4a and the similar slanted lines shown along the edges of Fig. 4b describe (69). The actual energy levels lie, therefore, at the intersection of the two families of lines.

The availability of many spectral levels and the large-redundancy in the unit square plot permit an accurate reduction of the experimental data to get the 3 key parameters θ, μ_I and μ_{II} . With these parameters and two dipole amplitudes d_α, QDT then provides a complete accounting of the spectrum over

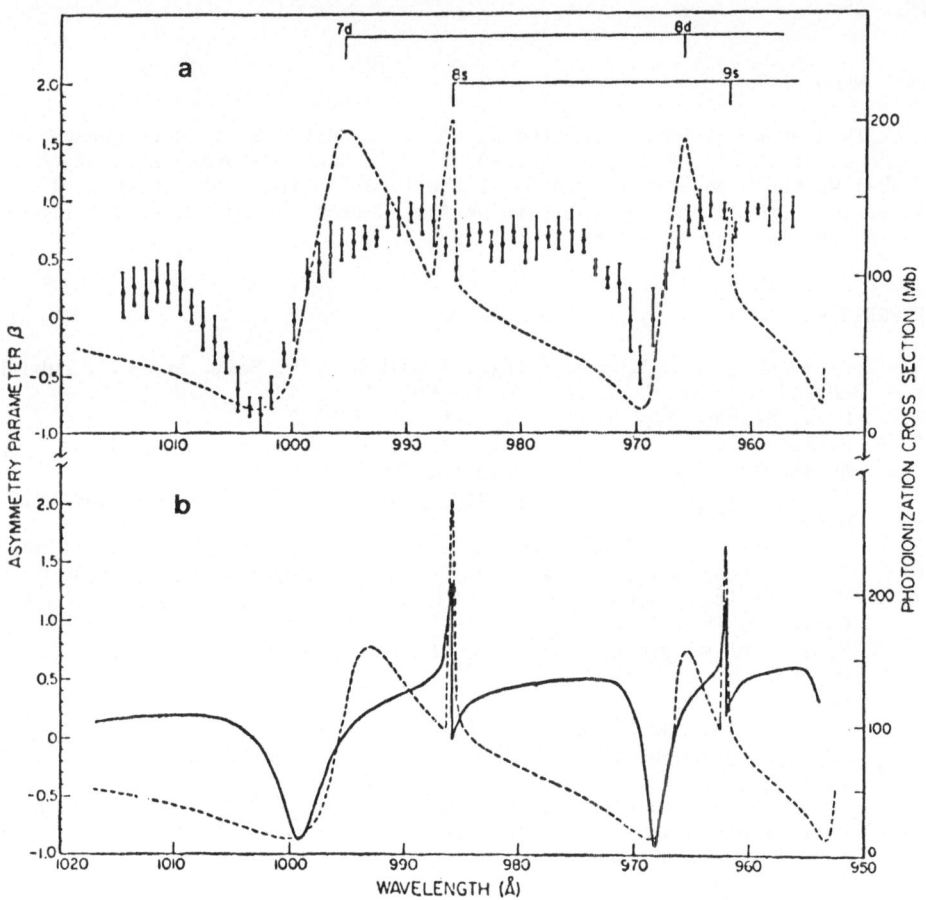

Fig. 5.   Photoionization cross-section and the asymmetry parameter β of the angular distribution $d\sigma/d\Omega = (\sigma/4\pi)\,[1+\beta\,P_2(\cos\theta)]$ for Xe between the $^2P_{3/2}$ and $^2P_{1/2}$ ionization thresholds.
a)   Experimental β values and photoabsorption resonances
b)   Theoretical β values (solid lines) and photoabsorption cross-sections (dashed lines) (from reference 23).

its entirety: 1) bound state energy positions and oscillator strengths (see Fig. 4c for the periodic variation of oscillator strengths along the spectrum); 2) profiles, positions and widths of autoionization levels lying between the two limits (Fig. 5 shows experiment and theory for photoionization cross-sections and the parameter β that governs the angular distribution of the photoelectrons)[23]; and 3) the branching ratios for the two groups of photoelectrons above the higher $I_{1/2}$ threshold. The full five-channel, two-limits analysis of a rare gas spectrum requires 5 μ_α's , 5 d_α's and the 10 parameters that define the 5x5 orthogonal matrix $U_{i\alpha}$.[24] Such multi-channel QDT analyses of complex atomic and molecular spectra are constantly being extended and refs. 2, 4 and 6 supply references to date.

In summary then, these lectures have presented a systematic introduction to the subject of quantum defect theory which views scattering and bound state phenomena in atoms and molecules in a unified way. The singling out of key aspects and variables permits a judicious application of analytical, numerical and semi-empirical techniques to bring about unity to diverse data. The theory continues to grow into new directions and the applications to specific spectra are also constantly expanding. Quantum defect theory has become a central element of modern spectroscopy and collision theory in atomic physics.

ACKNOWLEDGMENT

This work has been supported by the U. S. National Science Foundation under Contract PHY 84-01855 and 86-07721. I thank the organizers of the (1984) Workshop on Atomic Physics at the Tata Institute of Fundamental Research, Bombay, and the Laboratoire Aimé Cotton, Orsay, for invitations that led to the initial development of these lectures.

REFERENCES

1.   F. S. Ham, in Solid State Physics edited by F. Seitz and D. Turnbull, eds., Academic, New York (1955), Vol. 1, p. 127.

2.   M. J. Seaton, Rep. Prog. Phys. 46, 167 (1983).

3.   U. Fano, J. Opt. Soc. Am. 65, 979 (1975); J. J. Wynne and J. A. Armstrong, Comm. At. Mol. Phys. 8, 155 (1979).

4.   U. Fano and A. R. P. Rau, Atomic Collisions and Spectra, Academic, Orlando (1985).

5.   C. H. Greene, A. R. P. Rau and U. Fano, Phys. Rev. A 24, 2441 (1982).

6.   C. H. Greene and Ch. Jungen, Adv. At. Mol. Phys. 21, 51 (1985).

7.   V. A. Zilitis, Opt. Spectrosc. 43, 603 (1977); W. R. Johnson and K. T. Cheng, J. Phys. B12, 863 (1979).

8.   L. I. Schiff, Quantum Mechanics, McGraw Hill, New York (1968), 3rd edition, p. 40.

9.   ref. 8, p. 354.

10. L. D. Landau and E. M. Lifshitz, Quantum Mechanics: Non-Relativistic Theory, Pergamon, Oxford (1977), 3rd edition, p. 106.

11. E. P. Wigner, Phys. Rev. 73, 1002 (1948).

12. E. U. Condon and G. H. Shortley, The Theory of Atomic Spectra, Cambridge Univ. Press, Cambridge (1951), Sec. 7^4.

13. ref. 12, Sec. 6^{12}.

14. A. R. P. Rau and U. Fano, Phys. Rev. A4, 1751 (1971).

15. H. Hotop, T. A. Patterson and W. C. Lineberger, Phys. Rev. A8, 762 (1973).

16. M. Abramowitz and T. Stegun, Handbook of Mathematical Functions, Dover, New York (1965), chapter 14.

17. ref. 8, Sec. 39. See also R. G. Newton, Scattering Theory of Waves and Particles, Springer-Verlag, New York (1982).

18. U. Fano and J. W. Cooper, Rev. Mod. Phys. 40, 441 (1968), Fig. 1.

19. P. G. Burke and D. Robb, Adv. At. Mol. Phys. $\underline{11}$, 144 (1975); U. Fano and C. M. Lee, Phys. Rev. Lett. $\underline{31}$, 1573 (1973); C. M. Lee, Phys. Rev. A $\underline{10}$, 584 (1974).
20. U. Fano, Phys. Rev. A $\underline{2}$, 353 (1970); G. Herzberg and Ch. Jungen, J. Mol. Spectrosc. $\underline{41}$, 425 (1972).
21. A. F. Starace, J. Phys. B $\underline{6}$, 76 (1973).
22. K. T. Lu and U. Fano, Phys Rev. A $\underline{2}$, 81 (1970) , and ref. 3.
23. D. Dill, Phys. Rev. A $\underline{7}$, 1976 (1973); J. A. R. Samson and J. L. Gardner, Phys. Rev. Lett. $\underline{31}$, 1327 (1973).
24. K. T. Lu, Phys. Rev. A$\underline{4}$, 579 (1971); C. M. Lee and K. T. Lu, Phys. Rev. A$\underline{8}$, 1241 (1973).

QUANTUM DEFECT THEORY FOR MOLECULES

Ch. Jungen

Laboratoire de Photophysique Moléculaire du CNRS
Université de Paris-Sud, 91405 Orsay, France

1. Introduction

During the past 15 years, the developments of molecular quantum defect theory and of the high-resolution spectroscopy of molecular Rydberg states have run largely parallel and have mutually stimulated each other. Before 1970, only moderately high molecular Rydberg states had been observed whose effective principal quantum numbers ranged typically from 2 to 6. The most complete observations were made for H_2 and NO.[1,2] Mulliken [3] , in a famous series of papers which appeared in the sixties, summarized the knowledge of Rydberg states in molecules and gave a detailed conceptual description of their electronic and fine structures. He discussed in particular how the electronic wavefunction of a diatomic Rydberg system evolves through various stages when the nuclei move adiabatically farther and farther apart. While the structures of the low Rydberg states showed many interesting features reflecting the beginning uncoupling of the Rydberg electron from the nuclear framework as well as interactions with valence states, it was not obvious how these states would be related to the scattering or continuum states observed in collision or photoionization experiments.

A decisive step occurred in the development in 1969, when Herzberg [4] (and almost simultaneously Takezawa [5]) succeeded in obtaining highly resolved absorption spectra of H_2 which showed structure up to and beyond the ionization threshold. By using $para-H_2$ cooled to $liquid-N_2$ temperature, Herzberg was able to restrict the rotational-vibrational Boltzmann population distribution essentially to a single level, the $v'' = 0, J'' = 0$ ground level. (Double primes generally designate quantum numbers associated with the lower state of an optical transition). As a result every vibrational-electronic excited state is represented in the absorption spectrum by a single spectral line, $J = 1 \leftarrow v'' = 0, J'' = 0$ and the Rydberg structure therefore emerges clearly. Fig. 1 shows that the observed spectrum near threshold is nevertheless quite complex.

In a preliminary analysis we plot the total energy of each observed spectral line

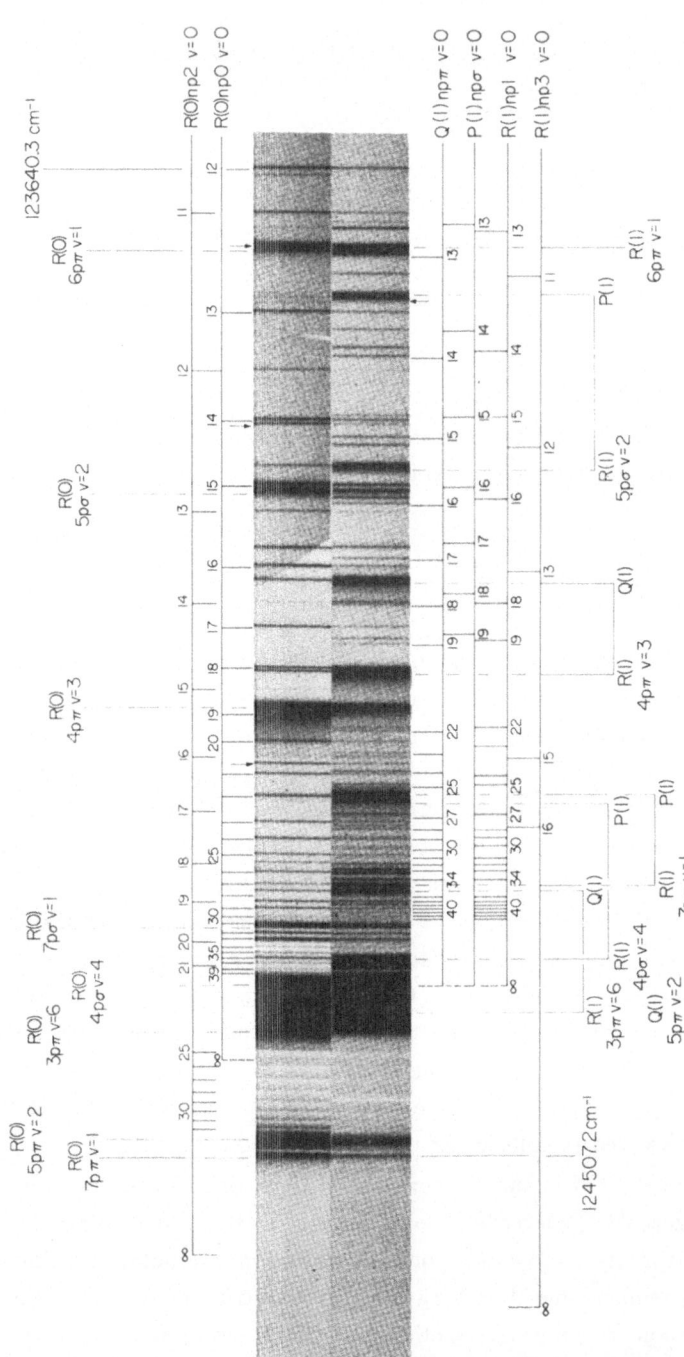

Fig. 1 Absorption spectrum of $p-H_2$ at liquid nitrogen temperature in the region 80.88–80.20 nm. After Ref. 16.

versus the part of the excited state energy which (very approximately, as we shall see later) is present in the vibrational and rotational degrees of freedom. The level diagram which emerges in this way is shown in Fig. 2. This diagram has prototype character in the sense that it represents the molecular analog of a single unperturbed atomic ionization channel (Rydberg series plus adjoining ionization continuum), as shown at the left of Fig. 2 for comparison. The tremendous complication arising through the presence of the nuclear degrees of freedom is evident.

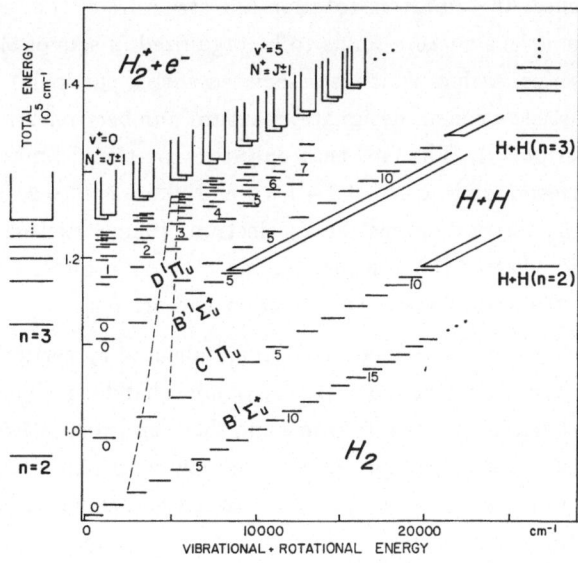

Fig. 2 Optically allowed levels in photoabsorption of H_2.

The H_2 molecule has played a prototype role in the development of molecular multichannel quantum defect theory (MQDT) in recent years - much as it did 50 years earlier when molecular electronic spectroscopy first began to bloom. We shall in this contribution concentrate on this system, our viewpoint being that the ideas developed here are significant for any molecular system. The aim of the present contribution is to describe how the familiar Rydberg equation must be generalized and extended such as to account for the complexities of the spectrum of Fig. 1, and how all of the levels and continua shown in Fig. 2 can be treated *as a unit*. The reader interested in the numerous applications of MQDT to other molecular systems is referred to a recent review [6].

2. Manifestations of channel interactions in H_2 spectra

A close inspection of the level pattern of Fig. 2 reveals a number of basic characteristics of a molecular ionization channel.

(i) The single atomic ionization threshold is split into a multitude of close-lying thresholds which reflect the vibration-rotation fine structure of the residual ion . Near these thresholds the level structure tends to be organized in sets of electronic progressions, or Rydberg series, each of which is associated with a particular threshold v^+, N^+. (Throughout this article we shall designate quantum numbers referring to the free ion core by the superscript +). The fact that only two rotational limits appear for each v^+ (cf. Fig. 1), corresponding to $N^+ = J - 1 = 0$ and $N^+ = J + 1 = 2$, indicates that the electron excited in H_2 absorption must be a p electron. (Singly excited $l = 0$ and 2 states are dipole-forbidden whereas the non-penetrating $l = 3, 5...$ orbitals have virtually zero radial overlap with the ground state wavefunction and are not coupled with $l = 1$).

(ii) In addition to the set of *ionization* continua, indicated by vertical hatching in Fig. 2, several *dissociation* continua are indicated by oblique hatching. These correspond to fragmentation of excited H_2 into two H atoms and they join on to *vibrational* progressions (variable vibrational quantum number v and fixed low electronic quantum number n). Each of these vibrational series is characterized by a well-defined electronic symmetry: there are $^1\Sigma_u^+$ and $^1\Pi_u$ electronic states. (The latter are identified experimentally in the spectrum of $ortho - H_2$ by the fact that the $J = 0$ level is absent).

(iii) The low and high energy portions of the pattern of Fig. 2 suggest two different ways of writing the wavefunction of the molecule.[3] At low energy a suitable form is a simple Born-Oppenheimer product

$$\Psi \approx \Psi_{el}\Psi_{vib}\Psi_{rot} = (A\Psi_{core}\psi_{Ryd})\Psi_{vib}\Psi_{rot}, \tag{1}$$

which, following Mulliken[3], we may call a Rydberg-coupled function. (The operator A in Eq.(1) makes Ψ antisymmetric in all the electrons). Eq.(1) indicates that the vibrational motion is governed by a potential $U(R)$ determined by the joint fields of the core electron, the Rydberg electron and the internuclear repulsion. Eq.(1) expresses the fact that the lower part of the level spectrum of Fig. 2 is dominated by *vibrational/rotational* level progressions. Near threshold a suitable Rydberg-uncoupled wavefunction has the form

$$\Psi \approx A[(\Psi_{core}\Psi_{rot}\Psi_{vib})\psi_{Ryd}], \tag{2}$$

i.e. ψ_{Ryd} is no longer included within the electronic factor of the B.O. approximation.

The B.O. approximation is still satisfied for the core, and the vibrational motion is now determined by the potential $U^+(R)$ of the core electron and the internuclear repulsion only. Eq.(2) expresses the fact that the upper part of the level spectrum of Fig. 2 is characterized by *electronic* progressions of levels.

A careful look at Fig. 2 convinces us that there is no one-to-one correspondence between the levels described approximately by Eq.(1) and those described by Eq.(2), respectively. Two broken lines connect all $\Lambda = 0(^1\Sigma_u^+)$ or $N^+ = 0$ levels with two vibrational quanta on the one hand, and all corresponding $\Lambda = 1(^1\Pi_u)$ or $N^+ = 2$ levels on the other hand. It can be seen that for low excitation these two lines are not parallel to each other. This tells us that the vibrational/rotational motion in the molecular complex is different from that of the free ion and depends on the molecular symmetry. In particular we see that the vibrational quantum in the lowest $^1\Sigma_u^+$ states is substantially reduced with respect to the ion value.

The preceding considerations show that the intra-molecular motion follows two regimes depending on whether the Rydberg electron is near the core or far away. We denote the two sets of states by $\{\alpha\} = \{R, \Lambda\}$ (because the Rydberg electron modifies the molecular potential $U(R)$ for each individual R value), and $\{i\} = \{v^+, N^+\}$. Now, even a Rydberg electron with high principal quantum number, or a continuum electron, will always have some amplitude near the core, and therefore a linear combination of several product functions of the type Eq.(2) will be required to describe its motion. In terms of Fig. 2 this means that the various channels v^+, N^+ near threshold are *coupled*, i.e. the system may pass from one channel to another. The important point at this stage is the realization that the low Rydberg states provide us with some information on how the dynamical regime changes when the outer electron approaches the core. In other words the lower part of the level diagram contains information on the degree of channel coupling occurring in the upper part. Similarly, small deviations from Born-Oppenheimer behavior in the lower states may be regarded as the beginning transition to the Rydberg-uncoupled regime, due to the fact that even a low energy Rydberg electron has some small amplitude far from the core.

We now examine the experimental manifestations of channel couplings in the different regions of the spectrum. In the lowest isolated states deviations from the B.O. approximations arise as the so-called adiabatic shifts [7] , which slightly displace the whole rovibrational fine structure of a state without altering its pattern. These shifts are inversely proportional to the reduced nuclear mass; they reflect the kinetic energy of the nuclei associated with the motion of the electrons (finite mass correction), and also the small additional electron kinetic energy arising because the electrons follow the slow nuclear motion. These corrections are shown in Fig. 3 for the $2p\pi\,^1\Pi_u$ states of H_2 and D_2 . At higher excitation energy spectral perturbations occur. This is illustrated by Fig. 4 which displays an enlarged section of the absorption spectrum of Fig. 1. The example shown involves the series converging towards the $v^+ = 0, N^+ = 0$ and $v^+ = 0, N^+ = 2$ thresholds. The presence of coupling between the two series is evidenced by the irregular spacings and intensities of successive Rydberg members. In the quantum defect plot shown at the far left of the diagram each perturbation shows up as a characteris-

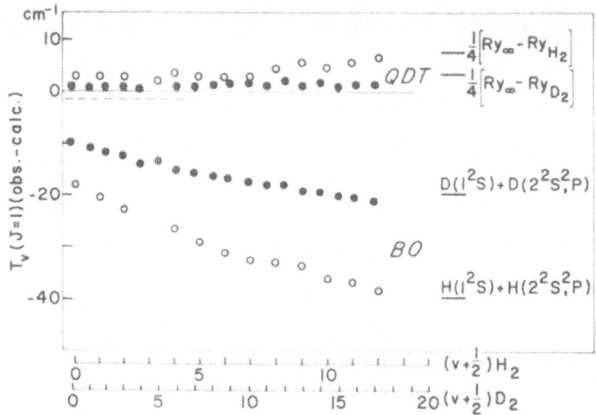

Fig. 3 Deviations of the observed $J = 1$ levels in the $2p\pi^1\Pi_u^-$ state from those derived *ab initio* for H_2 (circles) and D_2 (dots) in the Born-Oppenheimer approximation (BO) and using quantum defect theory (QDT). The difference between the two results represents the adiabatic and non-adiabatic corrections. The remaining deviations in QDT correspond to the specific isotope effect. After Ref. 18.

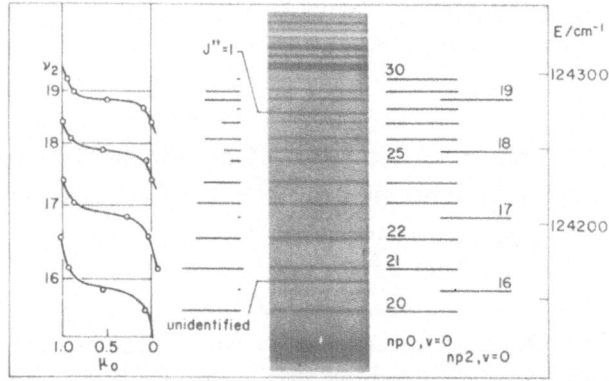

Fig. 4 Rotational perturbations between highly excited Rydberg levels in H_2. The spectrogram is adapted from Ref. 16 and represents absorption to $J = 1$, odd parity final-state levels. To its right are shown the (strong) np series converging to the $v^+ = 0, N^+ = 0$ level of H_2^+ and the perturbing (weak) series converging to the $v^+ = 0, N^+ = 2$ level. The stick spectrum to the left of the spectrogram has been calculated using quantum defect theory. Far left: observed (circles) and calculated (full lines) quantum defects of the individual levels evaluated with respect to the $v^+ = 0, N^+ = 0$ limit. After Ref. 6.

tic resonance pattern of the quantum defect, which is not a constant. When the total energy exceeds that of the lowest ionization threshold, interchannel coupling connects open channels (electronic continua) and closed channels (discrete levels). Preionization occurs, which synonymously is called autoionization. The channel interaction leads to a conversion of rotational-vibrational energy into electronic energy, and therefore the decay process is called rotational-vibrational preionization. Fig. 5 presents a section of the high-resolution photoionization spectrum of Dehmer and Chupka [8] just near the $v^+ = 0, N^+ = 0$ threshold. The resulting broad asymmetric autoionization line profiles can be discerned very clearly. The effect of preionization is also visible in Fig. 1 in many places. Further, the channel interactions may modify the product distributions

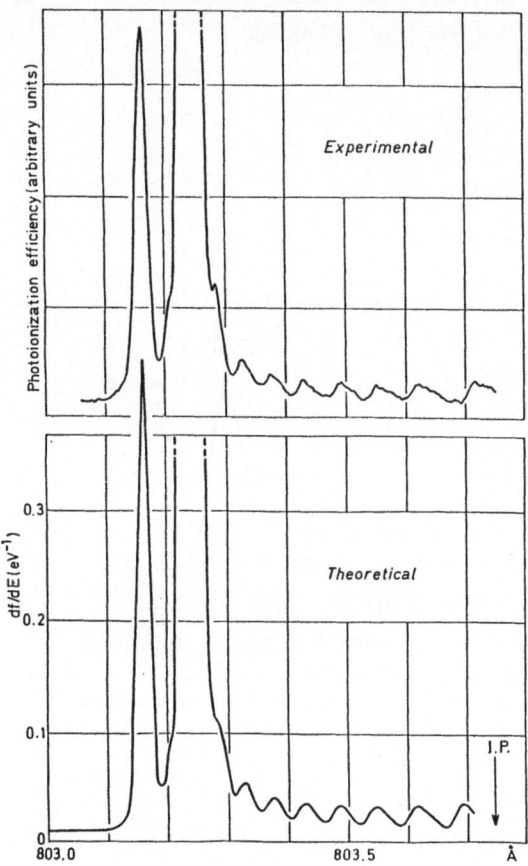

Fig. 5 Top: Relative photoionization cross section of $p-H_2$ in the region of the ionization threshold at $78K$ taken at a wavelength resolution of $0.0016\ nm$ (after Ref. 8). Bottom: MQDT calculation of the same spectrum from Ref. 14. The theoretical spectrum has been broadened to correspond to the experimental resolution.

in fragmentation processes. As an example, Fig. 6 shows a section of the rotationally resolved photoelectron spectrum of H_2 recorded by Pollard et al. [9] in excitation near the $v^+ = 6$ threshold. Two peaks are observed which correspond to rotational quantum numbers $N^+ = J''$ and $N^+ = J'' + 2$, and are the continuum analogs of the two Rydberg series observed in absorption near each vibrational threshold (cf. Figs. 1 and 2). The presence of only two peaks, i.e. the absence of the $N^+ = J'' + 4$ and higher peaks, once again point to the fact that the electronic transition has predominant $s \rightarrow p$ character. The fact that the $N^+ = J''$ peak dominates by far indicates that in first order the the bound-free transition leaves the rotational quantum number unchanged. This is characteristic of a so-called *direct* $s \rightarrow p$ bound-free transition, whereby the photoelectron carries away the angular momentum provided by the absorbed photon, with no possibility for exchange of angular momentum between the electron and the core. Fig. 5 shows that this selection rule is not exactly verified: the small photoelectron intensity of the $N^+ - J'' = 2$ group results from a higher order effect involving the coupling between the $N^+ = J''$ and $N^+ = J'' + 2$ $(J = 1)$ channels.

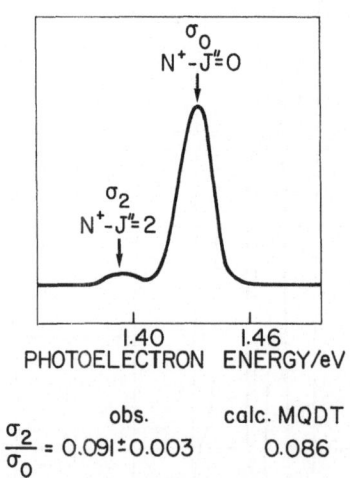

$$\frac{\sigma_2}{\sigma_0} = 0.091 \pm 0.003 \qquad 0.086$$

obs.   calc. MQDT

Fig. 6 Top: Rotationally resolved photoelectron spectrum of the transition $H_2(v'' = 0) \rightarrow H_2^+(v^+ = 0)$ for excitation at $73.6\ nm$ (electron collection at the magic angle). After Ref. 9. Bottom: observed (top spectrum) and calculated (MQDT, Ref. 23) partial cross section ratio σ_2/σ_0 .

3. The frame transformation approach

Although the preceding considerations suggest that the manifold manifestations of channel interaction in H_2 have a common physical origin, their unified treatment may at this point appear as a hopelessly complicated task - it was indeed perceived as such 20 years ago [10].

A decisive step was taken in 1970 when Fano [11] outlined how Seaton's [12] multichannel quantum defect theory should be combined with the concept of frame transformations to yield a manageable molecular theory. Fig. 7(a) is a plot of the radial Coulomb potential for a $l = 1$ electron in the asymptotic field of the molecular core. Since the core may be in one of several alternative vibration-rotation states, we must consider a whole set of such potentials, which are displaced from one another by the core level spacings. Fig. 7(a) illustrates two such potentials which are displaced by approximately the vibrational frequency of H_2^+. The shading indicates, roughly to scale, the core region where the

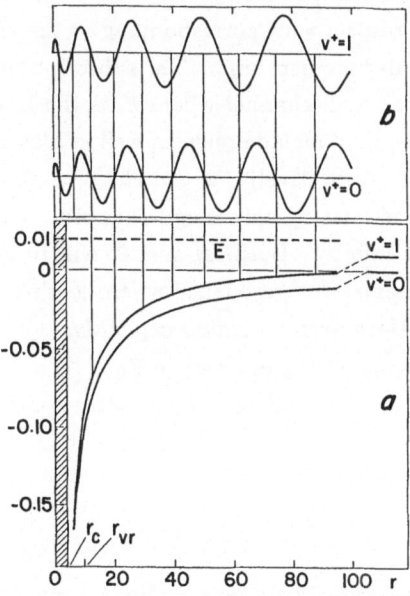

Fig. 7 (a) Potential and kinetic energies of an electron outside an ion core. The total energy E corresponds to a situation where the electron is in the continuum and has asymptotic energies $\epsilon = +0.02\ (+0.01)$ $a.u.$ above the $v^+ = 0$ ($v^+ = 1$) thresholds, respectively. The core region is indicated by hatching. (b) shows the corresponding radial electron wavefunctions for $l = 0$. Note how up to $r \approx 10\ a.u.$ the two wavefunctions remain in phase, whereas for $r \geq 50\ a.u.$ they are out of phase.

Coulomb potential turns into more complicated interactions involving all four particles individually. Fig. 7(b) shows two examples of radial Coulomb functions. Their energies have been chosen to differ by an amount corresponding to a vibrational quantum in such a way that the total energy is the same. Fig. 7(b) shows - and this is one of the main arguments in Fano's paper - that there is *a whole range of r values outside the core (* $r_c \leq r$) where the radial electron wavefunctions corresponding to the same total energy are *independent* of the internal vibrational state of the core.

We return at this point to Eqs.(1) and (2), which we rewrite in collisional form outside the core for a given total energy E. A superposition of Rydberg-uncoupled products (Eq.(2)) takes the form

$$\Psi_{r_c \leq r}(E) \approx \sum_i \Phi_i(\omega)[f_l(\epsilon_i, r)cos\pi\bar{\mu}_i(E) - g_l(\epsilon_i, r)sin\pi\bar{\mu}_i(E)]\bar{B}_i(E). \tag{3}$$

The factors $\Phi_i(\omega)$ include the core wavefunction $i = \{1\sigma_g, v^+, N^+\}$ and the angular part of the Rydberg electron wavefunction ψ_{Ryd} . f_l and g_l are the regular and irregular radial Coulomb functions, respectively, corresponding to the electron energy in channel i possessing an orbital angular momentum l. (The subscript l will be dropped from here on). One has $E = E_i + \epsilon_i$ for each channel where E_i is the internal energy of the core. It is understood that Eq.(3) may include open as well as closed channels corresponding to ϵ_i positive and negative, respectively. $\bar{\mu}_i$ are effective phase shifts and B_i channel amplitudes to be determined later by enforcing the appropriate boundary conditions at the core boundary and at infinity. Both, $\bar{\mu}_i$ and $\bar{B}_i$ will in general vary rapidly with energy E, e.g. near a preionization resonance where $\bar{\mu}_i$ rises sharply by unity. Their dependence on E has therefore been indicated explicitly.

The different arrangements of the brackets in Eqs. (1) and (2) indicate the different coupling conditions in the Rydberg-coupled and Rydberg-uncoupled situations, respectively. The states $i(= v^+, N^+)$ and $\alpha(= R, \Lambda)$ are linked by an orthogonal transformation with elements $U_{i\alpha}$:

$$\Phi_i(\omega) = \sum_\alpha U_{i\alpha} X_\alpha(\omega), \tag{4}$$

where X_α are the Rydberg-coupled core plus Rydberg angular functions. This transformation will be specified below.

The collisional form of the Rydberg-coupled (B.O.) wavefunction can be written down in the region where the vibration-rotation energy is negligible compared with the electronic energy, that is precisely in the range $r \leq r_{vr}$ displayed in Fig. 7. We have

$$\Psi_{r_c \leq r \leq r_{vr}}(E) \approx \sum_\alpha X_\alpha(\omega)[f(r)cos\pi\mu_\alpha - g(r)sin\pi\mu_\alpha]A_\alpha(E). \tag{5}$$

Since the dependence of f and g on the energy is small in this region on the scale of typical vibration-rotation energies, it has been suppressed in Eq.(5). Similarly, the $\pi\mu_\alpha$ are phase shifts reflecting the matching of the electron wavefunctions for $r \leq r_c$ and $r \geq r_c$ at the core boundary, and they should normally also vary only slowly with energy. On the other hand, the channel amplitudes $A_\alpha(E)$ which remain unspecified at this stage pending application of the boundary conditions at infinity, will vary quite rapidly with energy. For example, the $v = 1$ level of a low Rydberg state will be characterized by quite different coefficients A_α than the $v = 0$ level of the same state.

Obviously the two expressions Eqs.(3) and (5) are equivalent in the inner zone where $f(\epsilon_i, r) \approx f(\epsilon_{i'}, r) \approx f(r)$. By comparing the two expressions we see at once that we must have

$$\sum_i \Phi_i(\omega) cos\pi\bar{\mu}_i(E)\bar{B}_i(E) = \sum_\alpha X_\alpha(\omega)cos\pi\mu_\alpha A_\alpha(E) \tag{6a}$$

which implies

$$cos\pi\bar{\mu}_i(E)\bar{B}_i(E) = \sum_\alpha U_{i\alpha}cos\pi\mu_\alpha A_\alpha(E), \tag{6b}$$

with $U_{i\alpha} = \langle \Phi_i | X_\alpha \rangle$. Similar equations hold with cosines replaced by sines. Thus we can rewrite the wavefunction Eq.(3) as

$$\Psi_{r_c \leq r}(E) = \sum_i \Phi_i(\omega) \sum_\alpha U_{i\alpha}[f(\epsilon_i, r)cos\pi\mu_\alpha - g(\epsilon_i, r)sin\pi\mu_\alpha]A_\alpha(E). \tag{7}$$

This wavefunction is valid everywhere outside the core like the wavefunction Eq.(3), and it reduces to a superposition of the type Eq.(5) near r_{vr}. Finally, Eq.(7) is recast in a form which brings out the connection with scattering theory more directly. By introducing new channel amplitudes $Z_{i'} = \Sigma_\alpha U_{i'\alpha}cos\pi\mu_\alpha A_\alpha$ we rewrite Eq.(7) in the form

$$\Psi_{r_c \leq r}(E) = \sum_i \Phi_i(\omega) \sum_{i'} [f(\epsilon_i, r)\delta_{ii'} - g(\epsilon_i, r)K_{ii'}]Z_{i'}(E) \tag{8a}$$

where the real non-diagonal reaction matrix is given by

$$K_{ii'} = \sum_\alpha U_{i\alpha}tan\pi\mu_\alpha U_{\alpha i'}^{tr} . \tag{8b}$$

The significance of this development is the following . The electron motion is represented by Eq.(8a) as an electron-ion scattering process which proceeds in two spatially distinct stages. The scattering event properly speaking occurs separately in each B.O. channel α and is characterized by an elastic scattering phase shift $\pi\mu_\alpha$ defined at the core boundary r_c . As a short range parameter associated with the strongly attractive Coulomb field, this scattering phase should typically vary rather slowly with energy, just as this is well-known to be the case for Rydberg series in atoms. At larger r outside the core a recoupling occurs and is embodied in the frame transformation elements $U_{i\alpha}$. The whole approach leading to Eqs.(7) and (8) hinges on the predominance of the Coulomb field near r_{vr}, which dwarfs both the internal core level structure *and* the non-Coulombic interactions of the core. Therefore the various channels are effectively *degenerate* and the frame transformation reduces to a simple change of basis set. At the same time the preeminence of the Coulomb interaction over the rotational-vibrational fine structure ensures that the lower Rydberg states obey Born-Oppenheimer behaviour, in other words the α-channels are *directly visible* in the lower part of the spectrum (as is exemplified by Fig. 2). Finally we see from Eq.(8b) that the reaction matrix becomes diagonal in two circumstances, either when all μ_α 's are equal or when the frame trans-formation matrix $U_{i\alpha}$ is diagonal. Note finally that the boundary conditions at infinity have still not been applied to the multichannel wavefunction Eq.(8) (i.e. the coefficients $Z(E)$ still remain unspecified) , which therefore is valid indiscriminately for bound levels, autoionizing resonances, or pure continuum states.

The application of the boundary conditions at infinity to a wavefunction of the type of Eq.(8) is carried out by standard quantum defect procdures, such as described by Seaton [12], in the book of Fano and Rau [13] or in A.R.P. Rau's lecture. See Refs. 6 and 14 for molecular MQDT. One is led to an eigenvalue problem represented by the following algebraic linear system:

$$\sum_{i'} [\tan \beta_i(E) \, \delta_{ii'} + K_{ii'}] Z_{i'}(E) = 0 \tag{9}$$

for each channel i. Here the β_i express the condition imposed on each asymptotic channel component, closed or open, at large distance, while the elements $K_{ii'}$ reflect the conditions imposed on the wavefunction at the core boundary outlined above. Open-channel and closed-channel conditions are applied in the following way:

$$\beta_i = -\pi \tau_\rho, \quad i \ open \tag{10a}$$

$$\beta_i(E) = \pi[\nu_i(E) - l], \quad i \ closed. \tag{10b}$$

$\pi \tau_\rho$ are the asymptotic eigen-phases common to all open channels. These phases are labelled by ρ and their number equals the number of open channels. Correspondingly the number of eigenfunctions $\Psi^{(\rho)}$, Eq.(7) or (8), of course equals the number of open channels. $\nu_i(E) = (-2\epsilon_i)^{-1/2} = [-2(E - E_i)]^{-1/2}$ are the effective quantum numbers for the closed ionization channels. Note that in a calculation the effective quantum numbers ν_i for closed channels are inserted into the linear system Eq.(9) as appropriate for the given energy E, while the eigenphases $\pi \tau_\rho$ result from its resolution.

The conditions Eqs.(9) and (10) amount to matching the wavefunction Eq.(8) to the known asymptotic form (in terms of standing waves). This can be appreciated most easily by considering the one-channel case. Eq.(10a) then reduces to the condition that $\mu - \tau$ be an integer, which simply means that the asymptotic phase $\pi\tau$ added to the Coulomb phase equals the short-range phase shift $\pi\mu$ originating in the core. Similarly Eq.(10b) reduces to the condition that $\nu - l + \mu$ be an integer. The non-integral quantity $\nu - l$ represents [15] the number of half-wavelengths of the radial wavefunction f_l between $r = 0$ and $r = \infty$ for arbitrary negative energy. Therefore the above requirement for bound levels means that the *total* number of half-wavelengths, within and outside the core, must be an integer. [15]

Eq.(9) specifies the channel amplitudes Z only to within an overall normalization factor. This factor is recalled here for the sake of completeness. When all channels are closed (discrete range) and when the energy dependence of the reaction matrix K can be neglected, the normalization condition reads (see Refs. 11 and 12)

$$\sum_i \frac{1}{\pi} \frac{\partial \beta_i}{\partial \epsilon_i} \{ \sum_{i'} [cos\beta_i \delta_{ii'} - sin\beta_i K_{ii'}] Z_{i'}^{(\rho)}(E) \}^2 = 1, \tag{11a}$$

where β_i are given by Eq.(10b) and $\frac{\partial \beta_i}{\partial \epsilon_i} = \pi \nu_i^3$ (from Eq.(10b)) is the well-known Rydberg normalization factor. Above threshold the normalization condition for each open channel i reads (see e.g. Ref 6)

$$T_{i\rho} = \sum_{i'} [cos\beta_i \delta_{ii'} - sin\beta_i K_{ii'}] Z_{i'}^{(\rho)}(E), \tag{11b}$$

where β_i is now given by Eq.(10a) and the $\sum_{i'}$ runs over all channels, open or closed. The elements $T_{i\rho}$ form a unitary matrix which in fact is the eigenvector matrix of the scattering matrix (see Ref.6). The eigenvalues of the scattering matrix are $e^{2i\pi\tau_\rho}$. The calculation of partial and total cross sections for a given physical process is set up in terms of the scattering matrix and is not further discussed here (cf. e.g. Rau's article or Ref. 6).

4. Application to H_2 states with moderate vibrational excitation

The initial ideas to the approach outlined in the preceding section are contained in Fano's [11] paper which deals with the rotational aspects of the problem (rotational l-uncoupling). The subsequent extension to vibrational interactions was outlined in Refs. 16 and 17, and a rather comprehensive theory has been given in Ref. 18.

We must initially identify the frame transformation elements for our specific H_2 problem. Each element is a product of a rotational and a vibrational factor,

$$U_{i\alpha} = \langle N^+|\Lambda\rangle \langle v^+|R\rangle . \tag{12}$$

For integer l (united atom approximation) the rotational factor is implied by the work of Van Vleck[19] who considered the decoupling of l from the molecular axis (transition from Hund's case b towards Hund's case d). The elements are given by (see also Ref. (11))

$$\langle N^+|\Lambda\rangle^{(l.J)} = \sqrt{2N^+ + 1} \, (-1)^{2N^+ + J - \Lambda} \sqrt{\frac{2}{1 + \delta_{\Lambda 0}}} \begin{pmatrix} J & l & N^+ \\ \Lambda & -\Lambda & 0 \end{pmatrix} \tag{13a}$$

for a Σ^+ core. With regard to the vibrational factor we use an argument suggested by Fig. 7: owing to the strong Coulomb attraction the motion of the Rydberg electron is independent of vibrational motion near the core, i.e. much faster, so that in effect the latter can be neglected altogether during a single collision event while the Rydberg electron is inside r_{vr}. Fig. 8 shows this schematically for a Coulomb field and also for

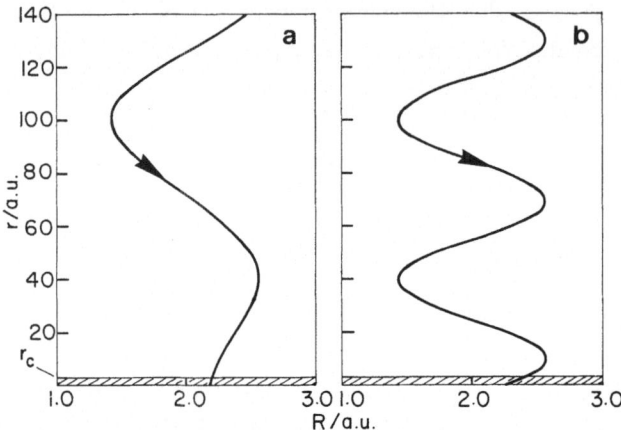

Fig. 8 Classical trajectories for a collision between an electron and a vibrating molecular target: (a) electron-ion, (b) electron-neutral. The trajectories correspond to a collision energy of 0.005 $a.u.$ and to a vibrational energy of 0.016 $a.u.$.

the zero field corresponding to an electron collision with a neutral molecule . We thus take the vibrational factor in Eq.(12) to be simply the vibrational wavefunction of the core,

$$\langle v^+|R\rangle^{(N^+)} = \chi_{v^+}^{(N^+)}(R). \tag{13b}$$

With Eqs.(12) and (13) the $\sum_\alpha$ in the reaction matrix of Eq.(8b) takes the specific form $\sum_\Lambda \int dR$,

$$K_{v^+N^+,v^{+\prime}N^{+\prime}}^{(J,l)} = \sum_\Lambda \langle N^+|\Lambda\rangle^{(J,l)} \int [\chi_{v^+}^{(N^+)}(R) tan\pi\mu_\Lambda(R) \chi_{v^{+\prime}}^{(N^{+\prime})}(R)] dR \, \langle \Lambda|N^{+\prime}\rangle^{(J,l)}. \tag{14}$$

The K matrix fully specifies the short-range part of a collision of an electron possessing a well-defined angular momentum l , with a positively charged core in a Σ^+ electronic state. The calculation of absorption cross sections requires in addition the appropriate transition matrix elements. The photoabsorption takes place in the core so that the transition to a state given by Eq.(7) is a simple linear combination

$$D = \sum_\alpha D_\alpha A_\alpha. \tag{15}$$

For unpolarized incident light and a dipole $s \to p$ transition one has

$$D_\alpha \equiv D_\Lambda^{(J)}(R) = \frac{d_\Lambda}{3} \sqrt{(2J+1)(2J''+1)} \, (-1)^\Lambda \sqrt{\frac{2}{1+\delta_{\Lambda 0}}} \begin{pmatrix} J & 1 & J'' \\ \Lambda & -\Lambda & 0 \end{pmatrix} \chi_{v''}^{J''}(R), \tag{16}$$

where $\chi_{v''}^{J''}(R)$ is the ground state vibrational wavefunction and J'' is the initial angular momentum. d_Λ is the radial dipole integral.

The only quantities which so far have not yet been specified are the short-range or eigen- quantum defects $\mu_\alpha = \mu_\Lambda(R)$. In principle the phase shifts $\pi\mu_\Lambda(R)$ should be calculated ab $initio$ in an fixed-nuclei electron-ion scattering calculation. Such calculations exist [20,21] for energies above threshold, but they do not quite have the accuracy required to reproduce the precisely measured bound state spectrum adequately. No *ab initio* calculations appear to have been carried out as yet for negative energy which would yield directly the quantum defects. An alternative less direct approach has been followed in Ref. 18 and consists in extracting the quantum defects from potential energy curves $U_{n\Lambda}(R)$ calculated by quantum-chemical methods. This is done by applying the Rydberg equation at each R value,

$$U_{n\Lambda}(R) = U^+(R) - \frac{1}{2[n - \mu_\Lambda(R)]^2}. \tag{17}$$

Note that at this point we make use of the assumption that the level system of Fig. 2 has basically a one-electron channel structure.

Fig. 9 shows the result of applying this procedure to the *ungerade* singlet states of H_2. In Fig. 9a are shown the high-quality *ab initio* potential energy curves $U_{n\Lambda}(R)$ which are available to date [22]. In Fig. 9b these curves have been transformed into a set of quantum defect *curves*, $\mu_{n\Lambda}(R)$, by means of Eq.(17). It can be seen that with the exception of the 2Σ curve for $R \geq 2.5 a.u.$, the quantum defect curves for each symmetry ($\Lambda = 0$ or 1, respectively) nearly coincide for different n values. This means that for each R the quantum defect curve exhibits only a weak energy-dependence, in other words,

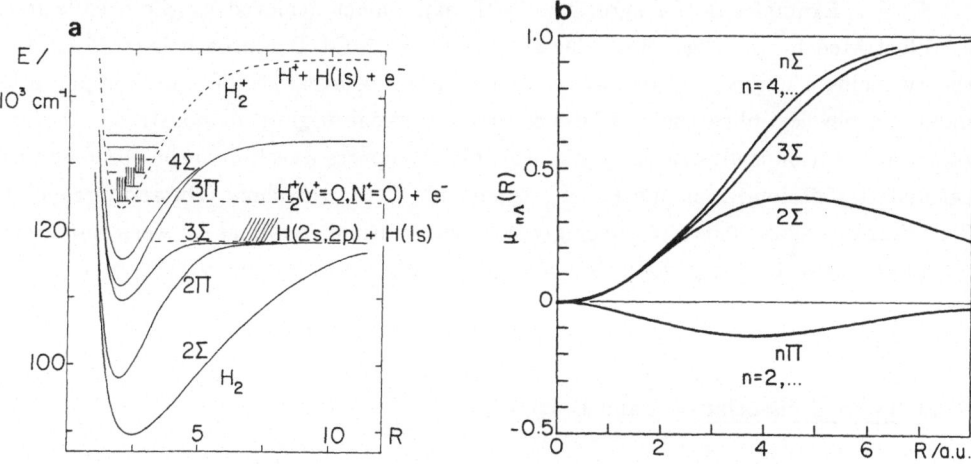

Fig. 9 Potential energy (a) and quantum defect (b) curves for the lowest *ungerade* excited singlet states of molecular hydrogen.

for any given value $\alpha = R, \Lambda$ a *single* quantum defect $\mu_\Lambda(R)$ represents the *whole* series of states reasonably well. From Eq.(14) it is clear that the $R-$ dependence of μ gives rise to the coupling between different *vibrational* channels, whereas its $\Lambda-$ dependence gives rise to the coupling between different *rotational* channels. We see in particular that the Σ^+ channel must be the primary origin of vibrational preionization. The strong $R-$ dependence of μ_Σ is characteristic of the *antibonding* nature of a $p\sigma$ electron: according to Mulliken [3] the $p\sigma$ orbital undergoes a *promotion* since its principal quantum number is by one unit larger in the united atom than for separated atoms. The promotion shows up in the quantum defect curves of Fig. 9 by the fact that μ *increases* by one unit upon separation of the constituent atoms. Conversely, the small variation of μ_Π with internuclear distance seen in Fig. 9 reflects the *non-bonding* character of the $p\pi$ orbitals.

Fig. 9b underscores another advantage of the frame transformation method which is crucial in the practical application. The ground state potential energy curve of H_2^+ supports 20 bound levels. Ignoring the vibrational continuum and including two rotational channels, we find that the K matrix to be considered has as many as 820 independent elements. The frame transformation expression Eq.(14) permits to generate these numerically on the basis of two smooth quantum defect curves (Fig.11b). Each of these can be represented by, say, 30 distinct points, and any additional required points can be obtained by straightforward interpolation procedures.

After the short-range K matrix and relevant transition elements have been determined, the appropriate boundary conditions have to be applied to the wavefunction at infinity as outlined in Sec. 3. The procedure yields the spectrum of bound levels as well as the total and partial photoionization cross sections including preionization profiles. The results obtained with this approach have been published in a series of papers [14.18.23-26]. Examples of the agreement with experiment achieved by the calculations are illustrated by the Figs. 3-6. The success of the MQDT approach is striking: the theory yields a unified and quantitative description of a variety of *a priori* apparently unrelated physical phenomena, whose common physical origin it demonstrates. A very recent application is illustrated by Fig. 10 which compares a section of the observed and calculated photoionization spectra of HD near the $v^+ = 1$ threshold (Du and Greene[27]). This example again shows convincingly how the multiple couplings between numerous open and closed Rydberg channels are handled successfully by the QDT approach.

5. Inclusion of dissociative channels

Dipole-allowed molecular dissociation sets in above the energy corresponding to a pair of separate H atoms in $n = 1$ and $n = 2$, respectively. Fig. 2 indicates that all discrete levels above this energy must to some extent be *predissociated*. Above the ionization threshold there occurs a *competition* between dissociation and ionization processes. Fig. 2 further indicates that predissociation may be viewed in some sense as the *inverse* process of preionization: in a preionization process the system evolves towards the left in Fig. 2, corresponding to a transfer of energy from the nuclei to the Rydberg electron. Conversely, in a predissociation process the system evolves towards the right in Fig. 2, because it is now the Rydberg electron which provides the nuclear kinetic energy necessary for the molecule to fragment. Both types of processes are driven by collisions between the Rydberg electron and the vibrating core - it is only the net result of the collisions which is different. This remark suggests that it should be possible

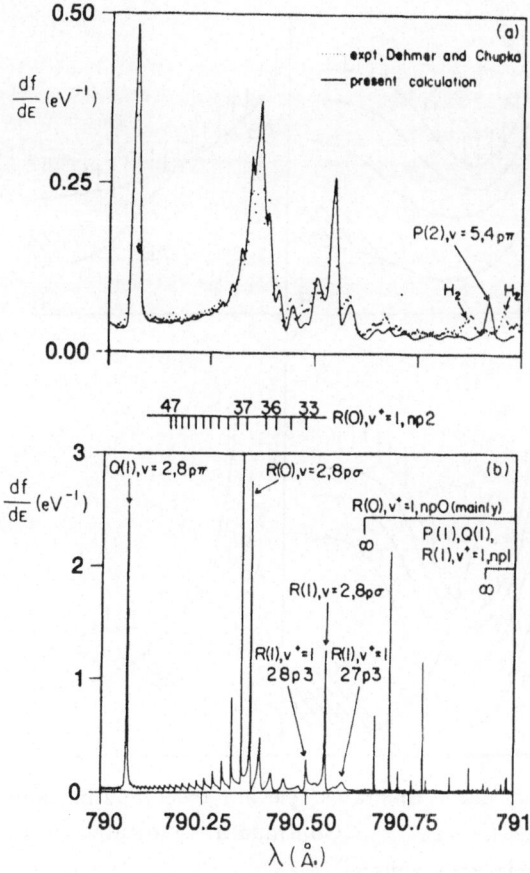

Fig.10 The oscillator strength distribution for photoionization of HD is compared with the quantum defect calculation. (a) The theoretical spectrum includes contributions from $J'' = 0, 1, and \, 2$ appropriately weighted, and is convoluted with the experimental resolution of $0.0016 \, nm$. (b) The same calculated spectrum is shown for perfect resolution, but omitting the contribution from $J'' = 2$. After Ref. 27.

to incorporate molecular dissociation processes into the quantum defect theory in an organic fashion. Indeed, Fig. 2 suggests that the dissociation continua involved in the process are the natural continuations of the vibrational progressions of low-n Rydberg states which have already been accounted for successfully in the preceding section.

Fig. 11 presents potential energy diagrams which illustrate different ways in which a competition between ionization and dissociation processes in a diatomic molecule may come about. The left-hand side pertains to neutral (electron-ion) systems whereas the right-hand side pertains to negative ions (electron-neutral molecule). The top-left diagram corresponds to the situation considered here. It constitutes the simplest situation giving rise to simultaneous ionization and dissociation, in the sense that both

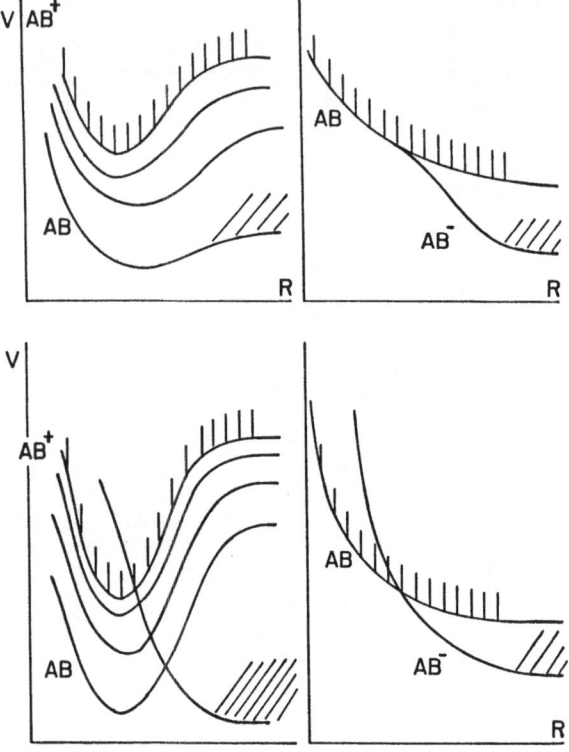

Fig.11 Potential energy curve diagrams relevant to competing ionization and dissociation processes (schematic). Continua are indicated by hatching (ionization - vertical, dissociation - oblique).

processes occur *within the same* Rydberg channel. In other words, it is the lower members of a Rydberg series which predissociate the higher members of the same series. The bottom-left diagram illustrates a situation where dissociation arises through interaction with a state which does not belong to the series. Unlike in the previous case such a process entails an electronic rearrangement within the core. The diagrams on the right-hand side illustrate analogous situations in negative ions; the main difference is that the strongly attractive Coulomb potential supporting an infinite series of levels is now lacking.

We consider photoabsorption by H_2 below $\sim 16.5 eV$ which can produce fragmentation $H_2 \rightarrow H(n = 1) + H(n = 2)$ in competition with ionization and with excitation to high

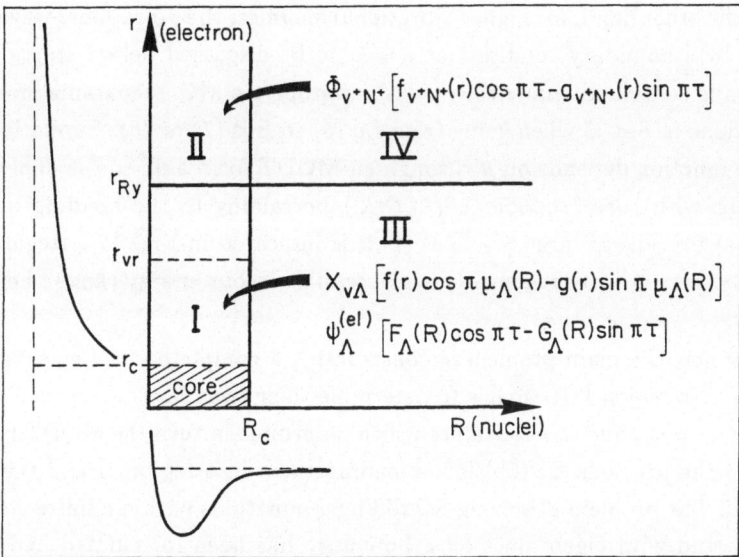

$$\Phi_{v^+N^+}\left[f_{v^+N^+}(r)\cos \pi\tau - g_{v^+N^+}(r)\sin \pi\tau\right]$$

$$X_{v\Lambda}\left[f(r)\cos \pi\,\mu_\Lambda(R) - g(r)\sin \pi\,\mu_\Lambda(R)\right]$$

$$\psi_\Lambda^{(el)}\left[F_\Lambda(R)\cos \pi\tau - G_\Lambda(R)\sin \pi\tau\right]$$

Fig.12 Configuration space relevant to correlated electronic and nuclear motions. In contrast to Fig. 8 it is assumed that the system may dissociate as well as ionize. The close-coupling expansion terms appropriate in different regions are indicated. The electron-ion and atom-atom interaction potentials are also sketched.

Rydberg states. The treatment of these alternative processes [28,6] extends the earlier formulation Eqs.(7)-(11) by combining the channels i of electron excitation with an *additional set* of dissociation channels. Whereas the wavefunction Eq.(7) and (8) had the single radial variable r, the treatment of dissociation involves also the internuclear distance extending to $R \to \infty$. Its wavefunctions must then be mapped over the two-dimensional diagram in Fig. 12.

This diagram is divided into four sections to be treated differently in the MQDT approach; region I is further subdivided into three zones. In the reaction zone, region I, energy eigenfunctions are to be calculated numerically as superpositions of products of functions of r, of R and of other variables, as discussed below. This zone has been extended beyond r_c, to $r \leq r_{Ry}$, to include all electron distances relevant to dissociation fragments. In the range $r_c \leq r \leq r_{Ry}$ each term depends on r according to Eq.(8), which now plays the role of Eq.(5) in Sec. 3. At the same time the construction of the function Eq.(8) by the frame transformation approach ensures that it takes the appropriate Born-Oppenheimer form for $r_c \leq r \leq r_{vr}$ as indicated in the diagram. R_c is chosen sufficiently large ($\sim 7a.u.$) so that all bound levels v^+ of H_2^+ have $\chi_{v^+}(R) \sim 0$ at R_c; the corresponding R-dependent factors of Eq.(8) in region I then practically coincide with the ordinary

$\chi_{v^+}(R)$. On the other hand, for higher vibrational energies, the vibrational wavefunctions are selected by a boundary condition at $R = R_c$ to be discussed below. In region II, the wavefunction is continued analytically , with appropriate MQDT parameters, similarly as this was done in Sec. 3 when going from Eq.(5) to Eqs.(7) or (8). Similarly in region III the wave function depends on R through an MQDT form akin to the dependence on r in region II, with base functions $\{F(R), G(R)\}$ pertaining to the bonding interatomic potential that prevails at large $R \geq 7 a.u.$. This is indicated in Fig. 12 . Region IV (and in fact the section $r \geq r_{vr}$ of region III) is inaccessible in our energy range ; hence we set $\Psi \approx 0$ there.

Consider now the main problem of constructing a complete set of eigenfunctions of given energy E in region I. Its goal is to determine their boundary values on *both* margins of this region , $r = r_{Ry}$ and $R = R_c$, values which determine in turn the MQDT parameters analogous to the previous Eq.(6b) for extending them into region II and their analogs for region III. The problem of solving Schrödinger equations within a finite volume with fixed energy and with eigenvalues on a boundary has been formulated and solved by Fano and Lee [29] and Lee [30] in a different context. As eigenvalue one uses in principle the normal logarithmic derivative , uniform over the *whole* boundary, and in practice equivalent phase shifts $\pi\tau_\rho$ which replace the MQDT parameters $\pi\mu_\alpha$ of Eqs.(6-8) and their analogs for fragmentation channels. Our specific problem is complicated by the fact that the boundary consists of two diverse sections. In addition the value of τ_ρ affects *both* the wavefunctions $\chi_{v^+}(R)$ (and hence the reactance matrix $K_{v^+v^{+\prime}}$ Eq.(14)), and the ionization thresholds E_{v^+} of the electron radial motion.

Transfer of the results of calculation in region I onto observable parameters is achieved as in Sec. 3 by matching the eigenfunctions of this region to those of regions II and III on the respective boundaries. At the boundary $r = r_{Ry}$ this is achieved analytically by means of the conditions Eq.(9) and (10a), whereas at $R = R_c$ numerical matching of vibrational wavefunctions is required. The reactance matrix $K_{ii'}$ is obtained from the eigenphase shifts $\pi\tau_\rho$ by means of Eq.(8b); the set of channels now includes dissociations and the matrix elements are no longer given by Eq.(14). In a final stage the calculation then proceeds as in Sec. 3, by application of the boundary conditions at infinity (Eqs.(9) and (10)). Thereby the open-channel condition Eq.(10a) is applied to the dissociation channels, and the influence of Rydberg levels in closed channels (whose ionization thresholds exceed the energy E and whose radial electron wavefunction extends beyond r_{Ry}) is taken into account through the condition Eq.(10b).

Fig. 13 illustrates the application of the method to photoabsorption by H_2. At the top of the figure is shown a section of the high-resolution photoionization efficiency curve of Dehmer and Chupka [8] from cold para-hydrogen. Complementary photodissociation excitation spectra have been recorded by Guyon, Breton and Glass-Maujean [31]. The spectrum is quite perturbed because of multiple interactions between the various Rydberg levels. This is revealed by the irregular intensity pattern and is confirmed by the calculations. In this range the $H_2^+(v^+ = 0$ and $1, N^+ = 0$ and $2) + e$ and $H(1s) + H(2s, 2p)$ channels are open. The multichannel quantum defect calculation yields directly all seven partial-cross section spectra evaluated point by point on a convenient energy mesh. In the lower part of the figure are given the total dissociation and ionization yields obtained by integration over each Rydberg resonance, and also the resonance width, along with the available experimental data from Refs. 8 and 31. Experiment and theory agree quite well.

Fig. 13 is a good illustration of the concept of propensity rules. [32,33]. According to the diagram, preionization is predominant in Rydberg levels with *high n* and *low v*

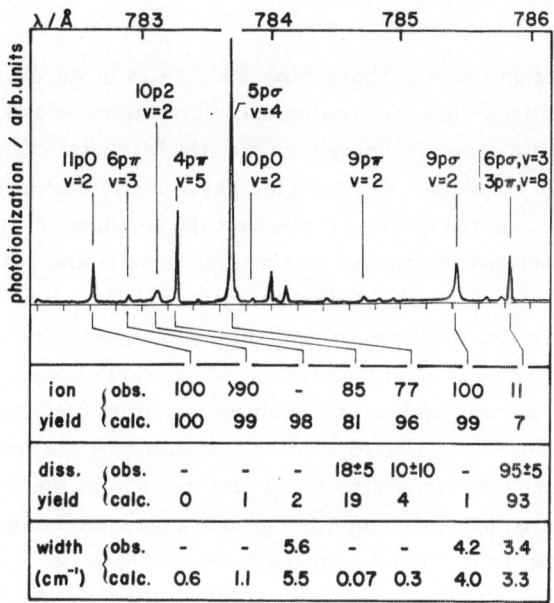

ion yield	obs.	100	>90	-	85	77	100	11
	calc.	100	99	98	81	96	99	7
diss. yield	obs.	-	-	-	18±5	10±10	-	95±5
	calc.	0	1	2	19	4	1	93
width (cm⁻¹)	obs.	-	-	5.6	-	-	4.2	3.4
	calc.	0.6	1.1	5.5	0.07	0.3	4.0	3.3

Fig.13 Top: Photoionization efficiency curve $(J'' = 0)$ of $p - H_2$ near $78.4\,nm$ (Ref. 8). The unassigned peaks correspond to $J'' = 1$ transitions. Bottom: Observed (Refs. 8 and 31) and calculated ionization and dissociation yields and resonance widths for the individual resonances. After Ref. 28.

values ($v = 2$ and 3 in the example shown), while predissociation is predominant, or at least significant, in Rydberg levels corresponding to *low n* and *high v* values ($v = 5$ and 8 in our example). Clearly, each type of decay is favored the more, the smaller is the required amount of energy conversion between the vibrational and the electronic degrees of freedom. This propensity rule has been spelled out more than 20 years ago by R.S. Berry [32,34]. One of the principal results of the approach described in this Section is that the rule emerges from quantitative calculations based on a comprehensive theoretical treatment. The present approach, however, applies also in circumstances where the simple propensity rule breaks down. This happens for instance when numerous Rydberg resonances corresponding to diverse degrees of vibrational excitation, clump together to form new entities, the so-called *'complex' resonances*. [26] The basic ingredient in the molecular MQDT which leads to the propensity rule (where it applies), is the vibrational integral in Eq.(14): this integral is significantly different from zero only for $v^{+'}$ not too different from v^+, owing to the fact that the electronic quantum defect function $\mu_\Lambda(R)$ (Fig. 9) varies only slowly on the scale of the characteristic vibrational wavelength.

6. Conclusions

The preceding sections have outlined how the physics of an electron attached to (or incident on) a positively charged rotating and vibrating molecular core may be approached in an efficient manner. We have seen how the frame transformation approach initiated by Fano takes advantage of simplifying physical circumstances in different regions of space. First of all, the treatment relies on the smallness of the rotational and vibrational energy spacings as compared to electronic kinetic and potential energies *in a zone near the core*. In this sense the molecular quantum defect theory appears as the natural generalization of the early approach of Born and Oppenheimer. Their approach was conceived in the framework of ground states and low excitations and is conceptually unsuited outside that frame. Inside a finite volume including the core, the characteristic wavelengths associated with electron and nuclear motions are comparable, so that the vast mass difference between electrons and protons ensures the Born-Oppenheimer separability. This means that the complicated short-range interactions embodied in the quantum defects μ_α can be evaluated with the nuclei *held fixed*. Secondly, at larger electron distances the favorably small mass ratio m_e/m_p is offset by the rapidly decreasing ratio of the electron and nuclear velocities, with the result that the Born-Oppenheimer approximation breaks down entirely. The frame transformation approach consequently switches to an entirely non-adiabatic description at large r values; all the attempts made in the past to extend the B.O. description beyond r_{vr} appear quite inadequate in retrospect.

In spite of the success of the MQDT treatment as applied to the dipole-allowed spectrum of the H_2 molecule, it should not be overlooked that the frame transformation matrix elements given in Eq.(13) reflect a basically crude level of approximation. There is therefore room for refinements of the theory if the need arises. For instance the united-atom values Eq.(13a) have been used for the rotational-electronic frame transformation matrix elements. More strictly speaking, these elements should be regarded as *functions* of the internuclear distance which reflect the progressive loss of spherical symmetry of the dissociating system [18]. Their evolution with internuclear distance should be evaluated *ab initio*. The specific form Eq.(13b) of the vibrational frame transformation elements also corresponds to an approximation. It is assumed that the Rydberg electron spends hardly any time inside the core in excess of its motion under the influence of the long-range field alone, so that the nuclei have no time to move during its presence. However, if the incident electron happens to be trapped for a time of the order of a molecular vibration, the use of Eq.(13b) will no longer be justified. The time delay associated with such a *resonant* scattering situation is reflected in an energy dependence of the scattering phase. In practical terms a problem arises when the energy dependence of $\mu(\epsilon, R)$ in Eq.(14) becomes significant, because it is not possible to define which value of the electron energy ϵ should be used in the vibrational integral. A solution to this problem has been outlined recently[35]. In this approach the assumption that the nuclei remain frozen during the scattering event is relaxed, but the Born-Oppenheimer separation of nuclear and electron motion is preserved within the core.

Another restriction of the approach outlined in this contribution is the fact that the physically relevant fragmentation channels were always assumed to correspond to the same electronic state of the core. In reality the conversion of core *electronic* energy during a collision event often plays a role much more important than the conversion of rotational-vibrational energy discussed here. In this event the quantum defect function $tan\pi\mu(R)$ in Eq.(14) must be expanded into a non-diagonal electronic matrix $K(R)$ which includes all the physically relevant electronic target states. At the same time, of course, the R-dependent Rydberg equation Eq.(17) must be replaced by the appropriate R-dependent multichannel expression. Extensions of molecular MQDT along these lines have been presented in a perturbative framework by Giusti-Suzor and Lefebvre-Brion [36] (see Ref. 6 for a review of numerous examples, and also Ref. 37), and in a non-perturbative approach more recently. [38]

A final point concerns the non-spherical symmetry of molecules. The asymptotic field even of a small ion like H_2^+ contains significant non-spherical components in addition to the dominating Coulomb field. In the present theory their influence is ignored by regarding them as effectively absorbed in the quantum defect. It is known that in fact the quadrupole field of H_2^+ accounts for a significant fraction of the quantum defects of the Rydberg states of H_2. [16,39] If one is to study increasingly larger systems and particularly polar polyatomic systems, one will have to deal with increasingly large effective core zones unless the pure Coulomb radial reference functions f and g are replaced by more realistic ones. The recently developed methods of generalized quantum defect theory, applicable to arbitrary asymptotic fields, [15,40,41], allow the farthest reaching non-spherical field components to be incorporated with the reference field. Thereby it should be possible to reduce the effective dimension of the core zone to that of the nuclear frame work itself.

7. Acknowledgement

I am grateful to Dr. U. Fano for his detailed criticism and numerous suggestions which have helped to improve the mansucript decisively. Comments on the manuscript by A. Giusti-Suzor are also appreciated.

8. References

[1] G. H. Dieke, J. Mol. Spectrosc. **2**, 494 (1958); *The Hydrogen Molecule Wavelength Tables of G. H. Dieke* , edited by H. M. Crosswhite (Wiley, New York, 1972).

[2] E. Miescher, J. Mol. Spectrosc. **20**, 130 (1966).

[3] R. S. Mulliken, J. Am. Chem. Soc. **86**, 3183 (1964); **88**, 1849 (1966); **91**, 4615 (1969).

[4] G. Herzberg, Phys. Rev. Lett. **23**, 1081 (1969).

[5] S. Takezawa, J. Chem. Phys. **52**, 2575, 5793 (1970).

[c] Chris H. Greene and Ch. Jungen, Adv. At. Mol. Phys. **21**, 51 (1985).

[7] J. H. Van Vleck, J. Chem. Phys. **4**, 327 (1936).

[8] P. M. Dehmer and W. A. Chupka, J. Chem. Phys. **65**, 2243 (1976).

[9] J. E. Pollard, D. J. Trevor, J. E. Reut, Y. T. Lee and D. A. Shirley, Chem. Phys. Lett. **88** , 434 (1982).

[10] F. H. Mies, Phys. Rev. **175** , 164 (1968).

[11] U. Fano, Phys. Rev. **A2**, 353 (1970).

[12] M. J. Seaton, Rep. Prog. Phys. **46** , 167 (1983).

[13] U. Fano and A.R.P. Rau, *Atomic Collisions and Spectra* (Academic, Orlando, FL, 1986).

[14] Ch. Jungen and Dan Dill, J. Chem. Phys. **73** , 3338 (1980).

[15] Chris. H. Greene, A. R. P. Rau and U. Fano, Phys. Rev. **A26** , 2441 (1982).

[16] G. Herzberg and Ch. Jungen, J. Mol. Spectrosc. **41** , 425 (1972).

[17] E. S. Chang, Dan Dill and U. Fano, Abstr. Proc. Int. Conf. Phys. Electron. At. Collisions 8th, Belgrade, p. 536.

[18] Ch. Jungen and O. Atabek, J. Chem. Phys. **66** , 5584 (1977).

[19] J. H. Van Vleck, Rev. Mod. Phys. **23** , 213 (1951).

[20] H. Takagi and H. Nakamura, Phys. Rev. **A27** , 691 (1983).

[21] G. Raseev, J. Phys. B **18** , 423 (1985).

[22] L. Wolniewicz and K. Dressler, J. Chem. Phys. **88** (1988) (in press), and earlier references therein.

[23] M. Raoult, Ch. Jungen and Dan Dill, J. Chim. Phys. Biol. **77** , 599 (1980).

[24] Dan Dill and Ch. Jungen, J. Phys. Chem. **84** , 2116 (1980).

[25] M. Raoult and Ch. Jungen, J. Chem. Phys. **74** , 3388 (1981).

[2c] Ch. Jungen and M. Raoult, Faraday Discuss. Chem. Soc. **71** , 253 (1981).

[27] N. Y. Du and C. H. Greene, J. Chem. Phys. **85** , 5430 (1986).

[28] Ch. Jungen, Phys. Rev. Lett. **53** , 2394 (1984).

[29] U. Fano and C. M. Lee, Phys. Rev. Lett. **31** , 1573 (1973).

[30] C. M. Lee, Phys. Rev. **A10** , 584 (1974).

[31] P. M. Guyon, J. Breton and M. Glass-Maujean, Chem. Phys. Lett. **68** , 314 (1979) and **69**, 591 (1979), and private communication.

[32] R. S. Berry, J. Chem. Phys. **45** , 1228 (1966).

[33] U. Fano, Phys. Rev. **A32** , 617 (1985).

[34] R. S. Berry and S. E. Nielsen, Phys. Rev. **A1** , 383,395 (1970).

[35] Chris H. Greene and Ch. Jungen, Phys. Rev. Lett. **55** , 1066 (1985).

[36] A. Giusti-Suzor and H. Lefebvre-Brion, Chem. Phys. Lett. **76** , 132 (1980).

[37] M. Raoult, J. Chem. Phys. **87** , 4736 (1987).

[38] S. Ross and Ch. Jungen, Phys. Rev. Lett. **59** , 1297 (1987).

[39] G. Herzberg and Ch. Jungen, J. Chem. Phys. **77** , 5876 (1982).

[40] F. H. Mies, J. Chem. Phys. **80** , 2514 (1984).

[41] Byungduk Yoo and Chris H. Greene, Phys. Rev. **A34** , 1635 (1986).

VARIATIONAL CALCULATION OF CHANNEL INTERACTION PARAMETERS

Chris H. Greene

Department of Physics and Astronomy
Louisiana State University
Baton Rouge, Louisiana 70803

ABSTRACT

The ability to calculate the short-range parameters used by multichannel quantum defect theory to describe electronic channel interactions has increased remarkably in the last few years. This improved capability derives more from simplifications in the underlying theoretical description than from the faster computational hardware that has become available. This lecture reviews some of these newer conceptual advances, and the manner in which they interface with more established tools such as quantum defect theory. As examples, recent small-scale calculations are described which have nearly eliminated the long-standing discrepancies between experimental and theoretical cross sections for photoabsorption by atomic calcium and strontium in the energy range up to ten electron volts. Prospects for extracting the dynamical origin of intriguing regularities observed in atomic valence-shell spectra throughout the periodic system are also discussed.

OVERVIEW

The original R-matrix theory of Wigner and Eisenbud[1] grew out of efforts to understand prolific resonance structures observed in low energy nuclear collisions. Prior to these efforts resonance features in cross sections had been described nicely by empirical Breit-Wigner-type formulas, rather than derived from a consistent theoretical framework. References 1 and 2, which I will refer to as the "Wigner-Eisenbud" formulation of R-matrix theory, provided such a framework, showing that confinement of all interacting nucleons within a finite "reaction volume" and with a specified boundary condition leads to a discrete spectrum of energy levels E_λ of the composite system. Moreover, the matrix of logarithmic derivatives (or R-matrix) of the total wavefunction on the surface of this reaction volume resonates at each energy level E_λ of the composite system according to the now-famous expression:

$$R_{ij}(E) = \frac{1}{2} \sum_\lambda \frac{\gamma_{i\lambda} \gamma_{j\lambda}}{E - E_\lambda} . \tag{1}$$

Each observed resonance was thus assigned as a single pole of Eq.(1), with parameters $\gamma_{i\lambda}$ related to partial widths of the λ'th resonance in the i'th channel. In Eq.(1) and in the following, greek indices like λ denote short-range eigenstates of the whole composite system while roman indices like i and j label the asymptotic fragmentation channels observable in a real scattering experiment.

A simple algebraic transformation determines the more familiar scattering matrix S_{ij} in terms of R_{ij}. Eq.(1) has been used widely to interpret resonant nuclear reactions phenomenologically both in nuclear physics and in high energy physics. Typically just a handful of terms with $E_\lambda \simeq E$ are included in the summation over energy eigenstates, thus permitting an empirical fit to experimental cross sections using only a small number of undetermined parameters $(E_\lambda, \gamma_{i\lambda})$.

Since the 1970's Burke, Taylor, and coworkers[3-5] have provided the driving force behind efforts to develop the Wigner-Eisenbud R-matrix method into an efficient framework for calculating numerous atomic and molecular properties. This goal places significantly more stringent demands on the theory than the semiempirical applications to nuclear physics. It may not be obvious that atomic physics, with its dominance of long range Coulombic forces, is at all appropriate to an R-matrix treatment focussed on short-range channel interactions. Because the exchange interaction is confined to a small volume of the order of 10 bohr radii, however, nearly all inelastic processes occur within this region and the R-matrix approach (or Wigner policy[6]) is still sensible. The motion at larger distances can then be described using either analytically known Coulomb functions coupled with a quantum defect theory treatment[7-9], or else by explicit close-coupling or perturbation analyses when long-range multipole effects are strong.[10,11]

A large effort has been expended in studies such as Refs.3-5 to develop general computer codes which can treat lighter closed- or open-shell atoms of almost arbitrary complexity. This Wigner-Eisenbud-type R-matrix program has now successfully treated numerous photoionization and electron-atom scattering cross sections for atoms such as C, N, O, Al, and Si^+. A very detailed and extensive review of progress achieved for such systems during the mid 1970's has been given by Le Dourneuf.[12]

On the other hand the great generality and complexity of the resulting computer codes (roughly 50,000 lines of Fortran programming in their current form) make it somewhat inflexible for extension to new types of problems or fo extraction of intermediate results. Desire for greater simplicity and for a more rapidly convergent physical description of channel interactions led Fano and Lee[13] to reformulate the Wigner-Eisenbud theory in terms of eigenstates of the R-matrix. More rapid convergence soon became apparent, but yet another reformulation, recasting the eigenchannel approach in noniterative form,[14,15] was needed to make it sufficiently fast and convenient to be truly competitive with the Wigner-Eisenbud treatment of Refs. 3-5. Recent applications to the photoabsorption spectra of calcium[16,17] and strontium[18,19], as well as the earlier studies of the lighter alkaline earths[20,21] and aluminum[22], have now shown conclusively that small-scale calculations can provide a simple and accurate description of the strong electron correlations in all of these atoms. In the following I will emphasize these more recent developments, and attempt an evaluation of the strengths and weaknesses of the eigenchannel and Wigner-Eisenbud versions of the R-matrix approach.

VARIATIONAL DETERMINATION OF SHORT-RANGE PARAMETERS

I turn next to a derivation of the R-matrix using a variational method which selects, at the outset, a set of eigenchannels within a short-range reaction volume V. A major characteristic of multichannel continuum wavefunctions, commonly described in terms of scattering or reaction matrices, is their large degeneracy. The large number of degenerate continuum solutions is associated with the large number of physically distinguishable scattering processes $i \rightarrow j$ accessible at each energy.

Any particular degenerate continuum solution has a logarithmic derivative that varies in a complicated fashion across the reaction surface S enclosing V. The so-called "eigenchannel R-matrix method" focusses on the special set of solutions with a *constant normal logarithmic derivative* $-b \equiv (\partial\psi/\partial n)/\psi$ at every point on the surface S. (The constant values b_β obtained amount to reciprocals of the eigenvalues of the conventional R-matrix.) One reason for using this particular criterion to identify the continuum solutions is that dynamical symmetries of the quantum mechanical system are in some cases naturally associated with the eigenchannels. A simple example is the description of scattering by a short-range, symmetric potential in one dimension, whose usual scattering solutions show no evidence of the reflection symmetry of the potential. But if the reaction volume is chosen as the symmetric region within the range x_0 of the potential, the R-matrix eigenchannels for this problem coincide with the even and odd parity eigenstates. This trivial example hardly describes a realistic atomic or molecular Hamiltonian. Nevertheless, R-matrix eigenstates have particular physical significance for many problems, as seen in lectures in this volume by Jungen and by Rau.

A variational expression for the R-matrix eigenvalues b_β can be derived simply by rearranging the familiar Ritz variational principle for Schrödinger energy eigenvalues calculated within the reaction volume V. The energy associated with a (real) eigenstate ψ is given by

$$E = \frac{\int_V \psi(-\tfrac{1}{2}\nabla^2 + U)\psi \; dV}{\int_V \psi^2 dV} . \tag{2}$$

Here dV is the differential volume element of configuration space, and the integrals extend only over the reaction volume V. The kinetic energy operator is $-\frac{1}{2}\nabla^2 \equiv -\frac{1}{2}\Sigma_i \nabla_i^2$ when all of the particles have unit mass, a result easily generalized to arbitrary mass. The potential energy operator U is assumed to be Hermitian, and without loss of generality the wavefunction is taken to be real. While the Ritz principle is common in the context of calculating bound state properties, its applications to continuum wavefunctions are less familiar. A major point of R-matrix methods is that even though the spectrum of energy levels becomes *discrete* after restricting the calculation to a finite volume, the wavefunction ψ still correctly describes a *continuum* state at the energy E, or at least that part of ψ which lies within the reaction volume.

To see this, the first variation δE of (2) can be evaluated to obtain

$$\delta E = N^{-1} \int_V [-\tfrac{1}{2}\delta\psi\nabla^2\psi - \tfrac{1}{2}\psi\nabla^2(\delta\psi) + 2\delta\psi U\psi - 2E\psi\delta\psi]dV , \tag{3}$$

where N is the normalization integral over the volume V. Application of Green's theorem brings Eq.(3) into the form

$$\delta E = 2N^{-1}\int_V \delta\psi(H-E)\psi \; dV + \tfrac{1}{2}N^{-1}\int_S \psi^2\delta[(\partial\psi/\partial n)/\psi] \; dS . \tag{4}$$

Of the two integrals in Eq.(4) the first is obviously zero since by definition ψ is an exact solution of the Schroedinger equation at energy E, $H\psi = E\psi$. On the other hand the second term, an integral over the reaction surface S involving the normal derivative $\partial\psi/\partial n$, does not necessarily vanish. We conclude that Eq.(2) is a variational principle only if $\delta[(\partial\psi/\partial n)/\psi]=0$, meaning that all trial functions used must be constrained to have the *same* logarithmic derivative (-b) on S as the exact solution at the energy E. Since b is not normally known in advance (indeed, the goal of our calculation is to find b), this makes the direct use of (2) somewhat inconvenient. Nevertheless Fano and Lee[13] showed how an eigenchannel R-matrix method could be based on Eq.(2), by guessing a trial value for b, followed by diagonalization of the Hamiltonian to obtain a set of energies $E_n(b)$. This process can then be repeated for different trial b values until one of the energy levels E_n coincides with the desired continuum energy E. This iterative procedure amounts to a variational determination of b(E). An application of the eigenchannel R-matrix method in this form has given a good description of argon photoabsorption spectra near the ionization threshold.[23] In another context Jungen has adapted the method to incorporate rovibrational predissociation into the frame transformation theory of interacting H_2 Rydberg channels.[24,9]

More recently, this eigenchannel formulation has been streamlined into a simpler and more efficient approach,[14,15] by recasting Eq.(2) in the form of a direct variational expression for the surface logarithmic derivative (-b) itself. The derivation begins by applying Green's identity to Eq.(2) in order to display b explicitly:

$$E\int_V \psi^2 dV = \int_V [\tfrac{1}{2}\nabla\psi\cdot\nabla\psi + \psi U\psi]dV + \tfrac{1}{2} b\int_S \psi^2 dS \ . \tag{5}$$

In the last (surface) term in (4), the condition that $b=-(\partial\psi/\partial n)/\psi$ is constant over the reaction surface for an R-matrix eigenchannel has been used. Solving now for b gives

$$b = \frac{\int_V [-\nabla\psi\cdot\nabla\psi + 2\psi(E-U)\psi]dV}{\int_S \psi^2 dS} \ . \tag{6}$$

This is easily demonstrated to be a stationary variational expression for the desired boundary parameter b, by evaluating δb and then applying Green's theorem. It should be stressed that δb=0 follows just from Eq.(6) and from the condition that the desired exact solution obeys

$$\frac{\partial\psi}{\partial n} + b\psi = 0 \quad \text{on S.} \tag{7}$$

In particular the trial functions to be used in Eq.(6) need not in general obey (7), nor must they have any specified behavior on the reaction surface.

The variational principle (6) can be reduced to a homogeneous system of linear algebraic equations by representing the trial function ψ in terms of a set of arbitrary (real) basis functions $\{y_k\}$ by:

$$\psi = \sum_k y_k c_k \ . \tag{8}$$

After inserting (8) into (6), stationary values of b emerge by stipulating that $\partial b/\partial c_k = 0$. This determines the logarithmic derivatives at each energy as the eigenvalues of a generalized eigensystem,

$$\underline{\Gamma} \vec{c} = b \underline{\Lambda} \vec{c} \ . \tag{9}$$

The matrix $\underline{\Gamma}$ can be evaluated using either of the expressions

$$\Gamma_{k\ell} = \int_V [-\nabla y_k\cdot\nabla y_\ell + 2y_k(E-U)y_\ell]dV \tag{10a}$$

or

$$\Gamma_{k\ell} = 2\int_V y_k(E-H)y_\ell dV - \int_S y_k(\partial y_\ell/\partial n) \ dS \ . \tag{10b}$$

The matrix $\underline{\Lambda}$ consists of a simple surface integral

$$\Lambda_{k\ell} = \int_S y_k y_\ell dS \ . \tag{11}$$

Equations (9)-(11) form the basis for a noniterative reformulation of the eigenchannel R-matrix method. A major advantage over the eigenchannel formulation as implemented by Fano and Lee[13,23] is that now *the logarithmic derivatives of all independent solutions (eigenchannels) at a desired energy E are determined by solving a single linear eigensystem.*

108

1. Multiplicity and Orthogonality of Eigenvectors

Eigenstates ψ_β and $\psi_{\beta'}$, corresponding to distinct eigenvalues b_β and $b_{\beta'}$ are orthogonal over the reaction surface, a condition guaranteed by Eq.(9) through the well-known result for such a generalized eigenvalue problem,

$$\vec{c}_\beta \cdot \underline{\Lambda}\, \vec{c}_{\beta'} = 0 \quad \text{when } b_\beta \neq b_{\beta'} \; . \tag{12}$$

This condition will be seen to guarantee that multichannel R-matrices are *exactly* symmetric, irrespective of deficiencies in the variational basis. Determination of the number of nontrivial solutions to Eq.(9) is somewhat more complicated. On physical grounds the number of eigensolutions must equal the total number of open and weakly-closed channels. (A strongly-closed channel, by way of contrast, is defined as one having negligible amplitude on the reaction surface.) The concept of *reaction channels* arises here by introducing a complete set of real, orthonormal surface harmonics ϕ_i which span the reaction surface S,

$$\int_S \phi_i \phi_j dS = \delta_{ij} \; . \tag{13}$$

Each of the basis functions y_k can be expanded on S in terms of the ϕ_i according to

$$y_k = \sum_i a_{ki} \phi_i \; , \quad \text{on S} \; , \tag{14}$$

and the matrix $\underline{\Lambda}$ simplifies to

$$\Lambda_{k\ell} = \sum_{i=1}^{N} a_{ki} a_{i\ell}^T \; . \tag{15}$$

Here N represents the number of channels in which an electron can escape from the reaction volume, or in other words N is the smallest number of surface harmonics needed to adequately represent each basis function on the surface in (14).

If $\underline{\Lambda}$ is a nonsingular n x n matrix, there will be n nontrivial solutions to Eq.(9). In practice, however, the number of channels N is much smaller than the total number of variational basis functions included in setting up Eq.(9), causing $\underline{\Lambda}$ to be highly singular. For the extreme case in which a single channel contributes to Eqs.(14) and (15), the matrix $\underline{\Lambda}$ is separable and has just one nonzero eigenvalue, whereby the system (9) has exactly one nontrivial solution as expected. Similarly, if a total of N channels are required to span the reaction surface in a particular energy range, the number of nontrivial eigensolutions will generally be N. The concept of surface harmonics should become clearer through an explicit application to two electron systems below.

At this point it is possible to complete the eigenchannel derivation of the R-matrix itself. With its N eigenvalues (b_β^{-1}) already determined by Eq.(9), the eigenvectors of the R-matrix are the only remaining quantities required. This set of vectors forms a matrix whose elements are simply the projections of the i-th surface harmonic onto the β-th eigensolution,

$$Z_{i\beta} = \int_S \phi_i \psi_\beta dS \,/\, N_\beta = \sum_i a_{ki} c_{k\beta} \,/\, N_\beta \; . \tag{16}$$

When N_β is chosen such that ψ_β / N_β is normalized over the reaction surface, the eigenvector matrix $\underline{Z}$ is orthogonal. The conventional R-matrix is now given by the manifestly symmetric expression

$$R_{ij} = \sum_{\beta} Z_{i\beta} b_{\beta}^{-1} Z_{j\beta} \; . \tag{17}$$

This is the inverse of the matrix denoted $\underline{R}$ in Refs. 14 and 20.

2. Explicit Solution of the One-Channel Problem

In order to gain some insight into the detailed workings of the eigenchannel R-matrix method, consider the case of a single channel, for which the summation in Eq. (15) includes one term only. The matrix $\underline{\Lambda}$ then becomes separable,

$$(\underline{\Lambda})_{k\ell} = (\underline{a}\,\underline{a}^T)_{k\ell} = a_k a_\ell \; . \tag{18}$$

In Eq. (18) the superscript T indicates the transpose of the column vector $\underline{a}$, with the i-th element of $\underline{a}$ given by the projection of the i-th basis function onto the lone relevant surface harmonic on S. The single eigenvalue in Eq. (9) can be shown by elementary means to be given by

$$b = 1/[\underline{a}^T \underline{\Gamma}^{-1} \underline{a}] \; , \tag{19}$$

a result derived previously by Nesbet and coworkers[25,26]. Since $\underline{\Gamma}$ is very nearly singular, however, there may be practical advantages to working with the eigensystem (9) directly, even in the one-channel problem.

It is quite simple to show how the one-channel analog of the Wigner-Eisenbud expression (1) can be derived by making a restrictive choice of the trial functions to be used in Eq. (9) or in the equivalent Eq. (19). In particular, suppose that the y_k are defined to be a complete orthonormal set of basis functions, all of whose logarithmic derivatives vanish on the reaction surface. Since the present example involves only one channel it will be simplest to imagine that the configuration space has just one dimension, say r, and that the reaction surface is just the point $r = r_0$, with $0 \leq r \leq r_0$ constituting the reaction volume. Then each of the Wigner-Eisenbud basis functions y_k has zero derivative at $r = r_0$. [It is not essential that the logarithmic derivatives of all basis functions vanish, but they must all be the same to recover a formula resembling Eq. (1).] If the finite-volume Hamiltonian matrix elements are constructed in this basis set and $\underline{H}$ is then diagonalized as in the usual Ritz method, an infinite set of energy levels E_λ are obtained along with the orthonormal eigenvectors $W_{k\lambda}$ of $\underline{H}$. The surface term in $\underline{\Gamma}$ vanishes for this choice of the y_k so the inverse needed in Eq. (19) is

$$\Gamma_{k\ell}^{-1} = \tfrac{1}{2} \sum_{\lambda} W_{k\lambda} (E - E_\lambda)^{-1} W_{\lambda\ell}^T \; . \tag{20}$$

Finally the single-element R-matrix is found to be

$$R = b^{-1} = \tfrac{1}{2} \sum_{\lambda} \frac{[\psi_\lambda(r_0)]^2}{E - E_\lambda} \; , \tag{21}$$

where ψ_λ is the value of the eigensolution evaluated at the reaction surface. The Wigner-Eisenbud formula showing the familiar resonance denominator has been influential largely because the dependence on the energy is shown explicitly. Owing to the restrictive type of basis set required to obtain this formula, the convergence of the variational expansion is much slower when a Wigner-Eisenbud-type basis set is used than can be achieved with alternative basis sets.

3. Comparison of Convergence Rates for a Single-Channel Problem

Let us turn to a model one-channel problem in which an s-wave electron experiences the following potential,

$$v(r) = -\exp(-r/2) \; , \tag{22}$$

110

and use the variational method to estimate the logarithmic derivative of the wavefunction at $r_0 = 5$ a.u. For simplicity we will work with a rescaled wavefunction $u(r) = r\psi(r)$ so as to eliminate first derivative terms in the Laplacian. A simple basis set to use in Eq.(9) is a set of free-particle solutions within the reaction volume $r \leq r_0$, namely

$$y_k(r) = \sin(\tfrac{1}{2}k\pi r/r_0) . \tag{23}$$

Each such basis function with odd k has zero derivative on the reaction surface, while for even values of $k \neq 0$ the basis function itself vanishes on the boundary. Two calculations will be compared, each of which uses five basis functions. For the first calculation, each of the basis functions used has zero derivative at $r = r_0$, with corresponding k-values k = 1, 3, 5, 7, and 9, so that this amounts to a standard five-state Wigner-Eisenbud analysis. For the second calculation the last basis function (k=9) used in the first calculation is replaced by a single basis function with k=2 which vanishes at $r = r_0$. This is representative of the typical basis set used in eigenchannel calculations in that basis functions with a range of surface logarithmic derivatives are included.

Figure 1(a) shows these simple basis functions used for the two calculations, illustrating the way in which a few basis functions calculated within a small volume have great flexibility for representing approximate wavefunctions. In Fig. 1(b) the variational b-values obtained in the two methods are plotted over a wide range of energy. Since the logarithmic derivative ranges from $-\infty$ to ∞, what is actually shown in Fig. 1(b) is a more convenient quantity $\tan^{-1}(b)/\pi$. The two calculations give similar qualitative results, with the apparent discontinuity near E = -0.15 a.u. related to a trivial branch change of the arctangent function at that point. This graph

<table>
<tr><td>(a)</td><td>(b)</td></tr>
</table>

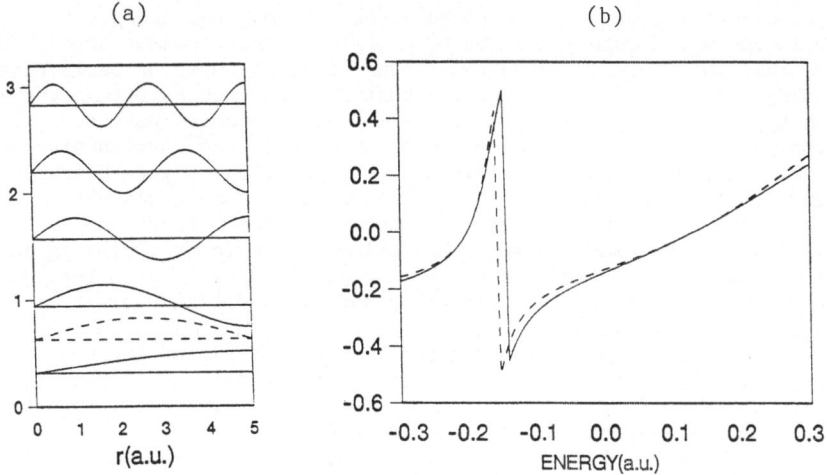

Fig. 1. (a) Basis set of free-electron radial functions used in a model problem. The ordinate gives the wavevector of each basis function. The solid curves show functions used in the Wigner-Eisenbud-type calculation, whose derivatives vanish at r_0. In the eigenchannel calculation the highest basis function is replaced by the dashed curve.
(b) Variational values of $\tan^{-1}b/\pi$, calculated as a function of energy with Wigner-Eisenbud (solid) and eigenchannel (dashed) basis sets.

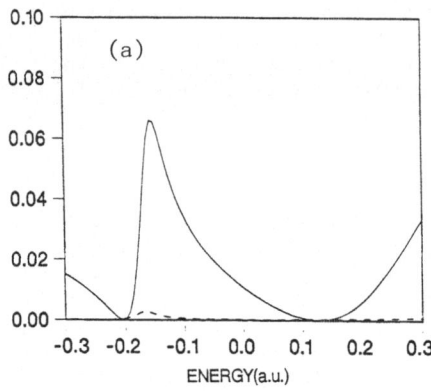
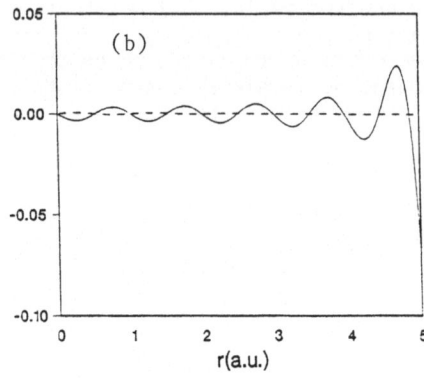

Fig. 2. (a) Difference between the exact and variational values
of $\tan^{-1}b/\pi$, for a Wigner-Eisenbud basis (solid) and for an
eigenchannel basis (dashed), as a function of energy.
(b) Difference between the exact and variational wavefunctions at
one energy as a function of r, using Wigner-Eisenbud (solid) and
eigenchannel (dashed) basis sets.

illustrates the fact that the b-value for the exact solution is a strictly
increasing function of the energy, i.e. db/dE > 0 at all energies.[1,2] Just as
the familiar Ritz functional provides an *upper* bound on exact energy
eigenvalues, likewise we see that Eq. (6) provides a *lower* bound on b (aside
from trivial branch changes of the type just mentioned). This criterion
already states that the dashed curve in Fig.1(b) is the more accurate since it
lies predominantly above the smooth curve except near two energies E≈-0.2
and E≈0.1 a.u. which both happen to be energies where b passes through 0. We
conclude that the typical eigenchannel-type calculation using basis functions
with a range of b-values gives results that are more accurate than the Wigner-
Eisenbud-type calculation using only basis functions which have b=0. The only
exception to this rule is at the energies where the true wavefunction has a
logarithmic derivative which coincides with that of the Wigner-Eisenbud basis
functions, in this case when b=0. This is documented more clearly in Fig. 2(a)
which shows the error itself, that is, the difference between the curves shown
in Fig. 1(b) and the exact curve. The maximum error in Fig. 2(a) is more than
an order of magnitude larger for the Wigner-Eisenbud-type basis set, showing
the greater flexibility of an arbitrary-boundary-condition basis.

The slower convergence of the fixed-boundary-condition basis expansion is
a general property, essentially equivalent to the Gibbs phenomenon in Fourier
analysis. Since no finite superposition of functions which all have b=0 can
adequately describe an exact solution with b≠0, the expansion is not uniformly
convergent. The connection with Gibbs phenomenon can be made still clearer by
considering Fig. 2(b), the error in the variational approximation to the
wavefunctions calculated with the two basis sets. (Actually, to better
illustrate the phenomenon, the Wigner-Eisenbud basis used to obtain the smooth
curve in Fig. 2(b) used ten basis functions whose derivatives vanish at r_0. To
determine the dashed curve in Fig. 2(b) the highest two Wigner-Eisenbud basis
functions with k=17 and k=19 were replaced by k=2 and k=4 basis functions.) In
Fig. 2(b) the "ringing" associated with nonuniform convergence is now clear,
with the error increasing sharply close to the R-matrix boundary. The
wavefunction has been normalized so that its first maximum is approximately
unity, so that Fig. 2(b) gives in essence the relative error.

The slow convergence properties of the Wigner-Eisenbud type of R-matrix treatment would seem to indicate that it is unsuitable for ab initio calculations. At least two methods have been devised, however, which approximate the effect of the highest terms in Eq.(1) which are neglected in any finite calculation. The one used by Burke is a correction to the logarithmic derivative developed by Buttle.[27] Another variational correction has been derived by Zvijac, Heller, and Light.[28] While either of these techniques will partially offset the damage caused by nonuniform convergence of the fixed-boundary-condition basis set, some degree of simplicity and generality appears to be lost when such corrections are needed. The noniterative eigenchannel formulation is simpler in the sense that no corrections are necessary since the basis functions have enough flexibility to give far more rapid convergence. One cost of the eigenchannel method is that the eigensystem (9) must be solved at each energy for which the R-matrix or the short-range wavefunction is desired, as opposed to the Wigner-Eisenbud approach where the Hamiltonian matrix needs to be diagonalizeα only once. As will be seen in the following section, the number of such diagonalizations required is rather modest, thanks to the slow variation of short-range channel interaction parameters with the energy.

APPLICATION TO ATOMS WITH TWO VALENCE ELECTRONS

1. Multichannel Formulation using a Close-Coupling-Type Basis

To see how the general theory of the preceding section is applied to a specific problem, and in particular to give a nontrivial example of the "reaction volume" and "reaction surface", consider the two-electron Hamiltonian

$$ H = - \frac{1}{2} \frac{\partial^2}{\partial r_1^2} - \frac{1}{2} \frac{\partial^2}{\partial r_2^2} + \frac{\vec{\ell}_1^{\,2}}{2r_1^2} + \frac{\vec{\ell}_2^{\,2}}{2r_2^2} - \frac{Z}{r_1} - \frac{Z}{r_2} + \frac{1}{r_{12}} \; . \tag{24} $$

(It is understood that H acts only on the *rescaled* wavefunction $\psi \equiv r_1 r_2 \Psi$ and that the volume element used for volume integrals is therefore $dr_1 dr_2 d\Omega_1 d\Omega_2$.) While it may prove desirable for some purposes to study this system using a hyperspherical reaction surface, as suggested by Fano[29], here I will use the more conventional "cartesian" reaction surface defined as the locus of all points in configuration space such that

$$ \max\{r_1, r_2\} = r_0 \; . \tag{25} $$

The spatial part of each antisymmetric close-coupling-type basis function has the form

$$ y_{ik}(r_1, r_2; \Omega_1, \Omega_2) = \phi_i(r_1; \Omega_1, \Omega_2) F_{ik}(r_2) + (-1)^S \phi_i(r_2; \Omega_2, \Omega_1) F_{ik}(r_1) \; . \tag{26} $$

Here the $F_{ik}(r)$ are arbitrary, nonorthogonal trial radial functions having ideally a variety of logarithmic derivatives at $r=r_0$, and vanishing at $r=0$. The "surface harmonics" ϕ_i are simply the states of the hydrogenic target ion of charge Z, including also the angular wavefunctions of both electrons coupled to a definite value of the total orbital angular momentum L:

$$ \phi_i(r_1; \Omega_1, \Omega_2) = r_1 R_{n_i \ell_{1i}}(r_1) \, Y_{\ell_{1i} \ell_{2i} L M_L}(\Omega_1, \Omega_2)/2^{\frac{1}{2}} \; . \tag{27} $$

Aside from the usual constants of the motion $\{L, M_L, S, M_S\}$, the channel index for this two-electron problem stands for $i \equiv \{n, \ell_1, \ell_2\}$. The $R_{n\ell}(r)$ are the usual radial eigenfunctions of a hydrogenic Hamiltonian of charge Z, except that they should vanish exactly on the reaction surface, $r=r_0$, rather than at $r \to \infty$. The derivatives will not vanish exactly on this boundary, but r_0 should be chosen large enough so that for all states to which excitation is to be considered, the radial derivative on the boundary is negligible. When this is satisfied, the radial functions with the lowest energies will for all practical purposes coincide with the standard Laguerre polynomial form, although high-lying channel functions are normally included in the basis which do have a nonzero derivative at r_0. These play the same role as the "pseudostates" incorporated into other methods of calculation, representing continuum-type functions. It is convenient (but not necessary) to choose the radial functions $F_{ik}(r_0)$ to be eigenfunctions of the hydrogenic Hamiltonian also, including functions that vanish and also some that do not vanish at r_0.

The calculation of Hamiltonian matrix elements in the matrix $\underline{\Gamma}$ of (9) is rather straightforward, but it may be worth showing the less familiar form for the surface integral involved in $\underline{\Lambda}$,

$$\Lambda_{ik,i'k'} = \int d\Omega_1 d\Omega_2 \int_0^{r_0} dr_1 [y_{ik}(r_1,r_0;\Omega_1\Omega_2) y_{i'k'}(r_1,r_0;\Omega_1\Omega_2)$$
$$+ \int_0^{r_0} dr_2 y_{ik}(r_0,r_2;\Omega_1\Omega_2) y_{i'k'}(r_0,r_2;\Omega_1\Omega_2)] . \tag{28}$$

This can be evaluated further giving a block-diagonal matrix, each block of which is separable as in (18):

$$\Lambda_{ik,i'k'} = \delta_{ii'} F_{ik}(r_0) F_{ik'}(r_0) . \tag{29}$$

Just as Eq.(18) was seen to have rank one, so (29) is seen to have rank N, where N is the number of blocks of $\underline{\Lambda}$ that are not identically zero. In more physical terms, we see that if a channel i is open or weakly-closed in the sense of quantum defect theory, then the radial basis set should include at least one function nonzero on S, i.e. $F_{ik}(r_0) \neq 0$. On the other hand, strongly-closed channels are those for which all radial functions vanish at $r=r_0$. These are the channels in which we can neglect the possibility of electron escape beyond $r=r_0$ owing to their (sufficiently rapid) exponential decay. It should be noted that the matrix elements in (9) can be evaluated once and for all independently of the energy, after which (9) can be solved at as many energies as are necessary.

After the N independent eigenstates ψ_β have been obtained by solving (9), the solutions on the boundary can be matched to a linear combination of Coulomb functions (or other appropriate long range solutions) in each open or weakly-closed channel i. Letting round brackets denote a surface projection, it is possible to write,

$$(\phi_i|\psi_\beta) = \sum_k F_{ik}(r_0) c_{ik,\beta} = f_i(r_0)I_{i\beta} - g_i(r_0)J_{i\beta} \tag{30}$$

$$(\phi_i|\partial\psi_\beta/\partial n) = -b_\beta(\phi_i|\psi_\beta) = f_i'(r_0)I_{i\beta} - g_i'(r_0)J_{i\beta} .$$

These equations determine the two NxN matrices $\underline{I}$ and $\underline{J}$, and the reaction matrix is simply $\underline{K} = \underline{J}\ \underline{I}^{-1}$. (This coincides with the smooth, short-range reaction matrix denoted $K^{(s)}$ in the lectures of Fano and of Rau.) At this point the eigenchannel formulation merges with standard quantum defect methods. These can now efficiently impose asymptotic boundary conditions on an arbitrarily fine energy mesh to describe any relevant scattering or photoabsorption process.

2. The Lighter Alkaline Earth Atoms

Ref. 20 used the preceding treatment to calculate the photoabsorption

spectra of Be and Mg, showing for the first time the simplicity and
effectiveness of the method. The main difference between this study and the
hydrogenic formulation in Sec.1 is that the one-electron potential $-Z/r$ in
Eq.(24) must be modified to the form $-Z(r)/r$. This form includes the effect of
the inner-shell electrons, to the extent that they just contribute an isotropic
r-dependent screening to the potential experienced by each valence electron.
This approximation is accurate as long as photon energies are not nearly
sufficient to excite an inner-shell electron. For any atom or ion in rows 1 or
2 of the periodic table with one valence electron, including e-Be^{++} and e-Mg^{++},
the screening function $Z(r)$ is given adequately by the Hartree-Slater method.[30]

The photoabsorption cross section up to the second threshold (np) of the
one-electron ion is determined by three interacting channels, namely
$ns\epsilon p$, $np\epsilon s$, and $np\epsilon d$. In this notation each channel labelled as "$\epsilon\ell$" is
treated as open or weakly-closed. The calculations of Ref. 20 were carried out
using the minimum number of two-electron basis functions, namely 10 of the
type $ns\epsilon p$, 9 of the type $np\epsilon s$, and 10 of the type $np\epsilon d$, with n=2 for Be and n=3
for Mg. Of the roughly 10 continuum-type orbitals used per channel,
approximately half were of the "closed-type", which vanish at r_0 and half were
of the "open-type" which by definition do not vanish at r_0. (In subsequent
calculations such as those for Ca and Sr described in Sec. 3 below, it was
observed that using at most two or three open-type functions per channel
normally gives somewhat better convergence.)

For Be the reaction surface was placed at $r_0=9$ a.u. which easily
accomodates the $Be^+(2p)$ radial wavefunction. Since Mg^+ is somewhat larger,
with weaker binding energy than Be^+, $r_0=11$ a.u. was used in the Mg
calculation.

Figure 3 shows the resulting photoionization cross sections in the energy
range between the ionic ns and np thresholds. This spectral range is dominated
by unusually strong and broad npn's autoionizing resonances. A narrow series
of npn'd resonances is also observed for both Be and Mg. Only the lower few
members of the infinite Rydberg series of autoionizing resonances are shown.
Since the final cross section is computed using quantum defect theory, however,
it is trivial to extend the calculation to arbitrarily high Rydberg states just
below either ionic threshold. Also shown for comparison are the (apparently)
best previous theoretical results of Dubau and Wells[31] and of Bates and
Altick[32]. Experimental measurements are difficult to compare quantitatively
with these calculations, but there appear to be no significant
discrepancies.[33,34] These npn's resonances are so broad that a Beutler-Fano
profile analysis of each autoionizing "state" would probably not characterize
the spectrum very well. In fact they are so broad that the doubly-excited
states decay after only half of a "Bohr-type orbit" is completed, thus
challenging the very notion of, for instance, 2p3s being considered a
quasistable state. As Refs. 35 and 20 have documented, the wavefunctions in
this region are approximately half "continuum-like" at small distances and half
"discrete-like", i.e. $ns\epsilon p \pm np\epsilon s$, almost exactly as demonstrated for doubly-
excited levels of the helium atom near the n=2 threshold by Cooper, Fano, and
Prats.[36]

Most features of the Be and Mg spectra are very similar. This similarity
is reflected to an amazing level in the channel mixing parameters as seen from
Fig. 4. Three parameters are shown as functions of energy for these two atoms,
namely two eigenquantum defects μ_α and a mixing angle θ. These three
independent quantities contain information equivalent to the 2x2 reaction
matrix obtained by ignoring the weak $np\epsilon d$ channel. In particular the
eigenvalues of the reaction matrix are $\tan\pi\mu_\alpha$ while θ is the rotation angle
needed to diagonalize K_{ij}. Note how the mixing angle goes to zero at energies
far below the lowest (ns) threshold, increasing to a maximum of $\approx\frac{1}{4}\pi$ at higher
energies where the two channels $ns\epsilon p$ and $np\epsilon s$ are essentially equally mixed.
One of the major features of all alkaline earth atoms is the occurrence of

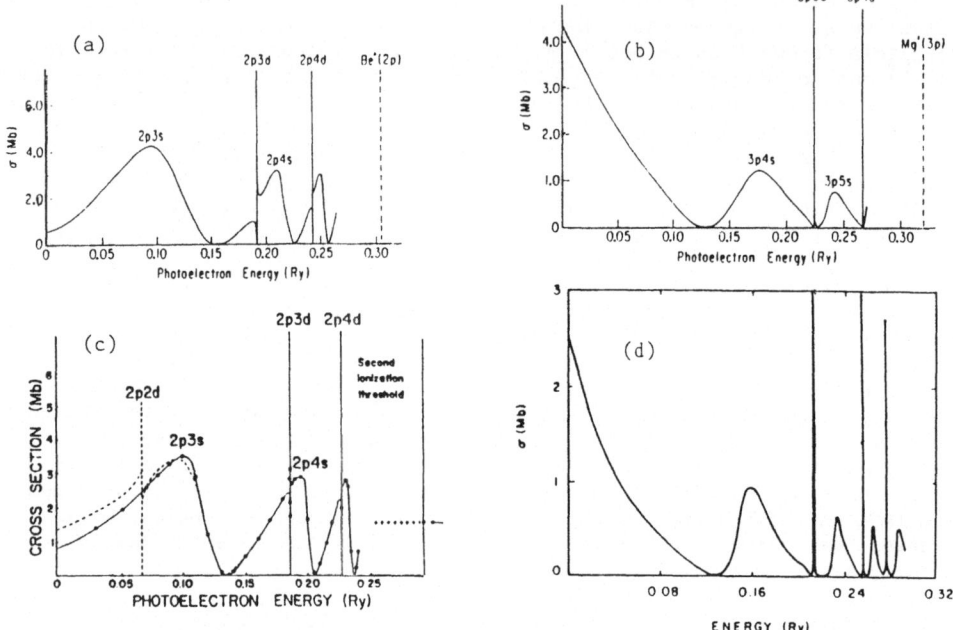

Fig. 3. Photoionization cross sections in Mb are shown as a function of
the photoelectron energy in Ry, between the lowest two ionization
thresholds. (a) and (b) show the eigenchannel R-matrix
calculations of Ref. 20 for Be and Mg, respectively, while (c)
and (d) are the corresponding calculations of Refs. 31 and 32.

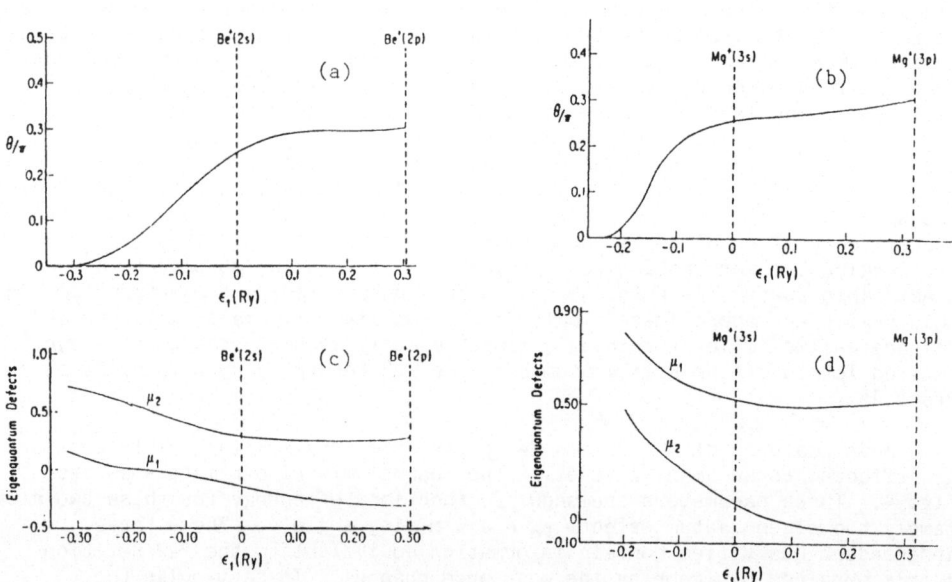

Fig. 4. Channel mixing parameters for $^1P^o$ Be and Mg are shown as
functions of the photoelectron energy. The channel mixing angle
is given in (a) and (b), the eigenquantum defects are given in
(c) and (d). Calculated by Ref. 20.

extremely strong channel interactions in a large number of symmetries, often approaching the limit of maximum interaction, i.e. maximum inelasticity.

The parallel between the dynamics of Be and Mg shows up even more dramatically in a plot of the two independent eigenstates of the reaction operator, shown in Fig.5. These eigenstates are almost indistinguishable for the two atoms, except for a slight expansion of the radial range for Mg and except for the region where either electron approaches the nucleus. (The Mg s and p wavefunctions have an extra node close to the nucleus.) The structure of the eigenchannel wavefunctions is instead greatly modified for the heavier alkaline earths Ca, Sr, Ba, and Ra because of the different ordering of ionic energy levels for those atoms. Nevertheless the characteristic feature of very strong, nonperturbative channel interactions continues to dominate their spectra.[37]

3. The Heavier Alkaline Earth Atoms

Beginning with calcium, the heavier alkaline earths become substantially more difficult to describe theoretically. The simplest independent-electron models of the the e-Ca^{++} interaction, like Hartree-Slater for instance, show large (≈ 10 %) errors in the lowest calculated energy levels 4s, 3d and 4p. The sensitivity to error of atoms located in this region of the periodic table should not be surprising, in view of the fact that even the ground state configuration is on the verge of changing from 4s (Ca) to 3d (Sc). The effective central potentials obtained in any standard independent-electron model reflect this instability, particularly for the $\ell=2$ partial wave.

While these considerations point to a difficulty in obtaining realistic *ab initio* potentials to describe the e-Ca^{++} interaction, they do not necessarily imply that the correlations between the valence electrons are any more difficult to describe in Ca than in Be or Mg. A good description of the e-Ca^{++} interaction is an absolute prerequisite for any calculation scheme attempting

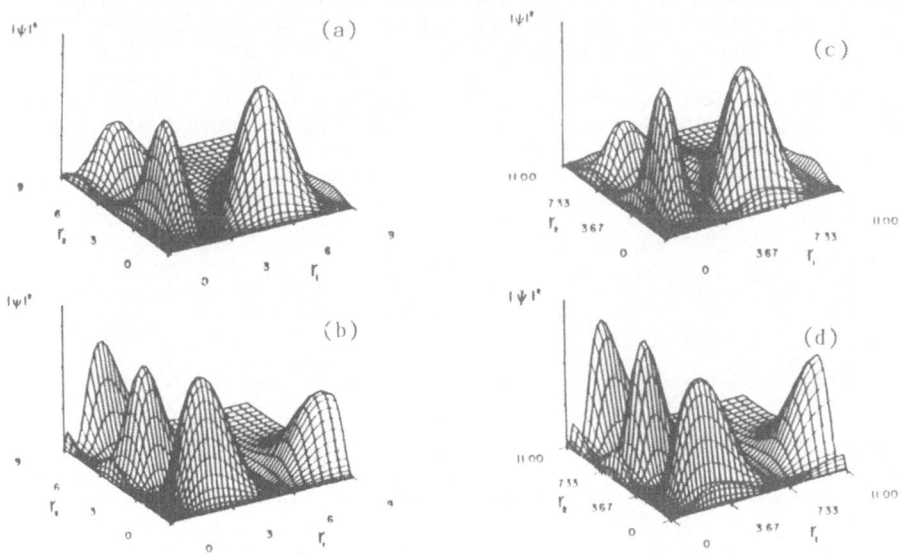

Fig. 5.   Radial sp-component of the $^1P^o$ probability densities for two independent eigensolutions at one energy (the second ionization threshold) for Be (a) and (b) and for Mg (c) and (d). These show the approximate validity of the +/- quantum number characterizing radial correlations in these eigenchannels. (From Ref. 20.)

to study these correlations, since the one-electron potential dictates the phase of a valence electron orbital when it emerges from the closed-shell ionic core. The detailed interactions of the two valence electrons outside this core depend sensitively on these phases, or equivalently on the one-electron quantum defects, of each emerging electron. It is thus extremely important to find a potential or screening function $Z(r)$ which can accurately reproduce the experimental quantum defects of the singly-charged alkaline earth ion.

With this motivation and reasoning, Refs. 16-19, 38 have abandoned for the time being attempts to determine the "best" one-electron potentials for Ca, Sr, and Ba from *ab initio* theory. Instead Ref.16 obtains, for example, a small empirical correction to the Hartree-Slater potential for e-Ca^{++}. The correction term takes the form of a cutoff polarization potential, but this is probably not a critical detail. Refs.18 and 38 apply empirical corrections in a somewhat different manner for e-Sr^{++} and e-Ba^{++}. In each case the potentials have been optimized solely on the basis of experimental information concerning the singly-charged ion, and *not* on the basis of improving agreement between the final R-matrix calculations for the *neutral* alkaline earth atom and the observed spectrum.

I turn now to the eigenchannel R-matrix calculation carried out for atomic calcium by Refs. 16 and 17, and to the very similar analysis developed by Refs. 18 and 19 for strontium. The initial calcium calculations were performed in LS-coupling for $^1P^O$ symmetry, which is the final channel relevant to dipole-allowed photoabsorption by the $4s^2$ ground state. The final R-matrix boundary was placed at $r_0=18$ a.u. although nearly identical results were found in trial calculations using $r_0=14$ a.u. and $r_0=22$ a.u. Several considerations are involved in choosing the radial basis functions to use in the variational calculation.

Radial orbitals $u_{n\ell}(r)$ of two distinct types were included in Refs. 16 and 17. The first, or "closed-type" $u_{n\ell}^c$ consists of eigenfunctions of the e-Ca^{++} radial Schroedinger equation which vanish at $r=r_0$. Figure 6(a) shows the set of seven closed-type functions used to represent s-wave electrons in the variational calculation. Note that each successive radial function in 6(a) has one additional node, and that the lowest has three nodes as expected for a 4s orbital. The second, or "open-type" orbitals $u_{n\ell}^O$ are also numerical eigenfunctions of the same differential equation, but they do not vanish at $r=r_0$. Three such orbitals are shown in Fig.6(b). They are called "open-type" because any open or weakly-closed channel involving an outer s electron must

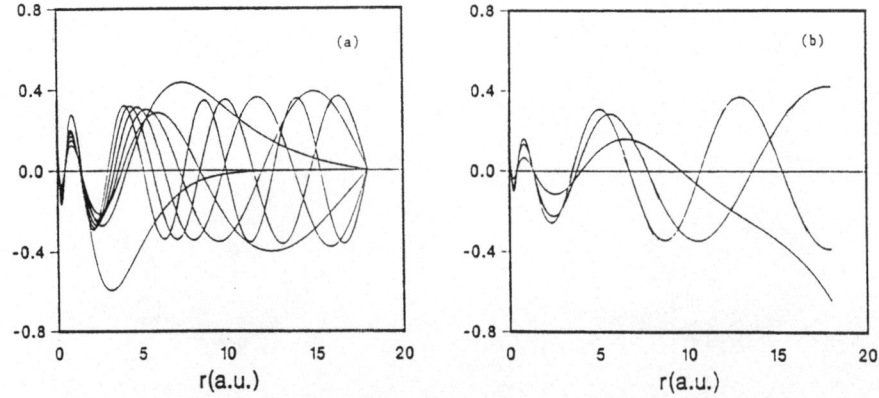

Fig. 6.   Radial one-electron ns-orbitals of Ca^+ used as basis functions in the eigenchannel R-matrix calculation for atomic calcium. Shown are both the closed-type orbitals in (a) and the open-type orbitals in (b).

include at least one such orbital. In practice these open-type orbitals are found by numerically solving the radial Schroedinger equation at energies midway between the successive eigenenergies found in the course of calculating the closed-type orbitals. The final results are not very sensitive to the choice of the open-type orbitals.

The choice of one-electron orbitals to combine into two-electron trial basis functions also requires some consideration. Since two-electron escape beyond the reaction surface is normally neglected in calculations of this type, each Slater-determinantal basis function utilizes at most a single open-type orbital. This guarantees that the final two-electron wavefunction vanishes at the point $r_1 = r_2 = r_0$.

To treat the energy range up to the 4p threshold of Ca^+, the "minimum" basis set would include two-electron basis functions of the type $4s\varepsilon p$ (of both open- and closed-type), as well as $3d\varepsilon p$, $3d\varepsilon f$, $4p\varepsilon s$, and $4p\varepsilon d$. This would give a total number of around 40 to 50 basis functions. The well-known slow convergence of such a "close-coupling" expansion suggests that some basis functions corresponding to "strongly-closed" channels such as 5s5p might be included as well. These should contain only closed-type orbitals for both electrons, of course. They play two roles in the variational expansion of the wavefunction: they may help to represent general electron correlation and

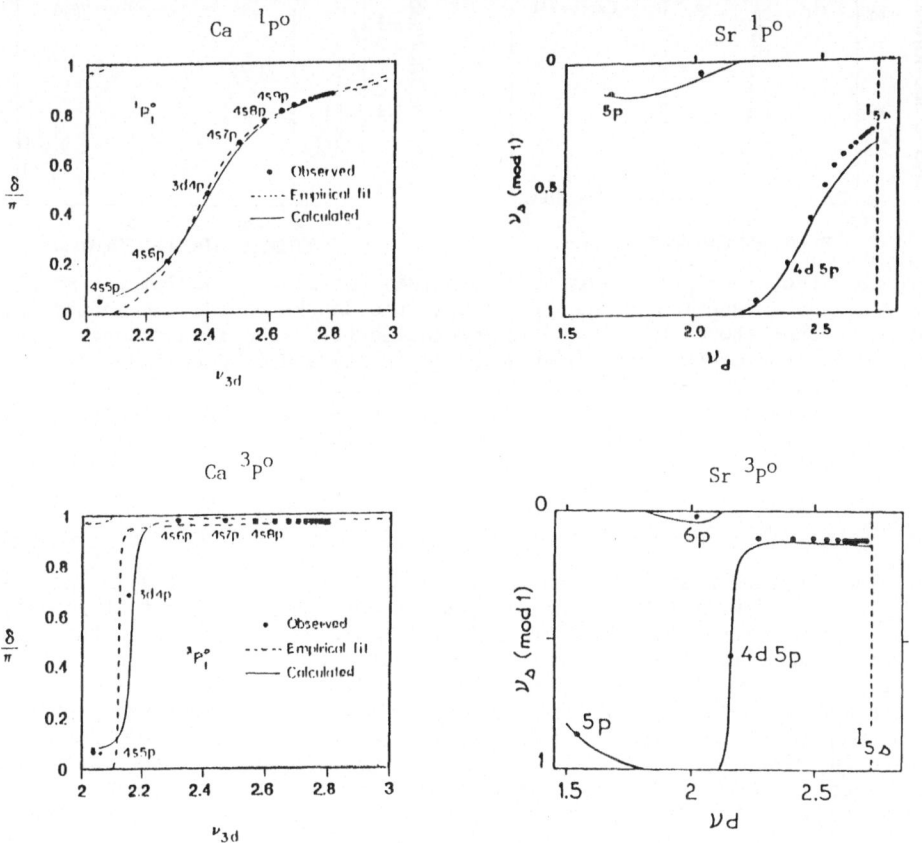

Fig. 7. Lu-Fano plots of the $^1P^o$ and $^3P^o$ discrete energy levels of Ca and Sr, comparing experimental values (solid points) with eigenchannel R-matrix calculations (solid curves) of Refs. 16 and 18. The dashed curves for calcium are empirical QDT fits of Ref. 37 using energy-independent short-range parameters.

relaxation-type features in a rather nonspecific manner, but a more clearly defined role is to describe the polarization of the inner (Ca$^+$) valence electron by the outer electron. Such polarization effects are usually neglected outside the reaction surface where Coulomb functions are utilized, but a basis set should be used which is sufficiently flexible to describe polarization inside the reaction volume. This latter criterion seems to quite useful and suggests that the strong polarizability of the 4p state will only be described correctly if some strongly-closed basis functions involving ns electrons and some involving nd electrons are included. In fact if all basis functions such as 5snp, 4dnp, and 4dnf are omitted, the apparent polarizability of the 4p state must be grossly incorrect as none of the normally dominant positive contributions are included.

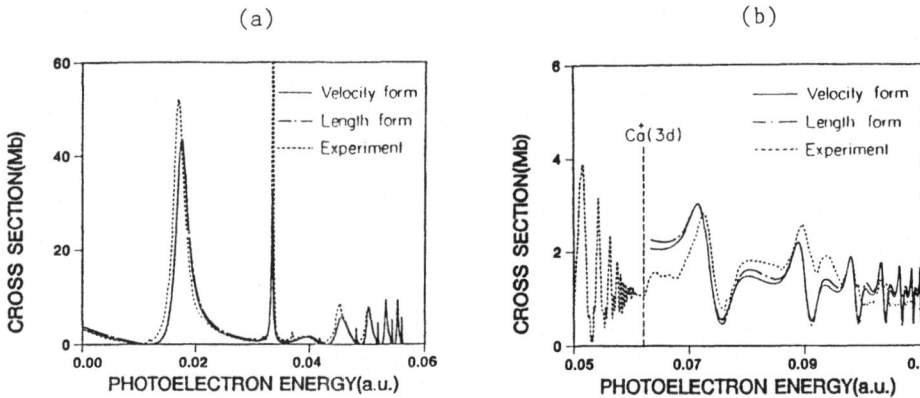

Fig. 8. Photoionization cross sections (Mb) for atomic calcium are shown as a function of energy, between the Ca$^+$ 4s and 3d thresholds (a) and between the 3d and 4p thresholds (b). The dashed curves give experimental results of Newsom[41] in (a) and of Connerade et al.[42] in (b). The solid and dash-dotted curves are velocity and length results for the cross section calculated by Greene and Kim[16] in LS-coupling.

The complete basis set used for calcium is given explicitly in Table II of Ref. 16, but it will suffice for present purposes to point out that roughly 50 such strongly-closed two-electron basis functions were included in the final calculations of that paper. Tests showed that such basis functions do not greatly change the qualitative shape of the calculated photoabsorption spectrum. In fact, the main results of Ref.16 can be obtained using a total of less than thirty two-electron basis functions. Nevertheless these strongly-closed functions are instrumental in improving the quality of the variational wavefunction, as judged by the agreement between length and velocity expressions for the dipole matrix element and as judged by other convergence criteria.[14] The strontium calculations of Aymar et al.[18,19] obtained good results without including any such strongly-closed basis functions, but show signs of somewhat poorer convergence in some energy ranges.

Some of the final LS-coupling calculations are compared with experimental results for calcium and strontium in Figs. 7 and 8. Fig. 7 shows the discrete energy level positions below the 4s and 5s ionization thresholds, respectively, in the form of a Lu-Fano plot[39] of the quantum defects for both ^{1}P^o and ^{3}P^o

symmetries. Since the Lu-Fano plot would consist only of vertical and horizontal lines in the absence of channel interactions, it is first of all clear that the interactions between calcium $4s\epsilon p$ and $3d\epsilon p$ channels are much stronger for $^1P^o$ than for $^3P^o$. The same conclusion applies to strontium. In fact the $^1P^o$ interactions approach the maximum strength allowed, much like the Be and Mg results described above. The calcium Lu-Fano plots also show the result of an empirical quantum defect theory fit with energy-independent short-range channel mixing. The better agreement of the eigenchannel R-matrix calculation confirms the very real need to incorporate energy-dependent mixing parameters into the quantum defect analysis. This energy dependence was often neglected in earlier semi-empirical MQDT studies treating narrower energy ranges.[40]

Fig. 8 compares the calculated photoionization spectrum[16] to a low-resolution plot of the spectra observed by Newsom[41] and by Connerade et al.[42], between the 4s and 3d thresholds in (a), and between the 3d and 4p thresholds in (b), respectively. The agreement with experiment in both ranges is quite good, although it deteriorates somewhat in the higher energy range. Two previous calculations in these energy ranges, one using many-body perturbation theory[43] and one using the Wigner-Eisenbud-type R-matrix formulation[44] (without adopting an empirical e-Ca^{++} potential), show distinctly poorer agreement with experiment. Perhaps as important as the agreement between theory and experiment in Fig. 8 is the agreement between length and velocity formulas for the cross section which is virtually exact in Fig. 8(a), as it should be if both the initial and final state wavefunctions are exact eigenfunctions of the Hamiltonian.[45] This indicates that the wavefunctions obtained in this small-scale eigenchannel R-matrix calculation are far more accurate than those obtained in the much larger R-matrix treatment by Scott et al.,[44] where length and velocity results generally differ by roughly a factor of two. Considering the excellent results found for other systems using the Wigner-Eisenbud treatment,[4,5,12,38] the discrepancies between Ref.44 and the experimental

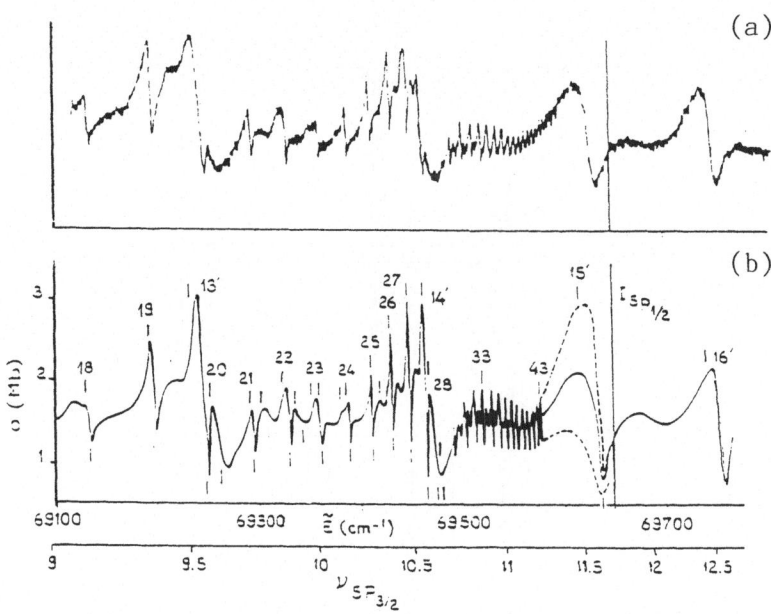

Fig. 9. Sr photoionization cross section (Mb) is shown as a function of the photon energy (cm^{-1}) near the Sr^+ $5p_{\frac{1}{2}}$ threshold. (a) is the experimental measurement of Brown et al.[46], while (b) is the eigenchannel R-matrix calculation of Aymar,[19] using an LS-jj frame transformation to account for fine-structure effects.

results can most likely be traced to their attempt to find an *ab initio* description of the e-Ca^{++} interaction, as opposed to the semi-empirical potentials used in Refs.16-19. (The calculation of Ref.44 was not strictly *ab initio*, however, since a small empirical correction was applied to the diagonal elements of calculated Hamiltonian matrices.)

Effects of the spin-orbit interaction on photoionization spectra of calcium[17] and strontium[19] have recently been described successfully using a geometric LS-jj frame transformation.[8] The idea behind the frame transformation is that the electron motion *within* the reaction zone is accurately represented by an LS-coupled wavefunction, i.e. ignoring spin-orbit effects altogether for $r < r_0$. Since the wavevector k_j of an escaping electron depends on the particular fine-structure state of the ionic residue, however, the wavefunction of a distant Rydberg electron must be described in jj-coupling. When spin-orbit effects are negligible for $r < r_0$ (that is, for Ca and Sr but not for Ba), LS-coupled reaction matrices can simply be recoupled into a larger jj-coupled reaction matrix. In the case of Sr J=1 odd-parity channels relevant to photoionization near the Sr$^+$(5p) thresholds, three LS-coupled reaction matrices are required: $^1P^O$ (5x5), $^3P^O$ (5x5), and $^3D^O$ (3x3). The angular momentum recoupling involves Wigner 9j-coefficients, and results in a single 13x13 reaction matrix and 13 corresponding dipole matrix elements. Once these are known, asymptotic boundary conditions at each energy E are efficiently imposed by multichannel quantum defect technology. The only point at which fine-structure actually enters this calculation is this last stage, where the experimental spin-orbit-split thresholds of Sr$^+$ are used in the MQDT calculation.

Aymar's recent 13-channel calculation[19] of Sr photoionization near the 5p$_{\frac{1}{2}}$ threshold is shown in Fig.9. Clearly, this simple procedure accounts for numerous subtle and complicated features in the experiment of Brown et

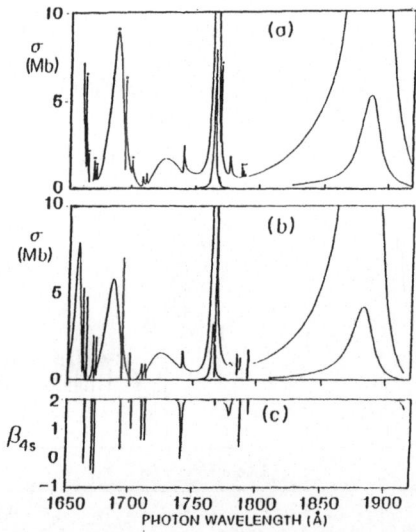

Fig. 10. Ca photoionization spectrum between the 4s and 3d thresholds. (a) is Newsom's high resolution measurement[41] and (b) is the calculation of Kim and Greene.[17] The insets in (a) and (b) show the 3d5p and 3d6p resonances reduced by a factor of 10. (c) is the predicted asymmetry parameter characterizing the angular distribution of photoelectrons.[17]

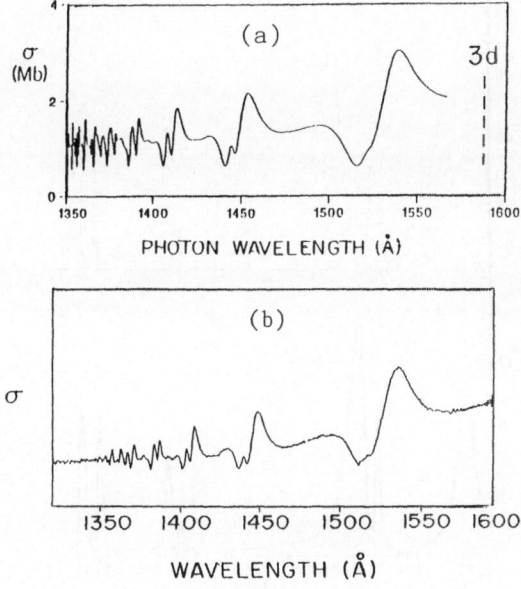

Fig. 11. Calcium photoionization spectrum between the 3d and 4p thresholds. (a) shows the spectrum calculated by Kim and Greene[17], while (b) gives the very recent measurement of Griesmann et al.[47]

al.[46] with great success. Figure 10 compares the experimental[41] and theoretical[17] calcium photoionization cross sections between the 4s and 3d thresholds. This is the same energy range shown in Fig.8(a), except at higher resolution. Most of the weaker and narrower features in Fig.10 are associated with resonances such as 3d4p $^3P^o$ whose excitation whould be forbidden for an electric dipole transition in LS-coupling. The position, width, and asymmetry of most of these weak lines emerge correctly from the calculation.

The spectral range between the 3d and 4p thresholds in calcium is shown in Fig.11, including the calculation of Ref.17 and also the more recent measurement of Griesmann et al.[47] The latter were only unveiled in the lecture of Connerade presented at this NATO school. Agreement is virtually exact, improving greatly over the older experimental results[42] depicted in Fig.8(b). Considering that the new experimental results were presented *after* the eigenchannel R-matrix study of Ref.17 was completed, this points toward the considerable *predictive* power of such calculations.

For heavier atoms, the effect of spin-orbit terms in the Hamiltonian is so strong that they must be included explicitly in the short-range calculation instead of by the simpler frame transformation method used for calcium and strontium. This has now been accomplished for barium[38] and for mercury[48] using the Wigner-Eisenbud-type formulation of R-matrix theory. The effects of all non-valence electrons are also treated in these calculations by a semi-empirical model potential. They have been quite remarkable in demonstrating that resonance features in such heavy atoms are accessible to detailed calculation. Figure 12 shows impressive agreement between calculated[38] and measured[49] photoionization cross sections of barium below the 5d threshold(s).

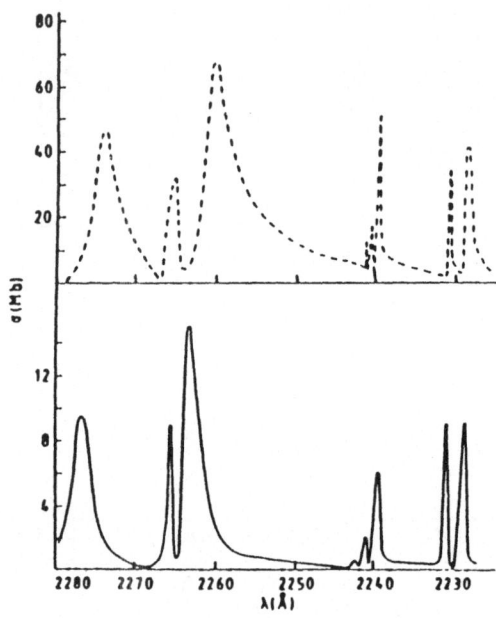

Fig. 12. Photoionization cross section of barium below the 5d thresholds,
comparing experimental results (dashed curve) of Hudson et
al.[49] with the Wigner-Eisenbud-type R-matrix calculation (solid
curve) of Bartschat and Scott[48] which includes the effect of
fine-structure on the electron pair inside the reaction volume as
well as outside.

PROSPECTS

The eigenchannel studies of Refs.16-22 have established that a very
realistic description of atoms having few valence electrons can be attained
using comparatively small computations. Except for strontium, each of these
calculations was carried out on a minicomputer. (The calcium calculations of
Ref.44, on the other hand, were carried out on a supercomputer.)

The aim of this line of research is not solely to attain impressive
agreement between theory and experiment, as desirable as that may be. The main
thrust is instead to elucidate global trends and systematics in the channel
interactions and electron correlations. The possibility of finding a small
number of smooth parameters controlling these interactions has now been firmly
established through the use of quantum defect theory. MQDT permits the
reduction of highly complicated spectra like those in Figs.3 and 7-12 down to a
few smooth quantities such as the eigenquantum defects μ_α shown in Fig.4 and
Fig.13. These characterize the short-range electron-ion scattering process and
should become, ultimately, the objects of greatest intrinsic interest.
Moreover, in moving from one atom to another through the periodic system, it is
these *smooth* quantities which should be compared directly, as in Ref. 37.

Refs. 16-22 and 38 have uncovered numerous relevant systematics, going
well beyond the early observation[37] of similarities among the alkaline earth
spectra. One striking aspect has been that photoionization of Ca, Sr, and Ba
near the np ionization threshold is most likely to produce ions in the first
excited state, (n-1)d. In calcium photoionization, for instance, the 3d
state(s) of Ca^+ is populated 64% of the time at this energy as opposed to only
17% of the time in the Ca^+(4s) ground state.[17] This is a dramatic example of
the complete inadequacy of any independent-electron model of the alkaline

earths, since these would predict that only Ca$^+$(4s) is produced. Another conclusion in the same vein can be reached by examining the smooth, short-range scattering matrix for $^1P^O$ calcium. When a d-wave electron collides with Ca$^+$(4p), 70% of the time it de-excites the ion into the Ca$^+$(3d) level and gains one unit of orbital angular momentum as it is accelerated,[16] again an indicator of nearly maximal channel mixing. Still another piece of information needed for a global understanding is the energy at which each channel "turns on" and becomes locally accessible for interaction as the energy is increased. This too emerges naturally from studies of this type, while being difficult to extract from empirical work alone.

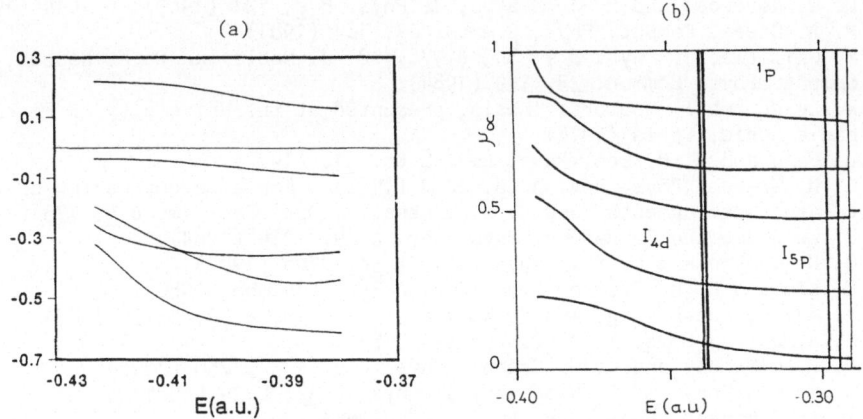

Fig. 13. Eigenquantum defects for $^1P^O$ calcium (a) and strontium (b) are shown to be smooth functions of energy. From Refs. 16 and 19.

Berkowitz has reviewed the systematics beginning to emerge from spectral observations throughout the periodic system.[50] Open shell atoms show strong electron correlations and broad autoionizing features almost universally owing to the low excitation energy needed for orbital hybridization, with several exceptions in the second period including fluorine and oxygen. The combined use of small-scale R-matrix calculations and of multichannel quantum defect technology currently seems poised to identify the key aspects in the evolution of channel interactions.

ACKNOWLEDGEMENT

This work could not have been completed without extensive collaboration with Longhuan Kim. Discussions with L. Kim, M. Aymar, and P. F. O'Mahony, along with access to their unpublished results have been most useful. Comments on the manuscript by U. Fano are appreciated. This work was supported in part by the National Science Foundation.

REFERENCES

1. E. P. Wigner, Phys. Rev. $\underline{70}$, 15 (1946); E. P. Wigner and L. Eisenbud, Phys. Rev. $\underline{72}$, 29 (1947); T. Teichmann and E. P. Wigner, Phys. Rev. $\underline{87}$, 123 (1952).
2. G. Breit, in Handbuch der Physik, Vol.41, Ed. S. Flugge (Springer-Verlag, Berlin, 1959), pp.107-231; A. M. Lane and R. G. Thomas, Rev. Mod. Phys. $\underline{30}$, 257 (1958).
3. P. G. Burke and W. D. Robb, Adv. Atom. Molec. Phys. $\underline{11}$, 143 (1975).
4. P. G. Burke and K. T. Taylor, J. Phys. B $\underline{8}$, 2620 (1975).
5. K. T. Taylor, C. J. Zeippen, and M. Le Dourneuf, J. Phys. B $\underline{17}$, L157 (1984).
6. M. Inokuti, Comments At. Mol. Phys. $\underline{10}$, 99 (1981).
7. M. J. Seaton Rep. Prog. Phys. $\underline{46}$, 167 (1983).
8. U. Fano and A. R. P. Rau, Atomic Collisions and Spectra, (Academic, Orlando, 1986).
9. C. H. Greene and Ch. Jungen, Adv. At. Molec. Phys. $\underline{21}$, 51 (1985).
10. D. W. Norcross and M. J. Seaton, J. Phys. B $\underline{2}$, 731 (1969); $\underline{6}$, 614 (1973); M. A. Crees, Comput. Phys. Commun. $\underline{23}$, 181 (1981)
11. M. Gailitis, J. Phys. B $\underline{9}$, 843 (1976); C. J. Noble and R. K. Nesbet, Comput. Phys. Commun. $\underline{33}$, 399 (1984).
12. M. Le Dourneuf, Doctoral Thesis, presented at the University of Pierre and Marie Curie, Paris (1976).
13. U. Fano and C. M. Lee, Phys. Rev. Lett. $\underline{31}$, (1973).
14. C. H. Greene, Phys. Rev. A $\underline{28}$, 2209 (1983). For a recent review of related developments, see C. H. Greene, J. Opt. Soc. Am. B $\underline{4}$, 775 (1987).
15. H. Le Rouzo and G. Raseev, Phys. Rev. A $\underline{29}$, 1214 (1984).
16. C. H. Greene and L. Kim, Phys. Rev. A $\underline{36}$, 2706 (1987).
17. L. Kim and C. H. Greene, Phys. Rev. A $\underline{36}$, (November 1987).
18. M. Aymar, Luc-Koenig, and S. Watanabe, J. Phys. B $\underline{20}$, 4325 (1987)
19. M. Aymar, J. Phys. B $\underline{21}$, (in press 1988).
20. P. F. O'Mahony and C. H. Greene, Phys. Rev. A $\underline{31}$, 250 (1985).
21. P. F. O'Mahony and S. Watanabe, J. Phys. B $\underline{32}$, L239 (1985)
22. P. F. O'Mahony, Phys. Rev. A $\underline{32}$, 908 (1985).
23. C. M. Lee, Phys. Rev. A $\underline{10}$, 584 (1974).
24. Ch. Jungen, Phys. Rev. Lett. $\underline{53}$, 2394 (1984).
25. R. K. Nesbet, Variational Methods in Electron-Atom Scattering Theory (Plenum, New York, 1980).
26. R. S. Oberoi and R. K. Nesbet, Phys. Rev. A $\underline{8}$, 215 (1973); $\underline{9}$, 2804 (1974).
27. P. J. A. Buttle, Phys. Rev. $\underline{160}$, 719 (1967).
28. D. J. Zvijac, E. J. Heller, and J. C. Light, J. Phys. B $\underline{8}$, 1016 (1975).
29. U. Fano, in Invited Papers and Progress Reports of the X ICPEAC, edited by G. Watel (North-Holland, Amsterdam, 1978), p.271.
30. J. P. Desclaux, Comput. Phys. Commun. $\underline{1}$, 216 (1969).
31. J. Dubau and J. Wells, J. Phys. B $\underline{6}$, 1452 (1973).
32. G. N. Bates and P. L. Altick, J. Phys. B $\underline{6}$, 653 (1973).
33. G. Mehlmann-Balloffet and J. M. Esteva, Astrophys. J. $\underline{157}$, 945 (1969).
34. J. M. Preses, C. E. Burkhardt, W. P. Garver, and J. J. Leventhal, Phys. Rev. A $\underline{29}$, 985 (1984).
35. C. H. Greene, Phys. Rev. A $\underline{23}$, 661 (1981).
36. J. W. Cooper, U. Fano, and F. Prats, Phys. Rev. Lett. $\underline{10}$, 518 (1963).
37. see, e.g. J. J. Wynne and J. A. Armstrong, Comments At. Mol. Phys. $\underline{8}$, 155 (1979), and references therein.
38. K. Bartschat, M. R. H. Rudge, and P. Scott, J. Phys. B $\underline{16}$, 2469 (1986).
39. K. T. Lu and U. Fano, Phys. Rev. A $\underline{2}$, 81 (1970).
40. C. M. Lee and K. T. Lu, Phys. Rev. A $\underline{8}$, 1241 (1973).
41. G. H. Newsom, Proc. Phys. Soc. $\underline{87}$, 975 (1966).
42. J. P. Connerade, M. A. Baig, W. R. S. Garton, and G. H. Newsom, Proc. Roy. Soc. A $\underline{371}$, 295 (1980).
43. Z. Altun, S. L. Carter, and H. P. Kelly, J. Phys. B $\underline{15}$, L709 (1982).
44. P. Scott, A. E. Kingston, and A. Hibbert, J. Phys. B $\underline{16}$, 3945 (1983).

45. see, e.g. A. F. Starace in Handbuch der Physik, vol.31, ed. W. Melhorn (Springer-Verlag, Berlin, 1982).

46. C. M. Brown, M. S. Longmire, and M. L. Ginter, J. Opt. Soc. Am. $\underline{73}$, 985 (1983).

47. U. Griesmann, Shen Ning, J. P. Connerade, K. Sommer, and J. Hormes, J. Phys. B (submitted 1987).

48. K. Bartschat, and P. Scott, J. Phys. B $\underline{18}$, L191 (1985); $\underline{18}$, 3725 (1985).

49. P. D. Hudson, V. L. Carter, and P. A. Young, Phys. Rev. A $\underline{2}$, 643 (1970).

50. J. Berkowitz, in Electronic and Atomic Collisions, eds. D. C. Lorents, W. E. Meyerhof, and J. R. Peterson (North-Holland, 1986), p.631.

ION-ATOM COLLISIONS

Joseph Macek

Department of Physics and Astronomy
University of Nebraska
Lincoln, Nebraska 68588-0111

ABSTRACT

Ion-atom collisions in the intermediate and high velocity regions typically employ the time evolution matrix $\underline{\underline{U}}(t,t')$. The matrix element $U(+\infty,0)_{j\alpha}$ represents a WKB-type approximation to the Jost matrix $J^\dagger_{j\alpha}$. Condensation channels α in this case represent channels appropriate near the untied atom limit. At low and intermediate velocities molecular bases for the channels α have proven fruitful. At high velocities multiple scattering states describe observed physical phenomena. Ion-atom collisions provide a model where experiment and theory combine to identify condensation channels, reaction coordinates, asymptotic channels and the propagation from condensation to asymptotic regions.

INTRODUCTION

In the introduction to the school program, I emphasized the central role played by fragmentation reactions. That is, fragmentation reactions form the natural bridge between collisions and spectroscopy in atomic and molecular physics. Photoionization was considered as a prototype of a fragmentation reaction. Similarly, photoexcitation was considered as a prototype for reactions where the emphasis is on the structure of bound excited states. Finally, we considered electron-atom scattering as an example of collisions and illustrated how such reactions are described in terms of the Jost matrices appropriate to fragmentation reactions. The next part of our school deals with ion-atom collisions and their relation to general issues of atomic and molecular physics.

Ion-atom collisions are part of the interdisciplinary field of ion-molecule collisions. We do not treat this broader subject at all, yet it is critical to physical chemistry applications. In addition, some progress in treating ion-molecule collisions has been achieved by building on insights that have proved fruitful in ion-atom collisions.[1,2] Our program is further narrowed by concentrating on energy ranges well above thermal, which again bypasses chemical applications. Even within the context of ion-atom collisions we do not treat the effect of electron correlations.[3] This area also has produced new results that point to fruitful future directions for the field.[4,5] Much experimental evidence exists indicating that these areas are ripe for development, but we concentrate on phenomena which fall within the scope of the independent particle approximation.[6] Most of the issues to be discussed emerge in collisions involving one electron and two positively charged particles.

It is convenient to classify ion-atom collisions according to the velocity of relative motion.[7] Actually, for ion-atom collisions there are several relevant parameters besides the relative velocity. To discuss these parameters let me introduce some notation. Let M denote the reduced mass of the target-projectile system, v the relative velocity, and m the electron mass. Also, let $\langle v_e \rangle$ denote the mean velocity of the active electrons in the initial or final states. Usually we use atomic units where $m=e=\hbar=1$, although it is often useful to exhibit the electron mass m explicitly.

The lowest velocity range corresponds to the region where $Mv \leq m\langle v_e \rangle$, i.e. the energy region where the wavelength of wavefunctions appropriate to relative motion and electron motion are of the same order of magnitude. Here one must use a wave treatment and the S-matrix is appropriately written as[8]

$$\underline{\underline{S}} = \underline{\underline{J}}^{+}(\underline{\underline{J}}^{-})^{-1} .$$

(1)

All of the issues identified for electron collisions, namely, the selection of condensation channels, reaction coordinates, frame transformations and asymptotic states are relevant here. This region is crucial to chemical reactions and for that reason has been extensively studied in a quantum chemistry context. This velocity region is not extensively treated here so let us move rapidly to the next velocity range.

In the next velocity range we consider that the velocity of relative motion is much greater than $m\langle v_e \rangle/M$, but that it is still less than $\langle v_e \rangle$. In this region one can make the semi-classical approximation wherein the relative motion of the heavy particles is treated classically or semi-classically and the electron motion treated quantally. Then the reaction coordinate is usually taken as the relative position of the heavy particles measured along a classical path. Often the classical path is taken as a straight line since particle deflection is generally minimal. In that case an appropriate reaction coordinate is the relative position of the target and projectile measured along a straight-line classical path. The projectile-target distance is then given by

$$R^2 = B^2 + Z^2$$

(2)

where B is the impact parameter and Z is a cartesian coordinate measured along the relative velocity vector $\vec{v}$. In this case Z represents the reaction coordinate. Usually one introduces an alternative parameter t, called time, defined according to $t=Z/v$ as is done in time-of-flight measurements. This definition and the semi-classical approximation means that one may employ (as an approximation) the time dependent-Schroedinger equation to describe the electronic motion

$$i\dot{\psi} = H\psi, \tag{3}$$

where H includes electron kinetic and potential energies only. The relative motion of the projectile and target is given by the Eq. (2) relating R and t.

In this approximation, appropriate when $\langle v_e \rangle m/M \ll v$, the reaction coordinate Z or t, is known. Furthermore, the S matrix takes the alternative form

$$\underline{\underline{S}} = \underline{\underline{U}}(\infty, -\infty), \tag{4}$$

where $\underline{\underline{U}}(t,t')$, with $t>t'$, is the time evolution matrix. The connection of this equation with the Jost matrices is intuitively obvious from the standard relation

$$\underline{\underline{U}}(t,t') = \underline{\underline{U}}(t,t'')\underline{\underline{U}}(t'',t'), \tag{5}$$

where t'' is any time coordinate. We need only select t''=0 to write the S matrix in the form

$$\underline{\underline{S}} = \underline{\underline{U}}(+\infty,0)\underline{\underline{U}}(0,-\infty) = \underline{\underline{U}}(+\infty,0)\underline{\underline{U}}(-\infty,0)^{-1}. \tag{6}$$

The factor $\underline{\underline{U}}(+\infty,0)$ which describes the propagation of the system from the distance of closest approach at t = 0 to the asymptotic region plays the role of $\underline{\underline{J}}^+$, and the factor $\underline{\underline{U}}(0,-\infty)$ which describes the propagation from the asymptotic region corresponding to negative times or, equivalently, to the initial state, plays the role of $\underline{\underline{J}}^-$. This connection between $\underline{\underline{J}}$ and $\underline{\underline{U}}$ will be made quantitively in the latter part of this lecture, but it is fairly obvious without a detailed derivation. What is important here is that Eq. (6) and ion-atom collisions generally serve as a model for interpreting collisions in terms of the Jost matrices. In essence, this is has long been the focus of ion-atom collision studies.

A great deal is known about ion-atom collisions in the intermediate velocity region. Here the semiclassical approximation and Eq. (5) give us a pictoral description of collisions where states at given instances of time are defined. Physical or mathematical intuition provides a guide to the selection of a set of instantaneous basis functions. In the velocity range where $v \ll \langle v_e \rangle$ a first approximation employs stationary states of the electrons at a fixed inter-nuclear distance R, i.e. molecular basis states. These are exact descriptions of the electron motion for fixed relative coordinate R, but a superposition of such states is required for non-zero velocities. In any event, the appropriate condensation channels are known; they are some sort of molecular channels.

Our understanding of molecular channels appropriate near the united atom region is still evolving. For example, for multiparticle systems

the diabatic states of Lichten[9] and Barat and Lichten[10] have been particularly fruitful.[11] In addition we know that purely molecular states
are not sufficient, rather one must allow for the translation of the
electrons with center of mass or charge of the system. Thus molecular
states must be supplimented by the much discussed translation factors.[4]
This becomes particularly important in the asymptotic region where molecular states naturally evolve into atomic states. In a reference frame
where the target is at rest the projectile moves and atomic states representing electrons around the projectile must incorporate the translational momentum of the electrons.

Some progress has been achieved by treating the trace of the moment
of inertia tensor of the ion-atom system, rather than the internuclear
distance, as an adiabatic variable and reaction coordinate.[11] This procedure keeps the conceptually useful aspects of the molecular picture,
such as molecular states and energy curves, but eliminates explicit and
somewhat ad. hoc. translation factors. Instead, electron translation in
the asymptotic region emerges naturally from the formulation.[12]

The conflicting requirements of stationary molecular states and non-
stationary dynamics complicates the simple molecular picture yet the
pictorial description of collisions in terms of molecular states, molecular energy curves, and transitions at crossings of molecular energy
curves is still fruitful. This picture is particularly appropriate for
the condensation channels α in the explicit expression for S_{ji}

$$S_{ji} = \sum_{\alpha} U_{j\alpha}(+\infty,0)U_{\alpha i}(0,-\infty). \tag{7}$$

We need both asymptotic channels i, j and condensation channels α. One
way to incorporate the two types of channels will be discussed in the
lecture later on this week by Lin. He will not discuss the wide variety
of states that have been used to build appropriate channel functions in
ion-atom collisions but will concentrate on a particular method, the AO+
method,[13] that he has developed.

A well-known example of a condensation channel in ion-atom collisions
is provided by the united-atom rotational coupling mechanism.[14] Here we
consider that the appropriate condensation channels for an ion-atom collision involving a one-electron target of nuclear charge Z_T and a projectile nucleus of charge Z_P are atomic states of a one-electron atom whose
nucleus has a charge $Z_T + Z_P$. Near the united-atom configuration the
electron energy levels are not sensitive to the spatial orientation of
the internuclear axis so that the molecular levels become degenerate in
this limit.[15] At points in coordinate space where molecular levels are
degenerate mixing of states, which leads to transitions between atomic
levels in the asymptotic region, occurs easily.[16] Our conceptual framework must incorporate just these condensation channels where level crossings occur, and these are frequently some sort of molecular channel.

A molecular-type picture can even be used to elucidate electron-atom
collisions.[17] Consider the example of s-p transitions in when the energy
of the s state is below that of the p state. Then the effective potentials $V_i(r)$ in a "molecular" energy diagram include the electron binding
energy ε_i plus the centrifugal potential;

$$'V_i(r)' \approx \varepsilon_i + \frac{\ell_p(\ell_p+1)}{2Mr^2} \tag{8}$$

where ℓ_p represents the angular momentum of relative motion. Let ℓ_T represent the angular momentum of the electron in the atomic state and L the total angular momentum $\vec{L} = \vec{\ell}_p + \vec{\ell}_T$. For the initial s-state channel ℓ_p=L but for the the p-state channel we may have either $\ell_p = L - 1$ or $\ell_p = L + 1$. Since $\varepsilon_s < \varepsilon_p$, the potential curve with $\ell_p = L - 1$ will cross the potential curve corresponding to the initial state, but the curve with $\ell_p = L + 1$ will not. It follows from the Landau-Zener theory that the amplitude for populating the p-state with $\ell_p = L - 1$ is much larger than the amplitude for populating states with $\ell_p = L + 1$. This implies that angular momentum of relative motion tends to decrease for an endothermic s-p transition. For ion-atom collisions we note that $\ell_p = mBv$, and that $v_i > v_f$ so that the decrease in ℓ_p is consistent with our curve-crossing argument.

For ion-atom collisions the argument is carried further. Because angular momentum is conserved one expects the average value of the angular momentum of the excited p state to point in the direction given by $j_t = \ell_{pi} - \ell_{pf} = MB(v_i - v_f)$. This provides a simple connection between the orientation $\langle L_T \rangle$ of the final state and the dynamics of the collision.[17] Of course, since one cannot simultaneously use momentum and angular momentum as good quantum numbers, the classical prediction just given for orientation represents a tendency, not a rigorous prediction. These tendencies, which relate to the anisotropies of collision-excited states, will be discussed by Andersen. He expresses these tendencies as "propensity rules" and discusses their observational implication.[18]

Professor Andersen's lectures also emphasize the characterization of final states in terms of state multipoles, and the connection between state multipoles and observable quantities. The state multipoles are expressed in terms of asymptotic eigenstates, and for bound systems the asymptotic eigenstates are well known. Secondary electrons which are bound neither to the target nucleus nor the projectile nucleus are also produced in ion-atom collisions. Since these electrons are not localized in coordinate space the asymptotic eigenstates of these continuum electrons are not known, indeed our understanding of such states is still evolving. Early theories used target continuum states or plane wave almost exclusively, but now we know that, for ion-atom collisions it is essential to consider eigenstates of the projectile also.[19]

We can see the relevance of continuum eigenstates of the projectile by a simple quantum defect argument.[20] Cross sections for electron capture to high Rydberg state $n'1'm'$ vary with n' according to the n'^{-3} law;

$$\sigma_{n'\ell'}^{(cap)} = \bar{\sigma}_{\ell'}^{(cap)} \left(\frac{1}{n'^3}\right), \quad n' \gg 1 \tag{9}$$

where the primes refer to quantities defined in the projectile frame. Since the energy of a Rydberg state is given by

$$\varepsilon_{n'} = -\frac{1}{2n'^2} \qquad (10)$$

one has

$$d\varepsilon'_n = \frac{dn'}{n'^3} \qquad (11)$$

so that Eq. (9) may be written

$$\sigma_{n'\ell'}^{(cap)} = \bar{\sigma}_{\ell'}^{(cap)} \frac{d\varepsilon'}{dn'} \; , \quad \varepsilon' < 0 \qquad (12)$$

valid for $\varepsilon' < 0$. But continuity requires that Eq. (12) hold also for positive energy. In this case, it is convenient to replace the discrete angular indices l'm' by the continuous index $d\Omega'/4\pi$, the differential of solid angle. Eq. (12) then becomes

$$\frac{d\sigma}{d\varepsilon'} = \frac{\bar{\sigma}^{(cap)}}{4\pi} \, d\Omega'. \qquad (13)$$

This equation is readily transformed to the laboratory frame;

$$d\varepsilon d\Omega' = \frac{v_e}{|\vec{v}_e - \vec{v}|} \, d\varepsilon \, d\Omega, \qquad (14)$$

so that

$$d\sigma^{(ion)} = \frac{v_e}{|\vec{v}_e - \vec{v}|} \, \frac{\bar{\sigma}}{4\pi} \, d\varepsilon \, d\Omega, \qquad (15)$$

where v_e represents the electron velocity.

The quantity $\vec{v}_e - \vec{v}$ vanishes when the electron's velocity equals the velocity of projectile, thus the cross section is infinite at that point. We conclude that capture extends into the continuum, and that the consequences of electron capture to continuum states are observable as a sharp peak in the secondary electron spectra centered at $\vec{v}_e = \vec{v}$. This peak manifests all of the collisional properties of bound states, in particular, the peak may exhibit anisotropies[21] to be discussed by Burgdorfer. He will emphasize the connection between anistropies of bound and continuum states.

In the high velocity region where $\langle V_e \rangle \ll v$, we regard the projectile as simply a source that supplies energy, momentum and perhaps angular momentum to a bound system. Then the focus is on the subsequent evolution of the system, just as in fragmentation by photons. The close correspondence between photon excitation and charged particle excitation will be discussed in the lecture by Lutz. He applies the virtual photon picture[22] (the Weizaker-Williams method) to excitation of atoms by fast charged particles. This virtual photon picture represents an old point of view which acquires new relevance with the current emphasis on collisions in presence of a laser field, i.e. in the presence of real as well as virtual photons.

Strong laser fields are another way to transfer energy and angular momentum to a bound system. Lectures by L. Agostini and A. Gusti-Suzor will consider the response of atoms to strong fields in the absence of collisions. Here the whole subject is in a state of rapid development. Questions concerning the appropriate condensation channels, the appropriate asymptotic channels and the propagation from the condensation region to the asymptotic region are to be examined.

In the high velocity region where $\langle v_e \rangle \ll v$ the first Born approximation represents a well accepted description of the most probable reactions. It is now known that the first Born approximation is inadequate for less probable process such as electron capture by charged particles, and a second or higher order approximation is needed. One second order process is known as the Thomas double-collision mechanism.[23] Here the projectile strikes a bound electron in the target which recoils and scatters once again from the target nucleus in a direction and with a velocity favorable for capture. Observations and related theoretical work establish that this mechanism does indeed operate at high velocity.

We want to set this particular result of Thomas in a more general context. To that end we interpret the Thomas mechanism in terms of condensation channels appropriate at high velocity. We will see that condensation channels α, which are high velocity extensions of the multiple scattering states discussed by Dehmer, incorporate the Thomas double-collision mechanism. Finally, these "multiple scattering" channels are amenable to experiment owing to characteristic structure in differential cross sections that they predict.

Early work by Thomas[23] and Drisko[24] considered that the electron traveled from the first to the second collision in a field free region. In reality the electron propagates in the field of both the projectile and the target nuclei. This more complicated motion is described by multiple scattering [25] states near t=0. Such states resemble the multiple scattering molecular states employed by Dehmer to describe final states in the photoionization of molecules.[26] Recall that for those states the atomic cores were fixed in space and the electron wave functions were represented by standing waves set up by multiple scatterings in the fields of the atomic cores. The multiple scattering wave functions for the Thomas mechanism are essentially the same except that the atomic cores move with velocities of the same order of magnitude as the mean electron velocities in the intermediate states. When the cores move so fast the number of multiple scatterings is small, in contrast to the infinite number implicit in the photoionization of molecules.[27] Even so, we see the close connection between a molecular description of the condensation channels at low velocities and the multiple scattering description at high velocities. Interpolating between the two extremes is the AO+ method described by Lin.[13]

Multiple scattering mechanisms have received new emphasis with the recognition that many of the theoretical details are amenable to rather direct experimental tests. The classic case is the Thomas double-collision mechanism for electron capture. Here one can show that the mechanism gives rise to a peak at a projectile scattering angle such that a struck free electron recoils at velocity equal to the incident velocity and making an angle of 60° with respect to the incident projectile velocity vector.[21] The corresponding scattering angle for the projectile θ_T is then given by momentum and energy conservation in the collision of the projectile with the bound electron;

$$\sin \theta_T = (m/M)\sin 60° \tag{16}$$

or

$$\theta_T \approx (m/M) \frac{\sqrt{3}}{2} . \tag{17}$$

To see how this emerges from a multiple scattering theory recall that the transition matrix element for electron capture is given by

$$T_{fi} = \langle f | V_{Te} + V_{Te}(E-H+i\eta)^{-1} | V_{Pe} \, i \rangle \tag{18}$$

where V_{Pe} and V_{Te} represent the electron-projectile and electron target nucleus interaction potentials respectively. The hamiltonian H represents the hamiltonian for an electron, projectile nucleus and target nucleus three-particle system in the center of mass system. To a good approximation we may replace the full hamiltonian H by H_T which omits the projectile-electron interaction in first approximation, as in the corresponding theories for excitation.[28] The resulting approximate amplitude,

$$T_{fi} \approx \langle f | V_{Te} + V_{Te}(E-H_T+i\eta)^{-1} V_{Pe} | i \rangle \tag{19}$$

then describes a second order mechanism with one scattering represented by V_{Pe}, propagation in the field of the target represented by the Green's function $(E-H_T + i\eta)^{-1}$, a second scattering from the target nucleus represented by V_{Te} and finally capture into a final state f.

Splitting the Green's function into its real and imaginary parts identifies a purely imaginary term where energy is conserved in intermediate states;[29]

$$T_{fi} = \langle f | V_{Pe} | i \rangle + \langle f | V_{Te} \frac{P}{E-H_T} V_{Pe} | i \rangle$$

(20)

$$- i\pi \langle f | V_{Te} \delta(E-H_T) V_{Pe} | i \rangle.$$

The term $P/(E-H_T)$ represents a principal part Green's function. Momentum is also conserved in intermediates states within the limits of the Compton profiles of initial and final states. Application of energy-momentum conservation for the third term in Eq. (20) predicts a unique angle for the scattered projectile that has picked up an electron.[21,30]

Eq. (20) also applies for charge transfer to continuum states where a secondary electron is produced. A close analysis Eq.(20) shows that if we detect a secondary electron with a specific velocity vector $\vec{v}_e$, then the double collision mechanism requires that the projectile scatter through a specific angle.

This is merely a consequence of applying energy-momentum conservation to a sequence of two scatterings. Quite generally multiple scattering states at the condensation limit give rise to observable structure in the asymptotic region.[31] This suggests that multiple scattering theories are a fruitful means of building in "molecular" type condensation channels where electrons move in the fields of both projectile and target. Burgdorfer will discuss one particular multiple scattering theory which describes capture to both bound and continuum states.[32]

This introduction to ion-atom collisions has emphasized the connection to electron collisions and the spectra of atoms and molecules. The similarity of Jost matrix and the $\underline{\underline{U}}(+\infty,0)$ matrix was central in this discussion. I will conclude this introductory lecture by sketching the precise nature of the approximations which are implied by the use of a classical trajectory for the heavy particle motion and a quantal treatment of the electron motion. The connection between $\underline{\underline{U}}(+\infty,0)$ and $\underline{\underline{J}}^+$ will emerge as a well defined WKB-type approximation for the Jost matrix;

$$\underline{\underline{J}}^+ \approx \underline{\underline{U}}(+\infty,0).$$

(21)

We can understand Eq. (21) by considering a wave traveling through an inhomogeneous medium with an index of refraction varying slowly with position. As the wave propagates through the medium only its phase, given by the WKB approximation, changes. The coefficient of the wave is just J^+ equal to $e^{i\delta}$WKB, where δ_{WKB} is the WKB phase shift. At points of rapid variation of the index of refraction, the wave may partially reflect. The coefficient of the reflected wave is J^- and is given by $e^{-i\delta}$WKB. We are able to identify the coefficients of these waves because the outgoing wave is uncoupled from the incoming wave in the WKB approximation. This decoupling of the two senses of propagation enables us to

employ an approximate first order Schroedinger equation. Since we now
deal with a first order equation in r it is possible to describe the
evolution of the system with a unitary matrix $\underline{\underline{U}}$. Further definition of a
new coordinate t according to

$$dt = dr/v(r)$$

where $v(r)$ is a local velocity, identifies the matrix $\underline{\underline{U}}$ as just the time
evolution matrix. We then have that for the simple model of a wave in an
inhomogeneous medium the result $U(+\infty,0) \approx J^+ \approx e^{i\delta}WKB$. Note that the
point r=0 of the one dimensional Schroedinger equation represents a reflection point where, to satisfy appropriate boundary conditions, we need
to superimpose a reflected wave with coefficient $J^- \approx e^{-i\delta}WKB$. This coefficient is identified as $U(0,-\infty) = U(+\infty,0)^+$ and since $U(t,t') =$
$U(t,0)U(0,t')$ we may employ the time evolution matrix for $0<t<+\infty$ rather
than the Jost matrices appropriate when incoming and outgoing waves are
strongly coupled.

To develop the WKB approximation in a way which emphasizes the decoupling of incoming and outgoing waves we write the second order
Schroedinger equation as two coupled first order equations. Neglect of
the coupling between incoming and outgoing waves gives the WKB approxima-
tion and enables us to identify $\underline{\underline{U}}(+\infty,0)$ as an approximate form of the
Jost matrix $\underline{\underline{J}}^+$.

The essential approximations are illustrated by the motion of a particle in an effective central potential $V(r)$, for which the radial
Schroedinger equation becomes,

$$\left[-\frac{1}{2M} \frac{d^2}{dr^2} + V(r) \right] f(r) = Ef(r) \tag{22}$$

We convert this second order differential equation into two coupled
first order differential equation by the standard definitions

$$y_1(r) = f(r)$$

$$\tag{23}$$

$$Y_2(r) = df(r)/dr = f'(r)$$

Since there are two linear independent functions $f(r)$ it is convenient to define the matrix of solutions $\underline{y}$

$$\underline{y} = \begin{pmatrix} y_{11} & y_{12} \\ y_{21} & y_{22} \end{pmatrix} \tag{24}$$

In addition we define the magnitude of the wave vector k and a local
wave vector $K(r)$ according to

$$k = \sqrt{2ME}$$

(25)

$$K(r) = \{2M\,[E - V(r)]\}^{1/2}$$

The Schroedinger equation is then equivalently written as two coupled first order differential equations in matrix form

$$\left[\frac{d}{dr} - \begin{pmatrix} 0 & 1 \\ -k^2(r) & 0 \end{pmatrix}\right]\, \underline{y}(r) = 0$$

(26)

This form is quite unsymmetric. A more symmetric form obtains by introducing the non-unitary transformation

$$\underline{y}(r) = \underline{\underline{A}}(r)\, \underline{g}(r)$$

where

$$\underline{\underline{A}}(r) = \begin{pmatrix} 1 & -1 \\ iK(r) & iK(r) \end{pmatrix}$$

(27)

so that $\underline{g}(r)$ satisfies

$$\left[\frac{d}{dr} + \frac{K^1(r)}{2K(r)}\begin{pmatrix} 1 & 1 \\ 1 & 1 \end{pmatrix} - K(r)\begin{pmatrix} 1 & 0 \\ 0 & -1 \end{pmatrix}\right]\, \underline{g}(r) = 0$$

(28)

The real diagonal terms $K'(r)/2K(r)$ are easily removed by a standard transformation $\underline{g}(r) = K(r)^{-1/2}\underline{\underline{u}}(r)$ to obtain;

$$\left[\frac{d}{dr} + \frac{K'(r)}{2K(r)}\begin{pmatrix} 0 & 1 \\ 1 & 0 \end{pmatrix} - iK(r)\begin{pmatrix} 1 & 0 \\ 0 & -1 \end{pmatrix}\right]\, \underline{\underline{u}}(r) = 0$$

(29)

In this form the exact Schroedinger equation appears as a coupled two channel equation for the non-unitary matrix $\underline{u}(r)$. Neglect of the coupling term gives the WKB approximation and $\underline{\underline{u}}$ becomes a diagonal unitary matrix. Upon defining a time coordinate according to

$$dt = 2Mdr/K(r),$$

(30)

we obtain a time-dependent Schroedinger equation.

More generally, in the multichannel case, we can recover the time dependent Schroedinger representation by performing the same transformations Eq. (23) - (29), introducing a time variable defined according to Eq. (30) and neglecting the coupling between incoming and outgoing waves. Corresponding to each channel in the multichannel case there is a local wave vector $K(r)$ defined according to

$$K_i(r) = \{2M[E - V_{ii}(r)]\}^{1/2}, \tag{31}$$

where $V_{ii}(r)$ is the diagonal element of the potential matrix. A common wave vector $K(r)$ is defined in terms of an average potential $V(r)$. The diagonal $K_{ii}(r)$ is then approximated by

$$K_i(r) \approx K(r) - M\left[V_{ii}(r) - V(r)\right]/K(r) \tag{32}$$

valid when the second term is small. The resulting approximate equations are just the usual time-dependent equations for the time evolution operators $U(t,0)$ and $U(-t,0)$, where t is positive. In this way we see that the time evolution operators $U(+\infty,0)$ and $U(-\infty,0)$ in the theory of ion-atom collisions represent approximate forms of the Jost matrices employed in our discussions of electron collisions and spectra.

Subsequent lectures will discuss ion-atom collisions from the standard evolution matrix point of view. This brief introduction shows the connection between these lectures and the broader subject of atomic collisions and spectra which forms the focus of the school.

ACKNOWLEDGEMENT

Support by the National Science Foundation under Grant PHY-8602988 is gratefully acknowledged.

REFERENCES

1. D. Dowek, D. Dhuic, V. Sidis and M. Barat, Phys. Rev. A 26, 746 (1982); O. Yenen and D.H. Jaecks, Phys. Rev. A 32, 836 (1985).

2. Arnold Russek in Electron and Atomic Collisions, edited by J. Eichler, I.V. Hertel and N. Stolterfoht (North-Holland, Amsterdam, 1983) p. 701.

3. J.F. Reading, A.L. Ford, G.L. Swafford and A. Fitchard, Phys. Rev. A 20, 130 (1979).

4. L.H. Andersen, P. Hvelplund, H. Knudsen, H.P. Moller, K. Elsner, K.G. Rensfelt and U.Y. Uggerhog, Phys. Rev. Lett. 57, 2147 (1986).

5. J.F. Reading and A.L. Ford, Phys. Rev. Lett. 58, 543 (1987).

6. J.H. McGuire, Phys. Rev. A 36, 1114 (1987).

7. Joseph Macek in Electronic and Atomic Collisions, edited by J. Eichler, I.V. Hertel and N. Stolterfoht (Elsevier Science Publishers B.V. 1984) p. 317.

8. U. Fano and A.R.P. Rau, Atomic Collisions and Spectra, (Academic Press Inc., New York, 1986) chapters 7 and 9.

9. Wm. Lichten, Phys. Rev. 139, 27 (1965).

10. M. Barat and Wm. Lichten, Phys. Rev. A 6, 211 (1972).

11. C.H. Greene, Phys. Rev. A 26, 2974 (1982); J.B. Delos, Rev. Mod. Phys. 53, 287 (1981); Joseph Macek and Khachig A. Jerjian, Phys. Rev. A 33, 233 (1986); A.V. Matveenko, and Y. Abe, Few-Body Systems, to be published.

12. J. Macek, M. Cavagerno, K. Jerjian and U. Fano, Phys. Rev. A 35, 3940 (1987).

13. W. Fritsch and C.D. Lin, Phys. Rev. A 26, 1255 (1982).

14. D.R. Bates and D.A. Williams, Proc. Phys. Soc. 83, 425 (1964).

15. J.S. Briggs and J.H. Macek, J. Phys. B 5, 579 (1972).

16. C. Zener, Proc. Roy. Soc. A 137, 696 (1932).

17. H.W. Hermann and I.V. Hertel, J. Phys. B 13, 4285 (1980).

18. N. Andersen and I.V. Hertel, Comments on Atomic and Molecular Phys. XIX, 1 (1986).

19. A. Salin, J. Phys. B 2, 631 (1969); J. Macek, Phys. Rev. A 1, 235 (1970); J. Crooks and M.E. Rudd, Phys. Rev. Lett. 25, 1599 (1970); K.G. Harrison and M.W. Lucas, Phys. Lett. 33A, 142 (1970).

20. M.E. Rudd and J.H. Macek, Case Stud. At. Phys. 3, 47 (1972).

21. R. Shakeshift and L. Spruch, Rev. Mod. Phys. 151, 369 (1979).

22. W. Heitler, The Quantum Theory of Radiation (Calrendon Press, Oxford, 1954) p. 414.

23. L.H. Thomas, Proc. Roy. Soc. 114, 561 (1927).

24. R.M. Drisko, Ph. D. Thesis, Carnegie Institute of Technology (1955).

25. J. Burgdorfer and K. Taulbjerg, Phys. Rev. A 33, 2959 (1986).

26. D. Dill and J.L. Dehmer, J. Chem. Phys. 61, 692 (1974).

27. Joseph Macek, Comments on Atomic and Molecular Physics VI, 169 (1977).

28. J. Macek and R. Shakeshaft, Phys. Rev. 22, 1441 (1980); P.A. Amundsen and D.H. Jakubassa, J. Phys. B 13, L467 (1980); J.S. Briggs, J. Phys. B 10, 3075 (1977); J. Macek and K. Taulbjerg, Phys. Rev. Lett. 46, 170 (1981).

29. J.H. McGuire, P.R. Simony, O.L Weaver and J. Macek, Phys. Rev. A 26, 1109 (1982).

30. J.S. Briggs, P.T. Greenland and L. Kocbach, J. Phys. B 15, 3085 (1982).

31. J.S. Briggs, J. Phys. B 19, 2703 (1986).

32. L.J. Dube and A. Salin, J. Phys. B 20, L499 (1987).

THEORETICAL STUDIES OF HEAVY PARTICLE COLLISIONS

C.D. Lin

Department of Physics, Cardwell Hall
Kansas State University
Manhattan, Kansas 66506

INTRODUCTION

Collisions between heavy ions and atoms alter the distribution of the electron charge cloud. Unlike electron–atom scattering where the collision complex is often limited in size (except for excitation to high Rydberg states and threshold impact ionization), in ion–atom collisions the electron cloud is pulled apart by the two collision centers and undergoes drastic change in size, shape and rotation. Experimental measurement of excitation, charge transfer or ionization processes reflects the selection of different pieces of the electron cloud after the collision. On the other hand, except for perturbative situations, theoretical descriptions have to follow the time evolution of the electron charge cloud. Not only has one to deal with the formation of the collision complex and the transfer of energy and angular momentum to the electron, (as in electron–atom collisions), but the existence of rearrangement channels (electron capture) also demands a proper account of the transfer of linear momentum as well.

Consider a one–electron collision system, the transfer of linear momentum to the electron plays an important role in determining the final outcome of an experiment. The charge cloud evolves quite differently as the collision speed varies. When the collision speed is much faster than the typical orbital speed of the electron, perturbative theories have been developed. This subject has been addressed in previous lectures in this series by Merzbacher, Briggs and Taulbjerg.[1] In this talk I will limit myself to collisions where the projectile speed is comparable to or slower than the orbital speed of the electron. This is a very active area of experimental study in recent years, in particular, in connection with the development of heavy ion sources.[2]

Theoretical description of ion–atom collisions is mathematically complicated by the need for treating the two-center character of the collision system. We will first limit ourselves to collisions where the motion of the heavy particles can be described classically. In this semi-classical model, the incident particle exerts a time-dependent field on the electron. Consider the collision of a bare projectile with a one-electron atom. The electronic motion is governed by the time-dependent Schrodinger equation,

$$(H-i \frac{\partial}{\partial t}) \psi(\vec{r},t) = (- \frac{1}{2} \nabla^2 - \frac{Z_T}{r_T} - \frac{Z_p}{r_p} - i \frac{\partial}{\partial t}) \psi(\vec{r},t) = 0 , \qquad (1)$$

where $Z_T(Z_p)$ is the charge of the target (projectile) and $r_T(r_p)$ is the distance of the electron from the target (projectile). Atomic units are used throughout unless otherwise indicated.

THE DESCRIPTION OF SCATTERING ATOMIC STATES

The solution of Eq.(1) requires the specification of initial and final states. Let us assume that the origin is chosen at the target center. The description of the initial target state and of its excitation channels is straightforward. For charge transfer channels, one has to take into account that the projectile center is moving with a velocity $\vec{v}$ with respect to the stationary target. In these channels the Galilean transformed projectile state wave functions are then represented by

$$\phi_j(\vec{r}_p)\exp [i(\vec{v}\cdot\vec{r} - \frac{1}{2} v^2 t - \varepsilon_j t)], \qquad (2)$$

where ϕ_j is the stationary wave function of the electron in the projectile frame and ε_j is the eigenvalue of state j. The velocity-dependent phase factor in (2) is called the plane-wave electron translational factor (ETF) and the orbitals (2) are called travelling atomic orbitals.

In considering final states one needs to include ionization channels as well. The continuum wave function of an electron in the field of two moving classical charged particles is not known analytically. Mathematically the ionization channels can be expanded in terms of continuum wave functions with respect to the target center. However, this is physically unattractive for describing several states of continuum electrons. For example, wave functions for the so-called electron capture to the continuum (ECC), that is, continuum electrons which have small linear momentum with respect to the projectile center, are not conveniently expanded in terms of target-centered wave functions.

THE COLLISION COMPLEX and THE MOLECULAR ORBITAL (MO) MODEL

The collision complex describing the time evolution of slow collisions was first introduced by Massey and Smith[3] under the so-called perturbed stationary state (PSS) approximation. In this model, they argued that when the relative velocity of the two heavy particles is small, the electron experiences the field from the two nuclei as if the two nuclei were stationary. The electronic wave function can then be expanded in terms of molecular orbitals (MO's),

$$\psi(\vec{r},t) = \sum_j a_j(t) \; \Phi_j(R(t);\vec{r}) \; \exp(-i \int_{-\infty}^t U_j(R(t'))dt'), \tag{3}$$

where Φ_j is the molecular wave function and U_j is the adiabatic molecular potential. In the PSS model, the collision complex is a quasi-molecule. Transition from one molecular state to another occurs only when the two curves undergo an avoided crossing. This simple model forms the basis for the interpretation of almost all low-energy atom-atom collision experiments since the 30's. It was adopted by Fano and Lichten[4] to explain the large inner-shell vacancy production probabilities in the 60's. Within the PSS model, the study of atom-atom and ion-atom collisions amounts to no more than the calculation of static molecular potential curves. This subject was reviewed by Barat[1] earlier for outer-shell processes.

The success of the PSS model in qualitatively interpreting experimental results has not deterred theorists from pointing out the major flaw in its mathematical formulation.[5] Although the molecular orbitals approach atomic orbitals on the target center (when the target is chosen as stationary) in the asymptotic limit, they do not approach the travelling atomic orbitals on the projectiles. To correct this situation, different forms of electron translational factors have been introduced such that each MO in (3) is replaced by

$$\Phi_j(R(t);\vec{r}) \; \exp[i \left(f_j(\vec{r},R)\vec{v}\cdot\vec{r} - \frac{1}{2} v^2 t - \int_{-\infty}^t U_j(R(t')) \; dt' \right)] . \tag{4}$$

In other words, travelling MO's are now used in the expansion so that each orbital reduces correctly to the travelling AO asymptotically. However, difficulties arise from the conceptual incompatibility of electron translational factors, which associate each electron with a specific fragment, and the molecular orbitals in which each electron moves throughout a fixed molecular complex. There is no obvious physical principle for the determination of the translational factors, or of the function $f(\vec{r},R)$, for finite values of R except in the asymptotic region. Different authors select different forms of ETF's based on physical arguments.[6-9] or on some mathematical variational principle.[10] The sensitivity of the final calculated results to ETF's, especially for the weaker channels, still has not been settled even today.[11]

THE ATOMIC-ORBITAL (AO) EXPANSION METHOD

To circumvent the difficulty of building ETF's in the PSS model and to provide a different way of representing the collision complex at higher collision energies, Bates[12] proposed a two-center atomic orbital expansion method where the time-dependent electronic wave function is expressed as

$$\psi(\vec{r},t) = \sum_i a_i(t)\phi_i(\vec{r}_T)e^{-i\varepsilon_i t} + \sum_j b_j(t)\phi_j(\vec{r}_p)e^{-i(\varepsilon_j t + \frac{1}{2}v^2 t - \vec{v}\cdot\vec{r})}.$$
(5)

In (5), ϕ_i and ϕ_j are the atomic states of the target and the projectile, with eigenenergies ε_i and ε_j, respectively. In the AO expansion, the excitation and charge transfer channels are correctly represented in the asymptotic region. The major question is whether such an expansion adequately incorporates the formation of the collision complex, or more precisely, whether the collision dynamics are well represented by such an expansion.

The wave functions (5) is similar to those of the close-coupling method used in electron-atom collisions in that the eigenstates used in the expansion are undistorted atomic wave functions. Actually this is true if the expansion includes the target states only. By including atomic orbitals at both centers, such an expansion can represent the formation of a molecular complex for collisions at large impact parameters (ρ), since it is known that static molecular orbitals at large inter-nuclear separations are easily expanded in terms of linear combinations of atomic orbitals (LCAO). Thus the AO expansion method is in fact capable of describing collisions occurring at large impact parameters, even at relatively low velocities.

THE AO+ MODEL

The LCAO model fails to describe molecular orbitals at small inter-nuclear separations; thus the two-center AO expansion method in (5) is not capable of representing the formation of molecular complexes at small impact parameters in slow collisions. For transitions occuring at small impact parameters, the AO method would fail.[13] This situation can be corrected by the AO+ model of Fritsch and Lin.[14] By recognizing that molecular orbitals in the R=0 limit are the united-atom (UA) orbitals, it

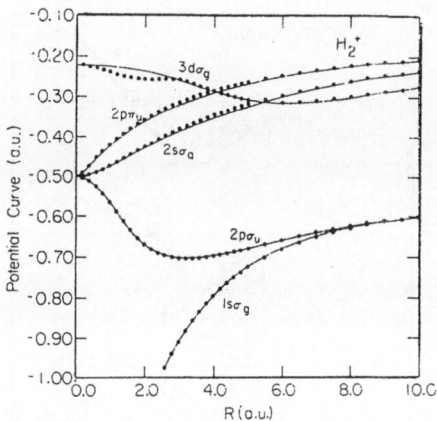

Fig. 1 Molecular potential curves of H_2^+. The solid lines are
exact results and the solid circles are from diagonalizing
the electronic Hamiltonian in the 22 AO+ basis set (from
Ref. 14).

was proposed that UA orbitals be incorporated in the expansion (5). To
avoid complexities in evaluating three-center integrals, these UA orbitals
were added at both collision centers. [In actual calculations these
orbitals were orthogonalized to the separated atom (SA) orbitals on each
center.] A useful check of the adequacy of these basis functions is to
see how they reproduce the molecular potential curves. This is
illustrated in Fig. 1 where the exact potential curves of H_2^+ are compared
with those calculated from a 22-state AO+ set. This set consists of
1s, 2s, $2p_0$, and $2p_1$ SA and UA orbitals (16 altogether), plus the $3d_0$,
$3d_1$, and $3d_2$ UA orbitals, at each center. We note that this AO+ set gives
accurate MO curves at all internuclear separations.

In the AO+ model for dynamical collisions, the SA and UA orbitals at
each center carry its plane wave ETF's. In collisions at low energies
where the collision complex is approximately represented by molecular
orbitals, we expect the basis functions in the AO+ model to be capable of
representing such a collision complex. On the other hand, the SA orbitals
do have the correct ETF's. The relative importance of UA and SA orbitals
is determined by the solution of the time-dependent Schrodinger equation
directly. We might remind the readers that the spirit of this method is
similar to the introduction of Hylleraas type correlated functions in the
close-coupling expansion of e-H scattering.[15]

The importance of UA orbitals in the AO+ model can be seen in the 2s
and 2p excitation cross sections of p+H collision from 1-60 keV, as shown
in Fig. 2. We note that good agreement with experimental data[16-19] is
achieved in the AO+ type calculation. Since rotational coupling between
the 2pσ and 2pπ orbitals is responsible for the population of 2p states at

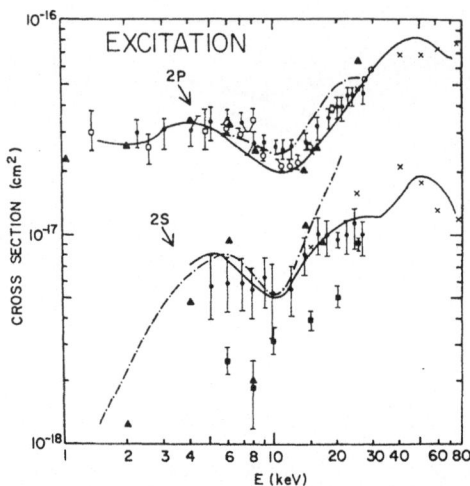

Fig. 2 Cross sections for excitation of atomic hydrogen to 2s and 2p
levels by proton impact. Experimental: ●, Ref. 16; ⬡ ,
Ref. 17; ◯ , Ref. 18; ⬛ , Ref. 19; Theoretical: solid line,
including pseudostates representing ionization, Ref. 22;
dash-dot curves, no pseudostate representation of ionization,
(from Ref. 22).

low energies, without the UA orbitals in the AO+ model the calculated
results would be wrong. This explains the earlier failure of the AO
expansion method.[13] We comment that the total resonant charge transfer
cross section in p+H collisions occurs at large impact parameters and thus
can be easily computed using the AO method. However, the famous 3°
oscillation[20] for the charge transfer probability in p+H collisions is due
to scattering at very small impact parameters[21] and can be explained only
when UA orbitals are included. (In these experiments, it was observed
that the capture probability at a fixed scattering angle of 3° displays
oscillatory structure when measured as a function of collision energies.)

In Fig. 2 we notice some discrepancies between the AO+ calculation
and experiments for E > 15 keV. At higher energies ionization becomes
more important. In a close-coupling type calculation, failure to include
states representing ionization channels results in ionization fluxes being
reflected and absorbed by the diffuse atomic orbitals included. This
explains why the calculated cross sections are too high at higher
energies. Since it is difficult to include continuum wave functions
directly in the calculation, pseudostates were introduced.[22] The precise
form of pseudostates is not important, but we need a few (4 or 5) states
in the energy band from threshold to about 1 or 2 au above it. The
diffuse pseudostates serve to 'absorb' the fluxes which would otherwise
populate 2s and 2p states. The flux to these pseudostates also provides
an estimate of ionization cross sections. We note from Fig. 2 that 2s and

2p excitation cross sections calculated with the inclusion of pseudostates compare quite well with experiments. Pseudostates have been similarly introduced in electron–atom scattering to account for ionization.[23] However, no fictitious resonances are associated with pseudostates in ion–atom collisions, in contrast to electron–atom scattering.

TRIPLY–CENTERED AO EXPANSION

It may be argued that the UA orbitals in the AO+ model should be placed at the center of charge, or more precisely, at the saddle point of the collision system. If the UA orbitals are placed at this third center in addition to the SA orbitals on the two collision centers, the result is a triply–centered AO expansion.[21] Such an expansion is more difficult to calculate since three–center integrals with plane–wave phase factors have to be evaluated. Nevertheless, this calculation has been carried out by Winter and Lin[24] for p+H system. Their results[24] for excitation and charge transfer cross sections to 2s and 2p states with comparable basis functions are basically identical to those of the two–centered AO+ calculations of Fritsch and Lin.[22] This finding indicates that the precise form of ETF's on the UA orbitals is not very important as long as UA orbitals are included. However, a significant charge concentration was observed at the third center, or the saddle point, in Ref. 24. The flux left at the third center was interpreted[25] as saddle point ionization, distinct from the direct Coulomb ionization and electron capture to the continuum which are referred to as continuum electrons near target and projectile centers, respectively. The saddle point ionization was described by Winter and Lin[25] as analogous to the Wannier mechanism for the breakup of three charged particles and ranks as the dominant ionization mechanism for slow collisions. From the PSS model viewpoint, saddle point ionization is a consequence of direct coupling of the $2p\pi$ orbital to the continuum states. The $2p\pi$ state was populated through rotational coupling with the $2p\sigma$ state. This is evidenced by the fact that the saddle point ionization probability and excitation and charge transfer probabilities to 2p states have the same impact parameter dependence.[25] Saddle point ionization has also been addressed in classical Monte–Carlo calculations[26] and in recent experiments[27,28] at higher collision energies.

THE AO–MO MATCHING METHOD

Although the AO+ model (in two– or three–center form) can adequately describe the collision complex and the asymptotic limits simultaneously, its application requires a fairly large basis set. For applications to many—electron collision systems, the calculation becomes very time consuming.[29] In slow collisions, one recognizes that the basic principle of the PSS model is qualitatively correct. Furthermore, in the asymptotic region, the states are represented by travelling atomic orbitals. These are obvious features in slow ion—atom collisions. In order to take advantage of these features, Kimura and Lin,[30,31] and Winter and Lane,[32] proposed the so—called AO—MO matching method. This method divides the internuclear separation into an inner region ($R<R_0$) and an outer region ($R>R_0$). The time—dependent Schrodinger equation is first integrated from large R at $t= -\infty$ to R_0 at $-t_0$ using travelling AO's. At R_0, the solution is expanded in terms of static MO's and then integrated within the MO basis from $-t_0$ to t_0 at $R_0=R(t_0)$ (assuming a straightline trajectory). At this second boundary, the solution is expanded again in travelling AO's and integrated to large R to extract scattering amplitudes.

This method is somewhat analogous to the R—matrix method used in time—independent problems.[33] It avoids many of the difficulties associated with the conventional AO and MO expansions. However, the matching radius R_0 is not well defined just as in the R—matrix method. The value of R_0 is usually chosen in the region where the dominant MO's are well represented by their LCAO expansion. Obviously a larger R_0 would require a larger MO set, and a smaller R_0 would require a larger AO set. For this method to be effective, the results should prove insensitive to the matching radius R_0.

This method has been tested for a number of one—electron systems and the results compared well with the AO+ calculations.[30-32] The dependence of the results on the matching radius has been found to be small if R_0 and basis sets are chosen reasonably. The most extensive application of this method thus far has been proton—helium collisions.[34] A schematic set of diabatic potential curves of the (H—He)$^+$ system is shown in Fig. 3, where the dominant radial and rotational couplings are indicated. For excitation or charge transfer to the n=2 states at low collision energies, the processes have to go through the 'pumping' mechanism via the radial coupling between the two lowest curves, as indicated by the rectangular box. By including the molecular orbitals and the atomic orbitals shown in Fig. 3 and choosing $R_0=2$ a.u., the method has served to calculate probabilities for capture and excitation to n=2 states.[34] As an example, in Fig. 4a we show the total capture probability vs reciprocal velocity at an impact parameter of 0.14 a.u., or $\Theta T=20$ keV deg. The measured[35] oscillatory structures (Fig. 3a) are well confirmed by the calculation using the matching method. Note that curved trajectories can be easily introduced in the inner region by the matching method and that the results at low energies favor a curved trajectory calculation. In Fig. 4b, a

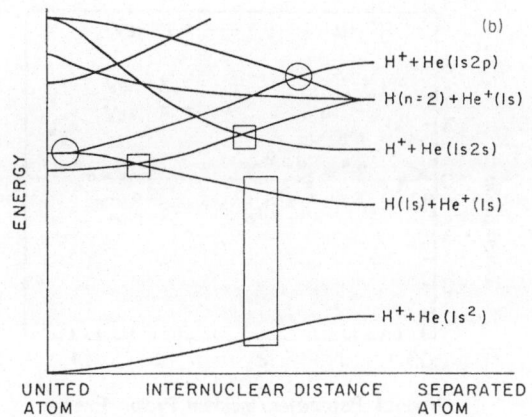

Fig. 3 Diabatic correlation diagram for the (H–He)$^+$ system.
Important radial and rotational couplings are indicated by
boxes and circles, respectively (from Ref. 34).

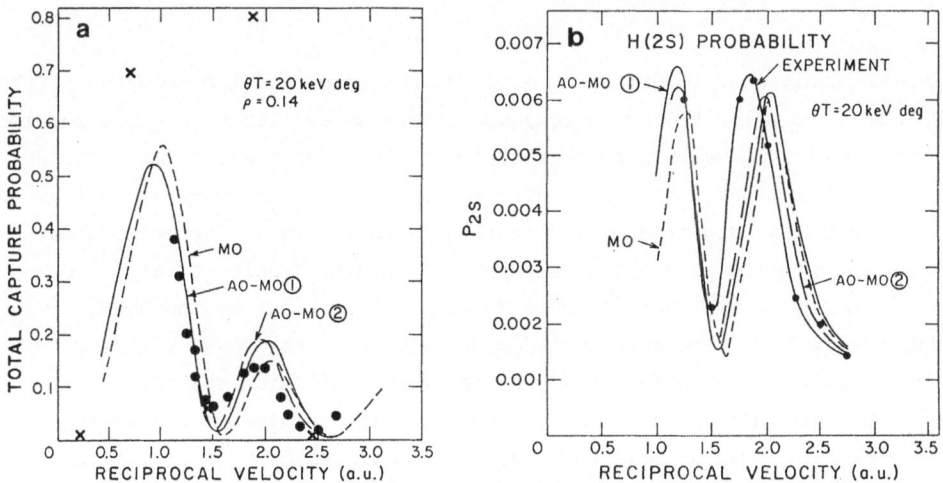

Fig. 4   (a) Total capture probability vs. reciprocal velocity for H$^+$+He
collision at ρ=0.14 a.u. or θT=20 keV.deg. Theory: curve 1,
AO–MO matching with linear trajectory; curve 2, AO–MO matching
with curved trajectory. (b) as in (a) but for capture to 2s
(from Ref. 34).

similar plot for the capture probability to 2s vs reciprocal velocity is
shown. The experimental data[35] are well reproduced by the calculation
also. Note that the 2s capture probability is only a fraction of one
percent. Transition probabilities to other states were discussed and
compared with experiments elsewhere.[34]

COHERENCE IN ION–ATOM COLLISIONS

Coherence parameters of excited states populated in ion–atom collisions have been studied experimentally for many systems. These parameters

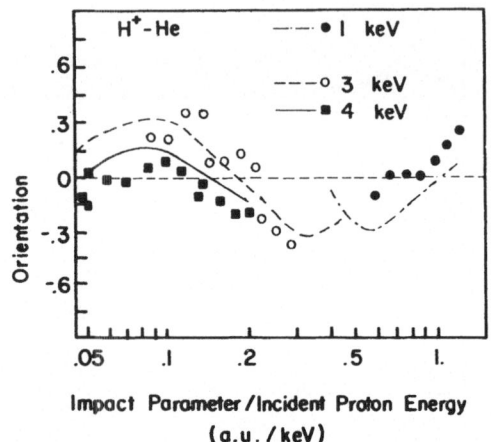

Fig. 5   Orientation parameter for capture to H(2p) in H$^+$–He
collisions: lines, theoretical calculations using AO–MO
matching method, Ref. 24; marks, experiments from Ref. 27.

provide additional information, e.g., on the shape and the rotation of the
charge cloud after the collision, not otherwise available in cross section
measurements, as well as further stringent tests of theoretical calculations. This subject has been addressed in previous lectures as well as in
the lecture by Burgdorfer. I only illustrate in Fig. 5 the calculated
H(2p) orientation parameters measured recently by Hippler et al.[37] for
p+He collisions and their comparison with calculations by the AO–MO
matching method. The agreement demonstrates that the AO–MO matching
method does provide a fair description of the collision system.

Another interesting coherence specific to the hydrogenic excited
states is the angular momentum coherence addressed by Burgdorfer[38] and
others.[39] Off–diagonal density matrix elements between degenerate excited
hydrogenic states can be measured by the polarization of the emitted
radiation in an external electric and/or magnetic field.[40] The AO+ model
has been applied to p+He collisions to extract the H(n=3) density
matrix,[41,42] and to compare with measurements by North Carolina State
and Harvard groups.[43] The comparison in Ref. 42 demonstrated that the AO+
model does give accurate off–diagonal density matrix elements.

From the density matrix several physical parameters can be extracted.
It was found that in general, for fast collisions where the projectile
speed is faster than the electron's orbital speed, the electron cloud
around the projectile center always lags behind the projectile. The
experiment on p–He collisions[40] is consistent with this picture. In a
recent study, the density matrices for the target and projectile excited

p+H (50 KeV)

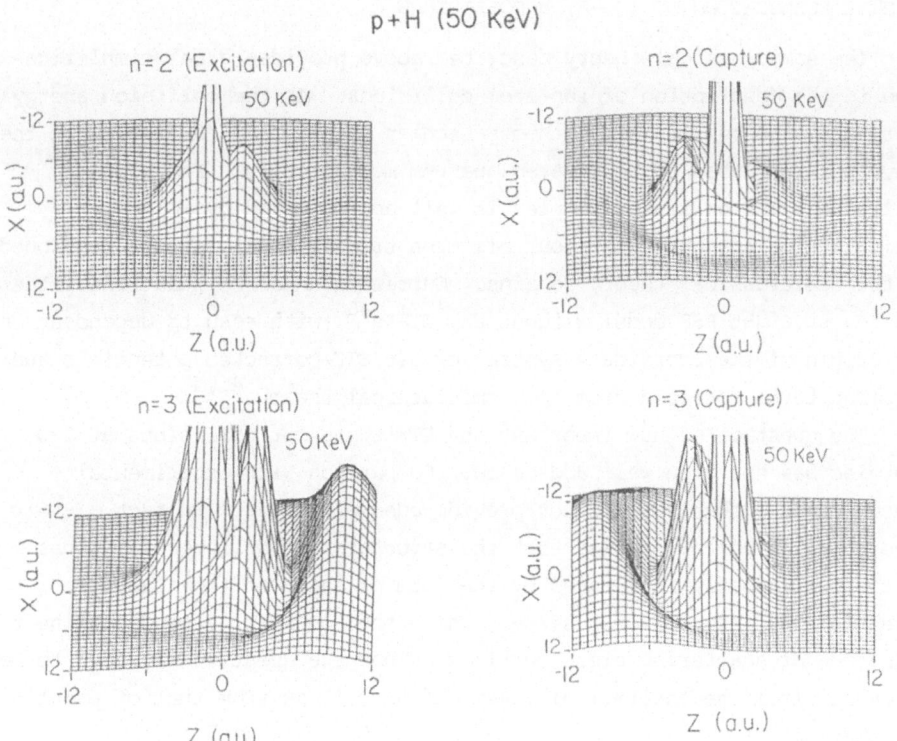

Fig. 6   Electron density plots for the H(n=2) and H(n=3) manifolds for
excitation and charge transfer in p-H collisions at 50 keV impact
energy.   Shown are the electron densities integrated over impact
parameters.   The beam direction is along +z, and xz is the
collisions plane.   Each plot is arbitrarily normalized.   Note that
the electron cloud is behind the projectile but ahead of the
target (from Ref. 44).

states in p-H collision have been calculated and analysed using the AO+
model.[44]   I only demonstrate one result here.   Fig. 6 shows the electron
density plots for the n=2 and n=3 manifolds for both the excitation and
charge transfer channels.   These plots are obtained after integration over
impact parameters and are normalized arbitrarily.   We note that at 50 keV
collision energy the electron cloud for the charge transfer channels lags
behind the projectile, while for excitation it is ahead of the target
center.   This is consistent with the intuitive picture that the electron
cloud lying between the two centers splits as the two centers separate
from each other.   We conclude this section by stating that detailed
analysis of the accurately calculated density matrix may provide deeper
insight into the excitation process.

The semiclassical theory described above provides great simplification in the description of ion-atom collisions. As the collision energy decreases, the semiclassical theory becomes invalid and the motion of the heavy particles has to be treated quantum mechanically. For slow collisions, the collision complex is well approximated by molecular orbitals, but the question about electron translational factors mentioned in the semiclassical theory remains. Current quantal calculations either use the straight PSS model without any ETF's[45] (with results dependent on the origin of the coordinate system) or use ETF-corrected potentials and coupling terms obtained from the semiclassical theory.[46,47]

The question of how important the ETF's are for collisions at low energies has not been well addressed. Comparison with experimental measurements probably would not provide adequate clues. In fact, theoretical studies[48] showed that the structures of differential cross sections are mostly influenced by the form or shape of the adiabatic potentials of the collision system, while the coupling terms determine the magnitude of scattering cross sections. Thus the question may have to be addressed from the theoretical viewpoint as well as from that of practical calculation.

REACTIVE SCATTERING AND HYPERSPHERICAL COORDINATES

It is well understood that the need for ETF's in the MO expansion arose from the incorrect representation of molecular orbitals for the asymptotic scattering states. Despite past intensive theoretical work aimed at amending the situation within the molecular orbital framework,[48] the situation remains unsettled. It becomes even more problematic when one treats muonic capture as in

$$d \ + \ (\mu t) \ \longrightarrow \ t \ + \ (\mu d). \tag{6}$$

This process was considered in connection with the investigation of muon-catalyzed fusion.[50,51] To calculate the cross section for the above process within the PSS formulation, the group at Dubna[52] had to couple up to about 300 equations in order to achieve converged results. Although a way to bypass the need for ETF's has been recognized recently[53,54] by formulating the problem in hyperspherical coordinates, no actual collision calculation has been performed as yet by this method.

Ion-atom collisions are a special case of general chemical reactive scattering involving three particles,

$$A + (BC) \longrightarrow (AB) + C$$

$$\longrightarrow B + (AC). \tag{7}$$

A proper formulation of excitation and charge transfer processes should incorporate the possibility of treating reactive scattering. To describe each dissociation channel, Jacobi coordinates of the respective dissociation limit are appropriate. For example, in the center-of-mass frame, the two Jacobi coordinate vectors which describe A + (BC) are $\vec{r}_{BC}$, the vector from B to C, and $\vec{r}_{A,BC}$, the vector from the center-of-mass of (BC) to A. Three sets of Jacobi coordinates are needed to describe the three dissociation limits indicated in (7) in actual calculations. Such a calculation has recently been formulated in the mass-weighted hyperspherical coordinates and some initial calculations have already been carried out for the bound states of three charged particles of arbitrary masses.[55] Within this approach, the methods for calculating scattering cross sections become identical to those for electron-atom scattering in the hyperspherical formulation, as reviewed by Starace.

CONCLUSION

I have attempted to summarize the various theoretical aspects of ion-atom collisions from the point of view of ab initio calculations. The wide variety of heavy ions and their broad velocity ranges in ion-atom collision studies pose a continuing challenge to theorists. Most ion-atom collisions can be considered as hard collisions in that electronic charge clouds undergo drastic changes in shape, size, and rotations. The concept of the collision complex is critical to any theoretical formulation, but its dependence on the collision velocity and impact parameter makes theoretical modeling difficult. For slow collisions, the collision complex is well approximated by the quasi-molecule of the system. We pointed out that an AO-MO matching method may be adopted for slow collisions when the motion of heavy particles can be described classically. For a quantal treatment of slow collisions, it appears that the remedy may come from a formulation in hyperspherical coordinates.

It is appropriate to end this article with a quotation from Fano.[56] In his speech presented at the American Physical Society, in Washington D.C. in April 1976, when he was awarded the Davisson-Germer Prize, he emphasized that 'the study of two-electron systems is a first step toward understanding the transitions that occur in chemical reactions'. The hyperspherical coordinates brought into atomic physics in the 60's by Fano not only did provide an understanding of doubly excited states in atoms after two decades, it appears that it may play a big role in the theoretical understanding of chemical reactions in years to come, thus achieving a great synthesis of atomic and molecular physics.

ACKNOWLEDGMENTS

The work reported here was performed in collaboration with Drs. W. Fritsch, A. Jain, M. Kimura and T.G. Winter. It was supported in part by the U.S. Department of Energy, Office of Energy Research, Division of Chemical Sciences.

REFERENCES

1. "Fundamental Processes in Energetic Atomic Collisions," eds. H.O. Lutz, J.S. Briggs and H. Kleinpoppen, NATO ASI Series B, Vol. 103, (1983) (New York, Plenum).
2. R.K. Janev and H. Winter Phys. Rept. 117, 265 (1985).
3. H.S.W. Massey and R.A. Smith, Proc. Roy. Soc. London Ser. A142, 142 (1933).
4. U. Fano and W. Lichten, Phys. Rev. Lett. 14, 627 (1965).
5. D.R. Bates and R. McCarroll, Proc. R. Soc. London Ser. A245, 175 (1958).
6. W.R. Thorson and J.B. Delos, Phys. Rev. A18, 117, 135 (1978).
7. S.B. Schneiderman and A. Russek, Phys. Rev. 181, 311 (1968).
8. J. Vaaben and K. Taulbjerg, J. Phys. B14, 1485 (1981).
9. L.F. Errea, J.M. Gomez-Llorente, L. Mendez and A. Riera, Phys. Rev. 32, 2158 (1985).
10. M.E. Riley and T.A. Green, Phys. Rev. A4, 619 (1971), see also the article by J. Briggs in Ref. 1.
11. M. Kimura and N.F. Lane, Phys. Rev. Lett. 56, 2160 (1986).
12. D.R. Bates, Proc. Roy. Soc. London, Ser. A274, 294 (1958).
13. D.F. Gallagher and L. Wilet, Phys. Rev. 169, 139 (1968).
14. W. Fritsch and C.D. Lin, Phys. Rev. A26, 762 (1982).
15. A.J. Taylor and P.G. Burke, Proc. Phys. Soc. 92, 336 (1967).
16. T.J. Morgan, J. Geddes and H.B. Gilbody, J. Phys. B6, 2118 (1973).
17. R.F. Stebbings, R.A. Young, C.L. Oxley and H. Ehrhardt, Phys. Rev. 138, A1312 (1965).
18. T. Kondow, R.J. Grinius, Y.P. Chong and W.L. Fite, Phys. Rev. A10, 1167 (1974).
19. Y.P. Chong and W. Fite, Phys. Rev. A16, 933 (1977).
20. H.F. Helbig and E. Everhart, Phys. Rev. 140, A715 (1965).
21. C. D. Lin, T. G. Winter and W. Fritsch, Phys. Rev. A25, 2395 (1982); M.J. Antal, D.G.M. Anderson and M.B. McElroy, J. Phys. B7, L118 (1974).
22. W. Fritsch and C.D. Lin, Phys. Rev. A27, 3361 (1983).
23. J. Callaway and D.H. Oza, Phys. Rev. A34, 965 (1986).
24. T.G. Winter and C.D. Lin, Phys. Rev. A29, 3071 (1984).
25. T.G. Winter and C.D. Lin, Phys. Rev. A29, 567 (1984).
26. R.E. Olson, Phys. Rev. A27, 1871 (1983).
27. W. Meckbach, P.J. Focke, A.R. Goni, S. Suarez, J. Macek, and M.G. Menendez, Phys. Rev. Lett. 57, 1587 (1986).
28. R.E. Olson, T.J. Gay, H.G. Berry, E.B. Hale and V.D. Irby, Phys. Rev. Lett. 59, 36 (1987).
29. W. Fritsch and C.D. Lin, J. Phys. B19, 2683 (1986).
30. M. Kimura and C.D. Lin, Phys. Rev. A31, 590 (1985).
31. M. Kimura and C. D. Lin, Phys. Rev. A32, 1357 (1986).

32. T. Winter and N.F. Lane, Phys. Rev. A31, 2698 (1985).
33. E.P. Wigner, Phys. Rev. 70, 15, 606 (1946); E.P. Wigner and
    L. Eisenbud, Phys. Rev. 72, 29 (1947).
34. M. Kimura and C.D. Lin, Phys. Rev. A34, 176 (1986).
35. H.F. Helbig and E. Everhart, Phys. Rev. 140, A674 (1964).
36. D.H. Crandall and D.H. Jaecks, Phys. Rev. A4, 2281 (1971).
37. R. Hippler, M. Faust, R. Wolf, H. Kleinpoppen and H.O. Lutz, submitted
    to Phys. Rev. (1987).
38. J. Burgdorfer, Z. Phys. A309, 285 (1983).
39. O. Scholler, J.S. Briggs and R.M. Dreizler, J. Phys. B19, 2505 (1986).
40. C.C. Havener, N. Rouze, W.B. Westerveld and J.S. Risley, Phys. Rev.
    Lett. 53, 1049 (1984); Phys. Rev. A33, 276 (1986).
41. A. Jain, C.D. Lin and W. Fritsch, Phys. Rev. A35, 3180 (1987).
42. A. Jain, C.D. Lin and W. Fritsch, Phys. Rev. A36, 2041 (1987).
43. M.C. Brower and F.M. Pipkin, Bull. Am. Phys. Soc. 31, 994 (1986).
44. A. Jain, C.D. Lin and W. Fritsch, submitted to J. Phys. B (1987).
45. T.G. Heil, S.E. Butler and A. Dalgarno, Phys. Rev. A23, 1100 (1981).
46. M. Kimura and R. Olson, J. Phys. B17, L713 (1984).
47. J. Tan, C.D. Lin and M. Kimura, J. Phys. B20, L91 (1987).
48. J. Tan and C.D. Lin, submitted to Phys. Rev. A (1987).
49. J.B. Delos, Rev. Mod. Phys. 53, 287 (1981).
50. L. Bracci and G. Fiorentini, Phys. Rept. 86, 169 (1982).
51. S.E. Jones, Nature 321, 127 (1986).
52. V.S. Melezhik, L.I. Ponomarev and M.P. Faifman, Sov. Phys. JETP 58, 254
    (1984).
53. J.H. Macek, M. Cavagnero, K. Jerjian and U. Fano, Phys. Rev. A35, 3940
    (1987).
54. A. Soloviev and I. Vinitzkn, J. Phys. B18, L557 (1985).
55. C.D. Lin and X.H. Liu, Phys. Rev. A submitted (1987).
56. U. Fano, Phys. Today, 29, 32 (1976).

COHERENCE AND CORRELATIONS IN FAST ION-ATOM COLLISIONS

Joachim Burgdörfer

Department of Physics and Astronomy
University of Tennessee, Knoxville, TN 37996-1200 and
Oak Ridge National Laboratory, Oak Ridge TN 37831-6377, USA

1.   Introduction

Inelastic ion-atom collisions are characterized by a rearrangement of
the electronic charge cloud of the collision partners. Electrons may be
lifted from the ground state to excited states ("direct excitation" of the
target or projectile) or may be lost during the collision ("ionization") or
may be transferred into orbits of the collision partner ("electron
capture"). In recent years the study of atomic collisions has tended toward
a more detailed understanding of the collision dynamics focussing on the
"shape" and the "circulation" properties of the electronic charge
distribution. The goal is to extract the maximum available information from
the collision process rather than just a few cross sections. Investigations
along these lines are closely connected with the notion of "coherence and
correlations" in atomic collisions. The work on this subject until the early
70's has been reviewed by Fano and Macek in their seminal paper[1] on "Impact
Excitation and Polarization of Emitted Light" which, in turn, has stimulated
a rapidly growing number of related studies in the fields of electron-atom
and ion-atom collisions and of beam-foil spectroscopy. Several reviews on
more recent developments have become available including those by Andrä,[2]
Blum,[3] Hermann and Hertel,[4] Hippler,[5] and Janev and Winter.[6]

In the following we focus on the description, classification and
interpretation of coherent excitation of atomic or ionic systems with
Coulombic two-body final-state interaction. The peculiarities of the
Coulomb interaction between the excited electron and the nucleus (or ionic
core) affords the opportunity to study a wealth of coherence phenomena
associated with the (near) degeneracy of different ℓ states. As is
well-known, the ℓ degeneracy in the non-relativistic Coulomb problem is a
consequence of a dynamical symmetry $O(4)$.[7,8] We will employ a

group-theoretical approach to classify and interpret coherent excitation. This method is a straightforward generalization of the Fano-Macek approach to describe coherent superposition of magnetic sublevels (Zeeman coherences) in terms of "state multipoles"[9] exploiting the geometric rotational symmetry SO(3) of free atoms.

The key point is that the maximum information on the state of excitation represented by a density operator can be mapped one-to-one onto expectation values of a set of conveniently chosen operators. The generators of the symmetry group (and functions of them) may provide a sensible and a physically meaningful representation of the density operator. "Identifying a state by means of such physical parameters brings out the operational basis of the theory and helps in forming a mental picture" (Fano, 1957).[9] Such an "operational basis" will provide us with an intuitive "classical" picture of the coherently excited charge cloud in terms of orbital parameters. The O(4) representation is particularly well suited for the study of n dependence of coherent states along a Rydberg series. The expectation values of certain O(4) operators turn out to be smooth functions of n and to possess a smooth continuation in the limit n→∞ in terms of the anisotropy coefficients of the angular distribution of low-energy continuum electrons.

We will illustrate with the help of a few selected examples what can be learned about the collision process from investigations of coherent excitation. We will restrict ourselves to charge transfer processes at high collision velocities ($v_p \gg 1$). Coherent excitation at low and intermediate energies is discussed in the lectures of N. Andersen[10] and C. D. Lin.[11] Atomic units are used throughout unless otherwise stated.

2. Basic properties of coherent excitation

An atomic collision may act as a sudden perturbation leaving the collision partners (projectile, target) in a final electronic state which can be represented as a linear combination of atomic eigenstates $|\phi_i\rangle$ as

$$|\psi(t=0)\rangle = \sum_i f_i |\phi_i\rangle. \tag{2.1}$$

The expansion coefficients f_i are, up to trivial factors, the scattering amplitudes for the electronic rearrangement process. The state (2.1) is a pure, fully coherent state. Coherence means the existence of correlations between any two specified pure states. Partial loss of coherence in the experiment stems from averaging over unresolved degrees of freedom in an ensemble of scattering events, i.e. from incomplete experimental determination of a full set of commuting observables in the preparation of

initial states and/or in the analysis of final states. Examples of
averaging processes frequently encountered in ion-atom collisions include
the average over scattering angles (or impact parameters), electronic and
nuclear spin quantum numbers, and the simultaneously excited electronic
states of the collision partner in complex collision systems. Such an
ensemble of scattering events is characterized by an ensemble of states
$|\psi^\alpha\rangle$ and of scattering amplitudes $f_i{}^\alpha$, where the index α denotes all degrees
of freedom to be averaged over. The description of ensembles of scattering
events is provided by a density matrix. Let us suppose we are interested in
the ensemble expectation values for the atomic observable O we find

$$\langle 0 \rangle = \sum_{i,j} \langle \phi_i | 0 | \phi_j \rangle \left[\frac{\sum_\alpha f_i{}^\alpha f_j{}^{\alpha*}}{\sum_{\alpha,k} |f_k{}^\alpha|^2} \right] \tag{2.2}$$

The quantity in brackets in (2.2) is called the density matrix element $\rho_{i,j}$.
With help of the density matrix ρ, Eq. (2.2) can be expressed as

$$\langle 0 \rangle = \text{Tr}(0\rho). \tag{2.3}$$

The notion of correlation comes into play as a means to prepare a
partially or fully coherent state. In this context, correlation means the
simultaneous determination of dynamical observables of both collision
fragments or of different observables of the same fragment. Examples
include the light emission by the target in coincidence with the scattering
angle of the projectile or the coincident detection of two subsequent
photons in a cascade. In any case, the goal is to observe a large number of
degrees of freedom, thereby reducing the number of averaging processes and
increasing the degree of coherence. Clearly, the ultimate goal is the
"ideal" experiment with all degrees of freedom observed resulting in a fully
coherent state ("state-to-state" processes). The production of fully
coherent states in electron-helium scattering is discussed by Kleinpoppen.[12]
For high-energy ion-atom collisions the resolution of the extremely small
scattering angles is a formidable task in view of the coincidence
requirements and the small cross sections. The experiments at high energies
discussed in the following therefore include an ensemble average over all
scattering angles.

The time evolution of the density matrix after a collision is given by

$$\rho_{i,j}(t) = e^{-i(\varepsilon_i - \varepsilon_j)t} \rho_{i,j}(0) \tag{2.4}$$

where $\varepsilon_{i,j}$ are the energies of stationary atomic states. Eq. (2.4) points
to an important practical limitation for the observation of coherences. In
principle, almost all pairs of excited states (i,j) may be coherently

excited, $\rho_{i,j}(0) \neq 0$. The range of (i,j) is limited only by the energy-(collision) time uncertainty principle and the geometric symmetry of the collision process. The latter is discussed in detail in the lecture by N. Andersen.[10] In practice, however, Eq. (2.4) implies that the observation of off-diagonal elements $\rho_{i,j}(t)$ requires a time resolution of the order of

$$\Delta t \leq |\varepsilon_i - \varepsilon_j|^{-1} . \tag{2.5}$$

Eq. (2.5) leads to the important conclusion that only coherences between almost degenerate pairs of states with small energy splittings are experimentally accessible. Averaging over the rapid oscillations ("quantum beats") leads to extinction of coherence effects. We note that Van der Straten, Morgenstern and Niehaus[13] have recently resolved quantum beats resulting from coherent excitation of the n=2 manifold in doubly-excited helium achieving thereby a time resolution of $\leq 10^{-14}$ sec.

3. Classification of hydrogenic bound-state coherences

Hydrogenic systems play the role of a "Rosetta stone" in the investigation of coherent excitation. This is due to both the simplicity of the atomic structure which promises a satisfactory theoretical treatment of the collision process as well as the large number of experimentally and theoretically accessible density matrix elements even for an angle-integrated collision geometry. The latter stems from the exact degeneracy of different ℓ states of the non-relativistic Coulomb problem. The quantum beats (Eq. (2.3)) due to the residual fine structure and Lamb shift splittings are easily resolvable in this case.

The degeneracy is consequence of a dynamical symmetry. The Coulomb problem possesses in addition to the angular momentum vector $\vec{L}$ a second constant of motion, the Runge-Lenz vector

$$\vec{A} = \frac{1}{2} (\vec{p} \times \vec{L} - \vec{L} \times \vec{p}) - Z\hat{r} \tag{3.1}$$

Here and in the following we set the reduced mass $m = 1$ (a.u.) The charge of the nucleus is denoted by Z. Thus we have

$$[H_a, \vec{L}] = [H_a, \vec{A}] = 0 \tag{3.2}$$

where

$$H_a = \frac{p^2}{2} - \frac{Z}{r} \tag{3.3}$$

denotes the Hamiltonian of the non-relativistic Coulomb problem. For simplicity we have omitted all relativistic corrections in (3.3). They play, however, an important role in the derivation of numerical coherence parameters from experimental quantum beat data. The vector $\vec{A}$ points to the

perihelion of the classical Kepler orbit (Fig.1). Its magnitude is
proportional to the eccentricity.

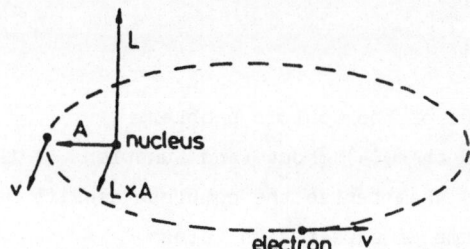

Fig. 1. Angular momentum $\vec{L}$ and Runge-Lenz vector $\vec{A}$ of a hydrogenic
         orbit. $\vec{L} \times \vec{A}$ points along the direction of the electron
         velocity at the perihelion

We consider in this section bound states ($\varepsilon < 0$). Acting on a subspace
of fixed n the renormalized Runge-Lenz vector

$$\vec{a} = \frac{n}{Z} \, \vec{A} \tag{3.4}$$

and the angular momentum vector $\vec{L}$ satisfy the commutation rules[8]

$$[L_i, L_j] = i \, \varepsilon_{ijk} L_k$$

$$[L_i, a_j] = i \, \varepsilon_{ijk} a_k \tag{3.5 a-c}$$

$$[a_i, a_j] = i \, \varepsilon_{ijk} L_k$$

where ε_{ijk} denotes the completely antisymmetric tensor in three dimensions.
Eqs. (3.5) correspond to the Lie algebra of the four-dimensional rotation
group O(4), the dynamical symmetry group of the hydrogenic bound-state
manifold for fixed n. The geometric symmetry group of three-dimensional
rotations, O(3), corresponding to the subalgebra (3.5a), pertains to the
degeneracy of different m states while the larger dynamical group O(4)
pertains to the "accidental" degeneracy of different ℓ states.

The symmetry group O(4) is homomorphic to the group SU(2) x SU(2).
This connection can be made explicit by introducing pseudospin operators

$$\vec{j}^{(1)} = \frac{1}{2} \, (\vec{L} - \vec{a}) \tag{3.6}$$

$$\vec{j}^{(2)} = \frac{1}{2} \, (\vec{L} + \vec{a})$$

which obey the commutation rules

$$[j_i^{(\gamma)}, j_j^{(\delta)}] = i \, \varepsilon_{ijk} \, j_k^{(\gamma)} \, \delta_{\gamma, \delta} \tag{3.7}$$

of two independent commuting angular momenta. In view of the orthogonality
requirement (Fig.1)

$$\vec{L} \cdot \vec{a} = 0 \tag{3.8}$$

only self-conjugate representations of SU(2) x SU(2) with identical quantum numbers

$$j^{(1)} = j^{(2)} = \frac{n-1}{2} \tag{3.9}$$

correspond to solutions of the Coulomb problem.

The generators of the O(4) group (and functions of them) provide a set of operators that can parameterize the complete density matrix of the nth principal shell in terms of expectation values

$$\langle O_i \rangle = Tr(O_i \rho) \quad (i=1, \ldots k) \tag{3.10}$$

where k is the number of independent parameters of the density matrix in the subspace of fixed n.[14] This set is denoted by $U_q^k(\pi, k_1, k_2)$. The indices (k,q) specify the rank and the component of the spherical tensor.[15,16] It has well-defined parity, $(-1)^\pi$ ($\pi = 0, 1$) and time reversal symmetry, $(-1)^{\pi+k_1+k_2}$, where k_1 and k_2 are the ranks of the multipoles of the pseudospins $\vec{j}_{1,2}$ involved. Explicit expressions for the operator subset pertaining to the n=2 shell will be given below.

The expectation values of the operator set $\{U_q^k\}$ provide a complete parameterization of the n shell density matrix. With help of some angular momentum algebra the expectation values can be expressed in terms of the standard representation of the density matrix as[14,17]

$$\langle U_0^k(\pi, k_1, k_2) \rangle = \frac{i^{k_1+k_2-k}}{[2(1+\delta_{k_1,k_2})]^{1/2}} \sum_{\ell,\ell'=0}^{n-1} [(2k+1)(2k_2+1)(2\ell+1)(2\ell'+1)]^{1/2} \cdot$$

$$\cdot [(-1)^\pi + (-1)^{\ell+\ell'}] \begin{Bmatrix} j & j & k_1 \\ j & j & k_2 \\ \ell & \ell' & k \end{Bmatrix} \rho_0^k(\ell,\ell')^*. \tag{3.11}$$

In (3.11)

$$\rho_q^k(\ell,\ell') = \sum_{m,m'} (-1)^{\ell-m} \sqrt{2k+1} \begin{pmatrix} \ell & k & \ell' \\ -m & q & m' \end{pmatrix} \langle n\ell m | \rho | n\ell'm' \rangle \tag{3.12}$$

are the well-known irreducible tensor components of the density matrix for fixed angular momenta (ℓ,ℓ'), namely, the state multipoles.[9] From (3.11) and (3.12) it is obvious that the O(4) parameterization is a straightforward generalization of the Fano-Macek description of magnetic substate coherences in terms of state multipoles to cases where the coherent excitation of different ℓ states is observed.

The physical significance of the O(4) operator set is illustrated in Table 1 where we list all q=0 operators for an axially symmetric density matrix for the n=2 manifold including their expansion in terms of the standard representation and parities with respect to coordinate and time inversion. The physical meaning of the two k=0 operators formed by L^2 and a^2 is obvious: The sum is the Casimir operator of the symmetry group O(4) and gives the n=2 excitation probability. The difference parametrizes the difference of the s- and p excitation probabilities. A large expectation value $\langle L^2 \rangle$ means a dominant p excitation whereas a large $\langle a^2 \rangle$ corresponds to a favoured s excitation. This is in accord with the Bohr-Sommerfeld theory of the hydrogen because $\langle a^2 \rangle$ gives classically the mean squared eccentricity of the orbit. The k=2 tensor describes the alignment parameter. In the special case of tensor components with $k=k_{max}$

Table 1. O(4) operator set for axially symmetric n=2 density matrix in hydrogen (Tr_x: trace over subspace x)

k	operator	expectation value	class
0	$U_0^0(0,0,0) = \frac{1}{2}(L^2+a^2)$	$\frac{1}{2} Tr_{n=2}(\rho)$	$P_e T_e$
0	$U_0^0(0,1,1) = \frac{1}{2\sqrt{3}}(L^2-a^2)$	$\frac{1}{2\sqrt{3}}(Tr_{2p}(\rho)-3\langle 200|\rho|200\rangle)$	
2	$U_0^2(0,1,1) = \frac{1}{2}(L_0^2-a_0^2)$	$\frac{2}{\sqrt{6}}(\langle 211|\rho|211\rangle-\langle 210|\rho|210\rangle)$	
1	$U_0^1(1,1,0) = -\frac{a_z}{\sqrt{2}}$	$-\sqrt{2}\ Re(\langle 200|\rho|210\rangle)$	$P_o T_e$
1	$U_0^1(1,1,1) = \frac{1}{2\sqrt{2}}(\vec{a}\times\vec{L}-\vec{L}\times\vec{a})_z$	$\sqrt{2}\ Im(\langle 200|\rho|210\rangle)$	$P_o T_o$

both operators $a^{k_{max}}$ and $L^{k_{max}}$ act on a single ℓ subspace with $\ell=\ell_{max}$. Consequently $L^{k_{max}}$ and $a^{k_{max}}$ are linearly dependent due to the Wigner-Eckart theorem. For n=2 this yields

$$(L_0^2) = -(a_0^2). \tag{3.13}$$

The only $P_e T_o$ parameter (even under parity, odd under time reversal) present in n=2 corresponding to the orientation parameter, $(L_q^1, q=\pm1)$, vanishes because of axial symmetry. The real part of the s-p coherence belonging to the class $P_o T_e$ can be represented by the component of the Runge-Lenz vector, a_z. This confirms the usual dipole interpretation[18] because the dipole moment, $\vec{d}$, is proportional to the eccentricity of the orbit (Fig.1),

$$d_z = \frac{3n}{2Z} a_z. \qquad\qquad (3.14)$$

Finally, the imaginary part of the s-p coherence ($P_O T_O$) are proportional to the z component of the vector $\vec{L} \times \vec{a}$. This vector points along the orbital velocity at the perihelion (see Fig.1). Several alternative representations have been proposed. Gabrielse and Band[19] parametrized the density matrix by expectation values of electric and magnetic multipoles and their time derivatives. Gabrielse[20] suggested to relate the hydrogenic density matrix to expectation values of a set of multipoles built by combinations of the position ($\vec{r}$) and angular momentum operators. More recently, Risley et al.[21] suggested to represent the coherently excited manifold in terms of local distributions of density and current density.

The choice of the O(4) operator set has several decisive advantages. All relevant matrix elements can be simply expressed in terms of angular momentum coupling coefficients. Their expectation values have a simple physical interpretation in terms of orbital parameters of the Kepler orbit and are constants of motion (or, slowly varying functions on an atomic time scale when relativistic corrections are taken into account). We note that the classical interpretation of these coherence parameters should be used with caution since for low n a classical orbital picture is, strictly speaking, not justified.

The determination of the density matrix, both experimentally and theoretically, has revealed new insights into collision processes, in particular into the perturbation series (Born series) for charge transfer. One key point is that the density matrix in first Born approximation obeys the multipole selection rule[22]

$$\langle U_0^k(\pi, k_1, k_2) \rangle = 0 \qquad \text{if } k_1 + k_2 \text{ odd.} \qquad\qquad (3.15)$$

Accordingly, the time reversal symmetry of non-vanishing multipoles is $(-1)^\pi$ and coincides with the parity. Eq. (3.15) is therefore equivalent to the statement that only $P_O T_O$ and $P_e T_e$ parameters are non-zero in Born approximation. The first example of this kind was discussed by Fano and Macek.[1] They pointed out that the orientation parameter $\sim \langle L_y \rangle$, which is $P_e T_O$, vanishes identically. The magnitude of $P_e T_O$ and $P_O T_e$ components is therefore a direct measure of the influence of high-order Born terms.

In the angle-integrated density matrix the only non-zero off-diagonal elements in the standard basis represent the ℓ coherences (e.g., the s-p-coherence). Following previous measurements of foil-excited hydrogen[18,23] first evidence has been found for ℓ coherences in the n=2 manifold produced in the charge transfer collisions

$$H^+ + He \rightarrow H(n) + He^+ \tag{3.16}$$

by a time-differential Lyα quantum beat experiment in external (anti) parallel electric fields by Sellin et al.[24] However, strong background contributions and field inhomogeneities had prevented an accurate determination of the density matrix.[24,25] An improved version of this experiment has been recently performed by DeSerio et al.[26] From the amplitude and the phase of the quantum beats the magnitude and the phase of the s-p coherence could be determined. We note that one remaining source of uncertainty in the Ly$_\alpha$ data lies in the "accidental" degeneracy of the H(2 → 1) transition with the Balmer-β transition in He$^+$(4 → 2).

The physical meaning of the multipole $\langle a_z \rangle > 0$ is that the electron lags behind the proton. Since $\langle a_z \rangle$ is of the class $P_o T_e$ this parameter vanishes in Born (or, Oppenheimer-Brinkman-Kramers (OBK))[27] approximation. Non-zero values are therefore a direct measure for the influence of higher-order Born terms. The continuum distorted wave (CDW) approximation[28] which includes only selected contributions from higher-order terms gives indeed positive values for $\langle a_z \rangle$ but underestimates the experimental data systematically. The

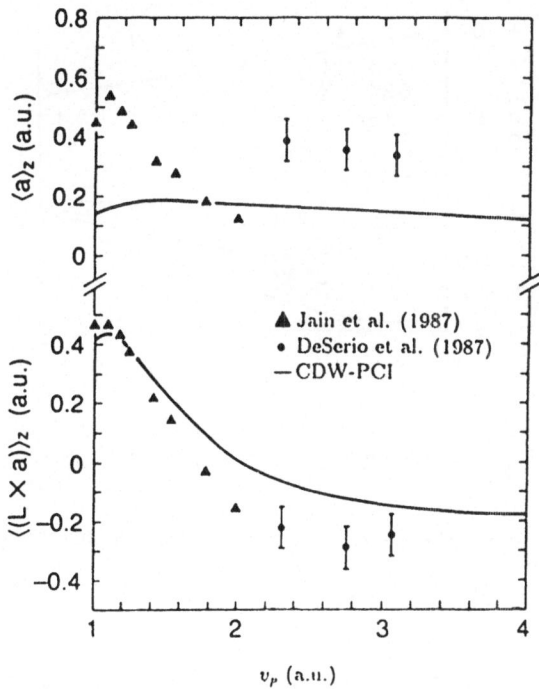

Fig. 2. Theoretical and experimental n=2 coherence parameters $\langle a_z \rangle$, and $\langle \vec{L} \times \vec{a} \rangle_z$. ▲, coupled-states calculation[29]; —, CDW-PCI; ●, DeSerio et al.[26]

close-coupling calculation of Jain et al.[29] gives large values of $\langle a_z \rangle$ at lower collision velocities which, however, fall rapidly below the CDW values near 80 keV and are not expected to be reliable at higher energies where contributions from intermediate continuum states become important. In the present CDW calculation referred to as CDW-PCI we have incorporated the long-range intrashell Stark mixing in the degenerate manifold in the exit channel using the post-collision interaction (PCI) model.[25] For the coherence parameter $\langle a_z \rangle$ the results of the standard CDW approximation and the CDW-PCI approximation are virtually indistinguishable. For the expectation value $\langle (\vec{L} \times \vec{a}) \rangle_z$ a completely different picture emerges. The CDW-PCI calculation reproduces the qualitative trend quite well while the CDW method without correction for Stark mixing (not shown in Fig. 2) fails completely. At lower velocities, the CDW-PCI model agrees with the close-coupling data of Jain et al.[29] surprisingly (and, in part, fortuitously) well. It appears that incorporation of Stark mixing improves the perturbation theory but the particular form chosen in the analytical PCI treatment may overestimate the effect.

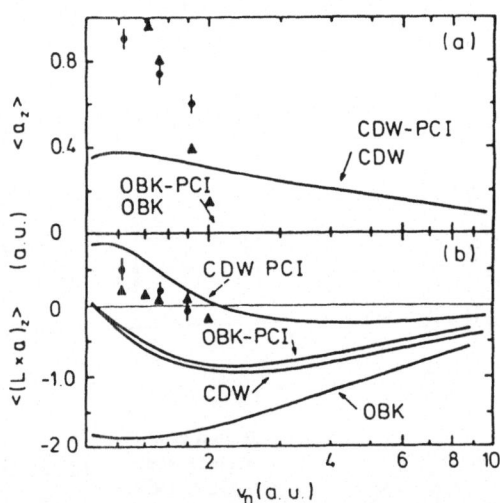

Fig. 3. (a) Expectation value of the z component of the Runge-Lenz
vector $\langle a_z \rangle$ following the charge transfer process (3.16) into
n=3 as a function of projectile velocity. Dots with error
bars, experimental data of Havener et al.[21]; OBK, OBK-PCI,
CDW and CDW-PCI.[22]; ▲, close coupling calculation, Jain et
al.[29] (b) Expectation value of the z component $\langle (\vec{L} \times \vec{a})_z \rangle$
as a function of the projectile velocity in different
approximations.

The n=3 density matrix following the reaction (3.16) has been investigated by Risley and coworkers[21]. Results for $\langle a_z \rangle$ for proton energies between 40 keV and 80 keV are in good agreement with the

close-coupling calculation by Jain et al.[29] (Fig. 3) while the CDW
approximation underestimates the charge cloud asymmetry similar to the n=2
result. The change of sign in $\langle(\vec{L}x\vec{a})_z\rangle$ near $v_p \cong 1.7$ a.u. indicated by the
CDW-PCI calculation for the n=2 level (Fig. 2) is indeed observed in the n=3
data. The change of sign in $\langle(\vec{L}x\vec{a})_z\rangle$ is the most direct evidence to date
for post- collisional Stark mixing which is incorporated in both the close
coupling calculation and the CDW-PCI model. We note that both intrashell and
intershell PCI effects have been proven to be important for direct
excitation as well.[30]

Comparison of Figs. 2 and 3 reveals a close overall similarity of the
coherence parameters for the n=2 and n=3 manifolds. This similarity
indicates a weak n dependence of coherence parameters. The consequences of
this observation will be analyzed in the following section.

4.   Extrapolation to high Rydberg and low-energy continuum states

Based on their correspondence to classical orbital parameters the O(4)
operator expectation values are expected to be particularly useful for
extrapolation to highly excited n manifolds. While the number of density-
matrix elements $\langle n\ell m|\rho|n\ell'm'\rangle$ increases dramatically as $n\to\infty$ ($\sim n^3$ for axial
symmetry) and a determination of all individual matrix elements becomes,
theoretically as well as experimentally, a formidable task, a few
statistical multipoles may provide an adequate description of the most
important and characteristic features of the coherently excited charge
cloud. This conjecture is supported by a classical orbital picture that
should become increasingly valid as n tends to infinity (more precisely, as
$(n, \ell) \to \infty$). In the classical limit, a few orbital parameters suffice to
characterize the Kepler orbit completely.

The extrapolation procedure can be motivated in a heuristic manner by
considering the eigenvalue equation

$$A_z |nn_1 n_2 m\rangle = Z \frac{n_2 - n_1}{n} |nn_1 n_2 m\rangle \qquad (4.1)$$

for hydrogenic states $|nn_1 n_2 m\rangle$ with parabolic quantum numbers n_1, n_2, and m.
For states with large dipole moments ($n_{1,2}/n \cong 1$, $n_{2,1}/n \cong 0$) the eigenvalue
becomes n independent for large n and equal to that for the zero-velocity
Coulomb continuum state

$$\lim_{v_z \to 0} A_z |v_z^{\pm}\rangle = \pm Z |v_z^{\pm}\rangle. \qquad (4.2)$$

This connection can be understood with the help of a simple classical
picture: the "perihelion" position of the Kepler orbit is almost identical
for negative energy eliptic orbits with large eccentricity and for zero-

energy parabolic orbits. One might therefore expect $\langle A_z \rangle$ to be a smooth function across threshold and furthermore weakly dependent on n along a Rydberg series.

We explore this conjecture by studying the Rydberg limit of the operator expectation values $\langle U_0^k(\pi, k_1, k_2) \rangle$. This limit involves both the n dependence of the density matrix elements, $\langle n\ell m | \rho | n\ell'm' \rangle$, and the expansion coefficients of the expectation values $\langle U_0^k \rangle$ in terms of the standard basis (through $j = (n-1)/2$ in the 9j symbol in (3.11)). Drastic simplifications can be achieved using that

$$\ell, \ell' \ll j \tag{4.3}$$

i.e., that only a "narrow" band of low final ℓ states significantly contributes to the excitation cross section. For most of the collisional excitation mechanisms at high energies known to date (direct excitation, electron capture, and ion-solid interaction), population of ℓ values is restricted to $\ell \leq 20$, and Eq. (4.3) is certainly satisfied for high n states (say, n > 100). In physical terms this means that the angular momentum of the electron becomes small compared to its dipole moment in the limit $n \to \infty$. Mathematically, this results in a contraction of the dynamical symmetry group in the Rydberg limit. Substituting (3.4) into (3.5) one finds

$$[A_i, A_j] = i\, \varepsilon_{ijk}\, \left(\frac{Z}{n}\right)^2 L_k \tag{4.4}$$

In the limit $n \to \infty$ (more precisely, $\langle L_k \rangle / n \to 0$) the components of the unnormalized Runge-Lenz vector commute. The resulting Lie algebra

$$[L_i, L_j] = i\, \varepsilon_{ijk}\, L_k$$
$$[L_i, A_j] = i\, \varepsilon_{ijk}\, A_k \tag{4.5}$$
$$[A_i, A_j] = 0$$

generates the dynamical symmetry group E(3) (Euclidean group) to which O(4) is contracted at threshold. We note that the continuation into the positive-energy continuum becomes the non-compact group O(3,1).

Using (4.5) and the fact that the properly normalized pseudospins become linearly dependent on each other in the limit $n \to \infty$,

$$\lim_{n \to \infty} \vec{J}(1,2)/n = \pm \frac{1}{2Z}\, \vec{A}, \tag{4.6}$$

the set of non-vanishing operator expectation values is given to leading order in n^{-1} by[17]

$$\frac{\langle U_0^k(\pi,k,0)\rangle_n}{\langle U_0^0(\pi,k,0)\rangle_n} \underset{n\to\infty}{=} \left(-\frac{1}{Z}\right)^k [(2k+1)(2-\delta_{k,o})]^{1/2}\langle A_0^k\rangle \tag{4.7}$$

with $k_1 = k$ and $k_2 = 0$ and $k+\pi$ even.

Without loss of generality we have chosen $k_2 = 0$ in (4.7) since different combinations (k_1,k_2) for given k become linearly dependent (see Eq. (4.6)). For later use we have displayed in (4.7) the n dependence explicitly. In terms of standard state multipoles the expectation values are given by

$$\frac{\langle U_0^k(\pi,k,0)\rangle_n}{\langle U_0^0(\pi,k,0)\rangle_n} \underset{n\to\infty}{=} (-1)^k(2-\delta_{k,0})^{1/2} \sum_{\ell,\ell'} (-1)^\ell((2\ell+1)(2\ell'+1))^{1/2}$$

$$\begin{pmatrix} \ell & k & \ell' \\ 0 & 0 & 0 \end{pmatrix} \rho_0^k(\ell,\ell') \tag{4.8}$$

The continuity of the density matrix[31] across the ionization limit implies that the Rydberg limit of the multipoles (4.7) also determines the anisotropy coefficients β_k of the angular distribution of continuum electrons near threshold. Low-energy electrons in the projectile rest frame give rise to a "cusp" in the forward emission in the laboratory frame.[31] The connection between (4.7) and β_k can be directly established by a partial-wave analysis of the doubly-differential cross section (DDCS) $d\sigma/d\vec{v}$. For axially symmetric excitation, the DDCS is independent of the azimuthal angle ϕ, i.e.

$$\frac{d\sigma}{d\vec{v}} = \frac{\sigma_0}{v} \sum_{k=0}^{\infty} \beta_k P_k(\cos\theta) \tag{4.9}$$

with[17]

$$\beta_k = \left[\frac{2k+1}{2-\delta_{k,0}}\right]^{1/2} \frac{\langle U_0^k(\pi,k,0)\rangle_v}{\langle U_0^0(0,0,0)\rangle_v} \tag{4.10}$$

and

$$\sigma_0 = \frac{v}{4\pi} \langle U_0^0(\pi,k,0)\rangle_v. \tag{4.11}$$

The multipoles for partial-wave coherences, $\langle U_0^k(\pi,k,o)\rangle_v$, in (4.10,11) are defined in complete analogy to corresponding Rydberg multipoles (4.7) as

$$\langle U_0^k(\pi,k,0)\rangle_v = (-1)^k(2-\delta_k,0)^{1/2} \sum_{\ell,\ell'=0}^{\infty} (-1)^\ell((2\ell+1)(2\ell'+1))^{1/2} \cdot$$

$$\begin{pmatrix} \ell & k & \ell' \\ 0 & 0 & 0 \end{pmatrix} \rho_0^k(\ell,\ell') \tag{4.12}$$

where

$$\rho^k(\ell,\ell') = \sum_{m,m'} (-1)^{\ell-m} \sqrt{2k+1} \begin{pmatrix} \ell & k & \ell' \\ -m & q & m' \end{pmatrix} \langle v\ell m|\rho|v\ell'm'\rangle \tag{4.13}$$

are the state multipoles for coherences between partial waves ($|v\ell m\rangle$, $|v\ell'm'\rangle$). The continuity of the density matrix weighted by the density of bound states and of continuum states

$$\lim_{n\to\infty} [\frac{n^3}{Z^2} \langle n\ell m|\rho|n\ell'm'\rangle] = \lim_{v\to o} [v\langle v\ell m|\rho|v\ell'm'\rangle] \tag{4.14}$$

yields

$$\lim_{v\to o} \beta_k = \lim_{n\to\infty} (-\frac{1}{Z})^k(2k+1)\langle A_0^k\rangle \tag{4.15}$$

and

$$\lim_{v\to o} \sigma_0 = \lim_{n\to\infty} [\frac{n^4}{4\pi Z^2}\langle U_0^0(0,0,0)\rangle_n]. \tag{4.16}$$

The latter expression allows one to relate absolute cross sections above with those below threshold provided the zeroth-order multipole is normalized to the n shell cross section rather than to unity. It should be noted that for large n coherences between different n states become observable, e.g. in photoemission experiments, since quantum beats associated with the energy difference $|\varepsilon_n-\varepsilon_n'|$ (see Eq. (2.4)) can be resolved for n,n'>>1. Inter -n coherences, however, do not contribute to $\langle A_0^k\rangle$ since $\vec{A}$ commutes with H_a nor do v-v' coherences contribute to the observed spectrum of continuum electrons which represents a projection onto energy eigenstates in the continuum.

The smooth behavior of $\langle A_0^k\rangle$ (or β_k) along the Rydberg series can be investigated numerically using the CDW approximation (Fig. 4). While the CDW results are not expected to be accurate, the relatively simple analytical structure of the CDW amplitude permits a systematic investigation of the n dependence of the density matrix over a wide range of n. Our numerical results serve primarily illustrative purposes. As noted above, differences between the CDW and the CDW-PCI results are negligible for $\langle A_0^k\rangle$.

All three low-order multipoles σ_0, β_1, and β_2 for proton-hydrogen charge transfer are remarkably weakly dependent on n quite similar to

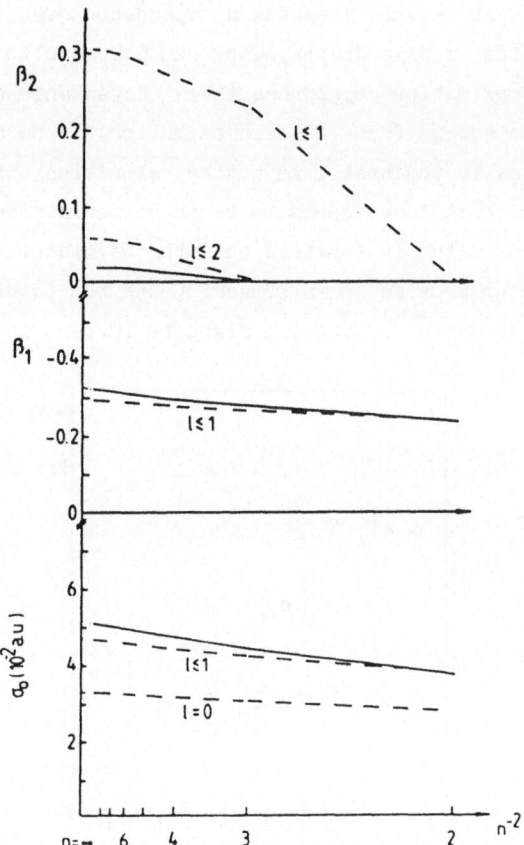

Fig. 4. Multipoles σ_0, β_1, and β_2 as functions of n^{-2} (binding energy of final state) for H^+ + $H(1s)$ →$H(n)$ + H^+ (v_p = 2 a.u.) in CDW approximation. ——, sum over all ℓ states; – – –, partial sums.

a quantum defect along a Rydberg series. The smooth behavior of the multipoles (4.7) is expected to be independent of the chosen approximation. The rapid convergence of partial sums over partial waves ($\ell \leq \ell_{max}$) indicates the dominance of low ℓ states. The limiting values ($n \to \infty$) give the cross section and the anisotropy parameters of the electron capture to continuum (ECC) cusp. The extrapolated values have been confirmed by a direct calculation for the ECC process.[32]

The smooth behavior of β_1 can also be verified by comparing experimental data for the charge-cloud asymmetry and the velocity asymmetry using (4.15). Fig. 5 displays $\langle A_z \rangle$, or equivalently, β_1, for reaction (3.16) for both bound states (n=2,3) and near-threshold continuum states (v≪1 or n=∞). We note the good agreement between the cusp data (n → ∞) of Andersen et al.[33] and Dahl[34] with bound state dipole data (n=3) of Havener et al.,[21] and (n=2) data of DeSerio et al.[26] supporting the conjecture of a weak n dependence of $\langle A_z \rangle$. The experimental data of Meckbach et al.[35] (not shown in Fig. 5) are in disagreement with all other data. The source of this discrepancy is not yet known. The weak n dependence over a wide range of projectile velocities is also displayed by the CDW results for n=3 and n=∞.

While CDW approximation reproduces the v_p dependence of the data better than the asymptotic second Born (B2) approximation,[36] the magnitude of $\beta_1(n)$ is systematically too small. At higher velocities, the impulse approximation[36] (IA-post form) seems to be in better agreement. However, since the post form of the IA is valid only for asymmetric systems with the projectile assumed to provide the stronger of the two potentials the agreement for proton-helium scattering might be in part fortuitous.

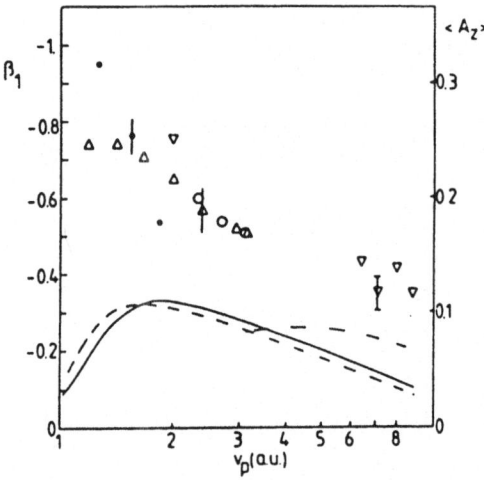

Fig. 5.   Dipole term β_1 for $H^+ + He \rightarrow H(n) + He^+$ as a function of projectile velocity; o, DeSerio et al.[26]; ●, Havener et al.[21] (n=3); ∇, Andersen et al.[33] (n=∞); Δ, Dahl[34] (n=∞); –·–·–, asymptotic B2 approximation[21] (n=∞); ····, IA[36] (n=∞); – – –, CDW[22] (n=3); ——, CDW (n=∞).

While various approximation schemes qualitatively reproduce the data, the quantitative agreement is still rather poor. At high projectile velocities a more advanced charge transfer theory like the DWB approximation suggested by Taulbjerg and Briggs[37] may provide the starting point for a more reliable description. At intermediate velocities close-coupling calculations are clearly superior. However, extension to the near threshold continuum seems to be a formidable task in view of the large number of coupled channels. The smooth n dependence of the O(4) multipoles suggests a practical method to circumvent this problem. Calculation of statistical multipoles for a few medium n states (above the most probable n shell) within a coupled-channel code and smooth extrapolation to threshold should suffice to reliably estimate the cusp cross section.

5. Concluding remarks

The operators of the O(4) group provide a convenient and physically meaningful representation of the density operator for (near) hydrogenic systems. Such an approach is particularly well-suited for cases where the determination of the complete density matrix is practically impossible and a selection of potentially important components becomes necessary. Due to dynamical selection rules for certain expectation values in Born approximation, investigation of coherent excitation provides a sensitive test for high-order contributions of the Born series to inelastic ion-atom collisions.

We close by mentioning future applications: The configuration mixing in doubly-excited helium-like systems can be approximately described by an O(4) basis states resulting from the broken O(4) x O(4) symmetry due to electron-electron interaction.[38-40] The application of the present representation of the density operator to the recently observed coherent excitation in doubly-excited states[13] may give new insights into the interplay between coherences and correlations in doubly excited states.

As the time resolution in the experiment may be further improved the observation of coherence between different n states becomes feasible. In this case the non-invariance O(4,2) group[8] may provide an adequate basis for the density operator in an extended Liouville space.

This work supported in part by the National Science Foundation and by the U.S. Department of Energy under contract No. DE AC05-84OR21400 with Martin Marietta Energy Systems, Inc.

REFERENCES

1. U. Fano and J. Macek, Rev. Mod. Phys. 45:553 (1973).
2. H. J. Andrä, in: "Progress in Atomic Spectroscopy," Part B, p. 825, W. Hanle and H. Kleinpoppen, eds., Plenum Press, New York (1979).
3. K. Blum, "Density Matrix Theory and Applications," Plenum Press, New York (1981).
4. H. W. Hermann and I. V. Hertel, Comments At. Mol. Phys. 12:61 and 127 (1982).
5. R. Hippler, Fundamental Processes in Atomic Collision Physics, in: "Proceedings of the Advanced Study Institute," Plenum Press, New York (1985).
6. R. K. Janev and H. Winter, Phys. Rep. 117:265 (1985).
7. V. Fock, Z. Phys. 98:145 (1935).
8. M. J. Englefield, "Group Theory and the Coulomb Problem," Wiley Interscience, New York (1972).
9. U. Fano, Rev. Mod. Phys. 29:74 (1957).
10. N. Andersen, these lectures.
11. C. D. Lin, these lectures.
12. H. Kleinpoppen, these lectures.
13. P. Van der Straten and R. Morgenstern, Comments At. Mol. Phys. 17:243 (1986).

14. J. Burgdörfer, Z. Phys. A 309:285 (1983).
15. U. Fano and G. Racah, "Irreducible Tensorial Sets," Academic Press, New York (1959).
16. D. Smith and J. Thornley, Proc. Phys. Soc. 89:779 (1966).
17. J. Burgdörfer, Phys. Rev. A 33:1578 (1986).
18. A. Gaupp, H. J. Andrä, and J. Macek, Phys. Rev. Lett. 32:268 (1974).
19. G. Gabrielse and Y. Band, Phys. Rev. Lett. 39:697 (1977).
20. G. Gabrielse, Phys. Rev. A 22:138 (1980).
21. C. C. Havener, N. Rouze, W. B. Westerveld, J. S. Risley, Phys. Rev. A 33:276 (1986).
22. J. Burgdörfer and L. Dubé, Phys. Rev. Lett. 52:2225 (1984); Nucl. Inst. and Meth. B10/11:198 (1985).
23. I. A. Sellin, J. R. Mowat, R. S. Peterson, P. M. Griffin, R. Laubert, and H. H. Haselton, Phys. Rev. Lett. 31:1335 (1973).
24. I. Sellin, L. Liljeby, S. Mannervik, and S. Hultberg, Phys. Rev. Lett. 42:570 (1979).
25. J. Burgdörfer, Phys. Rev. A 24:1756 (1981).
26. R. DeSerio, C. E. Gonzalez-Lepera, J. Gibbons, J. Burgdörfer, and I. A. Sellin, submitted for publication in Phys. Rev. A (1987).
27. M. R. C. McDowell and J. P. Coleman, "Introduction to the Theory of Ion-Atom Collisions," North Holland, Amsterdam (1970).
28. I. M. Cheshire, Proc. Phys. Soc. (London) 84:89 (1964).
29. A. Jain, C. D. Lin, and W. Fritsch, submitted for publication in Phys. Rev. A, and private communication, (1987).
30. O. Schöller, J. S. Briggs, and R. M. Dreizler, J. Phys. B 19:2505 (1986).
31. M. E. Rudd and J. Macek, Case Studies At. Mol. Phys. 3:47 (1972).
32. J. F. McCann and D. S. F. Crothers, J. Phys. B 20:L19 (1987).
33. L. Andersen, K. Jensen, and H. Knudsen, J. Phys. B 19:L161 (1986).
34. P. Dahl, J. Phys. B 18:1178 (1985).
35. W. Meckback, I. B. Nemirovsky, and C. Garibotti, Phys. Rev. A 24:1793 (1981).
36. D. H. Jakubassa-Amundsen, "Lecture Notes in Physics," Vol. 213, p. 17, K. O. Groeneveld, W. Meckbach, and I. A. Sellin, eds., Springer, Berlin (1984).
37. K. Taulbjerg, in: "Fundamental Processes in Energetic Atomic Collisions," NATO-ASI Series, p. 349, H. O. Lutz, J. S. Briggs, H. Kleinpoppen, eds., Plenum, New York, (1983).
38. C. Wulfman, Chem. Phys. Lett. 23:370 (1973).
39. D. R. Herrick, M. E. Kellman, and R. D. Poliak, Phys. Rev. A 22:1517 (1980).
40. C. D. Lin, these lectures.

THE PRODUCTION AND MEASUREMENT OF ANISOTROPY USING PHOTONS

Chris H. Greene

Department of Physics and Astronomy
Louisiana State University
Baton Rouge, Louisiana 70803

ABSTRACT

The alignment and orientation of a collision fragment can be determined by monitoring the polarized fluorescence emitted by a specific fragment state. Another way to determine fragment multipole moments is to observe the polarization of laser-induced fluorescence. The present paper reviews the relationship between the desired multipole moments and the intensity of laser-induced fluorescence, as derived from a viewpoint close to that of Fano and Macek. Time-resolved quantum beats in the fluorescence intensity have an amplitude which can often be directly expressed in terms of the fragment alignment and orientation. This relationship can be used to extract these moments or to correct for depolarization effects in time-unresolved experiments. Propensity rules for the angular distribution and the alignment of photofragments are discussed with reference both to examples and counterexamples. The photoelectron angular distribution and ionic alignment, generated by the simultaneous photoionization of helium accompanied by excitation of the hydrogenic ion, are analyzed with reference to predictions based on the hyperspherical potential curves of doubly-excited helium and are compared with recent experimental observations.

ANISOTROPY MEASUREMENTS

As discussed in Fano's first lecture, the analysis of anisotropic fragmentation can easily become cumbersome and impenetrable when numerous spatial directions and angular momenta are involved. A tremendous advantage is gained by systematically factorizing geometrical and dynamical aspects to the greatest possible extent. The next three sections illustrate how this factorization can be developed for two typical problems: simple fluorescence following a collision and laser-induced fluorescence following a collision. In each case the intensity expression is rearranged so as to isolate a small number of dynamical parameters multiplying standard factors of "geometry". The geometrical factors include angular functions containing the location and polarization sensitivity of each detector or monochromator, and also Wigner coefficients containing relevant angular momentum quantum numbers. Such a factorization can always be accomplished (to a greater or lesser extent) for any anisotropy measurement of interest, using the type of analysis developed below.

1. Impact Excitation of Fluorescence

The intensity of polarized fluorescence from a generic "collision fragment" depends on a small set of key dynamical quantities. These are frequently expressed in terms of irreducible components $\rho_q^{(k)}$ of the excited state density matrix. The k=0 component is proportional to the total excited state population, the k=1 components reflect the average orientation or magnetic dipole moment of the excited state probability density, and the k=2 components determine the net alignment or quadrupole moment of the density. Blum's book[1] offers an excellent introduction to this problem emphasizing thes irreducible components of the excited state density matrix.

A somewhat different viewpoint has been developed over the past thirty years by Fano and coworkers.[2-4] The central elements in this description are not density matrix elements, but rather mean values of a small set of operator relevant to the problem at hand. Often, as in the Fano-Macek treatment of fragment fluorescence, these operators are simple and familiar. The q=0 components of the orientation and alignment, for instance, are proportional to the mean values of irreducible angular momentum operators J_z and $3J_z^2 - J^2$, respectively. This philosophy of focussing on a small set of mean values of easily-interpreted operators has a practical advantage over dealing with the full (reducible) density matrix, in that the number of relevant dynamical parameters is greatly reduced. The same reduction in the number of dynamical parameters is obtained in any case by using the irreducible spherical tensor decomposition of the density matrix developed by numerous authors.[5-7]

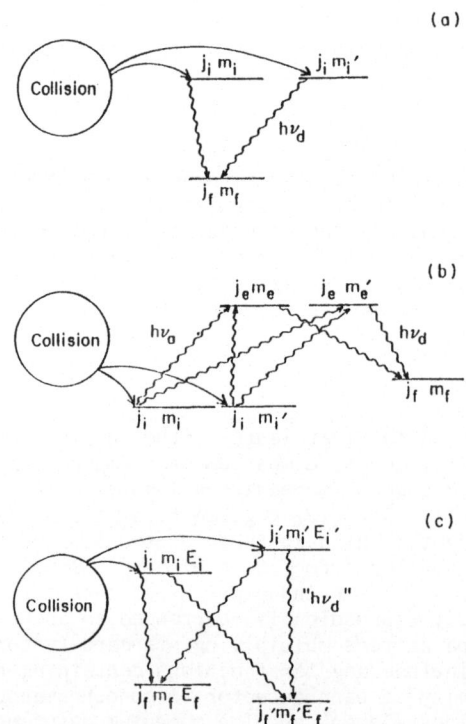

Fig. 1. Three experiments to measure the anisotropy of collision fragments: (a) collision-excited fluorescence; (b) laser-induced fluorescence (LIF) (c) quantum beats in fluorescence following coherent excitation of nearly degenerate states |i) and |i').

The intensity of emitted fragment fluorescence can derived using the simple ground rules enumerated by Breit in his classic 1933 review article.[8] Figure 1(a) shows a set of (indistinguishable) quantum mechanical pathways from the collision to the final detected photon $h\nu_d$, whose amplitudes must be added coherently. Processes leading to distinct final states $|j_f m_f\rangle$ are distinguishable, in principle, and their *probabilities* must therefore be added instead instead of their amplitudes. The resulting expression for the intensity of collision-excited fluorescence is given by

$$I = I_0 \sum_{m_f} | \sum_{m_i} (j_f m_f | \hat{\epsilon} \cdot \vec{r} | j_i m_i)(j_i m_i | C) |^2 . \qquad (1)$$

This expression involves an amplitude $(j_i m_i | C)$ to populate the fragment state $|j_i m_i\rangle$ in the course of the collision, and a matrix element for emission of an electric dipole photon of energy $h\nu_d$ and polarization $\hat{\epsilon}$. After expanding the absolute square into a double summation, Eq.(1) takes the form

$$I = \sum_{m_i m_i'} [(j_i m_i' | C)(C | j_i m_i)] \, Q_{m_i m_i'} . \qquad (2)$$

The quantity in square brackets in Eq.(2) is the density matrix $\rho_{m_i' m_i}$ of the collision-produced excited state and the quantity $Q_{m_i m_i'}$ is just

$$Q_{m_i m_i'} = I_0 \sum_{m_f} (j_i m_i | \hat{\epsilon}^* \cdot \vec{r} | j_f m_f)(j_f m_f | \hat{\epsilon} \cdot \vec{r} | j_i m_i') . \qquad (3)$$

Equations (2) and (3) have a less-than-transparent dependence on the initial and final state angular momenta, and on the detector geometry contained in $\hat{\epsilon}$. Fano and Macek realized that the intensity expression has the following tensorial structure:

$$I = I_0 \langle (j_i | \, (\hat{\epsilon}^* \cdot \vec{r}) \, P_f (\hat{\epsilon} \cdot \vec{r}') \, | j_i) \rangle , \qquad (4)$$

where

$$P_f \equiv \sum_{m_f} |j_f m_f)(j_f m_f| \qquad (5)$$

is a scalar operator contributing no multipolarity to the light intensity. The outermost brackets in Eq.(4) denote the averaging procedure (2), and the prime on the second dipole operator $\vec{r}'$ serves as a reminder that two independent integrations are involved in its evaluation.

The role of the fragment multipole moments emerges after recoupling the polarization and dipole operators in a different order. First of all the operator in Eq.(4) is written in spherical tensor notation as in Fano and Racah,[3]

$$(\hat{\epsilon}^* \cdot \vec{r})(\hat{\epsilon} \cdot \vec{r}') = 3\{[\epsilon^{*(1)} \times r^{(1)}]^{(0)} \times [\epsilon^{(1)} \times r'^{(1)}]^{(0)}\}_0^{(0)} . \qquad (6)$$

Next the constant geometrical factors ϵ are brought together to form a net k-th rank multipole of the detected photon through a recoupling transformation (see chapter 12 of ref. 3), replacing (6) by

$$\sum_{k=0,1,2} \{[\epsilon^{*(1)} \times \epsilon^{(1)}]^{(k)} \times [r^{(1)} \times r'^{(1)}]^{(k)}\}_0^{(0)} (2k+1)^{\frac{1}{2}} , \qquad (7)$$

in which the explicit recoupling coefficient $((11)k \, (11)k | (11)0 \, (11)0)^{(0)} = (2k+1)^{\frac{1}{2}}/3$ has been inserted. This recoupling is handled in a more rudimentary fashion by Fano and Macek, with the same result. Notice how the multipolarity k of the emitted radiation $[\epsilon^{*(1)} \times \epsilon^{(1)}]^{(k)}$ matches exactly the multipolarity of the excited fragment state,

$$\langle (j_i | [r^{(1)} \times r'^{(1)}]^{(k)}_q P_f | j_i) \rangle \ . \tag{8}$$

In other words, the "quadrupole moment" of the radiation field is proportional to the quadrupole moment of the fragment state $|j_i\rangle$, simply because the final state $|j_f\rangle$ is not observed following the light emission and hence contributes no multipolarity.

The last major element of the Fano-Macek analysis is the use of the Wigner-Eckart theorem to replace the complicated dipole operator in (8) by any other convenient operator of rank k, in particular by an angular momentum operator $J^{(k)}_q$ (see, e.g. Refs. 9-11),

$$\langle (j_i | J^{(k)}_q | j_i) \rangle \ (j_i \| [r^{(1)} \times r'^{(1)}]^{(k)} P_f \| j_i) \ / \ (j_i \| J^{(k)} \| j_i) \ . \tag{9}$$

Thus the collision dynamics influences the light intensity only by causing a few anisotropic angular momentum operators to have nonzero mean values. In the case of a cylindrically symmetric collision frame, only two such parameters are needed to fully determine the pattern of polarized light emission. These are explicitly the orientation parameter (normalized as in Ref.9 rather than as in Ref.4),

$$O^{(1)}_0 = \langle J_z \rangle / [j_i(j_i+1)]^{\frac{1}{2}} \tag{10}$$

and the quadrupole alignment parameter

$$A^{(2)}_0 = \langle 3J_z^2 - J^2 \rangle / j_i(j_i+1) \ . \tag{11}$$

Some further simplification of the reduced matrix elements in Eq.(9) gives a final expression for the fluorescence emitted after a cylindrically symmetric collision:

$$I = I'_0 \ \{1 + \frac{3}{2} h^{(1)} O^{(1)}_0 \cos\theta \ \sin 2\beta$$
$$- h^{(2)} A^{(2)}_0 [\frac{1}{2} P_2(\cos\theta) - \frac{3}{4} \sin^2\theta \ \cos 2\chi \ \cos 2\beta]\} \tag{12}$$

in which two factors $h^{(k)}(j_i, j_f)$ involving Wigner 6j-coefficients contain the only dependence of the anisotropic light emission on the final state quantum number j_f. These coefficients are given explicitly in Ref. 9, with $h^{(1)}$ also normalized differently from Ref. 4 to compensate for the different definition of $O^{(1)}_0$. In Eq.(12) the polar angle of the detector position relative to the collision symmetry axis is θ, and the orientation of a linear polarizer relative to the plane of the detector and the symmetry axis is χ. Lastly the angle β vanishes if the detector observes linearly polarized light only, but equals $\pi/4$ if the detector is sensitive to left-circularly polarized light.[4] For more general expressions of the light emission appropriate for more complicated collision geometries, the reader is referred to Refs. 1, 4, and 9.

2. Laser-Induced Fluorescence (LIF)

Figure 1(b) shows how two photons can be used to probe higher multipole moments (up to rank k=4) of a fragment state excited by a collision. In the standard laser-induced fluorescence experiment, a laser (or other source of photons $h\nu_a$) excites the fragment state $|i\rangle$ to an excited state $|e\rangle$, after which a detector observes a fluorescence photon $h\nu_d$ emitted in the course of a transition to a final state $|f\rangle$. By monitoring the fluorescence intensity as a function of the polarization of the absorbed (a) and detected (d) photons, the total number of fragments in state $|i\rangle$ excited by the collision can be extracted as well as several multipole moments of the fragment state $|i\rangle$. This process has been described by Case et al.[10] using a spherical tensor decomposition of the density matrix. I turn now to a Fano-Macek-type analysis developed by Greene and Zare.[11]

The starting point for the analysis is the addition of all quantum mechanical amplitudes shown in Fig 1(b) leading to the same final state $|j_f m_f\rangle$, followed by summation of all probabilities leading to different final states. Suppressing the explicit representation of the averaging procedure as in Eq.(4), the starting expression for the intensity of laser-induced fluorescence is

$$I = I_0 \sum_{m_f} \langle | \sum_{m_e} (f|\hat{\epsilon}_d \cdot \vec{r}|e)(e|\hat{\epsilon}_a \cdot \vec{r}|i)|^2 \rangle .$$ (12)

Writing out the absolute square explicitly, this can be put into the following form analogous to Eq.(4):

$$I = I_0 \langle (i| \hat{\epsilon}_a^* \cdot \vec{r}_1 \, P_{e'} \hat{\epsilon}_d^* \cdot \vec{r}_2 \, P_f \, \hat{\epsilon}_d \cdot \vec{r}_3 \, P_e \, \hat{\epsilon}_a \cdot \vec{r}_4 |i\rangle ,$$ (13)

where three scalar projection operators have been defined as

$$P_{e'} \equiv \sum_{m_{e'}} |e')(e'| , \quad P_f \equiv \sum_{m_f} |f)(f| , \quad P_e \equiv \sum_{m_e} |e)(e| .$$ (14)

In Eq.(13) $\hat{\epsilon}_a$ and $\hat{\epsilon}_d$ represent the polarization vectors of the absorbed and detected photons, while the subscripts (1-4) on the dipole operators serve as a reminder that *four* independent integrations are involved in Eq.(14).

At this point a recoupling is needed to isolate the multipole moments of the absorbed and detected photons, giving in exact analogy to Eq.(7),

$$(\hat{\epsilon}_d^* \cdot \vec{r}_2)(\hat{\epsilon}_d \cdot \vec{r}_3) = \sum_{k_d} \{[\epsilon_d^{*(1)} \times \epsilon_d^{(1)}]^{(k_d)} \times [r_2^{(1)} \times r_3^{(1)}]^{(k_d)}\}_0^{(0)} (2k_d+1)^{\frac{1}{2}}$$ (15)

The same equation also holds for the absorbed photon, with the replacements $d \rightarrow a$ throughout and also $2 \rightarrow 1$ and $3 \rightarrow 4$ for the subscripts. Next an additional recoupling combines the absorbed and detected photon multipole moments into a single multipole of rank k for the whole LIF process, replacing the operator in Eq.(13) by the following:

$$\sum_{k_d k_a k q} (-1)^{k-q} \epsilon_{-q}^{(k)}(k_d k_a) R_q^{(k)}(k_d k_a) ,$$ (16)

where a "polarization tensor" has been introduced,

$$\epsilon_{-q}^{(k)}(k_d k_a) \equiv \{[\epsilon_d^{*(1)} \times \epsilon_d^{(1)}]^{(k_d)} \times [\epsilon_a^{*(1)} \times \epsilon_a^{(1)}]^{(k_a)}\}_{-q}^{(k)} ,$$ (17)

and where a coupled electric dipole operator has been defined as

$$R_q^{(k)}(k_d k_a) \equiv \{[r_2^{(1)} \times r_3^{(1)}]^{(k_d)} \times [r_1^{(1)} \times r_4^{(1)}]^{(k_a)}\}_q^{(k)} P_{e'} P_f P_e .$$ (18)

The component q of each multipole is restricted to the range $-k \leq q \leq k$. For cylindrically symmetric collision systems only the q=0 components of each multipole fail to vanish. If in addition the collision partners initially have no net spin orientation or other helicity, then the odd rank (k=1,3) multipoles must vanish. Lastly the Wigner-Eckart theorem can be used as in (9) to replace the dipole operator in (18) by an angular momentum operator:

$$I = I_0 \sum_{k_d k_a kq} (-1)^{k-q} \, \varepsilon^{(k)}_{-q}(k_d k_a) \, \langle i | J^{(k)}_q | i \rangle$$

$$\times (j_i \| R^{(k)}(k_d k_a) \| j_i) \, / \, (j_i \| J^{(k)} \| j_i) \; . \tag{19}$$

The ratio of reduced matrix elements in (19) can be further broken down using Wigner-Racah algebra to give Wigner 6j- and 9j-coefficients. Explicit expressions can be extracted from Refs. 10 and 11. The main point is that the anisotropic laser-induced fluorescence has been expressed in terms of a few dynamical parameters as in Ref. 4, which are the mean values of angular momentum operators in the initial state of a fragment after the collision. In the case of cylindrical symmetry and no initial helicity, the two parameters consist of the quadrupole alignment parameter of Eq.(11), and of a hexadecapole alignment parameter:

$$A^{(4)}_0 = \langle [3\vec{J}^4 - 6\vec{J}^2 - 30J_z^2 \vec{J}^2 + 25J_z^2 + 35J_z^4]/8\vec{J}^4 \rangle \; . \tag{20}$$

This operator is normalized to coincide with the Legendre polynomial $P_4(\hat{J}\cdot\hat{z})$ in the high-j_i limit, though in fact the Legendre polynomial is a good approximation even at small values of $j_i \geq 2$. (Recall that $A^{(4)}_0 \equiv 0$ for $j_i < 2$.)

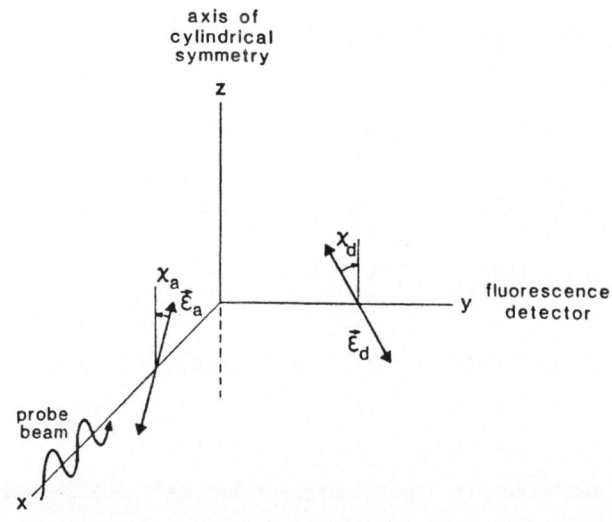

Fig. 2. Typical orthogonal excitation-detection geometry used in laser-induced fluorescence experiments. The incident laser photon is linearly polarized at an angle χ_a relative to the collisional symmetry axis. The fluorescence is detected with linear polarization at an angle χ_d relative to the symmetry axis.

The general expression for the intensity of laser-induced fluorescence in the case of a cylindrically symmetric collision frame with no helicity, for the orthogonal excitation-detection geometry shown in Fig.2, is given by

$$
\begin{aligned}
I = I_0' \{ 1 &+ \gamma_{220} t_{ie} t_{ef} [\tfrac{4}{3} P_2(\cos x_d) P_2(\cos x_a) - \sin^2 x_d \sin^2 x_a] \\
&+ A_0^{(2)} (-\gamma_{202}\, t_{ef} P_2(\cos x_a) - \gamma_{022}\, t_{ie} P_2(\cos x_a) \\
&\qquad + \gamma_{222} t_{ie} t_{ef} [\tfrac{4}{3} P_2(\cos x_d) P_2(\cos x_a) + \sin^2 x_d \sin^2 x_a]) \\
&+ A_0^{(4)} \gamma_{224} t_{ie} t_{ef} [8 P_2(\cos x_d) P_2(\cos x_a) - \sin^2 x_d \sin^2 x_a] \} \ .
\end{aligned} \tag{21}
$$

The factors t_{ie} equal unity for P and R absorption branches ($j_i = j_e \pm 1$) and are -2 for the Q absorption branch ($j_i = j_e$). The factors t_{ef} are defined similarly for the emission branches. Legendre polynomials of the second rank are denoted by $P_2()$. The angular momentum coupling factors γ are generally complicated Wigner coefficients, but in the high-j_i limit relevant to many molecular experiments they approach the following constants:

$$
\gamma_{220} \to 3/20, \ \gamma_{202} \to 1/2, \ \gamma_{022} \to 1/2, \ \gamma_{222} \to 3/28, \ \gamma_{224} \to 9/140 \ . \tag{22}
$$

Note that while the formulas for the intensity of laser-induced fluorescence are naturally more complicated than the Fano-Macek results for fluorescence excited by impact, this LIF treatment achieves the same *factorization of dynamics and geometry* which is characteristic of the Fano-Macek formulation. In particular the algebraic manipulations discussed in Eqs.(13)-(19) have reduced the combination of dipole absorption and emission processes to the mean value of an angular momentum operator in the initial state formed by the collision. Generalizations to more complicated collision geometries or to more complicated excitation-detection geometries are then quite straightforward.

Figure 3 shows two experimental examples in which laser-induced fluorescence was used to measure the population and alignment of diatomic CN fragments created by photodissociation of ICN.[12,13] Thus in this case three photons are involved, the first of which photodissociates the triatomic molecule, the second of which excites the diatomic fragment to an excited state, and the third of which comes from the excited state fluorescence. Since the collision system involves an electric dipole photon which creates the anisotropy, the hexadecapole alignment $A_0^{(4)}$ should vanish identically. Measurements of this type help to identify the dominant photodissociation pathways, such as whether the dissociation proceeds through bent or linear excited states of the ICN.

3. Quantum Beats and Depolarization Effects

Often two or more nearly degenerate levels are populated coherently by a sudden collision process. When this happens, as depicted in Fig. 1(c), it is often possible to observe this coherence in the fluorescence, which oscillates in time at frequencies corresponding to the energy differences between excited levels $E_i - E_{i'}$. The origin of these oscillatory time dependences or "quantum beats" can be understood by constructing the amplitude to observe a photon after the sudden collision occurs at time t=0. As before, recall that quantum mechanical *amplitudes* are added for indistinguishable processes, but for the distinguishable processes leading to different $|j_f m_f\rangle$ states the *probabilities* must instead be added. The intensity of fluorescence observed at time t following the collision is then

(a)

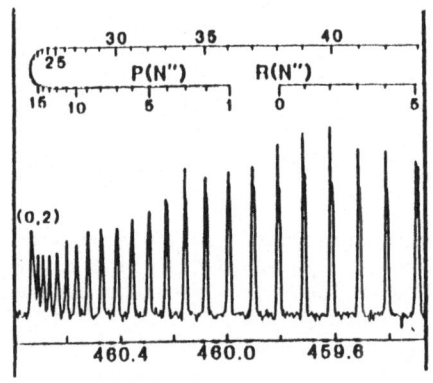

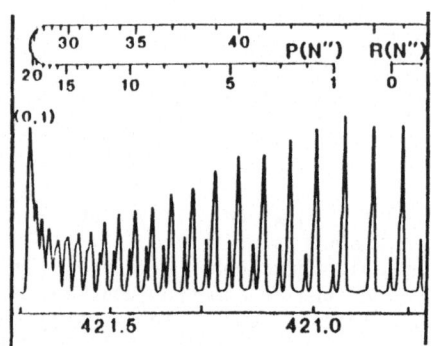

Wavelength(nm)

(b)   (c)

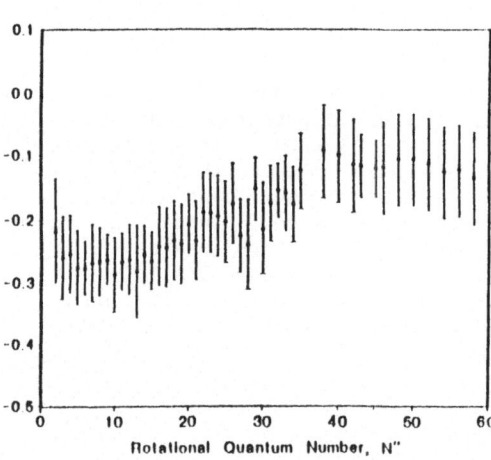

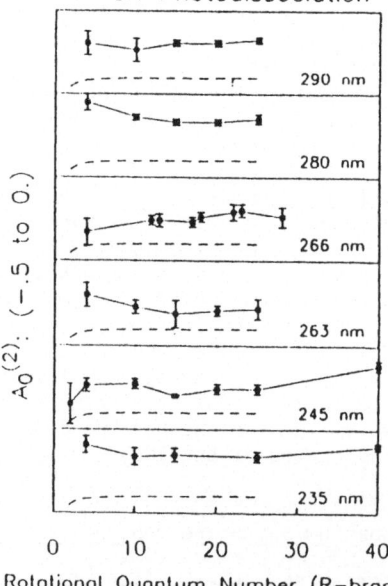

Fig. 3. Results from two experiments using laser-induced fluorescence to
measure the alignment of CN fragments produced by photolysis of ICN.
(a) Typical LIF spectrum for CN taken from O'Halloran et al.[12]
(b) Alignment of CN fragments measured by Ref. 12 is shown as a function
of the CN rotational quantum number N. (c) is the same as (b), except as
measured by Hall et al.[13] at six different wavelengths for the photolysis
laser (shown in the figures).

$$I = I_0 \sum_{m_f} |\sum_{j_i m_i} (j_f m_f |\hat{\epsilon} \cdot \vec{r}| j_i m_i) e^{iE_i t} (j_i m_i | C)|^2 , \qquad (23)$$

where E_i is the energy level of the stationary state $|j_i m_i)$. After expanding the absolute square in (23) there appear off-diagonal oscillatory terms like $\cos[(E_i - E_{i'})t]$ involving the excited state energy splittings, but notice that no final state energy splittings give rise to quantum beats. It should also be pointed out that the fluorescence detector must be incapable of distinguishing the different photon frequencies. For a high resolution detector the pathways become distinguishable and the quantum beats necessarily disappear.

Normally the frequency of the quantum beats is too fast to be observed, unless the energy splittings are very small, as is the case with hyperfine splittings or even fine structure splittings in small atoms. Alternatively the splittings can be induced by an external electric or magnetic field and chosen to lie in a convenient frequency domain. In the case of hyperfine quantum beats, for instance, Fano and Macek show that the derivation leading to Eq.(12) remains valid, except that the anisotropy itself acquires an oscillatory time dependence of the form

$$A_0^{(2)}(t) = A_0^{(2)}(0) \, g^{(2)}(t) \qquad (24)$$

where

$$g^{(2)}(t) = \sum_{FF'} \frac{(2F+1)(2F'+1)}{2I+1} \left\{ \begin{matrix} F & F' & 2 \\ j_i & j_i & I \end{matrix} \right\}^2 \cos(\omega_{FF'} t) . \qquad (25)$$

This expression assumes that the quantum beats arise from a single electronic angular momentum state $|j_i m_i)$ which has a small hyperfine splitting. The nuclear spin of the atom is taken to be I and the total angular momentum of the atom including nuclear spin and the electronic angular momenta is denoted F.

(a) (b)

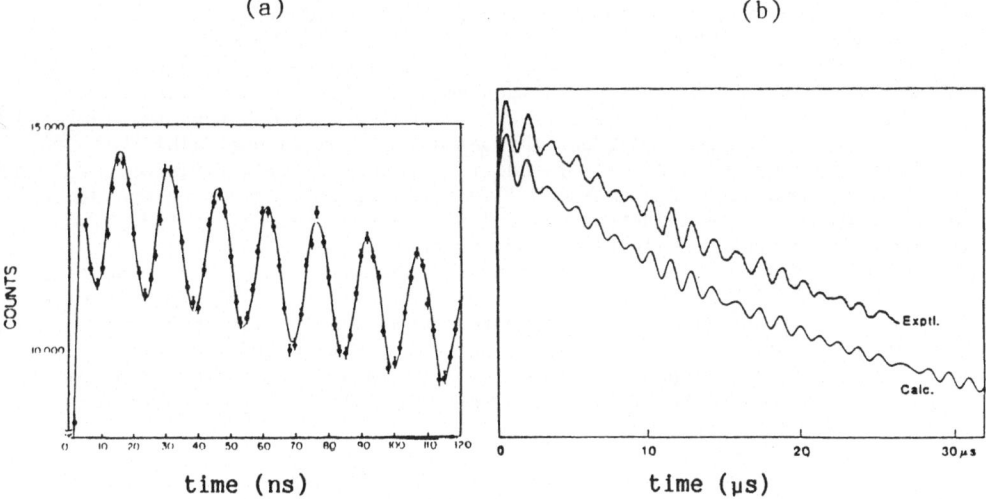

Fig. 4. Quantum beats in the time-resolved fluorescence intensity.
(a) Helium 3889 Å oscillations following beam-foil excitation (from Burns et al.[14]). (b) Zeeman quantum beats observed in NO_2 laser-induced fluorescence by Brucat and Zare.[15] Note that the time scales in (a) and (b) differ by almost three orders of magnitude.

Numerous experiments have been designed to detect these time dependent oscillations, and also to use them to determine the alignment and orientation. An example is provided by the experiment of Burns et al.[14] shown in Fig.4(a), which measured the oscillatory amplitudes for several different polarizer settings in a beam-foil spectroscopy experiment that monitored helium atoms which fluoresce after being excited by a thin tilted carbon foil. The analysis provided all three nonzero alignment parameters and a single orientation parameter relevant to a collision system having not an axis of symmetry but just a plane of symmetry. In this experiment the excited state energy differences derive from the fine structure splittings of the 3889Å helium transition, rather than from hyperfine splittings. Fig.4(b) shows quantum beats observed by Brucat and Zare[15] in the laser-induced fluorescence of NO_2. Here the excited state splittings are produced by a uniform magnetic field rather than by an internal molecular field. Fig.5 shows the time dependent modulation factor $g^{(2)}(t)$ caused by the hyperfine structure of HF, for four different molecular rotation states, N=1, 2, 3, and 10. In this case the analysis of Altkorn et al.[16] is complicated by the presence of two nonzero nuclear spins, but a reasonably straightforward derivation is presented by Ref.16 nevertheless. Note that in Fig.5 the depolarization effects become unimportant for higher values of N since the nuclear spin hardly affects the directional alignment of the angular momentum in that case. A last example of quantum beats is given by the recent electron-photon coincidence measurements made by Heck and Williams,[17] as is shown in Fig.6. In this experiment 350 eV electrons collide with atomic hydrogen and excite it to the n=2 states. The experiment is conducted in the presence of a weak electric field which permits an investigation of the 2s-2p coherence. By monitoring the scattered electron at one scattering angle and the photon emitted at a definite time t after the collision, an oscillatory time dependence is observed whose amplitudes and phases determine the entire n=2 density matrix of the collision-excited hydrogen atom.

Frequently the experimental time resolution is insufficient to detect the oscillatory behavior. In this case the factor $g^{(2)}$ in (25) is replaced by an average value including an exponential decay lifetime τ for the state $|j_i)$. This amounts to replacing each factor $\cos(\omega_{FF'}t)$ in Eq.(25) by $[1+(\omega_{FF'}\tau)^2]^{-1}$, which shows that the unobserved time dependence still affects the light emission. In fact it always has a "depolarizing effect" in that the apparent anisotropy is reduced in absolute value by such unobserved oscillations. In more physical terms, the oscillations are associated with the precession of the initially aligned vector j_i about the total angular momentum vector F. Accordingly when the lifetime τ is much smaller than the characteristic precession time ($\simeq \omega_{FF'}{}^{-1}$), the fluorescence is emitted before the vector j_i has time to precess and therefore the depolarization is negligible. In this limit the average value of $g^{(2)}$ is unity. In the opposite limit of very fast precession compared to the emission lifetime, the depolarization can reduce the apparent value of the anisotropy by a large factor since all off-diagonal terms with F ≠ F' are then negligible. Most theoretical studies of fragment alignment and orientation neglect this depolarization, making it essential to "correct" the experimental anisotropy before making detailed comparisons with theory. One of the cleanest experimental demonstrations of this depolarization effect is the strong isotope effect observed by Hafner et al.[18] in their measurements of polarization of the alkali resonance lines excited by electron impact at threshold. Their main results are displayed in the following table:

Table 1. Depolarization of alkali resonance lines.

Isotope	Nuclear Spin I	Observed P(threshold)[18]
^6Li	1	0.397 ± 0.038
^7Li	3/2	0.206 ± 0.030
^{23}Na	3/2	0.148 ± 0.018

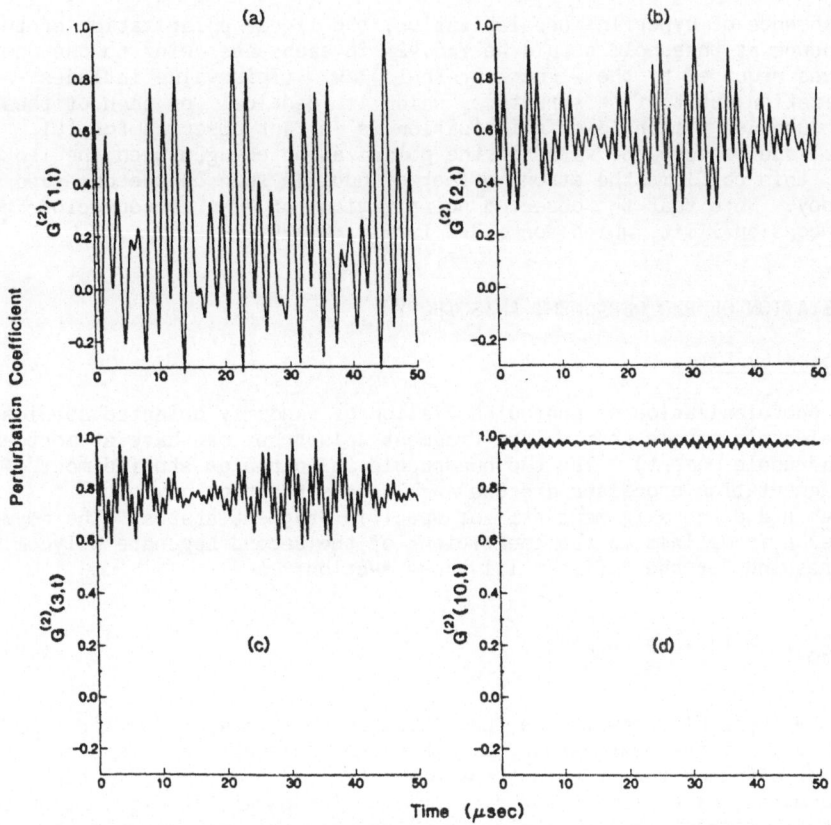

Fig. 5. Calculated hyperfine oscillations of the HF alignment are shown for four different rotational levels, N=1, 2, 3, and 10. These results, taken from Altkorn et al.,[16] demonstrate the decreasing importance of depolarization with increasing N.

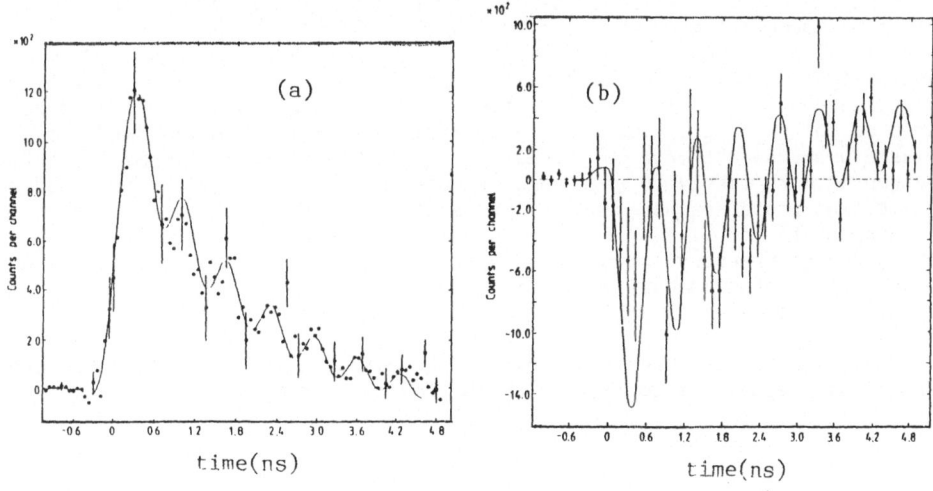

Fig. 6. Quantum beats observed in the electron-photon coincidence signal by Heck and Williams.[17] The electron and photon are detected with a time delay t, following electron impact excitation of H(2s,2p) in the presence of a weak electric field. The only difference between (a) and (b) is that the field direction has been changed by 180°.

In the absence of hyperfine depolarization, the linear polarization of the fluorescence at threshold should be P=0.429 in each case owing to the dominance of s-waves required by the Wigner threshold law. (This value includes depolarization due to fine structure, which is identical for each of these atoms.) Because a much lower polarization is in fact observed for ^{7}Li and ^{23}Na, and because the value of the polarization changes when the isotope is changed, this confirms the strong effect of nuclear spin on the observed anisotropy. Note that the observed polarization for ^{23}Na is very close to the fast precession limit, which for I=3/2 is $P_{min}=0.129$.

INTERPRETATION OF PHOTOFRAGMENT ANISOTROPY

1. Propensity Rules

In photoionization or photodissociation of randomly oriented species by a single electric dipole photon, the fragment anisotropy can have at most rank k=2 (quadrupole moment). The two quadrupole anisotropies studied most often in photofragmentation processes are the angular distribution asymmetry parameter β and the alignment $A_0^{(2)}$ of specific fragment states. The asymmetry parameter β is defined as the coefficient of the second Legendre polynomial in the expression for the differential cross section:

$$\frac{d\sigma}{d\Omega} = \frac{\sigma}{4\pi} \left[1 + \beta P_2(\cos\theta) \right] , \qquad (26)$$

where θ is the angle between the incident light polarization vector $\hat{\epsilon}$ and the relative axis of fragment escape, $\hat{k}$. When the incident light is circularly polarized, a third type of anisotropy parameter can be created in the photofragments, namely the first rank orientation parameter $O_0^{(1)}$. When the fragment of interest is a photoelectron, the orientation is simply the net electron spin polarization obtained after averaging over all photoelectron escape directions θ. It is also possible to observe fragment alignment and/or orientation in coincidence with the fragment escape direction θ, but such coincidence anisotropies will not be considered here.

A better qualitative understanding of anisotropy measurements can be derived by formulating approximate "propensity rules". Propensity rules are generally simple expectations for a particular observable which are often classically motivated. One of the more familiar propensity rules for the integrated photoabsorption cross section for an atomic $n_0\ell_0$ subshell can be stated as follows: *the electric dipole process $\ell_0 \to \ell_0+1$ normally tends to dominate over the alternative process $\ell_0 \to \ell_0-1$* . In classical terms the active electron is more likely to *gain* angular momentum when it acquires the energy of the incident photon rather than to lose angular momentum as it gains energy. This propensity for $\ell_0 \to \ell_0+1$ has been widely documented throughout the periodic system, though numerous exceptions can be identified. Fano has presented an alternative quantum mechanical argument for the success of this particular propensity rule.[19]

Returning to our three photofragment anisotropies of interest, they are first of all restricted to lie within the following ranges:

$$-1 \leq \beta \leq 2 , \qquad (27a)$$

$$-1 \leq A_0^{(2)} \leq 4/5 \qquad (27b)$$

$$-\tfrac{1}{2} \leq O_0^{(1)} \leq 1/\sqrt{2} \qquad (27c)$$

Here the limits on β and $A_0^{(2)}$ are appropriate for linearly polarized incident light. The limits on $O_0^{(1)}$ are instead appropriate for circularly polarized light with positive helicity (q=1) along the symmetry axis $\hat{z}$, and the limits (27c) change sign for the opposite helicity (q=-1).

The simplest propensity rules for these anisotropies can be deduced from fairly obvious arguments. Consider an incident photon which is linearly polarized along the z axis, the symmetry axis for this "collision". Any photodissociation or photoionization process is then caused by the oscillating electric field. It thus seems natural to expect the charge cloud to become elongated along the z axis, the direction of this applied external force. Translating this simple statement into a propensity rule can be shown to give the following expectations:

$$\beta \sim positive \tag{28a}$$

$$A_0^{(2)} \sim negative \ . \tag{28b}$$

The orientation parameter $O\binom{1}{0}$ is of course zero for symmetry reasons when the incident light is linearly polarized. For circularly polarized incident light the simplest expectation is that the net helicity or circulation of the final state probability density will have the same sign (or direction) as that of the incident photon. Thus if the photon has helicity q=1, this argument suggests a naive propensity rule for the orientation:

$$O_0^{(1)} \sim positive \ \ (for \ q=1) \ . \tag{28c}$$

Tests of the propensity rules for β are easily found since a large number of measurements have been made, especially for photoelectron angular distributions. These tests seem to indicate some sort of global validity for the propensity rule (28a), but it must be stressed that violations of the rule are hardly rare (occurring perhaps 20% to 30% of the time). Far fewer measurements of photofragment alignment have been made, despite increased activity in recent years. Nevertheless, the propensity rule (28b) seems thus far to predict the sign of the quadrupole alignment $A_0^{(2)}$ with much greater regularity than does the rule (28a) for β. As an example, the measured alignment of Zn^+ produced by photoionization of Zn is shown in Fig.7 to have the expected negative sign, as does the calculated alignment of Bartschat.[20] On the other hand the propensity rule says nothing about the interesting energy dependence of this alignment.

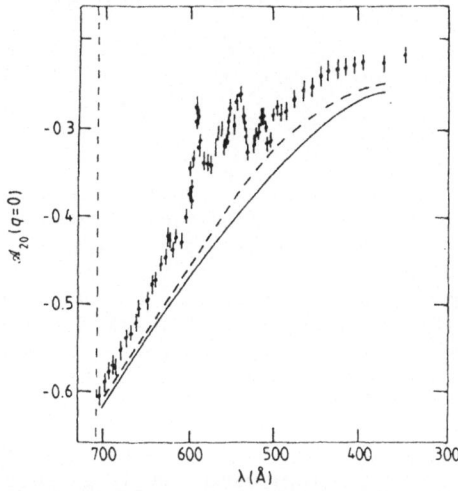

Fig. 7. Alignment of Zn^+ produced in Zn photoionization, as calculated by Bartschat (solid curves) and as observed by Kronast et al. (From Ref.20.) Alignment is negative as expected from the propensity rule.

Virtually no orientation measurements have been made to date, preventing an analysis of the success of the rule (28c). Photoelectron spin polarization measurements might be regarded as examples, except that the spin polarization is normally obtained as a function of the photoelectron escape direction θ. To obtain an average spin orientation $O\binom{1}{0}$ from these experiments, an integral over the θ-dependence is required. Some of the first measurements of orientation produced by photodissociation are now being studied by Zare and coworkers.[21]

The angular momentum transfer formalism developed by Dill and Fano[23] to describe photoelectron angular distributions can be used to give some more detailed expectations for specific systems. For instance, Dill and Fano show that a whole class of so-called parity-unfavored processes *always* lead to the opposite of Eq.(28a), namely $\beta = -1$, a consequence of little more than angular momentum conservation. A similar analysis of photofragment alignment[9] shows

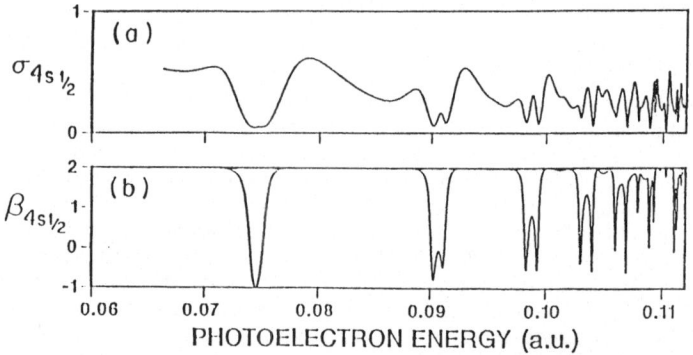

Fig. 8. Partial cross section $\sigma(4s)$ and asymmetry parameter $\beta(4s)$ characterizing the highest energy photoelectrons ejected from atomic calcium in its ground state. The energy range is between the 3d and 4p thresholds of Ca^+. Calculated by Kim and Greene.[23]

that departures from the propensity rule (28b) are *always* associated with parity-unfavored processes. On the other hand the opposite type, the parity-*favored* processes, can have either positive or negative contributions to the orientation. The positive contributions to the orientation dominate when the "observed" angular momentum quantum number j_i is larger than the "unobserved" angular momentum transfer j_t, whereas negative values of the orientation appear when $j_i > j_t$. This modified form of the orientation propensity rule should be more generally applicable than (28c). Ref. 9 should be consulted for more detailed considerations of this type, especially Figs.6 and 7 of Ref.9.

There can be many reasons why the propensity rules just described may fail to predict the sign of an anisotropy parameter in a specific instance. One of the most common reasons is that the dominant contribution to the photofragmentation cross section may happen to vanish or nearly vanish owing to some cancellation effect. Since this "dominant contribution" is the one which usually tends to obey a propensity rule, the "unfavored" contributions are more likely to be significant when the dominant part vanishes. One such example of this behavior shows up in a theoretical R-matrix calculation of the photoelectron asymmetry parameter β for the process,

$$h\nu + Ca(4s^2\ {}^1S_0) \rightarrow Ca^+(4s\ {}^2S_{\frac{1}{2}}) + e^- . \tag{29}$$

The partial cross section $\sigma(4s)$ and the asymmetry parameter calculated by Kim and Greene[23] are shown in Fig.8 as functions of energy between the 3d and 4p thresholds of Ca^+. At most energies in Fig.8, the asymmetry parameter β is equal to 2, the value it must have (in this case) if the final state is purely LS-coupled ${}^1P^0$. At energies where the cross section is small, however, owing to resonance effects, this dominant ${}^1P^0$ contribution is suppressed, and parity-unfavored contributions associated with coupling to ${}^3P^0$ symmetry cause β to approach the opposite value -1.

2. Anisotropies in the Photoexcitation of Doubly-Excited States

In some cases predictions can be made about photofragmentation anisotropy using rather elementary arguments,[24] without requiring elaborate variational calculations of the type shown in Fig.8. Such an example arises for the two-electron process

$$h\nu + He \rightarrow He^+(n) + e^- , \tag{30}$$

at photon energies just above the $He^+(n)$ threshold, thus producing very slow photoelectrons.

A prediction of the asymmetry parameter $\beta(n)$ (and other anisotropies) for this process is possible because of some unusual simplifications. First of all, the excited $n\ell$-states of the hydrogenic ion $He^+(n)$ with different ℓ can be regarded as degenerate. This "accidental degeneracy" permits the excited state $He^+(n)$ to form a permanent electric dipole moment. Gailitis and Damburg[25] and Seaton[26] have shown how to calculate this dipole moment, by using first-order degenerate perturbation theory to diagonalize all terms in the $e\text{-}He^+(n)$ Hamiltonian of order r_2^{-2} at large electron-ion distances r_2. The r_2^{-2} terms include the centrifugal potential of the outer electron and also the dipole portion of the electron-electron repulsion:

$$A = \frac{\vec{\ell}_2^{\,2} + 2\vec{r}_1 \cdot \hat{r}_2}{2r_2^2} , \tag{31}$$

where $\vec{r}_1$ is the position vector of the inner, hydrogenic electron. The degenerate basis functions used to diagonalize the operator A (at fixed r_2) are products of radial hydrogenic functions and spherical harmonics coupled to form a definite value of the total orbital angular momentum L and z-component M:

$$y_{n\ell_1\ell_2}(\vec{r}_1,\hat{r}_2) = R_{n\ell_1}(r_1)\, Y_{\ell_1\ell_2LM}(\hat{r}_1,\hat{r}_2) . \tag{32}$$

Since the outermost electron is taken here to lie outside the $n\ell_1$-orbital radius, it is not necessary to antisymmetrize these basis functions. The dimension of the resulting matrix of the operator A depends on n, L, and on the parity π. For the L=1, odd-parity final states excited after the He ground state absorbs an electric dipole photon, the matrix $\underline{A}$ has dimension (2n-1) x (2n-1).

The eigenvalues of the matrix $\underline{A}$ thus amount to 2n-1 dipolar potentials, each of which has a characteristic (constant) dipole moment a_α,

$$V_\alpha(r_2) \xrightarrow[r_2\to\infty]{} a_\alpha/2r_2^2 . \tag{33}$$

The corresponding eigenfunction

$$\psi_\alpha \underset{r_2 \to \infty}{\sim} \sum_{\ell_1 \ell_2} y_{n\ell_1 \ell_2} (\hat{r}_1, \hat{r}_2) \; c^{(nL)}_{\ell_1 \ell_2, \alpha} \qquad (34)$$

is determined by the α-th eigenvector $\hat{C}$ of $\underline{A}$, except for its dependence on the distance r_2.

Equations (33) and (34) identify the main features of a set of dipole channels appropriate for describing the escape of an electron from a hydrogenic ion $He^+(n)$ [or from excited hydrogen $H(n)$]. If only one such channel α is excited for some reason, then the photoelectron angular distribution asymmetry $\beta(n)$, as well as the orientation and alignment of the residual hydrogenic ion, are completely described by this wavefunction (34).[24,27] In general it need not be the case that a single such channel ψ_α is excited. In the more general case the final state wavefunction at $r_2 \to \infty$ will be a coherent superposition of all the ψ_α's in Eq.(34), each with a corresponding radial wavefunction $F_\alpha(r_2)$ to be determined numerically.

As pointed out by Refs. 24 and 28-30, the hyperspherical potential curves for $^1P^o$ He and H^- suggest that *a single dipole channel is predominantly excited in two-electron photoabsorption processes*. (See also the lectures by Lin and by Starace in this volume for elaboration of this point.) Specifically, these potential curves imply that out of all of the 2n-1 dipole channels which might be relevant at large-r_2, the dominantly excited channel will always be the one having the *second-most attractive dipole moment* a_α . Evidence for this conclusion was drawn from the $^1P^o$ potential curves of H^- shown in Fig.9,[28,29] and from those calculated by Macek[30] for He($^1P^o$). Note that in Fig.9 a single potential curve (labelled '+') is by far the most attractive at small-R and is consequently the channel which best overlaps the ground state and is predominantly excited. Moreover this '+' potential curve always undergoes a diabatic crossing with a '-' type potential curve, so that as $R \to r_2 \to \infty$, the '+' potential is the *second-lowest* rather than the lowest. Fano and Lin have presented arguments suggesting that such a +/- crossing must be present for all higher n as well.[31]

(a) (b)

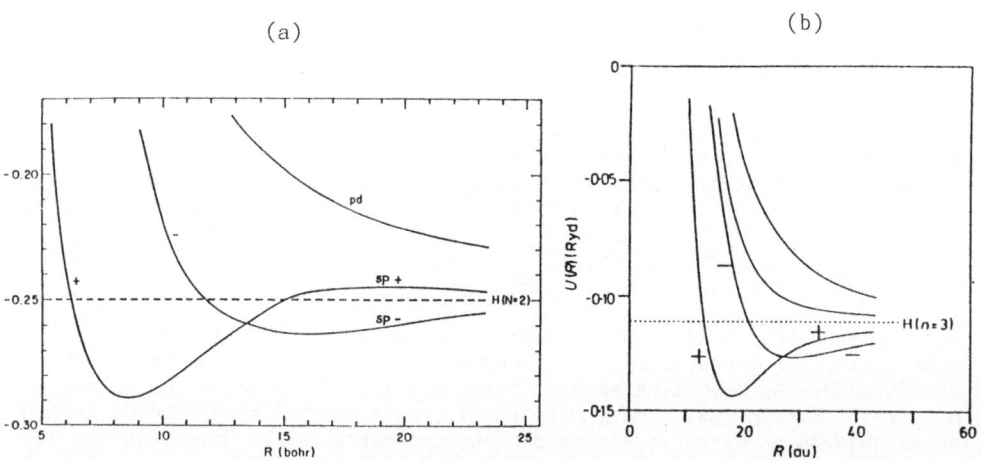

Fig. 9. Hyperspherical potential curves for H^- $^1P^o$ channels near the (a) N=2 and (b) N=3 thresholds of H, showing a +/- crossing in each case. Calculations are taken from Lin[28] and from Greene[29], respectively.

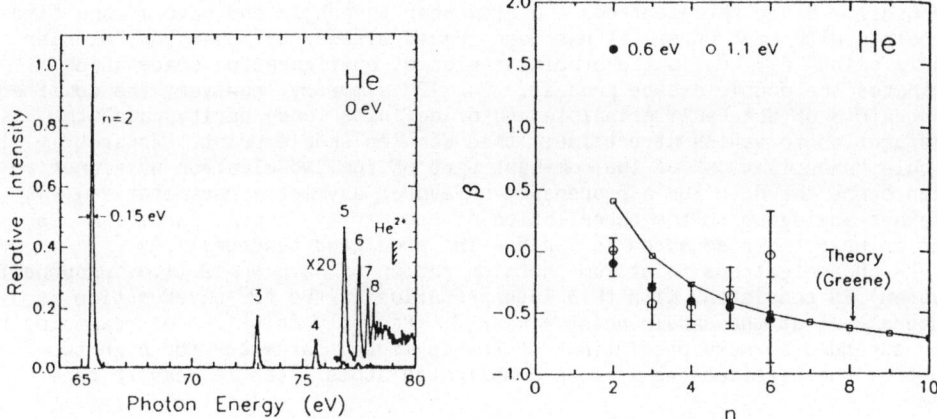

Fig. 10. Helium photoelectron spectra observed by Heimann et al.[32]
(a) Spectrum of zero-energy photoelectrons as a function of photon
energy, showing successive hydrogenic threshold peaks He$^+$(n);
(b) Observed asymmetry parameter β(n) for the lowest energy electrons at
successive thresholds are compared with the predicted values from Ref.24.

Consideration of the relevant $|\psi_\alpha|^2$ in Eq.(34) thus permits a simple
prediction of the anisotropy parameters. Figure 10 shows the values
of β(n) predicted by Ref. 24 for the slowest electrons ejected in helium
photoionization just above the n-th threshold. Also shown in Fig.10 are the
very difficult measurements made recently by Heimann et al.[32] The overall
agreement would seem to confirm the correctness of the preceding argument, and
equally, to confirm that the diabatic +/- crossing predicted by hyperspherical
coordinate analyses correctly describes a nontrivial aspect of the
photoionization process.

It might be noted that the analysis of Ref. 24 predicts that the *opposite*
limit β→2 at n→∞ would be relevant if there were no diabatic +/- crossings.
This should be the case if metastable helium atoms in the 1s2s
^{3}S state are photoionized instead of the singlet helium ground state, since
that process leads to the final state symmetry $^3P^O$ which has no +/-
crossings. This symmetry has not yet been studied experimentally.

This quantum mechanical system provides an unusual example in which the
propensity rule (28a) is violated in the extreme, despite the process being
parity-favored. The trend of negative β-values, approaching -1 in fact at n→∞,
is surprising since the application of an oscillating electric field along the
z-axis results here in ejection of an electron in a direction *orthogonal to the
applied force*. This is a very different situation from the β-values discussed
in the context of Fig.8, which become negative at resonances where the relative
parity-unfavored contribution was enhanced owing to cancellation of the parity-
favored component. Helium, on the other hand, has no parity-unfavored
component whatsoever (in LS-coupling).

A partial interpretation can nevertheless be given to this striking
violation of the propensity rule (28a). The high excitation of two electrons
simultaneously is closely related to the Wannier threshold problem in which
both electrons escape to infinity. Over a large region of configuration space,
the wavefunctions associated with these two processes must be essentially

identical. Since the electrons are slow near threshold and have a long time to correlate with each other, it has been argued elsewhere[33] that the "Wannier saddle point" $\vec{r}_1 = -\vec{r}_2$ is the crucial region of configuration space which dominates the double escape process. For $^1P^O$ symmetry, however, the combined constraints of the Pauli principle and of definite (odd) parity cause the wavefunction to vanish at precisely this Wannier saddle point. Apparently it is this "cancellation" of the dominant part of the two-electron wavefunction which opens the door for a propensity-unfavored asymmetry parameter $\beta \to -1$, somewhat analogous to the cancellation of the parity-favored partial cross section near the resonances in Fig.8. The predicted tendency[24],[34] of $^3P^O$ photoelectrons in helium photoionization to have $\beta \to 2$ (i.e. propensity-favored) is consistent with this interpretation as the $^3P^O$ wavefunction is nonzero[34],[35] at the saddle point $\vec{r}_1 = -\vec{r}_2$. Recently this line of reasoning has been extended to make predictions of the asymmetry parameter for high two-electron photoexcitations in more complicated atoms like the heavier rare gases.[37]

Finally, consider the orientation and alignment of the excited hydrogenic ions $He^+(np)$ produced by this photoionization process. The same wavefunctions have been used recently[27] to determine these anisotropy parameters which can be observed by monitoring the circular and linear polarization of the $np \to 1s$ fluorescence following the photoionization. Fig.11 shows the predictions as a function of n, for both the T=1 (i.e. $^1P^O$) symmetry and for the T=0 ($^3P^O$) symmetry. An interesting result in Fig.11 is the fact that the alignment of the levels is negative for both symmetries T=1 and T=0, in

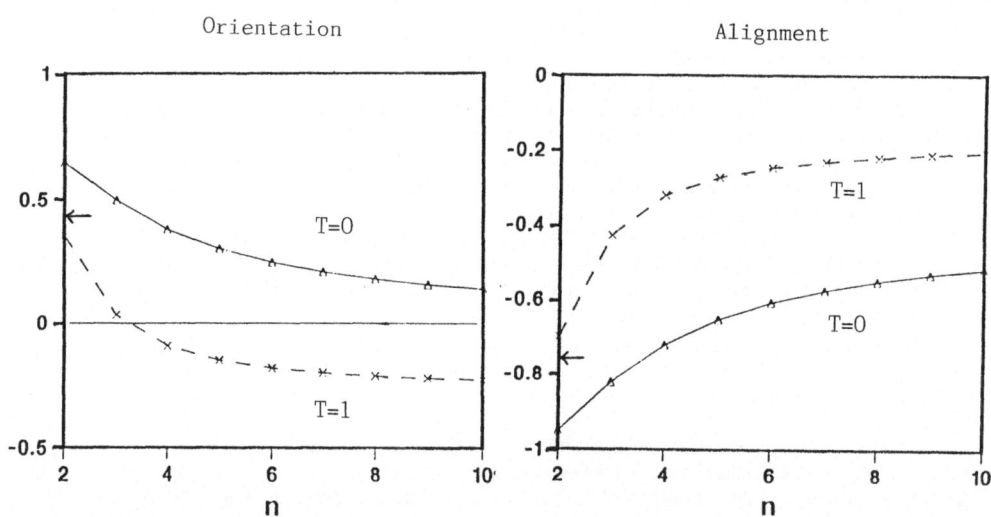

Fig. 11. Orientation (a) and alignment (b) of $He^+(np)$ levels excited in helium photoionization, as predicted by Sadeghpour and Greene,[27] ignoring fine-structure and the Lamb shift. Also shown are n=2 values deduced from the experiment of Jimenez-Mier et al.,[37] after correcting for depolarization effects. The T=1 curves are relevant for photoionization of ground state helium, while the T=0 curves describe photoionization of the 1s2s ^3S metastable state.

accordance with the expected range (28b). This shows how one anisotropy (in this case β) can lie in the propensity-unfavored range, while another ($A_0^{(2)}$) can lie in the propensity-favored range, *for the same physical process*. The predicted values for the orientation in Fig.11,on the other hand, have opposite signs at $n \to \infty$. The negative values obtained for T=1 thus violate the propensity rule (28c) whereas the T=0 values do not, somewhat analogous to the situation already discussed for the asymmetry parameter β. An observation of $He^+(2p)$ fluorescence by Jimenez-Mier et al.[37] are marked on Fig.11, and are seen to be in reasonable agreement with the T=1 theoretical values. (An earlier theoretical study by Ojha[38] based on an R-matrix calculation gives essentially exact agreement with the experimental polarization measurement.) Clearly, these results indicate that the dynamics determining photofragment anisotropies are more complicated and require more care in their interpretation than do total cross section measurements.

OUTLOOK

The use of photons to create and to observe anisotropies is becoming a well-established tool for studying the dynamics of fragmenting collisions. The widespread availability of lasers and synchrotron light sources has played a key role in this development. For theorists, photon collisions are advantageous because they involve a far smaller number of final state angular momenta owing to the electric-dipole nature of the radiation, as opposed to atom-atom or electron-atom collisions. Nevertheless our understanding of anisotropy must be viewed as extremely fragmentary at present. For many atoms and small molecules, R-matrix or other ab initio calculations are increasingly successful at determining anisotropy parameters when the number of relevant channels is less than about ten. The work of Refs.38 and 39 is especially encouraging in this regard.

Experimental progress has also been rapid,[40] as seen for instance in the large number of systems for which photoelectron angular distributions and spin polarizations have been mapped in detail. Photofragment alignment and orientation have been studied far less and are understood at a correspondingly lower level. In general, our intuitive or global understanding of systematics lags far behind current abilities to "calculate" or to observe individual systems experimentally. Improvement of this qualitative understanding should be one of the more urgent goals in this field.

ACKNOWLEDGEMENT

I am indebted to H. Sadeghpour for numerous discussions. This work was supported in part by the National Science Foundation.

REFERENCES

1.   K. Blum, <u>Density Matrix Theory and Applications</u>, (Plenum, New York, 1981), 217 pp.
2.   U. Fano, Rev. Mod. Phys. <u>29</u>, 74 (1957).
3.   U. Fano and G. Racah, <u>Irreducible Tensorial Sets</u>, (Academic, New York, 1959).

4.   U. Fano and J. H. Macek, Rev. Mod. Phys. 45, 553 (1973).
5.   I. V. Hertel and W. Stoll, Adv. At. Mol. Phys. 13, 113 (1977).
6.   A. Omont, Prog. Quant. Elect. 5, 69 (1977).
7.   K. Blum and H. Kleinpoppen, Phys. Rep. 52, 203 (1979).
8.   G. Breit, Rev. Mod. Phys. 5, 91 (1933).
9.   C. H. Greene and R. N. Zare, Ann. Rev. Phys. Chem. 33, 119 (1982).
10. D. A. Case, G. M. McClelland, and D. R. Hershbach, Mol. Phys. 35, 541 (1978).
11. C. H. Greene and R. N. Zare, J. Chem. Phys. 78, 6741 (1983).
12. M. A. O'Halloran, H. Joswig, and R. N. Zare, J. Chem. Phys. 87, 303 (1987).
13. G. E. Hall, N. Sivakumar, and P. L. Houston, J. Chem. Phys. 84, 2120 (1986).
14. D. J. Burns, R. D. Hight, and C. H. Greene, Phys. Rev. A 20, 404 (1979).
15. P. J. Brucat and R. N. Zare, J. Chem. Phys. 78, 100 (1983).
16. R. Altkorn, R. N. Zare, and C. H. Greene, Mol. Phys. 55, 1 (1985).
17. E. L. Heck and J. F. Williams, J. Phys. B 20, 2871 (1987).
18. H. Hafner, H. Kleinpoppen, and H. Kruger, Phys. Lett. 18, 270 (1965).
19. U. Fano, Phys. Rev. A 32, 617 (1985).
20. K. Bartschat, J. Phys. B 20, 5023 (1987); The experimental results in Fig.7 were obtained by W. Kronast, R. Huster, and W. Mehlhorn, Z. Phys. D 2, 285 (1987).
21. R. N. Zare et al., unpublished.
22. D. Dill and U. Fano, Phys. Rev. Lett. 29, 1203 (1972); U. Fano and Dill, Phys. Rev. A 6, 185 (1972).
23. L. Kim and C. H. Greene, Phys. Rev. A 36, (November 1987).
24. C. H. Greene, Phys. Rev. Lett. 44, 869 (1980).
25. M. Gailitis and R. Damburg, Proc. Phys. Soc. Lond. 82, 192 (1963).
26. M. J. Seaton, Proc. Phys. Soc. Lond. 77, 174 (1961).
27. H. R. Sadeghpour and C. H. Greene, to be published.
28. C. D. Lin, Phys. Rev. Lett. 35, 1150 (1975).
29. C. H. Greene, J. Phys. B 13, L39 (1980).
30. J. H. Macek, J. Phys. B 1, 831 (1968).
31. U. Fano and C. D. Lin, Atomic Physics 4, (Plenum, New York, 1975), p. 47,
32. P. A. Heimann, U. Becker, H. G. Kerkhoff, B. Langer, D. Szostak, R. Wehlitz, D. W. Lindle, T. A. Ferrett, and D. A. Shirley, Phys. Rev. A 34, 3782 (1986).
33. See, for instance, A. R. P. Rau, Phys. Rep. 110, 369 (1984), and references therein.
34. C. H. Greene and A. R. P. Rau, Phys. Rev. Lett. 48, 533 (1982); also J. Phys. B 16, 99 (1983).
35. A. D. Stauffer, Phys. Lett. 91A, 114 (1982).
36. C. H. Greene, J. Phys. B 20, L357 (1987).
37. J. Jimenez-Mier, C. D. Caldwell, and D. L. Ederer, Phys. Rev. Lett. 57, 2260 (1986).
38. P. C. Ojha, J. Phys. B 17, 1807 (1984).
39. W. R. Johnson, K. T. Cheng, K. N. Huang, and M. LeDourneuf, Phys. Rev. A 22, 989 (1980).
40. U. Heinzmann, in Fundamental Processes in Atomic Collision Physics, eds. H. Kleinpoppen, J. S. Briggs, and H. O. Lutz, (Plenum, New York, 1985), p.269.

ATOMS IN EXTERNAL FIELDS

P.F. O'Mahony

Department of Mathematics
Royal Holloway and Bedford New College
Egham, Surrey TW20 OEX, England

1. INTRODUCTION

The study of atoms in externally applied fields has progressed rapidly, experimentally and theoretically, in the last few years. (For a review up to 1984 see Clark et al. (1984); recent advances are documented in "Collisions and spectra in external fields II", Taylor et al. (1988).) We emphasise at the outset, that the imposition of an external field on an atom is not just a further complication to the already difficult zero-field problem. On the contrary the analysis of the motion of an electron in such fields serves as a simple physical prototype for problems where forces with disparate symmetries, e.g. spherical and cylindrical, combine and compete. The escape of two electrons from an atom near the double ionisation threshold is a more complicated example of such a problem (Fano (1983)). The insight gained from the study of atoms in fields should therefore be of more general significance.

The application of an external field, electric and/or magnetic, to an isolated atom or molecule in general serves to probe both the short range interactions of the constituent particles of the isolated atom or molecule and the effect of competition between the long range intrinsic fields and the externally applied field. Theoretically both of these aspects are analysed separately and then brought together in a final step using some novel techniques. The conceptual framework of quantum defect theory (Seaton (1983) and Rau's lecture), namely the partitioning of configuration space into several distinct regions where different forces dominate and the matching of solutions to the Schrödinger equation between these regions, will therefore play a key role in understanding the spectra of atoms in fields.

We will describe field effects in general but in particular we will deal with an isolated atom in (a) an electric field or (b) a magnetic field. There are important differences between the electric and magnetic cases as the Stark problem for hydrogen is separable in parabolic co-ordinates. Much of what will be said will hold for molecules also, as long as the fields are of laboratory strength.

The most obvious initial effect in applying an external field is to define a unique direction in space. This breaks the overall rotational symmetry of the atom. However the component of the total angular momentum about the axis defined by the field direction remains a conserved quantity. To investigate further the regions of the spectrum most affected by the potential due to the applied field, we write down the Schrödinger equation, in atomic units, for hydrogen in an electric or magnetic field

$$\left(-\frac{1}{2} \nabla^2 - \frac{1}{r} + V_{EX} \right) \Psi = E\Psi$$

The potential V_{EX}, in a.u. for a field applied along the z-axis is (apart from a linear Zeeman effect in the magnetic case).

$$V_{EX} = \begin{cases} F\ z & \text{ELECTRIC FIELD} \\ \frac{1}{2}\beta^2\ \rho^2 & \text{MAGNETIC FIELD} \end{cases}$$

For typical laboratory fields, the field strengths F and β are $\sim 10^{-5}$ a.u. thus at distance of the order $0 \leq r \leq 10, \frac{1}{r} \sim 10^{-1} \gg V_{EX}$. Therefore, for the ground or low-lying excited states, the fields are but a small perturbation compared to the internal Coulomb field, $-\frac{1}{r}$. However, when the radius of the electron's motion r becomes large the Coulomb potential and the potential due to the external field are comparable and have to be treated on an equal footing. Thus the external field's potential affects mainly the Rydberg states (since $r \sim n^2$ and for $r \sim 1000$, we obtain $n \approx 32$) or states excited to the continuum.

It is clear from the above that the motion of an electron in a <u>multi-electron atom</u>, which is excited to a Rydberg or continuum state in the presence of an external field of laboratory strength, can be treated in general by dividing space into different regions.

(I) a complicated multi-electron region where field effects are negligible, $0 \leq r \leq 10$

(II) Coulomb field dominates, $10 \leq r \leq 1000$

(III) Coulomb plus external field, $1000 \leq r \leq 3000$

(IV) external field dominates $3000 \leq r \leq \infty$

The radial ranges indicated of course can be varied for stronger fields. The motion of the electron in the first two regions can be analysed as in the zero-field case (Burke and Robb (1975), Greene and Rau lectures). The main task of the theory then is to describe the electron's motion in III and to understand how the solutions can be matched to those in II and IV. The Coulomb plus electric field, being separable, enables one to treat the problem semi-analytically (Harmin (1982)) and in section 2 we will describe this case in detail. Section 3 deals with the Coulomb plus magnetic field problem which in contrast is non-separable. This brings many new features not apparent in the Stark effect.

2. STARK EFFECT

2.1 Hydrogen in and Electric Field

The experimental observation of regular modulations in the photoionisation spectrum of rubidium in an external electric field stimulated renewed interest in the Stark effect (Freeman et al (1978)). These modulations were subsequently accounted for in a calculation of the photoionisation spectrum of hydrogen by Luc-Koenig and Bachelier (1980). They integrated numerically the one-dimensional equations obtained by separating the Schrödinger equation in parabolic coordinates. I will present here however the semi-analytic theory of Harmin (1982) as it is readily generalisable to non-hydrogenic atoms and also finds applications in problems other than the Stark effect (Greene (1987)). A main result of this theory is that the cross section can be factorised into a product of the <u>field</u> <u>free</u> cross section times an energy dependent term which contains the modulations.

The Schrödinger equation for the hydrogen atom in an external field F directed along the z-axis is

$$\left(-\frac{1}{2}\nabla^2 - \frac{1}{r} + F\,z\right)\Psi = E\Psi \qquad (1)$$

The potential in (1) is independent of φ so L_z is a constant of the motion. Eq. (1) is well known to be separable in parabolic coordinates (ζ,η) where $\zeta = r + z$ and $\eta = r - z$, (see for example Bethe and Salpeter (1957)). Letting

$$\Psi = (\zeta\eta)^{-\frac{1}{2}}\chi_1(\zeta)\chi_2(\eta)\,\frac{1}{\sqrt{2\pi}}\,e^{im\varphi} \qquad (2)$$

eq. (1) separates into

$$\frac{d^2\chi_1(\zeta)}{d\zeta^2} + \left[\frac{1}{2}\varepsilon - 2\left(\frac{m^2-1}{8\zeta^2} - \frac{\beta}{2\zeta} + \frac{F}{8}\zeta\right)\right]\chi_1(\zeta) = 0, \qquad (3a)$$

$$\frac{d^2\chi_2(\eta)}{d\eta^2} + \left[\frac{1}{2}\varepsilon - 2\left(\frac{m^2-1}{8\eta^2} - \frac{1-\beta}{2\eta} - \frac{F}{8}\eta\right)\right]\chi_2(\eta) = 0 \qquad (3b)$$

The m^2-1 part of eq. (3) is the centrifugal term. β is a separation parameter and as it enters as the coefficient of the Coulomb potential it can be thought of as an effective charge. Therefore a charge β in the ζ potential leaves a charge $1-\beta$ in the η potential. The linear terms due to the external field give rise to a bounded potential in ζ and a potential barrier in the η coordinate. Fig. 1 shows the 1-dimensional potentials in ζ and η for different values of β.

A solution of 3(a) and 3(b) can be labelled by m the magnetic quantum number and the parameters ε and β. i.e. $\Psi_{\varepsilon\beta m}(\zeta,\eta,\varphi)$. As we are interested primarily in Rydberg and continuum states we will look at WKB solutions to eq. (3)

$$\int_{\zeta_1}^{\zeta_2} k_\zeta d\zeta = \int_{\zeta_1}^{\zeta_2}\left(-\frac{m^2}{4\zeta^2} + \frac{\beta}{\zeta} + \frac{1}{2}\varepsilon - \frac{1}{4}F\zeta\right)^{\frac{1}{2}} d\zeta = \left(n_1 + \frac{1}{2}\right)\pi \qquad (4a)$$

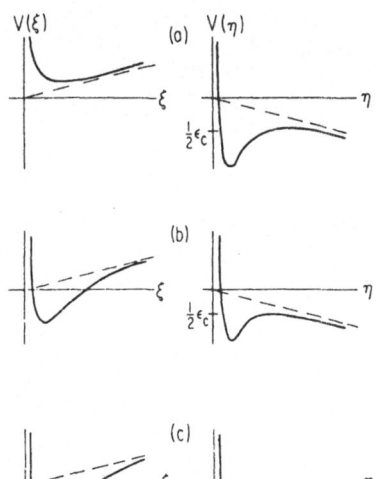

FIG. 1. Qualitative plots of the potentials $V(\xi)$ and $V(\eta)$ in Eqs. (3a) and (3b) for $m > 1$, $F \gtrsim 0$, and sample values of β : (a) $\beta \approx -0.1$, (b) $\beta \approx +0.4$, and (c) $\beta \approx 0.9$. $---$ pure-Stark potentials $+\frac{1}{4}F\xi$ and $-\frac{1}{4}F\eta$.

$$\int_{\eta_1}^{\eta_2} k_\eta \, d\eta = \int_{\eta_1}^{\eta_2} \left(-\frac{m^2}{4\eta^2} + \frac{(1-\beta)}{\eta} + \frac{1}{2}\varepsilon + \frac{1}{4}F\eta \right)^{\frac{1}{2}} d\eta = \left(n_2 + \frac{1}{2} \right)\pi \qquad (4b)$$

(A Langer correction is included in the centrifugal term.) We look initially at the familiar field-free F=0, Coulomb case. Given an energy ε we can perform the integral in eq. (4a) at a selection of values of β. However for integral n_1 the equation will only be satisfied for certain β's, $\beta(\varepsilon, n_1)$. This is shown in figure 2 where β is plotted as a function of ε for different n_1. The same is done for $1-\beta$ and n_2. A particular energy level exists when there is an integral number of nodes n_1 and n_2 in ζ and η respectively. The hydrogenic energy levels are thus obtained as $\varepsilon = -1/2n^2$ where $n = n_1 + n_2 + 1 + m$. Eigenstates can clearly be labelled by (n_1, n_2) or alternatively by (ε, β). We will use the two interchangeably.

When F is non zero a similar analysis can be performed yielding $\beta = \beta(\varepsilon, n_1)$ and WKB wavefunctions in ζ and η can thus be constructed. However to make a more general approach we note that in photoionisation from the ground or low-lying excited states of hydrogen the photon is absorbed by the electron at small radial distances. This being in region I, described in the introduction, the external field effects on photoabsorption proper are negligible. The energetic electron then moves over large radial distances where the potential is spherically symmetric and may reach regions III and IV where the appropriate coordinate system is parabolic. Therefore we need to transform the spherical Coulomb solutions in region II into solutions in parabolic coordinates so as to propagate the solution to infinity.

Since the F=0, Coulomb problem, is separable in spherical and parabolic coordinates there exists a transformation between their eigenfunctions.

$$\Psi^{F=0}_{\varepsilon\beta m}(\zeta, \eta, \varphi) = \sum_{\ell=m}^{\infty} a_{\beta\ell} \, \Psi_{\varepsilon\ell m}(r, \Theta, \Phi) \qquad (5)$$

e.g. in the n=2, m=0 manifold of hydrogen the parabolic states labelled by the quantum numbers (n_1, n_2, m) are equal to $(1,0,0)$ and $(0,1,0)$ and the transformation (5) is given by

$$\Psi_{100}(\zeta, \eta, \Phi) = \frac{1}{\sqrt{2}}\Psi_{200}(r, \Theta, \Phi) + \frac{1}{\sqrt{2}}\Psi_{210}(r, \Theta, \Phi) \qquad (6a)$$

$$\Psi_{010}(\zeta, \eta, \Phi) = \frac{1}{\sqrt{2}}\Psi_{200}(r, \Theta, \Phi) - \frac{1}{\sqrt{2}}\Psi_{210}(r, \Theta, \Phi) \qquad (6b)$$

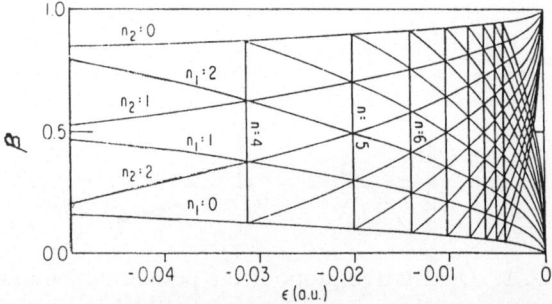

FIG.2. β as a function of ε for different n_1 and n_2. The principal quantum number for the Coulomb energy levels, n, is obtained when there are simultaneously an integral number of nodes in ζ and η. (Taken from Harmin (1981)).

$a_{\beta\ell}$ in (5) is essentially a Clebsch-Gordan coefficient when n is an integer and can be analytically continued for non-integral values (Holman and Biedenharn (1966)). To examine the transformation (5) further we consider the expansion of the wavefunctions in a volume surrounding the nucleus.

$$\Psi^{F=0}_{\varepsilon\beta m} = N^{F=0}_{\varepsilon\beta m}\left[(\zeta\eta)^{m/2}\frac{e^{im\Phi}}{\sqrt{2\pi}}\left[1 + 0(r)\right]\right] \tag{7}$$

$N^{F=0}_{\varepsilon\beta m}$ is the energy normalisation coefficient as detailed in Rau's lecture.

Similarly we have

$$\Psi_{\varepsilon\ell m} = N_{\varepsilon\ell m}\left[r^{\ell}P^m_{\ell}(\cos\Theta)\frac{e^{im\Phi}}{\sqrt{2\pi}}\left[1+0(r)\right]\right]; \tag{8}$$

and substituting (7) and (8) in (5) we get

$$\left(\frac{\Psi^{F=0}_{\varepsilon\beta m}}{N^{F=0}_{\varepsilon\beta m}}\right)N^{F=0}_{\varepsilon\beta m} = \sum_{\ell=m}^{\infty}a_{\beta\ell}N_{\varepsilon\ell m}\left(\frac{\Psi_{\varepsilon\ell m}}{N_{\varepsilon\ell m}}\right) \tag{9}$$

Thus

$$\left(\frac{\Psi^{F=0}_{\varepsilon\beta m}}{N^{F=0}_{\varepsilon\beta m}}\right) = \sum_{\ell=m}^{\infty}a'_{\beta\ell}\left(\frac{\Psi_{\varepsilon\ell m}}{N_{\varepsilon\ell m}}\right) \tag{10}$$

where $a'_{\beta\ell} = a_{\beta\ell}N_{\varepsilon\ell m}/N^{F=0}_{\varepsilon\beta m}$

So far in (10) we have only recast (5) in a form such that the behaviour of the wavefunction in parabolic coordinates on the L.H.S. of (10) is almost independent of energy near the nucleus. If we now look at the solution with a field F we can also expand it in a volume surrounding the nucleus as follows:

$$\Psi^F_{\varepsilon\beta m} = N^F_{\varepsilon\beta m}\left((\zeta\eta)^{m/2}\frac{e^{im\Phi}}{\sqrt{2\pi}}(1+0(r))\right); \tag{11}$$

as the applied field is negligible within the region $0<r\ll1/\sqrt{F}$, the coordinate dependence of the wavefunctions in (11) and in the field free case (7) are identical. The two solutions only differ in the energy normalisation factors, $N^F_{\varepsilon\beta m}$ and $N^{F=0}_{\varepsilon\beta m}$. Therefore from (7) and (11) we get

$$\frac{\Psi^F_{\varepsilon\beta m}}{N^F_{\varepsilon\beta m}} = \frac{\Psi^{F=0}_{\varepsilon\beta m}}{N^{F=0}_{\varepsilon\beta m}} \tag{12}$$

for $0 < r \ll \frac{1}{\sqrt{F}}$ and thus from (10) we get the basic equation

$$\Psi^F_{\varepsilon\beta m}(\zeta,\eta,\Phi) = \sum_i U_{\beta\ell}\Psi_{\varepsilon\ell m}(r,\Theta,\Phi) \tag{13}$$

where $U_{\beta\ell} = a'_{\beta\ell}N^F_{\varepsilon\beta m}/N_{\varepsilon\ell m}$.

This transformation between spherical and parabolic solutions is only valid when the external field is negligible, i.e. for $0<r\ll\frac{1}{\sqrt{F}}$, and is therefore a local frame transformation as opposed to the global transformation defined by eq. (5) for the Coulomb problem. The field in equation (13) acts indirectly through the energy normalisation coefficient $N^F_{\varepsilon\beta m}$. β depends on the quantisation conditions in eq. (4) and is determined by asymptotic boundary conditions together with $N^F_{\varepsilon\beta m}$ These quantities can be calculated semi-analytically in WKB (Harmin 1981).

The cross-section for photoionisation from the ground state of hydrogen in the presence of a field in atomic units can be written as

$$\sigma^F(\varepsilon) = \frac{4\pi^2}{137}\left(\varepsilon - \varepsilon_0\right)\left|<0\left|r_m\right|\Psi^F_{\varepsilon m}>\right|^2 \tag{14}$$

where $r_m = r\cos\Theta$ for $\Delta m = 0$ transitions and $r_m = r\sin\Theta e^{\pm i\Phi}/\sqrt{2}$ for $\Delta m = \pm 1$ transitions. ε_0 is the hydrogenic ground state energy. The wavefunction $\Psi^F_{\varepsilon m} = \Sigma_\beta \Psi^F_{\varepsilon\beta m}$ and on substituting for $\Psi^F_{\varepsilon\beta m}$ from (13) and using the dipole selection rules $\Delta\ell = \pm 1$ (we only access the $\ell = 1$ term) we get finally

$$\sigma^F(\varepsilon) = \frac{4\pi^2}{137}\left(\varepsilon - \varepsilon_0\right)\left|<0\left|r_m\right|\Psi_{\varepsilon 1 m}(r,\Theta,\Phi)>\right|^2\left(\sum_\beta U_{\beta 1}\right)^2 \tag{15}$$

therefore

$$\sigma^F = \sigma^{F=0}\left(\sum_\beta U_{\beta 1}\right)^2 \tag{16}$$

where $\sigma^{F=0}$ is just the 1s to εp cross section for hydrogen. The calculated cross section in a field of 77 kV/cm is shown in Fig. 3. The final state is m=0. The quasi-bound resonances in the ζ and η coordinates are labelled by (n, n_1, n_2, m). The smooth modulations about the zero field cross section are due, as indicated, to the quasi-bound resonances with $n_2 = 0$ and $n_1 = 12, 13, 14$ etc. . Theoretically this resonanct behaviour is contained in the energy variation of the normalisation factor $N^F_{\varepsilon\beta m}$ which enters in $U_{\beta 1}$.

2.2 Non-hydrogenic Stark Effect

In non-hydrogenic systems we have to describe the motion of the photoelectron in the complicated multi-electron core defined as region I in the introduction. However the effect of this region on the electron escaping into region II is embodied

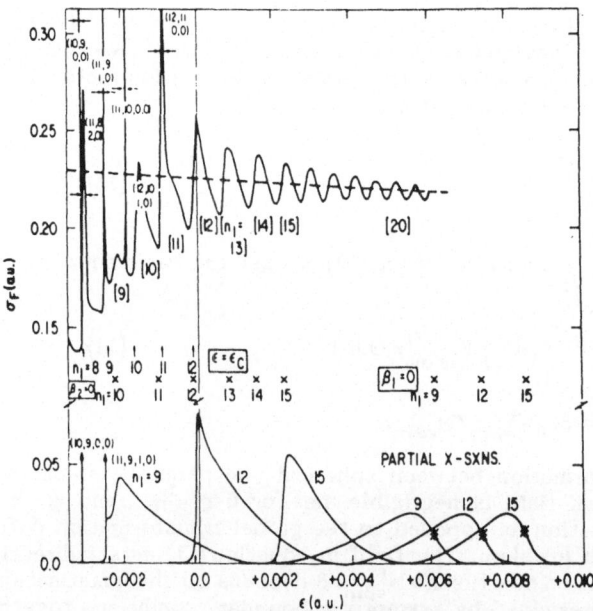

FIG. 3. The photoionisation cross section (in a.u) for hydrogen in a field of 77 kV/cm. The lower part of the diagram shows the partial cross sections. (Taken from Harmin (1982)).

in the phase shift δ_ℓ for the ℓth partial wave. Each spherical component of the wavefunction in region II can thus be written as a phase shifted Coulomb function as follows.

$$\Psi_{\varepsilon\ell m}(\vec{r}) = f_{\varepsilon\ell m}(\vec{r}) \cos\delta_\ell - g_{\varepsilon\ell m}(\vec{r}) \sin\delta_\ell \qquad (17)$$

where $f_\ell(\vec{r})$ and $g_\ell(\vec{r})$ are the regular and irregular Coulomb solutions respectively. The electric field is still negligible in this region so we use the transformations developed already for the Coulomb functions in (13) to recast eq. (17) in parabolic coordinates. We invert eq. (13) to obtain

$$f_\ell(\vec{r}) = \sum_\beta \left(U^{-1}\right)_{\ell\beta} \Psi_\beta^F(\vec{r}). \qquad (18)$$

Fano (1981) has shown how to transform between the irregular function $g_\ell(\vec{r})$ in spherical coordinates and that in parabolic $\chi_\beta^F(\vec{r})$ by comparing the Green's functions in both coordinate systems.

$$g_\ell(\vec{r}) = \sum_\beta \tilde{U}_{\ell\beta} \csc(\gamma_\beta) \chi_\beta^F(\vec{r}). \qquad (19)$$

The extra phase shift γ_β arises from the potential barrier in the η motion. Substituting (18) and (19) in (17) we obtain the solution in parabolic coordinates

$$\Psi_{\varepsilon\ell m}(\vec{r}) = \cos\delta_\ell \sum_\beta \left(U^{-1}\right)_{\ell\beta} \Psi_\beta^F - \sin\delta_\ell \sum_\beta \tilde{U}_{\ell\beta} \csc(\gamma_\beta) \chi_\beta^F. \qquad (20)$$

The solution (20) is not normalised however and thus by including the normalisation factor explicitly, denoted by $D^F = \langle \Psi' | \Psi \rangle^{-1}$, we can write the cross section as follows

$$\sigma^F = \frac{4\pi^2}{137} \left(\varepsilon - \varepsilon_0\right) \sum_{\ell\ell'} \langle 0|r_m|\Psi_{\varepsilon\ell'm}\rangle \left[\langle\Psi'|\Psi\rangle\right]_{\ell\ell'}^{-1} \langle\Psi_{\varepsilon\ell m}|r_m|0\rangle. \qquad (21)$$

The cross section can be written in the following very compact form

$$\sigma^F = \text{Trace}\left|\underline{D}^F \sigma^{F=0}\right|. \qquad (22)$$

This is the generalisation of the hydrogenic results (16). The quantities U, $\tilde{U}$ and γ_β that are involved in the calculation of D^F can all be determined in WKB approximation for hydrogen. The phase shifts δ_ℓ together with the fieldfree dipole matrix elements or cross section $\sigma^{F=0}$, which represent the non-hydrogenic character of the atom, can be calculated ab initio or taken from experiment. Harmin (1982) has used the above theory to calculate, ab initio, the photoionisation of sodium by two photons in the presence of a 3.59 kV/cm field. The experiment was performed by using a single photon at resonance with the excited Na $3^2P_{3/2}$ state and a second photon to subsequently ionise the atom (Luk et al. (1981)). Fig. 4 shows that the theoretical calculation reproduces all the resonance structure due to the quasi-bound levels indicated by (n_1, n_2) in the figure. Resonances associated with different m_ℓ are also accounted for by including the spin-orbit effects.

3. QUADRATIC ZEEMAN EFFECT

The study of the Quadratic Zeeman Effect was stimulated by a classic experiment performed by Garton and Tomkins (1969a). They found regular modulations in the photoionisation spectrum of barium in a magnetic field (see Figure 8). The spacing between the peaks in the modulations was found to be approximately 1.5 $\hbar\omega_c$, ω_c being the cyclotron frequency. This spacing was seen to be a novel

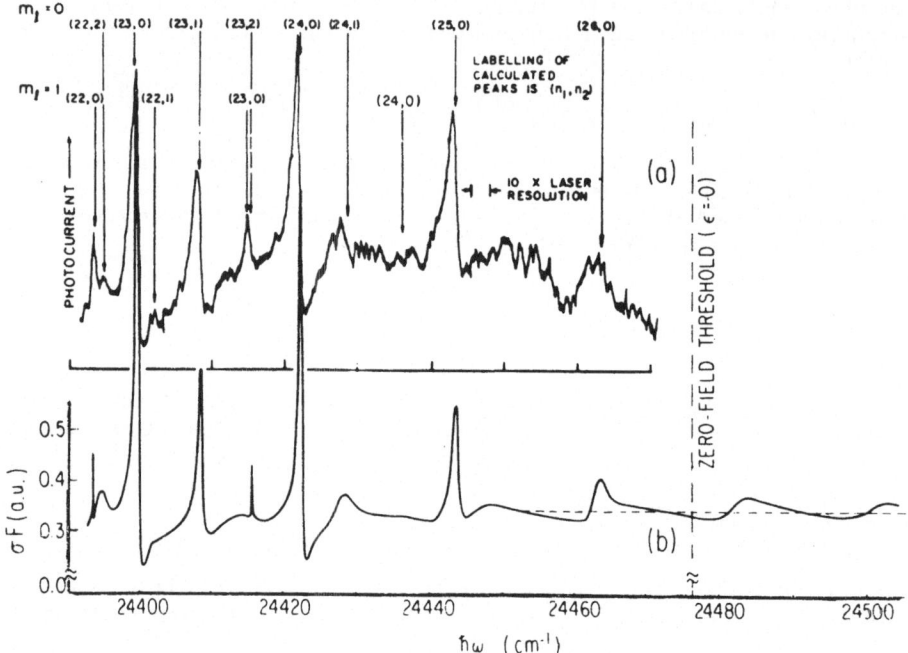

FIG. 4. (a) Experimental π photoionisation spectrum of Na $3^2P_{3/2}$ in a field F = 3.59 kV/cm, versus photon energy (b). Theoretical cross-section. ---$\sigma^{F=0}$ (Taken from Harmin (1982)).

effect due to competition between the intrinsic Coulomb field and the externally applied magnetic field. Recent higher resolution experiments on hydrogen (Main et al. (1986)) have shown the gross modulations seen in the Garton-Tomkins experiment to be full of sharp, long-lived resonances. This very rich spectrum, due to the multiplicity of possible modes of excitation for an electron moving in a two-dimensional potential, will be discussed in detail in the following sections. A detailed theoretical description of the photoionisation spectrum has yet to be found.

3.1 Discrete Spectrum for Hydrogen in the Inter-ℓ and Inter-n Mixing Regions

The hamiltonian in the center of mass frame for hydrogen in an external magnetic field B is

$$H = \frac{1}{2\mu}\left(\vec{P} + \frac{e}{c}\vec{A}\right)^2 - \frac{e^2}{r} \tag{23}$$

where μ is the reduced mass and $\vec{A}$ is the vector potential. Taking $\vec{A} = \frac{1}{2}\vec{r}\times\vec{B}$, where the magnetic field B is directed along the z-axis, we can write the Schröding equation in atomic units (Gasiorowicz (1974), p211).

$$\left[-\frac{1}{2}\nabla^2 - \frac{1}{r} + \beta L_z + \frac{1}{2}\beta^2 r^2 \sin^2\Theta\right]\Psi = E\Psi \tag{24}$$

L_z being the z-component of the orbital angular momentum. $\left(\beta = \frac{eB}{2\mu c}\right) = 2\omega_c$. (The above assumes that the centre of mass is at rest or is moving in the direction of the field otherwise a motional Stark effect is induced.) In addition to the Coulomb potential we have a term linear in β in (24), i.e. βL_z, which gives rise to the familiar linear Zeeman effect. The quadratic Zeeman effect stems from the $\frac{1}{2}\beta^2 r^2 \sin^2\Theta$ term. The potential in eq. (24) can be written in spherical or cylindrical coordinates as follows.

$$V = \frac{1}{2}\beta^2 r^2 \sin^2\Theta - \frac{1}{r} = \frac{1}{2}\beta^2\rho^2 - \frac{1}{\sqrt{\rho^2 + z^2}} = V_D + V_C \tag{25}$$

where V_C denotes the Coulomb potential. The Schrödinger equation with the above potential is not separable in any coordinate system as opposed to the Stark potential described in section I. However the potential does have the following properties

(a) V is independent of φ, so L_z is a constant of motion $L_z = m\hbar$.

(b) V is symmetric about the z = 0 plane, i.e. $V(z) = V(-z)$. Hence eigenstates of (24) have either even or odd parity, π, about the z=0 plane, denoted by π=0 or 1.

Eigenstates of (24) can thus be classified by m and π, which we denote by m^π. Field strengths accessible in the laboratory, namely B $\sim$ 5 Tesla or 50 kiloGauss, give a β in atomic units of $\sim 10^{-5}$ implying $\beta^2 \sim 10^{-10}$. Therefore, as pointed out in the introduction, $\left|\frac{1}{2}\beta^2 r^2 \sin^2\Theta\right| << \left|-\frac{1}{r}\right|$ for low lying excited states. In this, perturbative or inter-ℓ mixing region of the spectrum, we treat V_D as a perturbation. To diagonalise V_D within a constant-n hydrogenic manifold $|n\ell m\rangle$ we need to consider the matrix elements $\langle n\ell m|\frac{1}{2}\beta^2 r^2 \sin^2\Theta|n\ell'm\rangle$.

The $\sin^2\Theta$ term mixes different ℓ's namely $\ell' = \ell$, $\ell' = \ell$ -2, giving a tridiagonal matrix. The resultant perturbed energy levels are given by

$$E_{nKm} = -\frac{1}{2n^2} + E_{Km} \tag{26}$$

and the eigenvectors are denoted by $|nKm\rangle$. The oscillator strength from the ground state, $|1s\rangle$ to the $m^\pi = 1^0$ final state $|nK1\rangle$ is

$$f = 2\left(E_{nK1} - E_{1s}\right)\left|\langle 1s|r_1|nK1\rangle\right|^2 \tag{27}$$

where the dipole selection rules $\Delta\ell=\pm 1$ means that only the $\ell=1$ component of the final state is accessed as in the Stark effect. Figure 5 shows this oscillator strength as a function of binding energy for excitation to the n=23, 24 and 25 manifolds of hydrogen in a field of 4.7 Tesla. We use even K for even parity states and odd K for odd parity states.

Analytic expressions for the energy levels E_{nKm}(Herrick (1982), Soloviev (1982)) as well as numerical diagonalisation demonstrate that the energy spacings of the low K values, within an n-manifold, are characteristic of a rigid rotor where as the high K values have an oscillator type spacing. In addition the probability densities of the low K eigenvalues are localised about the z=0 plane whereas the high K eigenvalues are localised about the z-axis. This approximate separation in space of the high and low K is more pronounced when one looks at the probability distributions in momentum space (Herrick (1982) Delande and Gay (1988)). Fano et al. (1987) in a semi analytic treatment of degenerate perturbation theory have recently shown that the two types of solutions, i.e. for low K and high K can be viewed as conjugate to one another.

As one goes higher in excitation energy the different n-manifolds begin to overlap. This is referred to as the inter-n mixing region of the spectrum. It is necessary to go beyond perturbation theory to analyse the spectrum in the inter-n mixing region. A first step in this direction was taken by Clark and Taylor (1982) when

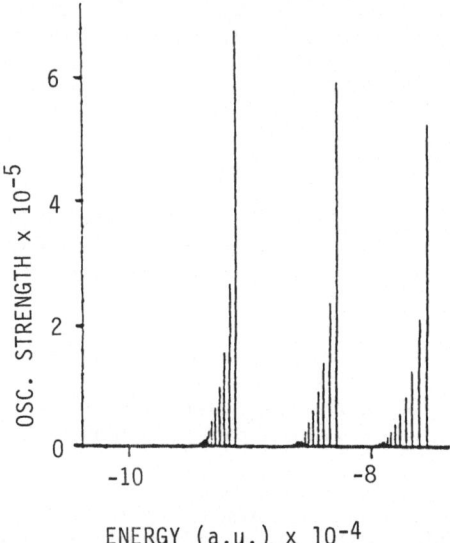

FIG. 5. The oscillator strength as a function of binding energy for the n = 23, 24, 25 manifolds of hydrogen in a field of 4.7T. Different levels are labelled by the quantum number K.

they extended the calculations into the inter-n mixing region by diagonalising the full Hamiltonian $H = H_0 + V_D$ in a basis set of Sturmian functions. They wrote the wavefunction as

$$\Psi = \sum_{n\ell} C_{n\ell} \frac{S_{n\ell}(\zeta r)}{r} Y_{\ell m}(\Theta, \varphi) \qquad (28)$$

where $Y_{\ell m}$ is a spherical harmonic. $C_{n\ell}$ are unknown coefficients to be determined by diagonalisation and $S_{n\ell}(\zeta r)$ is a Sturmian function.

$$S_{n\ell}(\zeta r) = \left[\frac{(n-\ell-1)}{2(n+\ell)}\right]^{1/2} (\zeta r)^{\ell+1} L_{n-\ell-1}^{2\ell+1}(\zeta r) \exp(-\zeta r/2) , \qquad (29)$$

and $L_{n-\ell-1}^{2\ell+1}$ is an associated Laguerre polynomial. The Sturmians have particular advantages over hydrogenic functions as a basis set in that for example they form a complete set for the discrete spectrum. ζ essentially determines the radial spread of the basis set as it enters in the exponential in (29). The parameter ζ can be varied also to check the convergence of the calculations. A further and more important aspect of the Sturmians is that the r^2 term in V_D gives non-zero matrix elements for $|n-n'| \le 3$ only. The resultant hamiltonian matrix is thus banded. A minor disadvantage is that the Sturmians are orthogonal over a weight function so that one ends up with a generalised eigenvalue problem

$$\underline{H}\vec{C} = E\underline{O}\vec{C} \qquad (30)$$

where $\underline{O}$ is the overlap matrix. The diagonalisation of hamiltonian matrices of dimension up to 1500 was possible on the Cray-1S computer. The oscillator strengths obtained using the eigenvalues and eigenvectors of these matrices are shown in fig. 6(e), the perturbative calculations shown in fig. 5 being now extended into the inter-n mixing region. It is clear from the diagram that different n manifolds almost overlap without perturbation and the assignment of a quantum number K in the inter-ℓ mixing regions remains valid in the inter-n mixing region (Zimmermann et al. (1980)). (The label K differs slightly from that used by Wintgen and Friedrich (1986) since here it refers to the exact eigenvalue, rather than as a label of the major component of the exact eigenvalue in the basis set). The individual K = 0, 2, 4, 6 are shown in fig. 6(a), (b), (c), (d) respectively. They decrease regularly even in the inter-n mixing region. The reason for the approximate conservation of the quantum number K is that the wavefunctions are localised in different regions of momentum space and in physical space and matrix elements between these states are therefore small.

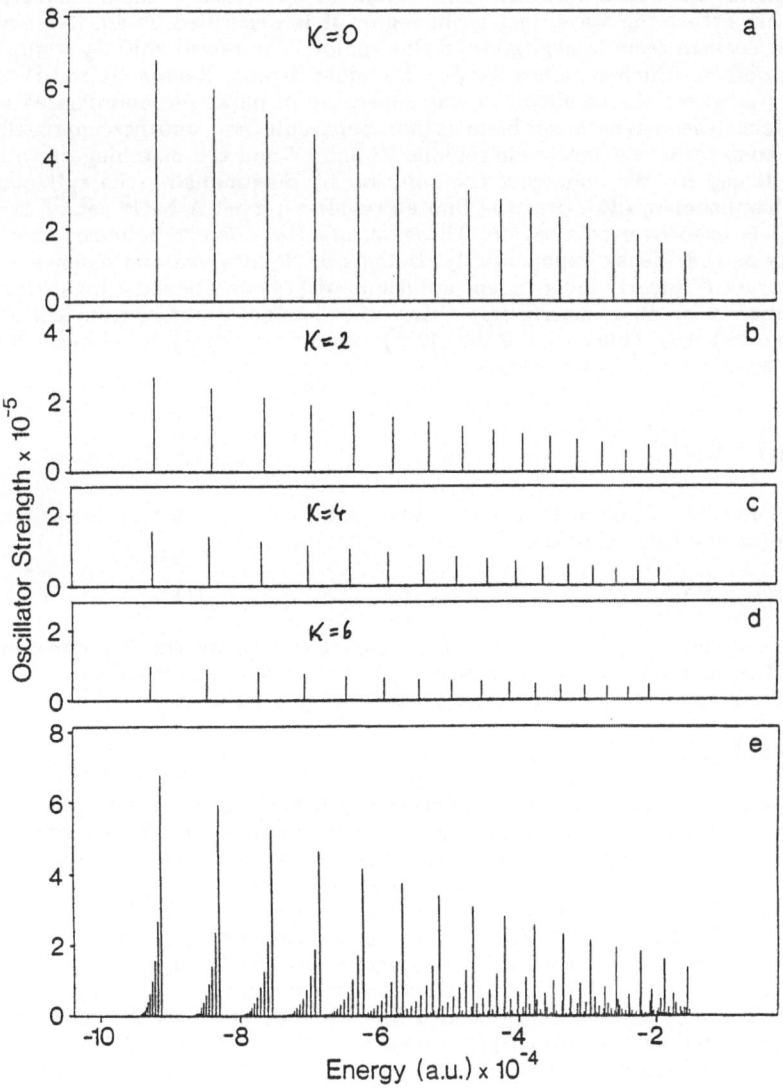

FIG. 6. (Oscillator strength as a function of energy for hydrogen in a field of 4.7T (a), (b), (c), (d) are the highest, second highest etc. energy lines plotted individually from each manifold. (Taken from Clark and Taylor (1982)).

The spacings of the low K values in Figure 6 were seen to approach 1.5 hω_c as the energy increased to threshold. This point I will come back to later when I describe the photoionisation spectrum. This type of basis set calculation has been extended to higher energies through the use of more powerful algorithms (Wintgen and Friedrich (1986), Delande and Gay (1986), Wunner et al. (1986)).

3.2 Non-hydrogenic Atoms in the Inter-ℓ and Inter-n Mixing Regimes

The basis set methods described in 3.1 are not readily applicable to non-hydrogenic atoms because of the multi-electron core potentials. Recently a method has been developed to extend the hydrogenic calculations to multi-electron atoms (O'Mahony and Taylor (1986)). The hydrogenic results are recovered when the multielectron potential vanishes.

We divide the space into different regions as described in the introduction. As in the Stark effect the wavefunction in region II is described by eq. (17) since the quadratic Zeeman term is negligible in this region. The phase shift $\delta_\ell = \pi\mu_\ell$ is the quantum defect, which vanishes for $\ell \gtrsim 4$ for most atoms. Region III and IV extend over $r_{II} \leq r \leq \infty$ where the hamiltonian was separable in parabolic coordinates for the Stark effect. The magnetic problem is non separable and another approach must be taken to describe the motion in regions III and IV and the matching of solutions between II and III. We construct the solution by diagonalising the full quadratic Zeeman hamiltonian, (24), over the limited region $r_{II} \leq r < \infty$. A basis set of Sturmian functions is used over this region. These satisfy the correct boundary conditions at infinity as they decay exponentially. Using the eigenvalues and eigenvectors we can construct ℓ linearly independent solutions of (24) and hence a logarithmic derivative matrix or an R-matrix $R_{\ell\ell'}$ (the orbital angular momentum acting as a channel index) at r_{II} (Burke and Robb (1975), Greene's lecture). $\underline{R}$ relates a function and its derivative at a given radius

$$\vec{F}(r_{II}) = \underline{R}(E) \frac{d\vec{F}}{dr}(r_{II}) \tag{31}$$

We know from region II that the wave function $\vec{F}$ must be a linear combination of the solutions (17) therefore

$$\vec{F}(r_{II}) = \underline{P}\vec{A} \tag{32}$$

where $\underline{P}$ is a diagonal matrix with elements given by the ℓ_{th} component in eq. (17). The coefficients, $\vec{A}$, are determined by matching the inner and outer logarithmic derivatives at r_{II}. Substituting (32) in (31) we obtain

$$\left(\underline{P} - R(E)\underline{P}'\right)\vec{A} = 0 \tag{33}$$

This equation has solutions, or in other words the logarithmic derivatives match, when the determinant of the term in brackets in (33) vanishes. This occurs only at certain discrete energies E_n. A search for the zeroes of the determinant gives the bound state energies E_n (Seaton (1985)). $\vec{A}$ is subsequently obtained from (33) and the nth eigenfunction is thus completely determined. The oscillator strength is again factored into a field free oscillator strength multiplied by a factor solely dependent on $\vec{A}$. This factorisation is identical to the Stark effect formula (16) except that the analog of $\vec{A}$, namely U, was determined semi-analytically. The μ_ℓ's and field free oscillator strength needed for the calculations with a magnetic field can be calculated ab initio for light atoms or taken from experiment for heavier atoms. The calculated spectra are compared with experiment for strontium and barium in figures 7 and 8. Figs. 7(a) and 8(a) are the ground state photoabsorption spectra in zero magnetic field showing a Rydberg series converging to the ionisation threshold. The experimental spectra in a field of 4.7 Tesla are shown in Figs. 7(b) and 8(b) together with the theoretical spectra in Figs. 7(c) and 8(c) using the above theory. The agreement between experiment and theory is excellent.

There is a big difference between the spectra of barium and strontium in this

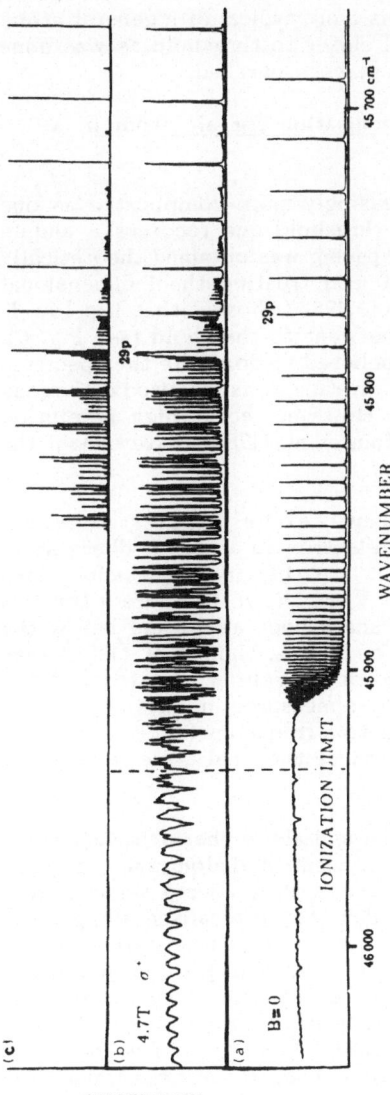

FIG. 7. Strontium absorption spectrum vs transition wave number from the ground state: (a) field-free spectrum; (b) experimental densitometer tracing for strontium in a magnetic field of 4.7 T; (c) theoretical photoabsorption spectrum in a field of 4.7 T. The theoretical results give the absolute oscillator strengths, but for n less than 29 the strongest lines have been reduced in size to facilitate comparison with the nonabsolute experimental measurements.

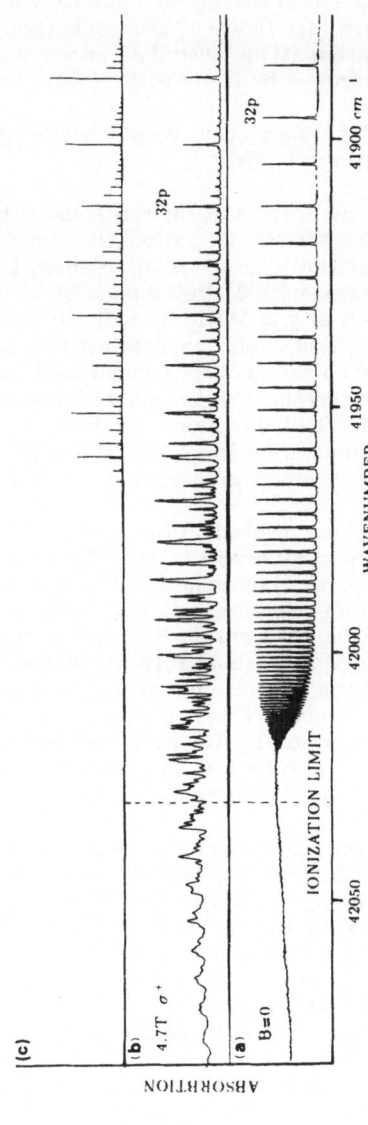

FIG. 8. Barium absorption spectrum vs transition wave number from the ground state: (a) field-free spectrum; (b) experimental densitometer tracing for barium in a magnetic field of 4.7 T; (c) theoretical photoabsorption spectrum in a field of 4.7 T.

energy range, due to the different behaviour of their significant quantum defects. In strontium the $\ell=1$ and $\ell=3$ quantum defects, μ_p and μ_f, are very different, $\mu_p \sim 0.72$ and $\mu_f \sim 0.08$ (Rubbmark and Borgstrom (1978)). The 'p' states therefore stand out from the manifold and hardly mix with the other states in the energy range below 29p. In fact the principal line in figure 7(c), below 29p, is much larger than indicated and is truncated to facilitate comparison with experiment. The significance of the K quantum number is lost in this region of the spectrum. Barium on the other hand, being strongly perturbed by the configuration 5d8p near threshold, has $\mu_p \sim 0.92$, $\mu_f \sim 1.00$ for $n \sim 32$. (Garton and Tomkins (1969b), Post et al. (1985)). Therefore the quantum defects are almost zero (modulo 1) which explains the 'hydrogenic' pattern (see figure 4) seen in barium. Strontium is more typical of a general atom. The calculations described here could be pushed closer to threshold as was done in hydrogen by Holle et al. (1987) but this has not been pursued.

3.3 Photoabsorption Near Threshold and Photoionisation for an Atom in a Magnetic Field

The spectra in figures 7 and 8 become increasingly more complicated as one approaches the ionisation threshold. Yet above threshold one recovers a simple modulation with a spacing of about 1.5 $\hbar\omega_c$. This spacing was obtained theoretically by Starace (1973) by taking the potential at z=0 and treating the 1-dimensional motion in ρ in WKB. In addition Clark and Taylor (1982) showed that the low K series, K=0, 2 etc. approached this spacing as one went to threshold (see Fig. 6). These so called quasi-Landau modulations were believed to dominate the spectrum in the threshold region and to be associated with wavefunctions localised orthogonal to the field direction i.e. about the z=0 plane. However recent high resolution experiments on hydrogen (Holle et al. (1985)) Main et al. (1986)) have shown the above interpretation to over simplify the problem.

Experimentally a first polarised photon resonantly excites the 2p state with a certain m value and a second photon excites the electron to a high Rydberg state or into the continuum. different m^π final states are therefore accessible. Two different experimental spectra for a field of 5.96 T are shown in figure 9 for two different final states. The spectra contain many sharp lines above and below the ionisation threshold with no obvious modulation of 1.5 $\hbar\omega_c$. Main et al. (1986) analysed their spectra by taking the real part squared of the Fourier transform of the cross section. This is shown in figure 10. The time is measured in units of the cyclotron period T_c. The peak marked (1) corresponds to a frequency $(=2\pi/T)$ of about 1.5 $\hbar\omega_c$. This peak is suppressed in figure 10(b) because the final state is $m^\pi=(-1)^1$ and is odd in parity about the z=0 plane.

The experiment covers an energy range below and above the ionisation threshold. Eigenstates, in this energy range, which are concentrated along the z-direction (Coulomb- Landau states) form a quasi-continuum as their energy separation is decreasing roughly as $1/n^3$. The low K, quasi-Landau states, localised along the ρ direction mix with these states giving the rise to many energy levels with varying admixtures of a particular eigenstates K. The assignment of a quantum number K to any particular energy level is no longer possible as K is no longer a conserved quantity. The resulting modulation of the absorption spectrum, however will preserve the characteristic quasi-Landau energy separation. (This type of phenomenon is familiar in for example a Rydberg series of autoionising resonances.) With the breakdown in the conservation of the quantum number K there is no conserved quantity left other than the energy and m^π. The energy level spacing distribution of the dense level structure seen in figure 9(a) will then follow a Wigner distribution (Wintgen and Friedrich (1986)), Delande and Gay (1986), Wunner et al. (1986)).

However many other frequencies are present in figure 10. Quantum mechanically these may be due to superpositions of a number of different K states. I will later describe a new quantum mechanical approach to the calculation of the superposition coefficients and the photoionisation spectrum but initially I will focus on some recent semi-classical calcualtions. The correspondence with and the applicability

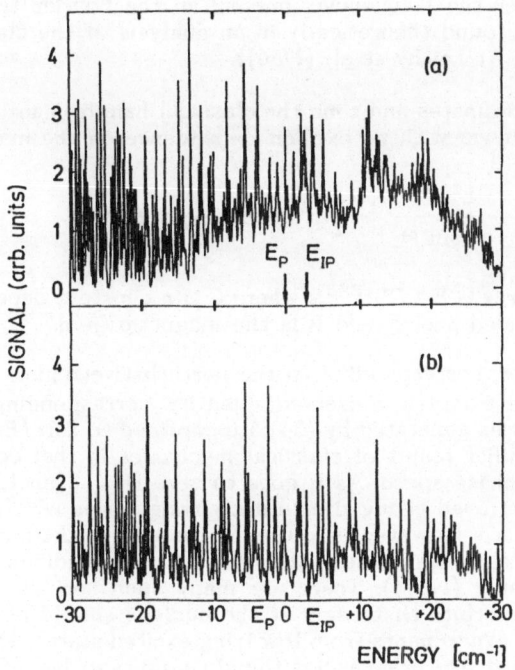

FIG. 9. Absorption signal as a function of energy for hydrogen in a field of 5.98T. The cross sections for two different final states are shown (a)$m^\pi = (0)^0$ (b)$m^\pi = (-1)^1$ (Taken from Main et al. (1986)).

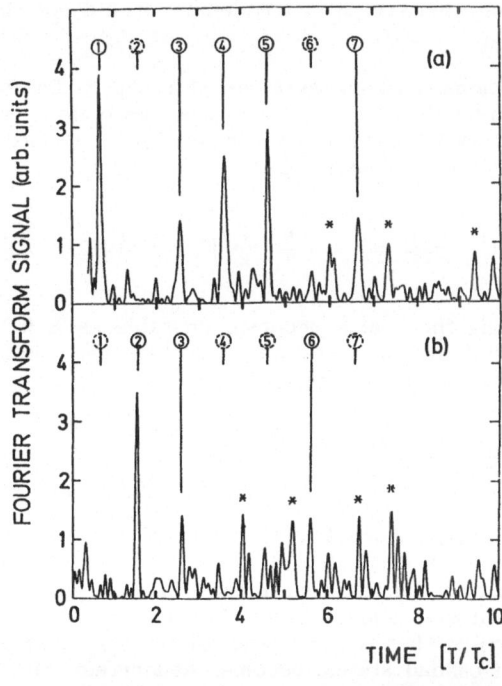

FIG. 10. Square of the real part of the Fourier transform of signals in Fig. 9 as a function of time measured in cyclotron periods (Taken from Main et al. (1986)).

of the semi-classical calculations to the quantum spectrum has yet to be fully understood, however the frequencies present in the Fourier transform spectrum (Fig. 10) have been found theoretically in an analysis of the classical hamiltonian (Main et al. (1986), Al-Laithy et al. (1986)).

Scaling the coordinates and time the classical hamiltonian can be written in a scaled form as follows, with no explicit dependence on the magnetic field

$$H = \frac{1}{2} \left[P_r^2 + \frac{P_\Theta^2}{r^2} + \frac{L_z^2}{r^2 \sin^2 \Theta} \right] + \frac{1}{2} r^2 \sin^2 \Theta - \frac{1}{2r} = E \qquad (34)$$

The scaled energy is $E = CB^{-2/3} \varepsilon$ where C is a constant depending on the units used, ε is the unscaled energy and B is the magnetic field.

At low energies, corresponding to the perturbative region or inter-ℓ mixing region where there exists a conserved quantity corresponding to the quantum number K, trajectories generated by (34) are confined to tori (Edmonds and Pullen 1980). This is a familiar result of classical mechanics in that conserved quantities define surfaces in phase space. As E goes to zero this is no longer true and the tori begin to be destroyed giving rise to many trajectories which wander over large regions of phase space. However, also the destruction of the old tori initially gives rise to new tori, localised in phase space, which have periodic orbits at their center (Al-Laithy and Farmer (1987)). There are many such tori but those trajectories which have periodic orbits that begin at the nucleus should be the most relevant to photoabsorption experiments from low lying excited states. These periodic orbits are shown in figure 11. The frequencies found (in units of $h\omega_c$) for these orbits are given by 0.64, 0.39, 0.28, 0.22, 0.18, 0.15 etc.. These are exactly the frequencies labelled 2, 3, 4, 5, 6, 7 in figure 10 and interpreted independently by Main et al. (1986). In addition we show in figure 12 the energy dependence of the frequencies of these orbits and at which energy each type of orbit begins (Al-Laithy et al. (1986)). The weak dependence on energy of the modulations in the absorption spectrum is important as otherwise such frequencies would not show up in a Fourier transform spectrum.

The energy positions at which these new tori or periodic orbits begin coincides with an instability in bundles of trajectories that move along the z-axis. Sumetskii (1982) showed such instabilities occur by linearising the equations of motion about the z-axis i.e. by expanding V as follows

$$V(\rho, z) = + \frac{1}{2} \beta^2 \rho^2 - \frac{1}{\sqrt{\rho^2 + z^2}} \approx - \frac{1}{z} + \frac{1}{2} \left(\beta^2 + \frac{1}{z^3} \right) \rho^2 \qquad (35)$$

The motion along the z-axis becomes unstable as E is increased and then oscillates between stability and instability until E=0. The energies at which the instabilities in the z-motion occur are given by

$$E_n = - \frac{1}{2} \frac{1}{2(n-\alpha)^{2/3}} \qquad (36)$$

Using some approximations Sumetskii found $\alpha = 2/3$ (Wintgen (1987) and Al-Laithy and Farmer (1987) have found $\alpha \approx 0.4$ numerically). The most important point however is that the instabilities in the z motion coincide exactly in energy with the generation of the periodic orbits shown in figures 11 and 12. These instabilities occur classically when the frequency of the motion in ρ, frequencies which are typical of quasi Landau states, becomes degenerate with frequencies of the Coulomb motion in z. Then one has coupling between the modes creating an instability in the z motion and the new periodic orbits. The 2/3 power in eq. (36) comes from the above resonance condition because the frequency of the Coulomb motion

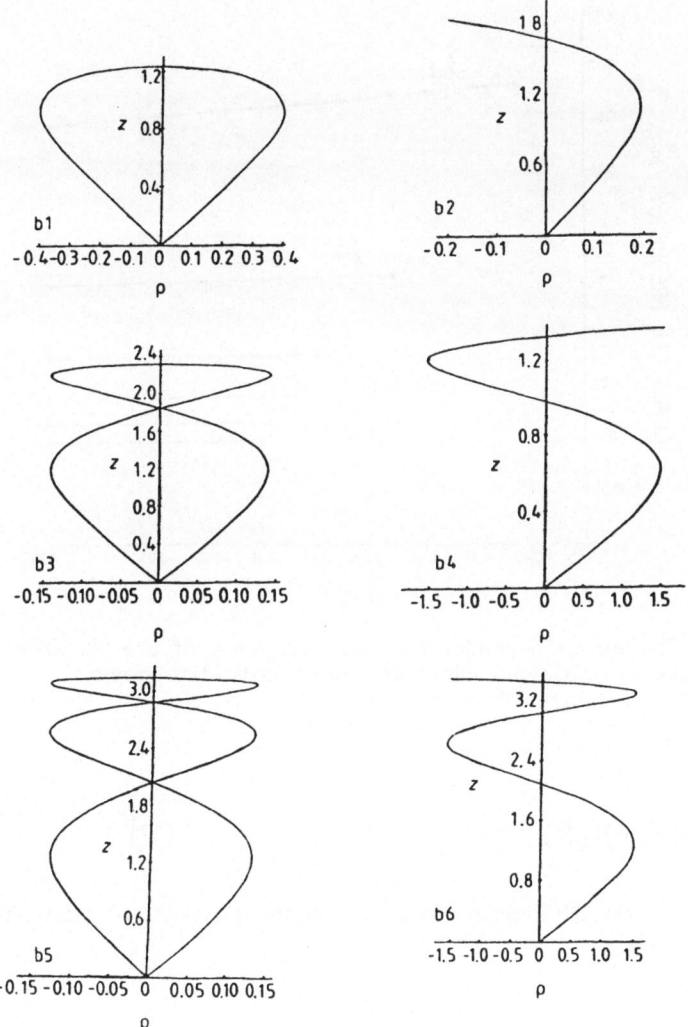

FIG. 11. Six classical periodic orbits plotted in the ρ-z plane. (Taken from Al-Laithy et al. (1986)).

in z, ω_z varies as $E^{3/2}$ whereas the frequency in ρ, ω_ρ, is constant and independent of energy. Equating multiples of the two frequencies one gets the energy dependence given in (36).

Quantum mechanically the energies at which these classical instabilities occur correspond to a breakdown in the quasi-conservation of the quantum number K. In analogy to the semi-classical description above, the breakdown in K occurs when the quasi-Landau and Coulomb-Landau states begin to mix.

To obtain a quantum mechanical solution to the problem, in order to calculate the above mentioned mixings, we extend the ideas of section 3.1. That is we use a fourth region $r_{III} \leq r \leq \infty$ where the potential can be expanded as in (35). The jth linearly independent solution in region IV is

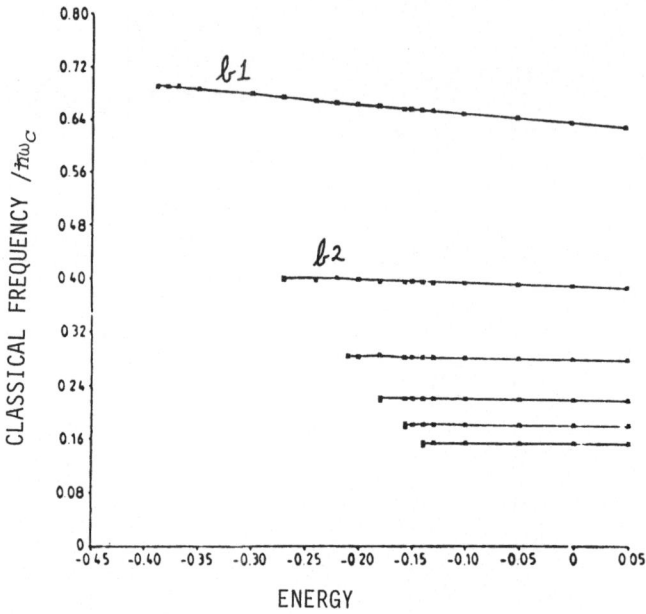

FIG. 12. The energy dependence of the frequency of the six orbits shown in Fig. 11. The energy at which each orbit originates is also shown.

$$\Psi_j = \sum_i \Phi_i(\rho,\varphi)F_{ij}(z) \qquad (37)$$

where Φ_i is the ith Landau channel, or harmonic oscillator state in ρ and

$$F_{ij}(z) \xrightarrow[z \to \infty]{} f_i - g_i K_{ij} \qquad (38)$$

where f_i and g_i are Coulomb functions in z and K_{ij} is the K-matrix. We construct ℓ linearly independent solutions in III as in section 3.2. These solutions are expressed in spherical coordinates. Projecting these solutions onto Landau channels at some z_0 and matching to eq. (37) we obtain the K-matrix. The solution is then determined over all space and the photoionisation cross section is easily calculated. Numerically the problem is complicated by a number of factors

(a) a unitary frame transformation has to be constructed between the spherical solutions in III and the solutions in cylindrical coordinates in IV,

(b) a large number of coupled channels may be required,

(c) the matching between III and IV is done at large radial distances.

However it certainly appears that the method is tractable.

In summary, the motion of an electron in a Rydberg or continuum state of an atom in a static external electric field seems now to be well understood. A lot of progress has been made, as outlined above, on the corresponding problem of an atom in an external magnetic field. However, we still lack a theoretical framework which would enable ab initio calculations of the rich and complex photoionisation spectra seen when an external magnetic field is applied to an atom. This problem should provide some new and interesting physics particularly in analysing the correspondence between the actual physical quantum solutions and the solutions obtained semi-classically.

ACKNOWLEDGEMENTS

I would like to thank M. Al-Laithy, C. Greene, K. Taylor and S. Watanabe for their conversations and collaboration in the above work This paper was written while the author held a Poste Rouge Position with E.R. 261 du C.N.R.S. at the Observatoire de Paris, Meudon.

REFERENCES

Al-Laithy, M. A. and Farmer, C., 1987, J. Phys. B, 20, L747.

Al-Laithy, M. A., O'Mahony, P.F. and Taylor K. T., 1986, J. Phys. B, 19, L773

Bethe, H. A. and Salpeter E. E., 1957, Secs. 6 and 53, in: "Quantum mechanics of one and two electron systems", Springer, Berlin.

Burke, P. G. and Robb W. D., 1975, Adv. Atom. Molec. Phys. 11 143.

Clark, C. W., Lu, K. T. and Starace, A. F., 1984, p. 247, in: "Progress in atomic spectroscopy, part C, H. J. Beyer and H. Kleinpoppen, eds. Plenum, London.

Clark, C. W. and Taylor, K. T., 1980, J. Phys. B, 13, L737.

Clark, C. W. and Taylor K. T., 1982, J. Phys. B, 15 1175.

Delande, D. and Gay, J. C., 1986, Phys. Rev. Lett., 57, 2006.

Delande, D. and Gay, J. C., 1988, in: "Collisions and spectra in external fields II", K. T. Taylor, C. W. Clark and M. Nayfeh, eds., Plenum, New York.

Edmonds, A. R. and Pullen, R. A., 1980, Preprints ICTP/79-801, Nos. 28, 29, 30 Imperial College London.

Fano, U., 1981, Phys. Rev. A, 24, 619.

Fano, U., 1983, p. 5, in: "Atomic physics 8", I. Lindgren, A. Rosen and S. Svanberg, eds., Plenum, London.

Fano, U., Robicheaux, F. and Rau, A. R. P., 1987, Submitted to Phys. Rev. A.

Freeman, R. R., Economu, N. P. Bjorklund, G. C. and Lu, K. T., 1978, Phys. Rev. Lett., 41, 1463.

Garton, W. R. S. and Tomkins,F. S., 1969a, Astrophys. J., 158, 839.

Garton, W. R. S. and Tomkins, F. S., 1969b, Astrophys. J., 158, 1219.

Garton, W. R. S., Connerade, J. P., Baig, M. A., Hormes, J. and Alexa, B., 1983, J. Phys. B, 16, 389.

Gasiorowicz, S., 1974, "Quantum Physics", Wiley, New York.

Greene, C. H., 1987, Phys. Rev. A, 36, 4236.

Harmin, D. A., 1981, Phys. Rev. A, 26, 2491.

Harmin, D. A., 1982, Phys. Rev. A, 26, 2656.

Herrick, D. R., 1982, Phys. Rev. A, 26, 323.

Holle, A., Wiebusch, G., Main, J., Rottke, H., Hager B. and Welge K. H., 1986, Phys. Rev. Letts., 56, 2594.

Holle, A., Wiebusch, G., Main, J., Welge, K. H., Zeller,G., Wunner, G, Ertl, T. and Ruder, H., 1987, Z. Phys. D, 5, 279.

Holman, W.J., III and Biedenharn, L. C., 1966, Ann. Phys., 39, 1.

Luc-Koenig, E. and Bachelier, A., 1980, J. Phys. B, 13, 1743.

Luk, T. S., Dimauro, L., Bergeman, T. and Metcalf, H., 1981, Phys. Rev. Lett. 47, 83.

Main, J., Wiebusch, G., Holle, A. and Welge, K. H., 1986, Phys. Rev. Lett. 57, 2789.

O'Mahony, P. F. and Taylor, K. T., 1986, Phys. Rev. Lett., 57, 2931.

Parkinson, W. H., Reeves, E. M., and Tomkins, F. S., 1976, J. Phys. B, 9, 157.

Post, B. H., Vassen, W., Hogervost, W., Aymar, M. and Robaux, O., 1985, J. Phys. B, 18, 187.

Rubbmark, I. R. and Borgstrom, S. A., 1978, Physica Scr., 18, 196.

Seaton, M. J., 1983, Rep. Prog. Phys., 46, 167.

Seaton, M. J., 1985, J. Phys. B, 18, 2111.

Soloviev, E. A., 1982, Sov. Phys. J.E.T.P., 55, 1017.

Starace, A. F., 1973, J. Phys. B., 6, 585.

Sumetskii, M. Y., 1982, Sov. Phys. J.E.T.P., 56 , 959.
Taylor, K. T., Clark, C. W. and Nayfeh, M., 1988:"Collisions and spectra in external fields II", Plenum, New York.
Wintgen, D. and Friedrich, H., 1986, J. Phys. B, 19 , 1261.
Wintgen, D. and Friedrich, H., 1986, Phys. Rev. Lett., 57 , 571.
Wintgen, D., 1987, J. Phys. B, 20 , L511.
Wunner, G., Woelk, U., Zech, I., Zeller, G., Ertl, T., Geyer, F., Schweizer, W. and Ruder, H., 1986, Phys. Rev. Lett., 57 , 3261.
Zimmermann, M. L., Kash, M. M. and Kleppner, D., Phys. Rev. Letts., 45 , 1980.

RYDBERG ELECTRONS IN LASER FIELDS :
A M.Q.D.T. TREATMENT WITH DRESSED CHANNELS

A. Giusti-Suzor[*]

Laboratoire de Photophysique Moléculaire, Bâtiment 213
Université Paris-Sud, 91405 Orsay   France

I.- INTRODUCTION

This lecture lies at the frontier between two domains usually disconnected : the domain of 'collisions and spectra' (from the title of the recent textbook by Fano and Rau[1]) and the domain of quantum electrodynamics (QED). In the first one the radiation field, when playing a role in the excitation process, is treated at first order of perturbation as a source term represented by a set of dipole matrix elements, at least when the wavelength is large enough to allow the dipole approximation. On the contrary, the detailed study of matter-radiation field interaction is the central problem of QED, starting from the basic model of the dressed two-level atom[2]. Extensions beyond this model have been based for a long time on state-by-state perturbative treatments, where the highest order of perturbation corresponds to the maximum number of photons involved in the process.

The need for a link between these two worlds stems from the recent vigorous development of powerful and tunable lasers, leading to a rapid increase of multiphoton experiments focused on various processes (see figure 1 and the lectures by P. Agostini). During the last decade several non-perturbative treatments have been proposed, generalizing the dressed atom model to include the complete set of bound and continuum states of atoms[3-8] and molecules[9-10]. These treatments rest either on a discretization of the continuum by the complex dilatation method ($r \rightarrow r\, e^{i\theta}$) with a L^2 representation of the continuum wave functions[3-7], or

---

[*]This lecture is based on work done in collaboration with P. Zoller (see Reference 11).

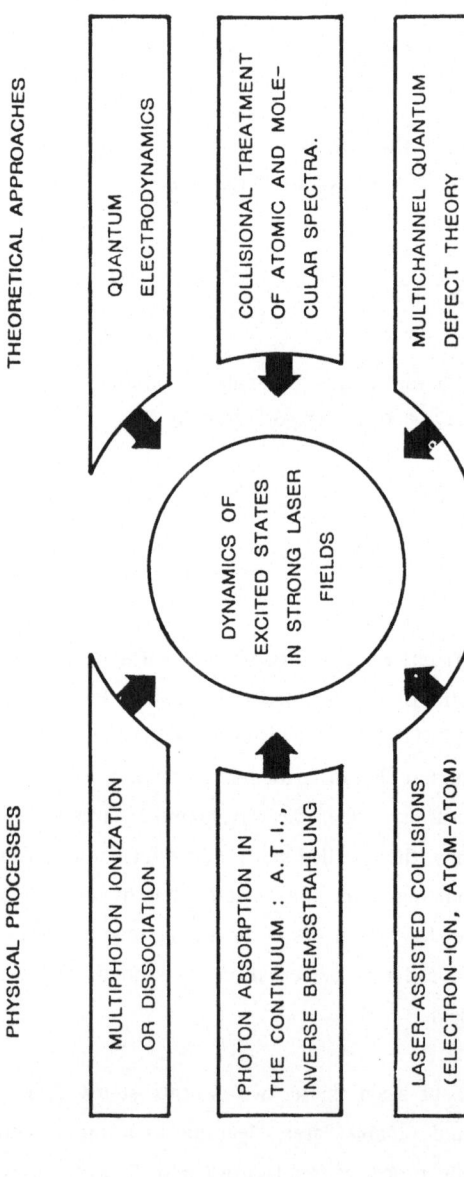

Figure 1 : Atomic and Molecular Processes
in Strong Laser Fields.

on close-coupling collisionnal calculations[8-10]. They become rapidly heavy when the number of photons involved is large (strong intensities) and in the energy regions of the spectra where many closely lying states, below or above a fragmentation limit, lead to complex resonance structures. This is typically the case of the Rydberg regions of atomic and molecular spectra.

We have thus been led to seek a more economical method to treat Rydberg and continuum electrons in laser fields, based on the observation that the main effect of laser radiation may be described as a finite volume interaction coupling Coulomb type fragmentation channels. The localization of radiative interaction in a finite volume opens the route for a theoretical approach which treats separately the short and long range processes and connects them in a further step. In particular, laser induced couplings may be incorporated in a multichannel quantum defect treatment (MQDT), leading us to define a set of 'dressed' channels associated with an intensity dependent reaction matrix, obtained by solving a system of close-coupled equations inside the reaction volume and matching it to uncoupled wavefunctions at the boundary. This reaction matrix varies slowly with the energy and may be calculated on a very coarse energy grid. The short range parameters (phase shifts, quantum defects and channel mixing coefficients) thus obtained are then incorporated in long range linear combinations of dressed Coulomb channels with appropriate asymptotic conditions, leading to ionization or collision cross sections.

Our aim in this lecture is mainly to sort out the physical ground of the limited range character of the radiative coupling and to indicate the main steps of the corresponding MQDT treatment. Technical details can be found in a recent paper[11] which is meant as the first step towards a unified treatment of the radiative and internal couplings in complex atoms and molecules.

II.- RANGE OF THE INTERACTION BETWEEN THE RYDBERG ELECTRON AND THE RADIATIVE FIELD

For weak radiation fields and sufficiently long wavelengths the radiative interaction is represented by dipole matrix elements which can be expressed in the length, velocity or acceleration forms. In particular the well known relation between the length form

$$D_L = \int_0^\infty dr \; f_l(\epsilon,r) \; r \; f_{l'}(\epsilon',r) \qquad (1)$$

and the acceleration form

$$D_A = \int_O^\infty dr \, f_l(\varepsilon,r) \, r^{-2} \, f_{l'}(\varepsilon',r) \qquad (2)$$

of the dipole matrix element between two continuum Coulomb radial wave functions f_l and $f_{l'}$ corresponding to positive energies ε and ε' (free-free transition) reads[12]

$$D_A = \frac{(\varepsilon' - \varepsilon)^2}{16} D_L \qquad (in \; a.u) \qquad\qquad (3)$$

where $\varepsilon' - \varepsilon = \hbar\omega$ is the transition energy.

It may appear surprising that the same interaction is measured by the matrix elements of two radial operators, r and r^{-2}, which lay the weight of the interaction at long and short ranges respectively. This apparent paradox disappears when one realizes that the two continuum wave functions correspond to different wavenumbers $k = \sqrt{2\varepsilon}$ and $k' = \sqrt{2\varepsilon'}$ (in a.u.) : asymptotically the integrands oscillate as $e^{i(k-k')r}$ and the long range contribution to both integrals is cancelled by destructive interferences[13]. Nevertheless, the two integrands (or accumulated integrals) behave very differently : the acceleration form leads to a rapid convergence as demonstrated on Figure 2, while the analogous figure for the length form would present at large distance tremendous oscillations with increasing amplitude and vanishing average contribution. Actually the integral in eq. 1 is only conditionally convergent and is usually computed by converting it to the acceleration form beyond a finite distance r_0, with a surface term calculated at $r = r_0$ [14-15].

These pedagogical remarks on dipole matrix elements lead to physical and methodological considerations on which our nonperturbative treatment will be based :

i) the effective limited range of the radiative interaction is physically grounded in the fact that at long range the Rydberg or continuum electron behaves almost like a free electron moving in a classical oscillating electric field, without actually absorbing or emitting photons.

ii) the relations between the alternative forms of the dipole matrix elements are usually obtained from commutator relations integrated on the whole space. They can also be derived by gauge transformations of the Schrödinger equation of an electron subject both to a Coulomb and a radiative field, that is by time-dependent unitary transformations which change the form of the radiative interaction operator without affecting its physical effects (see section 2 of ref. 11, summarized in Table 1). The length form of the dipole matrix element corresponds to the

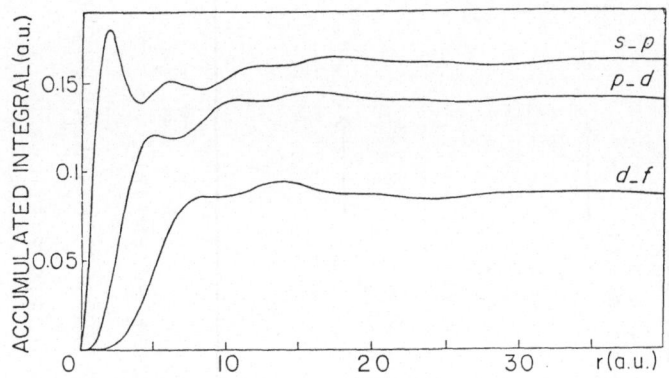

Figure 2 : Convergence of the radial integral in Eq.(2) for a $(\varepsilon = 0, \ell) \rightarrow (\varepsilon' = \hbar\omega, \ell + 1)$ dipolar transition, as a function of the upper limit of integration r and for $\hbar\omega = 0.3$ Ry.

interaction operator H_L and the acceleration one corresponds to the matrix element of H_A, which has a very clear physical meaning : the unitary transformation $e^{-(i/\hbar)p.\breve{\alpha}}$ represents a space translation into an accelerated frame[16] which follows the asymptotic oscillations $\breve{\alpha}(t)$ of the electron. In the oscillating frame the interaction operator $H_A{}^{int}$ (Table 1) vanishes at infinity $(r \gg |\alpha|)$ except for a translation energy term which gives only rise to a phase factor in the electron wave function. The r^{-2} form of the dipole operator is just proportional to the first term in the multipole expansion of the effective interaction potential

TABLE 1 : ALTERNATIVE REPRESENTATIONS OF THE RADIATIVE INTERACTION OPERATOR

ELECTRON + FIELD HAMILTONIAN	UNITARY TRANSFORMATION	INTERACTION HAMILTONIAN

$$H_V = \frac{1}{2m}(\vec{p} - e\vec{A})^2 + V(\vec{r})$$

$$= \underbrace{\frac{p^2}{2m} + V(\vec{r})}_{H_o} + H_V^{int}$$

VELOCITY FORM

$$e^{-\frac{i}{\hbar}\, e\vec{r}\cdot\vec{A}(0,t)}$$

$$e^{-\frac{i}{\hbar}\,\vec{p}\cdot\vec{\alpha}(t)}$$

$$H_L^{int} = -\, e\vec{r}\cdot\vec{\mathcal{E}}(0,t)$$

LENGTH FORM

$$H_A^{int} = V(\vec{r}+\vec{\alpha}) - V(\vec{r}) + \frac{1}{2}\,m\,\dot{\vec{\alpha}}^2(t)$$

ACCELERATION FORM

NOTATIONS :

$\vec{V}(\vec{r})$ e - CORE POTENTIAL

$\vec{A}(\vec{r},t)$ VECTOR POTENTIAL (COULOMB GAUGE)

$\vec{\mathcal{E}}(\vec{r},t)$ ELECTRIC FIELD

$\vec{\alpha}(t)$ ASYMPTOTIC OSCILLATIONS OF THE ELECTRON

$$m\,\ddot{\vec{\alpha}}(t) = e\,\vec{\mathcal{E}}(0,t)$$

$$V(\vec{r} + \vec{\alpha}) - V(\vec{r}) \approx \frac{e^2}{4 \pi \varepsilon_o} \frac{\vec{\alpha} \cdot \vec{r}}{r^3} \quad \alpha \quad \frac{\alpha_o}{r^2} \qquad (4)$$

The length α_0 is the amplitude of the electron asymptotic oscillations, related to the electric field modulus E_0 by

$$\alpha_0 = \sqrt{2} \ e \ E_0 \ / \ m\omega^2 \qquad (5a)$$

or in atomic unit

$$\bar{\alpha}_0 = \alpha_0 / a_0 = \sqrt{2} \ (2 \ Ry \ / \hbar\omega)^2 \ (I/I_0)^{1/2} \qquad (5b)$$

with $I = 2c \ \varepsilon_0 \ | \ E \ |^2$ the light intensity and $I_0 = 1.4 \times 10^{17}$ W/cm^2 the atomic unit of light intensity.

The physical picture which emerges from this discussion is that the electronic motion is governed by different forces in different parts of the space, an idea which guides most of the theoretical approaches described in this Symposium. Close to the core the atomic forces dominates ; in the asymptotic domain the electron vibrates rapidly in the time-dependent laser field, with a slow mean motion in the weak asymptotic Coulomb field. Transitions from Rydberg to different bound or free orbits by absorption or induced emission of laser photons occur in the transition zone between these two regions dominated by different forces.

III.- THE RADIATIVE REACTION MATRIX

The space-translated form of the Schrödinger equation allows us to define a set of reaction channels which are uncoupled at infinity, much as in laser-free collision problems. Going beyond the perturbation theory we can thus project the Schrödinger equation into this channel basis, obtaining a system of close-coupled equations which can be solved by standard numerical codes. Here we outline the main steps of this derivation in the simplest system, namely a Rydberg electron of a hydrogen atom subject to circularly polarized light :

i) We perform a Fourier expansion of the electron wave function Ψ_A represented in the accelerated frame

$$\Psi_A(\vec{r},t) = \sum_{N=-\infty}^{+\infty} \phi_N(\vec{r}) \, e^{-i\,N\omega t - i\,Et/\hbar} \qquad (6)$$

ii) We expand each Fourier component (corresponding to a given Floquet block in the language of QED) as a superposition of spherical harmonics :

$$\phi_N(\vec{r}) = \sum_{lm} F_{lm}^{(N)}[(r)/r] \, Y_{lm}(\theta,\varphi) \qquad (7)$$

where $F_{lm}^{(N)}(r)$ is a radial wave function. Each channel is thus identified by the parameters (N,l,m) with N the Floquet index (photon number) and (l,m) the angular momentum quantum numbers of the electron.

iii) Analysis of the Fourier expansion of the Schrödinger equation $i\hbar\, d\Psi_A/dt = H_A \,\Psi_A$ yields the time-independent coupled equations :

$$\left\{ E + N\,\hbar\omega - \frac{1}{2}\,m\,\omega^2\,\alpha_0^2 + \frac{\hbar^2}{2m}\left[\frac{d^2}{dr^2} - \frac{l(l+1)}{r^2} - V_{lm,lm}^{(0)}(\alpha_0,r) \right] \right\} F_{lm}^{(N)}(r)$$

$$= \sum_{N'l'm' \neq Nlm} V_{lm,l'm'}^{(N-N')}(\alpha_0,r)\, F_{l'm'}^{(N')}(r) \qquad (8)$$

The oscillation energy term $(1/2)\,m\omega^2\alpha_0^2$ affects equally each channel and may be considered as an overall shift of the thresholds, corresponding to a change of energy scale.

The potential terms $V_{lm,l'm'}{}^{(N-N')}$ are the Fourier coefficients of matrix elements of the potential $V(\vec{r} + \vec{\alpha}(t))$ between spherical harmonics functions. For circularly polarized light, the Fourier expansion of the translated Coulomb potential $|\vec{r} + \vec{a}(t)|^{-1}$ coincides with its multipole expansion and its successive components have the form :

$$
V_{lm,l'm'}^{(N-N')}(\alpha_o,r) = \frac{e^2}{4\pi\,\epsilon_o} \sum_{k=|N-N'|}^{\infty} <lm \,|C_{k,N-N'}|\, l'm'>(-1)^{N-N'}
$$

$$
\times \quad C_{k,-(N-N')}(\pi/2,\,0) \quad \frac{r_<^k}{r_>^{k+1}} \tag{9}
$$

where C_{kq} are spherical harmonics functions and $r_>$ ($r_<$) the largest (smallest) of α_o and r. In the optical range and for intensities well below the atomic unit, α_o is small (see eq. 5b) and the multipole terms behave as $\alpha_o{}^k/r^{k+1}$ already at short distances.

Therefore the parameter α_o, which depends on both the laser frequency and intensity (Eqs 5), determines the strength of the channel coupling and the number of equations to be retained in the close-coupled system. Except for the diagonal terms (k=0, l=l') for which the first contribution is just the Coulomb interaction, the terms with longest range (in r^{-2}) couple only channels differing by one photon (k=1), associated with wave functions whose asymptotic oscillations interfere destructively : the effective coupling is thus limited to a finite volume, as anticipated in Section 2.

Outside the interaction zone, the solutions of the close-coupling system may be written as linear combinations of Coulomb wave functions, as in usual MQDT treatments

$$
F_{ij}(r) = f_i\,\delta_{ij} \pm g_i\,K_{ij}{}^{(s)} \tag{10}
$$

where $j = (N_j,\, l_j,\, m_j)$ denotes the index of the solution and i the channel components. $f_i = f(\epsilon_i, l_i, r)$ and $g_i = g(\epsilon_i, l_i, r)$ are regular and irregular[17] energy normalized Coulomb wave functions for the energy $\epsilon_i = E + N_i\,\hbar\omega$, i.e. the electron energy referred to the threshold of channel i, dressed with N_i photons -see Fig. 3. The coefficients $K_{ij}{}^{(s)}$ are the elements of a symmetric matrix which is for the present problem the 'short-range' reaction matrix presented by Fano, in section 5 of the Introduction to this Theory Symposium. They are determined by matching the solutions of the close-coupled system to the analytical expressions (10) at a distance r_c beyond the range of the radiative interaction. An example of convergence with respect to the size of the close-coupled system is given in Table 2, for the photon energy $\epsilon = 0.35$ Ry. Convergence with respect to r_c is very good at 30 a.u. in this case.

Table 2 : Convergence of the reaction matrix element $K_{12}^{(s)}$ with the size of the radiative close coupling system ; (-2) means 10^{-2}. ω = 0.35 Ry.

$\overline{\alpha}_o$	$I\,[W/cm^2]$	perturb. (a)	Number of channels included: (b)						
			4 (l=2)	6 (l=3)	9 (l=4)	12 (l=5)	16 (l=6)	20 (l=7)	25 (l=8)
0.1	6.6(11)	·4.37(·2)	·4.37(·2)	·4.37(·2)	·4.37(·2)				
0.5	1.6(13)	·2.18(·1)	·2.24(·1)	·2.26(·1)	·2.27(·1)	·2.27(·1)	·2.27(·1)		
1.0	6.6(13)	·4.37(·1)	·4.77(·1)	·4.95(·1)	·5.00(·1)	·5.01(·1)	·5.02(·1)	·5.02(·1)	·5.03(·1)
1.5	1.5(14)	·6.55(·1)	·8.40(·1)	·8.85(·1)	·9.05(·1)	·9.13(·1)	·9.16(·1)	·9.18(·1)	·9.20(·1)
2.0	2.6(14)	·0.87	·1.80	·1.82	·1.84	·1.85	·1.86	·1.86	·1.86

(a) Golden rule calculation : $K_{12}^{(s)}$ proportional to α_0.
(b) The l-value in parentheses denotes the largest partial wave included in the calculation

The most important property of the radiative reaction matrix is its slow variation with the asymptotic energy of the electron, a result of being limited to short range where the electron is strongly accelerated. $K^{(s)}$ can thus be calculated on a coarse energy grid. In particular the calculations may be performed slightly above the highest threshold with all channels open (region I of Fig. 3), and then extrapolated below this threshold to study the effect of the laser field on the corresponding Rydberg series. Examples of such results are presented in the next section.

IV - FIRST RESULTS IN HYDROGEN

Once the smooth reaction matrix $K^{(s)}$ is obtained by solving the close-coupled system in a finite range and matching its solutions to Eq.(10), the physical observables (ac Stark shift and broadening, total and partial photoionization cross-sections) are easily calculated by analytical MQDT procedures in the external zone, where the Coulomb channels are decoupled. At this point only we have to specify the precise energy range we are interested in (region I or II on Figure 3) and to distinguish between open channels ($\varepsilon_i > 0$ in Eq. 10) and closed channels ($\varepsilon_i < 0$). In the open channels, outgoing waves will describe the ionization process while in closed channels the diverging part of the wave function must be eliminated, leading to a set of quasi-discrete levels or resonances.

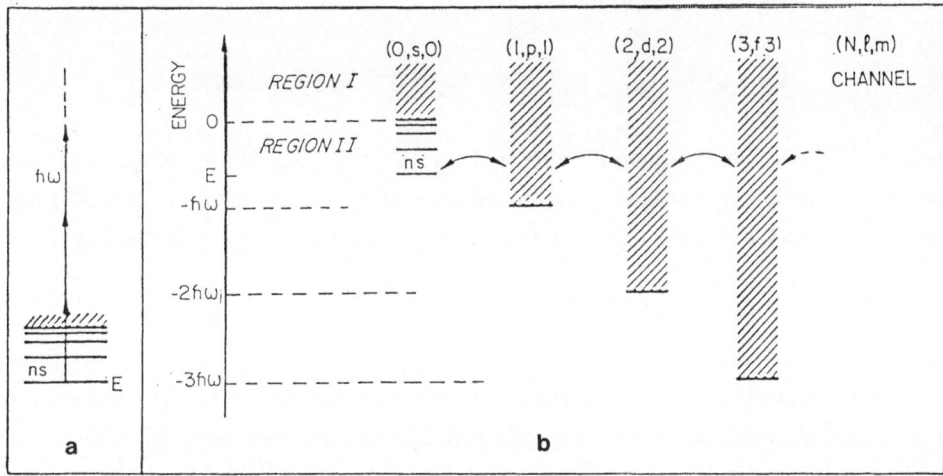

Figure 3 : Schematic representation of Rydberg state multiphoton ionization in H - (a) usual representation - (b) collisional representation. The arrows indicate the radiative transitions between dressed channels. For clarity stimulated emission is not indicated and only channels with $N = \ell$ are shown.

Let us first define the 'short-range' scattering matrix $S^{(s)}$ associated to the reaction matrix $K^{(s)}$ by the usual relation

$$S^{(s)} = [\, 1 + iK^{(s)} \,] \, [\, 1 - iK^{(s)} \,]^{-1}$$

This smooth scattering matrix, whose dimension equals the total number of channels, consists of 4 blocks

$$S^{(s)} = \begin{bmatrix} S_{oo}^{(s)} & S_{oc}^{(s)} \\ S_{co}^{(s)} & S_{cc}^{(s)} \end{bmatrix} \qquad (11)$$

with o and c referring to the subsets of open and closed channels, respectively.

The ordinary scattering matrix S, pertaining to the open channels only, is then obtained by eliminating the diverging parts of the closed channels from the total asymptotic wave function. This leads to the expression[18]

$$S = S_{oo}^{(s)} - S_{oc}^{(s)} [e^{-2i\pi\nu_c} - S_{cc}^{(s)}]^{-1} S_{co}^{(s)} \qquad (12)$$

where $e^{-2i\pi\nu_c}$ is a $N_c \times N_c$ diagonal matrix with $\nu_{ij} = \delta_{ij} (Ry/-\epsilon_i)^{1/2}$ the effective quantum number referred to the threshold of the closed channel i. In contrast to $K^{(s)}$ and $S^{(s)}$, this S matrix varies rapidly with the energy in the vicinity of the complex poles defined by

$$\det | S_{cc}^{(s)} - e^{-2i\pi\nu_c} | = 0 \qquad (13)$$

In the simplest case where all closed channels have the same threshold energy E_c, the roots of Eq.(13) correspond to the eigenvalues of $S_{cc}^{(s)}$, a complex submatrix which in the general case is not unitary. Its eigenvalues may thus be written as $e^{2i\pi\mu_\rho}$, where $\mu_\rho = \alpha_\rho + i\beta_\rho$ ($\beta_\rho > 0$) are _complex_ quantum defects ($\rho = 1, 2 ... N_c$). From the solutions $\nu_c = n - \mu_\rho$ of Eq.(13) one obtains the complex energies of the successive resonances in each series :

$$E_{n,\rho} = E_c - \frac{Ry}{(n - \mu_\rho)^2} = E_c - \frac{Ry}{n^2} + \Delta_{n,\rho} - \frac{1}{2} i \Gamma_{n,\rho} \qquad (14)$$

where $\Gamma_{n,\rho}$ and $\Delta_{n,\rho}$ denote the ac Stark broadening and shift (additional to the global energy shift $1/2 \, m\omega^2 \alpha_0^2$ due to the electron oscillations (see Eq.(8)).

Expanding the second term in Eq.(14) one gets the approximate expressions

$$\Delta_{n,\rho} \simeq -2 \, R_y \, \frac{\alpha_\rho}{n^3} \qquad (15a)$$

$$\Gamma_{n,\rho} \simeq 2 \, R_y \, \frac{\beta_\rho}{(n-\alpha_\rho)^3} \qquad (15b)$$

At moderate intensities the submatrix $S_{cc}^{(s)}$ is nearly diagonal and its resonances correspond approximately to s, d .... series, whereas strong radiative couplings cause l-mixing at higher intensities.

From the width (15b) one gets the <u>total</u> ionization rate $\Gamma_{n,\rho}/\hbar$ for each state of the Rydberg series. Our collisional approach also yields directly the <u>partial</u> ionization rates in terms of matrix elements of the short-range scattering matrix $S^{(s)}$: The rate $\gamma^{(k)}$ for ionization with k photons (k-1 being absorbed above the ionization threshold) is simply given by[11]

$$\gamma_{n,\rho}^{(k)} = (\Gamma_{n\rho}/\hbar) \sum_{i_k} S_{i_k \rho}^{(s)}$$

where the sum bears on all the open channels dressed with k photons.

These total and partial ionization rates are represented on Figure 4 for the 6s Rydberg state of H, and the field frequency $\omega = 0.35$ Ry. With increasing intensity the higher order processes become more noticeable, whereas the one-photon photoionization tends to saturate.

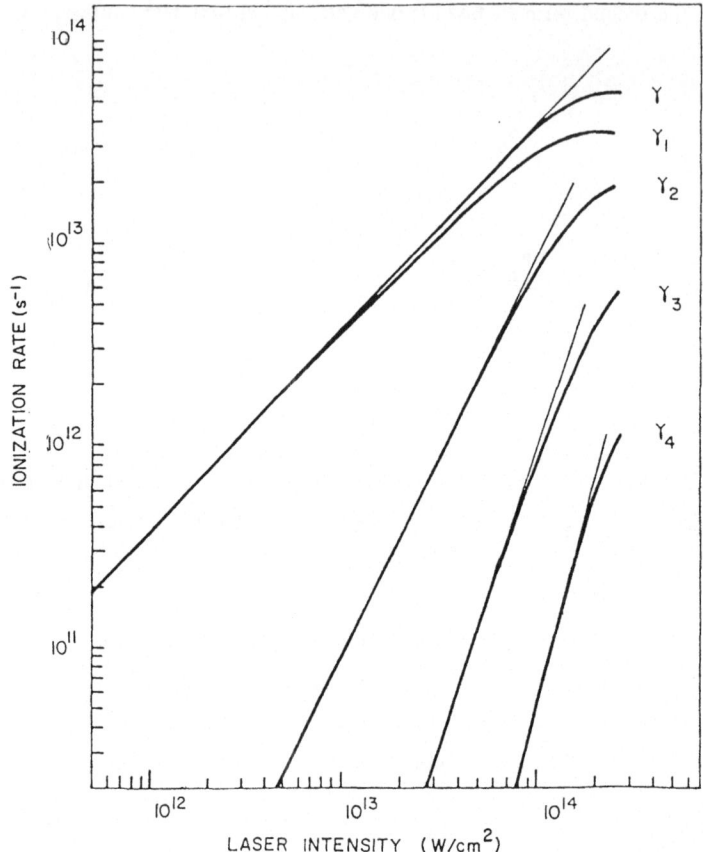

Figure 4 : Ionization rate of hydrogen 6s state by circularly polarized light (ω = 0.35 Ry). γ denotes the total ionization rate and γ_k the partial rate for k-photon ionization.

Another representation of the partial ionization cross sections is given by the photo-electron spectrum (Figure 5), which shows a succcession of peaks equally spaced by the photon energy $\hbar\omega$. This representation shows still more clearly the relative decrease of the first peak (one photon ionization) in favor of the following ones the intensity increases. It is interesting to note that the measurement of such photoelectron spectra in multiphoton experiments on Xe has been the first demonstration of the so-called "above threshold ionization" (ATI) process in atoms[19] and more recently in molecules[20].

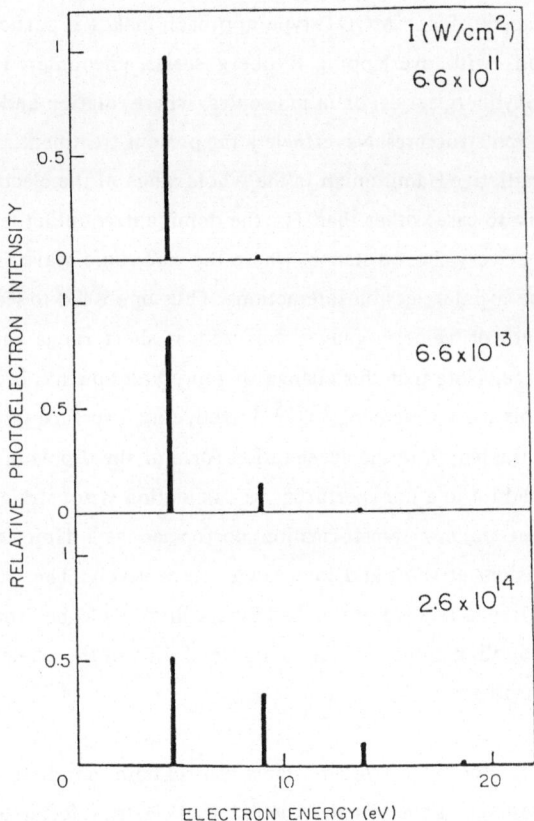

Figure 5 : Theoretical photoelectron energy distribution in the ionization process of hydrogen ns states, for various laser intensities I and circularly polarized light (ω = 0.35 Ry).

V.- CONCLUSION AND PERSPECTIVES

The most striking feature of the simple calculations described above is that all the results have been obtained by solving the close-coupling system (7) for <u>a single energy</u> value, slightly above threshold (ϵ = 0.01 eV). The resulting short-range reaction matrix $K^{(s)}$ has been verified to vary very smoothly with energy and thus may be used to calculate the physical observables in a limited energy range below and above the threshold. This consequence of the limited effective range of the radiative interaction reduces the computational effort considerably : only the analytical calculations in the external zone have to be performed on a narrow energy grid to account for resonant structures.

This economical aspect of our MQDT-type approach makes it particularly suitable for treating complex situations with overlapping Rydberg series, encountered in many-electron atoms (with core excited Rydberg series) or in molecules, where rotation and vibration give rise to complicated autoionization structures. Nevertheless the present treatment, performed with the acceleration form of the radiative Hamiltonian in the whole range of the electron-core distances, cannot be applied directly to cases other than H : the dominant coupling term in r^{-2} lays the weight of the interaction at very short distance, where the electronic wave functions are never accurately represented due to interelectron interactions. Thus one is led to seek a 'mixed-gauge' approach, where the length or velocity gauges are used at short range and the acceleration representation at long range. Note that this change of representation has been previously used for the calculation of dipole matrix elements[14-15], introducing a surface term evaluated at the radius of switching from the length to the acceleration form of the dipole operator (see Section 2). Extension of this procedure to a non-perturbative calculation is not straight forward since it requires a time-dependent unitary transformation performed at a finite distance, with two alternative Fourier expansions at short and long range, respectively. The variational R-matrix method described by C. Greene in a lecture of this Symposium could be more suitable for this purpose than the close-coupling approach due to its flexibility in the choice of basis orbitals. Work along this line is in progress.

Finally, extension to molecular Rydberg states subject both to radiative and rovibrational interactions will take advantage of the fact that the external electron decouples from the nuclear motion at large distances only : the radiative process may thus be studied in the Born-Oppenheimer framework (fixed nuclei) with rovibrational interactions accounted for in a later step by the frame-transformation technique used in the field-free molecular MQDT calculations[21].

In conclusion, we believe that calculations based on the identification of the region of the space where the effective radiative interactions take place will help considerably in the treatment of all kinds of atomic and molecular processes in strong laser fields, a domain where experiments grow at high rate and need tractable theoretical descriptions.

REFERENCES

1. U. Fano and R.P. Rau, Atomic Collisions Spectra (Ac. Press, 1986).

2. S. Autler and C.H. Townes, Phys. Rev. 100, 703 (1955); C. Cohen-Tannoudji and S. Haroche, J. Physique 30, 125 (1969).

3. Shi-I Chu and W.P. Reinhard, Phys. Rev. Lett. 39, 1195 (1977).

4. Y. Gontier, N.K. Rahman and M. Trahin, Phys. Rev. A24, 3102 (1981) and Nuovo. Cim. 4D, 1 (1984).

5. Shi-I Chu and J. Cooper, Phys. Rev. A32, 2769 (1985).

6. J.T. Broad, Phys. Rev. A31, 1494 (1985).

7. M. Crance and J. Sinzelle, in "Fundamentals of Laser Interactions", Lecture Notes in Physics 229 (Ed. F. Ehlotzky, Springer 1985), p. 290.

8. L. Dimou and F.H.M. Faisal, Phys. Rev. Lett. 59, 872 (1987).

9. A.D. Bandrauk and M.L. Sink, J. Chem. Phys. 74, 1110 (1981).

10. F.H. Mies and P. Julienne, in 'Spectral line shapes 3', Ed. By F. Rostas and W. de Gruyter, Berlin 1985.

11. A. Giusti-Suzor and P. Zoller, Phys. Rev. A (1987), in press.

12. see e.g. A.F. Starace, 'Theory of atomic photoionization', in Handbuch der Physik, vol. 31 (Ed. W. Mehlorn-Springer-Verlag-Berlin, 1979)

13. U. Fano, Phys. Rev. A32, 617 (1985).

14. G. Peach, Mon. Not. R. Astron. Soc. 130, 361 (1965).

15. M.J. Seaton, J. Phys. B14, 3827 (1981) and J. Phys. B19, 2601 (1986).

16. W.C. Henneberger, Phys. Rev. Lett. 21, 838 (1968).

17. see e.g. C.H. Greene, A.R.P. Rau and U. Fano, Phys. Rev. A26, 2441 (1982).

18. M.J. Seaton, Rep. Prog. Phys. 46, 167 (1983).

19. P.Agostini, F. Fabre, G. Mainfray and N.K. Rahman, Phys. Rev. Lett. 42, 1127 (1979) ; F. Fabre, G. Petite, P. Agostini and M. Clement, J. Phys. B15, 1353 (1982).

20. D. Normand, C. Cornaggia and J. Morellec, J. Phys. B19, 2881 (1986).

21. U. Fano, Phys. Rev. A353 (1970) ; Ch. Jungen and D. Dill, J. Chem. Phys. 73, 3338 (1980) ; see also the lecture by Ch. Jungen in this volume.

HYPERSPHERICAL DESCRIPTION OF TWO-ELECTRON SYSTEMS

Anthony F. Starace

Department of Physics and Astronomy
The University of Nebraska
Lincoln, Nebraska 68588-0111

INTRODUCTION

While the use of hyperspherical coordinates to describe two-electron systems is quite old,[1-8] it was Macek's[9] introduction of a quasi-separable approximation in hyperspherical coordinates which made possible a host of theoretical studies elucidating the symmetries of doubly-excited states and the dynamics of processes involving two-electron atoms and ions. This work, up to about mid 1982, has been reviewed by Fano.[10] However, much additional progress has been made since 1982, particularly on the symmetries of doubly excited electronic states, on the correspondences with molecular descriptions of two-electron systems, on the limitations of the quasi-separable approximation at small and at large distances, and on the hyperspherical descriptions of systems with an arbitrary number of electrons. It is the purpose of this and the next three papers, then, to present an updated review of our current theoretical understanding of electron correlations.

In this first paper we review the hyperspherical description of two electron systems, beginning with the quasi-separable approximation of Macek[9] and its applications. We next discuss the concept of two electron motion along a potential ridge as the "pathway" by which two-electron states of high excitation are realized. The need for alternative representations of two-electron wave functions near the nucleus and at distances far from the nucleus is then analyzed. Finally, a number of extensions of the quasi-separable approximation in hyperspherical coordinates to other three-particle systems are mentioned.

The following three papers in this series on the theory of electron correlations are concerned with these additional aspects: the paper of Lin discusses the symmetries of two-electron excited states, primarily based on the quasi-separable approximation in hyperspherical coordinates; the paper of Feagin discusses a molecular description of two-electron adiabatic potentials and symmetries; finally, the paper of Cavagnero discusses the hyperspherical description of N-electron atoms and ions, with $N > 2$.

THE HYPERSPHERICAL REPRESENTATION

General Orientation

A two electron wavefunction $\psi(\vec{r}_1, \vec{r}_2)$ is usually described by the six coordinates r_1, r_2, $\hat{r}_1$, and $\hat{r}_2$ of the two electrons. In hyperspherical coordinates the magnitudes of the individual radial coordinates, r_1 and r_2, are replaced by a hyperspherical radius, R, and a hyperspherical angle, α, where

$$R \equiv (r_1^2 + r_2^2)^{1/2} \tag{1}$$

and

$$\alpha \equiv \arctan(r_2/r_1). \tag{2}$$

The radius R measures the "size" of the two electron state, while the angle α measures the radial correlation of the two electrons. Note that when $\alpha = \pi/4$, $r_1 = r_2$; when $\alpha \approx 0$ or $\approx \pi/2$, one of the electrons is at a much larger distance from the nucleus than the other.

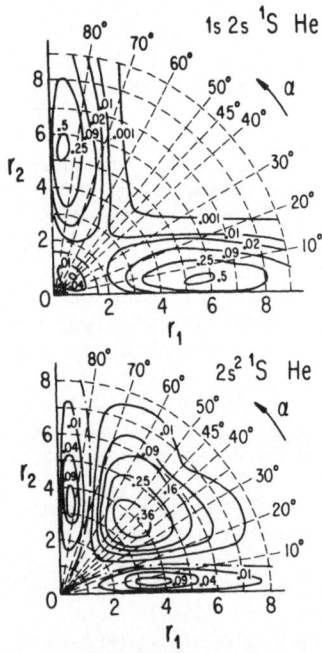

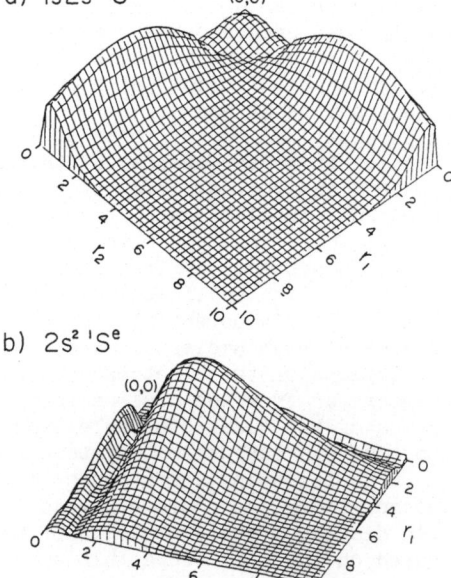

Fig. 1. Contour plot of the approximate probability distribution $|\psi(\vec{r}_1,\vec{r}_2)|^2$ for He. (a) 1s2s ^{1}S (b) 2s^2 ^{1}S. Solid Lines: lines of constant probability. Dot-Dash Lines: nodal lines. (From Ref. 12)

Fig. 2. Relief map of the approximate probability distribution $|\psi(\vec{r}_1,\vec{r}_2)|^2$ for He. (a) 1s2s ^{1}S (b) 2s^2 ^{1}S (From Ref. 13)

236

Before summarizing the features of the Schrödinger equation in these coordinates let us look first at plots of approximate two-electron probabilities, $|\psi(R,\alpha,\hat{r}_1,\hat{r}_2)|^2$, in these coordinates. Fig. 1 shows contour plots[11,12] and Fig. 2 shows relief maps[13] for the probability distributions of the singly-excited state 1s2s ^{1}S and the doubly-excited state 2s^2 ^{1}S of He. (Note that the wave functions are calculated in the approximation that each electron has an orbital angular momentum equal to zero in order to eliminate all dependence on the angular variables $\hat{r}_1$ and $\hat{r}_2$; since the angular dependence is trivial, these states are symmetric about $\alpha = \pi/4$, i.e., under interchange of r_1 and r_2.) The most obvious distinguishing features of the two probability distributions is that the one for the singly excited state is largest along $\alpha \approx 0$ and $\alpha \approx \pi/2$ (implying one electron is much further from the nucleus than the other) while the one for the doubly excited state is largest along $\alpha \approx \pi/4$ (implying both electrons are comparably excited, i.e., $\alpha = \pi/4$ when $r_1 = r_2$). A second important feature is the behavior of the nodal lines for the two probability distributions. The 1s2s ^{1}S state has a single nodal line along $R \approx 2$, while the 2s^2 ^{1}S state has two nodal lines along $\alpha \approx$ constant: one along $5° < \alpha < 30°$ and the other along $60° < \alpha < 85°$. The fact that the pattern of nodal lines is approximately along the orthonormal grid of constant R and constant α implies a quasi-separability of R and α coordinates.

The nodal line pattern for a particular state serves also to classify the state.[13] The ground state of He, 1s^2 ^{1}S, has a spherically symmetric probability distribution and is the first member of the singly-excited channel 1sns ^{1}S, which converges to the He$^+$(n=1) threshold. The single node in R for the state 1s2s ^{1}S, shown in Figs. 1(a) and 2(a), characterizes it as the second member of the 1sns ^{1}S channel. The state 2s^2 ^{1}S, shown in Figs. 1(b) and 2(b), has no radial nodes. It is the first member of the Rydberg series 2sns ^{1}S converging to the He$^+$(n=2) threshold. The two nodes approximately along constant α, symmetrical about $\alpha = \pi/4$, characterize 2s^2 ^{1}S as a member of this second Rydberg channel. Thus nodes in R characterize the excitation of a state within a channel while nodes in α characterize the various channels.[13]

Two-Electron Schrödinger Equation

In hyperspherical coordinates the non-relativistic two-electron Schrödinger equation becomes

$$\left(\frac{d^2}{dR^2} - \frac{1}{R^2}\left(-\frac{d^2}{d\alpha^2} - \frac{1}{4} + \frac{\ell_1^2}{\sin^2\alpha} + \frac{\ell_2^2}{\cos^2\alpha} \right) - \frac{C(\alpha,\theta_{12})}{R} + 2E \right)$$

$$\times \; (R^{5/2} \sin\alpha \cos\alpha \; \psi) = 0, \tag{3}$$

where the potential $-C(\alpha, \theta_{12})$ is proportional to the sum of the nuclear and electrostatic potentials,

$$-C(\alpha,\theta_{12}) = R\left(-\frac{2Z}{r_1} - \frac{2Z}{r_2} + \frac{2}{|\vec{r}_2 - \vec{r}_1|} \right)$$

$$= -\frac{2Z}{\cos\alpha} - \frac{2Z}{\sin\alpha} + \frac{2}{(1 - \sin2\alpha \; \cos\theta_{12})^{1/2}}, \tag{4}$$

$\vec{\ell}_1$ and $\vec{\ell}_2$ are the usual orbital angular momentum operators for the individual electrons, $\theta_{12} \equiv \cos^{-1} \hat{r}_1 \cdot \hat{r}_2$, and Z is the nuclear charge.

In the hyperspherical coordinate method of Macek,[9] the two-electron wavefunction $\psi_\nu(\vec{r}_1, \vec{r}_2)$ is expanded in terms of a complete set of adiabatic eigenfunctions $\phi_\mu(R; \alpha, \hat{r}_1, \hat{r}_2)$, which depend parametrically on the hyperspherical radius, R, and are functions of the five angular variables, α, $\hat{r}_1$, and $\hat{r}_2$. The form of ψ is thus:

$$\psi_\nu(R, \alpha, \hat{r}_1, \hat{r}_2) = (R^{5/2} \sin\alpha \cos\alpha)^{-1} \sum_\mu F_{\mu\nu}(R) \phi_\mu(R; \alpha; \hat{r}_1, \hat{r}_2) \tag{5}$$

The angular function ϕ_μ is defined to satisfy the following differential equation in atomic units ($\hbar = e = m = 1$):

$$\left(-\frac{d^2}{d\alpha^2} + \frac{\ell_1^2}{\cos^2\alpha} + \frac{\ell_2^2}{\sin^2\alpha} - RC(\alpha, \theta_{12}) \right) \phi_\mu = -U_\mu(R) \phi_\mu. \tag{6}$$

Here $-C(\alpha, \theta_{12})$ is defined in Eq. (4) and $U_\mu(R)$ is an eigenvalue which is parametrically dependent on R. Upon substituting Eq. (5) in the two-electron Schrödinger equation and using Eq. (6), one obtains the following set of coupled differential equations for the radial functions $F_{\mu\nu}(R)$:

$$\left(\frac{d^2}{dR^2} + \frac{U_\mu(R) + \frac{1}{4}}{R^2} + (\phi_\mu, \frac{\partial^2\phi_\mu}{\partial R^2}) + 2E \right) F_{\mu\nu}(R)$$

$$+ \sum_{\mu' \neq \mu} \left((\phi_\mu, \frac{\partial^2\phi_{\mu'}}{\partial R^2}) + 2(\phi_\mu, \frac{\partial\phi_{\mu'}}{\partial R}) \frac{\partial}{\partial R} \right) F_{\mu'\nu}(R) = 0 \tag{7}$$

In Eq. (7) the coupling matrix elements $(\phi_\mu, \partial^n\phi_\mu, /\partial R^n)$, $n = 1, 2$, involve integration over the five angular variables only and are thus parametrically dependent on R.

The Quasi-Separable Approximation

Each of the potentials $U_\mu(R)$ and its corresponding angular eigenfunction ϕ_μ define a hyperspherical channel μ. These channels are coupled through the radial derivative matrix elements in Eq. (7). In a quasi-separable hyperspherical (QSH) approximation,[9] one ignores the coupling terms in the second set of braces in Eq. (7). Then the wave function in Eq. (5) may be represented by a single term with $\mu = \nu$ in the summation on the right side, i.e.,

$$\psi_{\mu E}^{QSH} = (R^{5/2} \sin\alpha \cos\alpha)^{-1} F_{\mu\mu E}(R) \phi_\mu(R; \alpha, \hat{r}_1, \hat{r}_2). \tag{8}$$

For simplicity one usually sets $\mu = \nu$ and drops the double subscripts on F when referring to the quasi-separable approximation solutions. One sees from Eq. (8) that the quasi-separable approximation amounts to assuming that motion in R and motion in α are approximately independent of each other. This quasi-separability was inferred from Figs. 1 and 2, which show electron density plots obtained from quasi-separable approximation wave functions. This behavior may be confirmed by examining correlated two-electron wavefunctions and observing that the nodal lines of such wavefunctions also lie approximately along constant R and along constant α.[14]

It should be emphasized that although only single radial and angular functions are used to represent the two-electron wave function in Eq.(8), much electron correlation is implicitly included. This is illustrated in Fig. 3, which shows the $s^2(^1S)$, $p^2(^1S)$, $d^2(^1S)$, and $f^2(^1S)$ components of the numerically calculated $H^-(^1S)$ ground state angular function, ϕ_μ. One sees clearly that these higher angular momentum components are significant at small R, near $\alpha \approx \pi/4$ (i.e., $r_1 \approx r_2$). As R increases, however, only the $ss(^1S)$ component contributes significantly, in accordance with the independent electron model.

Notice also in Eq. (8) how all members of the channel μ have the same angular function ϕ_μ at any given R. Each state of excitation energy E within the channel μ is described by the radial function $F_{\mu E}$, which is calculated in the channel potential $U_\mu(R)$ using Eq. (7) and ignoring the off-diagonal coupling terms. Because each member of a Rydberg series of doubly excited states has the same angular function ϕ_μ and has a radial function $F_{\mu E}(R)$ that is calculated in the same potential $U_\mu(R)$, the

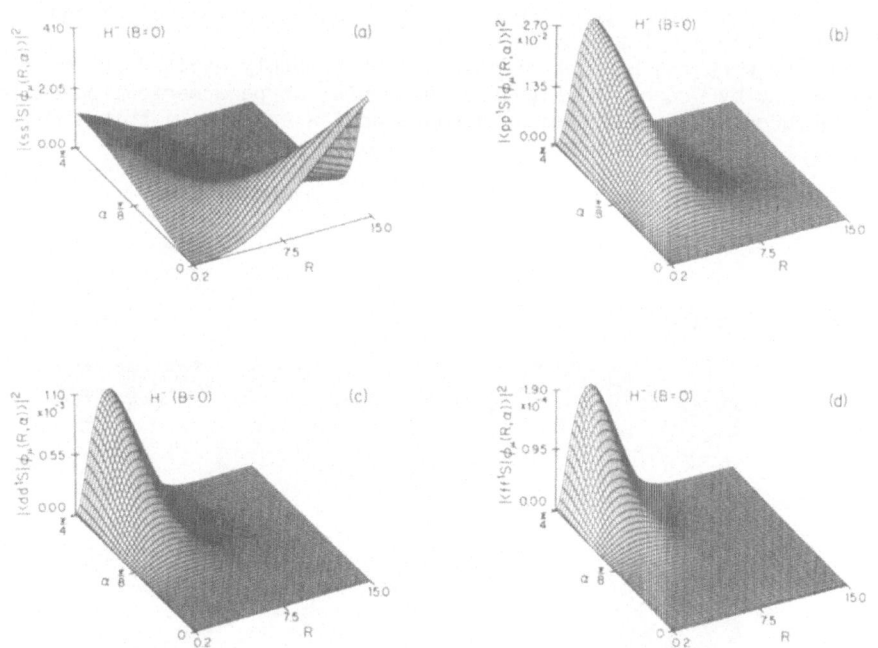

Fig. 3. Probability (per unit length in α) for the angular momentum state $\ell\ell^1S$ in the singlet ground state of H^- (a) $\ell=0$; (b) $\ell=1$; (c) $\ell=2$; (d) $\ell=3$. Note multiplication by factors of 10^2, 10^3, 10^4 in Figs. (b), (c), and (d), respectively. (From Ref. 15)

physical properties of states belonging to a particular channel μ are often immediately apparent upon examination of $U_\mu(R)$ and ϕ_μ. In what follows we illustrate the use of the potentials $U_\mu(R)$ to classify two-electron excitation channels. We then survey the accuracy of numerical predictions obtained using the quasi-separable approximation.

APPLICATIONS OF THE QUASI-SEPARABLE APPROXIMATION

Hyperspherical Classification of Two-Electron Excitation Channels

The first major success[9] of the quasi-separable approximation in hyperspherical-coordinates was the classification and interpretation of the photoabsorption spectrum of He in the region of the doubly excited Rydberg states converging to the n = 2 threshold. In the usual classification scheme there should be three Rydberg series of such levels of comparable intensity: 2snp ^{1}P, 2pnd ^{1}P, and 2pns ^{1}P. The experimental spectrum of Madden and Codling,[16] shown in Fig. 4, showed only one strong Rydberg series and one very weak Rydberg series. The third possible series was not observed. Cooper, Fano, and Prats[17] interpreted the relative intensities of the two observed series in terms of the so-called "+" and "−" series, (2snp ± 2pns)^{1}P. The "+" series members are more intense than those of the "−" series because the corresponding wavefunctions of the "+" members have a much larger amplitude near the origin, allowing therefore a much larger overlap with the ground state. This scheme, however, does not explain the weakness of the 2pnd ^{1}P channel. Fig. 5, however, shows Macek's hyperspherical potentials $U_\mu(R)$ for the three channels μ converging to the n=2 state of He$^+$. One sees immediately that the three channels have vastly different centrifugal barriers near the origin, explaining the large intensity differences of the three allowed channels. Furthermore, the first two hyperspherical channels have the "+" and "−" characteristics predicted by Cooper et al.[17]

Similar work has been carried out for the doubly excited states of H by Lin[18] and by Greene.[19] Fig. 6 shows Lin's[18] hyperspherical potentials for the three ^{1}P$^\circ$ doubly excited Rydberg series converging to the n=2

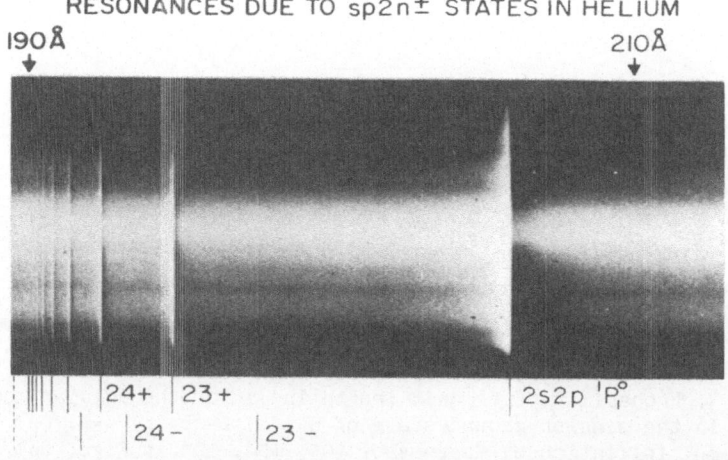

Fig. 4. Photoabsorption spectrum of He between 190 and 210 Å. The "+" and "−" series members are indicated below the spectrum. (From Ref. 16)

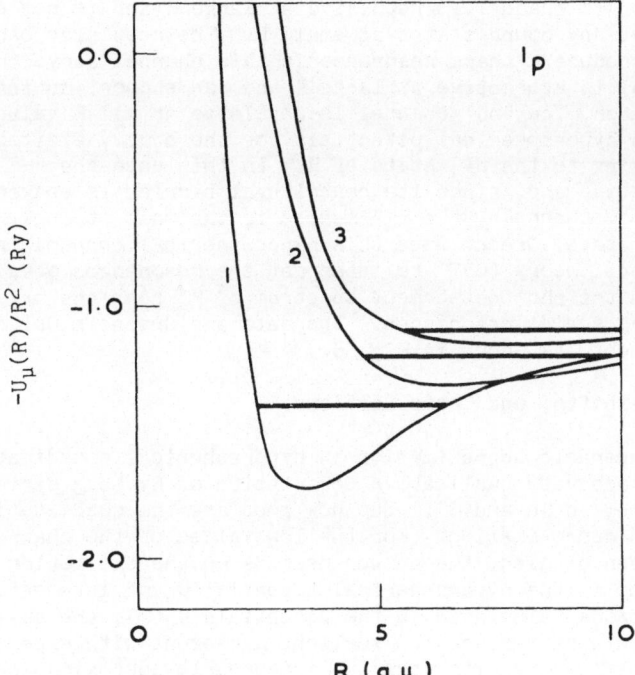

Fig. 5. Hyperspherical potential curves $-U_\mu/R^2$ vs. R for the three He doubly excited ^1P channels converging to the n=2 state of He$^+$. (From Ref. 9)

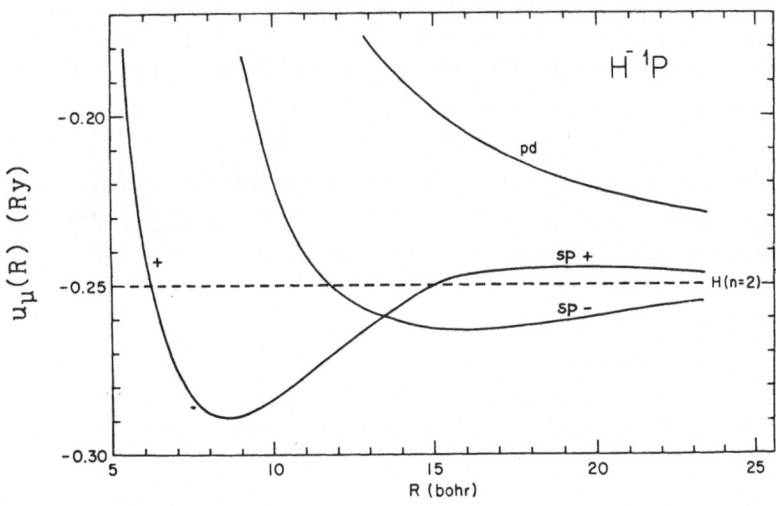

Fig. 6. Effective hyperspherical potential curves $u_\mu(R)$ vs. R for the

three H$^-$ doubly-excited ^1P channels converging to the n=2 state of H. Note

that $u_\mu(R) \equiv -[(U_\mu(R)+1/4)/R^2+(\phi_\mu, \dfrac{d^2\phi_\mu}{dR^2})]$ (Cf. Eq. 7). (From Ref. 18) state of H. The "+" channel is repulsive at large R and is not deep enough to support any bound states at small R. The repulsive barrier does, however, produce a shape resonance in this channel above threshold. The "−" potential is attractive at large R and can support an infinity of Feshbach resonances. The "pd" channel is repulsive at all R values. Fig. 7 shows Greene's hyperspherical potentials for the doubly excited $^1P^o$ channels converging to the n=3 state of H. In this case the "+" potential is always attractive and, since its centrifugal barrier is weaker than those of the other channels, the "+" series is the most strongly excited from the ground state. Greene used this hyperspherical channel calculation and quantum defect theory (QDT) to interpret the resonances obtained by Hamm et al.[20] in the photodetachment spectrum of H⁻ near the n=3 threshold as due to the "+" series resonances. The data and Greene's QDT fit are in excellent agreement, as shown in Fig. 8.

Energies, Phase Shifts, and Cross Sections

The quasi-separable approximation in hyperspherical coordinates thus provides a very accurate qualitative description of Rydberg series of doubly excited states in He and H⁻. But how good are the quantitative predictions in this approximation? For the low values of the channel index, μ, which have been studied, the answer depends on the excitation energy above the minimum in the hyperspherical potential U_μ of interest. For the lowest energy states calculated in the potentials $U_\mu(R)$, the quasi-separable approximation energies are in excellent agreement with experiment and with other theoretical results; the quasi-separable approximation wave function may also be used confidently. However, higher energy states of a particular channel μ calculated in the potential $U_\mu(R)$ are increasingly too high in energy,[9] if bound, or have too low phase shifts,[21-24] if unbound. This is not surprising since for higher excitation energies the coupling between the hyperspherical channels can no longer be ignored. Thus at present the quasi-separable approximation in the hyperspherical coordinate approach provides a very good initial approximation to the exact electron wave function, but its systematic improvement for states of moderate and high excitation energy to provide state-of-the-art numerical predictions has been carried out in only a few cases.[24-26] In what follows, we provide a few examples of the high level of accuracy to be expected from the quasi-separable approximation for the lowest excited

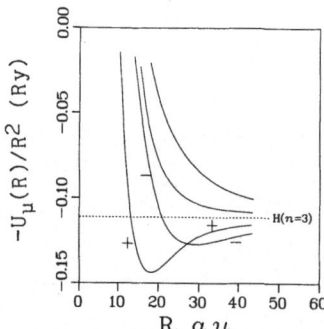

Fig. 7. Hyperspherical potential curves $-U_\mu/R^2$ vs. R for H⁻ doubly-excited ^1P channels converging to the n=3 threshold of H. (From Ref. 19)

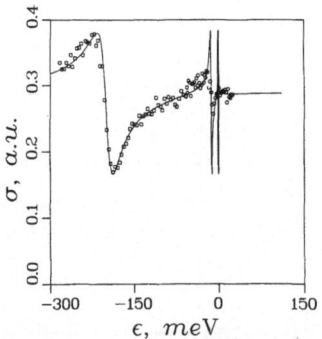

Fig. 8. Photodetachment cross section of H⁻ near the n=3 threshold of H at $\epsilon = 0$. Data: Hamm et al. (Ref. 20). Solid Line: QDT fit of Greene (Ref. 19). (From Ref. 19)

TABLE I. Elastic e-H ^{1}S Scattering Phase Shift (in radians)[a]

k(a.u.)	Three-Channel Close-Coupling[b]	Hyperspherical[c]		"Exact" Variational Result[d]
		One Channel	Two Channel	
0.1	2.491	2.513	2.521	2.553
0.2	1.974	1.983	2.023	2.067
0.3	1.596	1.568	1.659	1.696
0.4	1.302	1.242	1.380	1.415
0.5	1.092	0.989	1.142	1.202
0.6	0.93	0.784	0.926	1.041
0.7	0.82	0.618	0.622	0.930

[a] From Ref. 21, Table I.

[b] P.G. Burke and H.M. Schey, Ref. 30.

[c] C.D. Lin, Ref. 21.

[d] C. Schwartz, Ref. 29.

states in particular hyperspherical potentials.

Regarding level energies, Macek[9] calculated the He 2s2p(^{1}P) resonance excitation energy as 60.138 eV as compared with the experimental value[16] of 60.135 ± .015 eV. Similarly, Miller and Starace[23], calculated the ground state energy of He as ⊢2.895 a.u. as compared with the essentially exact non-relativistic theoretical value[27] of -2.904 a.u. Very recently Koyama et al.[25] and Fukuda et al.[26] have calculated energies for high-lying doubly-excited states up to the n=9 threshold in H⁻ and up to the n=7 threshold in He. They achieve agreement with results of other calculations within about 1% or so primarily by employing the diabatic[28] hyperspherical potentials in energy regions where the adiabatic (quasi-separable approximation) hyperspherical potentials have strongly avoided crossings. We shall discuss the use of diabatic hyperspherical potentials in the next section.

Regarding phase shifts, Lin[21] calculated the e-H ^{1}S elastic scattering phase shift for values of electron momenta, k, in the range, $0 \leq k$ (a.u.) ≤ 0.7. He used both the quasi-separable (one channel) hyperspherical approximation as well as a coupled two-channel hyperspherical approach. As shown in Table I for k = 0.1 a.u. the quasi-separable approximation phase shift of 2.513 rad. is closer to the "exact" variational result of 2.553 rad. of Schwartz[29] than is the three-channel close-coupling result of 2.491 rad. of Burke and Schey.[30] For $k \geq 0.3$ a.u., however, the quasi-separable approximation results[21] are lower than the three-channel close-coupling results[30] and both of these calculated predictions are lower than the variational results,[21] with the quasi-separable approximation results becoming increasingly too low as k increases. Coupling two hyperspherical adiabatic (quasi-separable) channels improves the hyperspherical predictions significantly, making them better than the three-channel close-coupling results up to k = 0.6. However, only calculations which include 7 coupled hyperspherical channels,[24] shown in Fig. 9, have provided results equivalent to those of the variational method.[29]

Lastly, regarding cross sections, a calculation[23] of the photoionization cross section of He using quasi-separable approximation hyperspherical coordinate wave functions demonstrates the strengths and weaknesses of the method. The initial and final wave functions for the process

$$He(^1S) + \gamma \quad \rightarrow \quad He^+ \; 1s(^2S) + e^-(^1P) \qquad\qquad (9)$$

both have the form of Eq. (8). For the initial state, μ corresponds to the lowest ^1S potential $U_\mu(R)$, and for the final state, μ corresponds to the lowest ^1P potential $U_\mu(R)$. The photoionization cross section obtained using the quasi-separable approximation wave functions is shown in Fig. 10. Fig. 10 also shows the revised experimental results of Samson,[31] which have error bars of ±3%. The hyperspherical results lie within these error limits near threshold (for kinetic energies $0.0 \leq \epsilon \leq 0.4$ a.u.) and in fact agree with experiment to within 1% at threshold. The hyper-

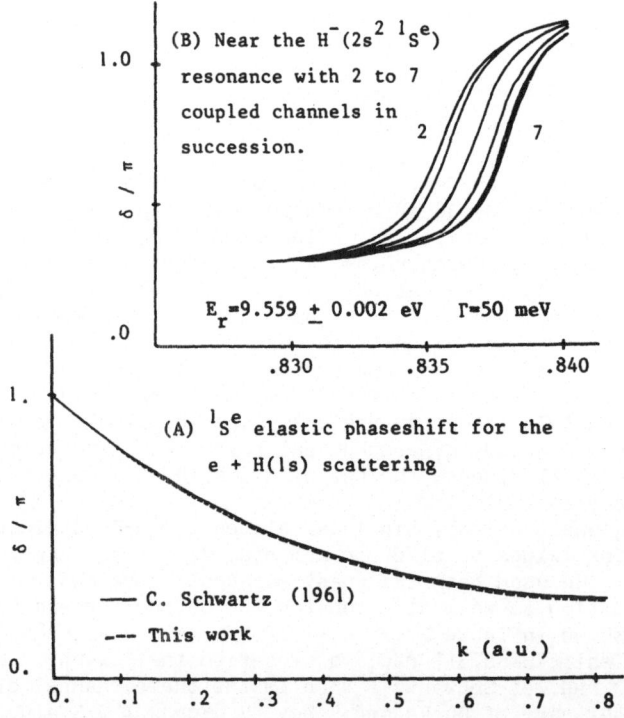

Fig. 9. Seven-channel hyperspherical calculation of the $^1S^e$ elastic scattering phase shift for electrons on H by L. Pelamourges, S. Watanabe, and M. LeDourneuf, Ref. 24. (A) Phase shifts for electron momenta in the range, $0 \leq k(a.u.) \leq 0.8$. Solid lines shows comparison with variational results of C. Schwartz, Ref. 29; (B) Phase shifts for electron momenta in the range, $0.83 \leq k(a.u.) \leq 0.84$. (From Ref. 24)

spherical results, however, are systematically lower than experiment above $\varepsilon = 0.4$ a.u. Of the many other theoretical calculations, we show one with very good overall agreement with experiment: the four channel $(1s\text{-}2\bar{s}\text{-}2\bar{p})$ close-coupling calculation of Jacobs.[32] In comparison with the close-coupling results, the single-channel hyperspherical results are in better agreement with experiment below $\varepsilon = 0.2$ a.u. and are systematically lower above $\varepsilon = 0.2$ a.u. Note that the length results of a more recent six-state R-matrix calculation by Berrington et al.[33] are in excellent agreement with the experimental results of West and Marr,[34] including those near threshold, thereby indicating the sophistication required to properly describe the threshold region by methods employing independent electron representations.

Discussion

We see from these numerous examples that the quasi-separable approximation in hyperspherical coordinates provides an excellent first approximation for the representation of two-electron states. Simply from the potentials U_μ one can obtain much qualitative understanding of entire Rydberg series of two-electron excitations. Furthermore, the lowest energy states in each potential U_μ are well represented by quasi-separable approximation wave functions, as confirmed quantitatively by calculations of energies, phase shifts, and cross sections. A simple improvement on the quasi-separable approximation in regions of sharply avoided crossings of the potentials U_μ is to switch to a diabatic representation in these regions, as discussed in the next section. However, for energies far above the minima of the potentials U_μ of interest, quantitative accuracy requires the solution of the coupled equations (7). The few calculations which have been carried out indicate that the convergence of the calculated results proceeds slowly as the number of channels increases. Further analysis of this slow convergence is presented in the section after next.

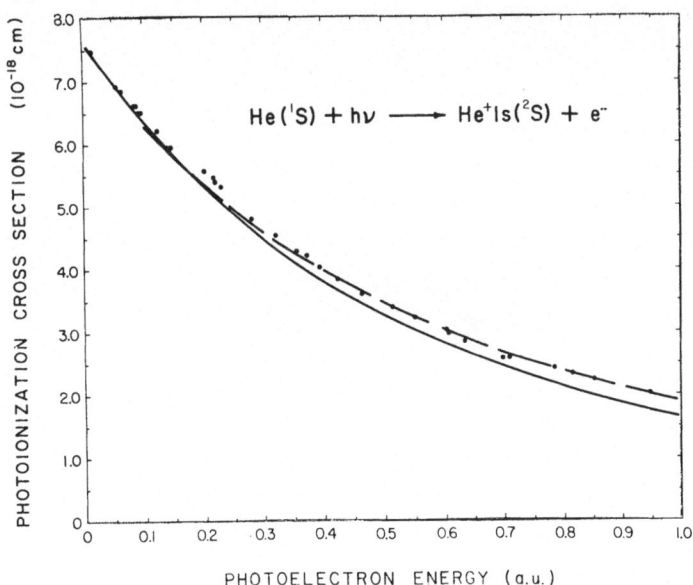

Fig. 10. Photoionization cross section for He. <u>Full Curve</u>: quasi-separable approximation (single channel) hyperspherical calculation of Miller and Starace (Ref. 23); <u>Dots</u>: Experimental results of Samson (Ref. 31); <u>Dashed Curve</u>: $1s\text{-}2\bar{s}\text{-}2\bar{p}$ (four channel) close-coupling calculation of Jacobs (Ref. 32). (From Ref. 23)

Analysis of the Angular Equation

The hyperspherical coordinate approach has not only been used to study stationary states, but also to understand qualitatively how a low-energy two electron state concentrated near the origin evolves to states of high excitation far from the origin upon receiving energy during a collision process. The key idea, stressed by Fano[35] and illustrated graphically by Lin,[13] is that such states describe motion along a potential ridge centered about the direction $\alpha = \pi/4$ (i.e., $r_1 = r_2$).

Consider Eq. (6) for the channel functions $\phi_\mu(R; \alpha, \hat{r}_1, \hat{r}_2)$. The potential $-C(\alpha, \theta_{12})$, defined in Eq. (4), is shown in Fig. 11 for $Z = 1$. States having one electron more excited than the other, i.e., $r_2 \gg r_1$ or $r_1 \gg r_2$, have an angle function ϕ_μ with maximum amplitude in the valleys of the potential in Fig. 11, near $\alpha = 0$ and $\alpha = \pi/2$. Doubly-excited states of comparable energy have $r_1 \approx r_2$ and thus the angle function ϕ_μ for these states has maximum amplitude on the ridge of the potential in Fig. 11, near $\alpha = \pi/4$, and preferably near $\cos\theta_{12} = -1$ (i.e., on opposite sides of the nucleus).

Consider now the R-dependence of the angle functions ϕ_μ. Eq. (6) shows that the potential $-C$ is multiplied by R. For large enough R, therefore, the potential $-RC$ on the ridge becomes equal to the eigenvalue $U_\mu(R)$. At this "classical turning point" the angle function ϕ_μ has no more "kinetic energy" of motion in α on the ridge. For larger R values, its amplitude on the ridge is exponentially damped and the probability amplitude in the channel μ must retreat to the valleys of the potential in Fig. 11, implying that for such large R values ϕ_μ describes states with one electron more highly excited than the other. Alternatively, the two-electron state on the ridge may "hop" to the next higher channel μ'. With a higher value of $-U_\mu(R)$, the two electron excitation could move to somewhat larger R along the ridge since the difference between $-U_{\mu'}$ and the top of the potential ridge of $-RC$ would restore some positive "kinetic energy" of motion in α. Actually the vicinity of the classical turning point is

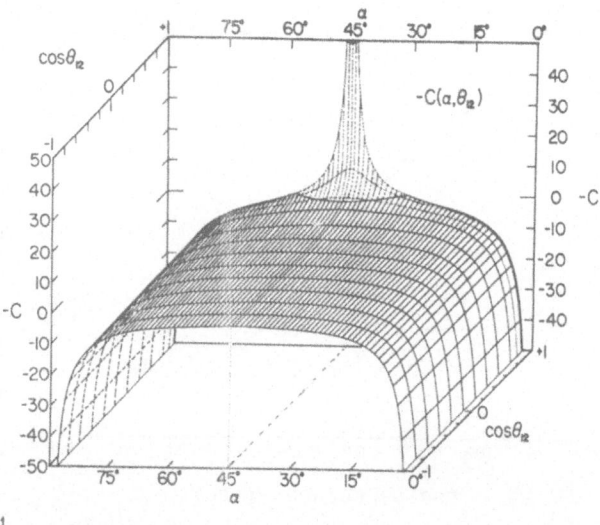

Fig. 11. Relief map of the potential $-C(\alpha, \theta_{12})$ defined in Eq. (4) for Z=1. (From Ref. 12)

propitious for such a transition to a higher channel μ' since the coupling matrix elements (cf. Eq. (7)) are largest precisely where the channel functions are changing most rapidly with R.

Lin[13] has shown graphically how the channel functions ϕ_μ behave as functions of R. In Figs. 12 and 13 we show the $H^-(^1S)$ channel functions $\phi_\mu(R;\alpha,\theta_{12})$ for μ = 1 and μ = 2 (i.e., the lowest two ^1S hyperspherical channels). In Fig. 12 one sees that at R = 1 the charge distribution in the first channel is peaked about α = π/4, lying on the potential ridge. At R = 4, however, the charge distribution is vacating the ridge and moving to the valleys near α = 0 and α = π/2. By R = 8, μ = 1 describes a channel with one electron much more highly excited than the other. Fig. 13 shows the next higher hyperspherical channel function. Note that at R = 4, precisely where μ = 1 has a depression along the ridge, the μ = 2 channel's charge distribution has a maximum. This peak in μ = 2 along the ridge progresses outward to larger R values until at R = 12 a depression appears along the ridge. If two-electron states in μ = 2 are to move to larger R <u>and</u> <u>remain</u> <u>comparably</u> <u>excited</u> they must hop again to the next higher hyperspherical channel, and so on.

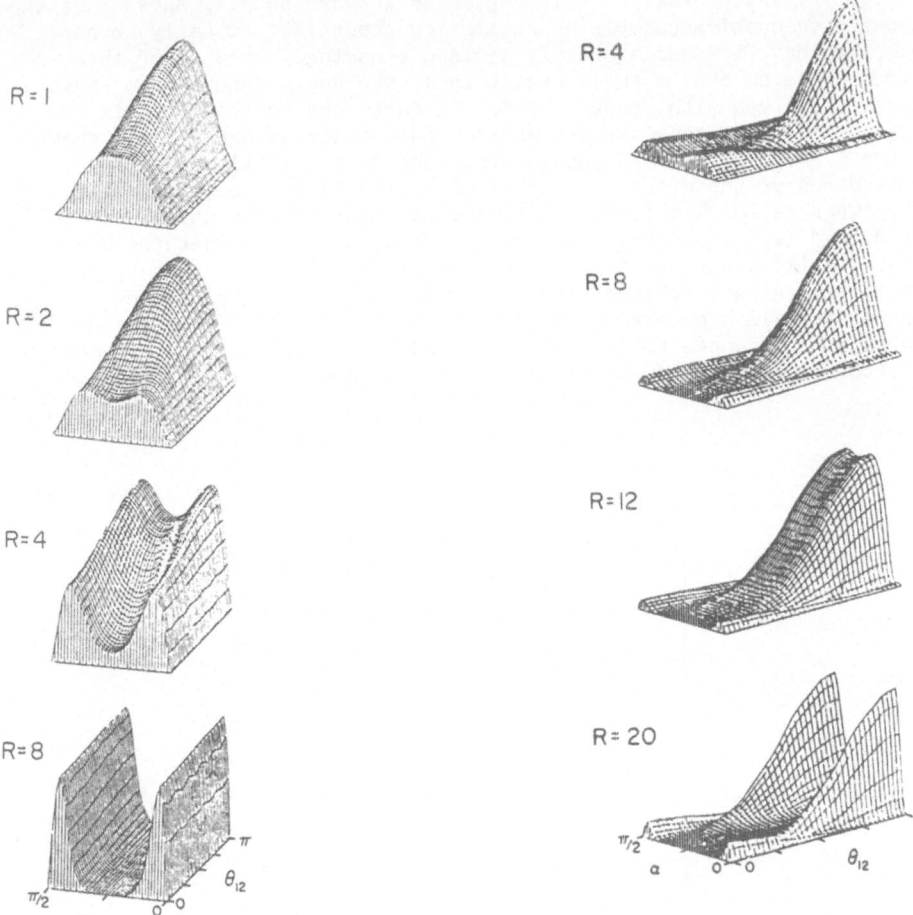

Fig. 12. Plot of $|\phi_\mu(R;\alpha,\theta_{12})|^2$ vs. α and θ_{12} for various R values for the first H^- ^1S hyperspherical channel, μ = 1. (From Ref. 13)

Fig. 13. Plot of $|\phi_\mu(R;\alpha,\theta_{12})|^2$ vs. α and θ_{12} for various R values for the second H^- ^1S hyperspherical channel, μ = 2. (From Ref. 13)

This perspective of two electron excitation states evolving toward large radii R along a potential ridge has its origins in the Wannier-Peterkop-Rau[36] analysis of electron impact ionization near threshold. Its application to quantitative predictions of excitation cross sections is just beginning. In particular, recent studies of photoionization of atoms having an outer s^2 subshell have shown some common features, which we now discuss.

Adiabatic Vs. Diabatic Potentials

Consider first the case of helium. As discussed above, the experimental observation of only a single intense Rydberg series converging to the n=2 threshold in the photoabsorption spectrum of He can be understood easily in terms of the hyperspherical potentials shown in Fig. 5. The potential labelled "1" is the so-called "+" channel, whose states overlap the ground state much more effectively than do states in either the "2" or "-" potential or the "3" or "d" potential. Note however, that the "1" and "2" potentials cross at R ≈ 7.64.

Fig. 14, which examines this region in greater detail, shows that the adiabatic or quasi-separable approximation potentials actually do not cross, but have instead a sharply avoided crossing. Because of this avoided crossing over a small region in R, the angle functions ϕ_μ have large derivatives with respect to R. In fact, the coupling matrix elements are so large that the "+" and "-" potentials exchange their character for R > 7.64. Fig. 15 shows this exchange by plotting the R-dependence of the overlap integral of $\phi_-(R=6.5)$ with $\phi_-(R)$ and with $\phi_+(R)$. Whereas for R < 7.64, $\langle\phi_-(6.5)|\phi_-(R)\rangle$ is close to unity, as expected, and $\langle\phi_-(6.5)|\phi_+(R)\rangle$ is close to zero, one finds that for R > 7.64 $\langle\phi_-(6.5)|\phi_+(R)\rangle$ is close to unity and $\langle\phi_-(6.5)|\phi_-(R)\rangle$ is close to zero. What is happening is that electronic excitations populated in the "+" channel at small R proceed outward at larger R and "hop" from the quasi-separable "+" channel to the "-" channel near R ≈ 7.64. For this reason one often employs the diabatic approximation shown in Fig. 14 in such

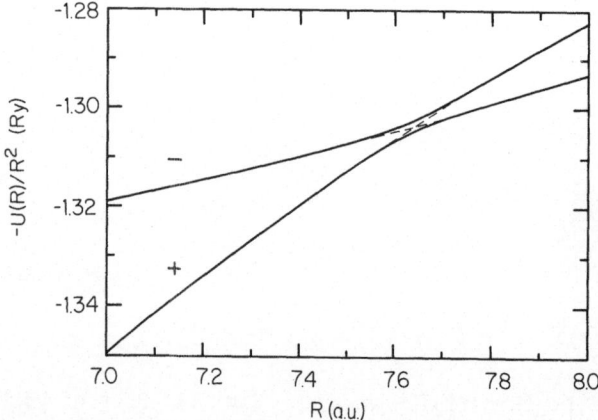

Fig. 14. The He*(n=2) ^{1}P "+"/"-" avoided crossing: a special case of transition from one $U_\mu(R)$ to another. Solid lines show the quasi-separable (adiabatic) hyperspherical potentials, which have an avoided crossing near R ≈ 7.64. The dashed lines show the diabatic potentials, which cross near R ≈ 7.64.

cases of sharply avoided crossings. That is, one connects the "+" potential and channel function below R ≈ 7.64 to the "–" potential and channel function above R ≈ 7.64 and vice versa. One then ignores the residual coupling between the new "+" and "–" diabatic potentials.

Photoionization of Be

Compare now this diabatic behavior observed in photoionization plus excitation of He leading to states converging to the n=2 threshold with photoionization of the outer $2s^2$ subshell of beryllium.

Greene[37] has calculated the photoionization cross section of Be including coupling between the lowest two hyperspherical channels, $\mu = 1$, corresponding to leaving the ion in its ground 2s state, and $\mu = 2$, corresponding to leaving the ion in its excited 2p state. (The inner $1s^2$ core was represented by a central potential so that only the correlation of the outer two electrons was treated.) Greene's procedure is to calculate the two adiabatic potentials $U_\mu(R)$ and angle functions ϕ_μ using the angular

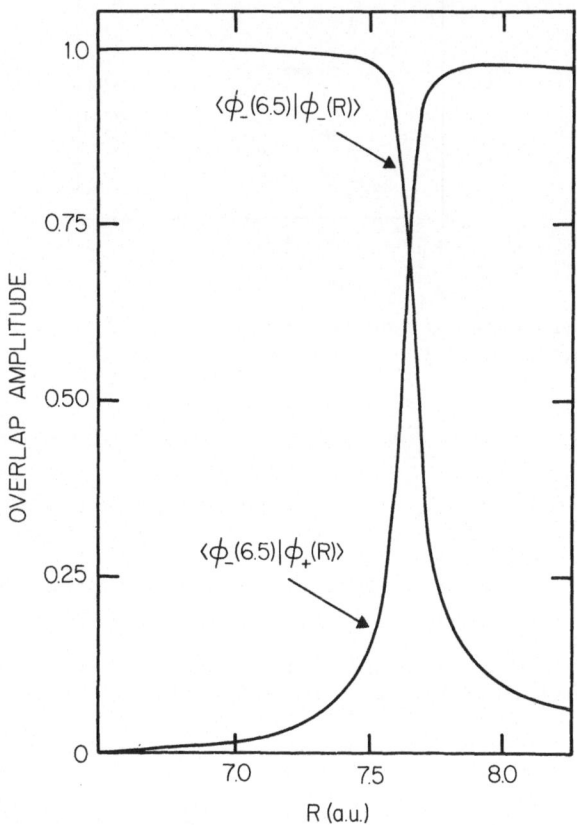

Fig. 15. Quasi-separable approximation channel function behavior at the "+"/"–" avoided crossing near R = 7.64 as exhibited by the R-dependence of the two overlap integrals $\langle\phi_-(R=6.5 \text{ a.u.})|\phi_-(R)\rangle$ and $\langle\phi_-(R=6.5 \text{ a.u.})|\phi_+(R)\rangle$. Integrations are over α, $\hat{r}_1$, and $\hat{r}_2$.

Eq. (6). The radial Eq. (7) is then solved including the first and second derivative coupling matrix elements connecting the channels $\mu = 1$ and $\mu = 2$. His results are in reasonable agreement with the close-coupling calculation of Dubau and Wells[38] and show a very large intensity for excitation of the ion to the 2p level.

The most interesting aspect of Greene's calculation[37] is the similarity his hyperspherical wave functions show to those in He, thereby indicating a similar behavior for He, Be, and all the alkaline earths. It is instructive first to compare the hyperspherical potentials $U_\mu(R)$ for the He ^1P levels converging to the $\mu=2$ threshold, shown in Fig. 5, to the corresponding potential curves in Be, shown in Fig. 16. One sees immediately from Fig. 5 why only one of the He$^+$ (n=2) excitation channels, $\mu = 1$, is strongly populated; it has a much less repulsive potential barrier than either the $\mu = 2$ or $\mu = 3$ channels. Furthermore, the channel function ϕ_μ for the "+" channel ($\mu=1$) is symmetric in α, having an antinode on

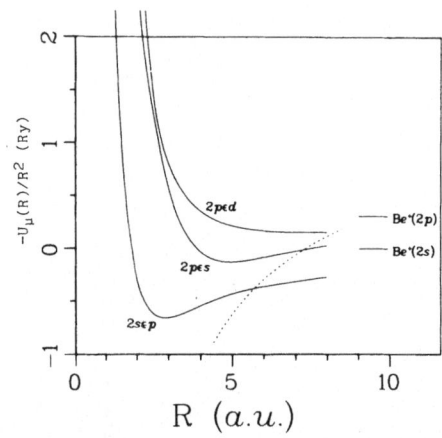

Fig. 16. Hyperspherical potential curves $-U_\mu(R)/R^2$ vs. R for the three Be ^1P channels converging to the n=2 state of Be$^+$. (From Ref. 37)

the potential ridge in Fig. 11, while the "−" channel ($\mu=2$) is antisymmetric in α, having a node on the potential ridge. The symmetry about $\alpha = \pi/4$ for the He wavefunctions holds for all R values due to the degeneracy in energy of these channels. Reiterating the discussion above, recall that although the "+" and "−" channels are shown to cross in Fig. 5, this crossing is actually avoided; in any case the channel functions ϕ_μ do not adjust to the crossing but proceed diabatically through it. For this reason the middle curve in Fig. 5 for R > 7.64 a.u. has "+" character while the lowest curve for R > 7.64 a.u. has "−" character.

Consider now the Be potentials in Fig. 16. Two differences from He are immediately apparent. First, the potential curves are non-degenerate for R → ∞. Secondly there is an avoided crossing between the first and second potential curves for 4 < R < 6. Otherwise, however, one expects most of the absorption strength, as in He, to go into the channel with the lowest potential curve. In his calculations Greene expanded the channel functions as follows:

$$\phi_\mu = \sum_{\ell_1 \ell_2} g_\mu^{\ell_1 \ell_2} (R;\alpha)\ Y_{\ell_1 \ell_2 LM}(\hat{r}_1,\hat{r}_2) \qquad (10)$$

The most important functions $g_\mu^{\ell_1 \ell_2}$ (those with $\ell_1 \ell_2$ = "sp") are shown in
Fig. 17 for the potential curves $\mu = 1$ and $\mu = 2$ for various R values.
For R = 2 one sees that the $\mu = 1$ function is approximately symmetric
about $\alpha = \pi/4$ while the $\mu = 2$ function is approximately antisymmetric,
just as for the the "+" and "-" channels in He. As R increases, however,
these adiabatic channel functions drop into one or the other of the potential valleys in Fig. 11, i.e., the $\mu = 1$ amplitude becomes concentrated
near $\alpha = 0$ while the $\mu = 2$ amplitude becomes concentrated near $\alpha = \pi/2$.
Thus, as R increases, the non-degeneracy of the thresholds in Be causes a
breakdown of the "+" and "-" symmetry about $\alpha = \pi/4$ observed at small R
values. Furthermore, this transition is seen to occur for R values
$4 \lesssim R \lesssim 6$.

What is remarkable about Greene's treatment of the coupled radial
equations (7) is the finding that the solution which at small R starts out
as the adiabatic wave function $F_{\mu=1}(R)\phi_{\mu=1}(R;\Omega)$ in the $\mu=1$ channel becomes
at R > 6 a <u>nearly equal superposition of the adiabatic wave functions for
$\mu = 1$ and $\mu = 2$ in such a way that the "+" symmetry is preserved through
the avoided potential crossing region.</u> In other words, just as in He, the
"+" solution proceeds diabatically through the avoided potential crossing.
This also explains the large excitation cross section observed in Be
since, unlike the case in He, the state having "+" character becomes at
R>6 a nearly equal superposition of the $\mu = 1$ and $\mu = 2$ channel functions.
Furthermore, it is expected that this diabatic behavior of the hyperspherical "+" solution will be a common feature of all alkaline earth and other
similar two electron systems.[10,37]

Indeed, R-matrix calculations[39] have found the eigenchannel functions
for Mg ^{1}P final states to be very similar in character to those for Be.
The heavier alkaline earth atoms Ca, Sr, Ba, and Ra require the treatment
of a still larger number of channels due to the proximity in energy of
bound "d" orbitals. However, even for these elements, a hyperspherical

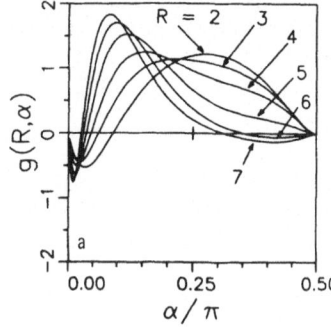
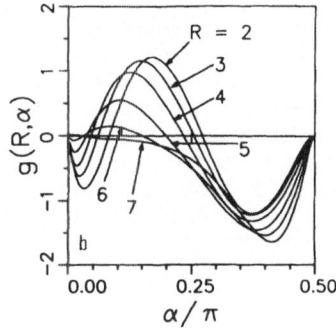

Fig. 17. Adiabatic "sp" channel wave functions associated with the lowest
two Be potential curves: (a) 2sϵp, $\mu = 1$; (b) 2pϵs, $\mu = 2$. (From Ref. 37)

analysis suggests that the diabatic character of a state populated at small R is preserved as the state evolves toward larger R,[10,37] although, of course, this idea must yet be tested by carrying out the calculations. Very recently, R-matrix calculations for calcium[40] and strontium[41] have found that at small radial distances the wave functions and channel interactions "look remarkably similar for all of these atoms [He, Be, Mg, Ca, Sr] including helium."[40(a)] In any case, the discovery of the common features of photoexcitation processes in He, Be, and Mg as well as in the heavier alkaline earths, despite vast differences in the coupling strength between the associated channels, is one of the new perspectives on two-electron correlations provided by the hyperspherical method. It indicates the usefulness of a combination of hyperspherical and R-matrix techniques to study electron correlations in these heavy atoms.[42]

HYPERSPHERICAL REPRESENTATIONS AT SMALL AND AT LARGE R

A number of recent studies of hyperspherical wave functions at both large and small radii, R, have shed light on the strengths and weaknesses of the quasi-separable approximation to two-electron wave functions. We review briefly here three such studies.

The Fock Expansion

The proper series representation for the wave function of two electrons moving in the Coulomb field of a nucleus has been a long-standing problem that is still not completely solved. Both Bartlett and Fock independently introduced logarithmic terms in R into the expansion, although nowadays such expressions have come to be known as "Fock expansions." In particular, Bartlett[43] showed that an acceptable series solution of the Schrödinger equation for a two-electron atom or ion is the following series representation near the nucleus,

$$\sum_{n,j} C_{nj} R^n (\ln R)^j, \tag{11}$$

where R (Cf.Eq(1)) is the hyperspherical radius introduced by Gronwall.[44] In particular, series representations in terms of only the variables r_1, r_2, $r_{12} \equiv |r_1 - r_2|$, and/or R--i.e., ignoring lnR terms - were shown not to exist.[43,45] Fock[3] independently demonstrated that Eq. (11) provides an acceptable series solution near the origin of the two-electron Schrödinger equation for ^1S states. He also discussed in detail the procedure for calculating the coefficients C_{nj}, which depend on five angular variables and hence obey rather complicated differential equations. Demkov and Ermolaev[4] showed that Fock's expansion applies generally to an N-electron wave function of arbitrary symmetry.

The usefulness of the series representation in Eq. (11) is dependent on its convergence properties. Macek[46] proved its convergence in the mean for values of R less than $(2|k|)^{-1}$, where the wave number, k, is related to the energy E (a.u.) of the two-electron system relative to the double ionization threshold by

$$E = k^2/2. \tag{12}$$

More recently, Morgan[47] has proved the convergence of the Fock expansion for S states pointwise for all values of R and has indicated that the expansion coefficients C_{nj} are analytic functions of E. Yet to be proved is whether for discrete energies one of the solutions of the Schrödinger equation having a convergent Fock expansion decays exponentially as either r_1 or r_2 tends to infinity, i.e., whether the "physical solutions" of the Schrödinger equation have Fock expansions. Leray[48] has begun to address this problem.

Although other studies of the Fock expansion have been carried out, the major application of Bartlett's and Fock's work has been to provide a better representation of the two-electron wave function in variational calculations.[7,49,50] In particular, Frankowski and Pekeris[7] have shown that a 101-term variational trial function including ℓnR terms gives better energies than a 1078-term variational trial function without ℓnR terms.

Very recently, theoretical interest has focused on determination of the exact two-electron wave function near the nucleus by explicit evaluation of the Fock coefficients, as opposed to the variational determination of a parameterized trial wave function including ℓnR terms. Feagin, Macek, and Starace[51] have presented a detailed numerical procedure for the evaluation of the Fock coefficients of arbitrary order as expansions in hyperspherical harmonics. Abbott, Maslen, and Gottschalk[52] have focused instead on analytic representations of the Fock coefficients, which they have obtained through second order in R for a state of arbitrary symmetry.

While the mathematical necessity of the logarithmic terms in Eq. (11) is acknowledged and the faster convergence their presence permits in variational calculations is known,[7] the physical consequences of these logarithmic terms remain unknown. Feagin, Macek, and Starace[51] attempted to determine this physical significance by matching numerically calculated Fock series solutions to quasi-separable approximation hyperspherical wave functions [cf. Eq. 8] at a boundary $R=R_0$ for ground and low-energy excited states of He and H⁻. Use of the Fock expansion for $R \leq R_0$ did not significantly change the energies, wave functions, or phase shifts of the states considered from those calculated using the simple the quasi-separable approximation wave functions in hyperspherical coordinates over the entire range in R.

These calculations,[51] then, seem to indicate that for small R the quasi-separable (adiabatic) approximation in hyperspherical coordinates is a very good representation of the lowest energy two-electron states. On the other hand, the quasi-separable radial wave functions do not have ℓnR terms in their series solution at $R=0$; an ordinary series expansion in powers of R applies. It is only when coupling between quasi-separable channels is considered that ℓnR terms are required to obtain a series solution of the full set of coupled equations (7). Future studies of processes for which the quasi-separable hyperspherical approximation is inadequate-such as excitation processes-are therefore required to identify the physical role played by the logarithmic terms in Eq. (11).

Matching to Independent Particle Wave Functions

In contrast to the apparent adequacy of the quasi-separable hyperspherical (QSH) wave function for two electron states at small R, as R tends to infinity, i.e., as the electrons become separated, the QSH wave functions tend too slowly to the independent particle (IP) wave functions. Christiansen-Dalsgaard[53] has tackled this problem by matching the QSH wave functions to IP wave functions at finite values of R. For the case of e⁻-H elastic scattering at low energies she has matched the QSH wave function corresponding to the H⁻ $1s\epsilon s(^1S)$ channel μ, i.e.,

$$\psi^{QSH}(\vec{r}_1,\vec{r}_2) \equiv \frac{F_\mu(R)\ \phi_\mu(R;\alpha,\hat{r}_1,\hat{r}_2)}{\left(R^{5/2}\ \sin\alpha\ \cos\alpha\right)}, \tag{13}$$

to the IP wave function for this channel, i.e.,

$$\psi^{IP}(\vec{r}_1, \vec{r}_2) \equiv \left(\frac{P_{1s}(r_1) P_{\epsilon s}(r_2) + P_{1s}(r_2) P_{\epsilon s}(r_1)}{2^{1/2} r_1 r_2} \right) Y_{00}(\hat{r}_1) Y_{00}(\hat{r}_2). \quad (14)$$

By requiring the logarithmic derivatives with respect to R for these two wave functions to be equal at $R = R_0$, one may obtain the scattering phase shift of the IP one electron wave function, $P_{\epsilon s}(r)$. Christensen-Dalsgaard chooses R_0 at each electron momentum k, defined by $\epsilon = k^2/2$, so that exchange terms are negligible relative to direct terms in the IP representation.

The results of this matching procedure for the e-H elastic scattering phase shift are shown in Fig. 18. One sees that they are equivalent to the "exact" variational results of Schwartz up to the vicinity of the $2s^2$ resonance energy. It achieves this accuracy by matching the best repre-

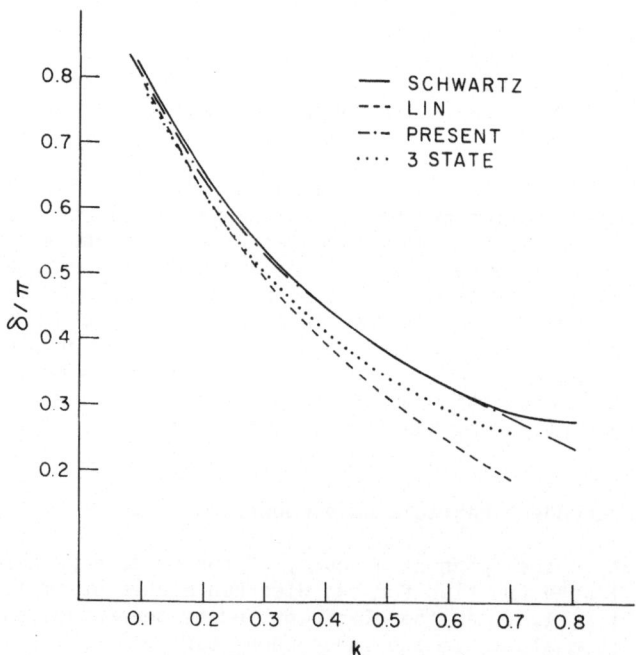

Fig. 18. Elastic scattering phase shifts, δ, for electrons on H plotted vs. electron momentum, k. Solid Line: variational results of Schwartz, Ref. 29. Dashed Line: one-channel quasi-separable hyperspherical (QSH) results of Lin, Ref. 21. Dotted Line: three-channel close-coupling results of Burke and Schey, Ref. 30. Dot-Dash Line: one-channel hyperspherical-independent particle matching results of Christensen-Dalsgaard, Ref. 53. (From Ref. 53).

sentation at short distances (the QSH wave functions) to the best representation at large distances (the IP wave functions). The matching is done for values of R in the range 6 - 8 a.u. As shown in Fig. 3, this range in R is transitional between electronic motion on the potential ridge (cf. Fig. 11) and electronic motion in the potential valleys (cf. Fig. 11).

The lessons provided by this calculation are that, firstly, the QSH wave functions are a very good representation at small distances where electronic motion is concentrated on the potential ridge; and secondly, when the wave function begins to depart from the potential ridge as R increases, it is best to switch to an IP representation of the electronic motion.

Asymptotic Expansions

As shown by Macek,[9] the QSH wave functions do tend to an IP form at asymptotically large values of R. The work of Christensen-Dalsgaard,[53] however, has shown that the numerically calculated QSH wave functions should be matched to IP wave functions at much shorter distances in order to obtain accurate phase shifts and energies. To facilitate an analytic match, Zhen and Macek[54] have obtained asymptotic expansions in powers of $(1/R)$ for QSH potentials and wave functions. The channel functions $\phi_\mu(R;\alpha,\hat{r}_1,\hat{r}_2)$ have been determined through terms of order R^{-2}, and the potentials $U_\mu(R)$ through terms of order R^{-4}. Among their findings is that the QSH wave functions evolve into polarized orbitals as $R \sim \infty$.

EXTENSIONS OF THE QUASI-SEPARABLE APPROXIMATION

While the majority of the three-body systems studied by means of the quasi-separable approximation in hyperspherical coordinates have been two-electron atomic and ionic systems in field-free space, a number of other three-body systems have been examined by this method in the last several years. These other three-body systems include the H_2^+ molecule by Greene,[55] the e^+-H system by Pelikan and Klar,[56] H- in an electric field by Lin,[57] H^- and He in a magnetic field $B \leq 10^9$G by Park and Starace,[15] the positronium negative ion ($e^+ - e^- - e^-$) by Botero and Greene,[58] and the HD$^+$ system by Macek and Jerjian.[59] In addition, Macek has used hyperspherical coordinates to study loosely bound states of three bosons interacting via short range two-body forces insufficient to bind any two of them.[60] One may expect further such extensions in the future.

CONCLUSIONS

We have surveyed broadly in this paper the use of the hyperspherical method to study three-body systems. In particular, we have shown that the quasi-separable (or adiabatic) approximation in hyperspherical coordinates provides a classification of doubly-excited electronic states, provides quantitatively accurate predictions for the lowest states in hyperspherical potentials and for processes connecting them, and provides at least qualitative insight into excitation processes. The evidence from several studies is that the quasi-separable wave functions, in general, provide an excellent representation for two electron wave functions at short distances near the nucleus, but tend too slowly to the more appropriate independent particle representation at large radial distances. Finally, we have mentioned a number of applications in recent years of the hyperspherical method to three-body systems other than two-electron atoms or ions in field-free space.

ACKNOWLEDGMENT

This work has been supported in part by the U.S. Department of Energy, Division of Chemical Sciences, under Grant No. DE-FG 02-85ER13440.

REFERENCES

1. E.C. Kemble, The Fundamental Principles of Quantum Mechanics with Elementary Applications, (Dover Publications, Inc., New York, 1937) p. 210.

2. P.M. Morse and H. Feshbach, Methods of Theoretical Physics (McGraw-Hill, New York, 1953) Vol. II, pp. 1730-1734.

3. V.A. Fock, Izv. Akad. Nauk USSR, Ser Fiz. 18, 161 (1954) [Eng. Transl.: Kong. Norske Videnskabers Selskabs Forh. 31, 138 (1958); 31, 145 (1958)].

4. Yu.N. Demkov and A.M. Ermolaev, Zh. Eksp. Teor. Fiz. 36, 896 (1959) [Sov. Phys.-JETP 36, 633 (1959)].

5. F.T. Smith, Phys. Rev. 120, 1058 (1960).

6. W. Zickendraht, Annals of Physics 35, 18 (1965).

7. (a) K. Frankowski and C.L. Pekeris, Phys. Rev. 146, 46 (1966);
    (b) K. Frankowski, Phys. Rev. 160, 1 (1967);
    (c) D.E. Freund, B.D. Huxtable, and J.D. Morgan, Phys. Rev. A 29, 980 (1984).

8. J.H. Macek, Phys. Rev. 160, 170 (1967).

9. J.H. Macek, J. Phys. B 2, 831 (1968).

10. U. Fano, Rep. Prog. Phys., 46, 97-165 (1983).

11. U. Fano and C.D. Lin, Atomic Physics 4 (Plenum, New York, 1975), pp. 47 - 70.

12. C.D. Lin, Phys. Rev. A 10, 1986 (1974).

13. C.D. Lin, Phys. Rev. A 25, 76 (1982).

14. C.D. Lin, Phys. Rev. A 27, 22 (1983).

15. C.H. Park and A.F. Starace, Phys. Rev. A 29, 442 (1984).

16. R.P. Madden, and K. Codling, Astrophys. J. 141, 364 (1965).

17. J.W. Cooper, U. Fano, and F. Prats, Phys. Rev. Lett. 10, 518 (1963).

18. C.D. Lin, Phys. Rev. Lett. 35, 1150 (1975); Phys. Rev. A 14, 30 (1976).

19. C.H. Greene, J. Phys. B 13, L39 (1980).

20. M.E. Hamm, R.W. Hamm, J. Donahue, P.A.M. Gram, J.C. Pratt, M.A. Yates, R.D. Bolton, D.A. Clark, H.C. Bryant, C.A. Frost, and W.W. Smith, Phys. Rev. Lett. **43**, 1715 (1979).

21. C.D. Lin, Phys. Rev. A **12**, 493 (1975).

22. H. Klar, and U. Fano, Phys. Rev. Lett. **37**, 1132 (1976); H. Klar, Phys. Rev. A **15**, 1452; H. Klar and M. Klar, Phys. Rev. A **17**, 1007 (1978).

23. D.L. Miller and A.F. Starace, J. Phys. B **13**, L525 (1980).

24. L. Pelamourges, S. Watanabe, and M. LeDourneuf, in <u>Electronic and Atomic Collisions: Abstracts of Contributed Papers, XIII ICPEAC</u>, Edited by J. Eichler, W. Fritsch, I.V. Hertel, N. Stolterfoht, and U. Wille (Berlin, 1983), p. 109.

25. N. Koyama, H. Fukuda, T. Motoyama, and M. Matsuzawa, J. Phys. B **19**, L331 (1986).

26. H. Fukuda, N. Koyama, and M. Matsuzawa, J. Phys. B **20**, 2959 (1987).

27. C.L. Pekeris, Phys. Rev. **112**, 1649 (1958).

28. See, e.g., B.L. Christiansen-Dalsgaard, Phys. Rev. A <u>**29**</u>, 470 (1984). See also the lecture by U. Fano, "General Features of Collisions and Spectra," in this volume.

29. C. Schwartz, Phys. Rev. **124**, 1468 (1961).

30. P.G. Burke, and H.M. Schey, Phys. Rev. **126**, 147 (1962).

31. J.A.R. Samson, Phys. Reports **28C**, 303 (1976).

32. V.L. Jacobs, Phys. Rev. A **3**, 289 (1971).

33. K.A. Berrington, P.G. Burke, W.C. Fon, and K.T. Taylor, J. Phys. B **15**, L603 (1982).

34. J.B. West and G.V. Marr, Proc. Roy. Soc. A **349**, 397 (1976).

35. U. Fano, Phys. Rev. A **22**, 2660 (1980).

36. G. Wannier, Phys. Rev. **90**, 817 (1953); R. Peterkop, J. Phys. B **4**, 513 (1971); A.R.P. Rau, Phys. Rev. A **4**, 207 (1971).

37. C.H. Greene, Phys. Rev. A **23**, 661 (1981).

38. J. Dubau and J. Wells, J. Phys. B **6**, 1452 (1973).

39. P.F. O'Mahony and C.H. Greene, Phys. Rev. A **31**, 250 (1985).

40. (a) C.H. Greene and L. Kim, Phys. Rev. A **36**, 2706 (1987).
    (b) L. Kim and C.H. Greene, Phys. Rev. A **36**, 4272 (1987).

41. M. Aymar, E. Luc-Koenig, and S. Watanabe, J. Phys. B **20**, 4325 (1987).

42. P.F. O'Mahony and S. Watanabe, J. Phys. B **18**, L239 (1985).

43. J.H. Bartlett, Phys. Rev. **51**, 661 (1937).

44. T.H. Gronwall, Phys. Rev. **51**, 655 (1937).

45. J.H. Bartlett, J.J. Gibbons, and C.G. Dunn, Phys. Rev. **47**, 679 (1935).

46. J.H. Macek, Phys. Rev. **160**, 170 (1967).

47. J.D. Morgan, Theor. Chim. Acta. **69**, 181-223 (1986).

48. J. Leray, in <u>Lecture Notes in Physics 195: Trends and Applications of Pure Mathematics to Mechanics</u>, Edited by P.G. Ciarlet and M. Roseau (Springer, Berlin, 1984), pp. 235-247.

49. A.M. Ermolaev and G.B. Sochilin, Dokl. Akad. Nauk SSSR **155**, 1050 (1964) [Sov. Phys.-Dokl. **9**, 292 (1964)].

50 G.O. Morrell and D.L. Knirk, Theor. Chim. Acta **37**, 345 (1975).

51. J.M. Feagin, J. Macek, and A.F. Starace, Phys. Rev. A **32**, 3219 (1985).

52. (a) P.C. Abbott and E.N. Maslen, J. Phys. A **20**, 2043 (1987).
    (b) J.E. Gottschalk, P.C. Abbott, and E.N. Maslen, J. Phys. A **20**, 2077 (1987).

53. B.L. Christensen-Dalsgaard, Phys. Rev. A **29**, 2242 (1984).

54. Z. Zhen and J. Macek, Phys. Rev. A **34**, 838 (1986).

55. C.H. Greene, Phys. Rev. A **26**, 2974 (1982).

56. E. Pelikan and H. Klar, Z. Phys. A **310**, 153 (1983).

57. C.D. Lin, Phys. Rev. A 28, 1876 (1983).

58. (a) J. Botero and C.H. Greene, Phys. Rev. A **32**, 1249 (1985).
    (b) J. Botero, Phys. Rev. A **35**, 36 (1987).

59. J. Macek and K.A. Jerjian, Phys. Rev. A **33**, 233 (1986).

60. J. Macek, Z. Phys. D **3**, 31 (1986).

CLASSIFICATION OF DOUBLY EXCITED STATES OF TWO-ELECTRON ATOMS

C. D. Lin

Department of Physics, Cardwell Hall
Kansas State University
Manhattan, Kansas 66506

INTRODUCTION

Doubly excited states of atoms were first detected and discussed in the 30's,[1] but their systematic study dates from the UV photoabsorption spectroscopy of helium in the early 60's by Madden and Codling.[2,3] In the spectra, two series of resonances, one intense and the other fainter and sharper, were observed. These resonances were attributed to the temporary formation of discrete states of the doubly excited intermediate, He^{**}, which eventually autoionize into $He^+(N=1)+e^-$. It took the great insight of Fano to recognize the important implication of the observed strong contrast in the strength of the two series. Cooper, Fano and Prats[3] argued that these doubly excited states cannot be adequately described by the independent electron model. The designation of states such as 2snp and 2pns $^1P^o$ would imply that the two series have nearly identical strength. They pointed out that the experimental data indicated that the two series are better approximated as 2snp+2pns and 2snp−2pns, and labelled as the '+' and the '−' series, respectively.

Double excitations may be regarded as the simplest and most easily studied examples of multiple excitations. The road to the understanding and the eventual complete classification of doubly excited states is by no means easy. Symmetry under interchange of the two electrons' radial distances was felt at once to underly the striking selection rule in the observed spectra.[3] Calculations by conventional atomic physics procedures confirmed the spectral evidence[4] without tracing the origin of the selection rules. A novel direct approach to the two-electron Schrodinger equation based on hyperspherical coordinates was designed by Fano and Macek and applied by Macek.[5] Macek's work pointed out how symmetry controls the intensity of different modes of excitations by invoking an adiabatic approximation in solving the wave equation for two-electron atoms in hyperspherical coordinates. Partial mappings of radial correlations were made in 1974 by Lin[6] which confirmed the + and - classification proposed by Cooper, Fano and Prats.[3] A complete mapping of electron correlations and a classification scheme of doubly excited states emerged only recently[7] by treating radial and angular correlations simultaneously.

Systematics of experimental data on doubly excited states are difficult to obtain. High resolution spectra of two-electron systems can be achieved through photoabsorption spectroscopy,[8,9] but only final $^1P^o$ states are studied in these experiments. Electron or ion collisions can populate many resonances in principle, but resolve only a few of them. Fortunately, for the simple two-electron systems, elaborate calculations[10] using traditional approaches have been performed and these calculations provide reliable numerical data for testing novel ideas.

Besides the hyperspherical approach, other methods have been developed for the understanding of doubly excited states. Among them are the group theoretical approach of Herrick[11-13] and others.[14] Individual doubly excited states have also been analyzed from a molecular viewpoint.[15,16] These studies explored the angular correlations of doubly excited states and pointed out the rovibrational pattern of their spectra. A complete classification scheme emerged[7,17] when the angular correlation quantum numbers from these studies are integrated with the +/- radial correlation quantum number. This eventual synthesis was carried out by mapping doubly excited state wave functions in hyperspherical coordinates from which correlation patterns characterized by angular and radial quantum numbers are displayed.[18-20] An analysis[21] of hyperspherical wave functions in the body frame of the atom further clarifies the meaning of these quantum numbers and the subtle limitation of the molecular model.

RADIAL AND ANGULAR CORRELATION QUANTUM NUMBERS

Correlation is a property of the relative motion of two particles. In this respect, hyperspherical coordinates provide a convenient basis for its description. Besides the three Euler angles for describing the rotation of the atom, one can choose the remaining three coordinates to be R, α, and θ_{12}. Since R is a measure of the size of the atom, it does not enter into the description of correlations. We regard radial correlation as represented by the dependence of the wave function on the angle α, and angular correlation as its dependence on the angle θ_{12}.

Both radial and angular correlations can be understood qualitatively within the CI approximation. For example, the wave functions for the strong and the weak series observed in the photoabsorption spectra of He can be represented approximately as 2snp+2pns and 2snp−2pns, respectively. This implies that the wave functions for '+' states have an antinode while '−' states have a node, near $\alpha \approx 45°$. Note that the '+' and '−' classification applies to every member of each Rydberg series.

Doubly excited states are known to have strong angular correlations. From the CI approach viewpoint, angular correlation is represented by the admixture of configurations of different (ℓ_1, ℓ_2) pairs. The approximate mixing coefficients for each state can be obtained using group approach. It was found by Herrick and Sinanoglu[22] that intrashell doubly excited states, i.e., states with two identical principal quantum numbers, are well approximated by 'Doubly Excited Symmetry Basis' (DESB) states. Their wave functions are the eigenstates of the SO(4) invariants, $\vec{B}^2$ and $(\vec{B} \cdot \vec{L})^2$, where $\vec{B} = \vec{b}_1 - \vec{b}_2$, and $\vec{b}_i$ is the Runge-Lenz vector of the ith electron. Two new quantum numbers K and T were used to characterize these states, replacing the ℓ_1 and ℓ_2 quantum numbers of the two electrons. The ranges of K and T are

$$T = 0, \ 1, 2 \ldots \ldots, \ \min(L, \ N-1)$$

$$K = (N-T-1), \ (N-T-3), \ldots \ldots, \ -(N-T-1) \tag{1}$$

where N is the principal quantum number of the intrashell states. For states with parity $\pi = (-1)^{L+1}$, T=0 is not allowed. Note that the assigned values of K and T do not depend on S. Roughly speaking, K and T can be understood as follows. For hydrogenic atoms, it is well known that the expectation values of $\vec{r}$ and $\vec{b}$ are proportional to each other within the subspace of constant principal quantum number N. Thus $\vec{B}^2$ is related to $\vec{b}_1 \cdot \vec{b}_2$, or $\vec{r}_1 \cdot \vec{r}_2$, i.e., to the angle θ_{12} between the two electrons. In fact, $\vec{B}$ is parallel to the interelectronic axis $(\vec{r}_1 - \vec{r}_2)$ and thus $(\vec{B} \cdot \vec{L})^2$ relates to the projection of the orbital angular momentum along the interelectronic axis.

The quantum numbers K and T can also be used to label asymptotic potential curves in the scattering of electrons by hydrogenic atoms.[23] This is achieved by extending the DESB functions to intershell states with N < n and then let n⟶ ∞ for fixed N. The quantum numbers K and T are then analogous to the quantum numbers labelling the linear Stark states, with K proportional to the average value of of $r_1 \cos\theta_{12}$ and T describing the magnitude of $\vec{L} \cdot \vec{r}_2$ (since $\vec{r}_{12} \approx \vec{r}_2$ for $r_1 \ll r_2$) where r_1 and r_2 are the radius vectors of the inner and the outer electrons, respectively. This interpretation of the quantum number T has interesting consequence on the values of the asymmetry parameter β in the angular distribution of the one-electron photoionization of helium accompanied by the excitation of the He^+ ion to the N-th excited level. It was predicted by Greene[25] that the values of β separate into two groups, depending on whether the two-electron complex is in a state of T=0 or T=1. The T=0 states will have β values approaching 2 as N$\rightarrow$∞ while the T=1 states will have β values approaching −1 as N$\rightarrow$∞. This prediction has now been confirmed experimentally[26] for N up to 6.

The integers K and T, especially when considered as describing linear Stark states in the asymptotic region, do not incorporate the +/− radial correlation. To treat radial and angular correlations simultaneously, an empirical rule was established to add a further quantum number A to each state.[7] Since not all two-electron states have approximate +/− radial symmetry, A was assigned to take one of the three values, +1, −1 or 0, with +1 and −1 corresponding to the + and −, respectively, and 0 to states without radial symmetry. This semiempirical classification was later 'justified' by analyzing the wave functions in the body-frame of the atom.[21]

The quantum number A is not independent of K and T for given L, S, N and π but follows the relation

$$A = \pi(-1)^{S+T} \quad \text{if } K > L-N$$
$$= 0 \quad \text{if } K \le L-N . \tag{2}$$

With the relations in Eqs. (1), and (2), all the correlation quantum numbers K, T, and A for states converging to a hydrogenic limit N can be assigned. All states with $L \ge 2N-1$ have A=0.

In terms of these quantum numbers, a Rydberg series is characterized by a channel index μ

$$\mu = (K, T)_N^A \, {}^{2S+1}L^\pi \tag{3}$$

and a state within the channel μ is labelled by $_n(K, T)_N^A \, {}^{2S+1}L^\pi$ where n is the principal quantum number of the outer (or Rydberg) electron. Using this notation, the 2snp+2pns and 2snp−2pns series of Cooper, Fano and Prats become $_n(0,1)_2^+$ and $_n(1,0)_2^-$, respectively. The so-called 'pd' series becomes $_n(-1,0)_2^0$. These rules can be applied to any state of a two-electron atom and each state is designated uniquely by the set of quantum numbers K, T, A, n, N, L, S and π.

RUDIMENTARY HYPERSPHERICAL APPROACH

The quantum numbers K, T and A provide convenient indices for labelling correlation properties of doubly excited states. To unravel the correlation pattern of each state represented by the set of $(K,T)^A$, it is necessary to examine its wave function. This will be done here for wave functions calculated in hyperspherical coordinates.

The basic formula for solving two-electron wave functions in hyperspherical coordinates have been reviewed in previous lectures. The notations used here are the same as in Starace's lecture. We only quote the essential equations here. We adopt the adiabatic approximation whose wave function is

$$\psi(R,\Omega) = F_\mu^n(R)\Phi_\mu(R;\Omega) \tag{4}$$

with

$$(-\frac{\Lambda^2}{R^2} + \frac{C}{R})\,\Phi_\mu(R;\Omega)=U_\mu(R)\Phi_\mu(R;\Omega). \tag{5}$$

The hyperradial wave function F(R) satisfies

$$\left[-\frac{d^2}{dR^2}+ U_\mu(R) - W_{\mu,\mu}(R) + 2E \right] F_\mu^n(R)=0 \tag{6}$$

where $W_{\mu,\mu} = (\Phi_\mu, d^2\Phi_\mu/dR^2)$ is included in the effective potential. In this approximation, μ identifies the channel and n denotes the degree of excitation in the hyperradial coordinate. Note that n corresponds here to the principal quantum number of the hydrogen atom and relates the number of nodes in the hyperradial wave function F(R) as well as to the radial extent of the atomic state. This part of the wave function is well understood from the basic quantum mechanics as well as from the independent particle picture. The main task of understanding electron correlations lies therefore in correlating the index μ with the behavior of Φ_μ. We characterize each channel function by displaying the surface charge density in the form of nodal patterns on the (α, θ_{12}) plane[18] at each R. As shown by Starace, the nodal patterns vary smoothly with R for each channel and the display of surface charge density for a single value of R suffices to characterize the correlation pattern of that channel. For states with L≠0, the rotational motion of the whole atom is averaged out so that only the relative radial (in α) and angular (in θ_{12}) motions are displayed.

ISOMORPHISM AND SUPERMULTIPLET STRUCTURE

To illustrate how the quantum numbers K, T and A indeed characterize the correlation properties of doubly excited states, we show in Fig. 1 a set of potential curves of He which converge to the N=3 limit of He^+. The

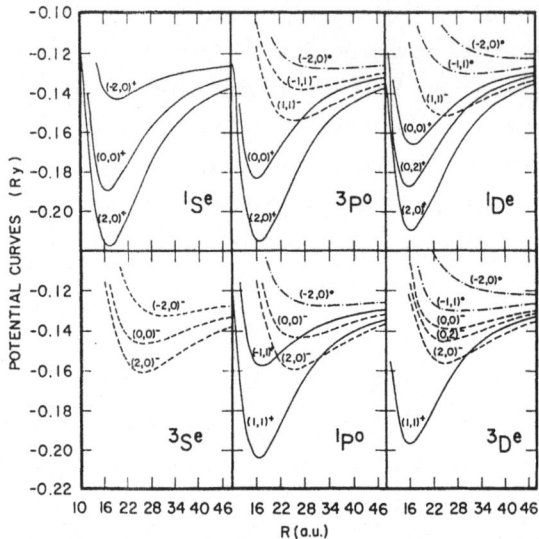

Fig. 1   Potential curves for $^{1,3}S^e$, $^{1,3}P^o$ and $^{1,3}D^e$ channels of He that
converge to the $He^+(N=3)$. Curves are labelled with correlation
quantum numbers K, T, and A. Reduced units with Z=1 are used.

curves are labelled with the K, T and A quantum numbers. The possible
sets of K, T and A for a given L, S, π and N follow the rules (1) and (2).
We first note that all the + curves have deep minima occuring at about the
same small values of R. Among the + curves, each curve is labelled from
below with the largest K and for a given K, with the largest T. The same
procedure also applies to the – curves and then to the A=0 curves,
respectively. This rule emerges from the fact that the magnitude of
correlation energies follow the hierachial order

$$U_A > U_K > U_T \tag{7}$$

where U_A, U_K and U_T are the local correlation energies associated with
each mode represented by the quantum numbers A, K, and T, respectively.
At large R, the +/– radial exchange symmetry is not important so that the
curves are ordered according to

$$U_K > U_T \tag{8}$$

only. In the asymptotic limit, curves are thus labelled from below with
the largest K and for a given K, with the largest T. By connecting the
curves in the two regions, some of the curves are allowed to cross each
other. Curve crossing is due to the change of relative importance of
radial and angular correlations with increasing R.

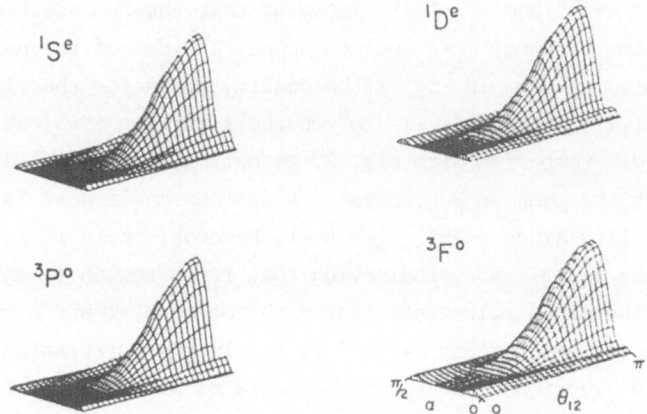

Fig. 2　Surface charge–density plots showing the isomorphic correlations for the $(2, 0)_3^+$ channels of $^1S^e$, $^3P^o$, $^1D^e$ and $^3F^o$ of He at R=20.

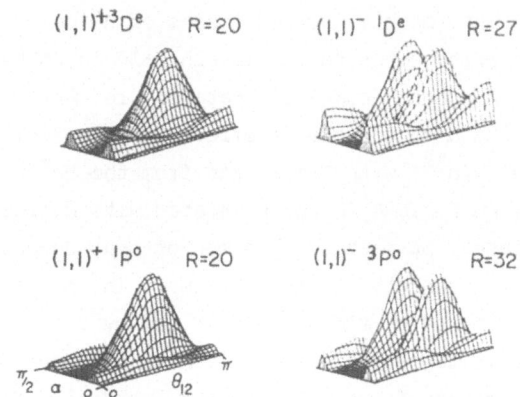

Fig. 3　As in Fig. 2 except for the $(1,1)^+$ D^e, $(1,1)^+$ $^1P^o$, $(1,1)^-$ $^1D^e$ and $(1,1)^-$ $^3P^o$ at the values of R indicated.　Note that all four plots have similar θ_{12} dependence as the K and T quantum numbers are identical, while A=+1 and A=−1 channels differ in the nodal structure at $\alpha=\pi/4$.

　　It is obvious from Fig. 1 that curves labelled with identical K, T and A are nearly degenerate, regardless of L, S and π.　This remark provides a first indication that the quantum numbers assigned have physical significance.　To see how the correlation quantum numbers are related to the correlation patterns, we show in Fig. 2 surface charge density plots for the $(2,0)_3^+$ channels of $^1S^e$, $^3P^o$, $^1D^e$ and $^3F^o$ at R=20 a.u.　We notice that the major features of these channels are the large densities near $\alpha=45°$ and $\theta_{12}=180°$, implying that the two electrons tend to stay at the same distance $r_1=r_2$ but on opposite sides of the nucleus. In another example, we show in Fig. 3 the density plots for the $(1,1)^+ \, ^1P^o$, $(1,1)^+ \, ^3D^e$, $(1,1)^- \, ^3P^o$ and $(1,1)^- \, ^1D^e$ channels of He** (N=3) at the values of R indicated. Comparing with Fig. 2, we notice that a smaller K value indicates that the peak in θ_{12} moves to a smaller angle near 135°. For A=1, the peak lies along $\alpha=45°$. For A=-1, however, there is an approximate nodal line along $\alpha=45°$, indicating that $r_1=r_2$ region is unfavored for − states. These plots illustrate that each set of K, T and A quantum numbers characterizes certain major features in the correlation patterns of two excited electrons. Doubly excited states with identical K, T and A quantum numbers have similar correlation patterns, i.e., they are isomorphic.

According to the hyperspherical method, approximate energy levels of doubly excited states can be calculated from Eq.(6). One important spectral regularity emerges obviously from the near degeneracy of potential curves with identical N, K, T and A quantum numbers. An example of large spectral regularity in the form of supermultiplet structure is shown in Fig. 4 for the doubly excited states of He below the He^+(N=3) threshold where we arrange the energy levels calculated by Lipsky et al.[27] using the K, T and A quantum numbers separately for A=+1 and A=-1. The energy levels are shown on a scale of effective principal quantum numbers n^*, defined by $E=-z^2/2n^{*2}$, with E measured from the He^+(N=3) limit. We note that for a given K, T, A, N and n, states with different L, S and π are nearly degenerate. In fact, each group of near−degenerate states forms an approximate <u>truncated</u> 'rotor' series. The number of levels in each series is determined by the value of N and by the rules (1) and (2). Note that the supermultiplet structure for A=-1 is obtained from that of A=+1 through interchange of the spin quantum number 1 and 3 and of parity. We further note the T-doubling in each multiplet for states with T≠0.

Only doubly excited states with A=+1 and A=-1 exhibit supermultiplet structure. States with A=0 do not display rovibrational normal modes and their spectra resemble those of singly excited states. For A=0, each pair of states with identical K, T, n, N, L and π, the energy of the triplet state is always lower than the energy of the singlet state, a behavior similar to that of singly excited states of two−electron atoms. An example of the spectral behavior of A=0 doubly excited states is shown in Fig. 5.

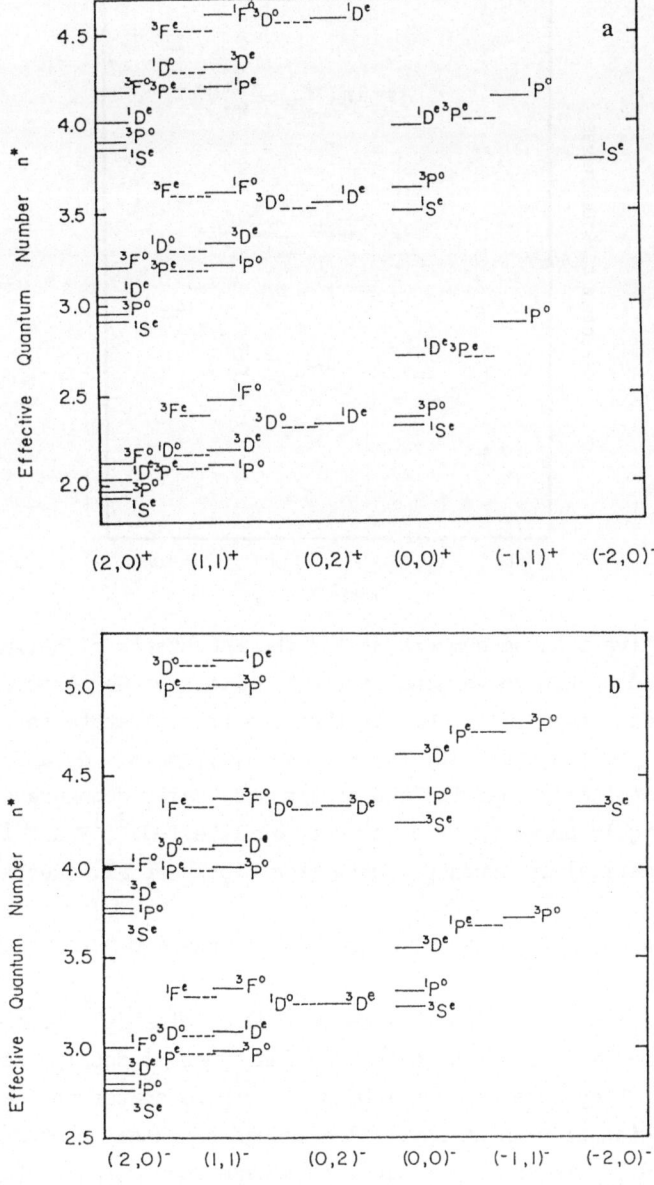

Fig. 4   Effective quantum number n* grouped according to the (a) $(K,T)^+$
and (b) $(K,T)^-$ quantum numbers for doubly excited states of
helium below the $He^+(N=3)$ limit. The rotorlike structure is
evident for each given $(K,T)^+$ and $(K,T)^-$. Energy levels taken
from the calculation of Lipsky et al. (Ref. 27).

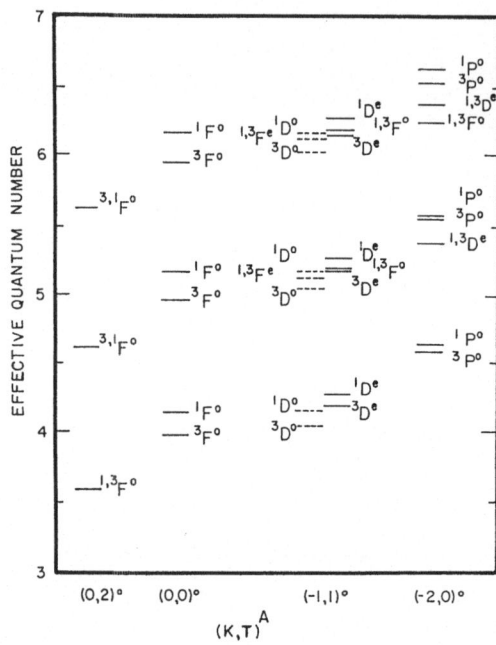

Fig. 5 Effective quantum numbers n* for the A=0 states of helium doubly exicted states converging to the N=3 limit of He$^+$ grouped according to $(K,T)^O$. Notice that the triplet state is always below the singlet state for a given K,T, and π. Notation like $^{1,3}F$ indicates that the two states are nearly degenerate, but ^1F is slighly above ^3F. The two states labelled $^{3,1}F$ are likely due to numerical inaccuracy. Data from Lipsky et al. (Ref. 27).

BODY–FRAME ANALYSIS OF CORRELATION QUANTUM NUMBERS AND MOLECULELIKE NORMAL MODES

The quantum numbers K and T were derived originally[22] from the group theoretical analysis of approximate wave functions of doubly excited states. Later they were used as labels for dipole states in the asymptotic region from which physical meanings of these quantum numbers were extracted.[23] In our classification scheme they were treated as approximate quantum numbers for describing electron correlations. To assess the 'purity' of these quantum numbers, one can analyze their wave functions in the body-frame of the atom.[21]

In the body-frame analysis we take the interelectronic axis as the internal axis of rotation. The two-electron orbital wave functions are transformed from the laboratory frame to the body frame through a rotation,

$$\mathcal{Y}_{l_1 l_2 LM}(\hat{r}_1, \hat{r}_2) = \sum_Q \mathcal{Y}_{l_1 l_2 LQ}(\hat{r}_1', \hat{r}_2') \, D_{QM}^{(L)}(\hat{\omega}) \tag{9}$$

where $(\hat{r}_1, \hat{r}_2)$ are defined in the laboratory frame and $(\hat{r}_1', \hat{r}_2')$ in the body frame, and D is the rotation matrix with Euler angles denoted by ω. Suppose the wave function is known in the laboratory frame,

$$\psi(\vec{r}_1,\vec{r}_2) = \sum_{l_1 l_2} \psi_{l_1 l_2}^L(r_1,r_2) \mathcal{Y}_{l_1 l_2 LM}(\hat{r}_1,\hat{r}_2), \tag{10}$$

transformation to the body frame changes (10) into

$$\psi(\vec{r}_1,\vec{r}_2) = \sum_Q \psi_Q^L(R,\alpha,\theta_{12}) D_{QM}^{(L)}(\hat{\omega}) \tag{11}$$

where

$$\psi_Q^L(R,\alpha,\theta_{12}) = \sum_{l_1 l_2} \psi_{l_1 l_2}^L(R\cos\alpha, R\sin\alpha) \mathcal{Y}_{l_1 l_2 LQ}(\hat{r}_1',\hat{r}_2') \tag{12}$$

and $-L \leq Q \leq L$.

From symmetry, one can show that under electron interchange

$$\psi_Q^L(R,\pi/2 - \alpha,\theta_{12}) = \pi(-1)^{S+Q} \psi_Q^L(R,\alpha,\theta_{12}) \tag{13}$$

where π is the parity of the state. We introduce a short hand notation for this phase factor:

$$A = \pi(-1)^{S+T} \tag{14}$$

where $T=|Q|$. The index A determines the symmetry of each component of the wave function projected along the interelectronic axis under particle interchange. In other words, if there is only one component in the projection in (11), T is an exact quantum number and hence A is an exact quantum number. On the other hand, if T=T' is the dominant component in (11), with small contribution from other T's, then we can assign the state to be labelled by T', with A' given by (14). In this case, both T' and A' are approximate quantum numbers.

To assess the purity of a quantum number T, we display in Fig. 6 the projection of the $(1,1)_3^{+\,1,3}P^o$ channel of He onto the body frame. The surface charge density plots for each T=0 and T=1 component are shown. Percentage represents the contribution to the normalization from each component. Note that for $^1P^o$ T=1 is the dominant component, thus A=+1. Admixture of T=0 represents a 10% effect here. This analysis shows the consistency in the assignment of T and A for this channel.

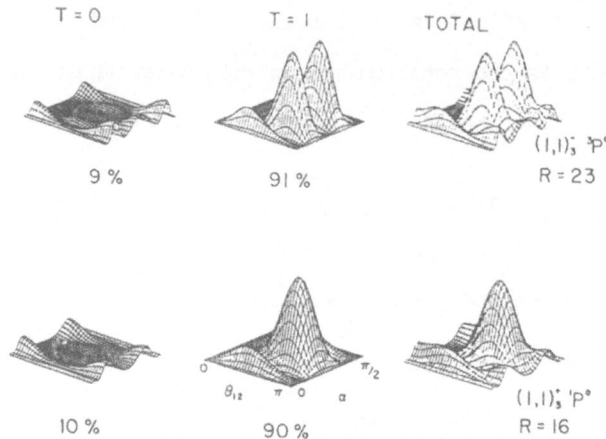

Fig. 6 Decomposition of the density plots into rotational components of the $(1,1)_3^-$ $^3P^O$ and $(1,1)_3^+$ $^1P^O$ channels of He at the values of R indicated. The percentages represent the contribution to the normalization form each T component.

The present classification scheme can also be viewed from the rovibrational model of molecules. For a given T (and A), K is related to the number of nodes, η_θ, in θ_{12}, or equivalently, to a vibrational quantum number v. A detailed analysis gives[21]

$$K = N - 2\eta_\theta - T - 1$$
$$= N - v - 1 \ . \tag{15}$$

When T is fixed, both K and v change in steps of 2.

The body-frame analysis permits us to interpret correlation quantum numbers in analogy to the rovibrational normal modes of molecules. On the other hand, there are subtle differences in doubly excited states:[21] (1) the dominant contribution to the rotational constant comes from the electron repulsion rather than from the kinetic energy; (2) the spectra show rotational contraction; (3) T-doubling results from atomic shell structure. Detailed discussion of these items is given elsewhere.[21]

We conclude this section by emphasizing that deviations of doubly excited states from the rovibrational model grow rapidly as the rovibrational energy rises. In particular, states labelled with A=0 fail to exhibit any rovibrational behavior and in fact resemble singly excited states. In this case, K and T (and A) serve to label the states without any relation to the rovibrational interpretation. This body-frame analysis also points out the limitation of treating doubly excited states purely from the molecular viewpoint.[28] Our conclusion is that there are

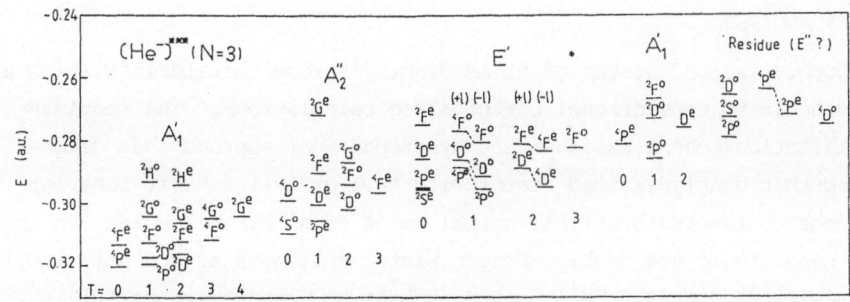

Fig. 7   Energy levels of triply exited states of He⁻ of the N=3 manifold
(3131'31") regrouped into moleculelike rotational manifolds.
States in the A_1' group have no nodes in their wave functions.
States in the A_2'' group have nodes when the nucleus lies on the
plane of the three electrons. States in the third group E'
correspond to the doubly degenerate bending mode.

two 'classes' of doubly excited states: those that exhibit rovibrational
normal modes, and those that resemble singly excited states. Only the
former can be handled approximately from a molecular viewpoint, while the
hyperspherical treatment is applicable to all doubly excited states.

TRIPLY EXCITED STATES

The body-frame analysis of doubly excited states provides guidance
for the classification of energy levels of multiply excited states from
the molecular viewpoint. Because of the complexity and the rich structure
of triply excited states, systematic calculations have only been performed
for a model problem by assuming that all three electrons are on the same
shell.[29] By carrying out a restricted CI calculation and then analyzing
the resulting wave function in the body-frame of the atom, it has been
shown that the correlation of the three electrons leads to normal modes
isomorphic to those of a molecule with D_{3h} symmetry. The energy levels
may thus be regrouped into manifolds, each revealing the rotational
structure of a symmetric top. An example of the energy levels of triply
excited states belonging to the N=3 manifold for He⁻ obtained from this
model calculation is displayed in Fig. 7. Each rotational manifold
represents a specific bending vibrational mode and is classified in the
D_{3h} group representation.

SUMMARY AND PERSPECTIVE

Doubly excited states of atoms display strong correlations which are not amenable to conventional perturbation calculations. The complete classification scheme based on the hyperspherical approach has demonstrated that doubly excited states with A=+1 or A=-1 exhibit behavior analogous to the rovibrational normal modes of an XYX molecule. On the other hand, there are doubly excited states analogous to the more familiar singly excited states. The quantum numbers K, T and A are useful in depicting how the electrons are correlated. Once the simple geometrical meanings of these quantum numbers are understood and the states are designated using the classification scheme discussed here, it is easy to get a mental picture of how the two electrons are correlated.

Despite the progress made in the classification of doubly excited states of two-electron atoms recently, there remain numerous problems to be explored. The systematics of decay widths is only partially known and its relation to K, T and A quantum numbers is not completely clear. There are limited studies of doubly excited states of multielectron atoms and ions. Knowledge of doubly excited states near the double ionization limit is practically nonexistent. However, the classification scheme developed for doubly excited states does point out new directions in atomic and molecular physics as well. The molecular rovibrational analysis has now been extended to a preliminary classification of triply excited states of three-electron atoms. The hypersherical coordinates methodology has also been extended to other Coulombic three-body systems[30] to provide new insights. Progress in these areas can be expected in the years to come.

REFERENCES

1. P. G. Krueger, Phys. Rev. 36, 85 (1930); E. Majorana, Nuovo Cimento, 8, 78 (1930); R. Widdington and H. Priestley, Proc. Roy. Soc. (London) A145, 462 (1934).

2. R. P. Madden and K. Codling, Phys. Rev. Lett. 10, 516 (1963); Astrophys. J. 141, 364 (1965); K. Codling and R. P. Madden, J. Appl. Phys. 36, 380 (1965).

3. J. W. Cooper, U. Fano and F. Prats, Phys. Rev. Lett. 10, 518 (1963).

4. U. Fano, "Doubly Excited States of Atoms," in Atomic Physics, Vol. 1 (Plenum, N.Y.) 209 (1969).

5. J. H. Macek, J. Phys. B1, 831 (1968).

6. C. D. Lin, Phys. Rev. A10, 1986 (1974).

7. C. D. Lin, Phys. Rev. A29, 1019 (1984); Phys. Rev. Lett. 51, 1348 (1983).

8. H. C. Bryant, et al. Phys. Rev. Lett. 38, 228 (1977).

9. P. R. Woodruff and J. A. Samson, Phys. Rev. A25, 848 (1982).

10. Y. K. Ho, Phys. Rept. $\underline{99}$, 1 (1983).

11. D. R. Herrick, Adv. Chem. Phys. $\underline{52}$, 1 (1983).

12. M. E. Kellman and D. R. Herrick, Phys. Rev. A$\underline{22}$, 1536 (1980).

13. D. R. Herrick, M. E. Kellman and R. D. Poliak, Phys. Rev. A22, 1517 (1980).

14. C. E. Wufman, Chem. Phys. Lett. $\underline{23}$, 370 (1973).

15. G. S. Ezra and R. S. Berry, Phys. Rev. A$\underline{28}$, 1974 (1983).

16. H. J. Yuh, G. Ezra, P. Rehmus and R. S. Berry, Phys. Rev. Lett. $\underline{47}$, 497 (1981).

17. C. D. Lin, Adv. At. Mol. Phys. $\underline{22}$, 77 (1986).

18. C. D. Lin, Phys. Rev. A$\underline{25}$, 76 (1982).

19. C. D. Lin, Phys. Rev. A$\underline{26}$, 2305 (1982).

20. C. D. Lin, Phys. Rev. A$\underline{27}$, 22 (1983).

21. S. Watanabe and C. D. Lin, Phys. Rev. A$\underline{34}$, 823 (1986).

22. D. R. Herrick and O. Sinanoglu, Phys. Rev. A$\underline{11}$, 97 (1975).

23. D. R. Herrick, Phys. Rev. A$\underline{12}$, 413 (1975).

24. M. Gailitis and R. Damburg, Proc. Phys. Soc. London $\underline{82}$, 192 (1963).

25. C. H. Greene, Phys. Rev. Lett. $\underline{44}$, 869 (1980).

26. P. A. Heimann et al. Phys. Rev. A$\underline{34}$, 3786 (1986).

27. L. Lipsky, R. Anania, and M.J. Connely, At. Data Nucl. Data Tables $\underline{20}$, 127 (1977).

28. J. Feagin and J. Briggs, Phys. Rev. Lett. $\underline{57}$, 984 (1987).

29. S. Watanabe and C. D. Lin, Phys. Rev. A$\underline{36}$, 511 (1987).

30. C. D. Lin and X. H. Liu, submitted to Phys. Rev. A (1987).

A MOLECULAR DESCRIPTION OF ELECTRON PAIR EXCITATION IN ATOMS

James M. Feagin

Department of Physics
California State University
Fullerton CA 92634 USA

INTRODUCTION

It has been thought for some time that the shape, depicted in Fig. 1, of the two-center Coulomb potential in two-electron atoms characterizes much of the atom's internal motions, particularly for double or "pair" excitations outside an ionic core. The idea is rooted in the familiar Wannier model of *threshold* double escape (Wannier, 1953; Rau, 1971; Peterkop, 1971). Small total energy E near threshold ensures that electron kinetic energies are small and hence that the escape is strongly dependent on the interplay of the interelectronic repulsion $1/R$ and the electron-ion attraction $-Z/r_1 - Z/r_2$. The breakup is thus characterized by a collinear configuration for the escaping pair along the "Wannier" saddle in the potential, clearly visible in Fig. 1. As the double escape develops and the two electrons and hence the two centers separate, one pictures the saddle as moving about the ion but with its center always located near the ion. This picture excludes the formation of one-electron bound states (single escape) when one or the other of the two-center wells orbits the ion, since the center of the saddle is located halfway between the two electrons at their center of charge.

For large enough electronic separations R the saddle is broad and flat, and the picture can also be applied to *bound* pair excitations for which both electrons are excited. Instead of separating, the two centers vibrate along the electron-pair axis. The Wannier saddle endows this axis with a rigidity, or what could be described as a sort of "false inertia," such that a correspondence with a homonuclear diatomic molecule arises. In fact, two-electron H^- has the same electrostatic energy as H_2^+ and thus Fig. 1 also depicts the two-center potential in H_2^+.

In this lecture a description of two-electron atoms based on a comparison with diatomic molecules is developed. The interelectronic axis is given the role of the internuclear axis in molecules. Symmetries of the electron pair are reclassified according to familiar molecular orbital (MO) symmetries of H_2^+ (Pack and Hirschfelder, 1968; Slater, 1977), and examples of both bound and continuum pair excitations are considered. The analogy is an old idea. It has been shown, for example, that H^- has a bound state by treating the state as a $1s\sigma_g$ molecular orbital (Gray and Pritchard, 1957; Hunter and Pritchard, 1967; see also Fröman, 1962). That the mass of the interelectronic axis is relatively small seems to be unimportant. In fact, it would appear that the structure of H_2^+ and to some extent the validity of the Born-Oppenheimer approximation in molecules may be more a consequence of the Wannier saddle in the three-particle potential than the inertia of the internuclear axis, as is generally thought.

What is the motivation for seeking such an unconventional description of atoms? Mostly, one would like to account for the extreme departure from independent-electron orbits that occurs in He, H^- and Ps^-. The three-body problem has no solution in terms of a finite number of elementary functions. If independent-electron states are used, large expansion lengths are needed and in turn large

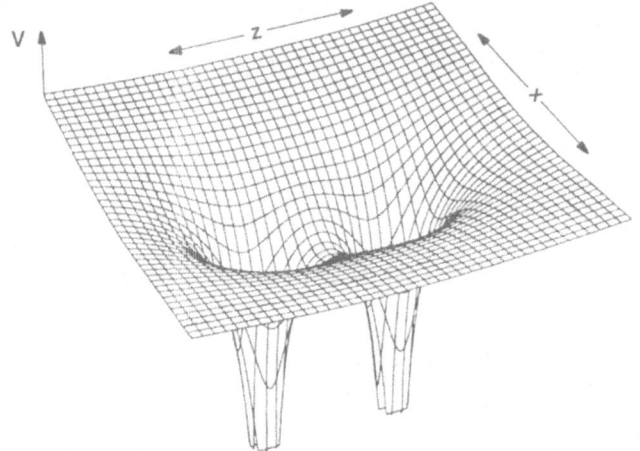

Figure 1. Two-center Coulomb potential $V = 1/R - Z/r_1 - Z/r_2$ showing the Wannier saddle with $Z=1$ for H^- or H_2^+. Here R is either the interelectronic separation in H^- or the internuclear separation in H_2^+. The z axis is taken along $\mathbf{R}$ and the electron-proton separations are calculated as $r_{i=1,2} = (x^2 + z^2 + R^2/4 \pm Rz)^{\frac{1}{2}}$. The quantity $r = (x^2 + z^2)^{\frac{1}{2}}$ determines the distance from the proton (in H^-) or from the electron (in H_2^+) to the saddle center at $z = x = 0$. The "two-center" wells are located at $x=0$, $z = \pm R/2$ with $r_{i=1,2} = 0$.

systems of coupled differential equations must be solved. From a computational viewpoint such an undertaking is often not a drawback, although for example Hartree-Fock methods give no bound state in H^- and Ps^-, and an independent-electron expansion applied to two-electron threshold escape does not appear to converge. The difficulty is instead more fundamental. Insight into approximate constants of the motion must be drawn "empirically" from say the numerical study of the nodal planes of the wavefunction. It has not been possible, in particular, to extract a complete set of commuting observables which fully define states of two-electron atoms. The analogy with diatomic molecules appears to improve this situation somewhat since the Schrödinger equation for two *fixed* Coulomb centers separates.

The molecular description of bound pair excitations given here extends the work of Hunter and Pritchard (1967) to include double excitations. The result is similar to the *adiabatic* hyperspherical description, familiar in the literature and described in this symposium, where the overall size of the atom — the hyperradius — is taken as the adiabatic variable (Macek, 1968; Fano, 1983). The molecular description given here, however, can also be used to interpret (Feagin, 1984a; Feagin et al., 1985; Feagin and Briggs, 1986) many features in hyperspherical calculations that are simply numerical results, as for example the strong similarities between He, H^- and Ps^- that hyperspherical computations have so accurately predicted. Any success of this molecular description represents an extreme breakdown of the independent-electron model and, like the adiabatic hyperspherical description, means a reduction in the number of coupled equations required to describe electron-pair excitations. One further notes that an analogy with H_2^+ is fundamentally different from one based on a linear triatomic molecule, introduced by Herrick and Kellman (1980) (Kellman, 1985), expanded upon by Ezra and Berry (1983), and compared to the adiabatic hyperspherical description by Watanabe and Lin (1986) (see also Lin's lecture in this symposium). These alternatives will be discussed again at the end of this lecture. A more detailed discussion is given elsewhere (Feagin and Briggs, 1988).

The plan of the lecture is as follows. Jacobi relative coordinates are first introduced and a general electron-pair wavefunction constructed, based on familiar molecular symmetries of H_2^+. As an application to *continuum* pair excitations, Wannier threshold angular distributions for electron impact ionization of helium are calculated. Emphasis is placed on the special nonadiabatic couplings required quantum mechanically in order to reproduce features Wannier derived classically. As an application to *bound* pair excitations, energies and eigenfunctions of H_2^+ for fixed internuclear separation are scaled to two-electron atoms. Adiabatic potential curves are obtained as a function of the interelectronic separation that bear a strong resemblance to potential curves as a function of hyperspherical radius. A molecular interpretation is given to features in the hyperspherical calculations, while the scaling is shown to account for observed similarities in He, H^-, and Ps^-. Finally, transitions between levels and other nonadiabatic effects and the connection of the H_2^+ scaling with other approaches are discussed. Atomic units are used throughout the text except where otherwise indicated.

ELECTRON PAIR SYMMETRIES

In order to describe the electron pair as a collinear unit it is most straightforward to introduce the relative coordinate vector $R=x_1-x_2$ of the two electrons as in Fig. 2, with $x_{i=1,2,3}$ the position vectors relative to the system center of mass (marked with an X in Fig. 2). Then only one other coordinate vector r is needed to describe the position of the interelectronic axis and specify completely the six degrees of freedom of this three-body problem in the center of mass system.

Jacobi Coordinates

This second vector r is fixed by requiring that both r and R and their conjugate momenta satisfy either Poisson bracket relations classically or commutation relations quantum mechanically. In this way one derives uniquely the *Jacobi coordinates* which take the form

$$P_R = \mu_{12} \frac{dR}{dt} \equiv -i \nabla_R \qquad\qquad \mu_{12} = \frac{m_1 m_2}{m_1 + m_2} \qquad (1a)$$

$$r = \frac{m_1 x_1 + m_2 x_2}{m_1 + m_2} - x_3 \qquad\qquad (1b)$$

$$P_r = \mu_{12,3} \frac{dr}{dt} \equiv -i \nabla_r \qquad\qquad \mu_{12,3} = \frac{(m_1 + m_2)m_3}{m_1 + m_2 + m_3} . \qquad (1c)$$

Thus, the proper choice of relative coordinates turns out to have the desirable feature that r and its conjugate momentum describe the motion of the center of charge and mass of the electron pair relative to the ion when R is the relative coordinate of the pair. As will become evident, these coordinates are useful in describing electron-pair symmetries, particularly for excitations in which the ion remains near the Wannier saddle in the two-electron potential (Feagin, 1984b).

In (1) μ_{12} and $\mu_{12,3}$ are the reduced masses of the electron pair and of the pair's center of mass relative to the ion, respectively. Thus, $\mu_{12} = \frac{1}{2}$, while in the cases of interest

$$He, H^-: \quad \mu_{12,3} = \frac{2 \cdot Am_p}{2 + Am_p} \simeq 2 \qquad\qquad (2a)$$

$$Ps^-: \quad \mu_{12,3} = \frac{2 \cdot 1}{2 + 1} = \frac{2}{3} \qquad\qquad (2b)$$

where A is the nuclear mass number and $m_p \simeq 1836$ is the mass of the proton. That these reduced

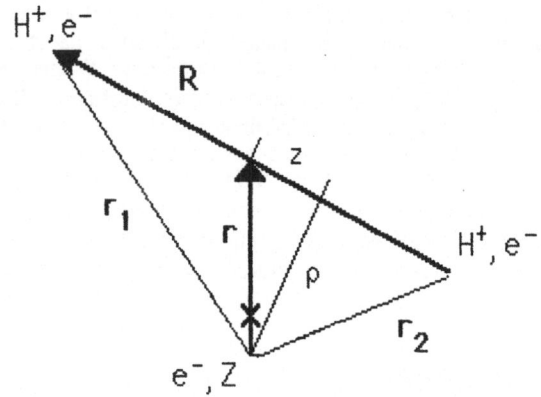

Figure 2. Jacobi coordinates for two electrons bound to an ion with charge Z or for H_2^+ with two protons and one electron. Here X indicates the center of mass of the atom and ρ and z are cylindrical components of $\mathbf{r}$ about a body-fixed z axis along $\mathbf{R}$.

masses are all of order unity helps to ensure the good analogy between two-electron systems and molecular H_2^+. In the molecule $m_1=m_2=m_p$ and the reduced mass of the internuclear axis is $\mu_{12} = \frac{1}{2}m_p$ while

$$H_2^+: \qquad \mu_{12,3} = \frac{2m_p \cdot 1}{2m_p + 1} \simeq 1 \; . \qquad\qquad (2c)$$

Finally, with these coordinates and momenta, one transforms the three-body hamiltonian to the center of mass system as

$$H = \frac{-\nabla_R^2}{2\mu_{12}} + \frac{-\nabla_r^2}{2\mu_{12,3}} + \frac{1}{R} - \frac{Z}{|\mathbf{r}+\frac{1}{2}\mathbf{R}|} - \frac{Z}{|\mathbf{r}-\frac{1}{2}\mathbf{R}|} \; . \qquad (3)$$

The Jacobi coordinates have the additional feature that no cross terms of the form $\nabla_R \cdot \nabla_r$ appear in the kinetic energy and are therefore sometimes referred to as "orthogonal coordinates."

Geometrical Symmetries

From the form of the hamiltonian in (3) it is evident that the system possesses a number of simple and familiar symmetries. For example, it is clear that H is invariant under coordinate inversion to simultaneous changes of both $\mathbf{R} \to -\mathbf{R}$ and $\mathbf{r} \to -\mathbf{r}$. This invariance leads, of course, to the classification of states with a definite even or odd *parity*.

The hamiltonian is also invariant to separate inversions of $\mathbf{R} \to -\mathbf{R}$ and of $\mathbf{r} \to -\mathbf{r}$. The former amounts to electron *exchange* and for two electrons (but no more!) leads to the familiar spin-S designation of singlet (symmetric) and triplet (antisymmetric) spatial wavefunctions. The latter symmetry is redundant in as much as it is a product of *inversion × exchange* but may in fact be more fundamental than either since pair correlations are well characterized, as will be demonstrated shortly, by orbital functions of internal coordinates $\mathbf{r}$ and R confined to the plane of the three particles. The $\mathbf{r} \to -\mathbf{r}$ symmetry is not easily recognized in more conventional descriptions involving functions of $\mathbf{r}_1$ and of $\mathbf{r}_2$, rather than functions of $\mathbf{r}=\frac{1}{2}(\mathbf{r}_1+\mathbf{r}_2)$ as shall be considered here. In any case, the evenness or oddness of the wavefunction under the operation $\mathbf{r} \to -\mathbf{r}$ leads to the familiar gerade/ungerade designation from homonuclear diatomic molecules (Slater, 1977).

Finally, it is clear that the hamiltonian is invariant to rotations of the system as a whole in which **R** and **r** are rotated together preserving their relative orientation. Overall *rotational* symmetry leads, of course, to the classification of states of definite total angular momentum **L** = **L**$_R$ + **ℓ_r** about the system center of mass and its projection along a fixed direction in the laboratory. These symmetries and their corresponding operators and eigenvalues are summarized in Table 1.

Approximate Dynamical Symmetries

Consider now the representation of a nearly rigid interelectronic axis. It is appropriate, as it is in molecules (Pack and Hirschfelder, 1968), to single out this axis explicitly by rotating to a body-fixed (BF) frame with z axis aligned along **R**. One thus expands the system wavefunction $\Psi(\mathbf{r},\mathbf{R})$ as a sum over rotation matrices $D_{MK}^{L}(\psi\theta\phi)$ determined by the spherical polar angles $\theta\psi$ of **R** in a laboratory-fixed frame and the rotation angle ϕ of the plane of the three particles about **R**. These angles constitute a set of Euler angles specifying the orientation of the plane of the three particles in a laboratory-fixed frame. Here M and K are the quantum numbers of the projections, $\mathbf{L}\cdot\hat{\mathbf{z}}=-i\partial/\partial\psi$ and $\mathbf{L}\cdot\hat{\mathbf{R}}=-i\partial/\partial\phi$, of the total angular momentum along the laboratory-fixed and the BF z axes, respectively. The expansion coefficients in the sum are then functions of the *internal* coordinates r and R in the plane of the three particles. In making the analogy with molecules, it is further useful to factor these coefficients as $\psi_{Kt}^{BF}(\mathbf{r},\mathbf{R})= F(R)\cdot\phi_{Kt}(\mathbf{r},\mathbf{R})$, defining a *channel function* $\phi_{Kt}(\mathbf{r},\mathbf{R})$ to describe the center of mass motion of the electron pair and a function F(R) to describe the two-electron relative motion. Here the quantum number t labels the g/u symmetry (from Table 1).

The transformation of the system wavefunction requires the simultaneous transformation of the hamiltonian (3) to the rotating BF frame. Thus, the usefulness of introducing BF wavefunctions and internal coordinates R and r is somewhat offset by the fact that centrifugal and Coriolis terms are introduced in the BF hamiltonian. These potentials arise in transforming the interelectronic kinetic energy, in particular, in transforming

$$\nabla_R^2 = \frac{1}{R^2}\frac{\partial}{\partial R}R^2\frac{\partial}{\partial R} - \frac{1}{R^2}\mathbf{L}_R^2 ,$$

where **L**$_R$=**L**-**ℓ_r** is the interelectronic angular momentum. The details are rather involved and the

Table 1. Summary of Symmetries, Their Commuting Operators and Eigenvalues

Parity	$R\rightarrow-R$	$\psi\rightarrow\pi+\psi$	=>	P:	$[P, H] = 0$
	$r\rightarrow-r$	$\theta\rightarrow\pi-\theta$		π	$\pi=0$ (even)
		$\phi\rightarrow\pi-\phi$			$\pi=1$ (odd)
Exchange	$R\rightarrow-R$	$\psi\rightarrow\pi+\psi$	=>	P_{12}:	$[P_{12}, H] = 0$
	$r\rightarrow+r$	$\theta\rightarrow\pi-\theta$		$(-1)^S$	$S=0$ (singlets)
		$\phi\rightarrow-\phi$			$S=1$ (triplets)
Inv/Exc	$R\rightarrow+R$	$\psi\rightarrow\psi$	=>	PP_{12}:	$[PP_{12}, H] = 0$
	$r\rightarrow-r$	$\theta\rightarrow\theta$		$(-1)^t=\pi(-1)^S$	$t=0$ (gerade)
		$\phi\rightarrow\pi+\phi$			$t=1$ (ungerade)
Rotations	$R\rightarrow R$	$\psi\rightarrow\psi'$	=>	L^2, L_z:	$[L^2, H] = [L_z, H] = 0$
	$r\rightarrow r$	$\theta\rightarrow\theta'$		$L(L+1), M\leq L$	
	$\mathbf{R}\cdot\mathbf{r}\rightarrow\mathbf{R}\cdot\mathbf{r}$	$\phi\rightarrow\phi'$			
Dynamical		$\phi\rightarrow\phi'$	=>	$\mathbf{L}\cdot\hat{\mathbf{R}}$:	$[\mathbf{L}\cdot\hat{\mathbf{R}}, H_{BF}] \sim 0$
				$K\leq L$	

result somewhat complicated (Pack and Hirschfelder, 1968). Instead, they can be summarized by writing $L_R{}^2 \rightarrow (L_R{}^2)_{BF}$, noting that the rotation to the BF frame affects only operators which depend on the Euler angles $\psi\theta\phi$ and not ones which depend on R and **r**. The transformed system hamiltonian (3) can then be written compactly as

$$H_{BF} = \frac{-1}{2\mu_{12}R^2} \frac{\partial}{\partial R} R^2 \frac{\partial}{\partial R} + \frac{(L_R^2)_{BF}}{2\mu_{12}R^2} + \frac{1}{R} + h \qquad (4a)$$

defining a hamiltonian for the electron-pair center of mass as

$$h = \frac{-\nabla_r^2}{2\mu_{12,3}} - \frac{Z}{|r+\frac{1}{2}R|} - \frac{Z}{|r-\frac{1}{2}R|} . \qquad (4b)$$

The BF hamiltonian *matrix* thus has block-diagonal elements in L from $(L_R{}^2)_{BF}/2\mu_{12}R^2$ that connect channel functions of different K and elements from $\partial/\partial R$ that connect channel functions of the same K.

The origins of approximate dynamical symmetries for electron-pair excitations of interest here can be now made clearer. First, one notes that if the interelectronic axis is long enough then the *off-diagonal rotational* couplings from the centrifugal and Coriolis terms are small compared to the interelectronic repulsion and can be ignored in zeroth approximation. That is in (4a) for large R

$$\frac{(L_R^2)_{BF}}{2\mu_{12}R^2} \ll \frac{1}{R} \qquad (5)$$

for the off-diagonal matrix elements in K. Then the projection of the total angular momentum along the BF z axis **R** is approximately conserved, such that $[L \cdot \hat{R}, H_{BF}] \sim 0$, and K becomes an "approximately good" quantum number. Note that this cylindrical symmetry is exact for the two-center Coulomb potential (Fig. 1) and is included in Table 1.

It also turns out in applications of interest that important features of the channel functions $\phi_{Kt}(r,R)$ remain unaltered over a wide range of interelectronic separations R. This *adiabatic invariance* is due in part to the rigidity or "false inertia" of the interelectronic axis and hence is a manifestation of the Wannier saddle. It means that *off-diagonal radial* couplings from $\partial/\partial R$ in (4a) connecting different adiabatic channels of the same K are small and can also be ignored. (The Wannier description of continuum excitations requires, however, modification of this approach; also isolated potential energy curve crossings can and do arise which lead to strong couplings but usually over a short range R only.) Then, along with the assumption that K is a good quantum number, the expansion of the system wavefunction can be reduced to a single adiabatic channel. From the transformation properties of the Euler angles (Table 1) one derives the transformation of the rotation matrices under parity and exchange. In this way, one constructs wavefunctions of definite LSΠ symmetry and approximate dynamical symmetry as

$$\psi_{NKt}^{L S \Pi} \sim \left[F(R) \cdot \phi_{NKt}(r,R) \right]_L \left[D_{MK}^L + \pi (-1)^{L+K} D_{M-K}^L \right] \chi_{SM_s} \qquad (6)$$

where χ_{SM_s} is a two-electron spin wavefunction. Here the notation $[...]_L$ is used to indicate an implicit dependence of the internal wavefunction on the total angular momentum quantum number. The pair of labels K and t give way to the more familiar molecular designations σ_g, σ_u and π_g, π_u etc. corresponding to K=0 (t=±1) and K=1 (t=±1) etc. (Slater, 1977).

The wavefunction (6) is an eigenfunction of parity P, exchange P_{12}, and the square of the total angular momentum L^2 and its projections, $\mathbf{L} \cdot \hat{\mathbf{z}}$ and $\mathbf{L} \cdot \hat{\mathbf{R}}$, along the laboratory–fixed and along the BF z axes. These operators form the basis of a complete set of commuting dynamical observables. Thus, two of the three quantum numbers S, π, or t (see Table 1) and L and the total energy E provide four of the six constants of the motion required to specify completely the six coordinate degrees of freedom. The remaining two constants of the motion are provided by the two nodal quantum numbers of the internal wavefunction in the plane of the three particles, at least one of which is related to K.

CONTINUUM PAIR EXCITATIONS

As a brief application consider a calculation of *threshold* angular distributions for electron impact ionization. Such an application enjoys the possibility of comparison with a recent detailed experiment. In particular, Selles, Huetz and Mazeau (1987a) at the Université Pierre et Marie Curie in Paris, in a sequence of remarkable measurements, bombarded helium with electrons at energies as low as one–half electron volt above the ionization threshold and detected escaping electrons in coincidence over a range of angles relative to each other and to the incident beam of electrons. They thus considered the reaction $e^- + He \rightarrow 2e^- + He^+$ to excite a pair of electrons simultaneously to the continuum with nearly zero total energy E. A sample of their data is shown in Fig. 3.

An incident beam of electrons means that the system retains only azimuthal symmetry with respect to the beam direction. Consequently, the wavefunction of the two escaping electrons must be taken as a superposition over all angular momentum components L, but with L=ℓ, where ℓ is the angular momentum of a component in the partial–wave decomposition of the incident beam (assuming that the initial atom and the final ion occupy s states). Thus, each component has parity $\pi = (-1)^\ell$ $= (-1)^L$ while the azimuthal symmetry allows one to set M≡0. With these restrictions, but summing over L, the electron–pair wavefunction (6) takes the simple form

$$\psi^S_{M=0} = \sum_{L, K \leq L} a_{LK} \left[F(R) \, \phi_{Kt}(\mathbf{r}, R) \right]_L \cos K\phi \, P^K_L(\cos\theta) \qquad (7)$$

where the a_{LK} are constant coefficients and the $P_L^K(\cos\theta)$ are associated Legendre functions.

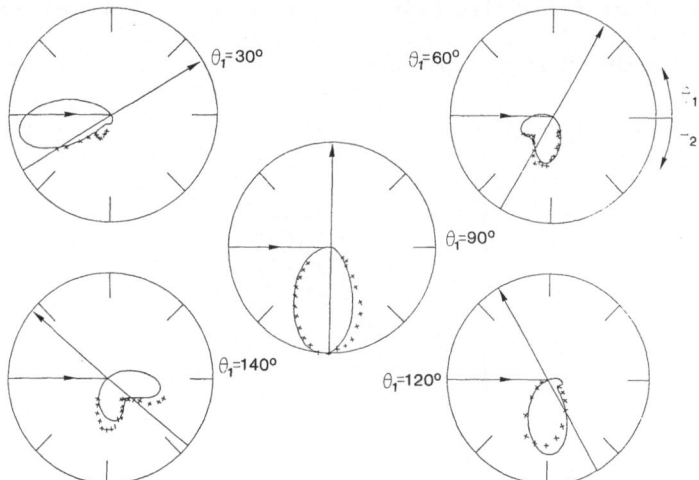

Figure 3. Polar plots of angular distributions for ionization of helium by electron impact at E=1eV above threshold, measured by Selles, Huetz, and Mazeau (1987a). Here θ_1 and θ_2 locate positions of a pair of coincidence detectors about the incident beam of electrons (incoming arrows).

Molecular Wannier Separation

To derive an energy dependence for the threshold cross section the internal wavefunction in (7) of the escaping electron pair must be constructed. The model of two-electron threshold escape that is generally believed to be the correct one was due originally to Wannier (1953). He argued that a classical analysis could perhaps be enlightening, since near threshold the deBroglie wavelengths of the escaping electrons are small compared to variations in the potential. He assumed that the two electrons, because of the shape of the "Wannier" saddle, would emerge in nearly opposite directions equidistant from the ion. He further argued that this configuration with an unstable equilibrium along the interelectronic axis would have to be maintained if the formation of bound systems and simple excitation were to be avoided. Although Wannier used hyperspherical coordinates in his analysis, it is clear that a description of the escape configuration is amenable to Jacobi coordinates.

The quantum mechanical formulation of Wannier's classical analysis uses an expansion of the internal wavefunction about the Wannier saddle in the potential for large separations of the particles, and only in the small E limit (Rau, 1971, Peterkop, 1971). This region of configuration space is illustrated in Fig. 4. The saddle is depicted here by equal separations $r_1=r_2$ of the two electrons from the ion ($r_{1,2}=|\mathbf{r}\pm\frac{1}{2}\mathbf{R}|$) (see Fig. 2). With Jacobi coordinates the Wannier configuration occurs for large R and small r; that is, for large electronic separations with the ion in the saddle near the center of mass of the two electrons. Near the saddle, $R \sim r_1+r_2$ and is tangent to a hyperradius $\mathcal{R}\equiv(r_1^2+r_2^2)^{\frac{1}{2}}$, as is evident in Fig. 4. Thus, the Wannier model expressed in Jacobi coordinates (Feagin, 1984b) is very similar to hyperspherical coordinate descriptions, although the Jacobi coordinates are well suited to the symmetries.

Crucial to the development of the Wannier formalism is the exclusion of the small r_1, r_2 region or Reaction zone depicted in Fig. 4. A solution valid there would be tantamount to a full solution of the three-body Coulomb problem. Instead, one usually attempts the less ambitious task of determining the analytic form of the electron-pair wavefunction outside the Reaction zone, and even there only near the Wannier saddle, forfeiting an evaluation of the expansion coefficients a_{LK} in (7). (See, however, Crothers (1986).)

Threshold Wavefunction

In practice, the Reaction zone is large enough that outside of it for large electron separations R the centrifugal and Coriolis couplings can be ignored and K taken to be a good quantum number. The arguments leading to eqs. (6) and (7) can be followed. It is then appropriate to introduce cylindrical-coordinate components of $\mathbf{r}$ about the BF z axis, i.e. $z\equiv\mathbf{r}\cdot\hat{\mathbf{R}}$ and $\rho\equiv\mathbf{r}\cdot[\hat{\mathbf{R}}\times(\hat{\mathbf{r}}\times\hat{\mathbf{R}})]$ (see Fig. 2). These coordinates and the scattering geometry appropriate to the experiment of Selles, Huetz and Mazeau are shown in Fig. 5.

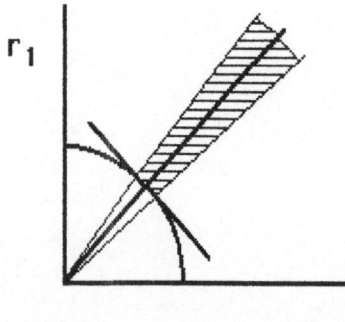

Figure 4. Schematic of the two-electron configuration space indicating the Wannier region and the Reaction zone bounded by the line $r_1+r_2 \sim R_0$ and by the hyperradius $\mathcal{R}_0 \equiv (r_1^2+r_2^2)^{\frac{1}{2}}$.

In order to imitate Wannier's classical analysis, the Schrödinger equation in the BF frame must nevertheless be solved as a pair of *coupled* equations (but diagonal in K)

$$\left[h - \frac{1}{2\mu_{12}} \left(\frac{1}{R} \frac{df}{dR} \frac{\partial}{\partial R} + \frac{\partial^2}{\partial R^2} \right) \right] \Phi_{Kt} = \mathcal{E}(R) \, \Phi_{Kt} \qquad (8a)$$

$$-\frac{d^2 f}{dR^2} + 2\mu_{12} \left(\frac{1}{R} + \mathcal{E}(R) \right) f = 2\mu_{12} E \, f \; . \qquad (8b)$$

where $f(R) = R \, F(R)$ has been introduced to simplify the radial derivatives from (4a). In addition, a large-R expansion of the three-body Coulomb potential about the Wannier saddle point $r = (\rho^2 + z^2)^{1/2} = 0$ is assumed, i.e.

$$V \underset{R\to\infty}{\sim} -\frac{C_0}{R} + \frac{C_1}{2R} \left(\frac{\rho}{R} \right)^2 - \frac{C_2}{2R} \left(\frac{z}{R} \right)^2 + \dots \qquad (9)$$

where the expansion coefficients depend on the ion's charge as $C_0 = 4Z-1$ and $C_2 = 2C_1 = 32Z$. The relative signs of C_1 and C_2 as they appear here are characteristic of a saddle. Thus, there is a restoring force along ρ giving rise to simple harmonic motion, while the motion along z is unstable, although for large R the saddle is nearly flat.

The distinguishing feature of the quantum mechanical formulation of the Wannier model is the need to include in (8a) in the electronic center of mass motion diagonal radial couplings to the two-electron relative motion. These nonadiabatic or *diabatic* couplings are required to concert the

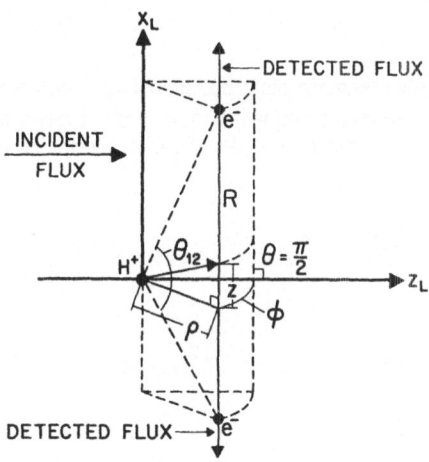

Figure 5. The scattering geometry and coordinates used to analyze threshold angular distributions. Here θ_{12} is the mutual angular separation of the two-electrons. Drawn for $\theta = \pi/2$ and for out-of-plane scattering $\phi \neq 0$.

electron-pair motion and continually redistribute the system energy, to ensure that double escape takes place (Feagin, 1984b). One thus obtains the quantities Wannier derived in his classical analysis. (Analogous radial couplings are also required if hyperspherical coordinates are used.)

In order to have the symmetries of Table 1, the wavefunction $\phi_{Kt}(z,\rho,R)$ of the electron pair center of mass must be proportional to $\rho^{K}\cdot z^{t+K} = \rho^{K}\cdot z^{L+S+K}$ since each scattering angular-momentum component has parity $\pi=(-1)^L$. From the definition $z=\mathbf{r}\cdot\hat{\mathbf{R}}$ it is clear that $z\rightarrow z$ under inversion while $z\rightarrow-z$ under exchange and under inversion×exchange. It is also clear that ρ is invariant to all symmetry operations since it measures the distance of the pair center of mass from the interelectronic axis. That no other nodes are given in the Wannier description is of course due to the approximate nature of the expansion of the internal wavefunction about the saddle, which remains a fundamental shortcoming of the model.

Threshold Cross Sections

Introducing a procedure for matching at $R=R_0$ the Wannier wavefunction onto the unknown one inside the Reaction zone (Peterkop,1971) allows one to infer the energy dependence Wannier obtained of the $L=S=0$ threshold cross section (but not its absolute value) as

$$
Q \sim E^m \qquad m = -\frac{1}{4} + \frac{3}{4}\left(1 + \frac{8\mu_{12}C_2}{9\mu_{12,3}C_0}\right)^{\frac{1}{2}}. \tag{10}
$$

As is evident, the Jacobi coordinates permit a clean extraction of the dependence of the Wannier index m on the reduced masses and Z. The occurrence of the square root in (10) — and thus the nonintegral magnitude of m — is a direct consequence of the diabatic couplings included in (8a). Values of the index are collected in Table 2.

Finally, the differential cross section for two-electron threshold escape can be derived from a probability current density obtained from the wavefunction (7) with the Wannier solutions of (8). One finds that

$$
\frac{dQ}{d\cos\theta\,d\cos\theta_{12}\,d\phi} = E^{m-\frac{1}{2}}\left[\frac{1}{4}|f_0|^2 + \frac{3}{4}|f_1|^2\right]\exp\left(-\frac{4\ln2\,(\pi-\theta_{12})^2}{w^2 E^{1/2}}\right) \tag{11a}
$$

where the width of the mutual angular distribution w is calculated for two-electron escape from helium to be $68°/(eV)^{1/4}$, in good agreement with a recent detailed analysis of independent sets of data (Cvejanović, 1987). Including only S, P, and D waves in the expansion, one derives for the *singlet* amplitude

$$
f_0 = b_0(^1S^e) + b_1(^1P^o)\,E^{-1/4}\,(\pi-\theta_{12})\cos\phi\,P_1^1(\cos\theta) + b_0(^1D^e)\,P_2(\cos\theta)
$$

$$
+ b_2(^1D^e)\,E^{-1/2}\,(\pi-\theta_{12})^2\cos2\phi\,P_2^2(\cos\theta) \tag{11b}
$$

and for the *triplet* amplitude

$$
f_1 = b_0(^3P^o)\,P_1(\cos\theta) + b_1(^3D^e)\,E^{-1/4}(\pi-\theta_{12})\cos\phi\,P_2^1(\cos\theta) \tag{11c}
$$

Table 2. Wannier Threshold Index for Some Systems of Interest

system	C_2/C_0	μ_{12}	$\mu_{12,3}$	m
He	64/7	½	2	1.056
H^-	32/3	½	2	1.127
Ps^-	32/3	½	2/3	1.886
H_2^+	32/3	$½m_p$	1	69.73

where the $\cos K\phi$ factors specify the out-of-scattering-plane contributions (see Fig. 5) while the $(\pi-\theta_{12})^K$ factors arise from the nodes ρ^K in the Wannier wavefunction; since $\rho \sim R|\pi-\theta_{12}|/4$ in the Wannier configuration. The expansion coefficients $b_K(LS\Pi)$, related to the a_{LK} in (7), are not determined by the matching procedure but are presumably energy independent.

In order to compare these expressions with experimental data it is useful to transform them to detector coordinates θ_1 and θ_2. This transformation has been made by Selles, Mazeau, and Huetz (1987b), who in fact have demonstrated the equivalence of the expressions in (11) to ones they derived using the detector coordinates. Their fits to the data, including only S, P, and D waves, are shown in Fig. 3. The cusps in the data are effects from the nodes in the Wannier wavefunction. The close fit indicates good convergence of the expansion in partial waves L of the electron-pair wavefunction and provides confidence in the Wannier model of continuum pair excitations near threshold.

BOUND PAIR EXCITATIONS

It has been emphasized in the foregoing that in order to reproduce in a quantum mechanical argument quantities Wannier obtained in a classical analysis, one must include diabatic couplings in (8a). It was further indicated that these radial couplings so severely complicate the solution of the pair of equations (8) that only a series expansion of the wavefunction about the Wannier saddle is known. The question then naturally arises: What kinds of pair excitations might be described when these diabatic couplings are dropped altogether from the electronic center of mass motion? At the same time one notes that the two-center hamiltonian defined in (4a) and included in (8a) can be separated for fixed R and in fact defines electron eigenfunctions and energies of *molecular* H_2^+ in the Born-Oppenheimer approximation for fixed internuclear separation. Although one could (and at some point should) investigate the question more systematically, e.g. by numerical integration, it is instructive to consider just a simple scaling of H_2^+ to two-electron atoms. Here only the lowest lying bound-pair excitations will be considered.

Molecular Adiabatic Separation

In order to obtain the simplest zeroth-order approximation to the internal wavefunction it is useful to assume that the interelectronic axis is rigid enough and long enough that at least the *off-diagonal* couplings from (4a) can be ignored. Then the factored wavefunction (6) is still a good approximation for the electron pair. Although the model must clearly work better for double excitations, one obtains in fact a fair qualitative description of ground and singly excited states of two-electron atoms. One thus works in zeroth order with a pair of equations, as in the Born-Oppenheimer approximation, with

$$h \, \phi_{nKt}(\mathbf{r},R) = \mathcal{E}_{nKt}(R) \, \phi_{nKt}(\mathbf{r},R) \qquad (12a)$$

describing the electronic center of mass motion and with

$$-\frac{df}{dR^2} + 2\mu_{12}\left(\frac{1}{R} + \varepsilon_{nKt}(R) + C_{nLK}(R)\right) f = 2\mu_{12}E\,f \qquad (12b)$$

describing the vibration along the interelectronic axis. Here the "diagonal coupling" matrix element (integrals over **r**)

$$C_{nLK}(R) = \left\langle \phi_{nKt} \left| \left(\frac{-\nabla_R^2}{2\mu_{12}}\right)_{BF} \right| \phi_{nKt} \right\rangle \qquad (12c)$$

is retained as a remnant of the Wannier diabatic couplings in (8a) and as a measure of the interelectronic kinetic energy in the BF frame for a given *adiabatic MO* channel $\phi_{nKt}(\mathbf{r},R)$.

As in molecular $H_2{}^+$ the two-center Coulomb problem (12a) for fixed R in the plane of the three particles separates in prolate spheroidal coordinates, often referred to as (confocal) elliptic coordinates, defined as (Bates and Reid, 1968; Slater, 1977)

$$\lambda = (r_1 + r_2)/R \qquad \mu = (r_1 - r_2)/R$$

$$1 \leq \lambda < \infty \qquad -1 \leq \mu \leq 1 \qquad (13)$$

$$\phi(\mathbf{r},R) = M(\mu)\,\Lambda(\lambda)$$

where r_1 and r_2 are electron-ion separations. Thus, points in the plane are located by intersecting ellipses (curves of constant λ) and hyperbolae (curves of constant μ). As $R{\rightarrow}0$ these coordinates go over smoothly into spherical polar coordinates r,θ appropriate to the united-atom or "united-electron" limit, and as $R{\rightarrow}\infty$ into parabolic Stark-state coordinates η,ξ appropriate to the separated-atom limit. In the elliptic coordinates the three-particle potential energy from (3) takes a particularly simple form

$$V(\mathbf{r},\,R) = \frac{1}{R} - \frac{Z}{r_1} - \frac{Z}{r_2} = \frac{1}{R}\left[1 - \frac{4Z\lambda}{\lambda^2 - \mu^2}\right].$$

The Wannier saddle is centered on $\lambda=1$ and $\mu=0$.

Associated with separability is a pair of molecular quantum numbers N_μ, N_λ that specify the number of nodes in the adiabatic MO $\phi_{nKt}(\mathbf{r},R)$. In particular, N_μ specifies the number of nodes along the interelectronic axis $\lambda=1$, and N_λ the number along its perpendicular bisector $\mu=0$. These nodes are conserved as R varies, and there exists a direct relation between the molecular ($N_\mu N_\lambda K$), the united-electron (nLK), and the separated-atom ($N_\eta N_\xi K$) quantum numbers (Bates and Reid, 1968). All three sets provide equivalent characterizations of the adiabatic MO, although it is usual to use the united-atom designation $1s\sigma_g$, $2p\sigma_u$, $2s\sigma_g$, etc. and write $\phi_{n\ell Kt}(\mathbf{r},R)$. It is readily seen that $(-1)^t = (-1)^\ell$, the united-atom parity. Finally, in the separated-atom limit one also uses the spherical-polar principal quantum number N of the single, bound electron where $N_\eta + N_\xi = N-K-1$. These points are summarized in Table 3.

Separability also implies an additional symmetry, which is present only for pure two-center Coulomb potentials. The dynamical operator which gives rise to this symmetry and which therefore commutes with the hamiltonian h in (12a) is known (Coulson and Joseph, 1967; Greenland, 1982). It is related to a one-electron Runge-Lenz vector but is conveniently expressed in the form

Table 3. Relations Between MO, United-Electron, and Separated-Atom Coordinates and Quantum Numbers

united-electron $R \to 0$	MO R finite	separated-atom $R \to \infty$
$\lambda \to 2r/R$	λ	$\lambda \to 1 + \eta/R$
$\mu \to \cos\theta$	μ	$\mu \to \pm(-1 + \xi/R)$
$N_\lambda = n - \ell - 1$	N_λ	$N_\lambda = N_\eta$
$N_\mu = \ell - K$	N_μ	$N_\mu = 2N_\xi \quad (N_\mu = \text{even})$
		$N_\mu = 2N_\xi + 1 \quad (N_\mu = \text{odd})$
$(-1)^\ell = (-1)^t$		$N_\eta + N_\xi = N - K - 1$

$$\Omega(R) = \tfrac{1}{2}(\boldsymbol{\ell}_1 \cdot \boldsymbol{\ell}_2 + \boldsymbol{\ell}_2 \cdot \boldsymbol{\ell}_1) + Z\,\mathbf{R} \cdot \left(\hat{r}_1 - \hat{r}_2\right) \tag{14}$$

where $\boldsymbol{\ell}_{j=1,2} = -i\mathbf{r}_j \times \nabla_j$ are one-electron angular momenta relative to the ion. The separation constant in the solution of (12a) is given by the MO expectation value of $-\Omega - \tfrac{1}{2}R^2 h$. Because of its dependence on R, $\Omega(R)$ does not commute with the full BF hamiltonian in (4a), although the applicability of the adiabatic approximation would indicate that it nearly does. If added to the list in Table 1, one obtains a complete set, albeit approximate, of commuting dynamical observables. This point will be taken up again at the end of this lecture along with a consideration of Herrick and Kellman's (1980) and of Lin's (1984) classification schemes.

Scaling H$_2{}^+$ Energies and Wavefunctions

Let $E(R)$ and $\varphi(\mathbf{r},R)$ denote the eigenenergies and eigenfunctions of H$_2{}^+$ obtained by solving (12a) for fixed internuclear separation with $Z=1$ and $\mu_{12,3}=1$. The methods for obtaining these solutions are well known and the results widely tabulated. Some of the low lying energies as a function of R are shown in Fig. 6.

Consider now scaling these solutions to two-electron atoms for which $\mu_{12,3} \neq 1$. The procedure will be to generalize somewhat the scaling of H$_2{}^+$ to $Z \neq 1$ diatomic molecules (Bates et al., 1953). The idea is an old one and has in fact been applied in some detail to the calculation of *single* excitations in He and H$^-$ (Hunter and Pritchard, 1967; see also Fröman, 1962). To simplify notation, subscripts shall be omitted when context allows. Scaled coordinates and wavefunctions are then defined as

$$\mathbf{x} = \mu Z \zeta\, \mathbf{r} \qquad \mathbf{S} = \mu Z \zeta\, \mathbf{R} \qquad \mu \equiv \mu_{12,3}$$

$$\phi(\mathbf{r}, R) = (\mu Z \zeta)^{3/2}\, \varphi(\mathbf{x}, S) \tag{15a}$$

where a R-dependent variational parameter $\zeta = \zeta(R)$ has been introduced (Slater, 1977; Rost and Briggs, 1987). The factor $(\mu Z \zeta)^{3/2}$ is required to preserve wavefunction normalization. The electron-pair energies from (12a) for fixed R can then be scaled from H$_2{}^+$ as follows:

$$\varepsilon(R) = \int \phi(\mathbf{r},R) \left[\frac{-\nabla_r^2}{2\mu} - \frac{Z}{r_1} - \frac{Z}{r_2} \right] \phi(\mathbf{r},R)\, d^3r$$

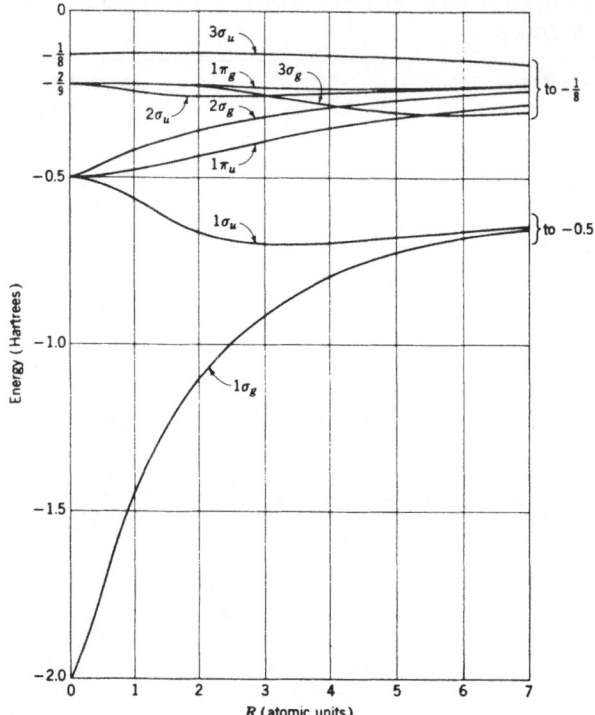

Figure 6. H_2^+ energy level diagram as a function of internuclear separation for $N=1$ and 2 with the internuclear repulsive energy $1/R$ not included. Reproduced from Slater (1977). Here $1\sigma_g \equiv 1s\sigma_g$, $1\sigma_u \equiv 2p\sigma_u$, etc.

$$= \mu Z^2 \int \phi(\mathbf{x},S) \left[\zeta^2 \left(\frac{-\nabla_x^2}{2} \right) - \zeta \left(\frac{1}{x_1} + \frac{1}{x_2} \right) \right] \phi(\mathbf{x},S) \, d^3x$$

$$= \mu Z^2 \left[\zeta^2 \, T(S) + \zeta \, V(S) \right] \tag{15b}$$

where $T(R)$ and $V(R)$ are $\mu=Z=1$ expectation values of the kinetic and potential energy operators, respectively, from H_2^+ with $E(R)=T(R)+V(R)$. These quantities are conveniently computed by invoking the following virial theorem (Slater, 1977)

$$T(R) = -E(R) - R \, E'(R)$$

$$V(R) = 2E(R) + R \, E'(R) \tag{16}$$

relating $T(R)$ and $V(R)$ to $E(R)$ and its derivative $E'(R)$ with respect to R. Likewise, the diagonal coupling matrix element in (12c) can be scaled to give

$$C(R) = \frac{(\mu Z \zeta)^2}{2\mu_{12}} \int \phi(\mathbf{x},S) \left(-\nabla_S^2 \right)_{BF} \phi(\mathbf{x},S) \, d^3x = \frac{(\mu Z \zeta)^2}{2\mu_{12}} \, \delta(S) \tag{17}$$

where $\delta(R)/2\mu_{12}$ is the corresponding quantity in H_2^+ which has no simple relation to $E(R)$ and has been computed for only a few low lying states (Hunter and Pritchard, 1967; Dalgarno and McCarroll, 1956,1957). In H_2^+ diagonal coupling is usually ignored because $1/2\mu_{12}=1/m_p$, while in two-electron atoms with $1/2\mu_{12}=1$ it plays an important role.

Collecting results, one can compute then from (12b) an effective one-dimensional potential as a function of R as

$$u(R) = \frac{1}{R} + \mathcal{E}(R) + C(R)$$

$$= \frac{1}{R} + \mu\zeta Z^2\left[(2-\zeta)E(S) + (1-\zeta)S\,E'(S) + \frac{\mu\zeta}{2\mu_{12}}\,\delta(S)\right]. \quad (18)$$

This potential determines the vibrations of the electron pair along the interelectronic axis and when plotted bears a strong resemblance to one as a function of hyperradius, that determines vibrations of the atom as a whole instead.

An optimum scale parameter ζ as a function of R is obtained by minimizing $u(R)$ with respect to ζ at each R. One finds, however, particularly for double excitations, that ζ varies little from its large-R separated-atom limit except near small R where the $1/R$ repulsion dominates the potential $u(R)$ anyway (Feagin et al., 1985). In the separated-atom limit in H_2^+ with $H^+ + H(N)$, one has with $\mu = Z = 1$ that

$$E(\infty) = -\frac{1}{2N^2} \qquad T(\infty) = \frac{1}{2N^2} \qquad V(\infty) = -\frac{1}{N^2} \qquad (19a)$$

$$\delta(R) \underset{R\to\infty}{\sim} \frac{1}{4N^2} + \frac{L(L+1)-\kappa^2}{R^2} \qquad (19b)$$

while $RE'(R) \to 0$ as $R \to\infty$. Inserting these limits into (18) and minimizing $u(\infty)$ with respect to ζ, one obtains (with $2\mu_{12}=1$) in the separated-atom limit

$$\zeta = \frac{2}{2+\mu}, \qquad (20a)$$

independent of Z and N, with the consequence that

$$u(\infty) = -\frac{\mu_{i3}Z^2}{2N^2} \qquad \mu_{i3} = \zeta\mu \qquad (20b)$$

the *exact* separated-atom energy and reduced mass for one electron ($i=1$ or 2) bound to the ion. Moreover, this scaling ensures that the adiabatic MO have the correct separated-atom Stark-state limits; for example, for the $N=1$ states,

$$\phi(r,R) \underset{R\to\infty}{\sim} e^{-x_1} \pm e^{-x_2} = e^{-\mu_{i3}Zr_1} \pm e^{-\mu_{i3}Zr_2} \qquad (20c)$$

where the $\pm$ determines the gerade/ungerade symmetry.

It might be noted that one obtains the correct separated-atom limits only if diagonal coupling $C(R)$ is included in the scaling. When omitted one finds that $\zeta(\infty)=1$ and $u(\infty)=-\mu Z^2/2N^2$, which is too deep by a factor of μ/μ_{13}. However, it is also noteworthy that the general shape of the electron-pair potential $u(R)$ can be determined without diagonal coupling (Feagin, 1984a). Thus, qualitative features in the potential curves, for example whether they support bound states or resonances or neither, are actually visible though less pronounced in H_2^+ potential curves $E(R)+1/R$. These curves for $N=1$ and 2, obtained from the diagram Fig. 6 by the addition of $1/R$, are shown in Fig. 7.

Physically, the scaling is required because the Jacobi coordinates $\mathbf{r}$ and $\mathbf{R}$ describe for large R a long, rigid rotor — the interelectronic axis — with mass $\mu=\mu_{12,3}$ and with one end bound to the ion. In reality, however, one must describe an electron at infinity and at rest relative to the center of mass of an electron-ion bound pair. Since scaling gives the correct large-R limits, it approximates the transformation, sometimes referred to as a "kinematic rotation," to the asymptotically correct set of Jacobi coordinates which connect the free electron to the center of mass of the bound pair. Exactly the same problem arises in H_2^+ in the Born-Oppenheimer approximation, only there the error is of order $1/m_p$. Thus, the separated-atom limits in eq. (19) and in Figs. 6 and 7 are off by a small amount but can also be corrected exactly by including diagonal coupling and scaling. It is interesting to note in passing that although the *adiabatic* hyperspherical potential curves have the correct separated-atom limits they provide accurate system eigenenergies E only when diagonal coupling has been added in, as has been stressed recently by Botero and Greene (1986).

An alternative to scaling would be to include a portion of the interelectronic kinetic energy in the solution of (12a). Such a procedure is known to improve the Born-Oppenheimer approximation and correct the separated-atom limits in H_2^+ (Pack, 1985). It also reminds one of the diabatic couplings in (8a) required in the quantum mechanical formulation of the Wannier model.

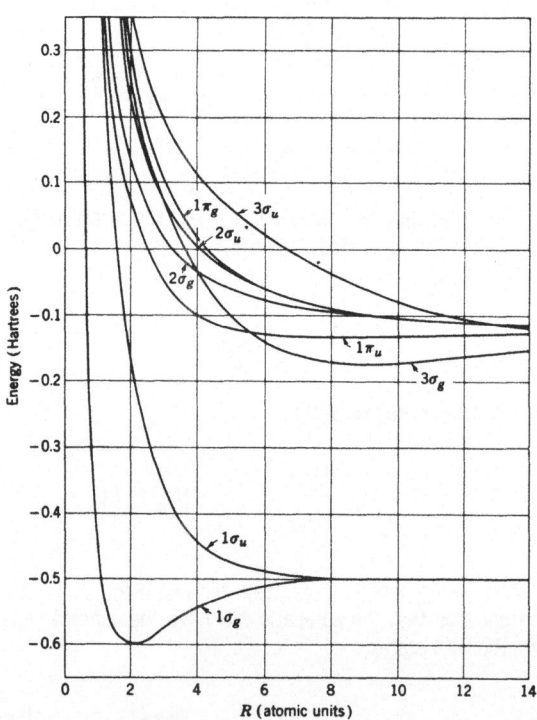

Figure 7. H_2^+ potential energy curves as a function of internuclear separation obtained from Fig. 6 by addition of the internuclear $1/R$ repulsion. Reproduced from Slater (1977).

Classification of Two-Electron States

The two-electron states and their associated symmetries can be obtained simply by a consideration of the ordering of the MO energy levels shown in Fig. 6. The lowest MO is the $1s\sigma_g$, correlating to the separated-atom ground state N=1, with quantum numbers K=t=0, and $N_\lambda = N_\mu = 0$ and therefore no nodal surfaces. Since K is zero, one sees from (6) that only states with $\pi(-1)^L = +1$ are allowed or, equivalently, since $\pi(-1)^S = (-1)^t = +1$, states with L+S= *even*. Hence the sequence of states $^1S^e$, $^3P^o$, $^1D^e$, $^3F^o$, etc. are allowed. Since these considerations depend only upon the K and t quantum numbers, the allowed states L+S=even and $\pi = (-1)^S$ are generated by all MO of σ_g character.

The next MO is the $2p\sigma_u$, which also correlates to the separated-atom ground state. Here K=0, t=1, and $N_\lambda = 0$, $N_\mu = 1$. The allowed states again have $\pi(-1)^L = +1$ from (6). However, one now has L+S= *odd*, since $\pi(-1)^S = (-1)^t = -1$, and hence the sequence $^3S^e$, $^1P^o$, $^3D^e$, $^1F^o$, etc. Again, this sequence will arise from all states of σ_u symmetry.

As a first application, one notes that the lowest singlet state $^1S^e$ is based on the *bonding* $1s\sigma_g$ MO, whereas the lowest triplet state $^3S^e$ is based on the promoted *antibonding* $2p\sigma_u$ MO. This classification explains immediately why the triplet state in helium lies so far above the singlet ground state and why the $^1S^e$ state of H^- is bound but the $^3S^e$ is not.

Quite generally, the MO appear in pairs of g,u symmetry separating to specific Stark states. For the N=2 separated-atom level there are $2s\sigma_g, 3p\sigma_u$ and $3d\sigma_g, 4f\sigma_u$ pairs analagous to $1s\sigma_g, 2p\sigma_u$ discussed above. Pairs with $K \neq 0$ have only the restriction that $L \geq K$ and that $\pi(-1)^S = (-1)^t$. Hence the pair $3d\pi_g, 2p\pi_u$ with K=1 give rise to the sequences $^1P^e$, $^3P^o$, $^1D^e$, $^3D^o$, etc. and $^3P^e$, $^1P^o$, $^3D^e$, $^1D^o$, etc., respectively. The states built upon pairs of MO of higher K value are the same but with the components L<K missing. These conclusions are summarized in Table 4, derived from Table 3.

Adiabatic Potential Curves

Scaled potential curves u(R) from (18) for electron-pair excitations in He, H^-, and Ps^- can now be computed. Because of the strong similarities between these curves and corresponding ones as a function of hyperspherical radius, much of the interpretation of physical phenomena in the present molecular description carries over from the literature and presentations in this symposium on *adiabatic* hyperspherical methods.

Two-Electron Ground States. For the $1s\sigma_g$ and the $2p\sigma_u$ MO converging to the N=1 limit, the diagonal coupling matrix element (17) has been evaluated in detail and tabulated as a function of R (Hunter and Pritchard, 1967), allowing u(R) and the scale parameter ζ to be optimized at each R (Feagin et al., 1985). These potential curves and $\zeta(R)$ are presented in Fig. 8 as a function of R for He and H^-. One notes the strong similarity of these curves in H^- with those in H_2^+, reproduced in Fig. 7. In He the ionic attraction is strong enough that the $2p\sigma_u$ ($^3S^e$) curve is pulled down below the N=1 threshold and crosses the $1s\sigma_g$ ($^1S^e$) curve, while in H^- it is completely repulsive. Thus, the $^3S^e$ state in H^- is not bound, as mentioned above.

The ground state system energy E_0 for both He and H^-, obtained by integrating (12b) numerically, is marked in Fig. 8 along with the adiabatic hyperspherical ground state energy. The discrepancy between the MO and the hyperspherical energies indicates the approximate nature of the scaling for these MO describing single excitations. Note, however, the general feature that $\zeta(R)$ deviates from its separated-atom limit (20a) only near the minimum in the $1s\sigma_g$ MO, whereas it remains almost constant over the range of variation of the $2p\sigma_u$ MO. Thus the scaling, and in particular the limit (20a), works better for doubly excited states where the Coulomb attraction of the ion is relatively weak compared to the 1/R repulsion, which thus pushes the potential wells to larger R.

Table 4. MO with N= 1 and 2, Their Quantum Numbers and Allowed LSΠ Sequences

$n\ell Kt$ united-electron	$N_\lambda N_\mu K$ MO	$N_\eta N_\xi$ N separated-atom	Sequence
$1s\sigma_g$	0 0 0	0 0 1	$^1S^e,\ ^3P^o,\ ^1D^e,\ ^3F^o,\dots$
$2p\sigma_u$	0 1 0	0 0 1	$^3S^e,\ ^1P^o,\ ^3D^e,\ ^1F^o,\dots$
$2s\sigma_g$	1 0 0	1 0 2	$^1S^e,\ ^3P^o,\ ^1D^e,\ ^3F^o,\dots$
$3p\sigma_u$	1 1 0	1 0 2	$^3S^e,\ ^1P^o,\ ^3D^e,\ ^1F^o,\dots$
$3d\sigma_g$	0 2 0	0 1 2	$^1S^e,\ ^3P^o,\ ^1D^e,\ ^3F^o,\dots$
$4f\sigma_u$	0 3 0	0 1 2	$^3S^e,\ ^1P^o,\ ^3D^e,\ ^1F^o,\dots$
$3d\pi_g$	0 1 1	0 0 2	$^1P^e,\ ^3P^o,\ ^1D^e,\ ^3D^o,\dots$
$2p\pi_u$	0 0 1	0 0 2	$^3P^e,\ ^1P^o,\ ^3D^e,\ ^1D^o,\dots$

Double Excitations. Scaled H^- potential curves of $^1S^e$, $^3S^e$, $^1P^o$, and $^3P^o$ symmetry converging to the N=2 level of hydrogen are shown in Fig. 9a. Because diagonal coupling matrix elements have not been evaluated for most of these MO, these curves have been computed using the asymptotic form of the diagonal coupling (19b) and the scale parameter (20a). For comparison, analogous hyperspherical potential curves computed by Lin (1976) as a function of the hyperradius $\mathcal{R}=(r_1^2+r_2^2)^{1/2}$ are reproduced in Fig. 9b. The similarity of the figures is striking. Among the most-discussed curves are those of $^1P^o$ symmetry, labelled "+", "−" and "pd" by Lin according to the earlier classification of Cooper, Fano, and Prats (1963) based on an independent-electron model. In the MO classification one sees from Table 4 that they are based upon the $2p\pi_u$, the $4f\sigma_u$, and the $3p\sigma_u$ MO. Their qualitative behavior can also be explained from the H_2^+ energy level diagram in Fig. 6.

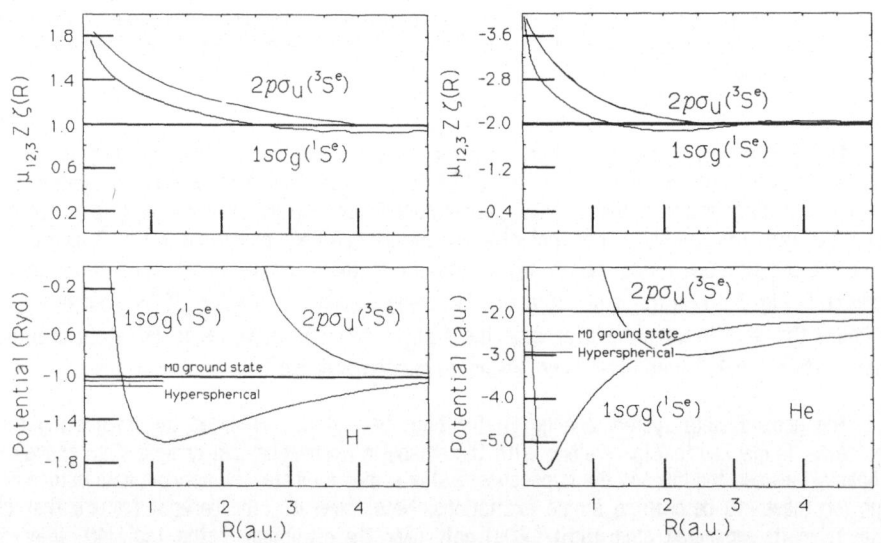

Figure 8. Scaled H^- and He potential energy curves and their optimized scale parameters as a function of the interelectronic separation converging to the N=1 limit. Ground-state energies E_0 obtained by integrating (12b) numerically are marked along with the adiabatic hyperspherical results. From Feagin et al. (1985).

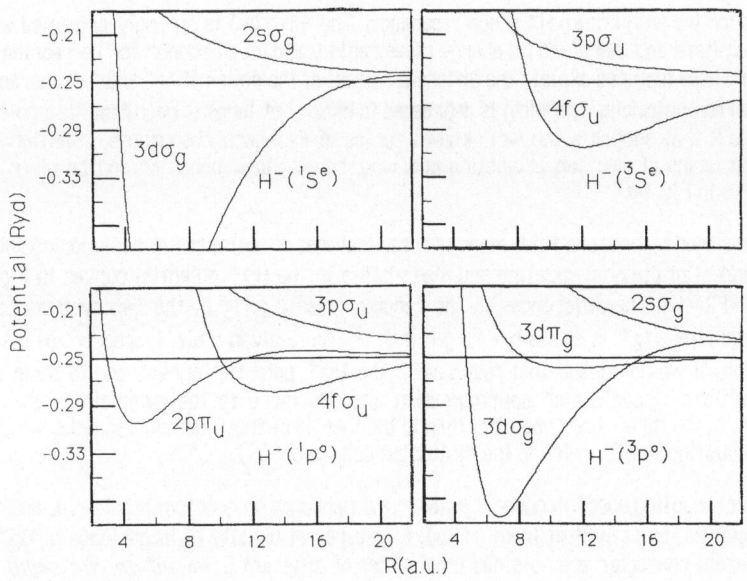

Figure 9a. Scaled H⁻ potential energy curves converging to the N=2 level of hydrogen.

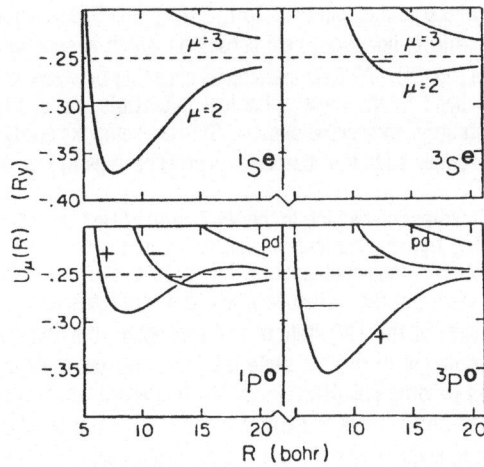

Figure 9b. H⁻ hyperspherical potential curves converging to the N=2 level of hydrogen, reproduced from Lin (1976).

The $3p\sigma_u$ MO is a promoted orbital and therefore its potential curve rises steeply for decreasing small R and, except for the 1/R interelectronic repulsion, would go to its united-electron limit n=3. Since it is also pushed up in energy for increasing large R, as in H_2^+, the result is a totally repulsive curve supporting no bound states. Similarly, the $2p\pi_u$ MO first rises in energy for decreasing large R, again as in H_2^+; however, in the united-electron limit it correlates to n=2 and therefore for decreasing intermediate R rapidly decreases in energy. Addition of the 1/R repulsion dominant at small R results in a potential well with a long range potential barrier such that this curve is responsible for the well known H$^-$ shape resonance. The 4fσ_u MO is strongly promoted with a n=4 united-electron limit and therefore its energy rises rapidly and much sooner for decreasing R than in the 2pπ_u case. These features explain the different "sizes" of the states of "+" and "−" character in the hyperspherical computations. The 4fσ_u is depressed in energy at large R resulting in a potential well at intermediate R that supports the well known series of Feshbach resonances. Similar molecular explanations in terms of electron promotion and long-range Stark behavior can be given of all the potential curves in Fig. 9a.

As mentioned before, one finds many of these features of scaled MO in two-electron atoms to be mostly independent of diagonal coupling and also visible in the H$_2{}^+$ potential curves in Fig. 7, when extended beyond R=14. The differences in the reduced mass $\mu_{12,3}$ in the two systems (see eq. 2) plays a role, such that H$_2{}^+$ is somewhat larger than H$^-$ for a given state. Moreover, as mentioned in the Introduction, it would appear that features in the H$_2{}^+$ potential curves and to some extent the validity of the Born-Oppenheimer approximation are due more to the symmetries and hence the Wannier saddle in the three-body potential than to the inertia of the internuclear axis, which renders only diagonal coupling small (refer to the discussion below eq. (5)).

Diagonal coupling does introduce, however, a dependence on quantum numbers L and K into the MO potential curves, as is evident from (19b), a feature not usually of importance in H$_2{}^+$. Hence, curves of different character are obtained from states of different L *based on the same MO*. For example, the close similarity of the "−" and "pd" curves of ^{1}P^0 symmetry and those of ^{3}S^e symmetry, another unexplained feature of hyperspherical computations, is explained here as curves built upon the same pair of MO, namely, the 4fσ_u and the 3pσ_u. The only difference is the extra unit of total angular momentum giving rise in the ^{1}P^0 to an additional repulsive centrifugal term in the diagonal coupling.

The curves of ^{3}P^0 symmetry, labelled "+", "−" and "pd" by Lin (1976), are seen to be clearly identified with the gerade counterparts of 2pπ_u, 4fσ_u, and 3pσ_u, namely, the 3dπ_g, 3dσ_g, and 2sσ_g MO. The MO curves exhibit, however, real crossings which appear as avoided crossings in the hyperspherical case. In fact, in the ^{1}P^0 case an avoided crossing between the "+" and "−" curves is actually drawn through by hand in the hyperspherical potentials, as in Fig. 9b. Since the two MO involved are of σ and π symmetry, they cross exactly. Similar remarks apply to the crossings seen in the ^{3}P^0 and ^{3}S^e MO curves compared to the absence of such crossings in Fig. 9b.

Scaled He potential curves converging to the N=2 level of He$^+$ are shown in Fig. 10a and can be compared with corresponding hyperspherical curves reproduced in Fig. 10b, which represent the first application of an *adiabatic* hyperspherical calculation to two-electron atoms (Macek, 1968). The similarities are again clear, but the influence of avoided crossings is more pronounced. The 2pπ_u MO again provides the deep well of the ^{1}P^0 state of "+" character, but the hyperspherical calculation shows a narrow avoided crossing with the "−" state, as in the H$^-$ case. Again the 2pπ_u and the 4fσ_u curves cross. The increased binding resulting from the increased ionic charge in helium compared with H$^-$ causes the promoted 3pσ_u curve to exhibit a shallow minimum (it is entirely repulsive in H$^-$) and to cross with the 4fσ_u around R=4. These two MO would couple via a radial coupling, and the corresponding hyperspherical curves reflect this by showing a strong avoidance.

Finally, scaled Ps$^-$ ^{1}P^0 potential curves converging to the N=2 level of Ps are shown in Fig. 11a and can be compared with hyperspherical curves as a function of $\mathcal{R}=(\mu_{12}R^2+\mu_{12,3}r^2)^{\frac{1}{2}}$ reproduced in Fig. 11b (Botero and Greene, 1986). The increase in size of Ps$^-$ (Z=1) from H$^-$(Z=1) is due principally to the smaller reduced mass $\mu_{12,3}$ in Ps$^-$ compared with that in H$^-$ (see eq. 2). As one would expect from the scaling, but which is another unexplained feature of hyperspherical computations, the curves in Ps$^-$ are very similar to the corresponding ones in H$^-$. Again, one should note that the avoided crossing between the "+" and "−" curves was simply drawn through in Fig. 11b. Also, as mentioned before, Botero and Greene have stressed the need to include diagonal coupling, particularly in the computation of Ps$^-$, if accurate potential curves are to be obtained.

294

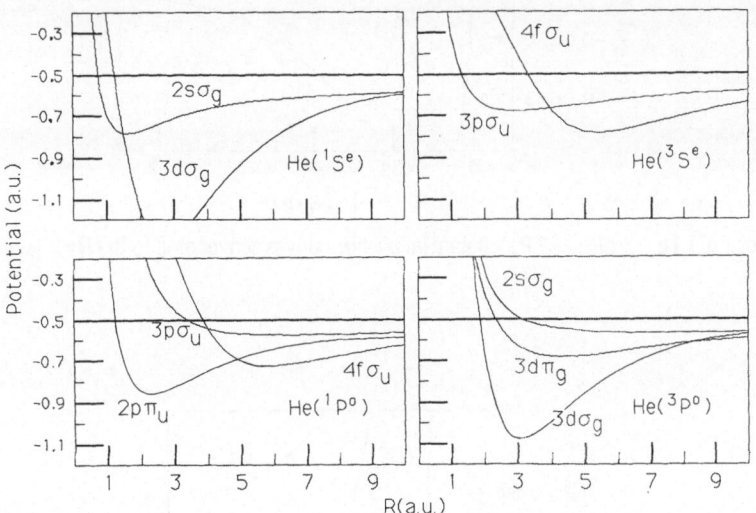

Figure 10a. Scaled He potential energy curves converging to the N=2 level of He$^+$.

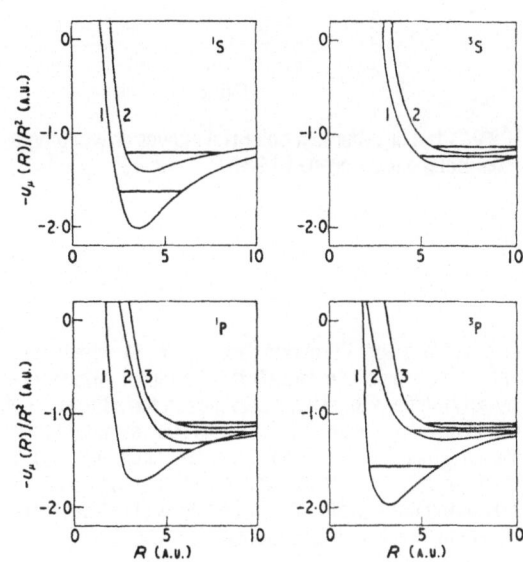

Figure 10b. He hyperspherical potential curves converging to the N=2 level of He$^+$, reproduced from Macek (1968).

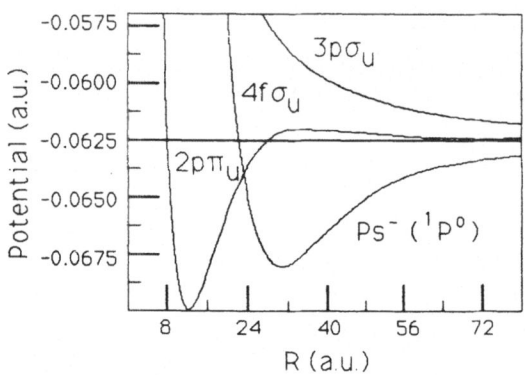

Figure 11a. Scaled $^1P^o$ Ps$^-$ potential energy curves converging to the N=2 level of Ps.

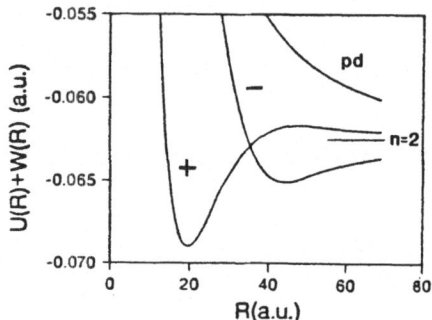

Figure 11b. $^1P^o$ Ps$^-$ hyperspherical potential curves converging to the N=2 level of Ps, reproduced from Botero and Greene (1986).

DISCUSSION

Nonadiabatic Effects

The adiabatic MO provide a zeroth-order description in which the symmetries of internal motion of states of two-electron atoms are specified. The model allows one to estimate bound and resonant state positions by solving for vibrational energies in the adiabatic potential curves. To locate energy levels accurately or to discuss dynamical processes such as autoionization, it is clear that off-diagonal, symmetry-breaking dynamical couplings must be taken into account. Such a nonadiabatic description can be achieved by solving directly the set of coupled equations obtained from the full body-fixed hamiltonian (4) using the wavefunctions (6) as basis states. The applicability of the MO description ensures that the number of coupled equations is minimized.

The states defined by the adiabatic potential curves are stationary. However, those states above the first ionization threshold lie in the N=1 continuum and therefore in principle will autoionize. In the independent-particle picture this autoionization is caused by the residual electron-electron interaction that is not diagonalized by the independent-particle basis (e.g. Hartree-Fock). In the MO description, however, the full electron-electron interaction is diagonalized by the MO basis, along with the electron-ion interaction. Hence, it is the off-diagonal *nonadiabatic* couplings, neglected by the assumption of a rigid interelectronic axis, that give rise to the decay of states lying energetically above the first ionization threshold.

The stationary states autoionize to a vibrationally unbound state of an N=1 MO resulting in the ejection of an electron to the continuum as the MO separates to the N=1 one-electron ground state. The principal selection rule is that the g, u MO symmetry is preserved. Two other selection rules arise from the requirement that radial couplings involving $\partial/\partial R$ have $\Delta K=0$ and centrifugal and Coriolis rotational couplings involving $\mathbf{L_R}$ have $\Delta K=\pm 1$. Physically, the radial coupling corresponds to squeezing or stretching of the interelectronic axis and changing the λ, μ nodal structure of the electronic center of mass motion about the ion without changing its alignment with respect to the interelectronic axis. For rotational coupling the alignment of the electronic center of mass motion changes due to a rapid rotation of the interelectronic axis which the electronic center of mass cannot follow. Specific examples of these assignments along with a more detailed discussion will be presented elsewhere (Feagin and Briggs, 1988).

Transitions between the stationary states may occur by the emission or absorption of photons, and the MO description provides a simple physical picture of the process. Moreover, a unified framework is provided in which to consider transitions to the special Wannier continuum levels. The recognition of the Wannier saddle and thus of a common set of Euler angles permits a straightforward evaluation of angular integrals and extraction of reduced matrix elements. The simplicity in the MO description comes about because the two-electron electric dipole operator $\hat{\mathbf{\varepsilon}}\cdot(\mathbf{r}_1+\mathbf{r}_2)=2\hat{\mathbf{\varepsilon}}\cdot\mathbf{r}$ depends only on the position coordinate of the electronic center of mass and is independent of the adiabatic coordinate R, as in a *one-electron* diatomic molecule. Dipole matrix elements connecting electron-pair states in (6) therefore become

$$\int dR \; f_v^{(L)}(R) \; f_{v'}^{(L+1)}(R) \; \langle\phi_{n\ell Kg}|(\hat{\mathbf{\varepsilon}}\cdot\mathbf{r})_{BF}|\phi_{n'\ell'K'u}\rangle \qquad (21)$$

where v is a vibrational quantum number of the pair. The angular integral over $\hat{\mathbf{R}}$ performed to obtain the transformed dipole operator $(\hat{\mathbf{\varepsilon}}\cdot\mathbf{r})_{BF}$ gives Clebsch-Gordan coefficients that lead to well known molecular selection rules (Pauling and Wilson, 1935). A further condition on the MO arises from the fact that, since the total parity must change while spin is conserved, the two MO involved must have different united-electron parity, $(-1)^\ell=\pi(-1)^S$, or what is equivalent, different gerade-ungerade symmetry (see Table 1).

The applicability of the adiabatic approximation means that the MO matrix element $\langle\phi|(\hat{\mathbf{\varepsilon}}\cdot\mathbf{r})_{BF}|\phi'\rangle$ is a slowly varying function of R and may be brought out of the integral in (21). Then the relative strengths of photon absorption are determined by the square of Franck-Condon integrals (Pauling and Wilson, 1935)

$$\int dR \; f_v^{(L)}(R) \; f_{v'}^{(L+1)}(R) \; . \qquad (22)$$

One thus sees that the language of molecular spectroscopy finds complete application in the MO description of atomic spectroscopy. Such an analysis helps explain for example why the $1s\sigma_g\ ^1S^e \rightarrow 2p\pi_u$ absorption to the $^1P^o$ series in helium is much stronger than to the $4f\sigma_u\ ^1P^o$ series. The potential minimum of the $4f\sigma_u\ ^1P^o$ curve in Fig. 10a lies much further out than that of the $2p\pi_u$ and hence the Franck-Condon factors with the $1s\sigma_g\ ^1S^e$ ground vibrational state are much reduced. Such an interpretation is a trademark of adiabatic hyperspherical descriptions (Fano,1983), although there less complete and more qualitative since the electric dipole operator depends on the adiabatic variable, i.e. on the hyperradius.

Connection With Other Classification Schemes

The set of approximate MO symmetries plays a role similar to the usual approximate symmetries represented by independent-electron product states labeled $n_1\ell_1 \cdot n_2\ell_2$. For L=0 the $1s\sigma_g$ MO gives rise to the $(1sns)^1S^e$ series, and its $(1s^2)\ ^1S^e$ ground state is the lowest vibrational state in the $1s\sigma_g$ potential well (marked in Fig. 8). The lowest independent-electron triplet state $(1s2s)^3S^e$ is the ground vibrational state of the $2p\sigma_u$ potential. However, in the two-electron atoms of interest here, configuration interaction strongly mixes the independent-electron orbits within these series. Thus, Cooper, Fano, and Prats (1963) considered the simplest linear combinations of independent-electron product states as new zeroth-order functions in order to classify the well known double-excitation spectra observed by Madden and Codling (1963) in helium. The scaling of H_2^+ which has been presented here was developed in this spirit while recognizing the need for an extreme departure from the independent-electron model.

Lin (1984) has extracted approximate two-electron symmetries empirically by studying numerical data on hyperspherical wavefunctions. He introduced a quantum number $A=\pm 1,0$ to label the in- and out-of-phase radial oscillations of the two electrons. Watanabe and Lin (1986) interpreted that analysis using a body-fixed frame with z axis also oriented along the interelectronic axis. (See also Lin's lecture in this symposium.) Thus, they included in their description the quantum number T to specify the projection of the total angular momentum along the interelectronic axis, i.e. the quantity denoted in this lecture by K. Using an approach akin to the one which has been used here (and in Feagin, 1984b), they resolved the two-electron wavefunction into rotational and body-fixed components and inferred that $A=\pi(-1)^{S+T}$ for $A=\pm 1$. ($A=0$ presumably describes single excitations only.) Following Herrick (1983, and references therein) and comparing two-electron configurations to a linear *triatomic* molecule, they analyzed a "bending-vibrational" quantum number K_H to describe the mutual angular correlation in θ_{12} of the two electrons (see Fig. 5). (Here K_H is used for Herrick's quantum number to distinguish it from the angular momentum projection quantum number K used in this lecture). Herrick and Sinanoglu (1975) had originally used a group theoretical reduction scheme to introduce K_H and to characterize the long-range electric-dipole form of the two-electron interaction. They showed that K_H assumes (at least for intrashell states having $n_1=n_2$ in the independent-electron model) the sequence of values $(N-T-1), (N-T-3), \ldots, -(N-T-1)$ with N the separated-atom principal quantum number. Their group theoretical approach gained strong support when it was demonstrated numerically that configuration-interaction mixing coefficients are well approximated by the recoupling coefficients characterized by K_H and T. The model of two-electron atoms based on a linear triatomic molecule was developed by Herrick and Kellman (1980) and Kellman (1985) to provide a physical interpretation of K_H and T. This model, which is fundamentally different from one based on diatomic H_2^+, has been expanded upon by Berry and coworkers (Ezra and Berry, 1983). Useful insight into the set of commuting observables which underlies the group theoretical approach has been recently provided by Mølmer and Taulbjerg (1987).

One readily draws a connection between Watanabe and Lin's identifications and the MO interpretation considered here. As pointed out already, their quantum number T is simply the projection quantum number K of the total angular momentum along the interelectronic axis. Since $\mathbf{L_R} \cdot \hat{\mathbf{R}} \equiv 0$, $\mathbf{L} \cdot \hat{\mathbf{R}} = \boldsymbol{\ell_r} \cdot \hat{\mathbf{R}}$ so that K also specifies the projection of the electronic center-of-mass angular momentum along the interelectronic axis and is thus equal to the united-electron and separated-atom magnetic quantum number. It has been shown (see Table 1) that the combination $\pi(-1)^S$ is equivalent to the MO quantity $(-1)^t$ and thus denotes the g or u character of the two-electron state. Furthermore, $(-1)^t$ is equal to $(-1)^\ell$, the united-electron parity (see Table 3). Hence, Lin's quantum number $A=\pi(-1)^{S+T}=\pm 1$ is seen (from Table 3 with $T\equiv K$) to be identical to $A=(-1)^{N\mu}$ and therefore determined by the number of hyperbolic MO nodes between the two electrons. The case $A=0$ does not arise in the MO description. One notes that the dichotomous classification $A=\pm 1$ is more limited than a specification of the number of nodes N_μ. The latter further leads to propensity rules for transitions between levels since the overlap of initial and final wavefunctions and hence transition matrix elements are strongly dependent on the relative nodal structure of the two wavefunctions.

The quantum number K_H is seen to be equivalent (again at least for intrashell states) to the sequence generated by $K_H=(N-K-1)-2N_\xi$ using the allowed values of the separated-atom parabolic quantum number N_ξ. Since $N-K-1=N_\eta+N_\xi$ (see Table 3), one also finds that K_H is equivalent to the sequence generated by $K_H=N_\eta-N_\xi$ using the allowed values of both parabolic quantum numbers. This extremely simple interpretation of K_H seems to go mostly unnoticed. Of course, K_H can be further expressed in terms of the MO quantum numbers N_λ and N_μ using the relations from Table 3. Thus, K_H

and A can be related and may not characterize truely independent degrees of freedom. It appears more significant that K_H is related to the *asymptotic* MO expectation value of the dynamical operator $\Omega(R)$ defined in (14), which commutes with the two-center hamiltionian in (12a) and accounts for separability of the Schrödinger equation for fixed R. Ones finds (Greenland, 1982) that

$$\frac{<\phi_{n\ell K\ell}|\Omega+\tfrac{1}{2}R^2 h|\phi_{n\ell K\ell}>}{R} + 1 \underset{R\to\infty}{\sim} -\frac{N_\eta - N_\xi}{N} = -\frac{K_H}{N}. \qquad (23)$$

Although the separated-atom parabolic quantum numbers are a possible means of classifying MO states, the simplicity of the MO picture and one's general familiarity with H_2^+ would suggest N_λ, N_μ and K as a more natural classification scheme of the lowest lying bound-pair excitations. In addition, the fundamental quantum number t (g/u symmetry) does not arise in the other analyses.

The discussion in these last paragraphs has emphasized the lowest bound-pair excitations. For higher excitations near the double-ionization threshold, adiabatic descriptions in general will lose their significance. This is the case for Wannier continuum pair excitations, because of the nonadiabatic couplings required, and even for Wannier *bound* pair excitations described by analytic continuation of the continuum-pair wavefunction below the double ionization threshold (Macek and Feagin, 1985). Precisely at what point and to what degree the adiabatic models will break down is still not known. Such puzzles lie in the way of a complete understanding of three-body Coulomb correlation. It is more encouraging (and perhaps more telling!) to note that while theoretical techniques remain much the same as they were some fifty years ago with the introduction of quantum mechanics an extensive catalog of precision laboratory and numerical experiments has developed on simple atomic systems. It is this resource, almost unique to atomic physics, that affords this field a decided advantage in seeking the means to solve the three-body problem.

ACKNOWLEDGEMENT

The author appreciates useful discussions with Roger Nanes and Richard Stevens and would like to thank them for their comments on the manuscript. Support of the U.S. Department of Energy, Division of Chemical Sciences, Office of Basic Energy Sciences and Energy Research is gratefully acknowledged.

REFERENCES

Bates, D.R., Ledsham, K., and Stewart, A.L., 1953, Phil. Trans. Roy. Soc. London A, **246**:215.
Bates, D.R., and Reid, R.H.G., 1968, Adv. At. Mol. Phys., **4**:13.
Botero, J., and Greene, C.H., 1986, Phys. Rev. Lett., **56**:1366.
Cooper, J.W., Fano, U., and Prats, F., 1963, Phys. Rev. Lett., **10**:518.
Coulson, C.A., and Joseph, A., 1967, Int. J. Quant. Chem., **1**:337.
Crothers, D.S.F., 1986, J. Phys. B, **19**:463.
Cvejanović, S., 1987, Institute of Physics, Belgrad, Yugoslavia, in preparation.
Dalgarno, A., and McCarroll, R., 1956, Proc. Roy. Soc. A, **237**:383, ibid., 1957, **239**:274, ibid., 1957, **239**:413.
Ezra, G.S., and Berry, R.S., 1983, Phys. Rev. A, **28**:1974.
Fano, U., 1983, Rep. Prog. Phys., **46**:97.
Feagin, J.M., 1984a, Bull. Am. Phys. Soc., **29**:801.
Feagin, J.M., 1984b, J. Phys. B, **17**:2433.
Feagin, J.M., Briggs, J.S., and Weissert, T.P., 1985, Abstracts of the 14th Int. Conf. on the Physics of Electronic and Atomic Collisions, Palo Alto, p. 147.
Feagin, J.M., and Briggs, J.S., 1986, Phys. Rev. Lett., **57**:984.
Feagin, J.M., and Briggs, J.S., 1988, submitted to Phys. Rev. A.
Fröman, A., 1962, J. Chem. Phys., **36**:1460.
Greene, C.H., 1980, J. Phys. B, **13**:L39.
Greenland, P.T., 1982, Phys. Rep., **81**:131.

Herrick, D.R., 1983, Adv. Chem. Phys., **52**:1.
Herrick, D.R., and Kellman, M.E., 1980, Phys.Rev. A, **21**:418.
Herrick, D.R., and Sinanoglu, O., 1975, Phys. Rev. A, **11**:97.
Hunter, G., and Pritchard, H.O., 1967, J. Chem. Phys., **46**:2153.
Kellman, M.E., 1985, Phys. Rev. Lett., **55**:1738.
Klar, H., and Klar, M., 1980, J. Phys. B, **13**:1057.
Lin, C.D., 1976, Phys. Rev. A, **14**:30.
Lin, C.D., 1984, Phys. Rev. A, **29**:1019.
Macek, J., 1968, J. Phys. B, **1**:831.
Macek, J., and Feagin, J.M., 1985, J. Phys. B, **18**:2161.
Madden, R.P., and Codling, K., 1963, Phys. Rev. Lett., **10**:516.
Mølmer, K., and Taulbjerg, K., 1987, Department of Physics, University of Aarhus, preprint.
Pack, R.T., 1985, Phys. Rev. A, **32**:2022.
Pack, R.T., and Hirschfelder, J.O., 1968, J. Chem. Phys., **49**:4009.
Pauling, L.P., and Wilson, E.B., 1935, *Introduction to Quantum Mechanics*, McGraw-Hill, New York; 1987, Dover Publications, New York.
Peterkop, R., 1971, J. Phys. B, **4**:513.
Rau, A.R.P., 1971, Phys. Rev. A, **4**:207.
Rost, J.M., and Briggs, J.S., 1987, Z. Phys. D, **5**:339.
Selles, P., Huetz, A., and Mazeau, J., 1987a, J. Phys. B, in press.
Selles, P., Mazeau, J., and Huetz, A., 1987b, J. Phys. B, in press.
Slater, J.C., 1977, *Quantum Theory of Matter*, Krieger, New York.
Wannier, G.H., 1953, Phys. Rev., **90**:817.
Watanabe, S., and Lin, C.D., 1986, Phys. Rev. A, **34**:823.

A HYPERSPHERICAL DESCRIPTION OF N-ELECTRON ATOMS

Michael Cavagnero

Department of Physics and Astronomy
University of Nebraska-Lincoln
Lincoln, Nebraska 68588-0111

ABSTRACT

General features of the hyperspherical adiabatic representation of
atomic structure and spectra are illustrated by means of a pair of simple
thought experiments. The thought experiments describe how correlated
atomic wavefunctions evolve under independent variations of the nuclear
charge and of the number density of electrons. Consequences of the
theory include systematic patterns of configuration mixing in both ground
state channels and multiply-excited states.

I. INTRODUCTION

In the second lecture of this symposium, Fano presented a framework
for the unified treatment of collisions and spectra.[1] Within this framework, a set of reaction channels describes the complex formed in a collision of electrons, photons, atoms and/or molecules. A complementary set
of fragmentation channels describes the dynamics after the complex has
separated into two (or more) of its constituents. Scattering or half-
scattering parameters are then expressed in terms of probability amplitudes connecting reaction and fragmentation channels; that is, in terms
of the Jost matrices described in Macek's opening lecture.

As is familiar from early applications of the quantum defect theory,
discussed herein by Rau, the reaction channels are generally specified by
a suitable set of quantum numbers which are approximately conserved while
the collision partners are confined within the reaction volume. (A familiar example is the projection of an electron's angular momentum along
the internuclear axis of a diatomic molecule, which is a good quantum
number for small radial distances of the electron.[2]) A highly localized
state, whether bound or resonant, contained within the reaction volume,
is described by the quantum numbers associated with one such reaction
channel. In this lecture, I will consider the reaction channels appropriate to the description of an atomic complex consisting of a number of
electrons, N, and a nucleus of charge Z, confined within a finite distance, R_O.

Macek's adiabatic hypothesis is one prescription for the calculation of reaction channels for two-electron processes.[3] This hypothesis states that the wavefunction of a pair of atomic electrons confined to a reaction volume is approximately separable in hyperspherical coordinates, defined as

$$R^2 = r_1^2 + r_2^2 \quad \text{and} \quad \alpha = \tan^{-1}(r_2/r_1). \tag{1}$$

The validity of this ansatz rests upon its demonstrated accuracy in representing doubly-excited states of H^- and He by a single adiabatic channel function (see the lecture by Starace). I will argue below that Macek's hypothesis is equally applicable to the description of atoms and ions with a larger number of electrons, and I will outline its implications for the theory of atomic structure and spectra. I will also describe a correspondence between the hyperspherical adiabatic representation of N-electron atoms and ions, and the "adiabatic" evolution of atomic structure along iso-electronic sequences.

One of the immediate results of this analysis relates to the study of perturbative correlations in one-electron excitation and ionization spectra. During the 1960's such correlation effects were initially misinterpreted as a collective response of whole subshells to incident radiation. It was later determined that these were indeed one-electron excitations, although their accurate reproduction by theory often required a massive superposition of configurations extending far up into the continuous spectrum.[4] The hyperspherical adiabatic representation treats such correlations as a residue of more pronounced collective modes, which are critically damped by the overwhelming dominance of the nuclear charge. This novel perspective, described in detail below, is quite useful for analyzing the systematics of such perturbative correlations along rows and columns of the periodic table.[5]

Non-perturbative correlations, on the other hand, have been observed only when a pair (or more) of electrons are excited out of the confines of filled atomic shells. These doubly-excited states emerged in the early sixties as the first such non-perturbative states. Their study has been evolving rapidly, as discussed by Lin in his lecture, and has been considerably aided by Macek's adiabatic hypothesis. Extension of these studies to 3-fold and multiple-excitations, as well as to multiple-ionization processes are only in their infancy. The pathways by which multiply-excited complexes proceed to fragmentation remain largely undescribed. The success of Macek's hypothesis in describing doubly-excited autoionizing states encourages us to attempt its application to these more complex circumstances.

II. TWO THOUGHT EXPERIMENTS

A few simple thought experiments can illustrate the dynamical evolution of correlations as described within the hyperspherical adiabatic representation. Consider a number of free electrons confined within a spherical "box" with a radius, R_0, of several atomic units. In the absence of electron-electron interactions, the spectrum of such a Fermi gas consists of densely packed nearly degenerate manifolds of states. These states are strongly coupled by electron-electron interactions, and the resultant interacting Fermi gas states are highly correlated.

Consider now that a nucleus of tunable charge Z is placed at the center of the spherical "box". This thought experiment consists of tuning the charge Z adiabatically from zero to N (the number of electrons). As Z is increased, the electrons slowly coalesce about the nuclear charge, and their wavefunction becomes progressively localized in deep valleys of the multi-dimensional potential surface. The prominent correlations of the Fermi gas are thus largely damped away by coupling to states spanning a wide range of the initial (Z=0) energy spectrum (which is clearly required to achieve localization). As mentioned above, the perturbative correlations which persist when Z=N are then viewed as a residue of the collective motion of the interacting Fermi gas.

This point of view is intimately connected with perturbation theory in an independent electron representation, the so-called 1/Z expansion.[6] We could, for example, continue our thought experiment by increasing Z further from N to infinity. In the Z=∞ limit, the scaled energy spectrum, E/Z^2, must approach the hydrogenic limit, in which electron-electron interactions are completely negligible. The thought experiment thus provides us with a "correlation diagram" connecting independent electron states with those of an interacting Fermi gas. It furthermore suggests that the correlations (configuration mixings, etc.) which exist at any fixed value of Z, say Z=N, should be viewed in the much larger context of an adiabatic evolution of states over the infinite range of Z.

The above thought experiment can be developed further to illustrate the kind of configuration mixing implied by a Z-dependent correlation diagram. The wavefunction of a free electron, confined to a spherical volume, $V = 4\pi R_0^3/3$, is

$$F(r) \quad \alpha \quad r^{-1/2} J_{\ell+1/2}(\sqrt{2E}\, r). \tag{2}$$

Using an approximate form for the zeroes of the Bessel's function, and imposing the boundary condition

$$F(R_0) = 0 \tag{3}$$

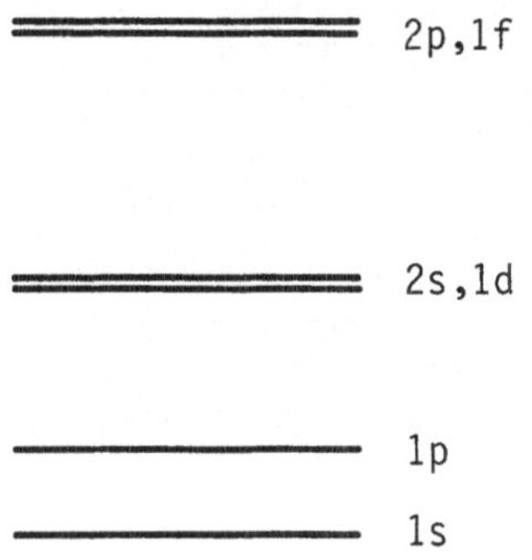

Fig. 1. Energy level spectrum of
a degenerate Fermi gas,
with nℓ-labels defined
in Eq. (4).

one obtains the energy spectrum depicted in Fig. (1), and given by

$$E_{n\ell} \approx (2n+\ell)^2 \pi^2 / 8R_0^2, \quad (\text{a.u.}) \tag{4}$$

where $n=1,2,3...\infty$. I have neglected the splitting of nearly degenerate levels, since these will be strongly mixed in the presence of electron-electron interactions. Note that this ordering of energy levels is pre-cisely that of a 3-dimensional harmonic oscillator (as in the theory of nuclear structure), and, furthermore, that the filling of such states yields (for large N)

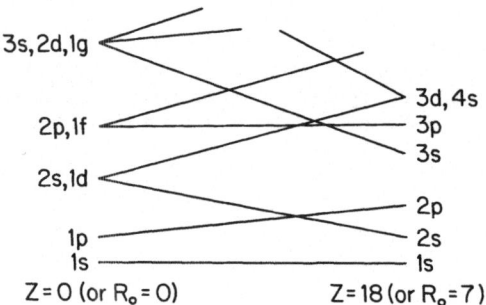

Fig. 2. Schematic correlation diagram connecting central field orbitals of the degenerate Fermi gas with those of the independent electron model.

$$E_{Fermi} \propto (N/V)^{2/3}, \quad E_{Total} \propto V(N/V)^{5/3}, \tag{5}$$

familiar results obtained by quantization in a box.

We now determine, within an effective central field approximation, how the orbital energies depicted in Fig. (1) vary with increasing Z. The resulting correlation diagram is given in Fig. (2). This schematic diagram correlates the central field orbitals of the electron gas with those of the independent electron model as Z is varied from zero to N. Owing to the incompatible symmetries governing the two limits, the dia-gram contains a series of crossings of the independent orbital energies. Clearly, the nature of the shell-structure obtained by filling the states depicted in Fig. (2) is highly Z-dependent. Energy level correlation diagrams which are essentially equivalent to Fig. (2) will emerge below in the hyperspherical adiabatic representation of atomic states.

In order to clarify this connection we can perform a second thought experiment designed to identify characteristics of the interacting Fermi gas that are preserved in atomic wavefunctions. Holding Z fixed (say Z=N), we adiabatically reduce the radius of the spherical box, R_0, from several atomic units to zero. (This is equivalent to varying the number density of electrons, as in solid matter under high pressure.) As R_0 is varied, we once again obtain a correlation diagram akin to Fig. (2). In the limit of high density ($R_0 \rightarrow 0$) the Coulomb interactions ($\sim 1/R_0$) are overwhelmed by the kinetic energy (see Eq.(4)), and the states approach those of a degenerate Fermi gas. This second thought experiment suggests that correlations of the Fermi gas electrons are preserved in the high-density portion of the configuration space of radial coordinates (r_i, $i \leq N$). It is this region of configuration space, characterized by smaller than average values of the hyperradius

$$R^2 = \sum_{i=1}^{N} r_i^2,$$ (6)

which the hyperspherical adiabatic representation is designed to treat accurately. Furthermore, such short distance properties are poorly represented by simple superpositions of independent electron wavefunctions.

III. THE HYPERSPHERICAL ADIABATIC REPRESENTATION

The hyperspherical coordinates used in describing atoms with a larger number of electrons are readily generalized from Eq. (1). For three electrons these are just spherical polar coordinates in the space of radial variables (r_1, r_2, r_3); that is,

$$r_1 = R\sin\alpha_1\cos\alpha_2$$

$$r_2 = R\sin\alpha_1\sin\alpha_2$$ (7)

$$r_3 = R\cos\alpha_1$$

More generally, the N-radial coordinates are replaced by the hyperspherical radius R of Eq. (6), and by N-1 angles $\{\alpha_i, i=1,2,3...N-1\}$. In terms of these coordinates Schroedinger's time-independent equation is (in atomic units)

$$\left[\frac{\partial^2}{\partial R^2} - \frac{4\Lambda_{3N}^2 + (3N-1)(3N-3)}{4R^2} - \frac{2C(\Omega_N)}{R} + 2E\right] R^{(3N-1)/2} \Psi(R, \Omega_N) = 0,$$ (8)

where Λ_{3N}^2 is a second order differential operator in each of the coordinates $\Omega_N = (\alpha_i, \hat{r}_j)$ and the potential energy takes the form

$$\frac{C(\Omega_N)}{R} = - \sum_{i=1}^{N} \frac{Z}{r_i} + \frac{1}{2} \sum_{i \neq j} \frac{1}{r_{ij}} \tag{9}$$

Equation (8) already suggests the scaling property which was the key to our second thought experiment. The kinetic energy terms clearly dominate the potential terms in the high-density limit, $R \to 0$. In fact, the exact expansion of $\Psi(R, \Omega_N)$ about $R=0$ (Fock's expansion[7]) has as its first term the eigenfunctions of Λ_{3N}^2, independent of Coulomb interactions. Reaction channels (RC) are thus classified by their behavior near $R=0$, i.e.

$$\lim_{R \to 0} \Psi_{RC}(R, \Omega_N) = R^\lambda \phi_\lambda(\Omega_N) \tag{10}$$

where $\lambda(\lambda+3N-2)$ is the eigenvalue of Λ_{3N}^2 corresponding to the hyperspherical harmonic $\phi_\lambda(\Omega_N)$. This classification scheme is a key element of the interpretation of observed oscillator strength "propensity rules" of doubly-excited states[8] (as discussed in Starace's lecture), for it determines the degree to which two (or more) excited electrons may be found in the vicinity of the highly localized ground state charge-density.

Upon recognizing the importance of this classification scheme, Macek[3] introduced a channel expansion

$$\Psi(R, \Omega_N) = \sum_\mu F_\mu(R) \, \Phi_\mu(R; \Omega_N) \tag{11}$$

with basis functions defined as adiabatic solutions of

$$\left[\frac{4\Lambda_{3N}^2 + (3N-1)(3N-3)}{4R^2} + \frac{2C(\Omega_N)}{R} \right] \Phi_\mu(R; \Omega_N) = 2U_\mu(R) \Phi_\mu(R; \Omega_N). \tag{12}$$

In the high-density limit the adiabatic potential curves, $U_\mu(R)$ (each of which supports a Rydberg series and its adjoining continuum), satisfy

$$\lim_{R \to 0} 2R^2 U_\mu(R) = \lambda(\lambda+3N-2) + \frac{1}{4}(3N-1)(3N-3)$$

and thus possess "centrifugal" barriers, the size of which is governed by the number of nodes, λ, of the relevant hyperspherical harmonic. Macek's hypothesis, alluded to above, states that a single term in Eq. (11) is sufficient to describe the lowest member of each autoionizing series of doubly-excited states. Higher members of each series are less localized, extending into the fragmentation region, and require a suitable frame transformation at large R.[9]

A notable aspect of this representation is that the $\Phi_\mu(R;\Omega_N)$ form a discrete infinity of orthonormal wavefunctions. This reflects the finite range of the α_i-coordinates $(o \leq \alpha_i \leq \pi/2)$, and implies that all continuum character is contained in the radial function, $F_\mu(R)$, of Eq. (11). The variable R is the only coordinate which spans an infinite range. As $R \to 0$, the $\Phi_\mu(R;\Omega_N)$ reduce to hyperspherical harmonics, which are simple oscillatory functions of the α_i, akin to the spherical harmonics representing angular wavefunctions.

That the antisymmetrized hyperspherical harmonics form a representation of a degenerate Fermi gas has been demonstrated in a previous paper,[5] and is in any event made obvious by setting $C(\Omega_N)=0$ in Eq. (8). These states can be labelled by specific term designations, $^{2S+1}L^\pi$, and by the individual orbital quantum numbers, ℓ_i (which are also conserved in the $R \to 0$ limit). The wavefunctions are highly correlated and are not

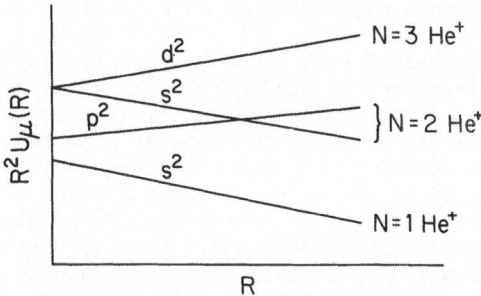

Fig. 3. Schematic diagram of the small-R dependence of the hyperspherical potential curves (drawn diabatically to conserve ℓ quantum-numbers) for the ground and doubly-excited $^1S^e$ channels of atomic He. Note the Fermi gas degeneracy of s^2 and d^2 levels at R=0.

separable in independent electron radial coordinates, but rather in the α_i coordinates defined, for example, by Eq. (7). The explicit construction of such wave functions, modeled after analogous work in nuclear theory,[10] is presented in Ref. (11).

The solutions of Eq. (12) can be expanded in terms of the hyperspherical harmonics, since this basis is complete and orthogonal with respect to the α_i-coordinates. This expansion converges rapidly only for small R, where the Coulomb interactions are perturbative, but it has the advantage that the matrix form of Eq. (12) can be cast in purely algebraic form. This last feature required a generalization of Racah's algebra to include the radial-coupling of the electrons motion, and was completed in Ref. (12). For the purpose of this lecture, I will concentrate on a few results of this analysis.

IV. EXAMPLES AND DISCUSSION

Perhaps the simplest example of the degenerate-gas correlations described above is the admixture of d^2 configurations in the $^1S^e$ autoionizing series of He converging to $He^+(N=2)$.[13] There are two series which converge to this threshold, and these consist of strong admixtures of $2s\epsilon s$ and $2p\epsilon p$ configurations. It is known that d^2 configurations admix strongly with the higher of the two series, leaving the lower series largely unperturbed.[14] This result can be traced to a degeneracy of the hyperspherical adiabatic curves at R=0; that is, to the admixture of d^2 and excited s^2 states of the Fermi gas, as sketched in Fig. (3). In this figure $R^2 U_\mu(R)$ is plotted so as to illustrate the small R limits of the potential curves. The curves are drawn diabatically (with ℓ^2 conserved) to indicate the angular momentum composition of the channel functions (note that the s^2 and p^2 curves will be repelled near their crossing in the actual adiabatic potential curves). A calculation of the hyperspherical adiabatic potential curves converging to N=2 indicates that the potential well supporting the higher series is substantially lowered ($\approx 1/3$ a.u.) at its minimum by the inclusion of d^2 states in this basis, while the lower curve is not noticeably effected, in substantial agreement with the results of configuration interaction calculations.[14] Note that the energy correlation diagram in Fig. (3) is topologically equivalent to the orbital correlation diagram in Fig. (2).

A systematic analysis of such Fermi gas correlations is presented in Ref. (5). As exercises for the student, two examples are recalled here. The ground state channel of Li is degenerate, at R=0, with a doubly-excited state channel with predominant angular momentum composition (s p^2), thus suggesting strong mixing of sp^2 configurations with the $1s^2ns$ $^2S^e$ series. This may be of substantial importance for the calculation of 2s-2p and $2s-\epsilon p$ oscillator strengths, now considered standard tests of the accuracy of correlated wavefunctions. Also, as pointed out in Ref. (5), the Be ground state channel potential curve avoids crossing with the intrashell channel curve designated $1s^2 2pnp$ $^1S^e$, indicating strong configuration mixing.

Finally, I should indicate that while the hyperspherical harmonics form a convenient basis for the analysis of small-R properties, an enormous number of them are required to represent the localization of the wavefunction at large R (this is akin to expanding Coulomb functions in terms of sine waves). A practical method for computing hyperspherical potential curves beyond a few atomic units has yet to be designed for atoms with more than two electrons. The algebraic method of Ref. (12) seems promising, if systematic ways of dealing with the extended matrix dimensions can be found. The alternative, a frame transformation to the independent electron representation at large distances, involves the computation of N-1 dimensional numerical integrals, and has yet to be attempted for atoms larger than Li.[15]

Examples of the extended algebraic calculations have been worked out for simple cases of atoms from helium to carbon.[12,15] Fig. (4) consists of hyperspherical potential curves (multiplied by R^2) for the $^1S^e$ channels of Be, within a restricted base of s^4 angular functions. These curves show interesting structure corresponding to valence excitations, K-shell holes, and even quadruply-excited states. Studies of the formation and fragmentation of such multiply-excited states lie well beyond the reach of current theory. While convergence at large radial distances is quite poor, roughly 95% of the total four-electron binding energy is obtained for the ground state channel. This serves as a typical example of the results obtained thus far for few electron atoms.

In conclusion, the hyperspherical approach provides a collective representation of atomic structure and spectra, which is capable of reproducing the predominately independent-electron character of low-energy excitations, while retaining the flexibility to describe the collective effects inherent at higher excitation energies.

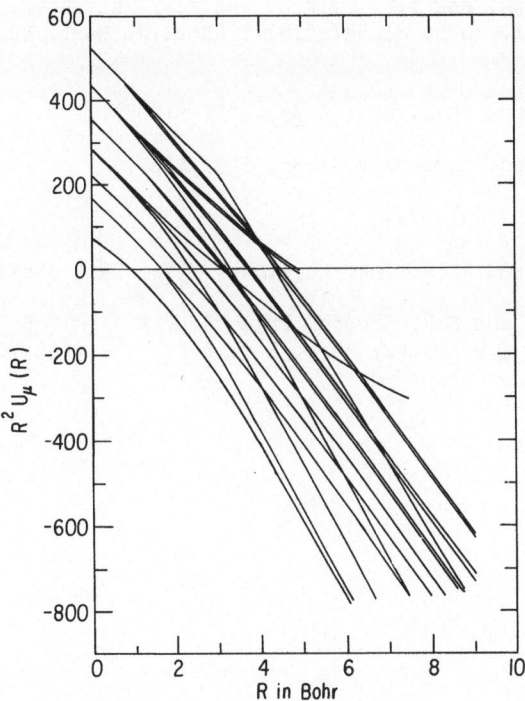

Fig. 4. Actual hyperspherical adiabatic potential curves (multiplied by R^2) for the $^1s^e$ channels of Be, calculated within a base of s^4 angular functions. Note the formation of families of curves with widely divergent ionization energies as $R \to \infty$.

ACKNOWLEDGEMENTS

The author gratefully acknowledges travel support from the National Science Foundation, Office of Postdoctoral Programs.

REFERENCES

1) U. Fano, Phys. Rev. A 24, 2402 (1981).

2) U. Fano, Phys. Rev. A 2, 353 (1970).

3) J.H. Macek, J. Phys. B 1, 831 (1968).

4) U. Fano and J.W. Cooper, Rev. Mod. Phys. 40, 441 (1968).

5) M. Cavagnero, Phys. Rev. A 30, 1169 (1983).

6) D. Layzer, Ann. Phys. $\underline{8}$, 271 (1959)

7) V. Fock, K. Nor. Vidensk. Selsk, Forh. $\underline{31}$, 138 (1958).

8) R.P. Madden and K. Codling, Astrophys. J. $\underline{141}$, 364 (1965).

9) B. Christensen-Dalsgaard, Phys. Rev. A $\underline{29}$, 2242 (1984).

10) For a review, see Y.F. Smirnov and K.V. Shitikova, Fiz. Elem. Chastits At. Yadra $\underline{8}$, 847 (1977) [Sov. J. Part. Nucl. $\underline{8}$, 344 (1977)].

11) M. Cavagnero, Phys. Rev. A $\underline{33}$, 2877 (1985).

12) M. Cavagnero, Phys. Rev. A $\underline{36}$, 523 (1987).

13) See Ref. (3), p. 839.

14) L. Lipsky and A. Russek, Phys. Rev. $\underline{142}$, 59 (1966).

15) C.W. Clark and C.H. Greene, Phys. Rev. A $\underline{21}$, 1786 (1980); C.H. Greene and C.W. Clark, Phys. Rev. A $\underline{29}$, 177 (1984).

COLLISION-PRODUCED ATOMIC STATES

Nils Andersen

Institute of Physics, University of Aarhus,
DK-8000 Aarhus C, and
Physics Laboratory, H.C. Ørsted Institute,
DK-2100 Copenhagen, Denmark

1. INTRODUCTION

The last 10-15 years have witnessed the development of a new, powerful class of experimental techniques for atomic collision studies, allowing partial or complete determination of the state of the atoms after a collision event, i.e. the full set of quantum-mechanical scattering amplitudes or - more generally - the density matrix describing the system. Evidently, such studies, involving determination of alignment and orientation parameters, provide much more severe tests of state-of-the-art scattering theories than do total or differential cross section measurements which depend on diagonal elements of the density matrix.[1] The off-diagonal elements give us detailed information about the shape and dynamics of the atomic states. Therefore, close studies of collision-produced atomic states are currently leading to deeper insights into the fundamental physical mechanisms governing the dynamics of atomic collision events.

The general theoretical framework for description and interpretation of these phenomena is excellently covered in two earlier sets of lecture notes from this series of summer schools by some of the pioneering theoreticians in the field, in 1982 by Joe Macek[2] and in 1984 by Karl Blum.[3] Selected experimental results were discussed by Rainer Hippler[4] and in several contributions to special symposia.

Thus, here I shall not duplicate these introductory reviews, but instead select a somewhat less general, but hopefully still tutorial approach, concentrating on simple, but important cases for which the basic physics is transparent without much recourse to the detailed mathematical machinery.

The first part of the lectures deals with the language used to describe atomic states, while the second part presents a selection of recent results for model systems which - to my taste - display fundamental aspects of the collision physics in particularly instructive ways. I shall here restrict myself to atom-atom collisions, since electron impact processes are extensively covered in lectures presented during the first week of this school.[5] The language used can, however, with little modification be applied also to electron-atom collisions. Indeed, that this is a fruitful point of view has become very clear in an extensive review of alignment and orientation studies of outer shell excitation in atomic collisions currently undertaken by NBS at the JILA Data Center.[6,7]

The discussion will be focused on states decaying by photon emission though most of the ideas can be easily modified to include electron emission as well.[8]

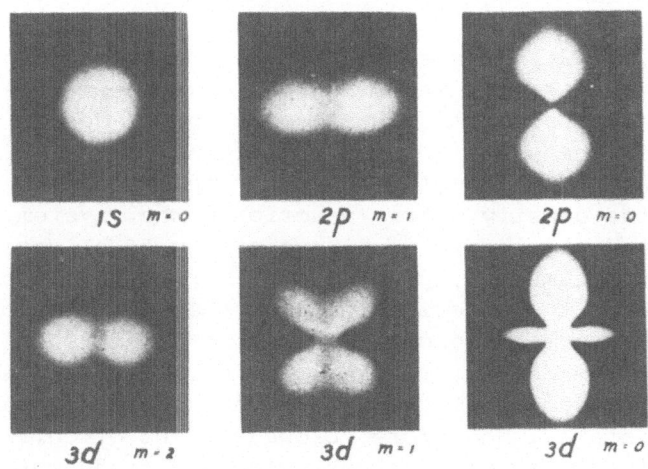

Figure 1

2. THE LANGUAGE

Figure 1 shows a selection of the famous pictures produced by Harvey White at Berkeley in 1931.[9] They represent the shapes of the electron clouds for the (non-relativistic) eigenstates of the hydrogen atom, in spherical coordinates. The upper left picture shows the spherically symmetric 1s ground state. The next one is a 2p electron with magnetic quantum number $M_L=1$, for which the electron circulates in the counterclockwise direction around the vertical quantization axis as seen from above. The 2p state with $M_L=-1$, with the electron circulating in the clockwise direction around the axis, has exactly the same shape, but one can distinguish the two states by the opposite circular polarisation of the photons that are emitted when the states decay. The third picture shows the 2p state with $M_L=0$, i.e. a p-orbital aligned along the z-axis. Similarly, the three following pictures display the shapes of the five 3d eigenstates.

As we all have learned, these shapes are, in Victor Weisskopf´s words,[10] *"the fundamental forms on which matter is built. These patterns <u>and their inherent symmetries</u> determine the behaviour of atoms"*. We shall explore this point of view in some detail, and in particular exploit their symmetry properties in space and time, as they apply to atoms in collisions. For simplicity we first restrict ourselves to the simple case of $^1S \rightarrow {}^1P$ excitation, thus neglecting the effects of electron and nuclear spin (We shall add their influence later on).

2.1 S → P Excitation

In a collision event a scattering plane is defined by the two directions of the incoming and outgoing particles. Furthermore, in a collision the total wavefunction conserves its symmetry with respect to this plane. We shall now derive some important consequences of this conservation law.

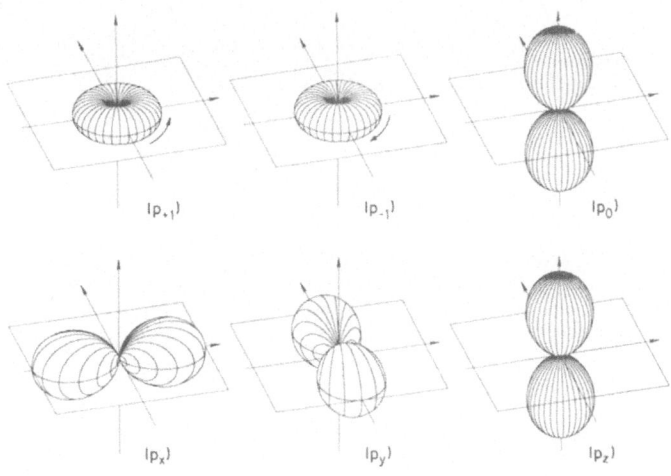

Figure 2

Figure 2 shows the shapes, i.e. the angular parts of the electron density of atomic p-states, cf. Figure 1. The upper row is the "Atomic Physics basis" corresponding to magnetic quantum numbers $M_L=+1$, -1, and 0. The lower row shows the "Molecular Physics basis", i.e. p-orbitals aligned along the three coordinate axes (x,y,z), corresponding to real-valued

313

wavefunctions, well-known from textbooks on Physical Chemistry. Recall that of these wavefunctions $|p_0\rangle = |p_z\rangle$ has negative reflection symmetry with respect to the (x,y) plane, while the remaining ones have positive reflection symmetry. The isotropic angular part of the wavefunction for an electron in an s-state has of course also positive reflection symmetry. Thus, in the simple case where an atom undergoes S → P excitation and the collision partner does not change, i.e. acts as a spinless, structureless particle, conservation of reflection symmetry with respect to the scattering plane implies that only the states $|p_{+1}\rangle, |p_{-1}\rangle$ (or $|p_x\rangle, |p_y\rangle$) are populated. "Out-of-plane" excitation of $|p_0\rangle$ is thus forbidden. The expectation value $\langle\bar{L}\rangle$ of the angular momentum vector of the p-state thus has a component along z only

$$\langle\bar{L}\rangle = (0,0,L_\perp) .\tag{1}$$

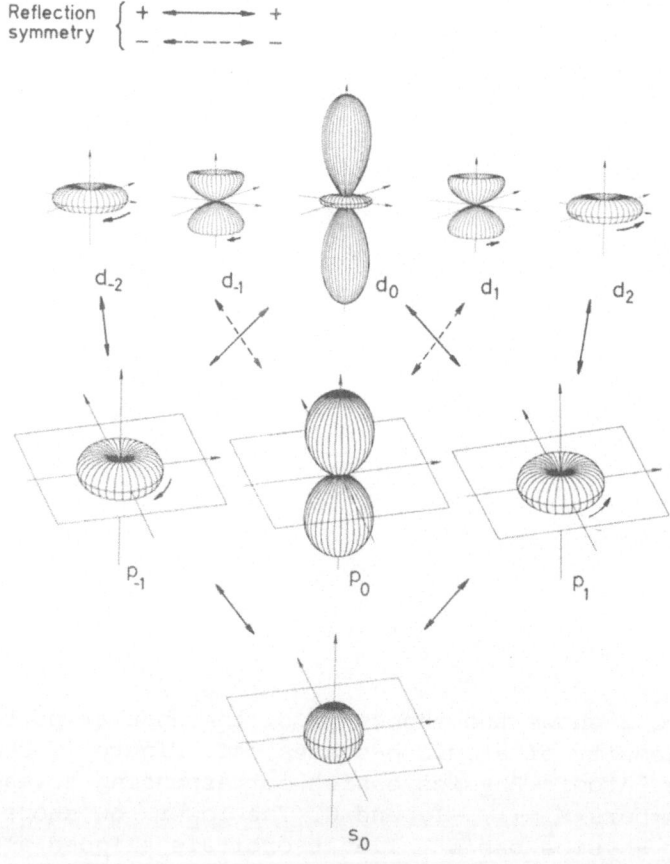

Figure 3

The actual size and sign of $L_\perp$ depends of course on the relative population of $|p_{+1}>$ and $|p_{-1}>$. Extension of these arguments to the general case of excitation-deexcitation processes gives a picture as outlined in Figure 3, which (for s, p, and d states) shows the two different families of states with positive and negative symmetry that - in the present case - can combine internally, but not with each other, in a collision event.

2.2 P → S Decay

Next, we consider the relation between the collision-produced atomic state and the radiation pattern of the subsequent decay. Figure 4(a) shows the shape of a p electron, having an *angular momentum* $L_\perp$ along the z axis; the major symmetry axis of the cloud is rotated an angle γ, the *alignment angle*, with respect to the x axis. The simplest expression for the wave-function in terms of the basis sets of Figure 2 is obtained in the so-called "atomic frame" (x_a, y_a, z_a) that is rotated by the angle γ with respect to the (x,y,z) frame (often termed "the natural frame").[11] We use the index "a" for the basis sets in the atomic frame, and obtain the relations[11]

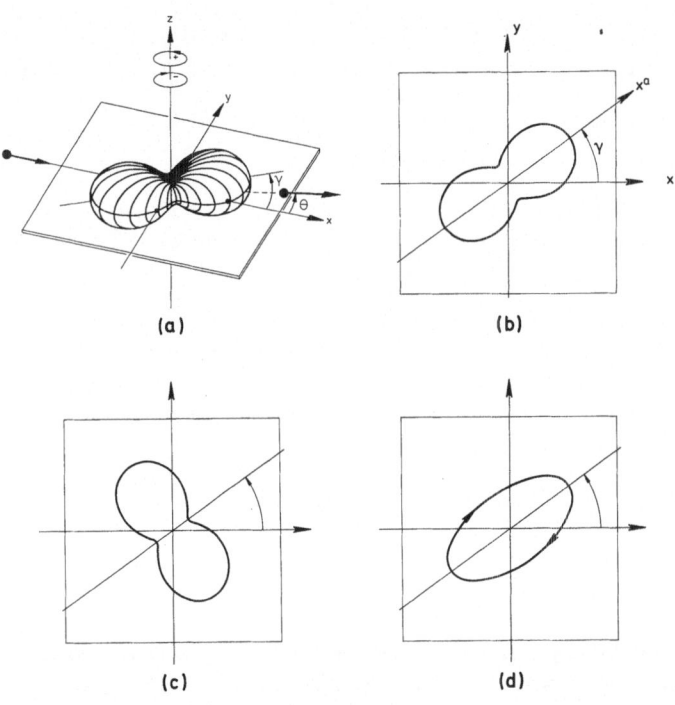

Figure 4

$$|p\rangle = a_x^a |p_x\rangle^a + a_y^a |p_y\rangle^a = a_{+1}^a |p_{+1}\rangle^a + a_{-1}^a |p_{-1}\rangle^a$$

$$= \{\sqrt{(1+P_1)}|p_x^a\rangle \pm i\sqrt{(1-P_1)}|p_y^a\rangle \}/\sqrt{2} \tag{2a}$$

$$= \{\sqrt{(1+L_\perp)}|p_{+1}^a\rangle - \sqrt{(1-L_\perp)}|p_{-1}^a\rangle \}/\sqrt{2} \tag{2b}$$

where the $+(-)$ sign refers to $L_\perp > 0 (< 0)$. The parameter P_1 the *linear polarisation*, will be defined below. From these equations the expression for the wavefunction in any other co-ordinate system may be found by suitable rotations.

The angular part T of the electron density in polar coordinates (θ, φ) in the natural frame may be written as (omitting a $3/4\pi$ normalisation factor)

$$T(\theta, \varphi) = \frac{1}{2} [1 + P_1 \cos 2(\varphi - \gamma)] \sin^2\theta \tag{3}$$

Figure 4(b) shows a cut through the charge cloud in the (x,y) plane $\theta = \pi/2$, which is proportional to the light intensity $I^z(\varphi)$ in the $+z$ direction transmitted through an ideal linear polariser with the direction of transmission tilted an angle φ with respect to the x axis

$$I^z(\varphi) = \frac{1}{2} [1 + P_1 \cos 2(\varphi - \gamma)] \tag{4}$$

Thus, the relative length l and width w of the charge cloud is given by

$$l = (1 + P_1)/2 \tag{5a}$$

$$w = (1 - P_1)/2 \tag{5b}$$

or

$$P_1 = (1-w)/(1+w). \tag{5c}$$

Recall that the radiation pattern from a classical dipole has its intensity maximum perpendicular to its direction. Therefore, a measurement of the angular distribution $I^{(x,y)}$ of the photons in the (x,y) plane is the same as (4), but rotated by $90°$ as in Figure 4(c), or

$$I^{(x,y)}(\varphi) = \frac{1}{2} [1 - P_1 \cos 2(\varphi - \gamma)] \tag{6}$$

This is the basic formula for a *photon angular correlation measurement*.

Alternatively one may determine the polarisation ellipse of the light emitted in the $+z$ direction, Figure 4(d), or, equivalently, the three Stokes parameters (P_1, P_2, P_3) defined by

$$I \cdot P_1 = I^z(\ 0^O) - I^z(90^O) \tag{7a}$$

$$I \cdot P_2 = I^z(45^O) - I^z(-45^O) \tag{7b}$$

$$I \cdot P_3 = I^z(RHC) - I^z(LHC) \tag{7c}$$

where I is the total intensity in the z direction. RHC and LHC means right-hand-circularly and left-hand-circularly polarised light, respectively, where we use the notation of classical optics[13] for which RHC (LHC) photons have helicity $\lambda = -1$ (+1). With this sign convention and the observation that decay takes place to an s state having zero angular momentum (such that the photon has to carry the angular momentum $L_\perp$ of the p state due to conservation of angular momentum), we get

$$L_\perp = -P_3. \tag{8a}$$

From (4) and (7) we get

$$P_1 = P_1 \cos 2\gamma \tag{8b}$$

$$P_2 = P_1 \sin 2\gamma \tag{8c}$$

or

$$P_1 + iP_2 = P_1 e^{2i\gamma} \tag{8d}$$

Eqs. (8) are the basic formulas for a *polarised photon coherence experiment*. For coherent light the degree of polarisation $P = |\bar{P}|$ is unity, or

$$P^2 = P_1^2 + P_2^2 + P_3^2 = P_1^2 + L_\perp^2 = 1. \tag{9}$$

For completeness we give the density matrix ϱ, defined by $\varrho_{mn} = a_m a_n^*$, where the a´s are the excitation amplitudes in the atomic physics basis in the natural frame

$$\varrho = \frac{1}{2} \begin{bmatrix} 1-P_3 & 0 & -P_1+iP_2 \\ 0 & 0 & 0 \\ -P_1-iP_2 & 0 & 1+P_3 \end{bmatrix} = \frac{1}{2} \begin{bmatrix} 1+L_\perp & 0 & -P_1 e^{-2i\gamma} \\ 0 & 0 & 0 \\ -P_1 e^{2i\gamma} & 0 & 1-L_\perp \end{bmatrix} \tag{10}$$

normalised so that tr $\varrho = 1$. Notice from ϱ_{-1+1} that the alignment angle γ is directly related to the difference between the phases δ of the amplitudes a_{+1} and a_{-1} ,

$$\delta_{+1} - \delta_{-1} = -2\gamma + \pi. \tag{11}$$

Expressions for ϱ in other basis sets or coordinate frames are easily derived from this expression.

In summary, the lesson from Figure 4 is that *coherence analysis*, i.e. a Stokes vector determination of the light in the z direction, may give $(L_\perp, \gamma)$ which completely determines the atomic state. Such an experiment has been termed a "perfect scattering experiment".[14] *Correlation analysis* yields the shape of the charge cloud, i.e. the parameters (P_1, γ), but is insensitive to the sign of $L_\perp$, which however is a crucial piece of information as far as collision dynamics is concerned. (The parameters used in Figure 4 are $\gamma=35^\circ$, $L_\perp=-0.8$ ($\hbar$), and $P_1=0.6$, corresponding to an amplitude ratio of 1:3 for $|a_{+1}\rangle, |a_{-1}\rangle$, or l=0.8, w=0.2.)

Finally, it deserves mentioning that complete information may also be obtained in the *time-reversed version* of a coherence experiment, i.e. collisional de-excitation of laser-excited atoms, as has been demonstrated in a series of studies by Hertel and coworkers and discussed elsewhere in this volume.[15]

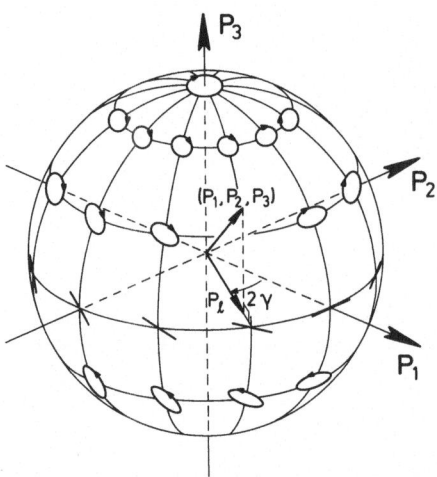

Figure 5.

A convenient framework for display of results of coherence analysis data is the so-called Poincaré sphere,[16] which represents the Stokes vector $\bar{P}$ as a point on the unit sphere. Figure 5 illustrates the polarisation ellipses corresponding to various points on the Poincaré sphere. Two opposite points on the sphere correspond to orthogonal states, e.g. the north pole ($P_3=1$) to $|p_{-1}\rangle$, the south pole to $|p_{+1}\rangle$. In this language a coherence experiment determines a point on the sphere, while a correlation experiment determines the projection only on the equatorial (P_1,P_2) plane, $P_1+iP_2 = P_1\exp(2i\gamma) = -2\cdot\varrho_{-1+1}$. How the shape of the electron cloud varies with latitude on the sphere is shown in Figure 6. It is evident from this graph that it is extremely difficult in a correlation experiment to detect a sign change of $L_\perp$, i.e. a crossing of the equator, while the circular polarisation, as determined by coherence analysis, is a direct probe of this event.

The parameters introduced above for the alignment and orientation of a p state are dimensionless numbers derived from relative measurements, and P_1 and $L_\perp$ are frame-independent. They are directly related to the experimental observables and enable at the same time easy visualisation of the shape and dynamics of the charge cloud of the excited electron.

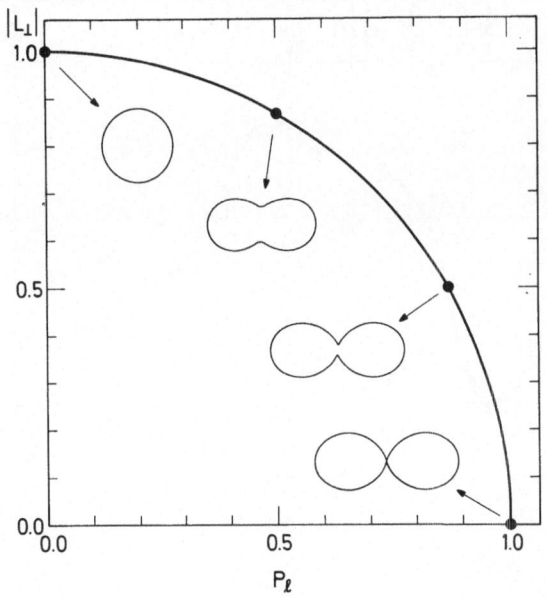

Figure 6

The following example serves to illustrate these points. Figure 7 shows a set of experimental $(L_\perp, \gamma)$ results for $He(2^1P)$ excitation by 80 eV electron impact as function of scattering angle. The dotted curve is the Born-prediction for γ (See Ref. 11 for further references). Figure 8 illustrates the shape and circulation of the radiating electron in the scattering plane at the angles corresponding to labels (a)-(j) in Figure 7. The two arrows show the direction of the incoming and scattered particle, respectively. The sign and magnitude of $L_\perp$ is indicated by the length and direction of the circular curves, a full circle corresponding to one unit of h. The upper row covers the counterclockwise circulation $(L_\perp > 0)$ and the lower row the clockwise circulation of the electron.

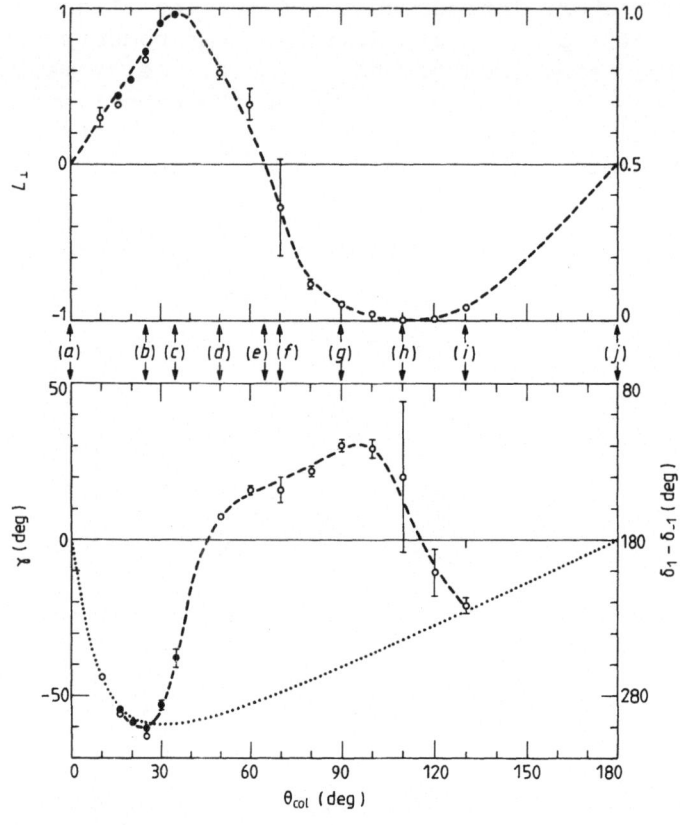

Figure 7

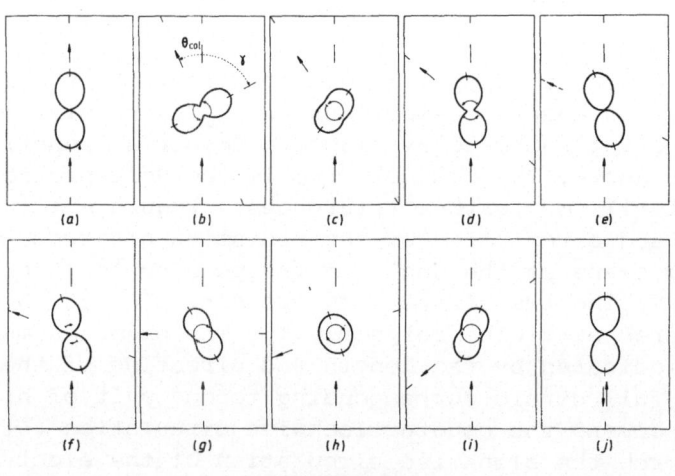

Figure 8

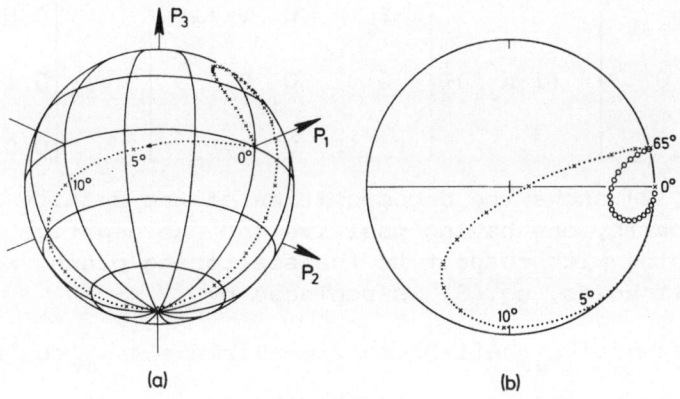

Figure 9

A theoretical prediction[17] close to the actual observations is shown on the Poincaré sphere in Figure 9(a). Crosses mark $5°$ steps of θ_{col}. The projection of this curve on the equatorial plane, which gives all information that can be determined in an angular correlation experiment, is shown in Figure 9(b), where x means $L_\perp > 0$ (i.e. the southern hemisphere), while o means $L_\perp < 0$ (the northern hemisphere).

2.3 Loss of Full Coherence

In cases where several processes which are in principle distinguishable contribute to the signal the corresponding density matrix elements have to be added, with weights given by their relative probabilities. Similarly, the resultant light polarisation vector is the sum of the weighted individual Stokes vectors. This situation can no longer be represented by a wavefunction. The light emitted in the z direction is no longer fully coherent, and P<1. P_1 and $L_\perp = -P_3$ are now independent parameters. Thus, even if positive reflection symmetry is still conserved, in this case *three* parameters are needed to specify the situation, namely $(\gamma, L_\perp, P_1)$. A correlation experiment only determines the shape(P_1) and the alignment angle γ, but gives no information about dynamics ($L_\perp$), while coherence analysis still provides complete information.

In the fully general case the assumption for the atomic wavefunction of positive reflection symmetry only cannot be maintained. This may happen, for example, for electron impact of heavy atoms if spin-orbit interaction causes the electron spin to flip. Equation (10) must then be replaced by

$$
\begin{bmatrix} \varrho_{11} & 0 & \varrho_{1-1} \\ 0 & \varrho_{00} & 0 \\ \varrho_{-11} & 0 & \varrho_{-1-1} \end{bmatrix} = (1-\varrho_{00})\frac{1}{2}\begin{bmatrix} 1-P_3 & 0 & -P_1+iP_2 \\ 0 & 0 & 0 \\ -P_1-iP_2 & 0 & 1+P_3 \end{bmatrix} + \varrho_{00}\begin{bmatrix} 0 & 0 & 0 \\ 0 & 1 & 0 \\ 0 & 0 & 0 \end{bmatrix} \quad (12)
$$

Equation (12) shows the decomposition of the density matrix into two parts, one having positive and one negative reflection symmetry with respect to the scattering plane. While eq(4) still holds, eq.(3) is replaced by

$$
T(\theta,\varphi) = (1-\varrho_{00})\tfrac{1}{2}[1+P_1\cos 2(\varphi-\gamma)]\sin^2\theta + \varrho_{00}\cos^2\theta \ . \quad (13)
$$

$h \equiv \varrho_{00}$ is thus a *height parameter*, the determination of which obviously requires observation from a direction different from z. It is particularly useful for analysis of e.g. electron impact excitation of the heavy rare gases for which spin-orbit coupling effects are pronounced near minima of the differential cross sections.[7] In the y direction the Stokes parameters are $(P_4,0,0)$ with $P_4<1$. Then the height is given by

$$
h = (1+P_1)(1-P_4)/[4-(1-P_1)(1-P_4)] = \varrho_{00} \quad (14a)
$$

while eqs. (5) are replaced by

$$
l = (1-\varrho_{00})(1+P_1)/2 \quad (14b)
$$

$$
w = (1-\varrho_{00})(1-P_1)/2 \quad (14c)
$$

The angular momentum is now given by

$$
L_\perp = -P_3(1-h) \ . \quad (15)
$$

Similarly, in a correlation experiment, determination of the height h requires observation from at least two θ-angles.

The fully general case is thus described by *four* parameters of which three can be determined from coherence analysis along z. The situation is summarized in Figure 10, which shows the shapes of the charge clouds and, below, in comparable scales, cuts along the principal axes in the atomic frame. In both cases (a) and (b) the alignment angle is 35° and the width parameter $P_1=0.6$, but the height parameter in (a) is h=0, i.e. positive reflection symmetry, while h=1/3 in (b). The angular momentum has no influence on the shape, and P_3 can be anywhere in the region $-0.8 \leq P_3 \leq 0.8$.

The parametrization suggested above is a natural development of the semiclassical model put forward by Macek and Jaecks in Sec. III of their fundamental paper.[18]

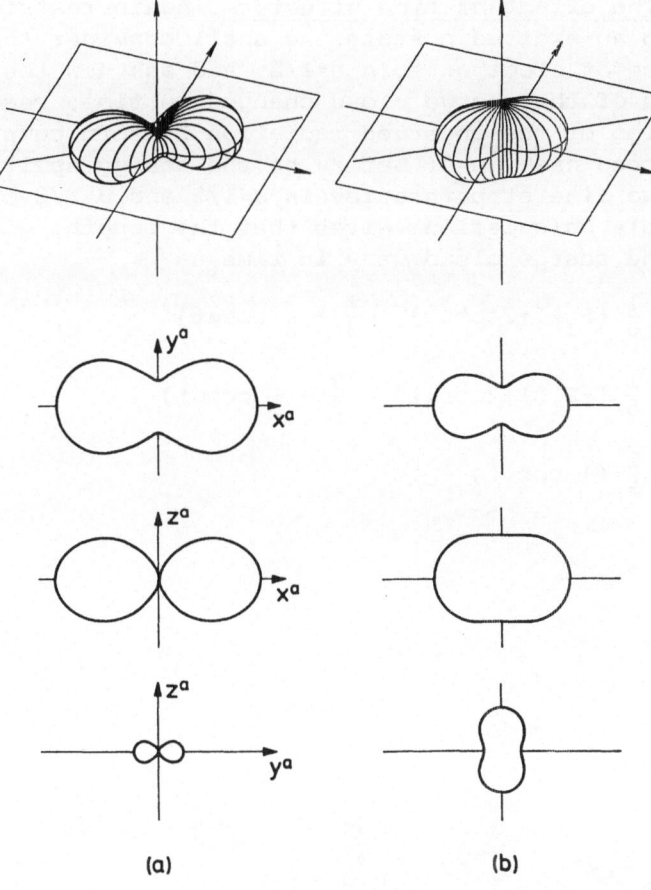

(a)   (b)

Figure 10

2.4 Time Evolution due to Internal Forces. Depolarisation

After the collision the isolated, excited atom may develop further under the influence of the fine structure (and
possibly hyperfine structure) force, which does not conserve
reflection symmetry of the spatial part of the wave function,
thereby allowing the charge cloud to change shape in time
until the optical decay takes place. However, even in this
case it is possible to reconstruct the nascent charge cloud
from an observation of the radiation pattern, as we now shall
see.

2.4.1 <u>The effect of fine structure</u>. Again restricting ourselves to an excited p state, we shall consider the simplest case of electron spin s=1/2, and ask how the shape and dynamics of the charge cloud changes in time, resulting in a modification of the observed radiation pattern compared to the unperturbed case, s=0. Let $\hbar\omega$ be the energy splitting between the two fine structure levels J=1/2 and J=3/2 of the P term. A simple calculation[7] gives that the length, width, and height of the charge cloud vary in time as

$$l(t) = \frac{1}{6} \{P_1(1+2\cos\omega t) + \frac{7}{3} + \frac{2}{3}\cos\omega t\} \qquad (16a)$$

$$w(t) = \frac{1}{6} \{-P_1(1+2\cos\omega t) + \frac{7}{3} + \frac{2}{3}\cos\omega t\} \qquad (16b)$$

$$h(t) = \frac{2}{9} (1-\cos\omega t) \qquad (16c)$$

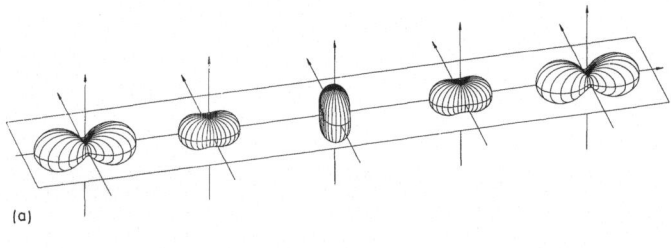

(a)

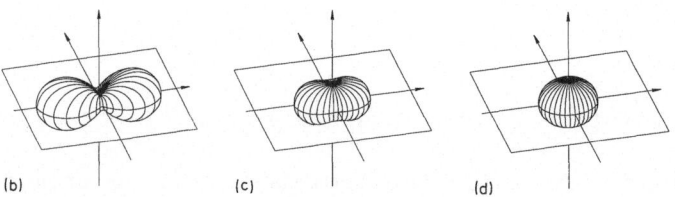

(b)   (c)   (d)

Figure 11

For the simple shape discussed previously in Figure 4 and Figure 10(a) having P_1=0.6, Figure 11(a) shows the corresponding (reversible) change in time of the shape for the five situations ωt=0, $\pi/2$, π, $3\pi/2$, 2π, respectively, showing the quantum-beat phenomenon well known in e.g. beam-foil spectroscopy. Since the symmetry axis stays fixed in space the plot has been made for γ=0°.

Most collision experiments only monitor the time average of these beats

$$\langle 1 \rangle_t = \frac{1}{6} (P_1 + \frac{7}{3}) \tag{17a}$$

$$\langle w \rangle_t = \frac{1}{6} (P_1 - \frac{7}{3}) \tag{17b}$$

$$\langle h \rangle_t = \frac{2}{9} \tag{17c}$$

thereby modifying the linear polarisation, eq. (5c),

$$P_1(s=1/2) = \frac{3}{7} \cdot P_1(s=0) \tag{18a}$$

provided that the sum of the two fine structure components is monitored.

An analogous calculation for the angular momentum gives

$$P_3(s=1/2) = P_3(s=0) . \tag{18b}$$

So, a measurement of the actual Stokes parameters for the $^2P \rightarrow {}^2S$ transition allows reconstruction of the Stokes vector corresponding to the nascent charge cloud created in the collision event, simply by multiplying the linear polarisations by a factor 7/3, while P_3 is unchanged.

2.4.2 The effect of hyperfine structure. In analogy with the fine structure effect, the presence of nuclear spin will cause further oscillatory behaviour. We shall not present details but refer to the literature[19] and here just give as an example the effect of further adding a nuclear spin I=3/2. The effect on the (time averaged) circular polarisation is a reduction to almost half the size (a factor 325/627), while the linear polarisation is reduced by almost a factor of eight (a factor 27/209). Figure 11 illustrates the effect on the (time-averaged) shape by subsequent adding an electron spin s=1/2, (b) → (c), and a nuclear spin I=3/2, (c) → (d), leading to an almost isotropic charge distribution. This situation is close to the case of the $^7Li(2^2P)$ and $^{23}Na(3^2P)$ states, where, however, effects due to a finite lifetime also show up.

Fine and hyperfine structure may thus cause a severe reduction in measured anisotropy compared to the nascent charge cloud, provided of course that they have time to develop, i.e. that the times characteristic for these effects are much shorter than the radiative lifetime.

2.5 S → D Excitation

S → D excitation has so far been studied in a few cases only. In the simple case where the excitation is fully coherent and the d-state posesses positive reflection symmetry,

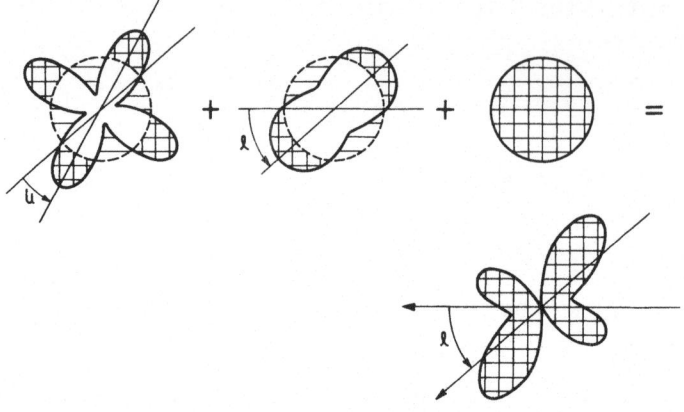

Figure 12

three amplitudes (a_{-2}, a_0, a_2) thus come into play, cf. Figure 3. The normalisation condition and an arbitrary phase factor leave four real parameters to be determined. The dipole radiation pattern for a subsequent $D \rightarrow P$ transition is completely determined by the four parameters P_1, P_2, P_3, P_4 introduced above. One might therefore think that a coherence analysis giving these four Stokes parameters determines the d state completely. Though characteristic parameters for the charge cloud, like γ, $L_\perp$, P_1, and h, still given by eq.(14) can be evaluated, analysis shows that in general *two* d-states exist having identical dipole radiation patterns, one charge cloud being the mirror of the other one in the (x^a, y^a) plane.[20] Here x^a is still the symmetry axis of the radiation pattern, but the charge cloud does not exhibit reflection symmetry with respect to this plane. Figure 12, which shows a cut through the cloud in the (x,y) plane and is based on an actual measurement,[20] indicates this duality. The ambiguity remaining corresponds to lack of knowledge of the sign of the angle η. Application of external fields, which break the symmetry even further and influence the time development of the charge cloud during the time from excitation to decay, is necessary in order to distinguish between the two alternatives.[21]

3. RESULTS FOR MODEL SYSTEMS

The class of model collision systems on which we will concentrate has the simple structure illustrated in Figure 13(a) and (b). The systems are characterized by the presence of a single, loosely bound valence electron outside two closed shells, such as the alkali atom-rare gas systems. As we shall see, the dominating inelastic process in these so-called qua-

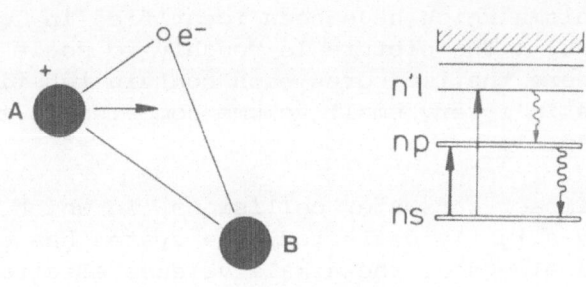

a   b

Figure 13

si-one electron systems is excitation of the ns → np resonance transition of the alkali atom. The np level of alkali-like atoms is - contrary to the case of a hydrogenic system[22] - relatively far from higher lying Rydberg levels and the continuum, and one may thus hope that a simple three state description, e.g. (ns, np_{-1}, np_{+1}), offers a reasonable starting point for a theoretical description. We shall now explore to which extent this expectation is fulfilled.

EXCITATION MECHANISMS

(a) Large impact parameters: DIRECT EXCITATION

One active electron. Cross section large when: $\dfrac{\Delta E\, a}{\hbar\, v} \simeq \pi$

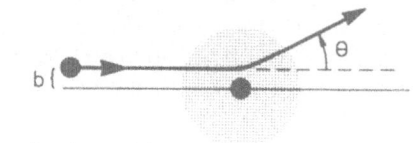

(Delocalized mech.)   1 a.u.

(b) Small impact parameters: also QUASI-MOLECULAR EXC.

Many-electron problem. Cross section small

(Localized mech.)

Figure 14

3.1 Excitation Mechanisms

Figure 14 illustrates schematically the two kinds of excitation mechanisms which have been identified in quasi-one electron systems.[23] The picture is roughly to scale for the Na-Ne system where the two cores each contain ten tightly bound electrons in a very small volume compared to the diffuse valence electron.

In large impact parameter collisions, in which the projectile is only slightly deflected, the system has essentially just one active electron, the alkali valence electron. It can undergo direct transitions to higher lying states, a process which may become very likely in a specific velocity region, as described below.

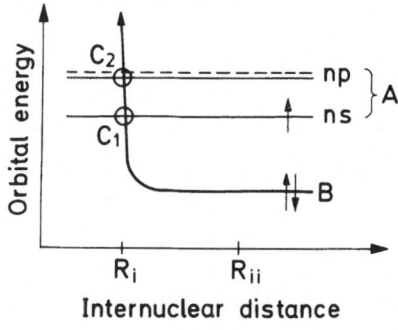

Figure 15

At small impact parameters, leading to large scattering angles, the two cores may interact significantly near the point of closest approach. This changes the problem to one with many active electrons, and it may be analysed in terms of the molecular orbital energy diagrams of the quasi-molecule, see Figure 15. Promotion of molecular orbitals may give rise to curve-crossings at specific internuclear distances, thereby allowing excitation of electrons to unoccupied orbitals. The outcome of such a collision is a fingerprint of the individual quasi-molecule and thus strongly system-dependent. For direct excitation, however, we expect to find some common trends when studying the various systems.

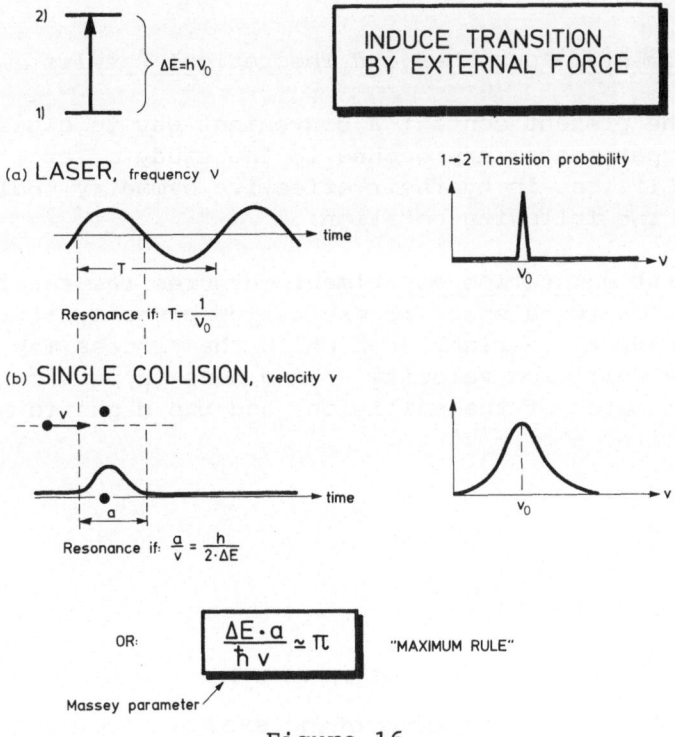

Figure 16

Figure 16 serves to place the direct mechanism in a more general context. Consider two atomic levels, 1) and 2), separated by an energy difference $\Delta E = h\nu_0$. Transitions can be induced by external forces in several ways. Consider first the case where the atom is illuminated by a laser with tunable frequency ν. As is well known, a strong increase of the transition probability is observed when the light frequency is such that the period T of the perturbation matches the natural frequency of the transition: $T = 1/\nu_0$. This effect is conveniently described by looking at (the square of) the Fourier Transform (FT) of the field. However, transitions can also be induced in a single collision, where another atom of velocity v passes by. Again, a peak is seen in the transition probability at the velocity where the collision time matches (half of) the natural frequency, $a/v = h/(2\Delta E)$, a being the interaction length, or

$$\frac{\Delta E a}{\hbar v} = \pi \ . \tag{19}$$

This is the so-called "*Maximum Rule*" for direct excitation. The quantity $\Delta E a / \hbar v$ is often termed the Massey parameter.

Because of the shape and short duration of the perturbation, the transition probability has a much broader peak here than in the previous case.

3.2 Experimental Approches and Theoretical Results

In the present context a convenient way to classify the various experimental approaches to the study of excitation in atomic collisions is by their effective symmetry. Below we shall use the following notation:

A first generation experiment measures *the total cross section* for a specific excitation process, i.e. it integrates over all variables on which the process may depend - except the collision velocity - and here in particular the impact parameter of the collision, and the magnetic sublevels of the excited atomic state.

EXPERIMENTAL SETUP

(1. Generation   exp.)

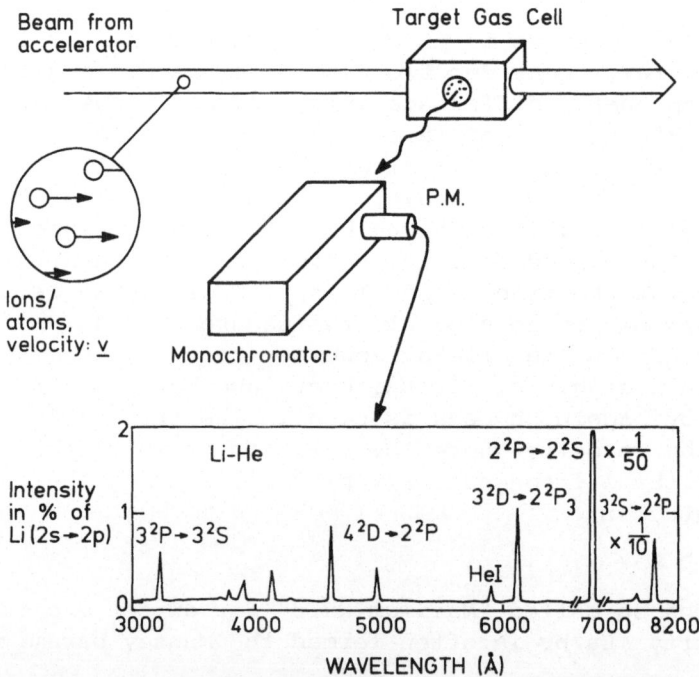

Figure 17

330

A second generation experiment makes use of *the rotational symmetry* of the typical experimental setup by determining e.g. the light polarisation with respect to the beam direction, or the impact parameter dependence by studying the variation with scattering angle in a differential cross section measurement.

A third generation experiment exploits fully *the planar symmetry* of the individual scattering events by coincidence techniques, as explained below. As we have seen in the previous chapter, this may allow a full determination of the *collision-produced atomic state*.

We shall now step by step discuss results obtained by the three generations of experiments in the light of corresponding theoretical results.

3.2.1 First generation. Total cross sections. Figure 17 shows schematically a typical setup for a first generation experiment. A beam - in our case e. g. alkali atoms of velocity v - from an accelerator is directed into a target cell with a rare gas. The light emitted from the cell is analysed in a monochromator. For the quasi-one electron systems the $n^2 S - n^2 P$ resonance line dominates completely the optical spectrum, as shown on the spectrogram for the Li-He collision. Total cross sections are then determined by monitoring line intensities versus collision energy.

Representative results are shown in Figure 18. We notice a resonance-shaped curve as expected from the arguments in Figure 16. For Li-He the lower experimental curve indicates the sum of the cross sections for higher excited Li levels, proving the dominance of the 2s-2p resonance transition. Spectral lines from He are even weaker. The full-drawn curve is the prediction of a simple theoretical model, based on the direct excitation picture, and it reproduces the experimental curve reasonably well. The maximum of the total cross section, about 4 Å^2, has a magnitude similar to the size of the Li atom, indicating that in this velocity region excitation probabilities may approach unity.

The criterion (19) will predict an upward shift of about a factor of four in impact energy for the cross section maximum when going from the Li-He to the Na-He system, as is indeed seen. For Na-Ne the maximum position should be as for Na-He since He and Ne atoms are of the same size. This too is observed, but a pronounced low-energy plateau also appears, which cannot be accounted for by a direct excitation model alone; a complicated quasi-molecular many-electron calculation (the dashed curve) was necessary to reproduce the experimental findings. That a different excitation mechanism is required to

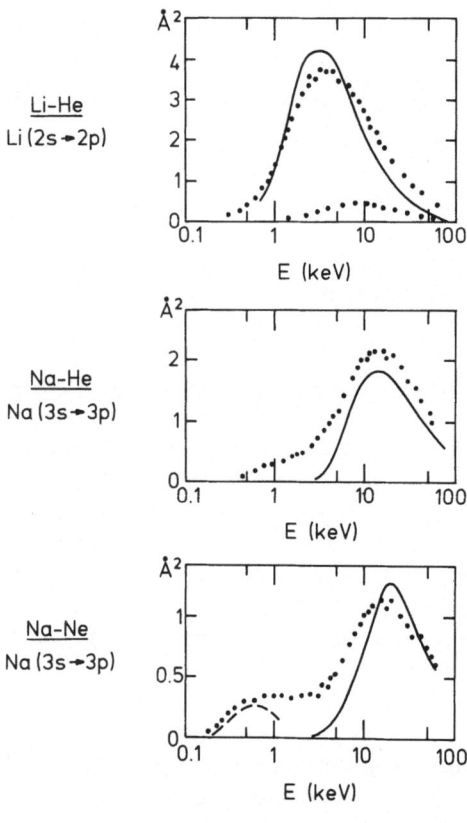

TOTAL CROSS SECTIONS
(1. Generation exp.)

Max. for: $\dfrac{\Delta E \cdot a}{\hbar \, v} \approx \pi$

Li-He
Li (2s → 2p)

Na-He
Na (3s → 3p)

Na-Ne
Na (3s → 3p)

Figure 18

explain the origin of this plateau is, however, much more evident from second generation experiments, to which we shall now proceed.

3.2.2 <u>Second generation. Differential cross sections.</u>
Figure 19 shows schematically a second generation experiment which determines the excitation probability (b) by electrostatic energy loss analysis of the ions scattered at a selected angle θ. For analysis of neutral particles one may alternatively use the time-of-flight technique. Since the interatomic potentials can be estimated from scaling laws or be calculated, the corresponding impact parameter b can be derived from the scattering angle.

Differential cross section:

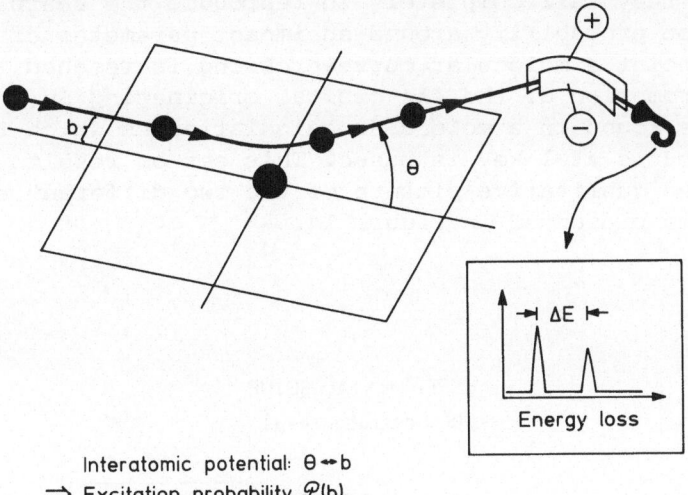

Interatomic potential: θ ⟷ b
⟹ Excitation probability $\mathscr{P}(b)$

Figure 19

EXCITATION PROBABILITIES
(2. Generation exp.)

Na-Ne

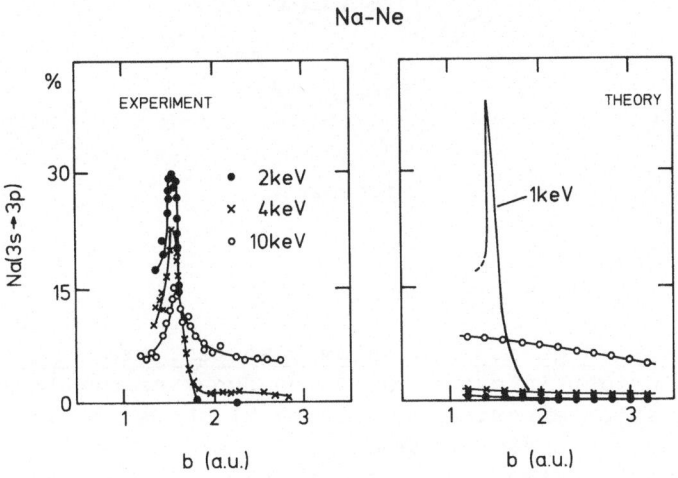

Figure 20

Results obtained in this way for the Na-Ne system are shown in Figure 20. Here, experimental results at the energies indicated are compared with corresponding theoretical results based on the direct excitation picture at 2, 4, and 10 keV impact energy (compare with the lower graph of Figure 18). As is seen, they fail completely to reproduce the sharp rise in excitation probability around an impact parameter of 1.4 a.u. At this point a molecular curve-crossing is reached, triggered by the promotion of the 4fσ orbital originating in the Ne L-shell, as found in a molecular calculation on which the theoretical curve at 1 keV is based. This set of results highlights the qualitative picture of the two different excitation mechanisms presented in Figure 14.

EXPERIMENTAL SETUP

(3. Generation exp.)

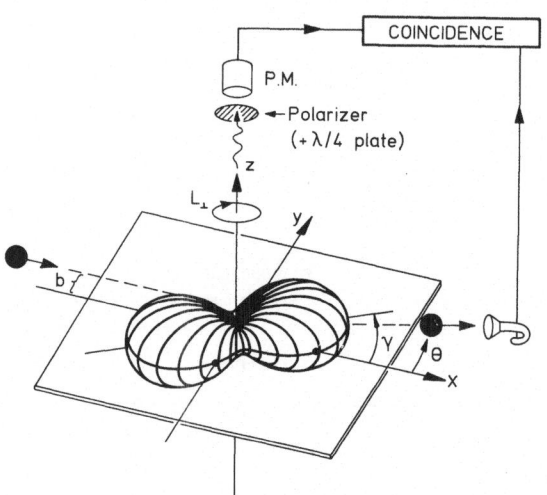

Figure 21.

3.2.3 Third generation. Atomic states. Propensity rules.

Even more insight into the collision dynamics may be gained in third generation experiments. Figure 21 shows the principle of such a setup. The scattering plane is defined by the directions of the incoming and scattered particles. Particles scattered at some angle θ are detected in coincidence with the photons emitted in the direction perpendicular to the scattering plane. A polariser, combined with a quarter-wave plate,

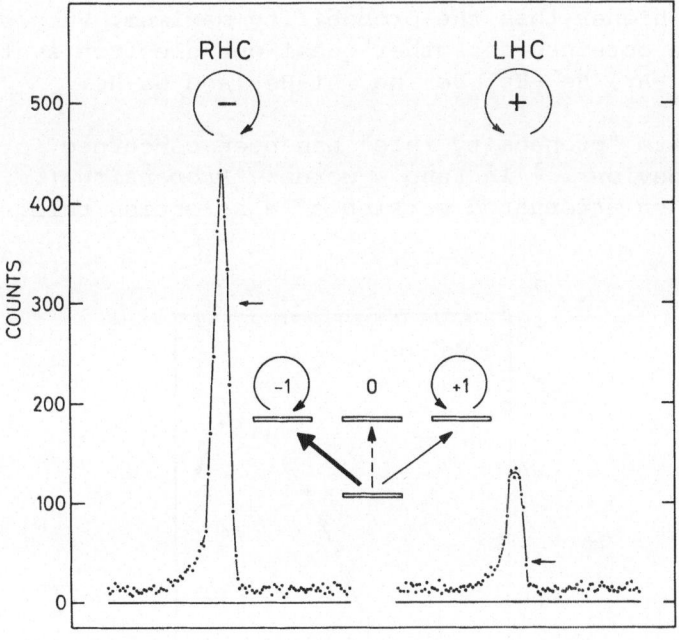

Figure 22

permits determination of the Stokes vector for the light. According to the considerations in Chapter 2, this setup may completely determine a collision-produced p-state.

Figure 22 shows results of such an experiment for the Mg^+-He collision at an impact parameter of 1 a.u. and an energy of 35 keV, which is the energy at which the maximum 3s-3p transition probability is observed. The number of coincidences is much larger for RHC- than for LHC-polarised photons, corresponding to a predominant population of the magnetic substate with M_L=-1, as shown. This difference is even larger than represented by the height of the peaks, since corrections (indicated by arrows) have to be done for polariser imperfections and cascade contributions from higher excited states.

Evidently, a *strong preference for clockwise orientation of the excited electron charge cloud is observed near the probability maximum* as seen also from Figure 23, which displays the variation with impact energy of the excitation probability and the angular momentum expectation value $L_\perp$.[24] In a region around the probability maximum $L_\perp$ stays in the $-(0.8-0.9)\hbar$ region corresponding to a M_L= -1:M_L=+1 ratio of the order of 10:1. When going below 10 keV, $L_\perp$ is reduced or even shows a transient sign rever-sal at 3-4 keV. Experimental limitations prevented a study above 60 keV, but studies of other systems show a reduction of $L_\perp$ also when going to velocities higher than the probability maximum. Very similar results are obtained for other quasi-one electron systems, such as Mg^+-Ar, Be^+-He, Be^+-Ne, Li-He, and Na-He.

The term "propensity rule" has been concerned for this kind of behaviour.[25] In Fano´s words: "*Propensity [...] amounts to an attenuated version of a selection rule. Selec-*

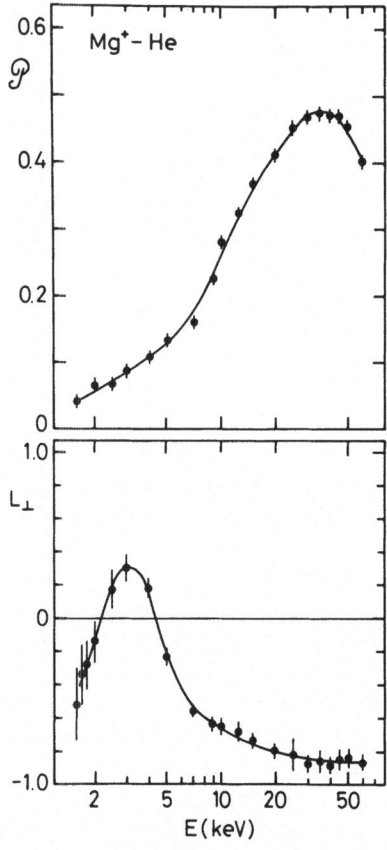

Figure 23

tion rules result from exact symmetries or other properties of a system".[26] The interpretation presented below is closely related to the analysis in recent papers by U. Fano and C.W. Lee,[27] and resembles the theory for Q-value spectroscopy in Nuclear Physics.[28]

We thus face the question of what is the physical origin of this orientation propensity rule, and what determines the range of validity in collision velocity and impact parameter. In the three state basis (s, p_{-1}, p_{+1}) we may write the time-dependent Schrödinger-equation determining the corresponding amplitudes (a_s, a_{-1}, a_{+1}) in the form[24]

$$i\hbar v \frac{d}{dx} \begin{bmatrix} a_s \\ a_{-1} \\ a_{+1} \end{bmatrix} = \underset{=}{A} \begin{bmatrix} a_s \\ a_{-1} \\ a_{+1} \end{bmatrix} \tag{20}$$

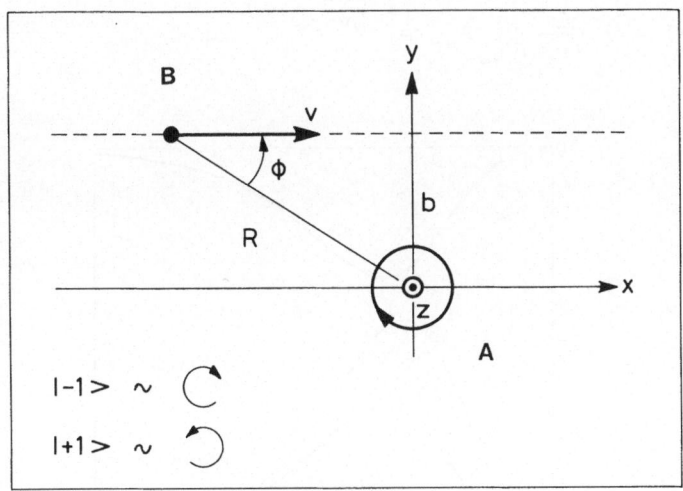

Figure 24

Referring to the collision geometry of Figure 24, which shows a collision between atoms A and B with an impact parameter b and velocity v, internuclear distance R determined by $R^2 = b^2 + v^2 t^2$ at time t (t=0 at x=0), and a rotation angle φ of the internuclear axis, the matrix $\underset{=}{A}$ has the form

$$\underset{=}{A} = F_{sp}(R) \begin{bmatrix} 0 & c.c. & c.c. \\ e^{i(\frac{\Delta Ex}{\hbar v}-\varphi)} & 0 & 0 \\ -e^{i(\frac{\Delta Ex}{\hbar v}+\varphi)} & 0 & 0 \end{bmatrix} + F_{+-}(R) \begin{bmatrix} 0 & 0 & 0 \\ 0 & 0 & e^{-i2\varphi} \\ 0 & e^{i2\varphi} & 0 \end{bmatrix} \tag{21}$$

The first term governs the s-p probability flow, with $F_{sp}(R) = \langle s|V|-1\rangle$, where V is the rare gas-valence electron interaction for which simple model potentials are available. The second term governs the $M_L=+1 \longleftrightarrow M_L=-1$ balance, with $F_{+-}(R) = \langle +1|V|-1\rangle$. The matrix elements F_{sp} and F_{+-} are conveniently calculated in the molecular A-B frame. In the interior of the matrices we recognize the two phase terms, one $\Delta Ex/\hbar v$, due to the s-p energy difference $\Delta E = \Delta E(R)$, the other one, $\pm\varphi$, due to the transformation properties under rotation of the two spherical harmonics $Y_{1\pm 1}$.

Starting the system in the ground state (1,0,0) we get to first order

$$a_{-1}(v) = \frac{1}{i\hbar v} \int_{-\infty}^{\infty} F_{sp}\, e^{i(\frac{\Delta Ex}{\hbar v} - \varphi)}\, dx \qquad (22a)$$

$$a_{+1}(v) = -\frac{1}{i\hbar v} \int_{-\infty}^{\infty} F_{sp}\, e^{i(\frac{\Delta Ex}{\hbar v} + \varphi)}\, dx \qquad (22b)$$

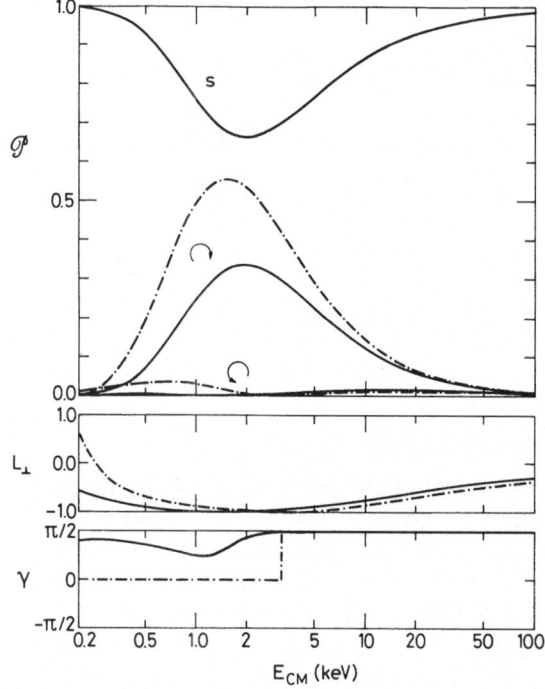

Figure 25

To understand the propensity for the $M_L = -1$ state near the excitation probability maximum we must analyse how to select the velocity v so that a_{-1} becomes large: The angle φ increases from zero to π during a collision. If v is selected so that the phase term $\Delta Ex/\hbar v$ also increases by π, i.e. eq. (19) is fulfilled, then the two phase terms in the a_{-1}-integral will cancel, and a_{-1} will approach its maximum value. In the a_{+1}-integral, on the other hand, the phase term will change by $\pi + \pi = 2\pi$ which leads to an almost cancellation of this amplitude. Thus, when the criterion (1) for maximum transition probability is fulfilled there is a strong propensity for population of the $M_L = -1$ state. Further theoretical analysis indicates that a reasonable estimate of the velocity range of validity of the propensity rule is $1/2 < v/v_{max} < 2$, where v_{max} is the velocity for maximum excitation probability.

Analysing next the impact parameter dependence, a rough estimate of a favourable choice of b can be obtained by the requirement that the time derivative of the phase term vanishes at x=0, where the interaction is strongest, or $\Delta E/\hbar = \dot{\phi}(0) = v/b$, which gives

$$b = \frac{\hbar v}{\Delta E} = \frac{a}{\pi} . \tag{23}$$

Thus, the propensity rule should be well fulfilled at impact parameters of the order of one third of the size of the atom.

Figure 25 shows numerical results for the Na-He system as function of energy at an impact parameter b=2.2 a.u. The dot-dashed lines are first-order perturbation calculations, eqs. (22). The orientation propensity $L_\perp \approx -1$ in the region of maximum excitation is strongly confirmed. However, since probabilities here are very large, one may question the reliability of first-order estimates. But solving instead the full set of equations in close coupling (the full-drawn lines) gives an even stronger propensity.

A fascinating aspect of the theoretical model is the possibility to follow in detail what goes on *during* the collision, i.e. to trace the growth of population of the excited state, the build-up of angular momentum, the changing shape of the valence electron charge cloud, and its alignment with respect to the internuclear axis along the trajectory. An example of this is shown in Figure 26, which displays the variation of these quantities in the interesting region near the point of closest approach. Again the system is Na-He with b=2.2 a.u. and an impact energy of 2 keV (c.m.), close to the energy of maximum excitation probability, cf. Figure 25. The atoms are seen to interact over a distance $a \approx 6-7$ a.u., according to which b was selected from the criterion (23). Note how the alignment angle develops: initially, the charge cloud is aligned along the internuclear axis, but as the collision

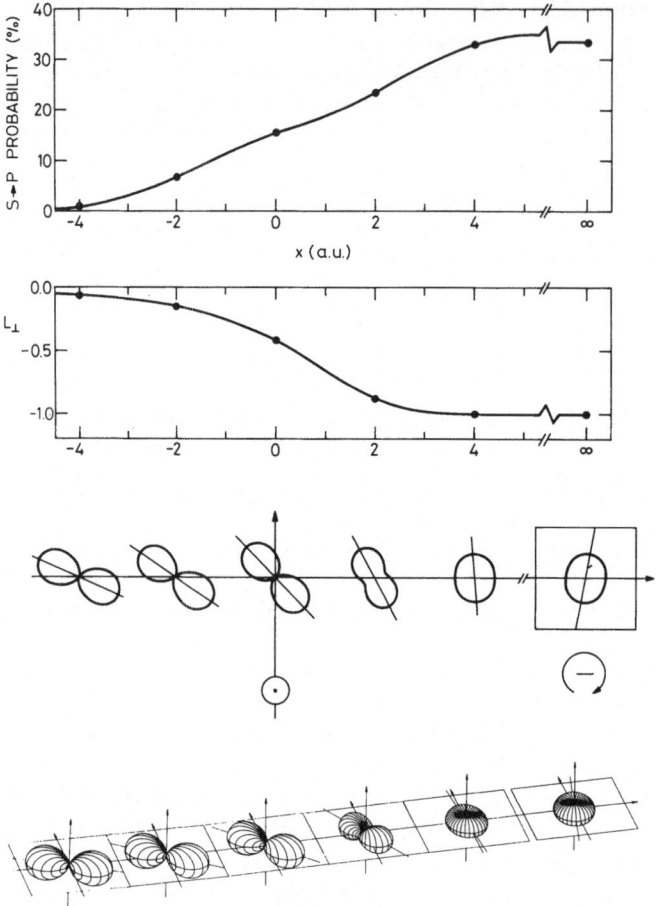

Figure 26

progresses, the cloud cannot follow completely the rotation of the axis and lags more and more behind. More pictures may be found in ref. 29.

One can generalize the analysis given above of the simple three-state model to the general problem with many states.[30] Using the sign conventions of Figure 27, the general criterion for orientation propensity takes the form

$$\frac{\Delta E a}{\hbar v} = - \Delta M \pi .$$

(24)

(In eq. (19) for s → p excitation $\Delta E > 0$ and $\Delta M = -1$). Note in particular that $\Delta E > 0 (<0)$ implies a change of angular momentum projection quantum number $\Delta M < 0 (>0)$.

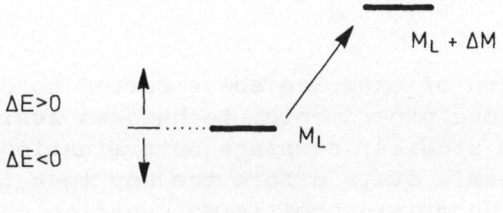

$\Delta E > 0$

$\Delta E < 0$

M_L

$M_L + \Delta M$

Figure 27

Some simple implications of this criterion concerning the collisional behaviour of other states may be seen, as illustrated in Figure 28. The upper part shows the opposite orientation obtained by excitation and deexcitation of an S state. Evidently, a sign change for ΔE will interchange the role of the two amplitudes a_{-1} and a_{+1}. The lower part of Figure 28 displays the predicted different behaviour of oppositely oriented P states. Referring again to the coordinate system above, the state with $M_L = -1$ is easy to deexcite, but hard to excite, while the one with $M_L = +1$ should behave in the opposite way. Experiments of this kind, where the dynamics of the active electron is completely controlled and probed, should be possible, but have not yet been done.

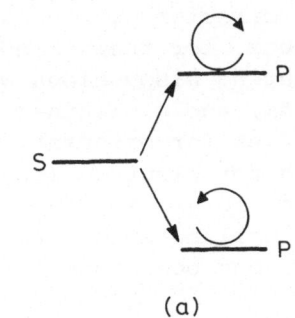

S

P

P

(a)

P

S,P,D

S

P

S,P,D

S

(b)

Figure 28

341

4 CONCLUSION

The selection of examples above served to demonstrate that of the various experimental techniques available for atomic collision studies, complete determination of the collision-produced atomic state offers the key to a full understanding of the underlying collision dynamics controlling the direct excitation process for atomic outer shells, as highlighted in the discovery of propensity rules for orientation.

Further interesting problems include the case of excitation by charge transfer from oriented atoms. Here one might expect right-left asymmetries in the electron capture probability from an oriented state in the range of collision velocities around the orbital velocity of the oriented electron. To which extent the propensity rule has to be modified in this case, where the active electron changes center, is a still unsettled question.

Whether the mechanisms discussed are of relevance to the case of orientation induced in grazing collisions with surfaces[31] may also deserve attention.

ACKNOWLEDGEMENTS

I wish to thank the directors of the school, J. Briggs, H. Kleinpoppen, and H. Lutz, for their kind invitation to lecture at Maratea. Illuminating discussions with U. Fano, I.V. Hertel, C.W. Lee, J. Macek, and A. Winther on the relationship between the propensity rules for orientation described above and analogous rules valid for electron impact, low-energy atom impact, photon impact, and in grazing nuclear collisions, are gratefully acknowledged. Collaboration over the years with T. Andersen and S.E. Nielsen has been essential.

REFERENCES

1. See J. Burgdörfer, this volume.
2. J. Macek, in: "*Fundamental Processes in Energetic Atomic Collisions*", H.O. Lutz, J.S. Briggs, and H. Kleinpoppen, eds., Plenum Press 1983, p.39.
3. K. Blum, in: "*Fundamental Processes in Atomic Collision Physics*", H. Kleinpoppen, J.S. Briggs, and H.O. Lutz, eds., Plenum Press 1985, p.103.
4. R. Hippler, in: "*Fundamental Processes in Atomic Collision Physics*", H. Kleinpoppen, J.S. Briggs, and H.O. Lutz, eds., Plenum Press 1985, p.181.
5. See in particular H. Kleinpoppen, this volume.

6. N. Andersen, J.W. Gallagher and I.V.Hertel, _in_:
    "_Electronic and Atomic Collisions_", D.C.Lorents, W.E.
    Meyerhof, and J.R. Peterson, eds. North-Holland 1986,
    p.57.

7. N. Andersen, J.W. Gallagher, and I.V. Hertel, Physics
    Reports, in print.

8. P. van der Straten and R. Morgenstern, Comm. At. Mol.
    Phys. 17 243 (1986).

9. H.E. White, Phys. Rev. 37 1416 (1931).

10. V.F. Weisskopf: "_Knowledge and Wonder. The Natural World
    as Man knows it._" MIT Press 1980.

11. N. Andersen, I.V. Hertel and H. Kleinpoppen, J. Phys. B
    17 L901 (1984).

12. N. Andersen, T. Andersen, H-P. Neitzke and E.H.
    Pedersen, J. Phys. B 18 2247 (1985).

13. M. Born and E. Wolf: "_Principles of Optics_", 4th Ed.
    Pergamon Press 1970.

14. B. Bederson, Comm. At. Mol. Phys. 1 41 (1969); ibid. 2
    7 (1970).

15. See E. Campbell, this volume.

16. H. Poincaré: "_Théorie Matematique de la Lumière_", G.
    Carré 1889, Chap. 12.

17. D.H. Madison, private communication. The theory is de-
    scribed in D.H. Madison and K.H. Winters, J. Phys. B
    16 4437 (1983).

18. J. Macek and D.H. Jaecks, Phys. Rev. A 4 2288 (1971).

19. K. Blum: "_Density Matrix Theory and Applications_",
    Plenum 1981.

20. N. Andersen, T. Andersen, J.S. Dahler, S.E. Nielsen, G.
    Nienhuis and K. Refsgaard, J. Phys. B 16 817 (1983).

21. U. Fano and J. Macek, Rev. Mod. Phys. 45 553 (1973).

22. See contributions by H. Lutz and C.D. Lin, this volume.

23. N. Andersen and S.E. Nielsen, Adv.At.Mol.Phys. 18 265
    (1982).

24. G.S. Panev, N. Andersen, T. Andersen, and P. Dalby,
    Z.Physik D 5 331 (1987).

25. R.S. Berry, J. Chem. Phys. 45 1288 (1966).

26. U. Fano, Phys. Rev. A 32 617 (1985).

27. C.W. Lee and U. Fano, Phys. Rev. A 36 66 (1987); ibid.
    p. 74.

28. A. Winther, _in_: "_Semiclassical Descriptions of Atomic
    and Nuclear Collisions_". Proceedings of the Niels Bohr
    Centennial Conferences, Copenhagen 1985, J. Bang and
    J. de Boer,eds. North-Holland 1985, p. 137.

29. N. Andersen and S.E. Nielsen, Z.Phys. D 5 309 (1987).

30. S.E. Nielsen and N. Andersen, Z.Phys. D 5 321 (1987).

31. See J. Andrä, this volume.

BASIC MECHANISMS IN ATOMIC COLLISION COMPLEXES

H.O. Lutz

Fakultät für Physik
Universität Bielefeld
4800 Bielefeld 1, F.R.G.

I. INTRODUCTION

The physics of ion-atom collisions is an old and venerable science.
Within the scope of these two lectures, an attempt at some degree of
completeness even in a rather cursory overview would be   entirely
impossible.   I will therefore discuss some basic aspects, thereby mostly
ignoring complications and sophistications which may arise along the way.
At appropriate positions, case studies will be presented which illustrate
some practical concepts and difficulties, and some connections to photon
and electron scattering processes will be pointed out.

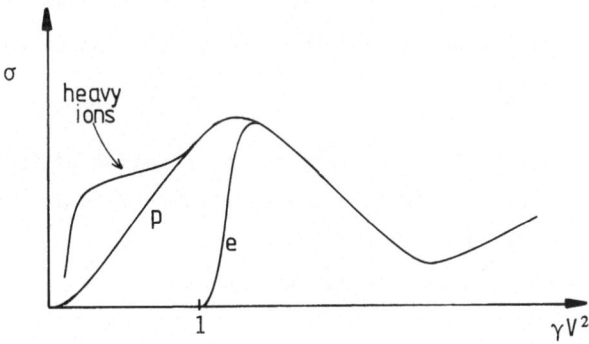

Fig. 1. Typical excitation or ionization cross section (schematic);
$(\gamma = (1-v^2/c^2)^{-1/2}$;   V = v/u with v the collision velocity, and u
the classical electron velocity).

Let us first have a look from the distance at the landscape we are going to walk around. Fig.1 shows schematically the behaviour of a typical cross section (e.g. excitation or ionization) for various types of projectiles. Characteristic features can be noted:

At low impact velocities v, even far below electron impact threshold, the large mass provides a heavy particle still with sufficient energy to excite or ionize the target. The maximum direct momentum transfer 2mv to a "free" electron (mass m) is, however, rather unfavorable; thus, the collision has to rely on the high-momentum components of the electron wave function. Therefore, the cross section rises with increasing v, until it reaches a maximum roughly around $V = v/u \approx 1$, with u the orbiting velocity of the active electron. In this low impact velocity regime ($V < 1$), the target electron wave function can be strongly perturbed by the presence of the projectile, particularly if the electrons experience forces of comparable strength from both the target and the projectile (i.e., $Z_1 \approx Z_2$); therefore, electronic processes proceeding through "quasimolecular" orbitals (whose binding energies can be drastically different from unperturbed atomic states) behave quite different from what would be expected from simple direct momentum transfer to electrons in unperturbed atomic states.

At high velocities, beyond the cross section maximum, all projectiles appear to behave roughly similar; this suggests that the target reacts mainly to the time-varying electromagnetic field created by the swift passing projectile charge. The cross section increase at relativistic velocities could then be understood in terms of the now Lorentz-contracted electromagnetic field pulse.

For a more quantitative description, the solution of the equation of motion for the particles involved can be attempted in various ways. For example, with high speed computers now available, the time-dependent equation of motion can be integrated directly as it stands; more customarily, however, one tries to find situations in which the process under study can be reduced to some kind of perturbative interaction. Very often, especially in the case of not too low energy ion scattering discussed here, semiclassical approximations (SCA) can be used: the projectile has a small de Broglie wave length, and moves along a classical path determined by an average (screened) ion-atom potential, and only the active electrons have to be treated quantum mechanically. Since this

approach also allows the most pictorial view of the collision process, we
will use it for the following discussion.

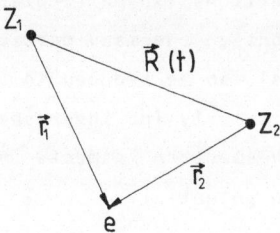

Fig. 2. The collision system (schematic)

The problem is sketched in Fig.2. For simplicity, we start from the
time-dependent Schrödinger equation

$$i\hbar\dot\Psi = H\Psi \tag{1}$$

where the (one-electron) Hamiltonian for the active electron is given by

$$H = -(\hbar^2/2m)\Delta - Z_1 e^2/r_1 - Z_2 e^2/r_2 \tag{2}$$

Expansion of the total electronic wave function into a set of basis states
yields

$$\Psi = \sum_s a_s\{R(t)\} * \psi_s * \exp(-i\int E_s \, dt/\hbar) \tag{3}$$

The ψ_s are eigenfunctions with energy E_s of a suitably chosen Hamiltonian
H_0, and depend parametrically on $\vec{R}(t) = \vec{r}_2 - \vec{r}_1$. Inserting equ.(3) into
(1), and using the orthonormality properties of the ψ_s, we obtain a set of
coupled equations for the transition amplitudes

$$i\hbar\dot a_k = \sum_s a_s(t) \, (\langle\psi_k|H-H_0|\psi_s\rangle - i\hbar \langle \psi_k|\dot\psi_s \rangle) *$$

$$\exp(i\int\Delta E_{ks} \, dt/\hbar) \tag{4}$$

with $\hbar\omega_{ks} = \Delta E_{ks} = E_k - E_s$. The summation includes continuum states.

The task is now to solve for a_k. So far, equ.(4) is equivalent to
(1) without any approximation, but in practice the basis set will be
incomplete, and this will limit the accuracy. This also shows that it is

347

quite important to adapt the Hamiltonian H_0 so that a minimum number of
basis states will suffice. Note that we consider a single active electron
only. Indeed, if the explicit electron-electron interaction is negligible
compared to the other interactions (in particular the usually dominating
electron-nucleus interactions), the many-particle situation (often found
in heavy ion-atom collisions) can be reduced to a summation over single
particle amplitudes.[1] Particularly for inner shells this is usually quite
well fulfilled, and the many-electron aspects very often reduce to a
screening of the interaction potential.

This is not necessarily so: the understanding of the single-electron
collision dynamics has now reached a rather high level of sophistication,
and the many-electron aspects become of increasing interest.

A simple look at the time evolution of a collision complex as e.g.
obtained from a numerical integration of the equations of motion (cf.
e.g. ref. 2) demonstrates that more than a total cross section is
required to characterize a collision. Evidently, the "shape" and the
internal motion ("rotation") of a collision complex is a significant
quality, being closely related to the population of substates having
certain angular momentum and magnetic quantum numbers. Such a discussion
has to consider two aspects: firstly, which states are produced in the
collision, and how do they evolve in the asymtotic region as the
projectile recedes to infinity; secondly, how can these states be
detected in an experiment and characterized as simply as possible. This
topic has received considerable attention in the past years. First
exploited in nuclear physics, these concepts have challenged theory as
well as experiment in electron and photon scattering, and lateron also
atomic collision studies. Rather detailed treatises can be found in the
literature and in particular in the proceedings of the two earlier Summer
Schools of this series.[3,4]

II. EXCITATION AND IONIZATION AT LOW COLLISION VELOCITIES (V<1)

1. Very Asymmetric Collision ($Z_1 << Z_2$)

The simplest case is obviously that of a very small perturbation of
the active electron wave function by the projectile; this occurs, e.g.,
in ionization (or excitation) of an inner shell electron by a light
projectile (note that most higher shells in a heavy atom are occupied,
i.e., ionization is here dominant). Therefore, the time-independent

348

atomic target wave functions offer themselves as a basis in equ.(3); thus $\dot{\psi}_s$, $\dot{E}_s$ = 0. Since usually in addition $|a_k| << |a_0|$, (s = 0 denoting the initial (ground) state) we obtain to first order ($\Delta E_{k0} \equiv \Delta E$)

$$i\hbar\dot{a}_k = \langle\psi_k|U|\psi_0\rangle \, a_0 \, \exp{(i\Delta E \, t/\hbar)} \tag{5}$$

with $U = H - H_0 = -Z_1 e^2/r_1 = -Z_1 e^2/|\vec{r}_2 - \vec{R}|$

Equ.(5) yields upon integration the final asymptotic amplitudes a_k (= $a_k(t \to \infty)$), the probabilities $P = |a_k|^2$, and the cross sections $\sigma = 2\pi \int P \, b \, db$. It is instructive to write $U(t)$ in terms of its Fourier transform $\tilde{U}$ (cf. E. Merzbacher in ref. 3).

$$U = 1/(2\pi)^3 \int \tilde{U}(q) \, \exp{[i\vec{q}(\vec{r}_2 - \vec{R})]} \, dq \tag{6}$$

(with $\tilde{U}(q) = -4\pi Z_1 e^2/q^2$).

Equ.(5) can then be written as

$$a_k \sim \int \tilde{U} \, F_{k0} \, T \, dq, \tag{7}$$

$$\text{with } F_{k0} = \langle k|\exp(i\vec{q}\vec{r}_2)|0\rangle \tag{8}$$

the inelastic form factor, and T a projectile trajectory function

$$T = \int \exp{i(\Delta E t/\hbar - \vec{q}\vec{R})} \, dt \quad ; \tag{9a}$$

in particular, for the specific case of a straight-line motion $\vec{R} = \vec{b} + \vec{v}t$ ("straight-line" SCA) T is given by (integration between $-\infty$ and $+\infty$)

$$T = 2\pi e^{-i\vec{b}\vec{q}} \, \delta(\Delta E/\hbar - \vec{q}\vec{v}), \tag{9b}$$

(and equ.(5) can be solved analytically).

We note several points; a few examples:

(a) This is a first order perturbation theory, and the cross sections scale with Z_1^2, as we would expect. From the structure of the above equations we might suspect that the straight-line SCA is equivalent to the PWBA; by closer inspection this is confirmed.[5]

(b) From equ.(9b) it follows that $\Delta E/\hbar = \vec{q}\vec{v}$, and the minimum momentum transfer $q_0 = \Delta E/\hbar v$. We see that the main contribution to T at small V comes from $\bar{b} \lesssim 1/q_0$ (cf. the "Massey criterion"), e.g. for ionization $\bar{b}/r_{shell} < 1$, i.e. from impact parameters much smaller than the shell radius. This can be clearly seen in the behaviour of the impact parameter dependent ionization probability (Fig.3a). This localization of the excitation region in slow collisions has several interesting consequences; for example, since the electronic interaction happens far within the electron cloud, it is mainly of the monopole type, populating low-l states. Another example is found in the structure of the total L_1-shell ionization cross section which shows a dip, for example at about 1.3 MeV impact energy in proton-Au collisions (Fig.3b). This is caused by the approximate coincidence of $\bar{b}$ with the location of the radial node in the 2s wave function.

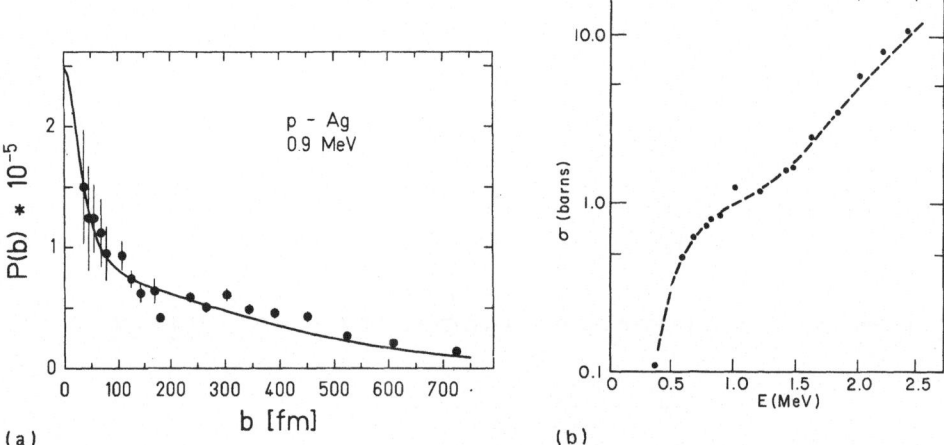

(a)   (b)

Fig. 3. (a) Impact parameter dependent ionization probability P(b) for 0.9
MeV p-Ag collisions (from ref. 6; $r_K \approx 1000$ fm); (b) total L_1
subshell ionization cross section for p-Au collisions (adapted
from ref. 7).

Other examples are the rather strong alignment effects which must be expected since different magnetic substates have different spatial distributions. Fig.4a shows as an example the integral alignment for $2p_{3/2}$ ionization in heavy atoms. First of all we note that the results for different targets (indicated by different symbols in Fig.4a) fall on two common curves for the two projectile ions if plotted vs. V. This is a consequence of the near-hydrogenic nature of inner shells. In addition, at small V strong negative alignment is predicted by theory and found experimentally. In a simplistic way, this is easily understood:

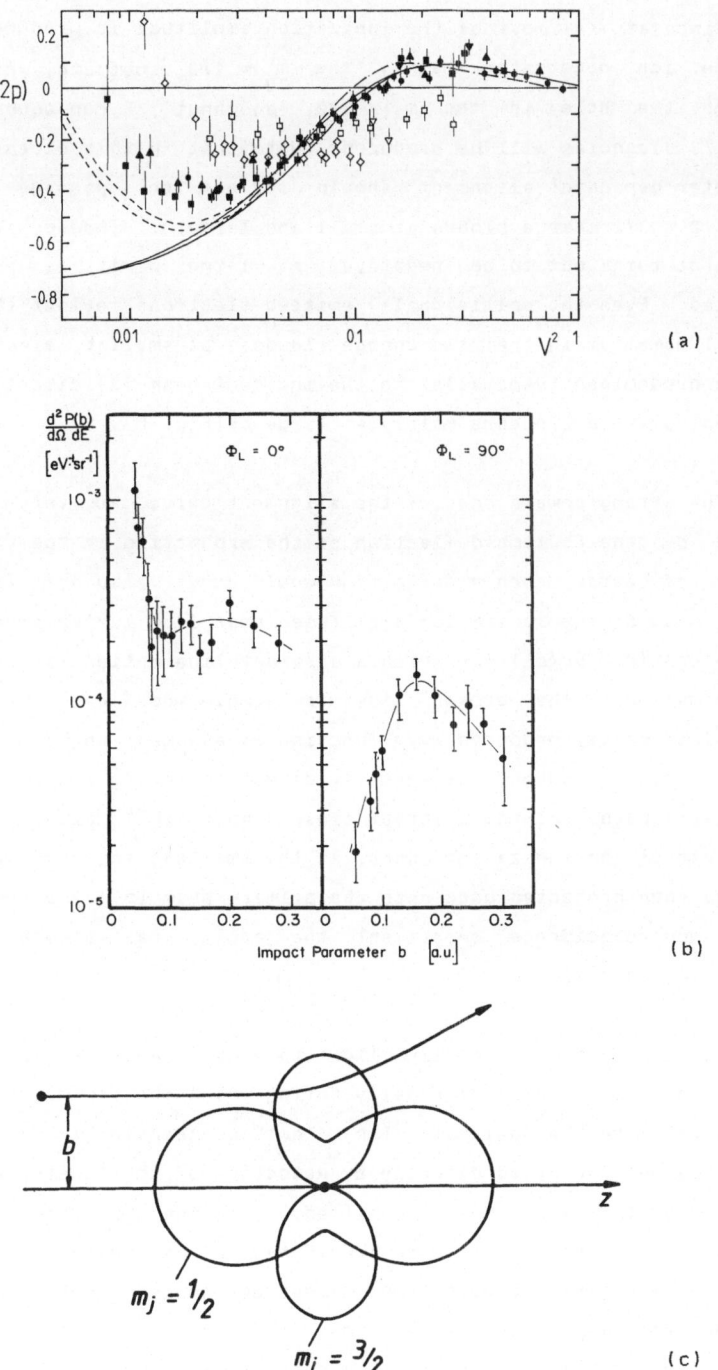

Fig. 4. (a) Integral alignment $A=[\sigma(3/2)-\sigma(1/2)]/[\sigma(3/2)+\sigma(1/2)]$ for $2p_{3/2}$ ionization by proton (closed symbols) and O ion (open symbols) impact;[8] (b) collisional electron ejection in 0.53 MeV/amu F^{8+}-Ne collisions.[10] (c) The $m_j = 1/2$ and $3/2$ electron clouds in a collision.

Since at V<<1 most of the ionization amplitude is produced at small R, the ion penetrates mainly the $m_j = 1/2$ substate, while it passes through the hole in the $m_j = 3/2$ "doughnut"; consequently, mostly $m_j = 1/2$ vacancies will be produced. Note that in this picture the impact parameter dependent alignment should reflect the passage through the $m_j = 1/2$, $3/2$ charge clouds at small and large b: indeed, at small b the alignment turns out to be negative, at large b it is positive,[9] as expected. Even the collisionally emitted electrons reflect the (preferred initial momemtum in the) two charge clouds: at small b, electron emission occurs predominantly parallel to the incident beam (z) direction, ejection perpendicular to z occurs mainly at large b (Fig.4b).

The strong upward bend of the alignment curve at very small V is caused by the Coulomb deflection of the projectile in the target nuclear field; of course, such a deflection would occur also at larger V but there most of the ionization amplitude comes from larger impact parameter collisions ($\bar{b} \approx \hbar v/\Delta E$) for which a straight-line motion is still a good approximation. The broken lines are simple model calculations in which the collisionally produced wave function is assumed to be aligned with respect to the outgoing trajectory direction (or to a direction parallel to the trajectory at the point of closest approach to allow for the fact that most of the ionization occurs at the smallest internuclear separation R), and then projected back onto the primary beam (z-)axis (remember that in a non-coincidence experiment the collisional ensemble is axially symmetric about z).

(c) This SCA treatment readily allows to break the trajectory halfway in order to allow for a time delay during which the ion may be absorbed by the target into the nucleus. The fleeting transiently formed "united atom" has been observed directly by detection of the "united atom x-rays", i.e., inner (K-) shell x-rays emitted from the "united atom" $Z_1 + Z_2$ during the very short time span 1/Γ (Γ the nuclear resonance width).[11] The change of ionization probability, caused by the corresponding phase delay between the amplitudes along the incoming and outgoing half of the trajectory, in principle offers the possibility to determine nuclear life times from experiment. A number of related studies have been performed; by far the largest effect has recently been seen by Spooner et al[12]: they found a pronounced resonance behaviour of the K-shell ionization cross section around the 10 MeV isobaric analog $^{138}Ba(p,p)^{138}Ba$ resonance. The effect is particularly strong here since the resonance width (Γ ≈ 68 keV) is comparable to the K-electron binding energy (I ≈ 37 keV), i.e., the delay of the projectile in the target nucleus is of the order of the K-electron orbiting time.

(d) Refinements: We see that this simple treatment contains many essential details of the collision process. Much refinement is possible, of course. Just to mention a few: account of the Coulomb deflection and retardation of the projectile in the field of the target nucleus (this is particularly important at small V where the effect can reduce the cross section by orders of magnitude due to the strong localization of the excitation region); inclusion of relativistic effects at high Z_2 which cause a compression of the wave function at the nucleus (the resulting increase of the high-momentum components results in a break-down of the simple hydrogenic scaling and shrink the P(b) dependence to smaller b); account of polarization and binding energy increase of the atomic electrons by the presence of the projectile; etc. Paul and Muhr[13] have compared a large number of experimental K-shell ionization cross sections to theory, to derive "reference cross sections" and to assess the limitations of the various corrections; the accuracy of cross sections for K-shell ionization by light projectiles is mostly approximately 5 to 10 %, although at certain impact velocities larger deviations from theory occur which are not completely understood yet.

2. Nearly Symmetric Collision Systems ($Z_1 \approx Z_2$)

If the collision becomes more symmetric, deviations from the simple picture presented above become evident. This concerns, for example, the behaviour of total cross sections (cf. Fig.1), the disagreement of experimental alignment data with SCA or BA model predictions (cf. O ion impact shown in Fig.4a), etc. These deviations from the simple expectations of section II.1 are, in the first place, caused by the particular choice of the basis states in that section, which is rather unpractical for more symmetric systems. Evidently, the states are strongly distorted by the presence of the comparatively high projectile charge, and a better picture is that of the transiently formed quasimolecule. Clearly, a natural choice for H_0 would now be the total Hamiltonian H itself, i.e., $H-H_0 = 0$. The wave functions and their energies are now parametrically a function of R(t); from equ.(4) we thus obtain

$$a_k = - \sum_s a_s \langle \psi_k | \psi_s \rangle \exp (i \int \Delta E_{ks}(t) dt / \hbar) \qquad (10)$$

Transitions are caused by the operator

$$\partial/\partial t \equiv R*\partial/\partial R + \theta*\partial/\partial\theta, \tag{11}$$

i.e., radial (Landau-Zener) and rotational (Coriolis) terms, connecting states of the same or different symmetry, respectively. During the collision the electrons move along their prescribed paths $\psi(t)$ of energy $E(t)$. Conventionally these paths are represented by correlation diagrams which display the electron energies as function of $R(t)$ (Fig.5).

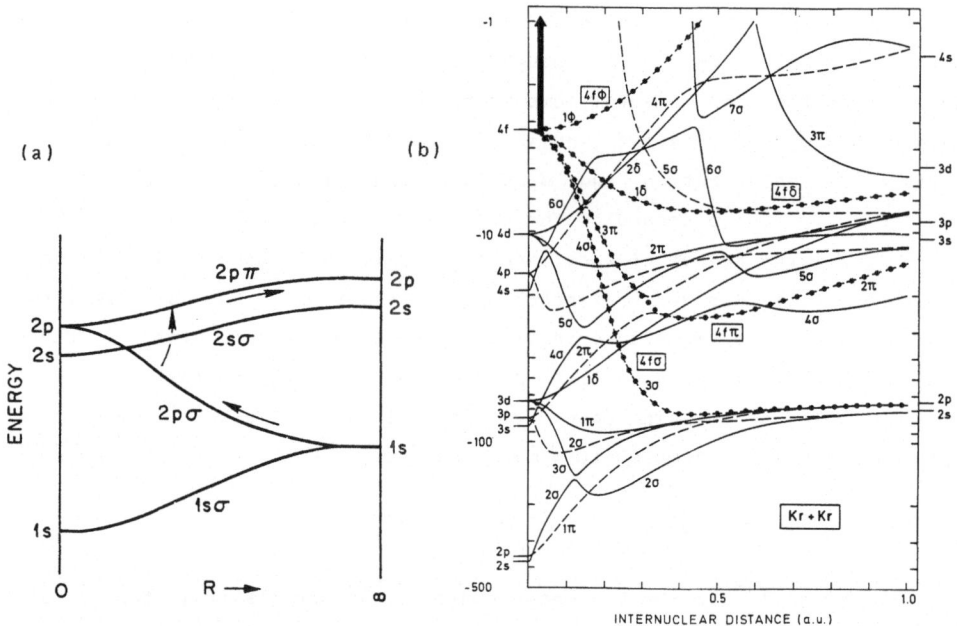

Fig. 5. (a) Schematic H_2^+ correlation diagram; (b) Kr-Kr correlation diagram (from ref. 19).

The practical choice of the Hamiltonian H_0 and, therefore, the corresponding correlation diagram, is largely a matter of convenience. Consider, for example, the H_2^+ collision system in which each state turns out to be uniquely characterized by a set of (three) different quantum numbers. According to the von Neumann - Wigner non-crossing rule, all states may therefore cross. In a more-electron collision system, on the other hand, the active electron moves in a two-center potential which is not quite Coulombic due to the screening by the other ("spectator") electrons. This deviation from a pure two-center Coulomb potential breaks the special symmetry of the H_2^+ quasimolecule: the "adiabatic" eigenstates are now characterized by one or two quantum numbers only (the angular momentum projection λ on the internuclear axis, and the parity in symmetric systems); as a consequence, certain states will have identical quantum numbers, and cannot cross. In the vicinity of such "avoided crossings", the adiabatic states rapidly change their symmetry properties. With some probability, this rapid change will not be followed (cf. the operator $\partial/\partial t$), causing transitions between the adiabatic states: the system shows a tendency to evolve along "diabatic" (H_2^+ like) states which thus incorporate to some extent the dynamical evolution of the system. For a more detailed discussion see, e.g., M. Barat in ref.3.

In passing we note that practical choices of basis states are frequently associated with more technical problems which nevertheless are often found to obscure the simple picture. Obvious is the increasing number of states generally necessary for increasing velocities, as a consequence of the increasing "motional width" of the quasimolecular states. Furthermore, in order to save computation effort, the MO basis is merged at some (finite) distance R into atomic states. The corresponding (diabatic, R-independent) energy levels do not cross, and the dynamical elements $\partial/\partial t$ vanish. However, since H_0 is not exact, the residual $H-H_0$ gives non-vanishing "potential coupling" elements. This model has been exploited in the treatment of "(vacancy) sharing", i.e., the distribution of vacancies or electrons among closely-spaced levels at large R. There is also another problem: So far, we have not differentiated between the active electron sitting finally at the target, or at the projectile. However, any state associated with the latter has asymptotically to be attached to a moving nucleus. Apparently, if this property is ignored, and a basis is used which properly describes the quasimolecule only in the (usually not too large) R-region of interest, it would have the wrong asymptotic behaviour: the consequences were again spurious couplings at large R. Just as an experimentalist has to correct his spectrometer setting for the Doppler shift in order to detect electrons emitted from a moving source, theory has to account for the asymptotic motion of target electrons which have changed position to the moving ion. This can be accomplished by so-called "translation factors". Clearly, in agreement with our intuitive feeling, it is particularly at large v where these tranlation factors are most significant. For more detailed discussion cf. e.g.,[14] and C.D. Lin in this volume.

We turn now back to the basic mechanism (cf. equ.(10)) with the aim to illustrate the breadth of this simple model by discussing a few examples.

It was more than 20 years ago when Fano and Lichten[15] proposed this quasimolecular model to explain discrete energy losses occurring in the scattering of Ar^+ ions on Ar atoms by the removal of one or two electrons from the Ar L shell via the steeply promoted $4f\sigma$ MO's, each removed electron causing an energy loss approximately corresponding to its binding energy. Since then, many special cases of MO level couplings have been identified and studied, particularly if they turned out to possess certain evident characteristics. The most famous example is the $2p\sigma-\pi$ rotational coupling in inner shells: in symmetric systems at not too high velocity, this coupling is rather isolated, and equ. (10) reduces to the two channels $2p\sigma$ and $2p\pi$ only (the $2s\sigma-2p\sigma$ radial coupling is then parity-forbidden; in $Z_1 \neq Z_2$, it is highly diabatic, and can be eliminated by using appropriate diabatic states[16]).

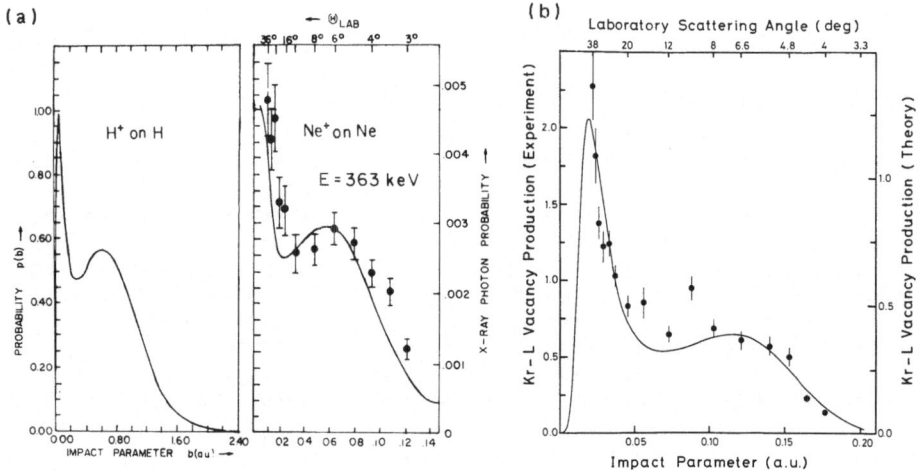

Fig. 6. Impact parameter dependent excitation: (a) K-shells in 250 eV H$^+$-H and 363 keV Ne$^+$-Ne,[17,18] (b) L-shell in 1.75 MeV Kr-Kr.[19]

The probability P for transfer of an electron originally in the 1s state into the 2p state shows (at not too low impact velocity) the characteristic doubly-peaked structure caused by the "slipping" of the electron cloud off the quickly rotating internuclear axis which has since been used many times to identify rotational interactions (cf. Fig.6); note that the independent electron model permits to scale P from the H$_2^+$ system, although in H$^+$-H collisions this behaviour has never been experimentally verified. Such rotational interactions are probably the best-studied individual couplings in slow atomic collisions; they have been quantitatively identified in a large number of systems with particular emphasis on inner shell processes where the independent electron model has a wide range of applicability (for a review cf. e.g. ref. 20).

The simple model contained in equ.(10) is able to describe a wide range of phenomena:

- rotational couplings as the 2pσ-π (which was first postulated by Bates and Williams[21] for H(2p) excitation in H$^+$-H collisions) have since been invoked many times particularly to explain inner (K, L, ...) shell processes;

- long-range couplings between asymptotic states determine the final destination of an electron (or vacancy) on the target or the projectile, particularly in somewhat asymmetric collision systems (vacancy sharing, cf. above);

- in not too heavy systems $Z_1 + Z_2$ at very short internuclear separations (i.e. close to the united atom limit) the MO energies hardly change with R, and the coherent motion of the two nuclei couples states which are practically atomic. The transition amplitude has a form quite similar to the one used for the direct amplitude (equ.(5)); the transition operator U is now given by the full molecular potential, U = $U_{proj.}$ + U_{target}.[22] This direct coupling of UA levels is (as in many other cases mentioned already) particularly conspicuous in the P(b) curves at small b (cf. refs. 9,20);

- in quite heavy collision systems $Z_1 + Z_2$, the atomic electrons are relativistic. As already mentioned in section II.1, this causes a strong compression of the wave functions around the nuclei, ψ_s and E_s still being strongly R-dependent even at R as small as the nuclear radius. The radial matrix elements for coupling of even the deepest states into the continuum become very large, with probabilities $P(b \approx 0)$ of the order of 10% (cf. e.g. G. Soff in ref.3). For MO binding energies I $\gtrsim 2m_0c^2$, also the negative Dirac continuum strongly couples to the internuclear motion, i.e., the ψ_s should now contain the appropriate negative energy solutions. Electrons can be coupled by the nuclear motion directly or in a two-step process out of the negative continuum, thus creating vacancies which are then ejected away from the UA nuclear charge as positrons ("stimulated positron emission"); or an empty quasimolecular state "dives" into the negative continuum, thus creating an unstable resonance in the positron continuum which can decay again by emitting rather sharp positron lines ("spontaneous positron decay") (cf. e.g. ref. 23). Experimentally, such positron structures have been found several years ago, typical positron energies lying around 350 keV, but recently new experimental information has been obtained: in particular, there appear to be coincident electrons of the same energy, emitted in the CMS system at 180° to the positrons. This seems to exclude the "atomic" positrons, but the problem still awaits a solution; speculations range from experimental artifacts, to a an e^+e^- resonance, to a new particle with a rest mass of about 2 * 350 keV + $2m_0c^2 \approx 1.8$ MeV (cf. ref. 24).

3. The Quantum Mechanical Phase

So far, our examples have mostly looked at aspects of the excitation process which, usually of course to a lesser degree of accuracy, could perhaps also be described classically: cross sections, ratio of total cross sections (e.g., alignment), and probabilities (cf. P(b)) are given by the square of the amplitude; the wave nature of the electron motion as expressed in the phase (cf. the factor $\exp(i\int \Delta E \, dt/\hbar)$ in equ. 4), did in most cases not appear explicitly. To bring out the full quantum mechanical complexity of the electron motion, we have to ask the right questions, i.e., do phase sensitive experiments.

One such example we have seen already: a nuclear reaction associated with some "sticking" of projectile and target nucleus introduces a phase delay and thus changes the total inner shell excitation or ionization amplitude.

Another example is the "rotation" of the excited state due to a phase between coherently excited magnetic substates (cf. e.g. H. Kleinpoppen and N. Andersen in this volume).

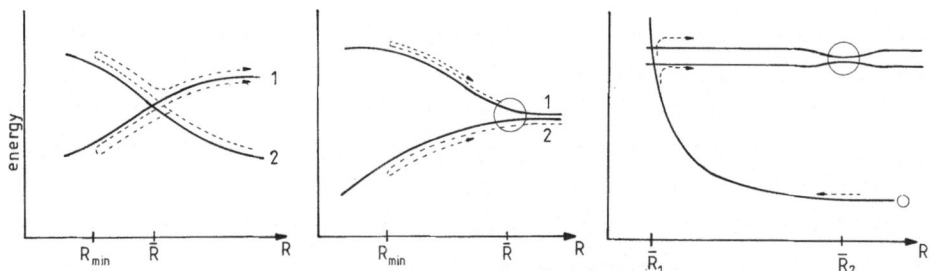

Fig. 7. Various types of interfering MO paths (schematic)

A sure way to introduce a phase difference is to force the system to evolve along different paths which are lateron brought together again. Consider e.g. Fig.7a: the system can pass an "avoided" crossing which on the ingoing part of the trajectory mixes states 1 and 2 at $R \approx \bar{R}$. On the outgoing part, they couple again, but have now suffered a phase difference $\chi = \chi_1 - \chi_2 = \int (E_1 - E_2) \, dt/\hbar$; Fig.7b shows a similar situation, in which the coupling region $\bar{R}$ is now a "long-range" coupling of the vacancy sharing type. Depending on the phase difference, the interference is constructive or destructive, causing oscillation in the excitation probability which is accessible to experiment ("Stueckelberg oscillations"). This situation is rather comparable to an optical two-slit interference experiment; the first passage through the coupling region prepares a coherent wave motion, the second passage effects the interference. Clearly, the appearance of this effect is connected to a well-defined trajectory: ill-defined trajectories will blur the coherence and thus the observed oscillation. As a consequence, the total cross section will in general not oscillate, due to a superposition of all impact parameters; it is only the impact parameter-selected probability P(b) which shows structure (Fig.8).

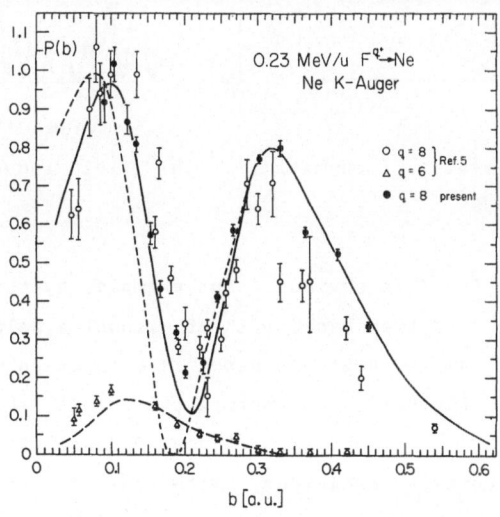

Fig. 8. Oscillatory K-K vacancy sharing in 0.23 MeV/amu F^{8+}-Ne collisions[25]

A special case has been pointed out by Rosenthal and Foley[26]: the specific trajectory with b≈0 can be isolated if one of the active couplings is located at very small R (Fig.7c). At $\bar{R}_1$, the coherent states 1 and 2 are prepared, at $\bar{R}_2$ they interfere on the outgoing part of the trajectory; if $\bar{R}_1 << \bar{R}_2$, only trajectories with small b≈0 are selected and can contribute, and the oscillations appear also in the total cross section; in particular, they have opposite phase in channel 1 and 2.

It is interesting to note that it does not need a collision to produce the behaviour described above if one recalls that the MO evolution in a quasimolecule is really caused by the R-varying (nuclear) Coulomb force. Such a Coulomb force can as well be simulated by an external electric field which distorts the atomic wave functions. Clearly, it would take extreme electric fields (order of 10^9 V/cm) to noticeably affect ground states. Rydberg states, however, exist in much weaker fields (order of few V/cm) and can readily be used for a demonstration, as has been shown by McAdam and coworkers.[27] Consider Stark states 1,2,3 which change their energy as a function of field strength E (Fig.9a). If state 2 and 3 have the proper symmetry, they interact at $E = \bar{E}$.

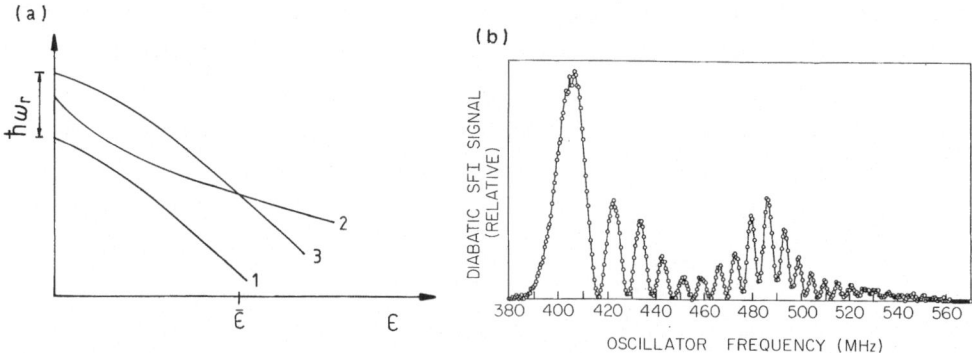

Fig. 9. (a) Stark states (schematic); (b) field-induced Stueckelberg oscillations[27]

The analogy to Fig.7 is obvious. For example, by running an E-ramp up to or across $\bar{E}$, and then back, with or without a delay at the maximum ramp field strength, all effects discussed in connection with Fig.7a can be obtained (cf. Fig.9b). To obtain the results shown in Fig.9b, an experimental trick has been applied: since Rydberg Stark states show usually quite a complex behaviour, with the location of interaction regions not well known, the interaction at $\bar{E}$ has been artificially produced by shining microwave radiation $\hbar\omega_r$ on the atom. This induces "dressed states", e.g., $|\psi_1 \pm n\hbar\omega_r\rangle$ etc., with the state 3 shown in Fig.9a being equivalent to $|\psi_1 + \hbar\omega_r\rangle$. By changing $\hbar\omega_r$, the value of $\bar{E}$ can thus be quantitatively controlled and changed at will, and the oscillations also appear as $f(\omega_r)$ in the signal of state 2 (produced, e.g., by field ionization). This is also a very nice example how the influence of static external fields could be studied experimentally. Theoretical work so far is quite sparse and has been concentrated on collisions involving ground state atoms or ions. Necessary fields to modify the collision are far too large to be produced in the laboratory; for magnetic fields cf. e.g. U. Wille in ref.4. By using Rydberg atoms, however, the field strengths could be reduced to quite manageable values. Theoretically, of course, this would introduce difficulties if a conventional quantum mechanical treatment were intended. A simple classical treatment might be quite adequate, though.

A peculiar type of coherence has recently been found in the spin exchange channel of an atomic collision.[28] Consider the collision of a spin polarized atom (e.g. Na) with an (unpolarized) open core ion (e.g. He^+ or Ne^+). Since the spin coupling to any other magnetic moment (spin-projectile orbit, spin-projectile spin, etc.) is small compared to the collision time, the spin is usually regarded as space-fixed during the interaction (Wigner's spin rule); in this picture, any spin change can be effected only long after the collision if some orbital angular momentum has been changed during the collision, and the spin then precesses (with a time constant much larger than the collision time) about the new total angular momentum j. Note, however, that the polarized electron can make exchange with an electron in the unpolarized open core. In the collision, parallel and antiparallel spin configurations will occur statistically. The antiparallel configuration is a coherent superposition of singlet and triplet states (the parallel configuration, being a pure triplet state, will not change the spin in an electron exchange), and their time evolution will again follow the $\exp(i\int\Delta E_{ex}/\hbar)$ behaviour: the resulting change in spin orientation of the originally polarized electron therefore happens with a frequency $\omega_{ex} = \Delta E_{ex}/\hbar$, where the exchange energy ΔE_{ex} is given by the singlet-triplet splitting. To observe this effect of spin-depolarization, the collision time can be used as a "clock" if it is of the order of $1/\omega_{ex}$; with ΔE_{ex} typically a few tenths of eV, the corresponding collision energies e.g. in Ne^+-Na collisions should thus lie around 1 keV. So far, only the first part of this oscillatory behaviour of the spin depolarization has been observed in He^+ and Ne^+-Na($\uparrow$) collisions (Fig.10).

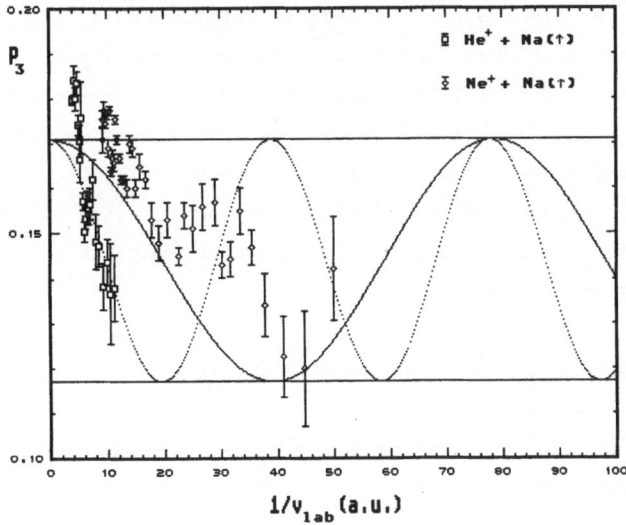

Fig. 10. Spin exchange in He^+-He and Ne^+-Ne collisions.[28] Upper and lower horinzontal lines indicate maximum and minimum spin polarization.

Note that this model is not at variance with the Wigner spin rule: the total spin is still conserved, a spin polarization is just transferred

<u>during the collision</u> from the outer electron to the open core, and back.

It might be mentioned that integration over all impact parameters (i.e. different collision times) would be expected to smear out this effect; however, observation of a process which is governed by a short-range interaction (as, e.g., Na 3p-excitation in Ne^+-Na collisions, which proceeds through a fairly short-range $4p\sigma$-$3p\pi$ coupling) selects specifically $b\approx0$ trajectories just as has been discussed above in case of the "Rosenthal-oscillations".

It is interesting to compare this effect to electron impact processes. Here, spin-changing effects constitute a major interaction particularly close to threshold, or around resonances, i.e., for sufficiently long "interaction times". Spin oscillations as a function of $1/\sqrt{E_0}$ as discussed above should in principle also be observable; in practice, however, tuning the impact energy E_0 with 10^{-2} eV resolution around the singlet threshold or through a resonance, and performing a spin-resolved experiment, appears to be very difficult indeed.

I want to conclude this section with a brief comment: evidently, the phase is a very sensitive quantity of a collision system. Phase changes of order π can cause drastic cross section variations. One should, however, not be mistaken as to regard the phase as the most critical quantity as such: if the phase change is not too rapid, its influence can be rather inconspicuous, and as you saw, phase measurements are usually difficult. You also recall that the "corrections" mentioned in section II.1 can change the cross sections by many orders of magnitude, too; e.g. the Coulomb deflection at low V has very strong influence on an inner shell cross section due to the strong localization of the ionization probability. Evidently, the phase is important, but there are other quite critical quantities as well.

4. Towards a More Complete Understanding of Few-Electron Systems

So far, we have concentrated on some basic aspects of atomic collisions. We want to use this knowledge now to ask how much do we really know about such a collision. As before, we concentrate on few-electron systems. The ideal case, of course, would be H^+-H; then there are the quasi-one electron situations, as e.g. H^+-Na (which at not too high impact energies has a practically inert core). A next step would be (quasi-)two electron situations, as H^+-He, or He^+-Na, and so forth. Note that inner shells, characterized to a high degree by an

362

independent-electron behaviour (as we have seen in the previous sections), are not a good test case. The main reason is that if we look deeper into the collision process, the other "spectator" electrons turn out to be not spectators at all; rather, they obscure much of the basic mechanisms. This can e.g. be seen when one tries to see alignment or orientation effects: as Fig.4 has shown, the integral alignment is smeared out even in a comparatively simple system as O^+-Z_2; no alignment is left e.g. in Ar^+-Ar collisions. This can also be (semi-)quantitatively demonstrated if a second vacancy is introduced into the collision system[29]: the angular momentum coupling between this "spectator" vacancy and the vacancy under study strongly reduces the measured alignment. An orientation has never been measured in inner shells since this would require a coincident circular polarization measurement in the x-ray regime, a practically impossible experiment. Therefore, real-few (or quasi-few) electron systems are a much better case; in particular, many interesting transitions are in the regime of visible radiation, or not too far away, and are therefore readily accessible to polarization measurements. For a case study we want to use the p-He system; in particular, let us concentrate on the reaction

$$H^+ + He \rightarrow H(2p) + He^+(1s) \tag{12}$$

$$\downarrow Ly\alpha$$

$$H(1s)$$

It is well suited because it is governed by several of the processes which we have discussed above. We want to work our way from one level of understanding to the next higher ones ("first, second, third generation studies" cf. N. Andersen in this volume), thereby meeting some typical techniques, results, and difficulties. To set the stage, we first have a look at the correlation diagram which gives an idea of the main reaction paths (Fig.11): at large distances R on the incoming part of the trajectory, one of the He-electrons is shared with the H(1s) state. Then the electron proceeds through the $2p\sigma$ and the well-known rotational coupling at small R into the $2p\pi$ state. Note, however, that the system is asymmetric and does not possess good parity; therefore, as already mentioned, radial coupling with $2s\sigma$ must also be taken into account, and it is not certain at all that the $2s\sigma-2p\sigma$ radial coupling is highly diabatic as in inner shells (indeed, as we shall see, it is not). On the outgoing part of the trajectory, the electron has the choice to proceed along various paths (some of them are indicated in the figure by dots),

363

until it is (coherently) shared again between various levels, one of them being the H(2p) channel which we finally detect by Lyman α emission. The residual He$^+$ ion at low energies is left predominantly in the ground state.

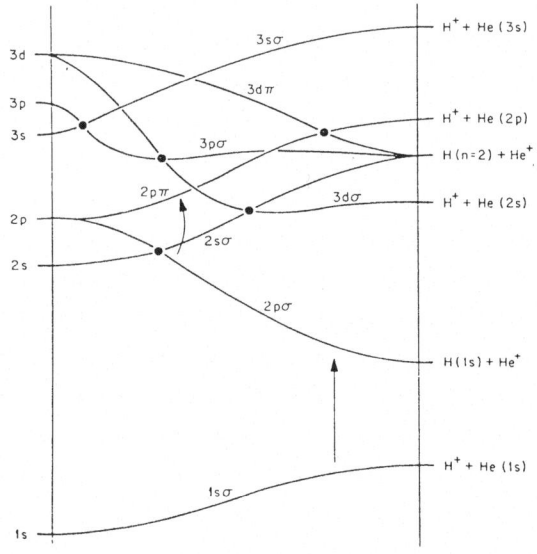

Fig. 11. H$^+$-He correlation diagram (schematic)

(a) A "first generation" study: the total cross section

The total cross section shows the typical behaviour mentioned already in section I. Some time ago it appeared as if rather advanced two-electron SCA calculations[30] are not able to correctly predict the total cross section, even though more sensitive details (as alignment and orientation) are well described (see below). This problem has recently been resolved[31]: the disagreement at impact energies below about 5 keV was apparently caused by experimental problems in earlier measurements. Since the charge exchange cross section at these energies is about two orders of magnitude smaller compared to H(2p) excitation in (neutral) H-He collisions, a small contamination of the proton beam by neutral H particles (created e.g. by charge exchange in the residual gas of the vacuum apparatus) can lead to drastically overestimated cross sections in the H$^+$-He experiment. Thus it can be concluded that the SCA treatment (as employed in ref. 30) is able to give a good account of the total cross section.

(b) A "second generation study": the integral alignment

The impact parameter integrated ("integral") alignment is rather disappointing: it is nearly zero at V < 1 (Fig.12). Note that this contrasts to the behaviour of neutral H+He → H(2p)+He collisions; this is already a first indication that the H^+-He system is not simply equivalent to a one-electron system, although the main reaction paths evident in the corresponding H-He correlation diagram look quite alike (of course, with the exception of the long-range charge exchange process on the incoming part of the trajectory). This point is emphasized by the behaviour of other (quasi-) one-electron systems, as H^+-H or H^+-Na (note, the H^+-H is particularly interesting, since the $2s\sigma$-$2p\sigma$ radial coupling is now parity-forbidden), and quasi-two-electron systems (as He^+-Na) (cf. ref. 33).

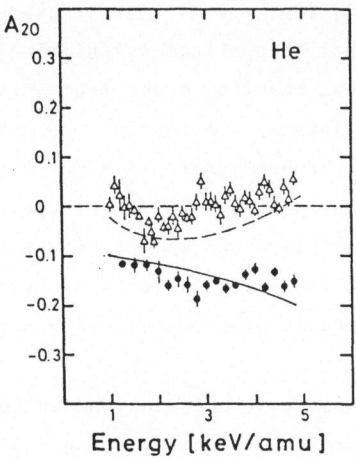

Fig. 12. Integral alignment of H(2p) in H^+,H-He collisions[32,33]; A_{20} = $(\sigma_1-\sigma_0)/(2\sigma_1+\sigma_0)$

The integral alignment displays a quite different behaviour in the different systems, and there seems to be no simple scaling rules. Of course, as in case of the inner L-shell alignment (section II.1) the vanishing integral alignment in H^+-He does not necessarily mean a vanishing impact parameter dependent alignment: quite possibly, the vanishing of the integral alignment is caused by the averaging over b. This problem is solved by coincidence experiments in which individual trajectories are identified; the axial symmetry of a non-coincidence experiment is broken in this way, allowing to determine the collision-transferred angular momenta $\langle L_y \rangle$. This leads to

(c) The "third generation" study: the complete experiment

Equ. (12) hints at an interesting possibility: three of the four particles (H^+, He, He^+) appear in the (quantum mechanically "trivial") isotropic ground state. Any "non-triviality" (anisotropic shape, rotation) must, therefore, appear in the H(2p) state; since H(2p) decays by Ly α photon emission into the again isotropic 1s ground state, the Ly α photons must contain all there is to know about the excitation process, i.e., this information is entirely contained in their linear and circular polarization which is detected in coincidence with the scattered projectiles (thus defining a scattering plane and an impact parameter).

As has been suspected above, the impact parameter dependent alignment by no means vanishes: at small b, mainly $|2p,m=0>$ ($=2p_0$) states are excited, at large b mainly $|2p, m = \pm1>$ ($=2p_1$) states (Fig.13). This is in agreement with recent theoretical calculation , but at variance with the simple 2pσ-π rotational coupling model from which one would naively expect mainly $2p_1$ population. According to Macek and Wang[34] this behaviour is caused by the non-adiabatic mixing of 2sσ and 2pσ over a wide R-range, with a particularly large radial transition strength at small b. Note that the differential alignment is not highly sensitive to details of the collision system: the calculations of Fritsch[35] (broken lines in Fig.13a) are made on the basis of a one-electron model.

Much more sensitive is the orientation vector T_{11} ($T_{10} \equiv 0$ by symmetry arguments). Fritsch's one-electron calculations would give positive and too large values while the experimental data show an oscillation-like structure as function of impact parameter. This is clearly in contrast to a simple "bouncing-ball" model; it makes evident that the orientation is really an expression of the quantum mechanical phase which is critically decided by the interplay of the paths and couplings through which the system evolves. This statement is emphasized by the actually accumulated phase χ which does not simply correspond to the adiabatic phase difference $\chi_{ad} = \int(E_1-E_0) \, dt/\hbar$ (E_1, E_0 the m=1,0 states, respectively). χ ranges approximately from π to 2π, while χ_{ad} lies between about π/2 and π.[36] The nice agreement between experiment and the two-electron calculations by Kimura and Lin[37] demonstrates that rather sophisticated models are necessary to properly describe such systems, taking into account the structure as well as the collision dynamics. This is also corroborated by the behaviour of the quasi-one electron (neutral) D + He → D(2p) + He system which again does not simply scale with the H^+-He results: in the neutral system χ_{ad} accounts fairly well for the accumulated phase, χ.[38]

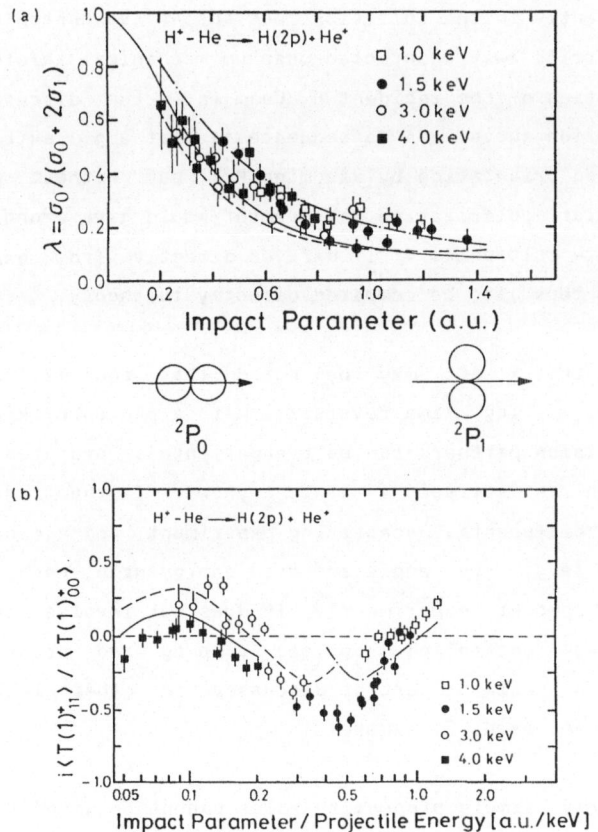

Fig. 13. H^+-He collision coherence parameters: (a) $\lambda = \sigma_0/(\sigma_0 + 2\sigma_1)$, and (b) orientation $\langle T_{11} \rangle$.[36]

With this data, we have now in principle all information which is quantum mechanically accessible. It should, therefore, be possible to construct the complete final state wave functions for all collision conditions (energy, impact parameter). In practice, however, one often faces the problem of incomplete coherence: whenever the investigated state is populated by different processes which are experimentally not distinguished even though they in principle could, then the wave function is not given by a coherent sum of amplitudes; the density matrix provides a better tool for describing such situations. This is also the case in the reaction equ.12: For complete coherence, the sum of all Stokes parameters, i.e., the "total polarization" P, must be unity (cf. N. Andersen, and H. Kleinpoppen in this volume). As it turns out, the total polarization P in our case study (equ.12) is not quite unity.[36] Reasons may be cascades populating H(2p), (i.e., higher states are populated coherently in the collision, but successive photon decay to the H(2p) state carries away undetected quantum mechanical information), or a H(1s) contamination of the incident H^+ beam (cf. the discussion of the total cross section above). A consequence is that e.g. extraction of the collision-induced orientation by assuming P=1, but without performing a separate circular polarization measurement would give wrong results. Of course, the state multipoles $\langle T_{KQ} \rangle$ derived directly from experiment are correct and can thus also be compared directly to theory.

As a final remark, it may be noted that equ.(12) can be read backwards, i.e., using time reversal: with a photon of known property, one of the collision partners can be brought into a "prepared state" (the other being in an isotropic 1s state anyhow). If the trajectory is now determined by a differential scattering experiment, and a final 1s state is identified (e.g. by angle-selected energy gain spectroscopy), this again is now a "complete experiment". It does not involve a coincidence experiment, but controlled optical pumping and projectile energy spectroscopy. This will be further discussed in other lectures during this School (N. Andersen, E. Campbell).

Unfortunately, simple propensity rules cannot be given yet for H^+-He and similar systems at small V, although one may hope to eventually develop a more general and intuitive picture of such basic collision systems. This is evidently a consequence of the action of many, often ill-separated interaction regions. This does not necessarily have to be so. There are many cases in which the separation of the critical interactions is much clearer. In particular this happens at very low impact velocities, and in situations in which direct coherent excitation of levels dominate (at intermediate and high velocities). The latter case is discussed by N. Andersen at this School. For very low velocities, the "collisional level broadening" is quite small, and different interaction regions can be much better separated. This makes it easier to follow the destiny of a rotation or alignment, induced at some small R, out to the asymptotic regime. Essentially, the procedure boils down to a (rather complicated) analysis of angular momentum coupling, as the system evolves through the different Hund's coupling schemes from molecular (small R) to atomic (large R) conditions. This problem shall not be discussed here any further, although it often allows to give a (in principle quite simple and intuitive) prediction of the final shape and orientation of a collision complex. Aside from much earlier work, a nice treatise has recently been published by Grosser[39] and the interested reader is referred to this paper and references cited therein.

(d) Remarks on the complete n=2 coherence

Although apparently the description of the reaction equ. (12) can be quantum mechanically complete, a significant part of the excitation amplitude is expected to go (coherently) into the practically degenerate H(2s) state. This state is metastable and cannot be observed without additional provisions. Usually, direct observation is achieved by applying an external electric field which Stark-mixes the s and p states. The field dependence of Ly α emission then varies in a characteristic manner which reflects the collisional 2s-p coherence. In our case study of H^+-He collisions, this has been done for the H(n=2) [40] and H(n=3) [41] state manifold in a non-coincidence experiment. Here, the collision ensemble is axially symmetric about the incident beam axis; therefore, the coherent admixture of an s state can manifest itself only in a forward-backward asymmetry of the H(n=2,3) electron (i.e., either leading or trailing the H after the collision). On the other hand, in any individual collision, i.e. a coincidence experiment, the 2p state is inclined ("alignment angle" γ) relative to the beam axis; consequently, also the leading-trailing effect does not necessarily have to be axially symmetric about the incident beam direction. In general, the electron distribution will be asymmetric with respect to the H-nucleus, and can even rotate around it.

Just to give an idea how complex the analysis of such an experiment can be, let us look at the simple quenching behaviour of e.g. a metastable H(2s) state which is exposed to a longitudinal $\vec{E}$-field. If we would assume a really degenerate n=2 manifold, the Stark effect would mix only 2s and $2p_0$, but not $2p_1$; the corresponding Stark states were $|2s> \underline{+} |2p_0>$ and $|2p_1>$. Naively, we would therefore expect quenching of H(2s) via $2p_0$ states, giving linearly polarized π-radiation. Instead, injection of a H(2s) beam in the keV regime into a longitudinal $\vec{E}$-field produces linearly polarized σ-radiation of nearly 100% polarization.[40] The difference between the naive expectation and the experimental finding is due to the non-degeneracy of the n=2 manifold, in combination with the almost adiabatic passage of the keV projectile through the fringe field region; to reach diabatic passage (in other words: to have a motional energy broadening in the ca. 1 mm long fringe field region corresponding to the fine structure splitting) and to observe the π quenching radiation, one had to have projectile energies of about 100 MeV! Therefore, an analysis of such s-p coherence experiments has very sensitively to take into account the actual form of the fringe field, and (if the collisional

s-p coherence is to be studied) also the gas target distribution in the
$\vec{E}$-fringe field region.

III. SOME REMARKS ON EXCITATION AND IONIZATION AT HIGH IMPACT ENERGY (V>1)

1. A General Picture

For increasing V the collisional width $\hbar v/a$ (a is some typical region
of space in which the excitation occurs) increases, until it becomes
comparable to (at V≈1) or larger than (V>1) the excitation (or ionization)
energy.   At low V, the projectile had to penetrate deeply into the target
(cf. section II.1); now at high V, ever increasing impact parameters
contribute, so that at V>>1 mostly distant collisions are significant.
Any perturbation is now so short that the perturbation scheme employed in
section II.1 should be quite satisfactory for all projectiles. In other
words, for $Z_1 << Z_2$, the straight-line SCA or PWBA should be a rather good
approximation at all except the lowest V (where the corrections mentioned
in section II.1 become important), and at V>>1 this is fulfilled to a
large extent even for $Z_1 \approx Z_2$.   Thus, for non-relativistic velocities
equ.(5) immediately leads to the well-known behaviour of such cross
sections, namely, a (1/E) ln E-behaviour for optically allowed (dipole)
transition, and a steeper decrease with energy for optically not allowed
terms.   Note that for heavy target atoms and (particularly) inner shells,
V>>1 (or even V>1) might mean relativistic projectile motion; the simple
non-relativistic treatment used so far would then break down. This is a
situation encountered now increasingly at high energy accelerators, even
with rather heavy ions where the corresponding impact energies then lie
far in the GeV regime. This has led to a wish for alternative methods
which may be less unwieldy than the full relativistic treatment which
would be called for in such a situation.   One method of comparative
simplicity is easy to see: in SCA or BA, the electronic motion is
governed by a form factor (equ.8) as in photoabsorption processes:
apparently, one can view photon and high-velocity particle impact somehow
as equivalent interactions. The concept seems simple: the time-varying
Coulomb field experienced by the target during the projectile passage
(Fig.14a) can be Fourier-transformed into a frequency spectrum of
"equivalent photons" from which the target then picks out those components
with which it can make a transition (cf. also N. Andersen in this
volume).

2. The "Equivalent Photon" Method

Let us put this idea on a more quantitative basis. Already in 1924, Fermi[42] treated excitation and ionization of atoms by fast non-relativistic α-particles, using this idea of equivalent photons. Lateron, the concept has been generalized by v. Weizsäcker[43] and Williams[44] (thus often "Weizsäcker-Williams" method, cf. also ref. 45). For very close collisions we immediately see a problem: if the interaction distance (e.g. the impact parameter b) becomes very small, the corresponding interaction Fourier components diverge with v/b (see also the discussion at the end of this section). Therefore, such "close collisions" (where the projectile actually penetrates through the target atom) have to be ignored; they are usually added as an (incoherent) correction by treating them as simple elastic scattering of the projectile by the (assumed quasi-free) atomic electrons. Neglecting these close collisions, the electric field "seen" by the target then consists of two components, one parallel and one perpendicular to the incident beam (z-) axis:

$$E_1 = -Z_1 e\gamma b/R_r^3 \tag{13a}$$

$$E_2 = -Z_1 e\gamma vt/R_r^3 \tag{13b}$$

with $R_r^2 = b^2 + \gamma^2 v^2 t^2$, and $\gamma = (1-v^2/c^2)^{-1/2}$. Fourier transformed, they correspond to two electromagnetic pulses P_1, P_2 incident on the target (Fig.14b); in dipole approximation, their intensities can easily be calculated[45]:

$$I_1(\omega,b) = const. x^2 * K_1^2(x) \tag{14a}$$

$$I_2(\omega,b) = const. x^2 * K_0^2(x)/\gamma^2 \tag{14b}$$

with $x = \omega b/\gamma v$. K_0, K_1 are the modified Bessel functions of order zero and one (Fig.14c). So far, everything is classical; quantum mechanics comes in by defining a photon number N, e.g. with

$$I \, d\omega = \hbar\omega * N(\hbar\omega) * d(\hbar\omega) \tag{15}$$

($I = I_1 + I_2$) which then gives the transition probability

$$P(b) = \int N\sigma_\gamma \, d\omega. \tag{16}$$

The photoabsorption cross section σ_γ can either be taken from experiment (as e.g. Fermi has done) or calculated separately.[46]

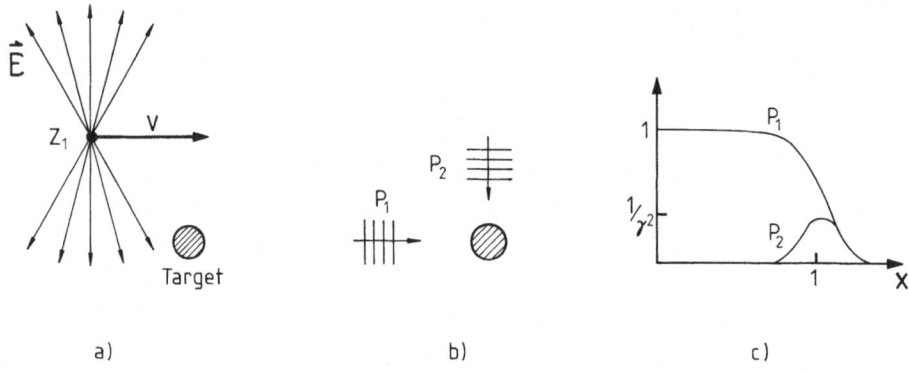

a)   b)   c)

Fig. 14. Two pulses of photons "equivalent" to the projectile-target Coulomb interaction

We see that at relativistic velocities ($\gamma>1$), the pulse P_2 becomes insignificant; at non-relativistic velocities ($\gamma\approx1$), both pulses contribute. This points to other difficulties of the method: by adding P_1 and P_2 incoherently one looses to some extent details of the Coulomb interaction; this is expected to be significant particularly in second- and third-generation experiments. Only at extremely relativistic velocities is this problem resolved as P_2 becomes negligible anyway. Nevertheless, even in the non-relativistic regime, the agreement with other methods (as e.g. BA) is good; for example, Kolbenstvedt[47] (calculating ionization cross sections for electron impact) and Komarov[48] (treating K-shell ionization by heavy particles) have found good agreement with BA and SCA, respectively, at non-relativistic velocities. At relativistic velocities, of course, the proper cross section increase is obtained. In the Weizsäcker-Williams picture this increase is due to the relativistic "compression" of the e.m. field lines in the pulse P_1 (cf. Fig.14a). In a proper (more elaborate) relativistic treatment the increase is caused by the increasing "transverse" magnetic interaction (i.e. the current density operator), while at non-relativistic velocities, the dominant interaction is due to the charge density operator (i.e. the "longitudinal" Coulomb interaction). If such "exact" calculations are compared to the Weizsäcker-Williams method conceptual problems of the latter become clearer. In the Weizsäcker-Williams picture, the photon exchange is broken into two separate events, namely, the calculation of a frequency spectrum, and the absorption of appropriate frequency components by the target with a cross section σ_γ. This apparently implies $(\hbar\omega/c)^2 = \vec{k}^2$ for the exchanged photon. The method thus relies on $\overset{*}{q}{}^2 = (\hbar\omega/c)^2 - \vec{k}{}^2$ being close to zero ($\overset{*}{q}$ the four-momentum of the exchange photon). Although this is very often true, it may in some instances be a poor approximation as has been shown by Eschwey and Manakos[49]: depending on the particular collision conditions, momenta $\overset{*}{q}{}^2$ far away from the "photon point" ($\overset{*}{q}{}^2 \equiv 0$) can be of importance.

Instead of following the above procedure (equs.14-16), one can also derive an "equivalent photon number" N from equ.16 by directly comparing it to an "exact" probability P. This "equivalent photon method" circumvents the problem mentioned above, and is being exploited frequently to calculate total cross sections and P(b) for excitation, ionization, bremsstrahlung production, etc. in electron and ion impact mainly in the relativistic regime (cf. ref. 50, and references therein).

IV. CONCLUDING REMARKS

Apparently, the basic mechanisms in ion-atom collisions can be understood quite often in simple and intuitive pictures. Much of the work is now concerned with test and refinements of these models. The tendency is clear: This often excellent knowledge about the very simplest systems is then applied to systems of increasing complexity. One such example is the increased interest in explicit electron-electron correlation in collision systems. Here, a rather broad range of phenomena attracts attention: among others there are transfer and excitation in ion-atom scattering; post-collision interaction in electron and ion scattering; correlated double ionisation in electron and ion scattering; the "45 degree" Thomas scattering peak in charge exchange; threshold effects in electron excitation; dielectronic recombination; etc. Other examples of even higher complexity may be found in the interaction of many-electron systems; or the hot topic of the interaction between very intense laser fields and atoms which can show striking parallels to energetic ion-atom collisions (cf. P. Agostini, this volume). While on the one hand, the quantitative understanding of such systems builds to a high degree on models developed for the more basic and elementary systems, a rapidly expanding computer potential also creates an increasing interest in the direct numerical solution of the many-particle equations of motion. It is clear, however, that mere computational descriptions would seriously limit the understanding of nature; it is the interplay between the "intuitive" models and concepts, and the more "tour de force" approach which will prove to be most fruitful.

ACKNOWLEDGEMENT

Valuable discussions with Drs. G. Baur and P. Eschwey are gratefully acknowledged. This work has been supported by the Deutsche Forschungsgemeinschaft (DFG) in Sonderforschungsbereich 216.

REFERENCES

1. J. F. Reading, Phys. Rev. $\underline{A8}$, 3262 (1973); J. F. Reading and A. L. Ford, Phys. Rev. $\underline{A21}$, 124 (1980).
2. R. Shakeshaft, Phys. Rev. $\underline{A18}$, 1930 (1978).
3. Fundamental Processes in Energetic Atomic Collisions (Proc. NATO ASI, Sept. 20 - Oct. 1, 1982, Maratea, Italy); eds.: H. O. Lutz, J. S. Briggs, H. Kleinpoppen; Plenum Press NATO ASI Series $\underline{B103}$, 1983.
4. Fundamental Processes in Atomic Collision Physics (Proc. NATO ASI, Sept. 10-21, 1984, Santa Flavia, Italy); eds.: H. Kleinpoppen, J. S. Briggs, H. O. Lutz, Plenum Press NATO ASI Series $\underline{B134}$, 1985.
5. J. Bang and J. M. Hansteen, Mat.-Fys. Medd. K. Dan. Vid. Selsk. $\underline{31}$, No.3 (1959).
6. L. Kocbach, J. M. Hansteen and R. Gundersen, Nucl. Instr. Meth. $\underline{169}$, 281 (1980).
7. M. H. Chen, B. Crasemann and H. Mark, Phys. Rev. $\underline{A26}$, 1243 (1982).
8. W. Jitschin, R. Hippler, R. Shanker, H. Kleinpoppen, R. Schuch, and H. O. Lutz, J. Phys. $\underline{B16}$, 1417 (1983).
9. e.g. G. Mehler, J. Reinhardt, B. Müller, W. Greiner, and G. Soff, Z. Phys. $\underline{D5}$, 143 (1987).
10. H. Schmidt-Böcking, A. Skutlartz, S. Hagmann, Phys. Lett. $\underline{A122}$ 421 (1987).
11. cf. W. E. Meyerhof, J.-F. Chemin, R. Anholt, and Ch. Stoller, Trans. New York Akad. Sci. $\underline{II40}$, 134 (1980).
12. D. W. Spooner, Ch. Stoller, J.-F. Chemin, W. E. Meyerhof, J.-N. Scheurer, and X.-Y. Xu, Phys. Rev. Lett. $\underline{58}$, 341 (1987).
13. H. Paul and J. Muhr, Phys. Rep. $\underline{135}$, 47 (1986).
14. R. McCarroll, in: Atomic and Molecular Collision Theory (Proc. NATO ASI, Sept. 15-26, 1980, Cortona, Italy); Ed.: F. Gianturco; Plenum Press NATO ASI Series $\underline{B71}$, 1982, pg. 191.
15. U. Fano and W. Lichten, Phys. Rev. Lett. $\underline{14}$, 627 (1965); W. Lichten, Phys. Rev. $\underline{164}$, 131 (1967).
16. J. S. Briggs, Rep. Prog. Phys. $\underline{39}$, 217 (1976).
17. W. Lichten, J. Phys. Chem. $\underline{84}$, 2102 (1980).
18. S. Sackmann, H. O. Lutz, and J. Briggs, Phys. Rev. Lett. $\underline{32}$, 805 (1974).
19. R. Shanker, U. Wille, R. Bilau, R. Hippler, W. R. McMurray, and H. O. Lutz, J. Phys. $\underline{B17}$, 1353 (1984).
20. U. Wille and R. Hippler, Phys. Rep. $\underline{132}$, 129 (1986).
21. D. R. Bates and D. A Williams, Proc. Phys. Soc. London $\underline{83}$, 425 (1965).
22. J. S. Briggs, J. Phys. $\underline{B8}$, L 485 (1975).
23. J. Reinhardt, T. de Reus, W. Greiner, B. Müller, U. Müller, A. Schäfer, P. Schlüter, S. Schramm, and G. Soff, in: Electronic and Atomic Collisions, eds.: D. C. Lorents, W. E. Meyerhof and J. R. Petersen, (Proc. XIV ICPEAC, Palo Alto, July 24-30, 1985), pg. 389.
24. G. Soff, J. Reinhard, T. de Reus, D. Jonescu, S. Schramm, B. Müller, W. Greiner, GSI-87-51 Preprint (August 1987).

25. S. Hagmann, S. Kelbch, H. Schmidt-Böcking, C. L. Cocke, P. Richard, R. Schuch, A. Skutlartz, J. Ullrich, B. Johnson, M. Meron, K. Jones, D. Trautmann, and F. Rösel, Phys. Rev. $\underline{A36}$, 2603 (1987).

26. H. Rosenthal and H. Foley, Phys. Rev. Lett. $\underline{23}$, 1480 (1963); H. Rosenthal, Phys. Rev. $\underline{A4}$, 1030 (1971).

27. J. Singh, X. Sun, and K. B. McAdam, 15th ICPEAC, Brighton, July 22-28, 1987, Book of Abstracts, pg. 764.

28. W. Jitschin, S. Osimitsch, H. Reihl, D. W. Mueller, H. Kleinpoppen, and H. O. Lutz, Phys. Rev. $\underline{A34}$, 3684 (1986); and to be publ.

29. U. Werner, W. Jitschin, and H. O. Lutz, J. Phys. $\underline{B18}$, 3111 (1985).

30. M. Kimura and C. D. Lin, Phys. Rev. $\underline{A34}$, 176 (1986).

31. R. Hippler, W. Harbich, H. Madeheim, H. Kleinpoppen, and H. O. Lutz, Phys. Rev. $\underline{A35}$, 3139 (1987).

32. R. Hippler, W. Harbich, M. Faust, H. O. Lutz, and   L. J. Dubé, J. Phys. $\underline{B19}$, 1507 (1986).

33. H. O. Lutz, Proc. 3rd Workshop on High-Energy Ion-Atom Collision Processes, August 3-5, 1987, Debrecen.

34. J. Macek and C. Wang, Phys. Rev. $\underline{A34}$, 1878 (1986).

35. W. Fritsch, private communication.

36. R. Hippler, H. Faust, R. Wolf, H. Kleinpoppen, and H. O. Lutz, to be published in Phys. Rev. (1987).

37. M. Kimura and C. D. Lin, ICAP-X, Book of Abstracts, eds.:   H. Narumi and T. Shimamura, Tokyo, 1986, pg. 487.

38. R. Hippler, W. Harbich, M. Faust, to be publ. in J. Phys. $\underline{B}$ (1988).

39. J. Grosser, Z. Phys. $\underline{D3}$, 39 (1986).

40. W. Harbich, R. Hippler, H. O. Lutz, to be publ.

41. cf. e.g. C. C. Havener, N. Rouze, W. B. Westerveld, J. S. Risley, Phys. Rev. $\underline{A33}$, 276 (1986).

42. E. Fermi, Z. Phys. $\underline{29}$, 315 (1924).

43. C. F. von Weizsäcker, Z. Physik $\underline{88}$, 612 (1934).

44. E. J. Williams, Mat. Fys. Medd. K. Dan. Vid. Selsk. $\underline{13}$, No. 4 (1935).

45. J. D. Jackson, Classical Electrodynamics, Wiley, N.Y. 1975.

46. One might note that this model also contains the adiabatic or Massey parameter: due to the strong decrease of σ_γ with increasing ω, the most important contributions come from $x \approx 1$, with $\hbar\omega \approx \Delta E$, thus (at non-relativistic velocities) $b \approx \hbar v / \Delta E$.

47. H. Kolbenstvedt, J. Appl. Phys. $\underline{38}$, 4785 (1967).

48. F. Komarov, Rad. Eff. $\underline{46}$, 39 (1980).

49. P. Eschwey and P. Manakos, Z. Phys. $\underline{A308}$, 199 (1982).

50. C. A. Bertulani and G. Baur, Nucl. Phys. $\underline{A458}$, 725 (1986).

SHAPES AND ORIENTATION OF ATOMS IN ENERGY AND CHARGE TRANSFER
COLLISIONS WITH LASER EXCITATION: RECENT DEVELOPMENTS

Eleanor E.B. Campbell and Ingolf V. Hertel

Fakultät für Physik
Universität Freiburg
Hermann-Herder-Straße 3
D-7800 Freiburg
F.R.G.

INTRODUCTION

In this contribution we wish to illustrate the insights into the underlying mechanisms of elementary molecular dynamical processes that can be obtained from alignment and orientation studies with laser excited atoms. We will do this with reference to two examples: 1) Resonant charge exchange in Na* + Na+ collisions and 2) Studies of energy transfer in Ca* – rare gas collisions. For a rather more comprehensive treatment of the subject the reader is referred to a recent review[1]. We will start however, with a few general comments on the nature of such experiments involving laser excited atoms.

EXPERIMENTS WITH LASER EXCITED ATOMS

By tuning polarised laser radiation to an atomic resonance transition it is possible to excite a sizable fraction of atoms travelling in a beam into an excited state with a well-defined alignment of its magnetic substates (i.e. with an anisotropic charge distribution) if the laser light is linearly polarised, or with a well-defined orientation (i.e. with a non-zero component of angular momentum) in the case of circular polarisation. Our emphasis here is on collision processes in which one atom A containing an np electron is involved, e.g. processes of the type

$$A(np) + B \longrightarrow AB^* \longrightarrow A(n'l') + B' \pm \Delta KE \qquad (1)$$

where ΔKE is the kinetic energy gained or lost in the collision.
We are mainly concerned with the molecular regime, i.e. the situation that arises in low energy collisions as long as the relative velocity of the interacting particles is small compared with the internal velocity of the atomic electrons involved in the process. Thus, the processes may be viewed as evolving through a quasi-molecule AB in various states between which relatively localised transitions may occur.

We are interested in polarisation effects observed in these collisions i.e. we want to know how the interaction process is influenced by the orbital alignment (or shape of the charge cloud) and the orientation of the electronic angular momentum prior to the scattering event. This type of question is closely related to the broad field of atomic collision studies in which the atomic charge cloud after an interaction is probed by detecting the photon re-emitted from a collisionally excited atom in coincidence with the scattered particle[2]. Fig. 1 illustrates that the two approaches yield essentially identical information as far as the collision dynamics are concerned and we can often describe a given process in either the forward or backward direction, depending on what appears more convenient.

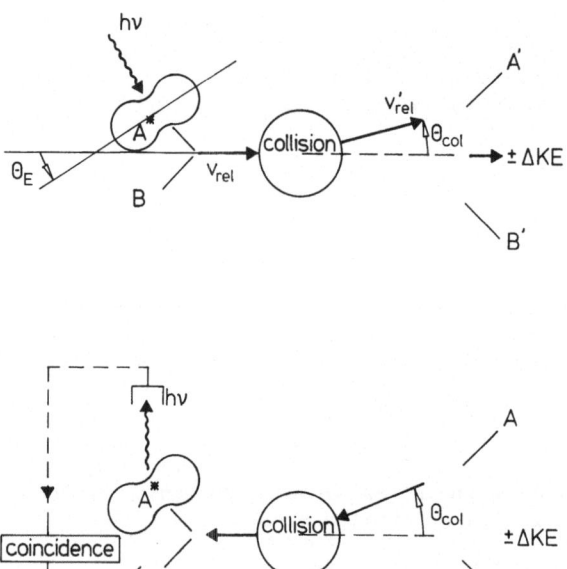

Fig. 1   Schematic view of a binary collision experiment (a) with a laser excited, polarised atom A* and collision partner B and (b) its relation to the inverse process studied in a particle–photon coincidence experiment. The initial and final relative velocities of the interactants define a scattering plane. Kinetic energy may be gained or lost in the process.

We use the density matrix formalism developed elsewhere[3] and described by N. Andersen[4] to interpret the experimental results in terms of the four independent alignment (3) and orientation (1) parameters for the collision-induced charge cloud, as illustrated in fig. 2. The differential cross section for scattering after excitation with linearly and circularly polarised light (incident perpendicular to the ion beam) can be written in terms of our four parameters as follows[5]:

$$I_{lin}(\theta_E, \theta_{c.m.}) = I_o/3 - I_o a_o(L)/24 \cdot \{(1 - 3\rho_{oo}) [2\cos^2\theta_E$$
$$- \sin^2\theta_E(1 + 3\cos2\phi)] + 3P_\ell^+(1 - \rho_{oo}) [\cos2\gamma(2\cos^2\theta_E$$
$$- \sin^2\theta_E(1 - \cos2\phi) + 2\sin2\gamma \cdot \sin2\theta_E \cdot \sin\phi]\} \qquad (2)$$

$$I_{cir}(\theta_{c.m.}) = I_o/3\{1 + 3o_o(L)/2(1 - \rho_{oo})L_\perp^+ \cdot \sin\phi$$
$$+ a_o(L)/8[1 - 3\rho_{oo} + 3(1 - \rho_{oo})P_\ell^+\cos2\gamma$$
$$- 3\cos2\phi(1 - 3\rho_{oo} - (1 - \rho_{oo})P_\ell^+\cos2\gamma)]\} \qquad (3)$$

where $a_o(L)$ and $o_o(L)$ are the optically prepared "alignment" and "orientation" multipole moments respectively[6], ϕ is the azimuthal scattering angle, I_o is the isotropic part of the differential cross section and θ_E is the polarisation angle of the linearly polarised light with respect to the relative velocity vector. The linear polarisation P_ℓ^+ indicates the shape of the charge cloud in the scattering plane with $(1+P_\ell^+)/2$ and $(1-P_\ell^+)/2$ determining the major (l) and minor (w) axes; the height of the charge cloud along the z-axis (h) is given by ρ_{oo} (see fig. 2).

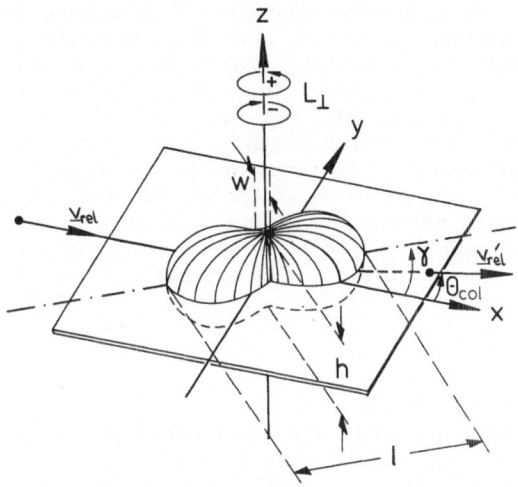

Fig. 2.   Charge cloud distribution of a collisionally excited p-atom in the "natural" frame, giving the definitions of the alignment angle γ and the angular momentum transfer L. The scattering plane is fixed by the directions of the incoming and outgoing particles and l, w and h refer to the length, width and height of the charge distribution measured at the intersections of the main axes.

The physical significance of these measurable parameters from the viewpoint of molecular collisions allows us to address three different aspects of the molecular collision dynamics. Firstly, we have the possibility of creating molecular states of specific symmetry by choosing the appropriate direction of the electric vector of the linearly polarised laser light. In this way we are able to "switch on" one of the molecular

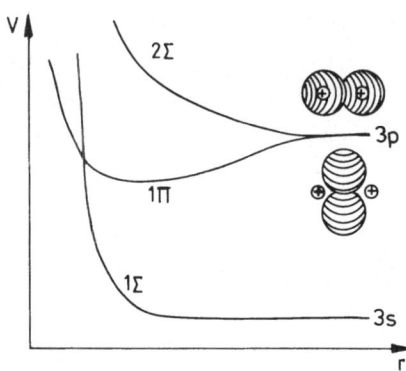

Fig. 3. Typical potentials for alkali – rare gas or alkaline earth
– rare gas interactions. The figure also shows how it is
possible to "switch on" either the Σ or the Π molecular
states.

Σ or Π states as they are schematically illustrated in fig. 3, in
connection with potentials typical in alkali – rare gas or alkaline earth
– rare gas interactions: the p-orbital may thus be aligned along (Σ) or
perpendicular (Π) to the internuclear axis of particles A and B.
Secondly we may consider the importance of reflection symmetry in
collisions with polarised atomic species. This can be studied explicitly
in experiments with planar symmetry. To illustrate this, fig. 4 shows the
motion of a dumbbell p-orbital where we have chosen a π (i.e.
perpendicular) preparation at large internuclear distances and display the
scattering plane explicitly. As indicated we can have either $\pi+$ or $\pi-$
symmetry. Although the molecular potentials are identical for both cases,
the dynamics may be vastly different as indicated in fig. 4. For large
impact parameters the same $\pi+$ atomic orbital prepared at large
internuclear distances may become either a Σ or a Π molecular orbital as
the particles approach. Space-fixed motion (large rotational i.e.
Coriolis coupling) corresponds to non-adiabatic transitions and yields a
molecular Σ state. On the other hand, "orbital following" i.e. rotation
of the atomic charge cloud with the molecular axis, corresponds to the
adiabatic case (large potential energy difference for the corresponding Σ
and Π states) and gives rise to a Π state in the quasi-molecule. What
happens in a real collision will depend on details of the adiabatic
molecular potentials, on the non-adiabatic coupling elements, on the
relative velocity of the interacting particles and on impact parameter.
The $\pi-$ atomic orbital, however, will stay space-fixed since reflection
symmetry of the wave function must be conserved. As we see from fig. 4
this implies that the molecular state definitely has Π character, in
contrast to the two possibilities occurring for states with positive
symmetry. It is thus very challenging to probe pure Π dynamics by $\pi-$
orbital preparation and to add the complications of $\Sigma-\Pi$ non-adiabatic
couplings by $\pi+$ and σ preparation of the atomic orbital.
Finally, we can consider the possible asymmetries of the scattering
signal which may arise for left or right-hand circularly polarised light
exciting the atomic target. Again, we require a planar symmetry for
observing such a phenomenon and scattering angles (different from zero
degrees) have to be well defined in the experiment. In a simple-minded
way we might then expect, as depicted in fig. 5, to find a different
scattering signal for particles scattered to the right or to the left from
a circular atomic state. Equivalently, the signal at a given scattering
angle may be different for left- or right-hand circular excitation of the
atom. Although fig. 5 intuitively suggests angular momentum

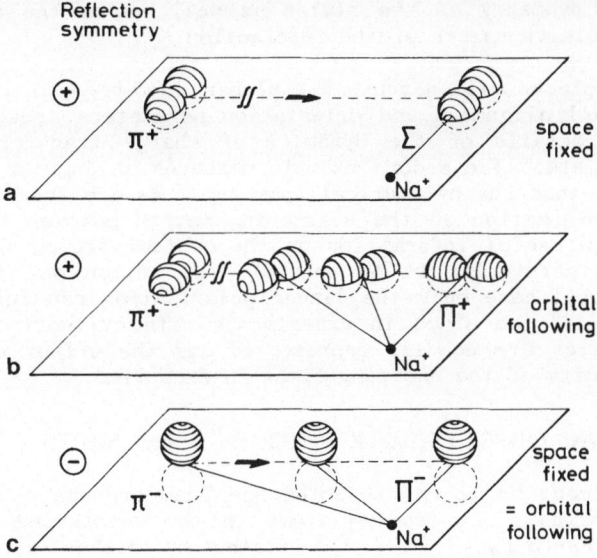

Reflection
symmetry

$\oplus$ π^+ Σ space fixed
   Na^+
a

$\oplus$ π^+ Π^+ orbital following
   Na^+
b

$\ominus$ π^- Π^- space fixed
     = orbital following
   Na^+
c

Fig. 4.   Motion of a dumbbell p–orbital for π (i.e. perpendicular) preparation at large internuclear distances. The $\pi+$ orbital may become (a) a Σ molecular state or (b) a $\Pi+$ molecular state at close internuclear distances depending upon whether the interaction is non–adiabatic or adiabatic respectively. The $\pi-$ orbital (c) remains space–fixed since reflection symmetry of the wave function is conserved and the molecular state must have $\Pi-$ character.

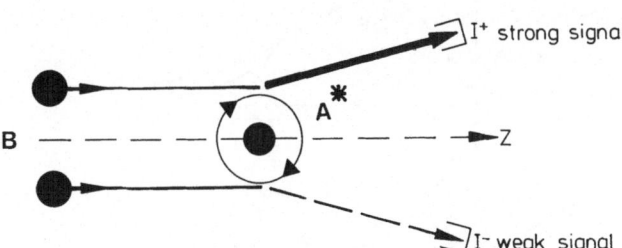

I^+ strong signal

B A^* z

I^- weak signal

Fig. 5.   A different scattering signal is expected for particles, B, scattered to the right or to the left, from an atom, A∗, in a circular atomic state.

considerations to determine the sign of such a left-right asymmetry (a picture which is applicable in the high energy regime), it turns out that for the molecular collision regime in which we are interested here, this kind of argument leads almost inevitably to the wrong conclusions[7]. Instead we find that the sign and magnitude of the asymmetry critically probes the long range interaction of the collision partners and also depends upon the symmetry of the states as well as on the effectively attractive or repulsive nature of the interaction.

The first example we will discuss has planar symmetry and allows us to fully determine all alignment and orientation parameters, revealing some very interesting details of the dynamics of charge transfer involving non-isotropic targets. The second example measures an angular integrated cross section and thus has cylindrical symmetry. We are therefore unable to obtain full information on the scattering matrix however it is still possible to obtain useful information on the orbital stereo-specifity of the process. In particular it is possible to "switch on" the Σ or Π potentials. In this case only the linear polarisation can influence the outcome of the collision since in experiments with cylindrical symmetry all azimuthal angles are equally represented and the effect of positive and negative helicity of the exciting light is cancelled.

EXAMPLE 1:   RESONANT CHARGE EXCHANGE IN THE Na^* + Na^+ SYSTEM

In this experiment an energy selected Na+ beam crosses a sodium atom beam at right angles. The sodium atoms in the scattering centre are excited to the $Na*(3^2P_{3/2}$, F=3, M_F) states by either circularly or linearly polarised light. A position sensitive microchannel plate

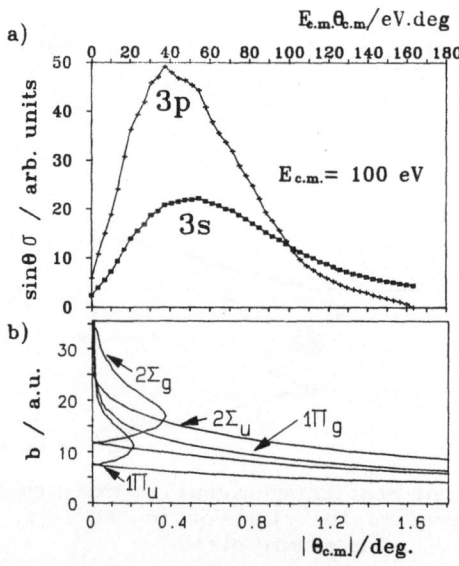

Fig. 6. a) Polarisation averaged differential cross sections for charge transfer from ground and excited state sodium.
b) Classical deflection functions for elastic scattering of Na+ from Na*(3p)

detector is used to measure both the polar ($\theta_{c.m.}$) and azimuthal (ϕ) angular distributions of the charge exchanged atoms simultaneously. Any residual ions are removed from the neutral beam by a deflecting electric field.

The isotropic part (i.e. integrated over the azimuthal angle ϕ and averaged over the polarisation state of the atom) of the differential cross section is shown in fig. 6 along with the classical elastic scattering deflection functions. Although we are unable to resolve any structure in the cross section and only see a broad peak at around 40 eV.deg we can nevertheless attribute this peak to rainbow scattering from the $1\Pi_u$ and $2\Sigma_g$ potentials with the maxima occurring at about 20 eV.deg and 40 eV.deg respectively (fig. 6b). This assignment has recently been confirmed by JWKB calculations of the differential cross sections for the four potentials involved in the resonant charge exchange process[8].

The alignment parameters (γ, P_ℓ^+, ρ_{00}), determined by linear polarisation studies are plotted in fig. 7 for $E_{c.m.} = 100$ eV. At small

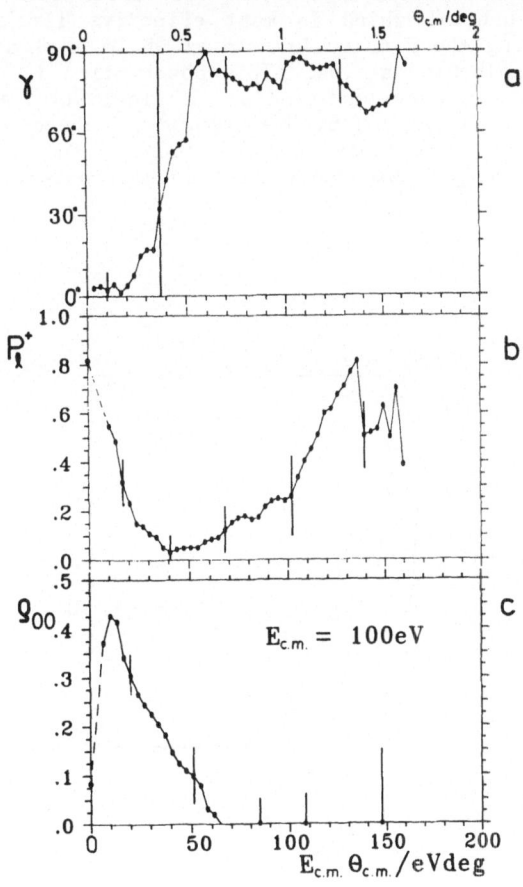

Fig. 7. Alignment parameters γ, P_ℓ^+ and ρ_{00} for $E_{c.m.} = 100$ eV. Upper and lower abscissae show the scattering angle $\theta_{c.m.}$ = $\theta_{lab}/2$ and the reduced scattering angle $\tau = E_{c.m.} \cdot \theta_{c.m.}$ respectively.

scattering angles, $\tau = E_{c.m.}\,\theta_{c.m.} < 20\text{eV.deg}$, the alignment angle, γ, is very small indicating that a pσ atomic orbital prepared at large internuclear separation of the interacting particles is most efficient for the charge transfer process. This is a substantial effect as shown in fig. 7b with $P_\ell^+ = 0.4$ at $\tau = 15$ eV.deg which implies that pσ is more than twice as effective for charge transfer than pπ+. As the scattering angle increases both orbitals become more and more equivalent until at around $\tau = 40$ eV.deg P_ℓ^+ almost vanishes and γ is no longer well defined as we can see from the large error bar computed in this range of scattering angle. The anisotropy increases again as the scattering angle is increased further, however, this time γ is nearly orthogonal to the situation at small scattering angles. For $\tau > 60$ eV.deg we find $\gamma = 80°$ and $P_\ell^+ = 0.8$ which means that a nearly pπ+ preparation of the atomic orbital at large internuclear distances is about nine times more effective for the process than the pσ orbital.

It would thus appear, when we consider the differential cross sections, that a σ atomic orbital which is most effective for small scattering angles is populating the Π molecular states while a π+ atomic orbital is populating the Σ molecular states. This observation is opposite to what was observed in our previous studies of inelastic and superelastic transitions in Na*(3p) induced by Na+ impact[9][10] where orbital following was found to be a reasonable approximation. In these rather less complicated collision processes only one molecular potential

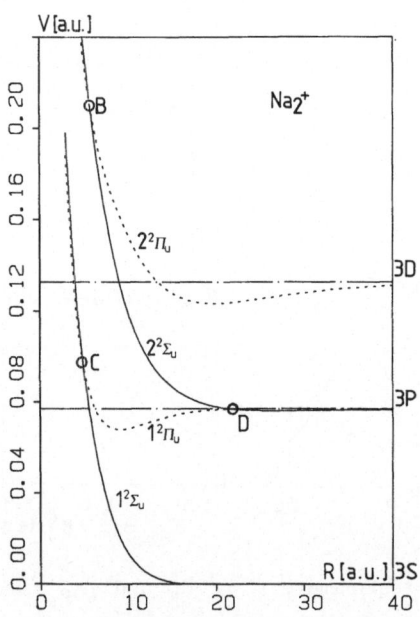

Fig. 8. Plot of the four potential curves contributing to p → s de-excitation through crossing C and p → d excitation through crossing B.

is responsible for the crossing between initial and final states as illustrated in fig. 8. It was found that the 3p → 3s de-excitation signal (progressing via the $1\Pi_u$ potential) was much stronger for initial preparation of a $\pi+$ atomic orbital whereas the 3p → 3d excitation process (progressing via the $2\Sigma_u$ potential) required the preparation of a σ atomic orbital. The alignment angles measured for the two processes were however found to be slightly less than the expected 90° or 180° predicted by the simple orbital following picture. This was explained by invoking the concept of a "locking radius": For internuclear separations less than this "locking radius" the charge cloud moves as depicted in fig. 4b i.e. orbital following while for larger separations space-fixed motion (fig. 4a) has to be assumed. For the small impact parameters contributing to these inelastic and superelastic processes involving only the u potentials, the "locking radius" was determined to be around 20 a.u., a number which was justified by theoretical calculations[7].

The additional complexities in the charge transfer study are such that
— both g and u potentials participate
— both Σ and Π states may be found in the exit channel
— as seen from fig. 6b several different impact parameters contribute to one scattering angle and may be very different in value
— the impact parameters of importance are typically much larger than in the inelastic (and superelastic) case and comparable in magnitude to the "locking radius".

Obviously when the impact parameter is about the same as the "locking radius" this is no longer a meaningful concept. Assume e.g. a pσ orbital is prepared at a large internuclear distance and we are interested in very small scattering angles $\approx 0.2°$. From fig. 6b we see that on the $2\Sigma_u$ curve this implies an impact parameter b $\approx$ 20 a.u., i.e. comparable to the "locking radius". Thus the orbital will move space-fixed until it reaches the distance of closest approach, R = b, and the system will find itself in a Π state so that we are left with the situation depicted in fig. 4a and can qualitatively understand the observed pσ dominance at small scattering angles in combination with the theoretically expected $1\Pi_u$ rainbow.

Additional support for this view is obtained from the angular dependence of the third alignment parameter, ρ_{00} (fig. 7c). As discussed previously this parameter gives an image of pure Π scattering and we can see that this is largest for the small scattering angles and reaches zero for $\tau > 60$ eV.deg

Figure 9 summarises the experimental findings derived from the studies with linearly polarised light showing the shape of the charge cloud at different values of $\tau = E_{c.m.} \theta_{c.m.}$.

The angular momentum transfer $L_\perp^+$, obtained from experiments with circularly polarised light, is plotted in fig. 10 for the same collision energy ($E_{c.m.}$ = 100 eV). The effect is fairly small but significant and shows an interesting structure. At small scattering angles $L_\perp^+$ is positive reaching a maximum of 0.08 $\hbar$ at around $\theta_{c.m.}$ = 1°. It then rapidly changes to become negative, crossing the axis at 1.1° and appears to reach a plateau with the value −0.08 $\hbar$.

In the inelastic and superelastic transitions in this same collision system[9][10] it was shown that for a Π final state the sign of $L_\perp^+$ was

positive but for the same conditions and a Σ final state the sign was
negative. In the charge transfer process there are contributions from
both Π and Σ final states and we could expect the two contributions to
cancel explaining the small value which we observe. Beyond this statement
any further conclusion would be pure speculation in view of the rather
more complicated situation encountered in the charge exchange process.

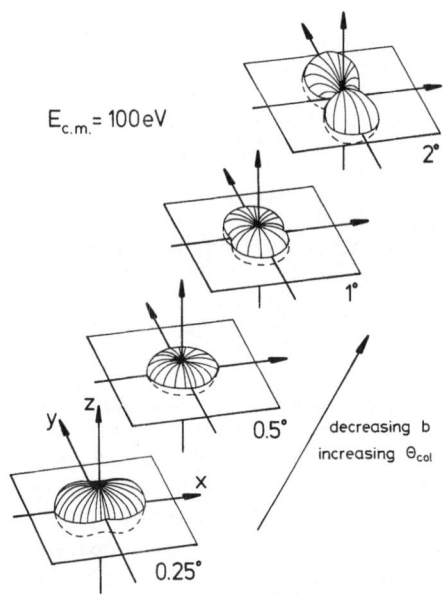

Fig. 9.   Charge distribution after resonant charge exchange from an
          isotropically populated Na*(3p) state as derived from the
          alignment parameters (fig. 7) for different scattering
          angles.

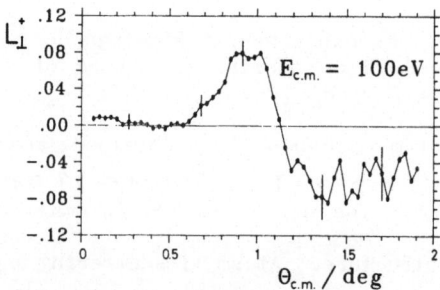

Fig. 10. Angular momentum transfer $L_\perp^+$ for resonant charge exchange
         at a collision energy of 100 eV.

386

EXAMPLE 2:   ENERGY TRANSFER IN Ca*-RARE GAS COLLISIONS

Calcium is particularly well suited for alignment studies since it has no nuclear spin and for the singlet states one can in principle prepare the electron orbital in a 100% aligned state.   This has been confirmed experimentally by observing the fluorescence from the laser excited atoms.

In these experiments carried out in the Leone group[11] [12] the effect of the alignment of the initially excited $Ca(4s5p^1P_1)$ state on the electronic energy transfer process

$$Ca(4s5p^1P_1) + Rg \longrightarrow Ca(4s5p^3P_j) + Rg + \Delta KE \qquad (4)$$

is studied where Rg stands for He, Ne, Kr or Xe.

A simplified energy diagram of the calcium atom is shown in fig. 11.   A pulsed laser is tuned to the $4s^2\ ^1S_0 \rightarrow 4s5p^1P_1$ transition at 272.1 nm. The product states, $3P_j$, are observed on the $4s5p^3P_j \rightarrow 4s3d^3D_j$ transitions at 616.9 nm and on the cascade transition $4s5p^3P_j \rightarrow 4s4p^3P_j$ at 616.2 nm.

The experiment consists of crossed beams of calcium and the rare gas, a pulsed laser for exciting and aligning the $Ca*(4s5p^1P_1)$ state and a photomultiplier tube to monitor the fluorescence from the product state. It measures an average over impact parameter, azimuthal angle and relative velocity.   To quantitatively interpret the experimental results in terms of the language discussed in the previous section, we have to average the evaluation formula (eq. 2) over all azimuthal angles.   This gives the

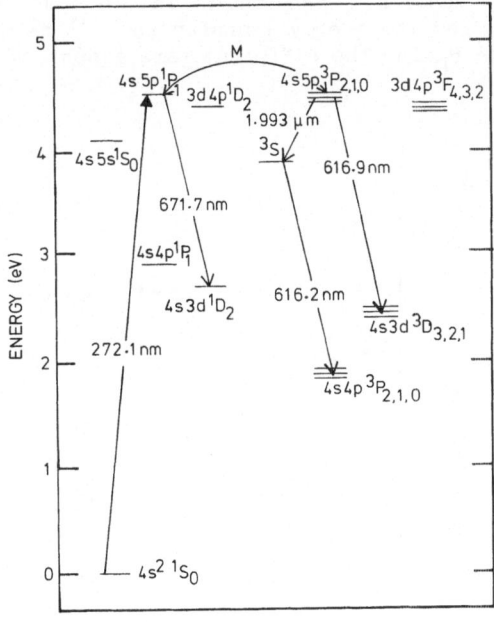

Fig. 11. Partial energy-level diagram for atomic calcium showing only the transitions used in the experiment

azimuthally averaged differential cross section, still as a function of the polarisation angle θ_E of the linearly polarised light used to excite the atoms:

$$I(\theta_E, \theta_{c.m.}) = \frac{1}{3}I_0 - I_0 a_0(L)/24 \; [1 - 3\rho_{oo} + 3P_\ell^+(1 - \rho_{oo})\cos 2\gamma] \cdot$$
$$(3\cos^2\theta_E - 1) \tag{5}$$

The experiment in addition sums $I(\theta_E, \theta_{c.m.})$ over all scattering angles, giving $I(\theta_E)$, and all that we can determine experimentally is an averaged "polarisation" (or sensitivity parameter) S which is usually defined as the relative difference of the scattering signal for $\theta_E = 0°$ and $90°$. From eq. 5 and with $a_0(L) = -2$ for optimal alignment which is possible for the $^1P_1$ case discussed here, we derive

$$S = \frac{I(0°) - I(90°)}{I(0°) + I(90°)} = \frac{1 - 3\langle\rho_{oo}\rangle + 3\langle(1-\rho_{oo})P_\ell^+\cos 2\gamma\rangle}{3 - \langle\rho_{oo}\rangle + \langle(1-\rho_{oo})P_\ell^+\cos 2\gamma\rangle} \tag{6}$$

The quantities in the brackets $\langle x \rangle$ are averaged over all scattering angles and weighted with the respective differential cross section $I_0(\theta_{c.m.})$.

The experimental results[12] are shown in fig. 12 which gives the relative probability for the energy transfer (eq. 4) as a function of the laser alignment angle θ_E for the different rare gases studied. For helium

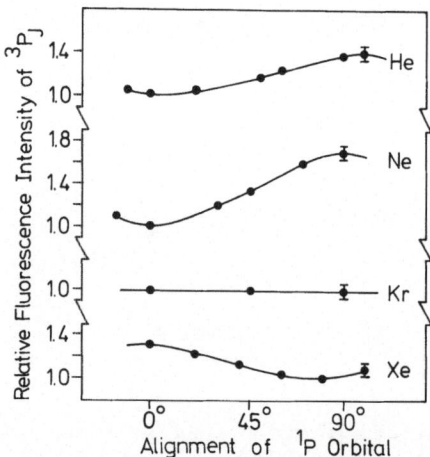

Fig. 12.   Relative fluorescence intensity as a function of alignment angle for the Ca*(4s5p$^1P_1$) + Rg → Ca*(4s5p^3P_j) + Rg processes.

and, especially, neon there is a clear maximum at an angle of approximately 90° corresponding to a perpendicular alignment of the Ca* p-orbital with respect to the initial relative velocity. Thus for Ca* + He, Ne the asymptotic preparation of a π orbital is preferred for the $^1P_1$ $\longrightarrow$ ^3P_j transition. With Xe we have a maximum at 0° indicating the dominance of the asymptotically prepared σ orbital. No dependence on laser alignment angle is observed for Kr.

The polarisation dependence for Ca* + He can be understood by considering the potential energy curves of the Ca*-He quasi-molecule formed in the collision. Figure 13 shows two sets of model potentials[13] both indicating the essential mechanism of the process to proceed via a $^1\Pi$ $\longrightarrow$ $^3\Sigma$ crossing at around 8.5 a.u. Thus we have the "orbital following" situation illustrated in the middle of fig. 4 with a π orbital prepared at large internuclear distances scattering as a Π molecular state. We can estimate the effective "locking radius", R_L, in this system from the observed averaged polarisation, S. We assume that we have a pure dumbbell ($P_\ell^+ = 1$), no spin-orbit effects ($\rho_{00} = 0$) and that only impact parameters close to the crossing radius R_c contribute to the process, as expected for

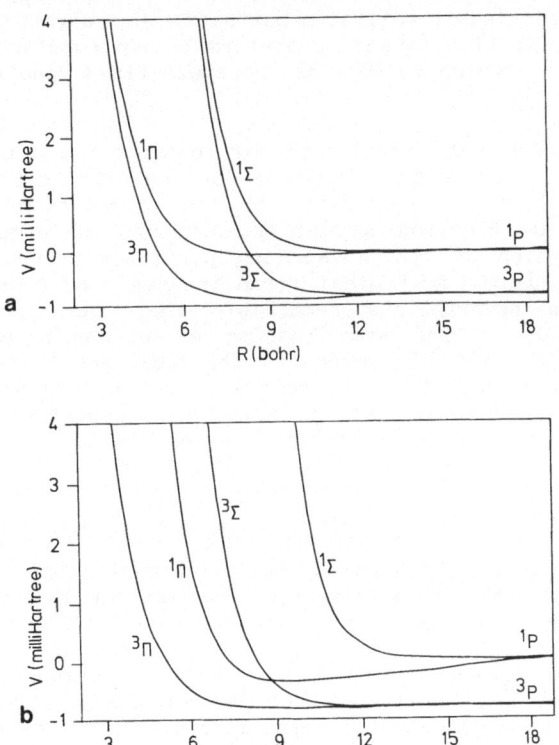

Fig. 13. Calculated molecular potentials for Ca* + He.
a) potentials which give best agreement with experimental data
b) potentials with increased $\Sigma-\Pi$ splitting for the ^1P state leading to enhanced polarisation effects.

rotational coupling, and can estimate the ratio of locking radius to crossing radius, $R_L/R_c \sim 1.95$ (making use of the relation $\cos\gamma = b/R_L$). If we take the crossing radius to be 8.5 a.u. (fig. 13) this gives a locking radius of about 17 a.u.

Alexander and Pouilly[13] performed quantum mechanical calculations with the model potentials chosen such that the total cross section is well reproduced by the calculation (i.e. the inner $^1\Pi - ^3\Sigma$ crossing is essentially kept at the same internuclear distance) while the $\Sigma-\Pi$ splitting was modified so that the polarisation effect could be simulated.

Under these conditions the calculations were able to reproduce the experimental results with the potentials shown in fig.13a giving an average polarisation, $S = -0.2$ (experimental value -0.17 ± 0.06). In contrast the potential curves depicted in fig. 13b give $S = -0.47$ which seems to illustrate that the observed polarisation effects are a direct probe of the dependence of the $\Sigma-\Pi$ splitting for the $Ca*(^1P_1)$ state on the internuclear distance. Although the crossing radius remains essentially unchanged at around 8.5 a.u., the different $\Sigma-\Pi$ splittings assumed lead to a greatly enhanced polarisation effect. This could be the reason for the larger ratio observed with Ne. Collisions with the heavier rare gases may have contributions from additional curve crossings or the Π and Σ potential curves may have a qualitatively different appearance for the heavier, more polarisable collision partners. One effect of this could be that the most effective impact parameters become substantially larger so that we have a situation resembling the space-fixed condition illustrated in fig. 4.

The quantum mechanical results appear to support our intuitive picture of the locking concept. Unfortunately for the time being this cannot be gleaned directly from the details of the calculation. The coupling between the atomic electron angular momentum and the angular momentum of the relative motion of the interacting particles prevents a one-to-one translation of quantum mechanical formalism and semi-classical intuitive model. It is nevertheless remarkable that the quantum mechanical calculation leads to the same results as we would expect from the intuitive concepts and also predicts the same trends as the splitting between Σ and Π potentials is increased. Thus we may well suspect that the quantum mechanical calculations contain the essential elements of the semi-classical picture even though the mechanism is not immediately obvious at the moment.

CONCLUSION

We have attempted here a brief illustration of the stimulating possibilities of orbital polarisation studies in the investigation of elementary molecular dynamical processes.

The situation appears to be fairly well understood in atom-atom or atom-ion collisions at collision energies in the range between a few eV and some hundred eV: At large internuclear distances the polarised atomic orbital stays essentially space-fixed while at small separations of the interacting particles a description in terms of the quasi-molecule formed during the collision is appropriate. Qualitatively the concept of a "locking radius" at which the change from space-fixed to body-fixed motion is thought to occur has proved to be very useful and intuitively appealing although certainly somewhat simplified.

REFERENCES

1. E.E.B. Campbell, H. Schmidt, I.V. Hertel, <u>Adv. Chem. Phys</u> 72:37 (1988)

2. M. Eminyan, K.B. MacAdam, J. Slevin, M.C. Standage, H. Kleinpoppen, <u>J. Phys. B</u> 7:1519 (1974)

3. K. Blum, Density Matrix Theory and Applications. New York:Plenum Press (1981)
   H.W. Hermann, I.V. Hertel, <u>Comm. At. Mol. Phys.</u> 12:61 and 127 (1982)

4. N. Andersen, this volume

5. R. Witte, E.E.B. Campbell, C. Richter, H. Schmidt and I.V. Hertel, <u>Z. Phys. D.</u> 5:101 (1987)

6. A. Fischer and I.V. Hertel, <u>Z. Phys. A.</u> 304:103 (1982)

7. I.V. Hertel, H. Schmidt and A. Bähring, <u>Rep. Prog. Phys.</u> 48:375 (1985)

8. R. Düren, private communication

9. A. Bähring, I.V. Hertel, E. Meyer, H. Schmidt, <u>Z. Phys. A.</u> 312:293 (1983)

10. A. Bähring, I.V. Hertel, E. Meyer, W. Meyer, N. Spies, H. Schmidt, <u>J. Phys. B.</u> 17:2859 (1984)

11. M.O. Hale, I.V. Hertel and S.R. Leone, <u>Phys. Rev. Lett.</u> 53:2296 (1984)

12. D. Neuschaefer, M.O. Hale, I.V. Hertel and S.R. Leone, in "Electronic and Atomic Collisions (invited papers of the XIV International Conference), D.C. Lorentz, W.E. Meyerhoff and J.R. Peterson, eds.,Elsevier (1986)
    W. Bussert, D. Neuschaefer, S.R. Leone, <u>J. Chem. Phys.</u>, 87:3833 (1987)

11. B. Pouilly and M. Alexander, <u>J. Chem. Phys</u> 86:4790 (1987)

ANALYSIS OF ELECTRON-ATOM COLLISIONS:

ELECTRON-PHOTON COINCIDENCE EXPERIMENTS

Hans Kleinpoppen

Atomic Physics Laboratory
University of Stirling
Stirling FK9 4LA, Scotland, U.K.

I. INTRODUCTION

Research on electron-atom scattering has progressed and developed to
a state by which detailed analysis may lead to a maximum complete information on the collision process. While cross section measurements of
elastic and inelastic electron scattering by atoms are of importance in
themselves and also for applications in other parts of physics and astrophysics recent advances particularly derived from angular and polarization
correlation experiments and from electron spin analysis of electron-atom
collisions have provided us with knowledge on scattering amplitudes and
their phases, atomic target parameter such as alignment and orientation
and shapes of electron charge distributions of states excited by electron
bombardment. These data on electron scattering by atoms are important
with regard to two aspects:

(1) The data provide a deep insight into physical mechanisms of electron-atom scattering processes. The data reveal what types of processes and
interactions occur or compete with each other in the collisional process.

(2) Scattering amplitudes and their phase differences and also atomic
target parameters extracted from the above experiments have successfully
been applied as most sensitive tests of modern collision theories. Data
from coincidence and electron spin experiments do not, in selected cases,
average or sum over partial cross sections for various sub-processes or
interactions of the collisional processes. Coincidence and spin experiments resulting in collision amplitudes, phases and target parameters have
been classified as "third generation" type of experiments going well beyond
the more limited kind of information obtained from differential ("second
generation" type of experiments) or total ("first generation" type of
experiments) cross section measurements.

In these series of lectures on analysis of electron-atom collisions
we are dividing the task into the various parts, namely on electron-photon
coincidence experiments, on electron spin experiments, and a comparison
between electron and positron scattering. Only very recently a start has
been made to combine spin and coincidence experiments; we will briefly
refer to this newest development.

This part of the lecture on analysis of electron-atoms collisions does not represent an overall review of the relevant field of research; emphasis rather is put on various possible types of data extracted from electron-photon coincidence experiments. The most comprehensive kind of information was so far obtained from the excitation of the singlet $^1P_1$ state of the helium atom. Accordingly this excitation process has served experimentalists as the most appropriate one so far to study most of the various physical effects revealed by the electron-photon coincidence technique.

We will also briefly refer to triplet ^3P excitation since in this excitation process the electron-photon coincidence experiment provides either direct information on spin-exchange amplitudes (e.g. for light atoms such as helium) or, by combining the coincidence technique with that applied in polarized electron-atom collisions, we obtain information on spin-orbit effects in electron impact excitation of atoms.

II. EXPERIMENTAL METHODS FOR ELECTRON-PHOTON-COINCIDENCE EXPERIMENTS

Electron-photon coincidence experiments have been carried out with and without optical polarizers (linear polarizers and $\lambda/4$ wave plates) in the detection channel of the photons. It has become common to refer to angular correlation experiments for schemes without optical polarizers and to polarization-correlation experiments for schemes including optical polarizers. Historically angular correlation experiments were first reported[1] followed soon by polarization-correlation experiments[2]. Polarization-correlation experiments may supplement angular correlation experiments due to the fact that the photon polarization reveals additional information compared to measurements without polarization analysis of photons.

a) Angular Correlation Experiments

When an electron of given energy E_o excites an atom the electron looses the energy $\Delta E = E_o - E_{thr}$ (E_{thr} threshold energy for exciting a higher atomic state). Both the inelastically scattered electron and the emitted photon from the transition of the excited state may have anisotropic distributions for their detection. However, the directional anisotropies or angular distributions of the electrons and photons may not be independent of each other; in general, the angular distributions may be correlated to each other. Accordingly, we would expect an angular correlation between the directions of detecting electrons and photons. Such correlation between angular distributions of electrons and photons can, however, only be recorded or detected if the two detection events from an electron and a photon are jointly detected in coincidence which means that the counting signal of the one "product particle" (e.g. the electron) is detected such that the counting signal of the second particle (e.g. the photon) occurs at a difference in time which would be expected to be fixed ("time correlation" for their joint detection) while "chance coincidences" between the electron and photon signal would have any arbitrary time difference for their joint detection. In practical coincidence experiments the time difference between the detection of the two signals is electronically transformed into a pulse height difference which can be recorded on a multichannel-analyzer.

The "true" coincidences occur as a peak on a background of chance coincidences. The integral of the counts in the coincidence peak minus those from the background is the coincidence signal of the joint detection from the inelastically scattered electron and the photon emitted from the excited atom. Each single coincidence count results from a single collision-excitation process between one electron and one atom. We can therefore refer the reaction process

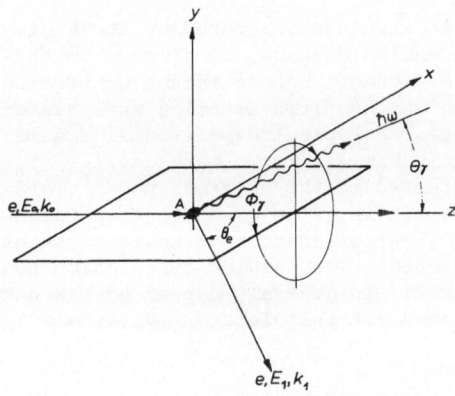

Fig. 1. Typical geometry of a coordinate frame (collision frame) for
electron-photon coincidence experiments. The incoming (e,E_0,k_0)
and inelastically (e,E_1,k_1) scattered electrons define the
scattering plane (X,Z). Angular correlations refer to the
electron scattering angle (θ_e), and the polar (θ_γ) and
azimuthal angle of the photon (Φ_γ).

$$e(E_0) + A \rightarrow A^* + e(E_0 - E_{thr})$$
$$ \rightarrow A + h\nu \qquad\qquad\qquad (1)$$

to a single electron-single atom collisional process.

An angular correlation experiment for the joint coincident detection
of the photon and electron underlies a geometry as revealed in Fig. 1. The
directions of the incoming and scattered electron define a scattering plane.
The photon $h\nu$ may be detected at a polar (θ_γ) and an azimuthal (Φ_γ) angle

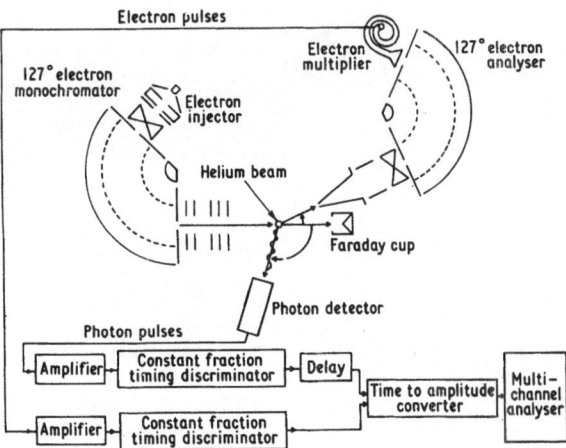

Fig. 2. Scheme of electron-photon coincidence apparatus for angular
correlation measurements including 127° cylindrical monochrom-
ators as electron injector and energy analyzer, a photomultiplier
as photon detector and fast coincidence electronics[1].

in coincidence with the electron scattered at the angle θ_e. The selection of the detection angles depends, of course, on the special case to be investigated. Detection of photons in the scattering plane with $\phi_\gamma = 180°$ has resulted in the first detailed analysis of electron-photon coincidence experiments. A typical experimental scheme for this special geometry is displayed in Fig.2. For a more detailed technical description the reader is referred to the original paper[1]. In these angular correlation experiments it is often usual to keep the initial energy of the electrons and the electron scattering angle constant while the photon detector is rotated. The angular correlation function for the coincident detection will, in general, depend on the excitation process and the joint coordinates for the electron ($\theta_e, \phi_e = 0°$) and the photon ($\theta_\gamma, \phi_\gamma$).

b) The Polarization-Correlation Method

The polarization-correlation method combines the angular-correlation experiment, as described in the previous sub-paragraph, with a polarization analysis of the coincident photon. A complete analysis of the photon polarization state requires in general the measurement of two linear and the circular polarization component (3 Stokes parameters, see chapter III.b.). A typical example[2] of an experimental scheme is shown in Fig.3 with a combination of a linear polarizer, a $\lambda/4$ wave plate and an isotope cell absorbing radiation from odd mercury isotopes and

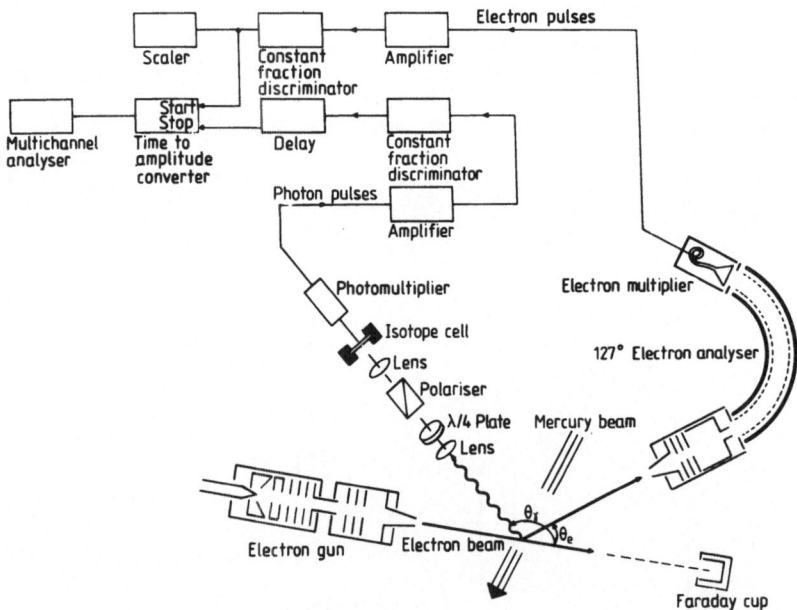

Fig. 3. Scheme of a typical polarization-correlation experiment[2] for electron-photon coincidences. The photon detection branch includes a linear polarizer, a $\lambda/4$-wave plate and an isotope cell for absorbing resonance radiation from odd mercury isotopes. Note that only the polar photon angle θ_γ is drawn in the Figure, the azimuthal angle may be chosen to any appropriate value.

transmitting radiation from even isotopes with good efficiency. Often the photons have been observed perpendicular to the scattering plane (i.e. $\phi_\gamma = 90°$) but special requirements may make it necessary to observe the polarized photons under any polar and azimuthal angle (interesting examples have been with the azimuthal angles $\phi_\gamma = 0°$ and 45° or 135°).

While photon-polarization correlation analysis has already been fully exploited in electron-photon coincidence experiments coincident electron spin analysis combined with polarized atom technique is beyond present range of applications. Nevertheless we display a future possible scheme (Fig.4) in which not only the photon polarization but also the spin of the coincident electron scattered from spin polarized atoms is analyzed.

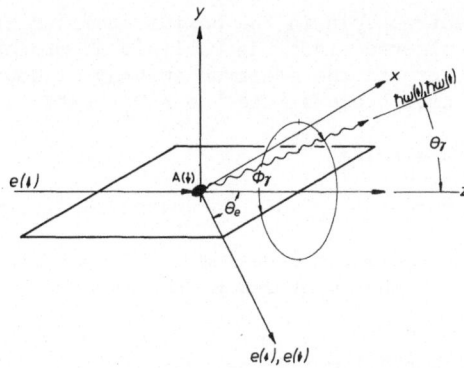

Fig. 4. Electron-photon angular correlations including coincident spin analysis for scattered electrons and photon polarization analysis with polarized incident electrons colliding with polarized atoms.

III. ANALYSIS OF ELECTRON-PHOTON COINCIDENCE EXPERIMENTS

In this main chapter we attempt to analyze the various types of information extracted from electron-photon coincidences in electron impact excitation of atoms. In a way it is very interesting or even amusing how angular correlation data can be interpreted in schemes which seem to differ from each other but they are not independent and can be "transferred" between each other without loss of information or obscuring in any way the physics involved.

We will not give a theoretical deviation of the most general case but restrict ourselves to the case of the P state excitation which has been most widely studied and first revealed the important new achievements in electron-impact excitation of atoms.

a) <u>The State Vector Description and the λ, χ Parameters</u>
   <u>for the $^1P_1$ Excitation</u>

Fano[3] suggested a new programme for collision experiments by which
colliding partners are first prepared in a pure quantum state before the
collisional process itself. After the collision the state vector of the
collisional process again represents a pure quantum mechanical state vector
since the Hamiltonian operator of the Schrödinger equation describing the
collision is a linear operator. The state vector of the final collisional
products then contains all the possible information obtainable, it may
then be a maximum kind of information extracted from the experiment.

The special case of impact excitation of the singlet ^1P state of
helium by monoenergetic, collimated electrons may follow the above scheme:
The initial state vector $|\Psi_{in}>$ represents the helium atom in the ground
state and the linear momentum $|P_e>$ of the collimated monoenergetic beam
of the electrons:

$$|\psi_{in}> = |He(1^1S_0)> |P_e>\qquad (2)$$

We note that the electron spin in the helium atom and the projectile electron
can be completely neglected since the collisional process of excitation will
be symmetric with regard to the electron spin up or down. The state vector
$|\Psi_{out}>$ of the excitation process into the n^1P_1 state is then

$$|\psi_{out}> = |He(n^1P_1)> |P_e\acute{}>\qquad (3)$$

with the new linear momentum $P_e\acute{}$ of the scattered electron.

We now represent the excited helium atom by a linear superposition of
the magnetic sub-states characterized by the magnetic quantum number $m_L = 0$
and ± 1. It follows

$$|\psi_{out}> = \underbrace{\sum_{m_L} f_{m_L} |n, L=1, m_L>}_{} |P'_e\acute{}>\qquad (4)$$

$$= |\psi(^1P_1)>$$

$$= |\psi(^1P_1)> |P_e\acute{}>\qquad , \text{ or }\qquad (5)$$

$$|\psi^1(P_1)> = f_0 |1\ 0> + f_1 |1\ 1> + f_{-1} |1-1>\qquad (6)$$

where the first number in the substates represent $L=1$ and the second number
the magnetic quantum number m_L; f_0, f_1 and f_{-1} are the amplitudes of the
magnetic substates. Note that our state vector as defined above shall only
represent the angular dependence of the excited state which determines the
polarization and angular distribution of the dipole matrix element for the
transition from the P state. Parity invariance requires reflection
invariance of the state vector with reference to the scattering plane.
It follows from the angular dependence of the substates that reflection
invariance only holds if $f_1 = -f_1$. Since the absolute phases of these
amplitudes cannot be determined only the relative phase difference χ between
f_0 and f_1 should be significant. By choosing f_0 real and positive we have

$$f_1 = |f_1| e^{i\chi}$$

and the differential excitation cross section for ^1P may be expressed in
terms of the bracket

$$<\psi(^1P_1)|\psi(^1P_1)> = \sigma(\theta_e, E) = |f_0|^2 + 2|f_1|^2\qquad (7)$$

398

with $|f_o|^2 = \sigma_o$ and $|f_1|^2 = \sigma_1$ as partial cross sections for magnetic sub-state excitation. With these definitions we can derive a joint angular correlation function of the coincidence count rate for given direction of the electron and photon by using a usual dipole matrix representation for the transition:

$$\frac{dN_c}{d\Omega_e d\Omega_\gamma} = nj \; C \left| \sum_{m_L} <0|\hat{\varepsilon} \cdot \underline{r}|\psi({}^1P_1)> \right|^2 \int_0^{\Delta t} e^{\frac{-\gamma t}{2}} dt \qquad (8)$$

with n as atomic density, j as electron current density, Δt the time interval over which the coincidence signal is recorded; $\hat{\varepsilon}$ is the polarization vector and the factor $e^{\frac{-\gamma t}{2}}$ accounts for the decay of the ${}^1P_1$ state ($\frac{1}{\gamma}$ mean lifetime of the excited state); $d\Omega_e$ and $d\Omega_\gamma$ are the aperture angles for the detectors.

A simple calculation gives for $\phi_\gamma = 180°$ (detection of photons in the scattering plane opposite to the scattered electrons)

$$\frac{dN_c}{d\Omega_\gamma} = \frac{3}{8\pi} N_{e-\gamma} \qquad (9)$$

with $N_{e-\gamma} = \lambda \sin^2\theta_\gamma + (1-\lambda)\cos\theta_\gamma - 2\sqrt{\lambda(1-\lambda)}\cos\theta_\gamma \sin\theta_\gamma \cos\chi$

$$= \frac{1}{\sigma}\left| f_o \sin\theta_\gamma - \sqrt{2} \; f_1 \cos\theta_\gamma \right|^2 \qquad (10)$$

and $\lambda = \frac{|f_o|^2}{\sigma} = \frac{\sigma_o}{\sigma}$, $1 - \lambda = 2\frac{\sigma_1}{\sigma}$, χ as relative phase between the excitation amplitudes f_o and f_1 for the magnetic substates. The physical significance of this angular correlation eq.(10) is as follows: Obviously the third term of the first of the two eqs. represents an <u>interference effect for the excitation of the two magnetic substates.</u> Such an interference term can neither be traced in the total nor in the differential cross section for the ${}^1P_1$ excitation process. The interference term vanishes for $\lambda = 0$, $\chi = \pm(2n + 1)\frac{\pi}{2}$ and if $\overline{\cos \chi}$ averages out. By considering special cases we may be able to find out from observed angular correlations between the coincident electrons and photons if such an interference term exists. The above dipole approximation suggests that the amplitudes and intensities of the electromagnetic field of the coincident photon are related to the excitation amplitudes and to the λ parameter as follows:

$$E_z \propto \frac{|f_o|}{\sqrt{\sigma}} \; , \quad E_x \propto \frac{\sqrt{2}|f_1|}{\sqrt{\sigma}}$$

$$|E_z|^2 \propto \lambda \; , \quad |E_x|^2 \propto 1 - \lambda \qquad (11)$$

These parameters allow us to interpret the angular correlation (and polarization, see chapter III.b) data as follows. The electric field vectors E_z and E_x trace out a typical optical polarization ellipse (Fig. 5a) while the photon intensities in the scattering plane represent a typical dipole dependence for two orthogonal dipoles coherently excited (Fig. 5b). However, for the special case that the interference term vanishes when all phase angles χ are equally distributed $\overline{\cos \chi} = 0$, incoherent excitation or $\chi = 90°$ the

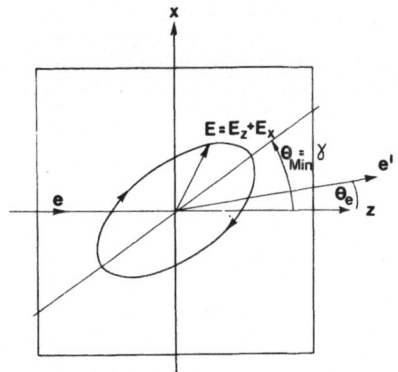

Fig. 5(a) Polarization ellipse for
the electric vectors of the
coincident photons in the
scattering plane. A finite
angle $\gamma = \theta_{Min}$ would be expected for a finite value of the
interference term of eq.10
(apart from $\chi = 90°$).

Fig. 5(b) Dipole intensity function
$I = |E_z + E_x|^2$ of coincident
photons observed in the
scattering plane.

principal axes of the polarization ellipse and the dipole intensity curve
are lying parallel to the z and x axes (Fig. 6a,b).

One might also contemplate about the prediction of the Born approximation for the angular correlation. It has already been shown by Bethe[4] in

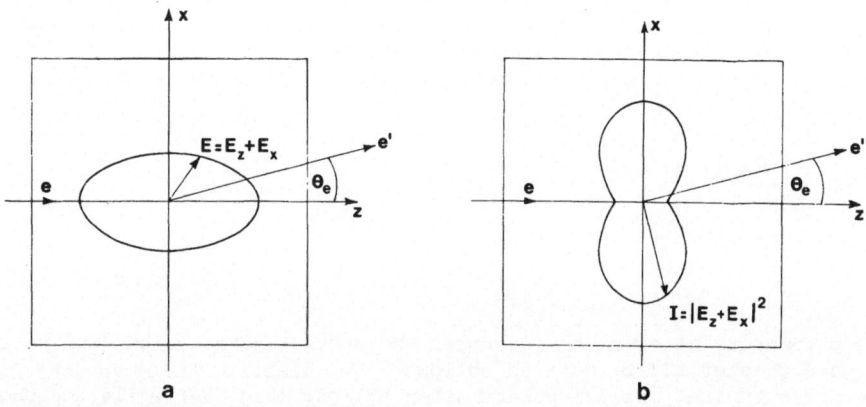

a   b

Fig. 6(a) and (b): same as in Fig. 5(a) and 5(b), respectively for
$\overline{\cos \chi} = 0$ (incoherent excitation) or $\chi = 90°$; θ_{Min} vanishes
for these cases.

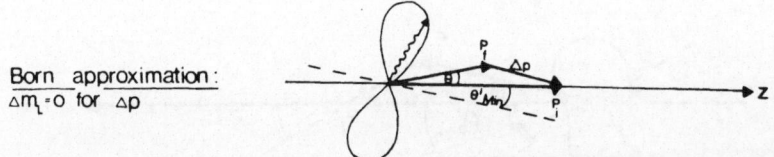

Born approximation:
$\Delta m_l = 0$ for Δp

Fig. 7. The momentum transfer ($\underline{\Delta p}$) direction as axis of quantization allows only excitation of the $m_L=0$ substate of the ^1P state. The transition from the ^1P state is a pure π transition with reference to $\underline{\Delta p}$ which gives a dipole intensity curve for a dipole oscillating parallel to $\underline{\Delta p}$.

the 30's that in the Born's approximation the excitation cross section is determined by the momentum transfer $\Delta p = p_i - p_f$ of the scattered electron (p_i linear momentum of the incoming, p_f linear momentum of the scattered electron) in the following way. The cross section for exciting the atom is given by the matrix element

$$\sigma_{Born}(\ldots, \Delta p, \ldots) \propto \left| \int \psi^* \psi_o e^{i/h(p_i - p_f)} \, d\tau \right|^2 \tag{12}$$

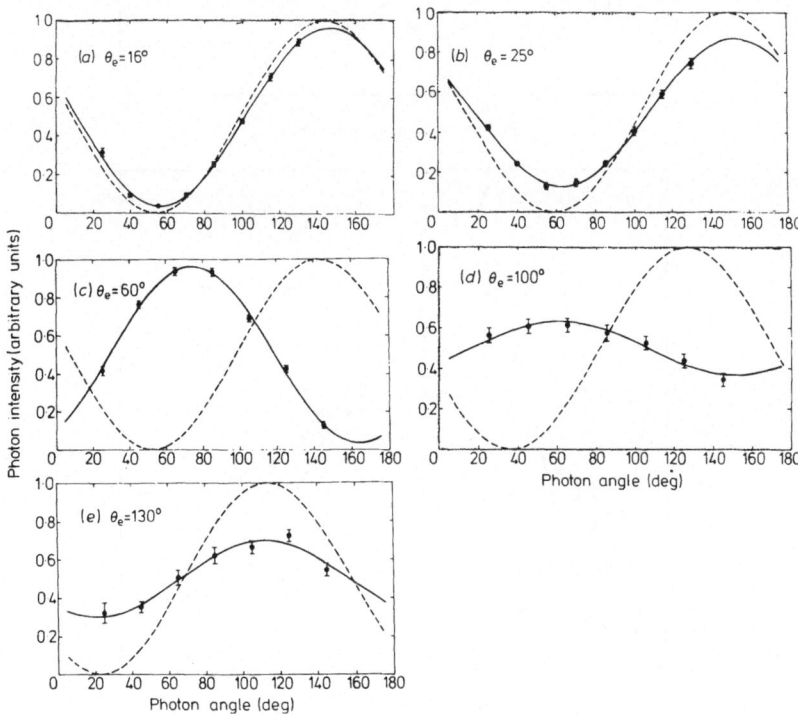

Fig. 8. Electron-photon angular correlations from He(2^1P) excitation at incident electron energy of 81.2eV for scattering angles (a) 16°, (b) 25°, (c) 60°, (d) 100° and (e) 130°. The full curve is a least squares fit of eq.(10) to the data. The broken curves are predictions by the first Born approximation (data by Hollywood, Crowe and Williams[5]).

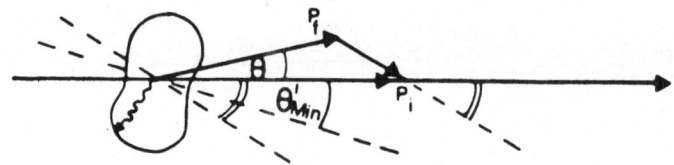

Fig. 9. Typical experimentally observed photon intensity curve as a
function of the photon observation angle θ_γ in the scattering
plane. Note that the curve is not symmetric with regard to the
momentum transfer $p_i-p_f = \Delta p$ of.the electron and θ_{Min} is not
identical to the angle between the z direction and the momentum
transfer.

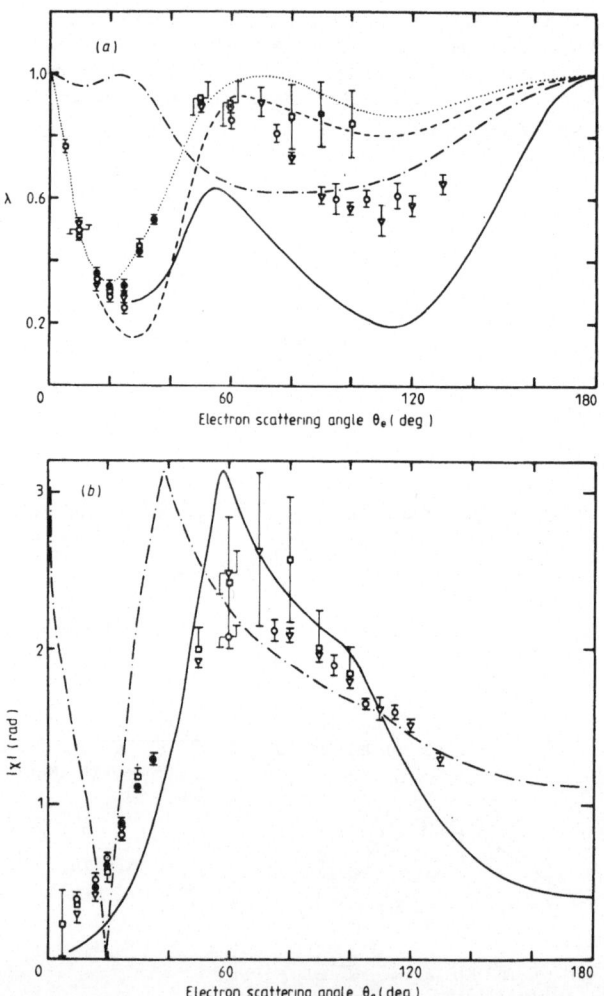

Fig. 10. λ and $|\chi|$ parameter for He(2^1P) excitation at incident electron
energy of $\sim$ 80 eV measured and calculated by different groups:
o Slevin et al.[6]; ∇ Hollywood et al.[5]; $\square$ Steph, Golden[7],
Theory: $\cdots$ Madison et al. (from Sutcliffe et al.[8]);
--- Thomas et al.[9] $-\cdot-\cdot-$ Catalan and Roberts[10];
_____ Scott, McDowell[11] (from Slevin et al.[6]).

with ψ as wavefunction of the excited and ψ_0 as wavefunction of the ground state. Taking, however, the momentum transfer direction as axis of quantization Bethe[4] showed that eq.12 remains finite only with $\Delta m_L = 0$ as a selection rule. For our excitation process $^1S \rightarrow {}^1P \rightarrow {}^1S$ this selection rule has the consequence that only the $m_L = 0$ is excited and the photon intensity curve is that of a dipole oscillating parallel to the momentum transfer as illustrated in Fig.7.

The predictions for electron-photon angular correlations by Born approximation or by incoherent excitation processes are very specific. The actual coincidence experiments have, however, shown data which are in general not consistent with such predictions as demonstrated in Figs. 8-10. Fig.8 shows a typical set of angular correlation data for various electron scattering angles which demonstrate the following. At small electron scattering angles the angular correlation data are quite close to the Born approximation. At larger electron scattering angles, however, there is no similarity between the actual angular correlations and both the Born approximation and the assumption of incoherent excitation according to Fig.6. A typical angular correlation intensity curve for the observed photons in the scattering plane is displayed in Fig.9 which is contrary to the shapes of Figs.6 and 7, which also demonstrates that the momentum transfer direction is not a symmetry axis of the photon distribution. It seems therefore justified that the angular correlation data can be interpreted in terms of eq.10 which requires inclusion of the interference term for the magnetic substate excitation. The state vector of eq.6 expresses linear or coherent superposition of the magnetic substates of the $^1P_1$ state. A more direct way of proving the state-vector Ansatz follows from polarization measurements (sub-section III.b).

Many investigations have been carried out to extract λ, χ and θ_{Min}* data from angular correlations of which Figs.10-12 give typical examples. These data have proved to be much more sensitive tests of theory than cross section measurements. Obviously the best agreement between theory and experiment is for He(2^1P) excitation for energies below the ionization threshold of helium (Fig.11 and 12). Many more improved theoretical models are required in order to achieve agreement between theory and experiment over a wide range of energies and scattering angles.

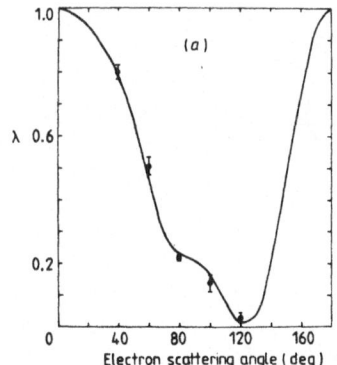 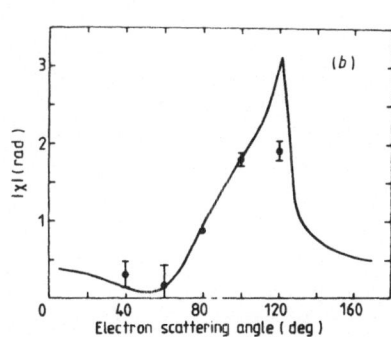

Fig. 11. λ and $|\chi|$ data for He(2^1P) excitation at 24 eV incident electron energy: ● experimental data of Crowe, Nogueria and Liew[12]; full curve is a five-state R-matrix calculation by these authors.

---

*The minimum value of $N_{e-\gamma}$ (eq.10) occurs at $\theta_\gamma = \theta_{Min}$ where

$$\tan 2\theta_{Min} = \frac{2[\lambda(1-\lambda)]^{\frac{1}{2}}\cos\chi}{(2\lambda-1)}$$

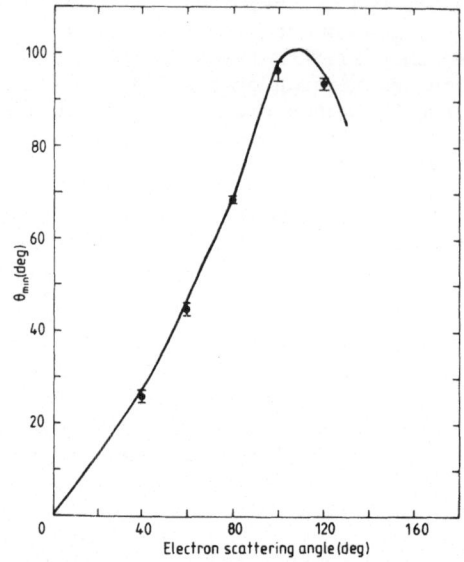

Fig. 12. θ_{Min} as extracted from the λ and $|\chi|$ data of Fig.11 in comparison to the same theory[12] of Fig.11.

b) <u>Polarization and Coherence Analysis</u>

The above analysis of electron-photon angular correlations has, in a way, given evidence for the <u>"model of coherent sublevel excitation"</u> of the helium $^1P_1$ state by electron impact. However, as will be seen in this sub-chapter, a more direct way of experimentally confirming the model of coherent sub-level excitation will follow from a polarization and coherence analysis of the coincident photon radiation. The argument is connected to

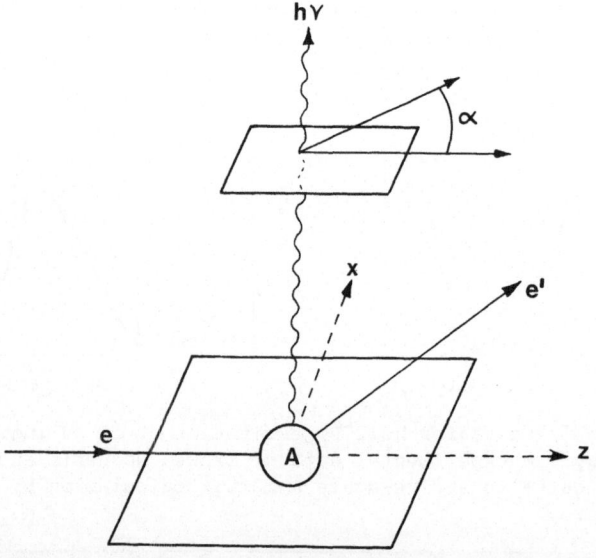

Fig. 13. Coincident photons hν polarized with their light vector at an angle α and observed perpendicular to the scattering plane.

the question if the coherence of the sublevel excitation can be "seen" or detected in the "collisional products", the inelastically scattered electron and the emitted photon of the excitation process . Obviously a direct connection exists between the statevector (eq.6) of the coherent sublevels and the photon radiation intensity described by the dipole matrix element for the optical transition. Following Born and Wolf[12] we can carry out a coherence analysis of photons by measuring all polarization Stokes parameters of the photon. By providing clear experimental evidence that the coincident photons are completely coherent one can then trace it back to completely coherent excitation of the atom.

In order to associate our statevector (eq.6) $|\psi(^1P)>$ for the $^1S \rightarrow {}^1P$ excitation with the coordinate frame (Fig.1, collisional frame) we introduce the following basis set for the angular parts of the state vector $|L_{m_L}>$, L = 1, m_L = 0, ±1:

$$|\psi_z> = |10>$$

$$i|\psi_y> = -\frac{1}{\sqrt{2}} \{|11> + |1 - 1> \tag{13}$$

$$|\psi_x> = -\frac{1}{\sqrt{2}} \{|11> - |1 - 1>\}$$

Adding and subtracting these functions gives

$$|\psi_x> + i|\psi_y> = - \sqrt{2}|11>$$

$$|\psi_x> - i|\psi_y> = + \sqrt{2}|1-1> \tag{14}$$

Multiplying eqs.(14) with the amplitude $f_1 = - f_{-1}$ and adding them gives

$$- \sqrt{2} f_1|\psi_x> = f_1|11> + f_{-1}|1-1> .$$

Our statevector can then be written as

$$|\psi(t=o)> = f_0|\psi_z> - \sqrt{2} f_1|\psi_x> , \tag{15}$$

it does not contain a y- component which is compatible with the fact that the $|\psi_y>$ angular part has negative symmetry with regard to reflection at the scattering plane (violation of parity invariance!).

We will use this statevector for deriving the coherence parameters and the photon Stokes parameters. We already discussed the physical significance of the amplitudes f_0 and $\sqrt{2} f_1$ for two coherently excited (Hertzian or classical) oscillators vibrating parallel to the z or x-axis, respectively. We will observe the photon radiation perpendicular to the scattering plane and will introduce the electric vectors and the photon density matrix with reference to the z- and x-axis of the collision frame (Fig.13):

$$E_z = |E_z|e^{-i(\omega t - \varphi_z)} , E_x = |E_x|e^{-i(\omega t - \varphi_x)} \tag{16}$$

$$S = \begin{pmatrix} E_z E_z^* & E_z E_x^* \\ E_x E_z^* & E_x E_x^* \end{pmatrix} = \begin{pmatrix} \rho_{zz} & \rho_{zx} \\ \rho_{xz} & \rho_{xx} \end{pmatrix} . \tag{17}$$

We normalize the photon intensity to unity $\text{Tr}\rho = I_0 = \rho_{zz}+\rho_{xx}=1$, and define the first Stokes parameter P_1 with reference to the linearly polarized components parallel to the z-direction ($\alpha = 0°$) and x-direction ($\alpha = 90°$):

$$P_1 = \rho_{11} - \rho_{22} = I(\alpha = 0°) - I(90°) \tag{18}$$

Generally the intensity component polarized at the angle α is given by

$$I(\alpha, t) = \langle E(t,\alpha) E^*(t,\alpha) \rangle \tag{19}$$

We may assume now that the x-component is subjected to an additional retardation ε with respect to the z-component (e.g. by means of an optical compensator such as an $\lambda/4$ wave plate):

$$E(t,\alpha,\varepsilon) = E_z \cos\alpha + E_x e^{i\varepsilon} \sin\alpha \tag{20}$$

Eq. (19) then becomes:

$$I(\alpha, t, \varepsilon) = \rho_{zz} \cos^2\alpha + \rho_{xx} \sin^2\alpha + \rho_{zx} e^{-i\varepsilon} \cos\alpha \sin\alpha + \rho_{xz} e^{i\varepsilon} \sin\alpha \cos\alpha \tag{21}$$

The second Stokes parameter P_2 follows from $\varepsilon = 0$, $\alpha = 45°$, $\alpha = 135°$ with $\cos 45° = \sin 45° = \frac{1}{\sqrt{2}}$, $\cos 135° = -\frac{1}{\sqrt{2}}$ and $\sin 135° = \frac{1}{\sqrt{2}}$:

$$P_2 = \rho_{xz} + \rho_{zx} = I(\alpha = 45°) - I(\alpha = 135°) \tag{22}$$

The third Stokes parameter is related to the retardation $\varepsilon = 90°$ measured with the two positions of $\alpha = 45°$ and $135°$ for a combination of a $\lambda/4$-wave plate with a linear polarizer:

$$P_3 = i(\rho_{xz} - \rho_{zx}) = I_{RHC} - I_{LHC} \tag{23}$$

with the right-(RHC) and left-(LHC) handed polarized intensity components. The density components of the photon matrix can then be expressed by I_0, P_1, P_2 and P_3.

$$\rho_{zz} = \tfrac{1}{2}(I_0 + P_1) \quad , \quad \rho_{xx} = \tfrac{1}{2}(I_0 - P_1)$$
$$\rho_{xz} = \tfrac{1}{2}(P_2 - iP_3) \quad , \quad \rho_{zx} = \tfrac{1}{2}(P_2 + iP_3) \tag{24}$$

Born and Wolf defined the "complex correlation factor" by means of the coherence matrix as follows,

$$\mu_{zx} = |\mu_{zx}| e^{i\beta_{zx}} = \frac{\rho_{xz}}{\sqrt{\rho_{zz} \rho_{xx}}}$$
$$|\mu_{zx}| = \frac{P_2 - iP_3}{\sqrt{1 - P_1^2}} \tag{26}$$

Completely coherent photon radiation is characterized by $|\mu_{zx}| = 1$ (or $\rho_{xz}\rho_{zx} = \rho_{zz}\rho_{xx}$) and a fixed phase different β_{zx} between the two orthogonal, linearly polarized light vectors vibrating parallel to z and x. Incoherent photon radiation ($|\mu_{zx}| = 0$) would result from statistically varying phase differences β_{zx}.

Alternatively, we can characterize completely coherent photon radiation by the "vector polarization"

$$|P| = \sqrt{|P_1|^2 + |P_2|^2 + |P_3|^2}$$
$$= \rho_{zz}^2 + \rho_{xx}^2 - 2\rho_{zz}\rho_{xx} + 4\rho_{xz}\rho_{zx} \tag{27}$$

For completely coherent photon radiation with $\rho_{zx}\rho_{xz} = \rho_{zz}\rho_{xx}$ this equation becomes

$$|P| = \sqrt{\rho_{zz}^2 + \rho_{xx}^2 + 2\rho_{zz}\rho_{xx}} = 1 \tag{28}$$

by applying eqs.(24) and $I_0 = 1$. In other words unity of $|P|$ and $|\mu|$ are characteristic tests for completely coherent photon radiation.

We now apply the coherence and polarization analysis to the $^1S_0 \rightarrow {}^1P_1 \rightarrow {}^1S_0$ excitation/de-excitation in an electron-photon coincidence experiment by which polarized radiation is observed perpendicular to the scattering plane. The first Stokes parameter P_1 follows immediately from the amplitudes f_0 and f_1 and the λ parameter of the ^1P excitation, which are related to the linearly polarized components parallel to the z-axis $(\alpha = 0°)$ and the x-axis $(\alpha = 90°)$:

$$I(0°) \propto |f_0|^2 \propto \frac{|f_0|^2}{\sigma} = \lambda \tag{29}$$

$$I(90°) \propto 2|f_1|^2 \propto \frac{2|f_1|^2}{\sigma} = 1 - \lambda \tag{30}$$

$$P_1 = \frac{I(0°) - I(90°)}{I(0°) + I(90°)} = \frac{\lambda - (1-\lambda)}{\lambda + (1-\lambda)} = 2\lambda - 1 \tag{31}$$

The second Stokes parameter P_2 follows from the intensity components for $\alpha = 45°$ and $\alpha = 135°$ in the relation

$$I(\alpha) \propto |f_0 \cos\alpha - \sqrt{2} f_1 \sin\alpha|^2$$

$$I(45°) = |f_0 \frac{1}{\sqrt{2}} - \sqrt{2}|f_1|e^{i\chi}\frac{1}{\sqrt{2}}|^2 = \tfrac{1}{2}|f_0 - \sqrt{2}|f_1|e^{i\chi}|^2$$

$$I(135°) = -\tfrac{1}{2}|f_0 + \sqrt{2}|f_1|e^{i\chi}|^2$$

$$P_2 = \frac{I(45°) - I(135°)}{I(45°) + I(135°)} = \frac{-\tfrac{1}{2}(\sqrt{2}|f_0||f_1|2\cos\chi)}{\underbrace{\tfrac{1}{2}(|f_0|^2 + 2|f_1|^2)}_{\sigma(E,\theta_e)}}$$

with $\sqrt{\lambda} = \sqrt{\dfrac{|f_0|^2}{\sigma}}$, $\sqrt{\dfrac{|f_1|^2}{\sigma}} = \sqrt{\tfrac{1}{2}(1-\lambda)}$,

$$P_2 = -2\sqrt{\lambda(1-\lambda)} \cos\chi \tag{32}$$

The third Stokes parameter P_3 can easily be extracted from a transformation of the statevector $|\psi(^1P)\rangle$ describing the excitation process (eq.15):

$$|\psi(^1P)\rangle = f_0\psi_z - \sqrt{2} f_1\psi_x$$

$$= \tfrac{1}{2}\underbrace{(f_0 - i\sqrt{2} f_1)}_{a_r}(\psi_z - i\psi_x) + \tfrac{1}{2}\underbrace{(f_0 + i\sqrt{2} f_1)}_{a_\ell}(\psi_z + i\psi_x)$$

$$P_3 = \frac{I(RHC) - I(LHC)}{I(RHC) + I(LHC)} = \frac{|a_r|^2 - |a_\ell|^2}{|a_r|^2 + |a_\ell|^2}$$

with

$$|a_r|^2 = |f_0|^2 + 2|f_1|^2 - 2\sqrt{2}|f_0||f_1| \sin\chi$$

$$|a_\ell|^2 = |f_0|^2 + 2|f_1|^2 + 2\sqrt{2}|f_0||f_1| \sin\chi$$

$$P_3 = \frac{2\sqrt{2}|f_0||f_1| \sin\chi}{|f_0|^2 + 2|f_1|^2} = 2\sqrt{\lambda(1-\lambda)} \sin\chi \tag{33}$$

Eqs. (31-33) are correct for positive electron scattering angles (i.e. $\varphi_e = 0°$). As shown by Beyer, Blum and Standage[13] P_2 and P_3 change their signs for negative scattering angles ($\varphi_e = 180°$).

By applying the eqs. 30-32 to the coherence analysis we obtain

$$\left| \mu_{zx} \right| = \left| \frac{P_2 - i\,P_3}{\sqrt{1 - P_1^2}} \right| = \left| \frac{-2\sqrt{\lambda(1-\lambda)}\,\{\,\cos\chi + i\,\sin\chi\,\}}{2\sqrt{\lambda(1-\ )}} \right|$$

$$= 1 \qquad\qquad (100\%\ \text{coherence of } {}^1P_1 \text{ excitation})$$

$$\mu_{zx} = \left| \mu_{zx} \right| e^{i\beta}{}_{zx} = \left| \frac{-2\sqrt{\lambda(1-\lambda)}}{2\sqrt{\lambda(1-\lambda)}} \right| e^{i\chi} \qquad\qquad (34)$$

It follows that the quantum mechanical phase difference χ between the two amplitudes f_0 and f_1 for the sublevel excitation with the quantum numbers $m_L = 0$ and $m_L = \pm 1$ is directly identical to the phase difference of the light vectors of the photons polarized parallel to the z and x axis and observed perpendicular to the scattering plane. In other words an <u>atomic phase difference is direct measurable as a macroscopical phase difference</u>.

Of course experiments on the full coherence and the total vector polarization are the crucial test for the model of coherent excitation of the ${}^1P_1$ state. Such measurement has first been reported for the $3{}^1P$ excitation of helium by Standage and Kleinpoppen[14]. Their data for $\left|\mu\right|$ and $\left|P\right|$ are in agreement with 100% coherence and unity of the magnitude of the vector polarization.

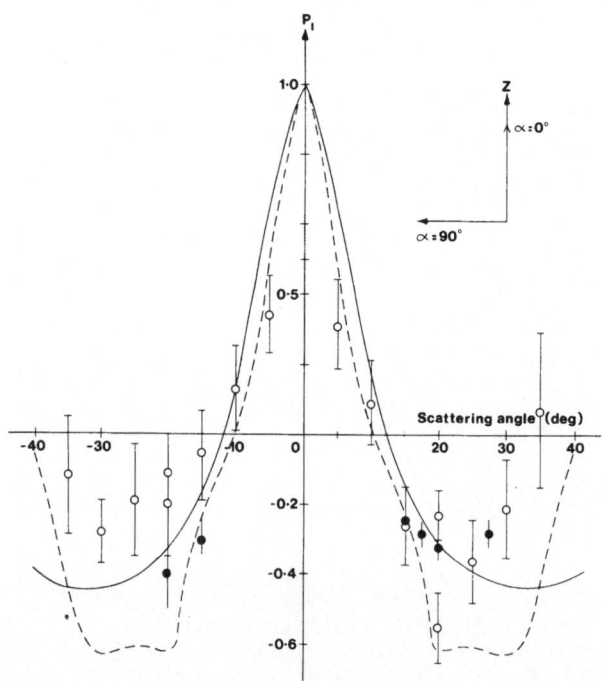

Fig. 14. Stokes parameter P_1 for $1{}^1S_0 \to 3\ {}^1P_1 \to 2{}^1S_0$ excitation/deexcitation (5016Å) of helium at 80eV electron impact energy.
φ Silim et al.[15], $\bullet$ Standage and Kleinpoppen[14]. Full curve first Born approximation, broken curve theory of Flannery, McCann[16].

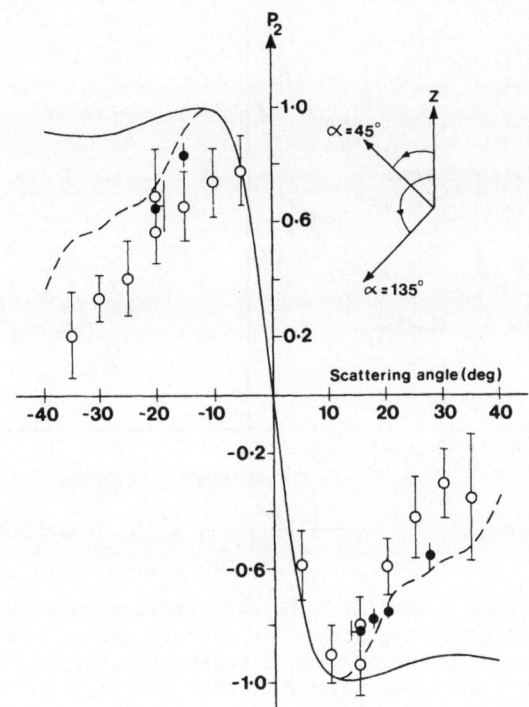

Fig. 15. Stokes parameter P_2 with legend as in Fig. 14.

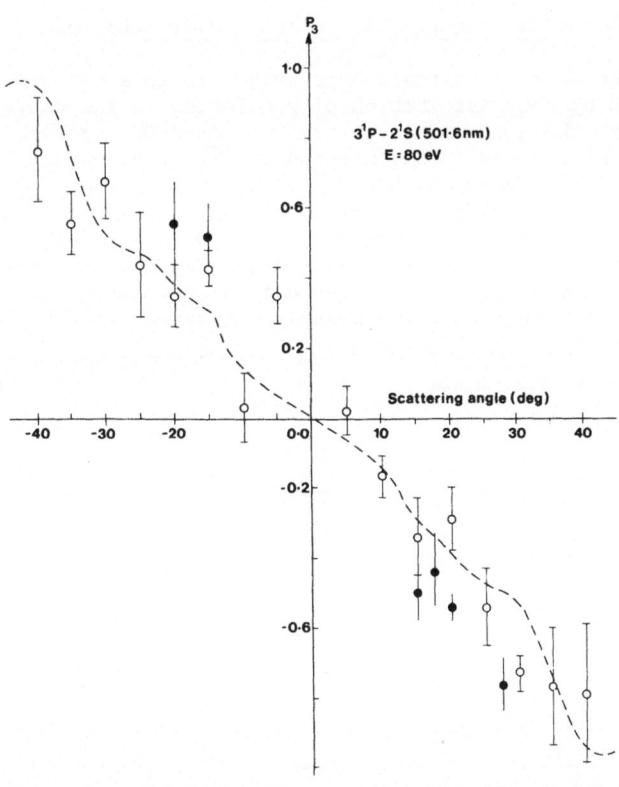

Fig. 16. Stokes parameter P_3 (circular polarization) with legend as in Fig. 14.

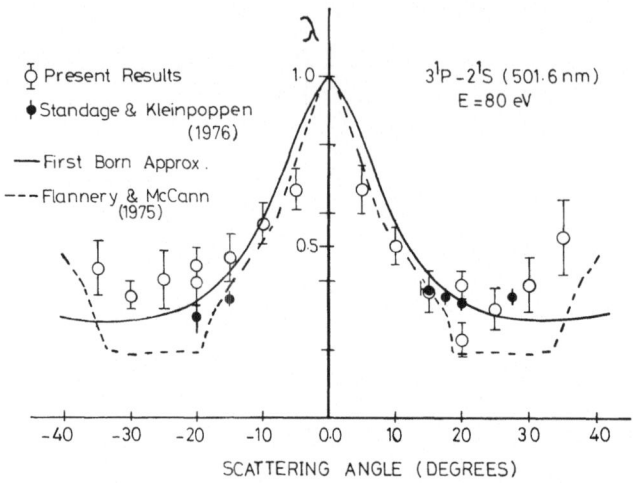

Fig. 17. Parameter $\lambda = \dfrac{|f_0|^2}{|f_0|^2 + 2|f_1|^2}$ with legend as in Fig. 14.

The clear evidence for $|\mu| = |P| = 1$ implies that angular correlation data such as λ and χ and the Stokes parameters can be used as important tests of theories for electron impact excitation. Fig. 14-18 show some typical recent data of these parameters for positive and negative electron scattering angles. The Stokes parameters or the λ and χ-parameter can also be related to various other physical interpretations of the electron-photon coincidences which are discussed in the following sub-sections.

c) Angular Momentum Transfer, Alignment and Orientation

After the excitation the atomic target is in a state which can be characterized by the distribution of population in the various magnetic substates with the quantum number m_L. This distribution can be described by the state multipoles (Blum, Kleinpoppen[17]) or alignment tensors and an orientation vector (Fano and Macek[18]). The excitation of P states can particularly be expressed in terms of orbital angular momentum transfer which is related to the orientation and alignment as follows. During the collisional excitation and because of angular momentum conservation the impinging electron can only transfer orbital angular momentum to the atom perpendicular to the scattering plane. The expectation value $\langle L_y \rangle$ of this

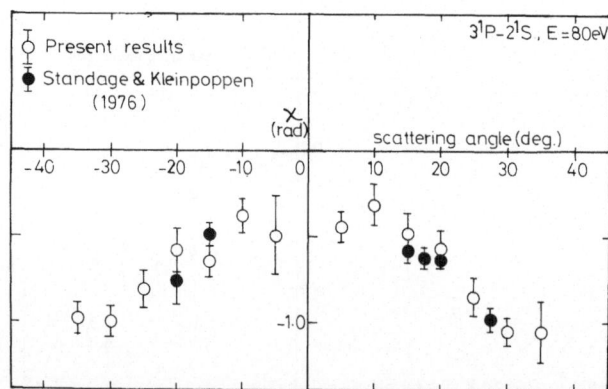

Fig. 18. Phase shift χ between the amplitudes $f_o(m_L=0)$ and $f_1(m_L=\pm1)$ versus electron scattering angle for the $3\,^1P \to 2\,^1S$ - transition at 80 eV electron impact energy. Experimental data ● by Standage and Kleinpoppen[14] and ○ by Silim et al.[15].

angular momentum transfer is the relative difference between the number of atoms with positive and negative orbital angular momentum and can be calculated from λ and χ data for the $^1P_1$ excitation process:

$$\langle L_y \rangle = \frac{N(L_y) - N(-L_y)}{N(L_y) + N(-L_y)} = -2\sqrt{\lambda(1-\lambda)}\,\sin\chi \tag{35}$$

This orbital angular momentum transfer can be related to the "orientation vector" O^{col}_{-1} as introduced by Fano and Macek[18]

$$O^{col}_{-1} = \frac{\langle L_y \rangle}{L(L+1)} \tag{36}$$

By observing photons from the de-excitation $^1P_1 \to {}^1S_0$ perpendicular to the scattering plane orbital angular momentum of the excited atomic state is "transferred" to photon radiation which should have the same expectation value for the photon spin transfer as the above orbital angular momentum

$$\frac{N(h\nu,\uparrow) - N(h\nu,\downarrow)}{N(h\nu,\uparrow) + N(h\nu,\downarrow)} = \langle L_y \rangle = -P_3 \tag{37}$$

$N(h\nu,\uparrow)$, $N(h\nu,\downarrow)$ are the numbers of photons with spin up (normal of the scattering plane) and spin down with regard to the y-direction, respectively.

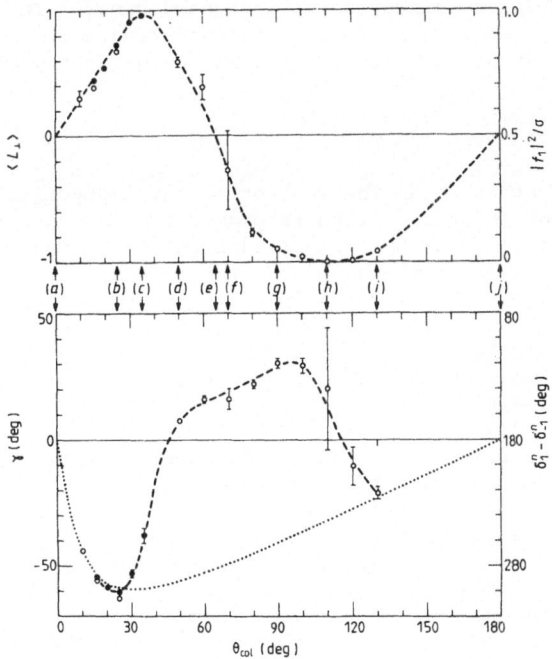

Fig. 19. Top: Orbital angular momentum $\langle L_y \rangle$ transferred to the excited 2^1P_1 state of helium at 80 eV electron impact energy versus the electron scattering angle $\theta_{col} = \theta_e$. Data from Eminyan et al.[1] (o) and Hollywood et al.[5] (o); dotted line first Born approximation. Broken line guides the eye.

Bottom: The corresponding alignment angle $\gamma = \theta_{Min}$ of the electron cloud (see sub-chapter III.e).

Because of the traditional difference between atomic physics and classical optics the sign for the measured optical circular polarization is opposite to the sign for spin polarization of photons or particles in atomic and elementary particle physics.

Eq.(37) provides a striking interpretation of observed circular polarization as directly related to the orbital angular momentum or the orientation of the $^1P_1$ state. Fig.19 gives an example for the collision-induced angular momentum transfer for the He(2^1P) state as a function of the scattering angle ($\langle L_y \rangle = \langle L_\perp \rangle$, $\perp$ refers to the normal of the scattering plane). Note the interesting oscillation of the transferred orbital angular momentum between positive and negative values. We will further discuss this phenomenon in the sub-sections III.g.

(d) Dynamical Asymmetries in Electron Scattering

In electron-photon coincidence experiments one can apply optical polarizers in a certain predetermined way. An interesting example is the choice of given circular polarization for the coincident photon radiation. For the $^1S_o \rightarrow \, ^1P_1 \rightarrow \, ^1S_o$ excitation/ de-excitation process of helium the coincident detection of completely circularly polarized photons of either right or left-handedness means that the photons were emitted from oriented atoms with orbital angular momentum $\langle L_y \rangle$ = either +1 or -1 (following the previous sub-chapter). The question can now be asked what happens with the number of scattered electrons for a given scattering angle θ_e if we measure them in coincidence with photons of given circular polarization, say right hand circular polarization first and afterwards when we have switched the circular polarization to left handedness. In sub-section III.b we derived already the relevant electron scattering amplitudes a_ℓ and a_r for right and left-hand circularly polarized coincident photons which are proportional to

$$f_o + i\sqrt{2}\ f_1 \text{ and } f_o - i\sqrt{2}\ f_1 \quad , \tag{38}$$

respectively. In other words the scattering intensity I_e^R for electrons coincident with RHC polarized light is proportional to $|a_r|^2$ while the scattering intensity I_e^L for LHC polarized light is proportional to $|a_\ell|^2$. Both intensities differ from each other. The relative electron scattering asymmetry becomes then identical to the circular polarization

$$\frac{I_e^R(+\theta_e) - I_e^L(+\theta_e)}{I_e^R(+\theta_e) + I_e^L(+\theta_e)} = P_3(\theta_e) \tag{39}$$

Another way of looking at electron scattering asymmetries with regard to circularly polarized light is to measure a left-right electron scattering asymmetry when the circular polarization of the coincident photons is kept fixed. In other words the electron detector is to be switched from a positive scattering angle ($+\theta_e$) to a negative one ($-\theta_e$) while the optical circular polarization filter for the coincident photons (observed perpendicular to the scattering plane) is set for only one of the kind. It has been shown by Beyer, Blum and Standage[13] that the excitation amplitudes f_{m_L} depend on the choice of the azimuthal electron scattering angle ϕ_e given by the equation

$$f_{m_L} = \left| f_{m_L}(E_o, \theta_e\ , \Phi_e = 0°) \right| e^{i(X + m_L \Phi_e)} \quad ; \tag{40}$$

moving the electron detector from a positive polar angle θ_e with $\phi = 0°$ to a negative one ($-\theta_e$) is equivalent to a changing of the azimuthal angle from $\Phi_e = 0°$ to $\Phi = 180°$. In other words

$$f_1(-\theta_e, \Phi_e = 180°) = -f_1(\theta_e, \Phi_e = 0°) \tag{41}$$

while f_O is not affected by this change. By measuring coincidences with given circular polarized light the excitation amplitudes for positive and negative electron scattering angles are of the same structure as in eqs.(40). We should therefore expect an electron scattering asymmetry for measuring the electron intensities $I_e(+\theta_e)$ and $I_e(-\theta_e)$ in coincidence with given circular polarization of the coincident photons. The relative right-left electron scattering asymmetries are then defined by

$$A^R = \frac{I_e(+\theta_e) - I_e(-\theta_e)}{I_e(+\theta_e) + I_e(-\theta_e)} = P_3(\theta_e)$$

$$A^L = \frac{I_e(\theta_e) - I(-\theta_e)}{I_e(\theta_e) + I(-\theta_e)} = -P_3(\theta_e)$$

(42)

with A^R and A^L as relative electron scattering asymmetries for fixed right (R) or left (L) hand circularly polarized coincident photons, respectively. Together with $P_3(\theta_e) = -P_3(-\theta_e)$ the above electron scattering asymmetries have been tested for the $1^1S_O \rightarrow 3^1P_1 \rightarrow 2^1S_O$ process by Silim et al.[20] The data of Figs. 20-22 do indeed confirm the equivalence of the relative electron scattering asymmetries of eqs.(42) with the circular polarization measurement of the coincident photons.

It is interesting to remark that the inelastic scattering amplitude and cross section is directly related and dependent on the sign of the orientation of the excited atomic state. In other words there is a direct connection or coupling between the produced orientation of the atom and the scattered electron having excited the atom.

Several authors have made an attempt to find a mechanism or a special interaction responsible for the sign of the orientation of the excited state in the electron-atom excitation process. Of particular interest have been classical grazing models which imply that repulsive and attractive potentials in the electron-atom scattering should directly be related

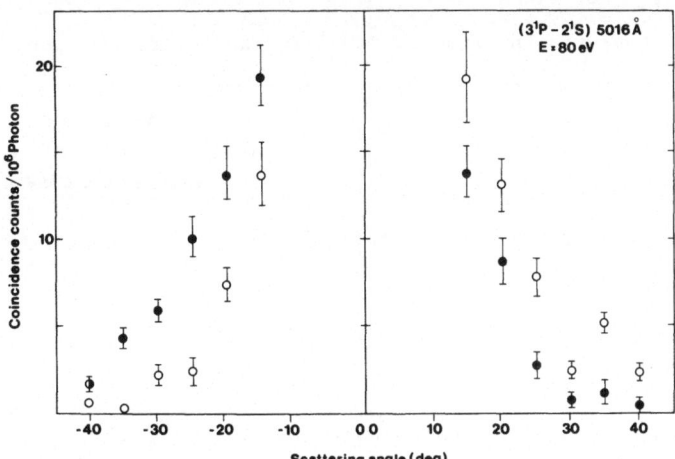

Fig. 20. Electron-photon coincidence count rates of $1^1S_O \rightarrow 3^1P_1 \rightarrow 2^1S_O$ helium transitions normalized to the photon counts as a function of the electron scattering angle. Note that for positive θ_e the results for right hand circularly polarized light (●) are lower than for those for left hand circularly polarized light (○). The situation is reversed for negative electron scattering angles (Silim et al.[19]). Electron impact energy 80 eV.

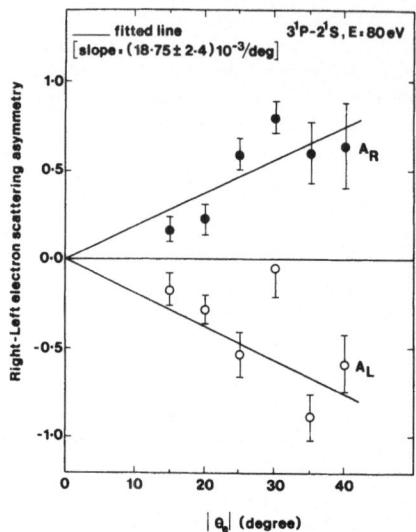

Fig 21. Right-left electron scattering asymmetry as a function of the
scattering angle for detecting right (●) or left (○) hand
circularly polarized coincident photons of $1\,^1S_O \rightarrow 3\,^1P_1 \rightarrow 2\,^1S_O$
helium transitions at 80 eV (Silim et al.[19]). The full lines
are fitted to the experimental data.

to either sign of the orientation of the excited state. Beyer et al.[20]
showed that the amplitudes for left- and right-hand circularly polarized
coincident photons can be expressed in terms of amplitudes resulting from
repulsive (f_R) and attractive potentials (f_A) as follows

$$f_R = \frac{1}{\sqrt{2}} (f_O - i\sqrt{2}\, f_1)$$

$$f_A = -\frac{1}{\sqrt{2}} (f_O + i\sqrt{2}\, f_1)$$

(43)

Results of an analysis in terms of repulsive and attractive amplitudes has

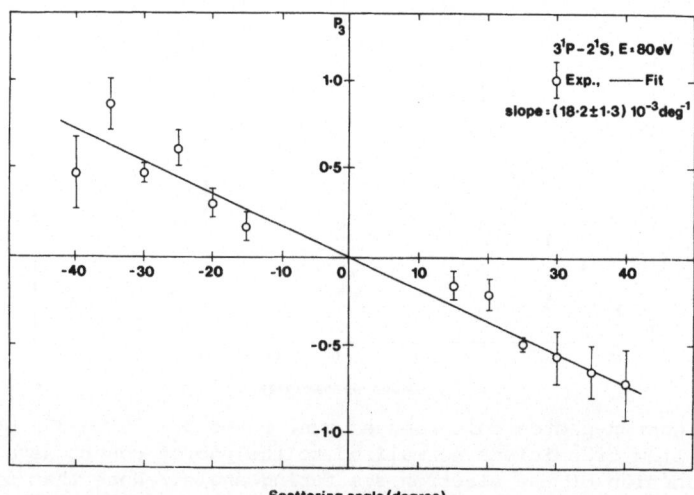

Fig. 22. Circular polarization Stokes parameter P_3 as a function
of the electron scattering angle[20] (helium $1\,^1S_O \rightarrow 3\,^1P_1 \rightarrow 2\,^1S_O$
transition at 80 eV). Note that the slope of the fitted line
agrees with that of the electron scattering asymmetry in the
previous figure proving the identities of the eqs. 42.

been reported at the previous Maratea School by Beyer and Kleinpoppen[21]. However, no direct experimental evidence has come forward for the actual relation between the sign of orientation of the excited state and the sign of the scattering potential so far.

e) Angular Dependence of Electron Charge Distribution of Excited States

The angular and polarization correlation data can also be applied to determine the angular distribution of the electron of the excited state. The relative charge contribution in the volume element $d\tau$ is $-e\psi^*\psi d\tau$ with the $\psi = f_0\psi_z - \sqrt{2} f_1\psi_x$ (eq.15) as statevector for the ^1P excitation. The angular dependence of the electron charge distribution accordingly depends on f_0 and f_1 or λ and χ and can be calculated and graphically represented for a given electron impact energy and scattering angle. Fig.23 gives a typical picture of the shape of the electron charge cloud of the excited ^1P state. The shape of the electron cloud in this Fig. corresponds to $\gamma = \theta_{Min} = 45°$ and $< L_\perp > = 0.75$ (in units of $\hbar$).

By observing the coincident photons of the $^1S_0 \to {}^1P_1 \to {}^1S_0$ transition we introduced already in sub-section III.b the three relevant Stokes parameters (eqs. 31, 32, 33). The linear polarized intensity $I(\alpha)$ of the coincident photons observed normal to the scattering plane with the polarizer angle α is given by

$$I(\alpha) \propto \lambda\cos^2\alpha + (1-\lambda)\sin^2\alpha - 2\sqrt{\lambda(1-\lambda)} \sin\alpha \cos2 \cos\chi \qquad (44)$$

By eliminating λ and $\cos\chi$ from eqs.31 and 32 we obtain

$$I(\alpha) \propto 1 + P_1\cos2\alpha + P_2\sin2\alpha = 1 + P_{lin}\cos2(\alpha-\gamma) \qquad (45)$$

with

$$P_{lin} = \sqrt{(P_1^2 + P_2^2)}, \quad \cos2\gamma = \frac{P_1}{P_{lin}}, \quad \sin2\gamma = \frac{P_2}{P_{lin}} \qquad (46)$$

On the other hand and since the polarization characteristics is directly depending on $<\psi(^1P)|\psi(^1P>$ of the dipole transition $^1P_0 \to {}^1S_0$ $I(\alpha)$ is proportional to the square of the electric light vector of the polarization ellipse (Fig.5a) and the density $e|\psi|^2$ of the charge cloud of the excited state in the scattering plane.

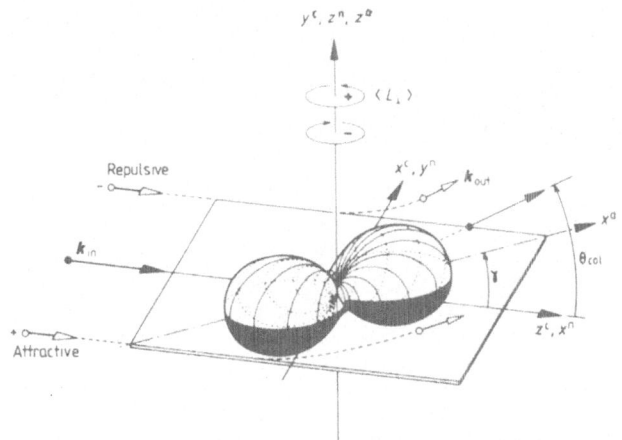

Fig. 23. Charge distribution of a coherently excited ^1P state for $\gamma = \theta_{Min} = 45°$ and $< L_\perp > = 0.75$ h. The coordinate indices c, n and a refer to the possible coordinate systems: collision, natural and atomic frame, respectively.

Observing the coincident photons without any polarization analysis is described by the angular correlation eq.10 as discussed in sub-section III.a. Again substituting λ and $\cos\chi$ by eqs.31 and 32 links the angular dependence of the observed photon intensity $I(\theta)$ in the scattering plane with the Stokes parameters P_1, P_2 and the minimum angle $\theta_{Min} = \gamma$ or the alignment angle γ of the electron cloud:

$$I(\theta_\gamma) \propto 1 - P_1\cos2\theta - P_2\sin2\theta = 1 - P_{lin}\cos2(\theta-\gamma) \qquad (47)$$

This angular distribution is rotated by 90° compared to the angular dependence of the electron charge distribution (eq.45) or the alignment angle γ is just identical to the minimum angle θ_{Min} of the photon intensity for an angular correlation experiment in the scattering plane.

f) <u>Spin-Exchange Amplitudes in the Impact Excitation of Triplet P States of Helium</u>

Electron impact excitation of helium triplet states from singlet states can only occur through spin exchange interaction. This is contrary to the singlet excitation from singlet states which can be achieved by either Coulomb direct and exchange interaction. In singlet-singlet transitions both types of interactions cannot be separated from each other, e.g., the amplitudes f_0 and f_1 for the $^1S_0 \rightarrow {}^1P_1$ excitation are indeed indistinguishably superposed by the two amplitudes describing the direct and the exchange interaction. In singlet-triplet transitions, however, we should be able

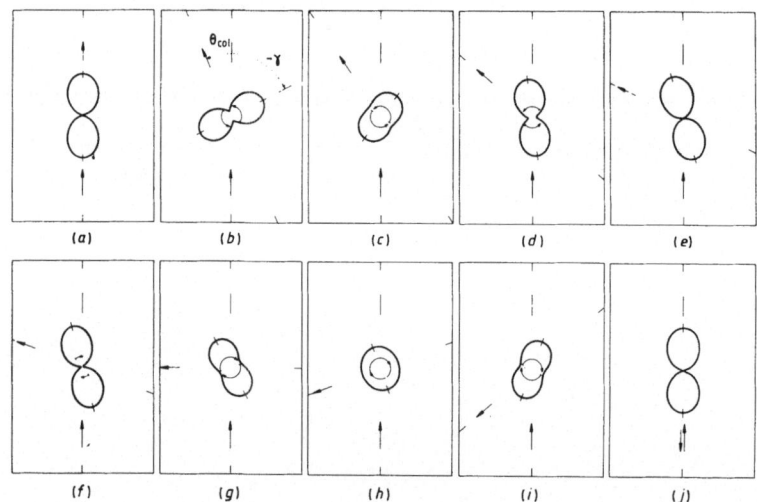

Fig. 24. Shape and dynamics of the electron charge distribution in the scattering plane for the He(2^1P) state at electron scattering angles $\theta_e=\theta_{col}$ with reference to (a) $\theta_e=0°$, (b) $\theta_e=25°$, (c) $\theta_e=35°$, (d) $\theta_e=50°$, (e) $\theta_e=65°$, (f) $\theta_e=70°$, (g) $\theta_e=90°$, (h) $\theta_e=110°$, (i) $\theta_e=130°$ and (j) $\theta_e=180°$. The labels (a), (b), ··· refer to the θ_e values indicated in Fig.19. Ingoing and outgoing electrons are indicated by the arrows. The sign and magnitude of the transferred orbital angular momentum L of Fig.19 is indicated by the length and direction of the circular arrows, a full circle corresponding to the unit of $\hbar$ (positive sign in counterclockwise direction, from ref.28).

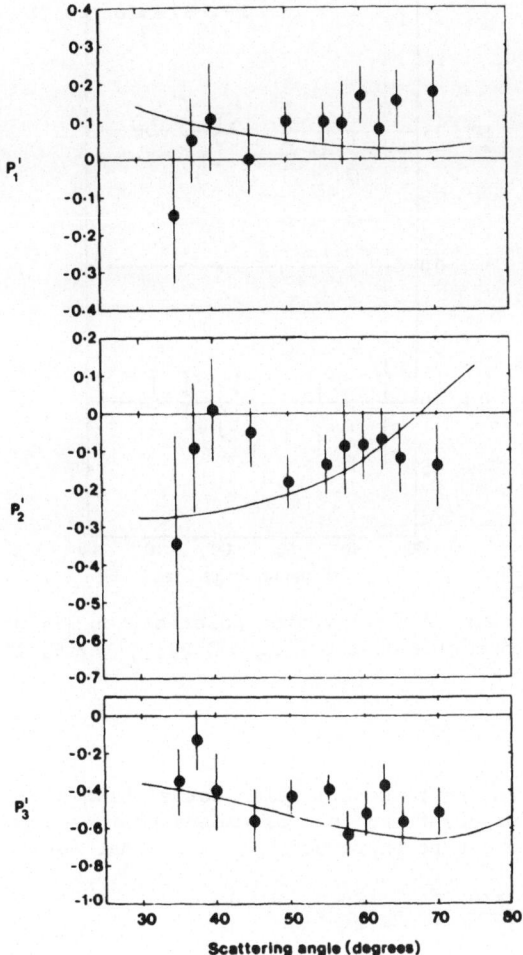

Fig. 25. Measured Stokes parameters of the excitation/de-excitation process $1\,^1S_0 \rightarrow 3^3P_{2,1,0} \rightarrow 2^3S_1$ of helium[22]. Electron impact energy 60 eV.
Full curves theoretical predictions by Cartwright and Csanak[23].

to describe the excitation process by "pure" spin exchange amplitudes (or often simply called exchange amplitudes). For $^1S \rightarrow {}^3P$ excitation with $L = 1$ and $m_L = 0, \pm1$ exchange amplitudes g_{m_L} with g_0 and g_1 would be expected to be appropriate as long as we can prove experimentally full coherence for the excitation process and account for the effect of the fine-structure splitting. While we expect that the amplitudes for the excitation of the orbital angular states are sufficient (we assume that the LS coupling during the very short time of the electron impact interaction is broken up) the influence of the fine structure coupling on the emission process of the photons should affect the electron photon angular and polarization correlation.

Symbolically the excitation/de-excitation process can be written as

$$e + He(1\,^1S_0) \rightarrow He(n^3P) + e$$
$$\longrightarrow He(n^3P_{2,1,0}) \rightarrow He(m^3S_1) + e \qquad (48)$$

After the initial excitation of the ^3P state the fine structure interaction will couple the orbital and spin momenta to the total angular momenta

417

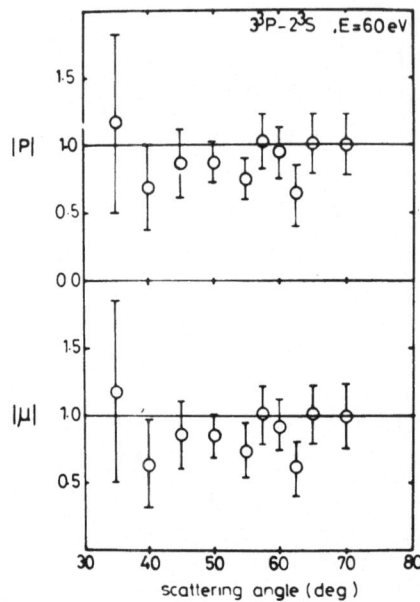

Fig. 26. Magnitude of total vector polarization $|P|$ and degree of coherence $|\mu|$ of the $1^1S_0 \rightarrow 3^3P_{2,1,0} \rightarrow 2^3S_1$ transition[22].

$J = 2, 1, 0$. With regard to polarization correlation experiments the Stokes parameters for observing the coincident photons perpendicular to the scattering plane can be expressed by λ and χ parameters as follows:

$$P_1' = \frac{I(0°) - I(90°)}{I(0°) + I(90°)} = \frac{15}{41}P_1 = \frac{15}{4}(2\lambda - 1) \tag{49}$$

$$P_2' = \frac{I(45°) - I(135°)}{I(45°) + I(135°)} = \frac{15}{41}P_2 = -\frac{30}{41}\sqrt{\lambda(1-\ )}\cos\chi \tag{50}$$

$$P_3' = \frac{I(RHC) - I(LHC)}{I(RHC) + I(RHC)} = \frac{27}{43}P_3 = \frac{54}{41}\sqrt{\lambda(1-\lambda)}\sin\chi \tag{51}$$

In these equations the measured Stokes parameters P_1', P_2', P_3' can be linked to the Stokes parameters P_1, P_2, P_3 corrected for the depolarization caused by the fine structure coupling, and to the angular correlation parameters λ and χ defined in analogy to the case for the singlet excitation $^1S \rightarrow {}^1P$ (eqs. 10 and 11).

$$\lambda = \frac{|g_0|^2}{|g_0|^2 - 2|g_1|^2}, g_1 = |g_1|e^{i\chi}, g_0 \text{ be real and the}$$

differential cross section is $\sigma = |g_0|^2 + 2|g_1|^2$ (52)

Fig. 25 gives an example of the measured Stokes parameters for the helium $3^3P_{2,1,0} \rightarrow 2^3S$ excitation. The magnitude of the vector polarization $|P| = \sqrt{|P_1|^2 + |P_2|^2 + |P_3|^2}$ and the coherence correlation factor $|\mu| = \frac{P_2 - iP_3}{\sqrt{1 - P^2}}$ are shown in Fig. 26; these data are compatible with the unity value and accordingly confirm that the exchange excitation of the 3^3P state is completely coherent and the state vector for the ^3P excitation can be represented by a linear superposition of the magnetic sublevel states:

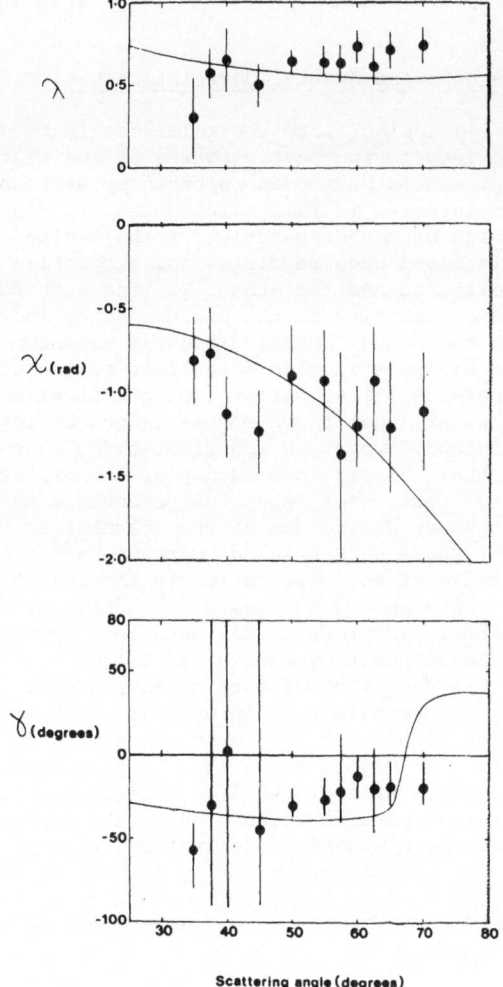

Fig. 27. Angular parameters λ, χ and γ extracted from the measured
Stokes parameters of Fig.25. Full curves theoretical
predictions by Cartwright and Csanak[23].

$$|\psi(^3P)\rangle \;=\; g_O|L = 1, m_L = O\rangle + g_1|L = 1, m_L = \pm 1\rangle$$

$$+ \; g_{-1}|L = 1, m_L = -1\rangle$$

$$=\; g_O|\psi_z\rangle - \sqrt{2}\, g_1|\psi_x\rangle \tag{52}$$

with $g_1 = -g_{-1}$ because of reflection invariance as in the case of singlet
(^{1}P) excitation (see sub-chapter III.a). Accordingly we can extract typical
angular correlation data such as λ, χ and the alignment angle γ of the
charge distribution (Fig.27).

It is obvious that these angular correlation data for exchange excitation are expected to be the most crucial and sensitive test quantities for
theoretical approximations. The accuracy of the experimental data for the
^{3}P excitation are inferior to the ^{1}P excitation process. The reason for
this being the fact that the lifetime of the ^{3}P states are in general
almost by two orders of magnitude larger ($\approx 10^{-7}$ sec for the 3^3P state!)
compared to the ^{1}P states. Accordingly the coincidence peak from the

$3^3P \rightarrow 2^3S$ transition is spread over a much larger time interval than for $^1P \rightarrow {}^1S$ transitions.

(g) Partial Wave Interference and Angular Correlation Data

Recently an interesting insight into the relationship between angular correlation data and the interfering partial waves of the electrons involved in excitation processes have been reported by Madison, Csanak and Cartwright[24]. Their interest in this relation mainly arose through the question about the sign of the orientation of the excited $^1P_1$ state in connection with models based upon repulsive and attractive potentials between the projectile electron and the atom. Various authors[25] have discussed models for the orientation of the excited state induced by the electron collision which attribute positive angular momentum transfer to long-range attractive atomic polarization potentials and negative transfer to electron-electron repulsion (classical or semi-classical models). The above authors argued, however, that these models cannot be correct since quantal calculations (distorted-wave, DW and first-order many-body theories, FOMBT) reveal that the sign dependence of the orientation of the excited state on either attractive or repulsive potentials should not be valid. Instead of the angular dependence of the orientation vector on the sign of the potential Madison, Csanak and Cartwright[24] showed that the characteristic structure of the orientation is the result of a quantum mechanical interference phenomenon with regard to partial waves of different orbital angular momenta. In order to get detailed physical insight into this interference phenomenon they made a partial wave study of the orientation vector $O_{1-}^{col} = \frac{1}{2} < L_\perp >$ by distorting an increasing number of incident electron waves, ℓ_p, up to a certain maximum (i.e. $\ell_p = 0, 1_1 \ldots, \ell_p^{max}$ incident orbitals) and the associated scattered waves with $\ell_q = (\ell_p \pm 1)$. Results of $< L_\perp >$ are given in Fig. 28 which shows for incident electron energy of 80 eV that the distortion of the $\ell_{p=0,1}$ incident electron waves (and the associated scattered partial waves) contribute most significantly to the "positive hump" and by disturbing the incident $\ell_p = 2$ partial wave a

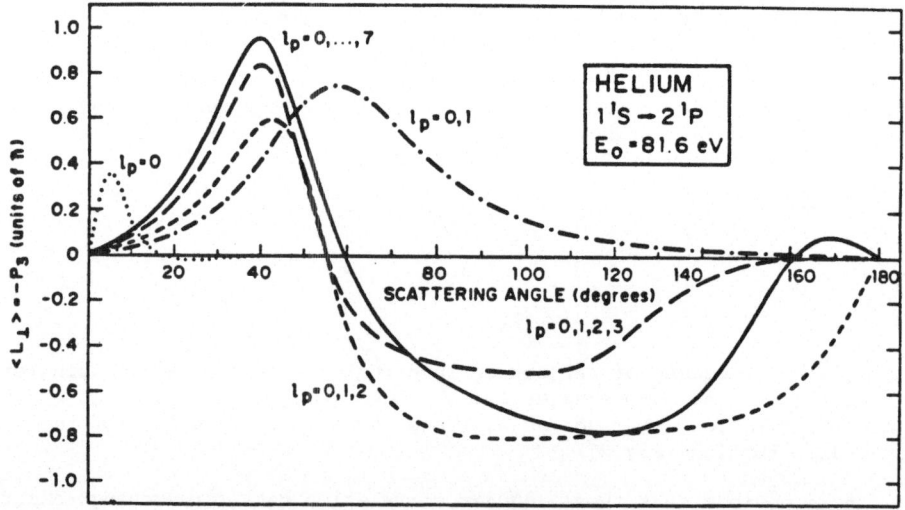

Fig. 28. Transferred orbital angular momentum $< L_\perp >$ for electron impact excitation of the helium 2^1P_1 state at 81.6 eV impact energy in first-order many-body theory, FOMBT (Csanak, Cartwright[26] $\ell_p = 0, 1, \ldots$) ℓ_{max} refer to the incident-electron partial waves that were distorted in the calculation.

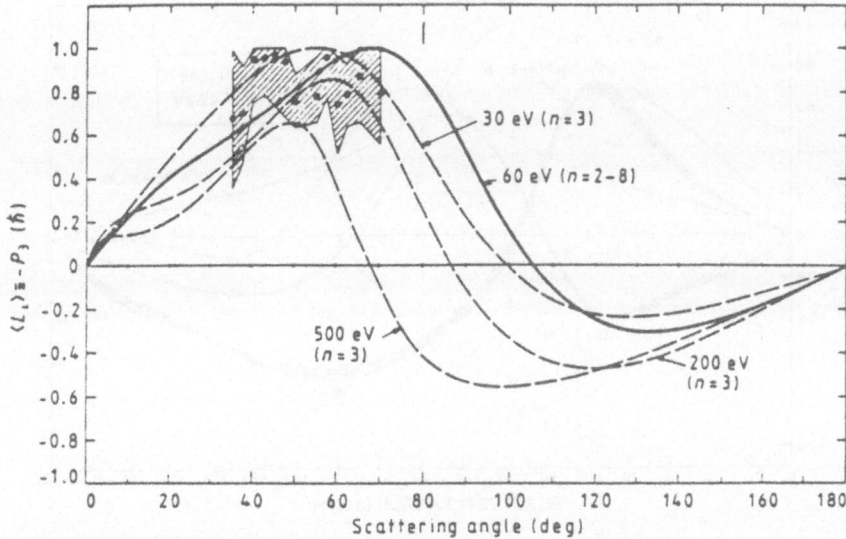

Fig. 29. Transferred orbital angular momentum $< L_\perp >$ for $n^3 P$ states of helium as a function of the scattering angle for various energies and principal quantum numbers n. The broken and full curves are the FOMBT data of Csanak and Cartwright[26]. The shaded region are the experimental data of Silim et al.[22] for the $3^3 P \rightarrow 2^3 S$ transition.

large "negative hump" suddenly occurs. Further additional distortions do not drastically change this interpretation. This partial wave analysis indicates that the positive hump is principally due to small impact parameter collisions and that the typical behaviour of the angular characteristics of the orientation vector O^{col}_{1-} or the transferred angular momentum $< L_\perp >$ of the ^1P state excitation is the result of a quantum mechanical interference phenomenon. It is notable that this observation essentially holds to energies investigated as high as 500 eV (Fig.29). At this incident electron energy, angular momentum transfer is still possible at small scattering angles and is associated with the distortion of $\ell_p = 0,1$ incident electron partial waves; FOMBT predicts a strong interference effect for 500 eV. Contrary to this the Born and Glauber approximation predict zero orientation $O^{col}_{1-} = \frac{1}{2}< L_\perp >$ for all scattering angles and energies. Following this it was suggested by Csanak to interpret $< L_\perp >$ as a physical quantity extracted from the angular correlation experiments which does not have any "background", i.e. the whole angular behaviour of $< L_\perp >$ is "structure" due to quantum mechanical interference between the various free-electron partial waves. This behaviour appears to be different from the Ramsauer-Townsend effect in which the quantum mechanical interference structure shows itself on a background for the differential or total scattering cross section.

Cartwright and Csanak[26] have extended their calculations even to n^1P, n^3P, n^1D and n^3D of which Fig.30 gives an example. The interesting result that can be observed is that the angular dependence of $< L_\perp >$ (and also the alignment angle γ of the electron cloud) were found to be essentially independent of the principal quantum number n. Csanak and Cartwright showed that this observation can be explained on the basis of the quantum defect theory which in turn is based on the conclusion by Hartree[27] that, close to a nucleus, a bound state orbital is essentially independent of the principal quantum number n except for a normalization constant.

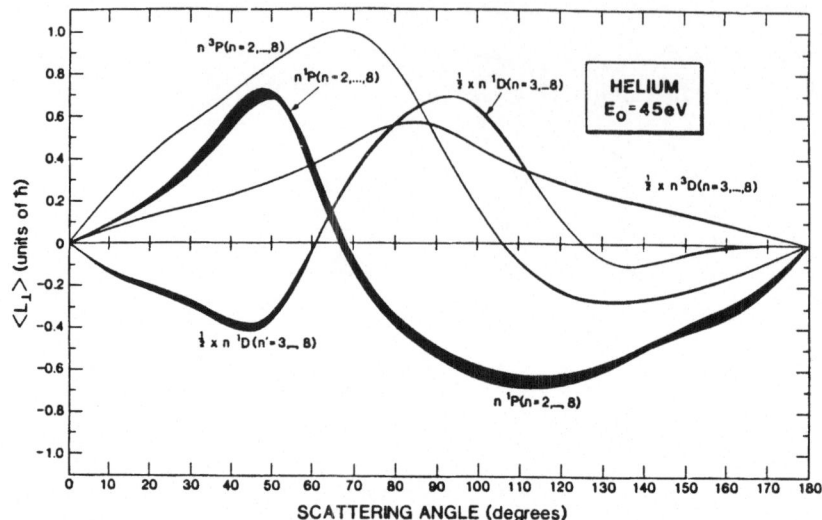

Fig. 30. Transferred orbital angular momentum $< L_\perp >$ for $n^{1,3}P(n=2,\cdots,8)$ and $^{1,3}D(n=3,\cdots,8)$ states of helium at 45 eV calculated by Csanak and Cartwright[26] FOMBT). Note the negative sign of $< L_\perp >$ for ^1D states for $\theta_e < 50°$ and the positive sign for ^3D states over the whole angular range.

We return now to the semiclassical model attributing the positive hump in the $< L_\perp >$ data of ^1P excitation to attractive polarization forces and the negative hump (approximately for $60° < \theta_e < 180°$) to repulsive forces (presumably electron-electron correlations). In a normal distorted-wave calculation the atomic distorting potential is a sum of the nuclear attraction term and the electron repulsion term from the charge distribution of the atomic electrons. Madison, Csanak and Cartwright[24] have consequently carried out two kinds of calculations. The first was a usual distorted-wave (DW) approximation in which both the repulsive and attrac-

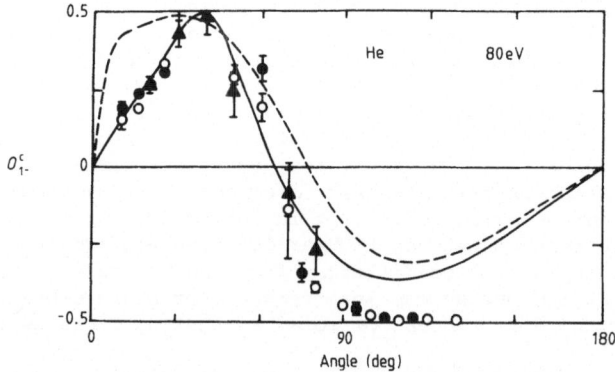

Fig. 31. Orientation $O_{1-}^{col}=\frac{1}{2}< L_\perp >$ of the He2^1P state at 80 eV as a function of the scattering angle. Theoretical curves: ——— normal distorted wave theory, - - - - distorted wave theory using a purely attractive distorting potential (Madison, Csanak, Cartwright[24]). Experimental data: o Hollywood et al.[5], ● Slevin et al.[6], ▲ Khakoo et al.[30]. The Hollywood et al. and the Slevin et al. data were set negative in accordance with the trend implied by theory.

tive terms were included. In the second calculation the authors have dropped the electron-electron repulsion term and retained only the nuclear attraction (effectively a Coulomb wave approximation, CWA). All these calculations included first-order exchange interactions. The results of both kinds of calculations for O_{1-}^{col} are shown in Fig.31. The difference of the DW-curves and the CWA-curves of Fig.31 represents the contribution of the electron-electron repulsion. It is obvious from the data of this figure that the electron repulsion does not have a significant effect on the orientation of the 2^1P state as a function of the scattering angle. Surprisingly enough electron repulsion appears to be more important at small angles than at larger angles.

In a semiclassical model in which the projectile is treated classically and the interaction with the atom quantally, Fargher and Roberts[29] obtained in a first-order time dependent perturbation approximation the characteristic shape of positive orientation of the 2P excitation of atomic hydrogen at small scattering angles and negative orientation at large angles although only the repulsive force between the projectile and the target was included. Madison, Csanak and Cartwright[24] have carried out a similar calculation for this case in the distorted-wave approximation, in which the distorting potential was chosen to be only the electron-electron repulsion term. This distorted-wave calculation, however, predicted orientation values of O_{1-}^{col} which were negative for all scattering angles, contrary to the semiclassical results of Fargher and Roberts.

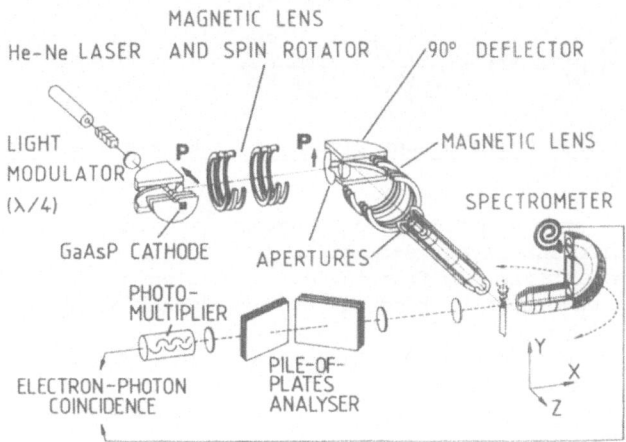

Fig. 32 (a). Schematic diagram of the Münster apparatus[31] for electron photon coincidences with primarily partially polarized electrons. The polarized electrons are produced by photo-ionization at the GaAsP cathode with circularly polarized light. The spin direction of the electron is rotated by a magnetic focussing spin rotator from the z direction to the y direction. After deflection by 90° the electron beam is focussed on to the mercury beam. The scattered electrons having passed an electron analyzer are detected in coincidence with the λ=2537Å photons. The photon polarization is analyzed by a pile of plate polarization analyzer.

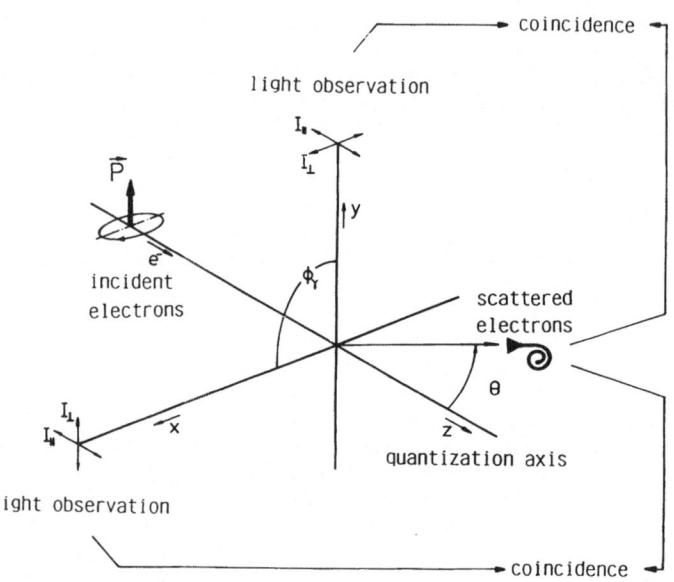

Fig. 32 (b). Geometry for the spin direction of incident electrons ($\vec{P}$) for the coincident light observation with polarized intensity components $I_{\parallel}$ and $I_{\perp}$ (with reference to quantization axis) and the detection angle of the scattered electrons. With these notations spin asymmetries of eqs. can be written as

$$A(M_y = 0) = \frac{I_{\parallel}(P_y) - I_{\parallel}(-P_y)}{I_{\parallel}(P_y) + I_{\parallel}(-P_y)} \text{ , and}$$

$$A(M_y = \pm 1) = \frac{I_{\perp}(P_y) - I_{\perp}(-P_y)}{I_{\perp}(P_y) + I_{\perp}(-P_y)} \text{ .} \tag{53}$$

(h) Electron-Photon Coincidences with Spin-polarized Electron Impact Excitation

We are now at the beginning of a new area of investigations, namely the study of electron impact excitation with polarized electrons detected in coincidence with photons from the de-excitation process. Two types of experiments have been carried out by the Münster group.

(1) Detection of inelastically scattered polarized electrons (without spin analysis) in the forward direction and in coincidence with the polarized 2537Å photons from the $6^3P_1 \rightarrow 6^1S_0$ de-excitation in mercury.

(2) The same process for non-zero scattering angles.

Contrary to the two exchange amplitudes required for the ^3P excitation of helium (sub-chapter III.f) six amplitudes are required for the above mercury excitation taking into account exchange, spin-orbit interactions and their interference terms. Although it is not feasible to extract all these amplitudes from the above combined coincidence-spin experiment, spin-up-down asymmetries have been measured by observing σ and π-components of the coincident 2537Å photons thus relating the excitation process to the $M_y=0$ and $M_y=\pm1$ magnetic sublevels of the 6^3P_1 state of Hg. Fig.32 shows and explains the complicated scheme and detection geometry of the Münster experiment. Fig.33 gives results for the spin asymmetries of the coincidence count rates with electron spins up ($A^\uparrow$) or down ($A^\downarrow$) with reference to the scattering plane.

The spin asymmetry connected to π light is

$$A(M_y = 0) = \frac{A^\uparrow(M_y = 0) - A^\downarrow(M_y = 0)}{A^\uparrow(M_y = 0) + A^\downarrow(M_y = 0)} \tag{54}$$

and that for σ light to

$$A(M_y = \pm 1) = \frac{A^\uparrow(M_y = \pm 1) - A^\downarrow(M_y = \pm 1)}{A^\uparrow(M_y = \pm 1) + A^\downarrow(M_y = \pm 1)} \tag{55}$$

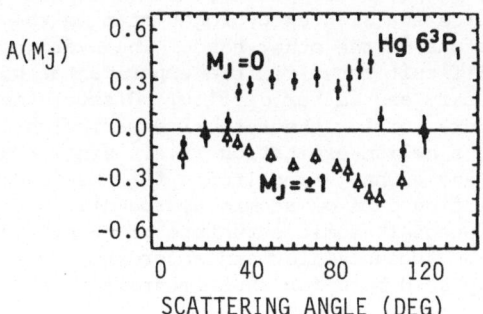

Fig. 33. Spin asymmetries for excitation of magnetic sublevels $M_y=0$ and $M_y=\pm 1$ of Hg 6^3P_1 state at incident electron energy of 9 eV (Münster group[31]).

Previous non-coincidence experiments on spin asymmetries did not allow distinction between the magnetic substates in the spin up/down asymmetries. However, the non-averaging coincidence data of Fig. 33 will certainly be of substantial help for more detailed understanding of electron impact excitation of heavy atoms. These novel data from combined spin-coincidence experiments will also be of guidance for theoretical approximations. Certainly these experiments do show the way into future developments by which full sets of excitation amplitudes might be forthcoming.

IV CONCLUSIONS AND FUTURE DEVELOPMENTS

Since the beginning of the seventies the experiments on electron-photon coincidences have provided us with new kinds of information and physical insight in electron impact excitation of atoms. In this lecture on the Analysis of Electron-Atom Collisions we primarily concentrated on the various types of information extracted from the angular correlation and polarization correlation of the $S \rightarrow P \rightarrow S$ transitions of helium. Excitation amplitudes and their relative phases, alignment and orientation, left-right scattering asymmetries, spin-exchange amplitudes, the electron charge cloud of excited states and spin up/down asymmetries represent such a comprehensive, beautiful collection of information and deep insight in the physics of electron impact excitation which is a basic quantum mechanical scattering process. Recent theoretical investigations suggest that the phenomenon of quantum mechanical interference of electron partial waves shows a strikingly strong and direct influence on the structure of correlation data such as the orientation of the excited state. This kind of Ramsauer-Townsend structure effect in correlation data of electron impact excitation appears to be detectable without background and is existent and noticeable even at high impact energies. It remains, however, an open problem whether the sign of the orientation of the excited state is directly correlated to either repulsive or attractive interaction potentials.

These lecture notes do not represent a current review or summary on the subject, we therefore refer the reader to a collection of various reviews already published[17,32-37] or to be published in the near future[38-41]. Although the helium excitation of nP states played a dominant role in the field of coincidence experiments, very impressive results have been reported with atomic hydrogen, sodium, magnesium, rare gases and mercury as target atoms. The degree of complexity with these atoms increases in both directions, towards the lighter hydrogen atom on the one hand and towards the heavier atoms on the other hand. The hydrogen atom as target appears to be more difficult both from the experimental point of view (because of the necessary and rather difficult dissociation of H_2 into 2H) as well as from the more complex theoretical description due to spin effects compared to the helium excitation (i.e., singlet and triplet amplitudes or direct and exchange amplitudes for the description of the excitation and correlation data of atomic hydrogen). With heavier atoms the influence of the special atomic structure of relativistic and of spin-orbit effects has to be accounted for and accordingly a lot of surprises and challenges are expected both for experimentalists and theoreticians in the future.

Combining the electron-photon coincidence technique with that of electron-spin and atomic spin analysis (see also following lecture by W. Raith) will be the ultimate goal for determining a full set of excitation amplitudes f_0, f_1, (direct) and g_0, g_1 (exchange) of the nP excitation of light one-electron atoms. The complicated situation of excitation of heavy atoms will further lead to applications of spin and coincidence methods which will depend on the special cases.

Finally it is fitting to note that the electron-photon coincidence technique applied first in electron-atom collisions has paved the way for similar investigations in heavy-particle impact excitation of atoms and molecules[42] and electron-molecule excitation[43].

Acknowledgements

The author gratefully acknowledges clarifying and stimulating discussions with Dr H J Beyer, Dr G Csanak and Dr H A Silim. Thanks are also due to Dr J Goeke, Professor G F Hanne and Professor J Kessler for reporting on their research results prior to publication. The research reported on the Stirling work was supported by the Science and Engineering Research Council, U.K.

REFERENCES

1. H. Kleinpoppen, Columbia University, unpublished report, 1967;
    M. Eminyan, K. MacAdam, J. Slevin and H. Kleinpoppen, Phys. Rev.Lett. 31 (1973) 576; and J.Phys.B7 (1974) 1519.

2. A.A. Zaidi, S.M. Khalid, I. McGregor, H. Kleinpoppen, J.Phys.B14, L503 (1981), and A.A. Zaidi, I. McGregor, H. Kleinpoppen, Phys.Rev.Lett. 45, 1168 (1980).

3. U. Fano, Rev.Mod.Phys. 29, 74 (1957).

4. H.A Bethe, Handbuch der Physik, Vol.XXVI/1, 2.Aufl. 1933.

5. M.T. Hollywood, A. Crowe, J.F. Williams, J.Phys. B12, 819 (1979).

6. J. Slevin, H.Q. Porter, M. Eminyan, A. Defrance, G. Vassilev, J.Phys. B13, 3009 (1980).

7. N.C. Steph, D.E. Golden, Phys.Rev. A21, 759 (1980).

8. V.C. Sutcliffe, G.N. Haddad, N.C. Steph, D.E. Golden, Phys.Rev. A17, 100 (1978).

9. L.D. Thomas, G. Csanak, H.S. Taylor, G.S. Yarlagadda, J.Phys. B10, 1073 (1977).

10. G. Catalan, M.J. Roberts, J.Phys. B12, 3947 (1979).

11. T. Scott, M.R.C. McDowell, J.Phys. B8, 1851 (1975).

12. M. Born, E. Wolf, "Principles of Optics", 3rd edition, Pergamon Press, 1964 and 1965.

13. H.J. Beyer, K. Blum, M.C. Standage, in "Fundamental Processes in Atomic Collision Physics", Nato-ASI series B, Vol.134, 573, Plenum Press (1985).

14. M.C. Standage, H. Kleinpoppen, Phys.Rev.Lett. 36, 577 (1976).

15. H.A. Silim, H.J. Beyer, H. Kleinpoppen, to be published and H.A. Silim, PhD thesis, Mansoura University, Egypt, 1985.

16. M.R. Flannery, K.J. McCann, J.Phys. B8, 1716 (1975).

17. K. Blum, H. Kleinpoppen, Physics Report 52, 203 (1979).

18. J. Macek and D.H. Jaecks, Phys.Rev.A4 1288 (1971).
    U. Fano and J. Macek, Rev.Mod.Phys. 45, 553 (1973).

19. H.A. Silim, H.J. Beyer, H. Kleinpoppen, to be published, and PhD thesis by H.A. Silim, Mansoura University, Egypt, 1985.

20. H.J. Beyer, H. Kleinpoppen, I. McGregor, I.C. McIntyre, J.Phys.B15, L545, (1982).

21. H.J. Beyer, H. Kleinpoppen, in "Fundamental Processes in Energetic Atomic Collisions", Nato-ASI series B, Vol.103, 531, Plenum Press (1983).

22. H.A. Silim, H.J. Beyer, A. El-Sheikh, H. Kleinpoppen, Phys.Rev. 35. 4454 (1987).

23. D.C. Cartwright, G. Csanak, J.Phys.B19, L485 (1986).

24. D.H. Madison, G. Csanak, D.C. Cartwright, J.Phys.B19, 3361 (1986).

25. (a) M. Kohmoto and U. Fano, J.Phys.B14, L447 (1981).
    (b) H.W. Hermann and I.V. Hertel, J.Phys.B13, 4285 (1980).
    (c) C.W. Lee and U. Fano, Phys.Rev.A33, 921 (1986).
    (d) N.C. Steph and D.E. Golden, Phys.Rev.A21, 1848 (1980).
    (e) N. Andersen, I.V. Hertel, H. Kleinpoppen, J.Phys.B17, L901 (1984).
    (f) N. Andersen, J.W. Gallagher, I.V. Hertel, Electronic and Atomic Collisions, D.C. Lorents, W.E. Meyerhof, J.R. Peterson (eds) Elsevier Science Publishers B.V., 57 (1986).
    (g) N. Andersen and I.V. Hertel, Comm.At.Mol.Phys. 19, 1 (1986).

26. G. Csanak, D.C. Cartwright, Inv. Progress Report for XVth Int.Conf. on the Physics of Electronic and Atomic Collisions, Brighton, U.K. July 1987.

27. D.R. Hartree, Proc.Cambr.Phil.Soc. 24, 426 (1928).

28. N. Andersen, I.V. Hertel, H. Kleinpoppen, J.Phys.B17, L901 (1984).

29. H.E. Fargher, M.J. Roberts, J.Phys.B19, 3361 (1986).

30. D.H. Khakoo, K. Becker, J.L. Forand, J.W. McConkey, J.Phys.B12, L209 (1986).

31. G.F. Hanne, Physics Report $\underline{95}$, 95 (1983); J. Goeke, PhD thesis 1987,
    Münster; J. Kessler, $\overline{\text{Inv}}$. Paper at XVth ICPEAC, Brighton, U.K.,
    July 1987; J. Goeke, G.F. Hanne, J. Kessler, Abstracts of
    Contributed Papers, XVth ICPEAC, Brighton, J. Geddes, H.B. Gilbody,
    A.E. Kingston, C.J. Latimer, H.J.R. Walters (eds.), p.175 (1987).
32. H. Kleinpoppen in "Atomic Physics 4", G. zu Putlitz, E.W. Weber,
    A. Winnacker, eds. p.449, Plenum Press (1975).
33. M. Eminyan, H. Kleinpoppen, J. Slevin, M.C. Standage in "Electron and
    Photon Interactions with Atoms", H. Kleinpoppen, M.R.C. McDowell,
    eds., p.455, Plenum Press (1976).
34. H. Kleinpoppen, K. Blum, M.C. Standage, IXth ICPEAC, Invited Lectures,
    J.S. Risley, R. Geballe (eds.), p.641, University of Washington
    Press (1976).
35. H. Kleinpoppen, I. McGregor, in "Atomic Physics 8", I. Lindgren,
    A. Rosen, S. Svanberg, eds., p.431, Plenum Press (1983).
36. J. Slevin, Rep.Progr. in Physics $\underline{47}$, 461 (1984).
37. N.O. Andersen, J. Gallagher, I.V. Hertel, Invited Papers XIVth ICPEAC,
    D. Lorents, W. Meyerhof, J.R. Peterson, (eds.), p.57, North-
    Holland (1986).
38. J.W. McConkey, P. Hammond, M.A. Khakoo, Inv. Paper, XVth ICPEAC, to
    be published.
39. G. Csanak, D.C. Cartwright, Inv. paper XVth ICPEAC, to be published.
40. H.J. Beyer, Inv. paper, International Symposium on Polarization and
    Correlation in Atomic Collisions, to be published.
41. N.O. Andersen, J.W. Gallagher, I.V. Hertel, to be published in
    Physics Report.
42. H. Kleinpoppen, J.F. Williams (eds.), "Coherence and Correlation in
    Atomic Collisions", Plenum Press (1980); R. Hippler in
    "Fundamental Processes in Atomic Collisions", Nato-ASI series
    Vol.134, p.181 (1985) and N.O. Andersen and H.O. Lutz in these
    Proceedings.
43. K.Blum, H. Jackubowicz, J.Phys.B$\underline{11}$, 909 (1978); M.A. Khakoo, J.W.
    McConkey, J.Phys.B$\underline{20}$, L175 (1987); J.W. McConkey, S. Trajmar,
    T.C. Nickel, G. Csanak, J.Phys.B$\underline{19}$, 2377 (1986).

ANALYSIS OF ELECTRON-ATOM COLLISIONS:

SPIN-DEPENDENT EFFECTS

Wilhelm Raith

Fakultät für Physik
Universität Bielefeld
D-4800 Bielefeld
Fed.Rep. of Germany

INTRODUCTION

At low energies, particularly below 100 eV, the principal spin-dependent interaction is exchange between the free electron and the bound atomic electrons. Electron exchange is a quantum phenomenon for which no classical model exists. Exchange occurs regardless of spin direction. Its spin dependence is rooted in the Pauli principle: Two electrons with antiparallel spins are distinguishable (one can tell whether exchange took place), whereas with parallel spins they are indistinguishable (exchange can happen but its occurrence cannot be determined experimentally). In quantum theory distinguishable and indistinguishable processes are treated differently; in the latter case the amplitudes for "direct" and "exchange" scattering can interfere. This makes exchange spin dependent.

The interaction between the spin-associated magnetic moments (and also the interaction between the magnetic moments of the free electron and the atomic nucleus) is completely negligible at low energies.

Spin-orbit interaction is a relativistic effect. In low-energy collisions spin-orbit effects due to the coupling of the spin of the free electron and its orbital motion are observable only in special circumstances, namely in scattering from high-Z atoms near a deep minimum of the differential cross section. The spin-orbit interaction of bound electrons in states with non-zero angular momentum can also influence the collision in some cases ("fine-structure effect").

We want to give an introduction into this highly developed field of atomic collision physics and discuss some recent experiments in order to exemplify the state of the art. More information and references can be found in Kessler's book.[1]

Most scattering experiments on exchange effects employ crossed polarized beams. Therefore we start this analysis with a survey of the experimental techniques relevant for polarized atomic beams and polarized electron beams.

POLARIZED ATOMIC BEAMS

Definition of Polarization

At first we consider only beams of one-valence-electron atoms from effusive or jet sources (in contrast to fast atoms from neutralized ion beams). The interesting spin polarization of such an atom is that of the valence electron; the nuclear polarization is irrelevant in low-energy scattering. This atomic polarization is defined with respect to an axis of quantization (usually a very weak magnetic field) as

$$P_a = \langle m_s \rangle / (½) ,$$

where m_s is the magnetic spin quantum number of the atomic electron in the ground state. All one-valence-electron atoms, however, have a non-zero nuclear spin I which couples with the electron spin s to form the total angular momentum F. In a weak magnetic field the atom is described by the eigenstates $\langle F, m_F \rangle$ with $F = I±½$ and $m_F = -F, ..., +F$. As an example, the hyperfine structure (HFS) Zeeman levels of the ground state of lithium-6 are shown in Fig. 1, together with the P_a-values for each level.

Polarizing Methods

In order to polarize an atomic beam one has to create an unequal distribution over the HFS Zeeman states favoring high values of $\langle m_s \rangle$. Two methods are employed: high-field state separation in a hexapole magnet and optical pumping.

In high magnetic fields the HFS coupling is destroyed. Atoms in high-field eigenstates $\langle m_s, m_I \rangle$ with $m_s = +½$ are transmitted through a hexapole magnet on sinusoidal trajectories which can even be utilized for some focusing of the atomic beam. Atoms in states with $m_s = -½$ [which are the atoms with ($F=I-½$, all m_F) and ($F=I+½$, $m_F=-F$)] are deflected away from the axis on hyperbolic trajectories and disappear inside the hexapole magnet. In the fringe field at the magnet exit the atoms turn their spins adiabatically into the direction of the external field. In a strong magnetic field the polarization of a perfectly state-selected atomic beam is unity. In a weak external field, as employed in most electron scattering experiments, the nuclear spin couples in again and the electronic polarization is reduced accordingly. The low-field polarization of a high-field state-selected atomic beam is

$$P_{atom} \leq \frac{1}{2I + 1}$$

with a maximum value of 0.5 for ^1H ($I = ½$) but only 0.125 for ^{133}Cs ($I = 7/2$).

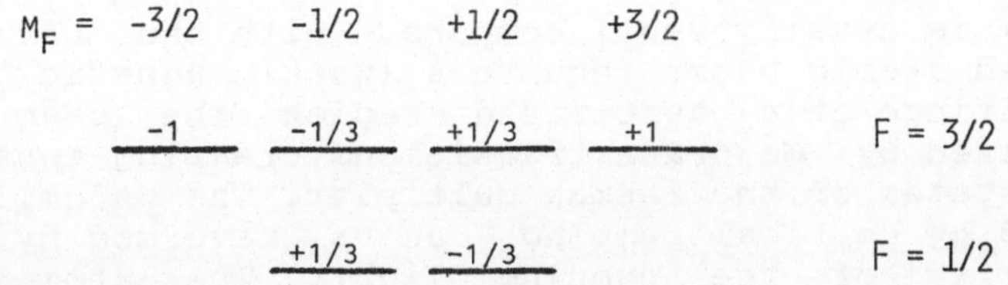

$$M_F = \quad -3/2 \quad -1/2 \quad +1/2 \quad +3/2$$

$$P_a = \quad \quad \underline{-1} \quad \underline{-1/3} \quad \underline{+1/3} \quad \underline{+1} \quad \quad F = 3/2$$

$$P_a = \quad \quad \quad \quad \underline{+1/3} \quad \underline{-1/3} \quad \quad \quad \quad F = 1/2$$

Fig. 1. Hyperfine-structure Zeeman states of a $^2S_{\frac{1}{2}}$ state with nuclear spin I=1 (lithium-6). Independent of the specific I-value the following can be stated: In the $F=I+\frac{1}{2}$ Zeeman multiplet P_a varies from −1 (for the state $m_F=-F$) to +1 (for $m_F=+F$) with equal increments between adjacent states. In the $F=I-\frac{1}{2}$ multiplet the states have the same $|P_a|$ as the state directly above <u>but with opposite sign</u>. State selection in a hexapole magnet leads to an atomic beam which contains only atoms in the states $F=I+\frac{1}{2}$ from $m_F=-F+1$ to $m_F=+F$ (but not $m_F=-F$), the other atoms are discarded. Optical pumping in its most efficient form causes all atoms to go into the state $F=I+\frac{1}{2}$, $m_F=+F$ or $-F$.

Optical pumping of atomic beams has become feasible with tunable lasers. The light intensity usually suffices for many photon absorptions per atom, which leads to a significant enhancement of the states with maximum m_F-values. Complications arise due to the fact that the HFS splitting of the ground state is much larger than the laser line width. If only one HFS state is optically pumped, then the atoms in the other state are not pumped at all. Furthermore, the ones which are pumped can fall back into the other HFS state and be excluded from further pumping. Cures for this problem are the following:

(1) Pumping the $F=I+\frac{1}{2}$ ground state with light which allows only absorption transitions to an excited state with $F'=F+1$. Then the spontaneous emission which obeys $\Delta F=\pm1$ can only go back to $F=I+\frac{1}{2}$ and the maximum polarization achievable in this way is $(I+1)/(2I+1)$.

(2) Combining cure (1) with a prior high-field state selection can lead to a polarization up to unity.[2]

(3) Pumping both HFS ground states with two different wavelengths will eventually empty the Zeeman states of $F=I-\frac{1}{2}$ and preferentially populate the state $\langle F=I+\frac{1}{2}, m_F=+F \rangle$. If the two wavelengths are produced with one laser and an electro-acoustical coupler,[3] the achievement of high polarization is impaired by coherence population trapping.[4] Better is the employment of two different sources, e.g., two GaAs laser diodes for the pumping of cesium atoms.

Polarization Reversal

The atoms of effusive or jet beams move with velocities of less than 1 km/s and go adiabatically through most laboratory magnetic fields because the angular velocity with which the direction of the magnetic field changes in the rest frame of the atom is usually small compared with the Larmor frequency. Polarized atomic beams require a (weak) magnetic guiding field. In the fringe of a zero-field region the beam would become depolarized by Majorana transitions creating equal populations of all states of the Zeeman multiplet. The atomic polarization produced by optical pumping can be reversed by switching the circularity of the pumping light. The atomic polarization produced with a hexapole magnet, however, is always positive; a reversal of the guiding field would also change the adiabatic spin motion in the fringe field of the magnet exit. For such beams a "spin flipper" is essential (Fig. 2).

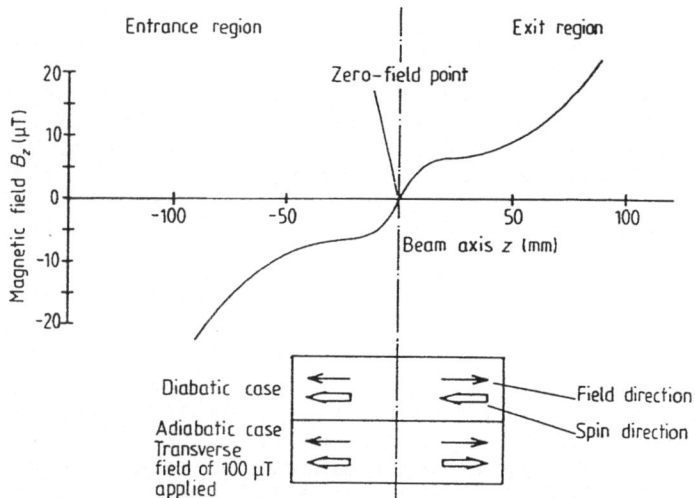

Fig. 2.   Magnetic field along the axis of a spin flipper for polarized atoms. The arrows symbolize the directions of the field and the spins in entrance and exit region for the two modes of operation. In the adiabatic mode a transverse field is applied in the center which facilitates spin rotation through 180°. (From ref. 5)

Measurement of the Atomic Polarization

Measuring the atomic polarization is important but not trivial. For high-field state-selected beams this problem has been solved by the following approaches:
(1) A nearly perfect state selection is achieved, e.g., by inserting a center stop which blocks the low-field-strength region of the magnet. Then the theoretical value of $P_a = (2I-1)^{-1}$ is taken for granted.
(2) Atomic trajectory calculations through the hexapole magnet are used to derive how much the achieved atomic polarization differs from the theoretical value.
(3) It is a reasonabe approximation to assume that all the states with high-field quantum number $m_s = +\frac{1}{2}$ have equal high population probabilities and those with $m_s = -\frac{1}{2}$ have equal low population probabilities. This allows the determination of P_a by means of a spin flipper and an analyzing hexapole magnet. (The reversal efficiency of the spin flipper can be determined by using two such devices.)

For atomic beams polarized by optical pumping the measurement of a <u>partial</u> polarization would be rather difficult as the
HFS Zeeman states have, in general, unequal populations. Fortunately, laser pumping is sufficiently effective to achieve
polarizations very close to $P_a=\pm 1$. This result can be verified
by optical absorption measurements with a weak "probe" laser
beam. A measurement of absorption by the $F=I-\frac{1}{2}$ states can prove
that these are nearly empty. Measurement of the absorption of
left and right circular polarized light by the $F=I+\frac{1}{2}$ states can
give an upper limit for $1-|P_a|$.[6]

Atomic-beam polarimeters are useful for measuring the
primary-beam polarization. Polarization measurements on scattered atoms are more difficult not just for intensity reasons
but because of the small scattering angles of the recoil atoms.
The atomic-beam recoil technique development by Bederson and
collaborators for measuring scattered-atom polarization will be
utilized by Jaduszliwer.[7]

Polarized-Beam Quality

Polarized atomic beams are comparable in flux and density
to unpolarized beams. The density decreases with increasing
distance between source and scattering region. Hexapole
magnets, which take up some space, provide, however, advantageous beam focusing. Optical pumping with a laser can be
accomplished in a short illumination region.

Atomic Beams with Spin 1

Aside from one-electron atoms, thus far only beams of
metastable $He(2^3S_1)$ have been polarized. A hexapole magnet with
center stop yields a beam of atoms in the state $\langle S=1, m_s=+1\rangle$
which is almost free of atoms in the other Zeeman states $m_s=0,-
1$ as well as free of $He(2^1S_0)$ and $He(1^1S_0)$ atoms.[8]

POLARIZED ELECTRON BEAMS

Definition of Polarization

For <u>atomic</u> beams it was sufficient to define P_a with
respect to the axis of quantization as $\langle m_s\rangle/(\frac{1}{2})$, since the
atomic polarization is produced in this way and atoms of
effusive or jet beams move slowly enough for adiabatic transitions into new field directions (exception: the special
conditions in a "spin flipper"). Although it is possible, in
principle, to produce an atomic beam with a polarization
orthogonal to the magnetic guiding field as a coherent superposition of the eigenstates with $m_s = +\frac{1}{2}$ and $-\frac{1}{2}$, it has so far
had no significance in scattering experiments.

For <u>electron</u> beams, however, this simplification cannot be
made. The electrons travel with velocities larger than 1000
km/s and usually move diabatically through regions of changing
magnetic field. It is, therefore, more useful to describe the

electron-beam polarization by a vector P_e which defines the spin orientation in space, regardless of any existing magnetic field. Obviously, in a scattering experiment involving polarized atoms P_e has to be turned parallel or anti-parallel to the magnetic field which guides the atom polarization. Quantum-mechanically P_e is properly defined as the trace of the product of the density matrix and the Pauli spin-matrix vector, but for understanding the behavior of electron polarization in an experiment it is most often sufficient to visualize the polarization of an ensemble of N electrons classically as

$$P_e = \frac{1}{N} \sum_{i=1}^{N} s_i / (\tfrac{1}{2}),$$

where s_i is the spin of the ith electron. In principle, one can measure all three components of P_e by performing three spin-sensitive measurements with their axis of quantization in x, y and z direction, respectively: $P_{e,i} = \langle m_s \rangle_i / (\tfrac{1}{2})$, i = x,y,z. For low-energy experiments it is irrelevant that P_e is not a co-variant quantity (relativistically P_e is only defined in the rest frame of the beam electrons).

Change of Polarization Direction

The direction of P_e is either specified in space (x,y,z) or in reference to the direction of propagation (transverse, with statement of the azimuth angle, or longitudinal).

Non-relativistically, electric fields do not affect the direction of P_e in space. Therefore, a 90° electrostatic beam deflector is a common device for transforming longitudinal into transverse polarization or vice versa.

A longitudinal magnetic field changes the azimuth angle of a transverse polarization with the Larmor frequency. A transverse magnetic field turns the beam direction with the cyclotron frequency. If P_e is orthogonal to the field, then for a non-relativistic beam the velocity and the polarization are turned in the same way as Larmor and cyclotron frequencies are equal, except for the very small g-factor anomaly.

For a rotation of P_e, without changing the beam direction in space, a Wien filter is employed. This has a magnetic field for turning P_e around the field direction (usually through 90°, a quarter of a Larmor cycle) and an electric field which provides a Coulomb force for compensation of the Lorentz force.

Production of Polarized Electron Beams

Nowadays the one and only satisfactory polarized-electron source is based on photoemission from cesiated and oxidized GaAs or GaAsP using circularly polarized light. With such a source the electron polarization can easily be reversed by changing the circularity of the light. The source technique is well described in the literature.[9] Potential users should know the following facts:

- The photocathode must be operated in ultra-high vacuum, of about 10^{-10} Torr. The crystal has to be finally cleaned by heating close to the melting point (unless one can cleave the crystal inside the UHV system).
- During the surface treatment with cesium and oxygen the photoemission has to be monitored.
- The emitted electrons are longitudinally polarized. A 90° electrostatic deflector (with a straight-through hole for the light) turns the electron beam away from the light and makes P_e transverse.
- For good photocathodes P_e lies in the range of 0.3 to 0.45; it must be measured and monitored!
- A Kr^+ laser is adequate for GaAs as well as GaAsP photocathodes. A He-Ne laser is only compatible with GaAsP.

Measurement of Electron Polarization

Several electron polarimeters described in the literature are only suitable for relative measurements; they can be used for monitoring a polarized electron source but for absolute measurements they must be calibrated against an "absolute" polarimeter and care has to be taken in order to keep the operation conditions sufficiently constant between calibration and measurement. Such relative polarimeters are the "absorbed current" devices based on spin-dependent secondary electron emission and consisting either of a tungsten crystal[10] or a polycrystalline gold surface[11] with non-orthogonal beam incidence (utilizing spin-orbit interaction) or a ferromagnetic conducting glass[12] with normal beam incidence (utilizing spin exchange). Other relative polarimeters are low-energy versions of the the high-energy Mott polarimeter employing reflective scattering from solid (or a thick layer of evaporated) gold.[13]

There are only two ways of determining the electron polarization absolutely, the helium polarimeter (utilizing electron-photon polarization transfer) proposed by Gay[14] and the old-fashioned high-energy Mott scattering. The former is being developed in at least three laboratories (Münster, Mainz, and Rolla/Missouri), the latter is receiving considerable attention since it was pointed out that all electron-polarization measurements performed thus far have a systematic uncertainty of $\Delta P/P = \pm 5\%$.[15] For wide-angle single scattering of high-energy electrons on a high-Z atom [(usually $\Theta = 120°$, E = 100-120 keV, Z = 79 (gold)] the Mott scattering can be reliably calculated using relativistic Hartree-Fock potentials (Fig. 3). The most serious technical problem is the elimination of multiple scattering by means of foil-thickness extrapolation. Stringent energy discrimination of the scattered electrons is essential. The approach of discriminating against electrons which suffer energy losses of, say, more than 100 eV and trusting the most refined calculations on Mott scattering might soon lead to an accurate absolute polarimeter. The execution of a genuine Mott double-scattering experiment, utilizing improved technology, is under way in Münster.[16]

A Mott polarimeter exhibits a left-right intensity asymmetry for incident electrons polarized orthogonal to the scattering plane. Two intensity measurements made sequentially with one scattered-electron detector for P_e "up" and "down" give the ratio

$$I\uparrow / I\downarrow = (1 + \delta)/(1 - \delta)$$

where $\delta = P_e \cdot S$ and S is the so-called "Sherman function", that is, the analyzing power of the polarimeter. By using two detectors at opposite azimuth angles the measurement becomes insensitive to intensity fluctuations. By using four detectors, 90° apart in azimuth, both transverse polarization components, P_x and P_y, can be determined simultaneously.

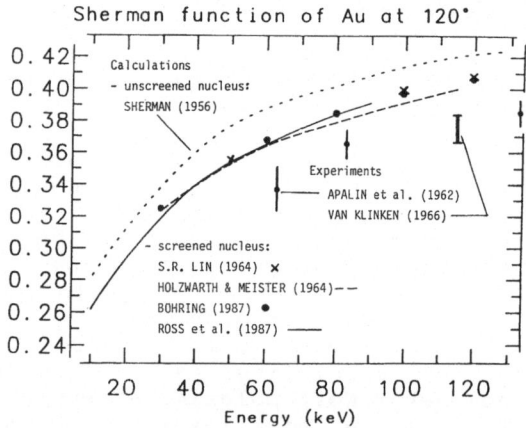

Fig. 3. Theoretical and experimental values for the analyzing power of 120° Mott scattering as a function of the electron energy. These values apply to infinitely thin gold foils. (Courtesy of M. Fink, Data from ref. 17)

Mott polarimeters are most conveniently used for electron beams of 10^{-9} to 10^{-11}A because such intensity leads to acceptable counting rates for reasonable foil thicknesses and detector solid angles. In order to measure the polarization of smaller electron currents one has to go to long counting times, thicker foils (with problems of calibration) and large detector solid angles (with problems of averaging the Sherman-function value). Measuring the polarization of a typical primary beam of 10^{-6}A creates other problems: Either the beam must be artificially weakend until the counting rates become tolerable or scattered-electron currents must be measured instead of counting rates.

Polarized-Beam Quality

Polarized electron beams from GaAs(P) sources are equal to unpolarized electron beams from thermionic sources in current and brightness, - both are usually space-charge limited. The inherent energy width of these polarized beams is about 100-200 meV, which is considerably smaller than that of thermionic beams.

ELASTIC SCATTERING

Exchange in Scattering from One-Electron Atoms

We consider the exchange between the beam electron (also referred to as the "free" or "continuum" electron) and the bound atomic electron (the "valence" electron). It is customary to refer to the spin orientation or the polarization "of the atom" while meaning only that of the atomic electron and disregarding the nuclear spin.

In the absence of spin-orbit coupling the total spin is conserved. The spin orientations of the two electrons are described in reference to an axis of quantization which is usually given by a weak magnetic field. The orientation of this axis with respect to the scattering plane is irrelevant.

The scattering is described by two amplitudes, f for "direct" and g for "exchange" scattering. The initial and final spin states can be symbolized by two arrows, the first one referring to the free electron and the second to the valence electron of the atom, with "up" and "down" corresponding to $m_{1,2} = +\frac{1}{2}$ and $-\frac{1}{2}$, respectively. For parallel spins it is impossible to distinguish whether exchange occurred or not. Therefore both amplitudes interfere (Table 1). The minus sign is required in order to get an asymmetric form of the spatial

Table 1. Scattering matrix for the elastic electron scattering from a one-valence-electron atom without spin-orbit interaction.

$m'_{1,2}$ \ $m_{1,2}$	↑ ↑	↓ ↑	↑ ↓	↓ ↓
↑ ↑	f−g	0	0	0
↓ ↑	0	f	g	0
↑ ↓	0	g	f	0
↓ ↓	0	0	0	f−g

wave function (to go with the symmetric spin states). The states ($\uparrow\uparrow$) and ($\downarrow\downarrow$) are triplet states of the two-electron scattering complex with total spin S=1 and magnetic quantum numbers m_s=+1 and -1, respectively. The states ($\downarrow\uparrow$) and ($\uparrow\downarrow$) are superpositions of the triplet state S=1, M=0 and the singlet state S=0, M=0. For initially parallel spins the cross section is given by

$$\sigma^{\uparrow\uparrow} = |f-g|^2$$

and for initially antiparallel spins by

$$\sigma^{\downarrow\uparrow} = |f|^2 + |g|^2$$
$$= \tfrac{1}{2}|f-g|^2 + \tfrac{1}{2}|f+g|^2.$$

With reference to the total spin states the term $|f-g|^2$ is called the "triplet", $|f+g|^2$ the singlet cross section. A survey of the different cross-section terminologies is given in Table 2.

In a crossed-beam experiment on electron-atom scattering the beam particles usually do not have perfect spin orientation. A partially polarized beam of intensity I with the degree of polarization P can be viewed as the incoherent superposition of a completely polarized beam with the intensity I·P and an unpolarized beam the the intensity I(1-P). We use the indices "e" and "a" to symbolize the polarization of electron and atomic beam, respectively.

In order to determine the partial cross sections for the scattering of partially polarized particles from the collision matrix it is useful first to consider completely polarized particles and to correct for the partial polarization later. It is customary to use the symbols $\sigma^{\downarrow\uparrow}$ and $\sigma^{\uparrow\uparrow}$ not only for anti-parallel and parallel spins, respectively, but also for antiparallel and parallel polarization vectors of the crossed beams, the later is obviously meant if the formula contains P_e and P_a.

Table 2. Different expressions for the partial cross sections

| Spin states | | Partial cross sections | | |
initial	final	direct/ exchange	parallel/ antiparallel	triplet/ singlet
$\uparrow$ $\uparrow$	$\uparrow$ $\uparrow$	$|f-g|^2$	$\sigma^{\uparrow\uparrow}$	σ_t
$\downarrow$ $\uparrow$	$\downarrow$ $\uparrow$	$|f|^2$	$\sigma^{\downarrow\uparrow}$	$\tfrac{1}{2}\sigma_t + \tfrac{1}{2}\sigma_s$
	$\uparrow$ $\downarrow$	$|g|^2$		

If either one or both beams are unpolarized, the resulting cross section is the average of the "up" and "down" partial cross sections:

$$\bar{\sigma} = \tfrac{1}{2}\sigma^{\uparrow\uparrow} + \tfrac{1}{2}\sigma^{\downarrow\uparrow}$$

$$= (3/4)\sigma_t + (1/4)\sigma_s$$

$$= (3/4)|f-g|^2 + (1/4)|f+g|^2$$

$$= \tfrac{1}{2}[|f|^2 + |g|^2 + |f-g|^2]$$

$$= |f|^2 + |g|^2 - \mathrm{Re}(f^*g)$$

The last form of writing $\bar{\sigma}$ suggests the partitioning in a "direct", an "exchange" and an "interference" cross section as follows:

$$\bar{\sigma} = \sigma_{dir} + \sigma_{ex} - \sigma_{int}.$$

$$\sigma_{dir} = |f|^2$$

$$\sigma_{ex} = |g|^2$$

$$\sigma_{int} = |f||g|\cos\Theta$$

Note that the "interference cross section" can become negative; "interference term" would be a more appropriate name for it. The angle Θ is the phase difference of direct and exchange scattering amplitude.

The simplest crossed-beam polarization experiments on exchange are those in which only two devices for production and/or measurement of polarization are employed. This class of experiments is schematically shown in Fig. 4. Type (1) can be viewed as an experiment in which a polarized electron beam is analyzed by using the scattering from a polarized atomic beam, or vice versa. Type (2) yields the trivial result of $P_e = P_a = 0$. Type (3) and (4) are experiments in which a polarized beam as the target acts as a polarizer for the other initially unpolarized beam. In experiments of type (5) and (6) the unpolarized target acts as a depolarizer for the other initially polarized beam. The asymmetries which can be determined in these experiments are A, A', and A":

$$A_{(1)} = \frac{1}{P_e P_a} \; \frac{\sigma^{\downarrow\uparrow} - \sigma^{\uparrow\uparrow}}{\sigma^{\downarrow\uparrow} + \sigma^{\uparrow\uparrow}}$$

$$A_{(3)}' = P_e'/P_e \quad \bigg| \quad A_{(4)}' = P_a'/P_a$$
$$A_{(5)}'' = P_a'/P_e \quad \bigg| \quad A_{(6)}'' = P_e'/P_a$$

With the help of the collision matrix it can be shown that

$$A = \sigma_{int}/\bar{\sigma},$$

$$A' = 1 - \sigma_{ex}/\bar{\sigma},$$

$$A'' = 1 - \sigma_{dir}/\bar{\sigma},$$

and

$$A + A' + A'' = 1.$$

Therefore, the measurements of any two of those asymmetries, in combination with a measurement of $\bar{\sigma}$ will give all three partial cross sections σ_{dir}, σ_{ex} and σ_{int}. In order to determine the sign of Θ, a more elaborate polarization experiment has to be performed.[18]

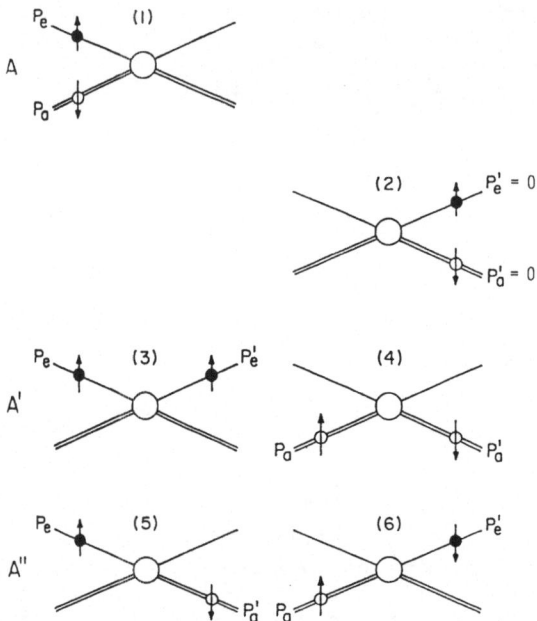

Fig. 4. Typology of the simplest polarization experiments on
         exchange in electron-atom scattering. Single lines
         represent incoming and outgoing electron beams, double
         lines, atomic beams. Black-dot arrows on left and
         right side symbolize electron polarizers and polari-
         meters, respectively, and open-circle arrows, atomic
         polarizers and polarimeters. At least two polarization
         devices are required. Present technologies favor
         experiments of type 1 in which the polarization
         asymmetry A is measured. Type 2 is not useful. Types 3
         and 4 permit the determination of A', types 5 and 6 of
         A". The three asymmetries are related, only two have
         to be measured. (From ref. 18, corrected)

     Experiments of type 1 are most feasible, thanks to the
efficient ways of producing polarized electrons and atoms. Such
an experiment alone (without measuring $\bar{\sigma}$ and one of the other
asymmetries) suffices to determine the ratio

$$r = \sigma_t / \sigma_s = \frac{1 - A}{1 + 3A}$$

which allows meaningful interpretations.

Since $0 \leq r \leq \infty$,

$$A = \frac{1 - r}{1 + 3r}$$

can vary between unity ($r = 0$, pure singlet) and $-1/3$ ($r = \infty$,
pure triplet).

     So far, there have been three different measurements of A

for selected scattering angles or energies and different atoms: hydrogen at Yale (Fig. 5), lithium at Bielefeld (Figs. 6,7), and sodium at NBS (shown later in Fig. 10). The agreement with elaborate scattering theories is satisfactory, particularly with close-coupling approximations. The polarized-orbital approximation, on the other hand, seems unable to describe the exchange effects properly.

Spin-Orbit Interaction

In elastic scattering from an unpolarized atom, only spin-orbit coupling can lead to polarization effects. The Mott scattering can be derived from Dirac theory and applies also to low-energy scattering, provided that the very important screening of the Coulomb field by the atomic electrons is properly taken into account.

Mott scattering (as for any spin-orbit interaction) is only sensitive to polarization orthogonal to the plane of scattering. It can be described by two scattering amplitudes f and g, called "direct" and "spin-flip". The basic formulas are

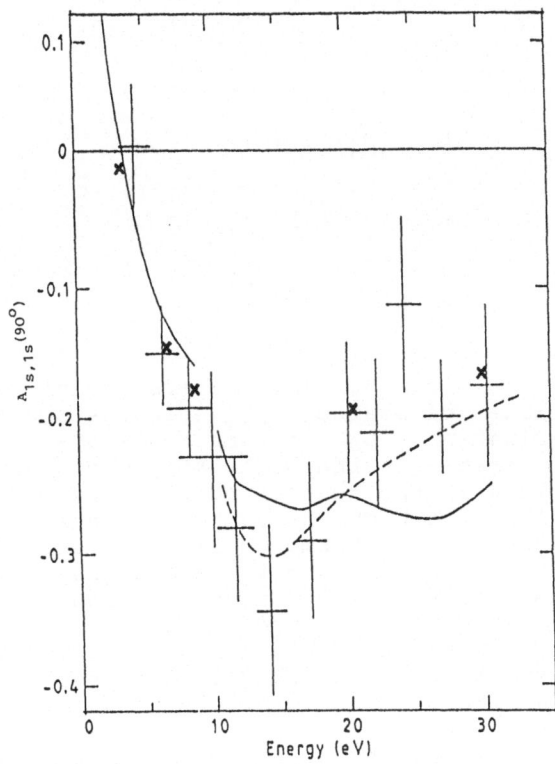

Fig. 5.   Polarization asymmetry A for the elastic 90° scattering on atomic hydrogen. Error bars - experimental results of the Yale group; crosses, solid and dashed curves - recent theoretical results. (From ref. 19)

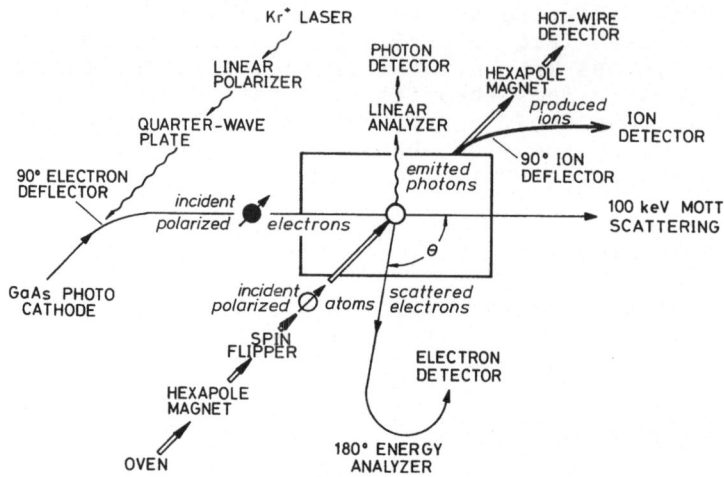

Fig. 6.   Layout of the Bielefeld crossed-beam experiment with
polarized electrons and polarized atoms, utilized for
studying differential elastic and inelastic scattering
(detection of the scattered, energy-analyzed electrons), integral inelastic scattering (detection of
the fluorescence photons), and ionization (detection
of the ions).

$$\bar{\sigma} = |f|^2 + |g|^2,$$
$$S = -2|f||g| \sin(\gamma_1-\gamma_2)/\bar{\sigma},$$
$$T = (|f|^2 - |g|^2)/\bar{\sigma},$$
$$U = 2|f||g| \cos(\gamma_1 - \gamma_2) /\bar{\sigma},$$

and
$$S^2 + T^2 + U^2 = 1.$$

If the initial beam has a transverse polarization P which
lies in the plane of scattering, then the scattered beam has
aquired a polarization P' whose components are given by S (for
the direction normal to the scattering plane) and T·P and U·P
for the directions in the scattering plane as shown in Fig. 8.
By measuring S, T and U and by using $\bar{\sigma}$-values from the literature, the Münster group determined all parameters of the theory
(Fig. 9). For low-Z atoms and for scattering angles away from
minima of the differential cross section the spin-orbit
interaction of the free electron in the field of the screened
nucleus is too weak for producing observable polarization
effects. An exception is the elastic resonance scattering, for
example, on the low-Z atom neon.   In this case the scattered
electron is temporarily held in a Ne$^-$ compound P-state with
fine-structure (FS) splitting. If the small energy width of the
electron beam allows the $P_{1/2}/P_{3/2}$ FS-doublet to be resolved,
then polarization effects can be observed.[22]

442

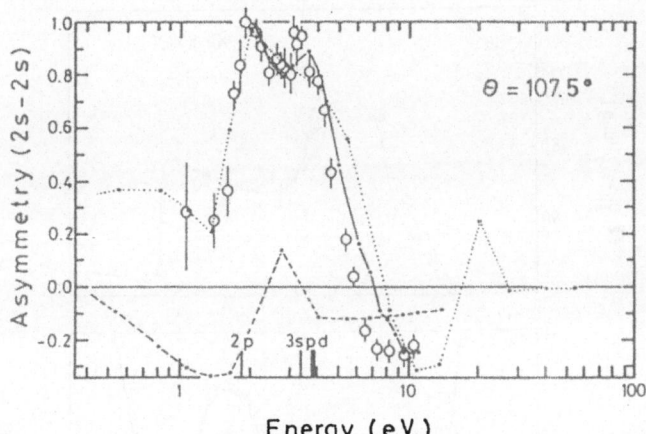

Fig. 7. Polarization asymmetry A measured for the elastic
107.5° scattering of polarized electrons from polarized lithium atoms. Open circles - experimental
results of the Bielefeld group; dashed curve - modified polarized-orbital approximation; dotted curve -
two-state close-coupling approximation; solid curve -
recent five-state close-coupling results of Moores.
(From ref. 20)

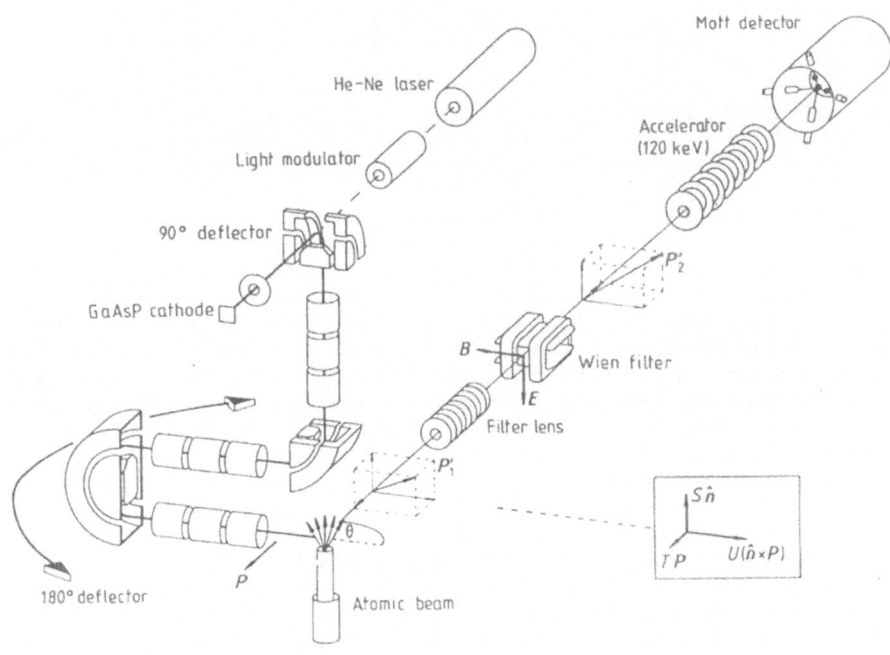

Fig. 8. Layout of the "complete" experiment on elastic electron scattering from Hg and Xe performed at Münster.
(From ref. 21)

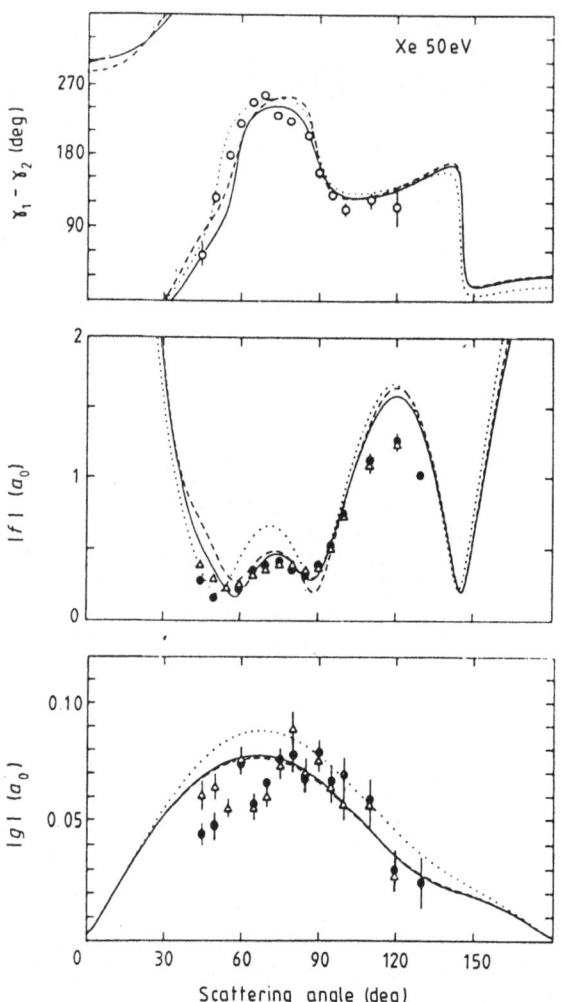

Fig. 9. The three parameters of Mott scattering (phase difference and moduli of direct and spin-flip amplitude)
from theory (dotted, dashed and solid curves) and
experiment (circles, dots and triangles). The "complete" polarization experiment of Fig. 8 yields the
phase difference and the cross section ratios $|f|^2/\bar{\sigma}$
and $|g|^2/\bar{\sigma}$. For evaluating $|f|$ and $|g|$ literature
values of the differential cross section $\bar{\sigma}$ were used.
(From ref. 21)

Both, exchange and spin-orbit interaction

If both, P_a and P_e are orthogonal to the scattering plane,
exchange and spin-orbit interaction can both be observed in an
experiment of type 1 (Fig. 1). By taking appropriate averages
of detector counting rates I, it is possible to observe the
effects of each interaction separately.[23] In the following, the
first and second arrows refers to P_e and P_a, respectively.

The exchange asymmetry (called A above)

$$A^{exch} = \frac{\sigma^{\downarrow\uparrow} - \sigma^{\uparrow\uparrow}}{\sigma^{\downarrow\uparrow} + \sigma^{\uparrow\uparrow}}$$

can be obtained from

$$A^{exch} = \frac{1}{P_e \; P_a} \frac{(I^{\downarrow\uparrow} + I^{\uparrow\downarrow}) - (I^{\uparrow\uparrow} + I^{\downarrow\downarrow})}{(I^{\downarrow\uparrow} + I^{\uparrow\downarrow}) + (I^{\uparrow\uparrow} + I^{\downarrow\downarrow})} \; ,$$

in which the Mott asymmetry cancels because the averages taken correspond to unpolarized incident electrons. The spin-orbit asymmetry

$$A^{s-o} = \frac{\sigma^{\uparrow} - \sigma^{\downarrow}}{\sigma^{\uparrow} - \sigma^{\downarrow}}$$

follows from

$$A^{s-o} = \frac{1}{P_e} \frac{(I^{\uparrow\uparrow} + I^{\uparrow\downarrow}) - (I^{\downarrow\uparrow} + I^{\downarrow\downarrow})}{(I^{\uparrow\uparrow} + I^{\uparrow\downarrow}) + (I^{\downarrow\uparrow} + I^{\downarrow\downarrow})} \; .$$

Here $\sigma^{\uparrow}$ and $\sigma^{\downarrow}$ refer to electron spins "up" and "down" the scattering plane ("up" means in the direction of $\mathbf{k} \times \mathbf{k'}$ where $\mathbf{k}$ and $\mathbf{k'}$ are the electron momenta before and after the collision) and the signal averages correspond to unpolarized atoms and, therefore, cancelled exchange asymmetry. Both asymmetries were measured by the NBS group (Fig. 10).

But there is also the possibility of interference of both effects, to which Farago drew attention.[24] If unpolarized electrons are scattered from polarized atoms neither exchange nor spin-orbit action alone would lead to an intensity asymmetry to be observed with a scattered-electron detector at some scattering angle Θ. In the terminology introduced above this asymmetry could be observed in a type 1 experiment by combining the intensity measurements as follows:

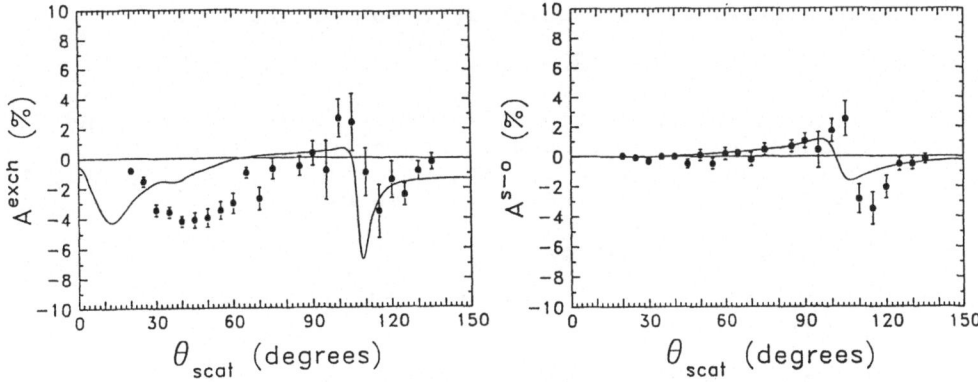

Fig. 10. Exchange and spin-orbit asymmetry for the scattering of 54.4 eV electrons from sodium atoms. Dots – experimental results of the NBS group, curves – theoretical estimates. (From Ref. 23)

$$A^{both} = \frac{1}{P_a} \frac{(I^{\uparrow\downarrow} + I^{\downarrow\downarrow}) - (I^{\uparrow\uparrow} + I^{\downarrow\uparrow})}{(I^{\uparrow\downarrow} + I^{\downarrow\downarrow}) + (I^{\uparrow\uparrow} + I^{\downarrow\uparrow})}$$

In the case of cesium, for which spin-orbit interaction is expected to be much stronger than for sodium, and for the much lower energy of only a few eV, for which exchange is expected to be stronger than at the 54.4 eV used in the Na measurements, Walker[25] estimated that this interference should be observable near the minimum of the differential cross section at about 90°. Qualitatively, this interference effect can be understood as due to electrons which first become polarized by spin-orbit interaction (Mott scattering) and then produce an intensity asymmetry by exchange, or vice versa. Other imaginable mechanisms, such as the interaction of the spin of one electron with the orbit of the other, were also estimated by Walker[26] and were all found to be negligible at low energies. The sum and the difference of A^{s-o} and A^{both} give asymmetries which can be related to spin-conserving and spin-nonconserving processes.[27]

Burke and Mitchell[28] showed that, in general, six complex amplitudes are required to describe the electron scattering from a one-electron atom with both exchange and spin-orbit interaction taken into account. However, it is not yet known how many of these amplitudes are significantly different from zero at low energies and how many polarization experiments will be worthwhile.

EXCITATION

Selection of Initial Spin States

The excitation of a one-electron atom from its S ground state to a P state is governed by four amplitudes: f_0, g_0, f_1 and g_1 where f and g mean again "direct" and "exchange" and the indices refer to the associated change in magnetic angular-momentum quantum number $\Delta M_L = 0$ and $\Delta M_L = \pm 1$, respectively.

Each partial cross section is a sum of the form

$$\bar{\sigma} = \sigma_0 + 2\sigma_1.$$

The excitation cross section [which we write $\sigma(S,P)$, and not with an index "exc" which could be confused with "exchange"] for parallel and antiparallel initial spins is (analogously to the elastic-scattering formalism) given by:

$$\sigma(S,P)^{\uparrow\uparrow} = |f_0 - g_0|^2 + 2|f_1 - g_1|^2$$
$$\sigma(S,P)^{\downarrow\uparrow} = |f_0|^2 + |g_0|^2 + 2|f_1|^2 + 2|g_1|^2$$

And the excitation asymmetry follows accordingly as

$$A(S,P) = \frac{\sigma_{int}}{\bar{\sigma}} \quad \text{with } \sigma_{int} = \text{Re}(f_0{}^*g_0) + 2\,\text{Re}(f_1{}^*g_1)$$

or $\quad A(S,P) = \frac{1-r}{1+3r} \quad$ with

$$r = \frac{|f_0 - g_0|^2 + 2|f_1 - g_1|^2}{|f_0 + g_0|^2 + 2|f_1 + g_1|^2} \ .$$

Such measurements for lithium are shown in Fig. 11. Here only those electrons were detected which had suffered the corresponding energy loss in the scattering through the angle Θ. Again, the 5-state close-coupling approximation gives a rather satisfactory description of the experimental results.

By measuring the differential excitation at all angles, the integral excitation cross section and the integral polarization asymmetry could, in principle, be obtained. In practice, however, one studies the photon emission instead. Detecting the fluorescence photons automatically provides an integration over all electron scattering angles. But this method also has distinct disadvantages:

- For incident-electron energies above the next inelastic threshold the photon signal is affected by cascading from higher states.
- The spontaneous-emission transitions with $M_L = 0$ and $M_L = \pm 1$ have different angular distributions for the photons. There fore, a photon-intensity measurement at a certain angular position might not include these two types of transition with the same weight 1:2 as the excitation cross section.
- Electron-impact excitation is a fast process of about 10^{-15}s duration. The spontaneous photon emission occurs typically 10^{-8}s after excitation. This time delay is more than sufficient for establishing the coupling of spin and angular momentum of the excited atomic electron (fine-structure relaxation time $\approx 10^{-12}$s) and also the coupling of the electronic angular momentum and nuclear spin (hyperfine-

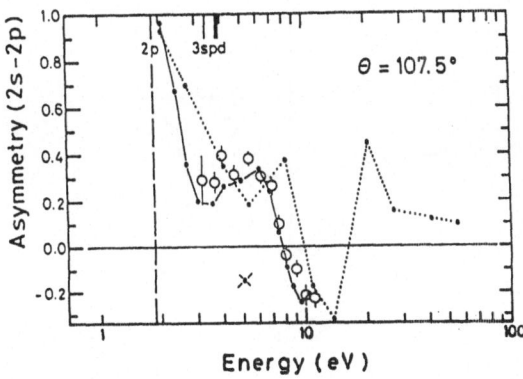

Fig. 11. Polarization asymmetry A measured for the inelastic 107.5° scattering with 2s-2p excitation of lithium atoms. Open circles – experimental results of the Bielefeld group; cross – unitarized distorted-wave approximation, dotted curve – two-state close-coupling approximation; solid curve – recent five-state close-coupling results of Moores. (From ref. 29)

structure relaxation time $\approx 10^{-10}$ s). The consequences of FS and HFS coupling can be calculated, but the fact remains that the photon emission is not in a simple way related to the excitation process and its spin dependence.

Measuring the Polarization Transfer to the Photon

If polarized electrons excite unpolarized atoms, some polarization is transferred to the excited state and, subsequently to the emitted photon. Here we consider experiments in which the scattered electrons are <u>not</u> observed.

The most obvious polarization transfer is that from a longitudinally polarized incident electron to a circularly polarized photon emitted in forward direction. For the alkali atoms Na, K, Rb and Cs this effect was measured by the Mainz group. For Cs, for example, $P_{circ}/P_e \approx 0.43$ was found near threshold.[30] Earlier attempts to utilize this phenomenon for an absolute electron polarimeter concentrated on Hg but failed because of a resonance 30 meV above the 6^3P_1 threshold which causes significant deviations from the theoretically predicted polarization transfer.[31] Helium, however, seems to be well suited. Gay[14] suggested the utilization of exchange excitation of the He 3^3P_3 states and measurement of the circular polarization of the unresolved 2^3S-3^3P multiplet at the convenient wavelength of 388.9 nm. The analyzing power $A = P_{circ}/P_e$ is predicted to be 0.49. In order to avoid cascade effects, however, it is necessary to stay within 0.59 eV of the 23.0 eV threshold of the 3^3P excitation.

A more general discussion of the observable photon-polarization parameters uses the coordinate system of Fig. 12. The polarization transfer discussed above is now called a measurement of the Stokes parameter

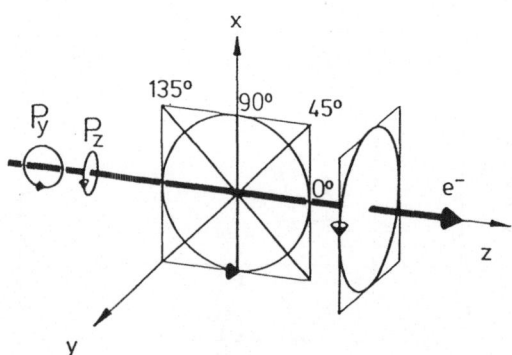

Fig. 12. Diagram for explaining the definitions of the Stokes parameters. P_y and P_z symbolize polarization components of the incoming electron beam. Measured is the circular polarization of the light emitted in the y and z directions and also the linear polarization of the light emitted in the y direction with E-fields in planes which form angles of 0, 45, 90 and 135° with the z axis. (From ref. 32)

$$\eta_{22} = \frac{I(+) - I(-)}{I(+) - I(-)} \qquad \alpha \ P_z$$

where $I(+)$ and $I(-)$ are the light intensities for positive and negative helicity, respectively. For light observation in y-direction three Stokes parameter are defined:

$$\eta_{1y} = \frac{I(45°) - I(135°)}{I(45°) + I(135°)} \qquad \alpha \ P_y$$

$$\eta_{2y} = \frac{I(+) - I(-)}{I(+) - I(-)} \qquad \alpha \ P_y$$

$$\eta_{3y} = \frac{I(0°) - I(90°)}{I(0°) + I(90°)} \ , \ \text{independent of } P_y.$$

The intensities I with degree arguments refer to light linearly polarized in those planes (cf. Fig. 12). Symmetry arguments show that η_{1z} and η_{3z} must vanish and that the above equations completely describe the dependence on the incident electron polarization. Theory also shows that $\eta_{1y} \neq 0$ only if there is spin-orbit interaction in the excitation process, whereas $\eta_{2y} \neq 0$ only if exchange effects play a role. Because the scattered electron is not observed in these measurements, these η-parameters are called integrated Stokes parameters. They are related to "integrated state multipoles" which characterize the excited atomic state.[33]

Coincidence Experiments

More detailed information is obtained if the scattered electron is detected in coincidence with the photon. Such polarization-coincidence experiments will be covered in Professor Kleinpoppen's lecture on correlations.

Photon detection is possible with very high resolution. It is much easier to resolve the fine-structure splitting opti-cally than to restrict the electron energy distribution sufficiently for selective excitation of only one multiplet component. Hanne showed that this fact can be utilized in polarization experiments.[34] The scattering of an electron, po-larized transverse to the scattering plane, from an unpolarized atom (say, in an $S_{\frac{1}{2}}$ ground state) through an experimentally defined angle Θ leads to a polarization of the orbital angular momentum L transferred to the excited atoms (Fig. 13). If only one of the fine-structure states is observed, say, $^2P_{\frac{1}{2}}$, one knows that the bound electron must have its spin preferentially oriented with $S \downarrow \uparrow L$. Thus, utilization of the "fine-structure" effect is equivalent (at least qualitatively) to electron spin-state selection in the final state. Therefore, polarization effects can be expected in the scattering of polarized elec-trons on unpolarized atoms, even if the spin-orbit interaction of the continuum electron is negligibly small.

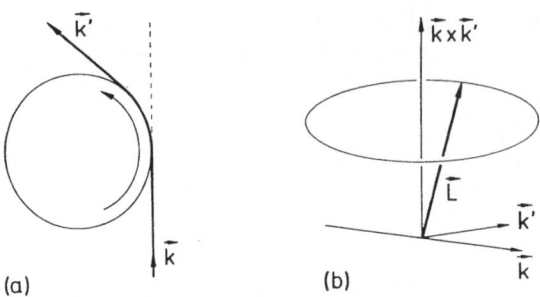

(a)　　　　　　　　　　　(b)

Fig. 13.   Semiclassical model of orientation by collision. (a)
Orientation in the scattering plane caused by a
grazing collision. (b) Orientation described by the
vector model, exemplified by a pure orbital angular
momentum state with a nonvanishing component of
orbital angular momentum normal to the scattering
plane. (From ref. 34)

SUPERELASTIC COLLISIONS

Such experiments are the time-reversal equivalents of the
electron-photon coincidence experiments discussed above:

(a) electron-photon coincidence

$$e^{\uparrow}(E,\Theta=0) + A \longrightarrow A*^{\uparrow} + e(E-\Delta E,\Theta)$$
$$\phantom{e^{\uparrow}(E,\Theta=0) + A \longrightarrow A*} \hookrightarrow A + \gamma^{\uparrow}$$

(b) superelastic collision

$$A + \gamma^{\uparrow}$$
$$e^{\uparrow}(E,\Theta=0) + \quad A*^{\uparrow} \longrightarrow A + e(E+\Delta E,\Theta)$$

The excited state of A* is experimentally defined either (a) by
observing a polarized photon of well-defined wavelength in
coincidence with the scattered electron or (b) by optical
excitation with laser light of well-defined wavelength and
polarization. With incident polarized electrons (a) is an
electron-photon polarization transfer process measured in differential inelastic scattering, (b) is a polarization-asymmetry
measurement for the differential superelastic scattering on a
polarized excited atom. Technically, the coincidence experiment
is difficult for intensity reasons: Only those events which
lead to photon emission into the small solid angle of the
polarization-sensitive photon detector are recorded. The
superelastic experiment is "wasteful" in the sense that many
more atoms have to be excited than actually serve in a collision. An optical photon-energy selection makes energy analysis
of the coincident electron superfluous. But in (b) not all
electrons scattered through Θ undergo an inelastic collision
and, therefore, electron energy analysis is required. If the
process studied involves a wavelength for which sufficiently
powerful tunable lasers are available, then the superelastic

scattering experiment[35] is probably more feasible than the coincidence experiment.

In a recent experiment[36] the NBS group studied the super-elastic scattering process

$$e^\uparrow(E) \ + Na^\uparrow(3^2P_{3/2},F=3) \qquad Na(3^2S_{1/2}) + e(E + 2.1 \text{ eV}).$$

With incident electrons polarized orthogonal to the scattering plane four intensities of superelastically scattered electrons were measured; $I^{R\uparrow}$, $I^{R\downarrow}$, $I^{L\uparrow}$ and $I^{L\downarrow}$, where the arrows refer to incident-electron spin directions and R, L to the circularity of the laser light. As shown in a theoretical analysis,[37] these four quantities (whose absolute values have no significance, thus giving only three independent ratios) are best transformed into the following physically meaningful parameters: (1) The ratio of triplet-to-singlet excitation cross sections, r; (2) the angular momentum component, orthogonal to the scattering plane, which has been transferred to the atom in the collision, for singlet excitation; (3) same as (2) for triplet excitation. Fig. 14 shows the results.

IONIZATION

In order to point out the pertinent theoretical facts for electron-impact ionization of atomic hydrogen we follow Peterkop[38] and start with the (so-called "single-differential") cross section for ionization leading to two electrons of energy ε and $E-\varepsilon$, where $E = E_0 - E_I$ (E_0 = initial energy, E_I = ionization energy). Thus we integrate over all angles of the outgoing electrons but not yet over all the partitions of energy.

$$d\sigma(\varepsilon,E-\varepsilon)/d\varepsilon = \frac{k_1 \, k_2}{k_0} \iint [(1/4)\,|f(k_1,k_2)+g(k_1,k_2|^2$$

$$+ (3/4)\,|f(k_1,k_2)-g(k_1,k_2)|^2]\; dk_1 dk_2$$

where k_0, k_1 and k_2 are the momenta of incident, outgoing, low-energy and outgoing high-energy electron, respectively. Therefore,

$$\varepsilon = \tfrac{1}{2}k_1^2, \quad E = \varepsilon + \tfrac{1}{2}k_2^2.$$

Unlike elastic scattering, in ionization both the initially free and the initially bound electron go out as free indistinguishable electrons. Therefore the following identities hold:

$$g(k_1,k_2) = f(k_2,k_1)$$

$$d\sigma(\varepsilon,E-\varepsilon)/d\varepsilon = d\sigma(E-\varepsilon,\varepsilon)/d\varepsilon$$

Since $d\sigma(\varepsilon,E-\varepsilon)$ is the transition cross section into a state in which the energy of either one of the electrons is within the interval $d\varepsilon$, the total ionization cross section for unpolarized particles is obtained as

$$\sigma = \int_{\varepsilon=0}^{E/2} d\sigma(\varepsilon,E-\varepsilon) = \tfrac{1}{2} \int_{\varepsilon=0}^{E} d\sigma(\varepsilon,E-\varepsilon).$$

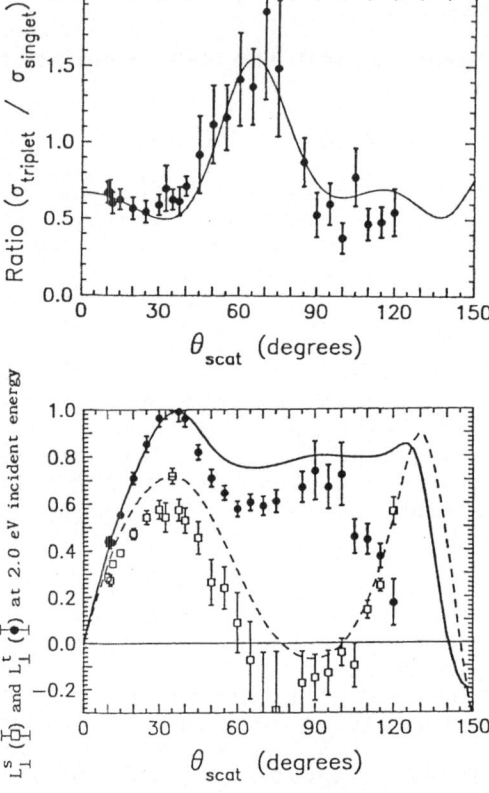

Fig. 14. The three physical parameters of a superlelastic collision which can be determined from relative cross-section measurements with polarized electrons and polarized excited atoms (here sodium atoms in the $3^3P_{3/2}(F=3)$ state). The data points are the experimental results of the NBS group, the curves are theoretical results of a four-state close-coupling approximation. (From ref. 36)

With the equations above, which allow the substitution

$$\int_0^{E/2} (|f|^2 + |g|^2)d\varepsilon = \int_0^{E/2} |f|^2d\varepsilon + \int_{E/2}^{E} |f|^2d\varepsilon = \int_0^{E} |f|^2d\varepsilon$$

the following result is obtained:

$$\sigma = \int_0^{E} \frac{k_1k_2}{k_0} d\varepsilon \quad |f(k_1,k_2)|^2 dk_1dk_2$$

$$- \int_0^{E/2} \frac{k_1k_2}{k_0} d\varepsilon \quad Re[f(k_1,k_2)f^*(k_2,k_1)]dk_1dk_2 .$$

452

The first term is the full-range cross section "without exchange", the second term describes the interference of direct and exchange amplitudes, which we can call σ_{int} in analogy to the elastic-scattering formalism.

The ionization asymmetry is given by

$$A_{ion} = \frac{\sigma_{ion}^{\downarrow\uparrow} - \sigma_{ion}^{\uparrow\uparrow}}{\sigma_{ion}^{\downarrow\uparrow} + \sigma_{ion}^{\uparrow\uparrow}} = \sigma_{int}/\sigma \; ,$$

with

σ_{int} = "full-range cross section <u>without exchange</u>"

     −   "σ, properly calculated <u>with</u> exchange".

Therefore, from ionization calculations "with" and "without" exchange one can obtain the polarization asymmetry. This fact allows the extraction of theoretical information about A_{ion} from older papers whose authors had not anticipated polarization experiments on ionization.

Fig. 15 shows the experimental results of A_{ion} for atomic hydrogen compared with various calculations. The Bielefeld results for Li, Na and K are given in Fig. 16. The light one-electron atoms H, Li and Na show a rather similar behavior of $A_{ion}(E)$ starting at threshold near 0.5 and decreasing monotonically with increasing energy, approaching zero at above 20 times the ionization energy. The abnormal form of $A_{ion}(E)$ for potassium is not yet explained. An experiment with cesium is underway at Bielefeld.

The threshold behavior of $A_{ion}(E)$ has been of considerable interest. It is now understood that the following partial waves can contribute at threshold:

$$^1S^e, \; ^3P^o, \; ^1D^e, \; ^3F^o \; ...$$

with A = 1 for the singlet waves and A = −1/3 for the triplet waves.[40] Temkin's threshold theory predicts oscillations in $\sigma(E)$ and also non-monotonic behavior of A_{ion}.[41] Careful measurements have not revealed any such features.[42]

New types of polarization experiments on ionization are under way at Bielefeld: With polarized beams of He(2^3S) the spin dependence of single and double ionization will be investigated. In an Ehrhard-type (e,e)-coincidence experiment with polarized electrons and polarized lithium atoms it will be studied whether the electrons within the "binary peak" of the angular distribution have a different spin dependence than those within the "recoil peak". There are as yet no theoretical predictions.

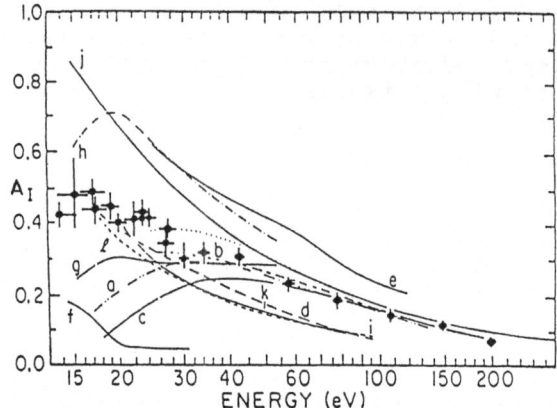

Fig. 15. Ionization asymmetry versus electron energy for atomic hydrogen. The data points are the experimental results of the Yale group, the curves labled a - k are various theoretical estimates. (From ref. 18)

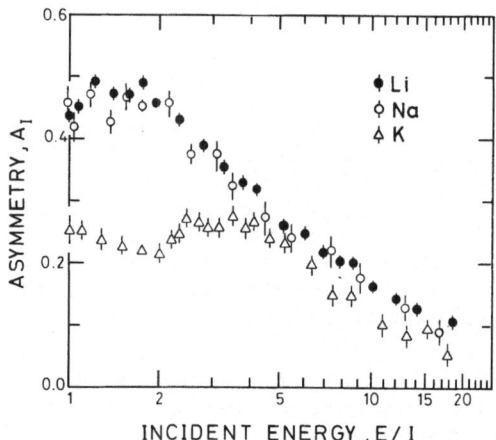

Fig. 16. Ionization asymmetries versus electron energy divided by the ionization potential for the light alkali atoms, measured by the Bielefeld group. (From ref. 39)

CONCLUSIONS

The existing techniques and the foreseeable developments allow the measurement of a vast variety of polarization parameters, some more accurately than others. Theoretically, these parameters are of different significance. Experimentalists and theorists together should assess the values of accuracy and interpretability in order to select the most interesting and fruitful experiments and not to waste resources with less important measurements.

ACKNOWLEDGEMENTS

The experiments at Bielefeld have been supported by the Deutsche Forschungsgemeinschaft in Sonderforschungsbereich 216 "Polarization and Correlation in Atomic Scattering Complexes" and by the University of Bielefeld. For valuable suggestions and careful reading of the manuscript the author is indebted to Professor M. Fink, Dr. L. Frost and B. Granitza.

REFERENCES

1. J. Kessler, "Polarized Electrons", 2nd ed., Springer, Berlin (1985).
2. D. Hils, W. Jitschin, and H. Kleinpoppen, Production of a Highly Polarized Sodium Atomic Beam, Appl.Phys. 25:39 (1981).
3. G. Baum, C.D. Caldwell, and W. Schröder, Dual-Frequency Optical Pumping for Spin-Polarizing a Lithium Atomic Beam, Appl.Phys. 21:121 (1980).
4. H. R. Gray, R. M. Whitley, and C. R. Stroud, Jr., Coherent trapping of atomic populations, Opt.Lett. 3:218 (1978).
5. W. Schröder and G. Baum, A spin flipper for reversal of polarization in a thermal atomic beam, J.Phys.E 16:52 (1983).
6. R. N. Watts and C. E. Wieman, The production of a highly polarized atomic cesium beam, Optics Com. 57:45 (1986).
7. B. Bederson, Crossed-Beam Electron-Neutral Experiments, in: "Methods of Experimental Physics" Vol. 7A, B. Bederson and W.L. Fite, eds., Academic Press, New York (1968), p. 57; B. Jaduszliwer, Optical State Preparation and Analysis for Spin-Effects Studies in Electron-Cesium Collision, Abstract, XV ICPEAC (Brighton, 1987), p. 811.
8. G. Baum et al., to be published.
9. D. T. Pierce and F. Meier, Photoemission of spin-polarized electrons from GaAs, Phys.Rev.B 13:5484 (1976); D. T. Pierce, R. J. Celotta, G.-C. Wang, W. N. Unertl, A. Galejs, C.E. Kuyatt, and S. Mielczarek, GaAs spin polarized electron source, Rev.Sci.Instrum. 51:478 (1980); E. Reichert and K. Zähringer, Electron Spin Polarization in the Photoemission of NEA $GaAs_{1-x}P_x$, Appl.Phys.A 29:191 (1982); J. Kirschner, H.P. Oepen, and H. Ibach, Energy- and Spin-Analysis of Polarized Photoelectrons from NEA GaAsP, Appl.Phys.A 30:177 (1983); F.C. Tang, M.S. Lubell, K. Rubin, and A. Vasilakis, Operating experiences with a GaAs photo-emission electron source, Rev.Sci.Instrum. 57:3004 (1986).
10. D. T. Pierce, S. M. Girvin, J. Unguris, and R. J. Celotta, Absorbed current electron spin polarization detector, Rev.Sci.Instrum. 52:1437 (1981).
11. M. Erbudak and G. Ravano, Spin dependent electron absorption in gold, J.Appl.Phys. 52:5032 (1981).
12. H. C. Siegmann, D. T. Pierce, and R. J. Celotta, Spin-dependent absorption of electrons in a ferromagnetic metal, Phys.Rev.Lett. 46:452 (1981).

13. H. Boersch, R. Schliepe, and K.E. Schriefl, Polarized electrons by Mott scattering at bulk targets, Nucl.Phys.A 163:625 (1971); J. Unguris, D.T. Pierce, and R. J. Celotta, Low-energy diffuse scattering electron-spin polarization analyzer, Rev.Sci.Instrum. 57:1314 (1986).

14. T. J. Gay, A simple optical electron polarimeter, J.Phys.B 16:L553(1983).

15. G. D. Fletcher, T. J. Gay, and M. S. Lubell, New insights into Mott-scattering electron polarimetry, Phys.Rev.A 34:911 (1986).

16. A. Gellrich, J. Jost, and J. Kessler, Elimination of instrumental asymmetries in Mott analyzers, Abstract XV ICPEAC (Brighton, 1987), p. 818.

17. N. Sherman, Coulomb Scattering of Relativistic Electrons by Point Nuclei; Phys.Rev. 103:1601 (1956); S.-R. Lin, Elastic Electron Scattering by Screened Nuclei, Phys.Rev. 133:A965 (1964); G. Holzwarth and H. J. Meister, Elastic Scattering of Relativistic Electrons by Screened Gold Nuclei, Nucl. Phys. 59:56 (1964); A.W. Ross, M. Fink, and H. Hilderbrandt, Complex Scattering Factors for the Diffraction of Electrons by Gases, in: "International tables of x-rays crystallography", to be published; W. Bühring (Heidelberg), unpublished; V.A. Apalin, I. Ye. Kutikov, I.I. Lukashevich, L.A. Mikaelyan, G.V. Smirnov, and P.Ye. Spivak, Asymmetry in Double Mott Electron Scattering for Energies from 45 to 245 keV, Nucl.Phys. 31:657 (1962); J. Van Klinken, Double Scattering of Electrons, Nucl.Phys. 75:161 (1966).

18. G. D. Fletcher, M. J. Alguard, T. J. Gay, V. W. Hughes, P. F. Wainwright, M. S. Lubell, and W. Raith, Experimental study of spin-exchange effects in elastic and ionizing collisons of polarized electrons with polarized hydrogen atoms, Phys.Rev.A 31:2854 (1985).

19. W. L. van Wyngaarden and H. R. Walters, Spin asymmetry parameter for electron-hydrogen scattering: I Elastic Scattering, J.Phys.B 19:1817 (1986).

20. G. Baum, M. Moede, W. Raith, and U. Sillmen, Measurement of Spin Dependence in Low-Energy Elastic Scattering of Electrons from Lithium Atoms, Phys.Rev.Lett. 57:1855 (1986).

21. O. Berger and J. Kessler, Elastic scattering of polarized electrons from mercury and xenon, J.Phys.B 19:3539 (1986).

22. E. Reichert and H. Deichsel, Spinpolarisation durch Elektronen-Resonanzstreuung an Neon, Phys.Lett. 25A:560 (1967).

23. J. J. McClelland, M. H. Kelley, and R. J. Celotta, Spin-Orbit and Exchange Effects in Elastic Scattering of Spin-Polarized Electrons from Spin-Polarized Na Atoms, Phys.Rev.Lett. 58:2198 (1987).

24. P. S. Farago, On the detection of spin-orbit interaction in the elastic scattering of electrons from one-electron atoms, J.Phys.B 7:L28 (1974).

25. D. W. Walker, On the asymmetry in single scattering of electrons from one-electron atoms, J.Phys.B 7:L489 (1974).

26. D. W. Walker, Conservation of Total Spin in Electron-Atom Collisions, in: "Electron and Photon Interactions with

Atoms", H. Kleinpoppen and M.R.C. McDowell, eds., Plenum, New York (1976), p. 203.

27. V. D. Ob'edkov and M. J. Van der Wiel, Asymmetry and spin polarisation in scattering of electrons by one-electron atoms, J.Phys.B 11:L329 (1978).

28. P. G. Burke and J. F. B. Mitchell, Spin-polarization in the elastic scattering of electrons by one-electron atoms, J.Phys.B 7:214 (1974).

29. G. Baum, W. Raith, U. Sillmen, and H. Steidl, Measurement of relative singlet and triplet contributions in inelastic collisions of electrons with low-Z atoms, Abstracts XV ICPEAC (Brighton, 1987) p. 153.

30. M. Eller, P. Naß, and E. Reichert, Polarization transfer in the impact excitation of rubidium by polarized electrons, Abstracts XV ICPEAC (Brighton, 1987), p. 157.

31. A. Wolke, K. Bartschat, K. Blum, G. F. Hanne, and J. Kessler, Investigation of Stokes parameters for studying resonances near threshold for $6^1S_0-6^3P_1$ excitation of mercury by polarized electrons, J.Phys.B 16:639 (1983).

32. G. F. Hanne, Spin Effects in Inelastic Electron-Atom Collisions, Physics Reports 95:95 (1983).

33. K. Bartschat and K. Blum, Theory and Physical Importance of Integrated State Multipoles, J. Phys.A 304:85 (1982); K. Blum and H. Kleinpoppen, Spin-dependent phenomena in inelastic electron-atom collisions, Adv. in At.Mol.Phys. 19:187 (1983).

34. G. F. Hanne, Orientation and Spin Exchange in Electron-Atom Collisions, Comments At.Mol.Phys. 14:163 (1984).

35. I. V. Hertel and W. Stoll, Collision experiments with laser excited atoms in crossed beams, Adv. At.Mol.Phys. 13:113 (1977).

36. M. H. Kelley, J. J. McClelland, and R. J. Celotta, Superelastic scattering of spin-polarized electrons from optically pumped sodium, Abstracts, XV ICPEAC Brighton, 1987), p. 152.

37. I. V. Hertel, M .H. Kelley, and J. J. McClelland, Analysis of Collisional Alignment and Orientation Studied by Scattering of Spin-Polarized Electrons from Laser-Excited Atoms, Z.Phys.D 6:163 (1987).

38. R. Peterkop, Consideration of Exchange in Ionization, Proc.Phys.Soc. 77:1220 (1961).

39. G. Baum, M. Moede, W. Raith, and W. Schröder, Measurement of spin asymmetries in the electron impact ionization of alkali atoms, J.Phys.B 18:531 (1985).

40. A. D. Stauffer, The nodal structure of two-electron wave functions on the Wannier ridge, Phys.Lett. 91A:114 (1982); C.H. Greene and A.R.P. Rau, Effects of symmetry on two-electron escape at threshold, J.Phys.B 16:99 (1983).

41. A. Temkin, The energy distribution cross section in threshold electron-atom impact ionization, J.Phys.B 7:L450 (1974).

42. M. H. Kelley, W. T. Rogers, R. J. Celotta, and S.R. Mielczarek, Near-Threshold Measurements of Spin Dependence of Electron-Impact Ionization, Phys.Rev. Lett. 51:2191 (1983).

ANALYSIS OF ELECTRON-ATOM SCATTERING:

POSITRON-ATOM SCATTERING, A VARIATION OF ELECTRON-ATOM SCATTERING

Wilhelm Raith

Fakultät für Physik
Universität Bielefeld
D-4800 Bielefeld
Fed.Rep. of Germany

INTRODUCTION

The interaction of an electron with a (screened) nucleus is attractive, of a positron, repulsive. Classically, hyperbolic trajectories result for the Coulomb scattering of electrons and positrons, but the nucleus is located in the near (inside) focal point for electrons, in the far (outside) one for positrons. Despite this difference the Rutherford cross section is identical for electrons and positrons as only the square of the charge enters into the formula. Similarly, the scattering in First Born Approximation (FBA), without accounting for exchange in the electron case, is identical. Since we consider here only the low-energy scattering below, say, several keV, we can disregard relativistic effects. In this energy range the FBA is the appropriate "high-energy" approximation. In fact, the experimentally determined convergence of electron and positron scattering cross sections can be taken as a criterion for the validity of the FBA.

The main reason why electron and positron scattering differ at low energies is the interaction of the charge of the projectile with the induced electric dipole moment of the atom which is attractive in both cases. This interaction is often called "polarization interaction" and asymptotically described by a $1/r^4$ "polarization potential". In order to avoid confusion with <u>spin</u> polarization we will use the terminology of the positron-physics community and speak of

and
static Coulomb interaction between the projectile's charge and the static field of the screened nucleus

dynamic Coulomb interaction between the projectile's charge and the time-dependent induced dipole moment of the atom.

Low-energy positron-atom scattering is a good testing ground for theoretical approximations dealing with the dynamic Coulomb interaction. It has the advantage over electron-atom scattering of the absence of exchange.

In the positron-atom interaction there are two reaction channels which do not have analogies in the electron-atom interaction. One is the free annihilation with one of the atomic electrons and the other is the formation of positronium (Ps = e^+e^-), the "lightest hydrogen isotope" with a mass of 1.097×10^{-3} amu. The free-annihilation cross section, given by

$$\sigma_{annih} = \pi \, r_0^2 \, c/v$$

($r_0 = 2.8 \times 10^{-13}$ cm = classical electron radius, c = velocity of light, v = positron velocity), is several orders of magnitude smaller than the angle-integrated elastic cross section and, therefore completely negligible in atomic-physics scattering experiments. (Free annihilation is investigated by means of positron-lifetime measurements in very dense gases. Positrons which survive the slow-down below the Ps-formation threshold can decay only via free annihilation.)

Positronium formation is a very promiment reaction channel which opens at an energy

$$E_{Ps} = E_{ion}(atom) - E_{ion}(Ps)$$

with $\quad E_{ion}(Ps) = \frac{1}{2} E_{ion}(H) = 6.8$ eV.

Since Ps formation can be measured, for example, by utilizing its decay into 2 or 3 photons as a signature, it can be deducted before the comparison with electron scattering is made.

Because it can be measured, Ps formation in positron scattering is easier to deal with than exchange in electron scattering. Electron scattering experiments with spin-polarized particles can only serve to learn about the exchange with the polarized atomic electrons. But there is also exchange with the electrons of closed shells. Therefore, positron scattering should lead to satisfactory approximations for the dynamic Coulomb interaction which, in turn, can be applied to electron-atom scattering (with sign reversal of course). Once this is understood, the comparison of experiment and theory in electron scattering can be devoted to developing satisfactory approximations for exchange.

The Ps formation is a theoretically demanding rearrangement process. Also of interest is the comparison with H formation in proton-atom scattering.

Positron-atom scattering, especially on atomic hydrogen, is of importance in astrophysics. During the 1970's gamma ray telescopes in balloons detected the 511 keV annihilation radiation ($e^+e^- \rightarrow 2\gamma$) from the direction of the galactic center. Although the origin of the positrons is uncertain, their annihilation is certainly related to collisions with atoms or molecules in the neighborhood (within about 1 light year) of the positron source.

Atomic physics with positrons is a rapidly expanding field of research. The most comprehensive articles can be found in the proceedings of positron workshops held biannually since 1981.[1-4]

SOURCES AND MODERATORS

Positrons are produced with energies on the order of 1 MeV in radioactive ß-decay and in e^+e^- pair creation by the bremsstrahlung of high-energy electrons. It is not feasible to produce a low-energy positron beam by decelerating a portion of those high-energy positrons because the remaining intensity would be negligibly small. More accurately, the brightness of a charged-particle beam – which is "current per source area and solid angle" –, divided by the particles' energy, is a constant for beam-transport devices based on conservative forces (Liouville's theorem). Low-energy positron beams of appreciable brightness can be obtained by employing dissipative forces in a solid-state moderator in which high-energy positrons are decelerated down to thermal energies and expelled into the vacuum if the moderator surface has a negative work function for positrons (Fig. 1).

For electron emission from a metal the work function can be viewed as the sum of three different potentials:

$$\phi(e^-) = \text{(image potential)} + \text{(dipole-layer potential)}$$
$$- \text{(Fermi potential)}$$

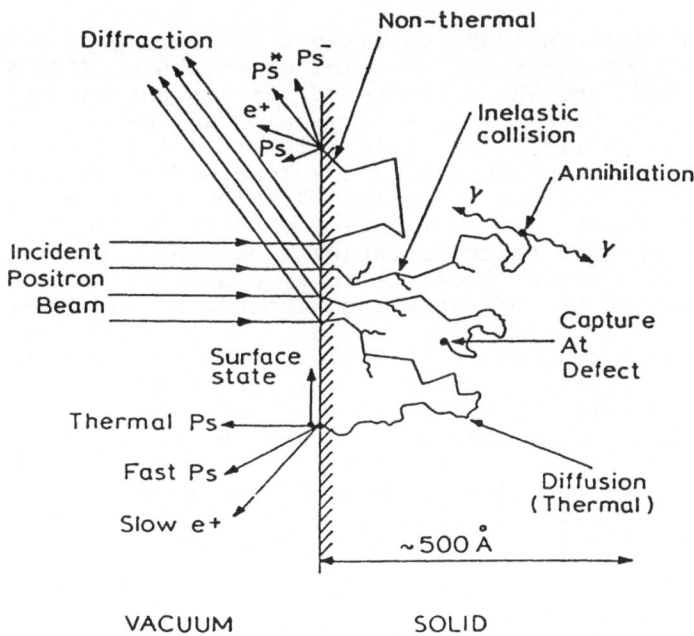

Fig. 1. Schematics of positron interactions in a solid. (Adapted from A.P. Mills, Jr., by R.M. Nieminen, ref. 2)

The Fermi potential enters with a minus sign because the electrons come most probably from the vicinity of the Fermi level. All three potentials amount to several volt and the resulting sum, therefore, is of the same magnitude. For positron emission from a metal the dipole-layer potential has the opposite sign and the Fermi potential is zero:

$$\Phi(e^+) = \text{(image potential)} - \text{(dipole-layer potential)}$$

Depending on which term is larger the positron work function is either positive or negative and is always small ($\leq$ 3 V).

Positron sources are of two types:

(1) Portable medium-current sources for use in the laboratory

(2) Stationary high-current sources at special research facilities

Type (1) is usually a long-lived sodium-22 source (half life 2.6 years), commercially available with activities of up to 100 mCi. Type (2) is either a pulsed source utilizing an electron accelerator, operational at Lawrence Livermore National Laboratory and the University of Gießen (under construction at Oak Ridge and Tsukuba), or a dc source consisting of a high-activity short-lived copper-64 source (half life 12.8 hours) activated in a high-flux reactor, operational at Brookhaven National Laboratory (planned at Munich-Garching). Another way of producing a positron beam of high brightness is to use ISOLDE, the ion-beam mass separator of fission products at CERN; a promising nuclide is rubidium-81 with a half life of 4.7 hours.[5]

For use under non-UHV conditions the only good moderator is polycrystalline tungsten, annealed up to about 2200 K (between annealing and use an exposure to air is not harmful). Better moderators are clean single crystals in UHV, e.g., W(111), Ni(100) and Cu(111). The moderation in a solid can be applied more than once. Re-moderation of a low-energy positron beam is useful for "brightness enhancement" which was first proposed by Mills.[6] Moderators can be surfaces of solids used in "reflection mode" or thin foils (often held between supporting meshes) used in "transmission mode". A cheap quasi-transmission moderator consists of two or three sandwiched tungsten meshes.

Moderator conversion efficiencies are quoted in different ways. At accelerator sources one obtains several moderated positrons per 10^6 high-energy electrons. The first moderation of radioactive positrons has at best an efficiency of 10^{-3}. Remoderation of 5-10 keV positrons is much more efficient; efficiencies of more than 10^{-1} have been obtained. The newest and supposedly best moderator is solid neon which also features a high reflectivity for moderated positrons.[7]

Moderated positrons have thermal energies plus the energy acquired by the work-function acceleration. Positrons from a tungsten-mesh moderator have an energy distribution of about 1.5 eV width. Positrons from a plane single-crystal moderator all experience the work-function acceleration in forward direction. The width of their energy distribution can be quite small, close to kT.

Typical <u>electron</u>-atom scattering experiments are performed with currents of about 10 μA, usually limited by space charge. <u>Positron</u>-atom scattering experiments started 15 years ago with about 1 e⁺/s and steadily progressed to 10^3-10^5 e⁺/s in laboratory experiments. High-brightness beams of about 10 pA will soon be available at Brookhaven, but even that will still be only 10^{-6} of typical electron beams.

The positrons from ß-decay are partially spin-polarized due to parity violation. Moderation does not necessarily destroy the polarization. If source and moderator are designed to minimize backscattering and to suppress the low-energy positrons of the radioactive decay, moderated positron beams with polarizations as high as 0.7 can be obtained.[8]

ELASTIC SCATTERING

Experimental information on elastic scattering is available from total cross section measurements below the lowest inelastic threshold (which for all atoms is E_{Ps}) and, only very recently, from crossed-beam differential-scattering experiments with retarding-potential energy discrimination.

By and large the positron cross sections are smaller than the electron cross sections because the repulsive static and the attractive dynamic Coulomb interaction subtract whereas in the electron case both are attractive and add.

Transmission experiments yield the angle-integrated elastic cross section below E_{Ps}. The most distinctive features in this low-energy region are Ramsauer minima (Fig. 2). Theoretically, the Ramsauer effect is related to a zero (modulo π) transition of the s-wave phase shift. In electron scattering from noble-gas atoms the s-wave phase shift η_0 starts at nπ (n = 1,2...) for E=0. With increasing E it decreases monotonically for He and Ne which do not exhibit the Ramsauer effect; for Ar, Kr and Xe $\eta_0(E)$ starts with a positive slope but soon changes to a negative slope and goes through nπ at the small but finite energy of the Ramsauer minimum. For positron scattering the Ramsauer effect is more straightforward: For all noble gases $\eta_0(0)$ starts at zero with a positive slope because at low energies the attractive dynamic Coulomb interaction dominates. At higher energies, the repulsive static Coulomb interaction dominates and, consequently, $\eta_0(E)$ becomes negative. In between, $\eta_0(E)$ goes through zero and a Ramsauer minimum is observed for He and Ne. In positron scattering from the heavier noble gases the Ramsauer effect is masked by substantial contributions from higher partial waves and, therefore, the cross section does not show a minimum although the s-wave phase shift goes through zero.

The first data on the angular dependence of forward elastic scattering came from a gas-target time-of-flight experiment with a longitudinal magnetic guiding field (Coleman and McNutt).[36] Now, crossed-beam experiments are feasible. So

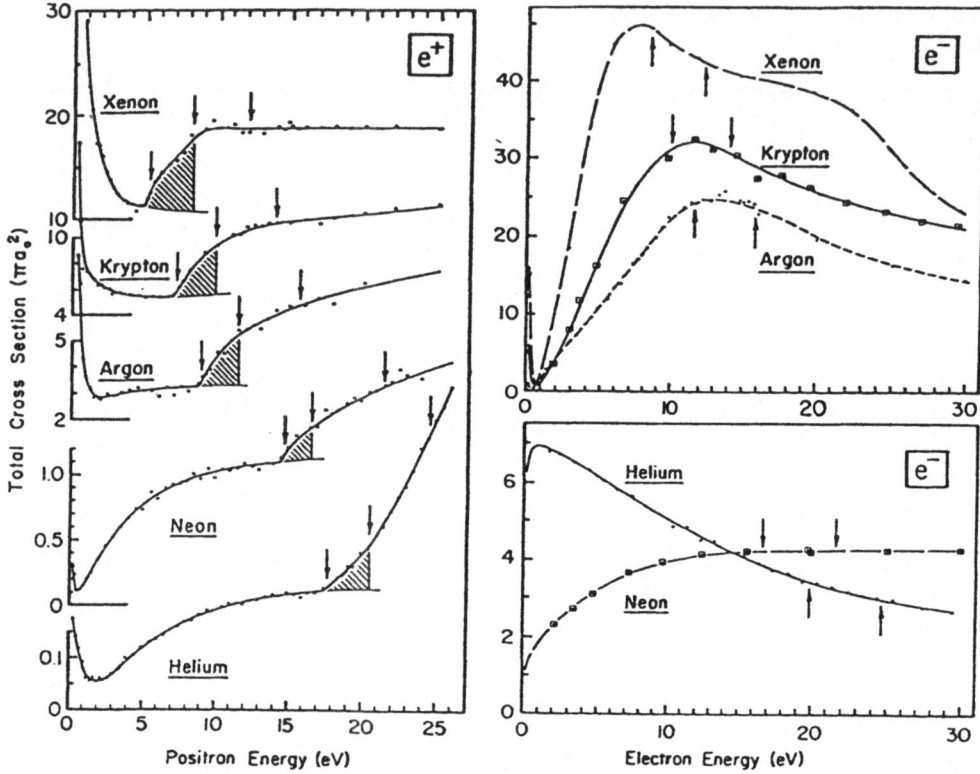

Fig. 2. Low-energy positron and electron scattering on noble-gas atoms. The three arrows shown in the diagram on the left indicate the positronium (Ps)-formation threshold, E_{Ps}, the first excitation energy, E_{exc}, and the ionization energy, E_{ion}; the two arrows in the diagrams on the right indicate E_{exc} and E_{ion}. At energies below the first arrow these total cross sections, $\sigma_t(E)$, (measured by the Detroit group) are the angle-integrated elastic cross sections, $\sigma_{el}(E)$. The first estimates for the Ps formation cross section were obtained by extrapolating $\sigma_{el}(E)$ to energies between E_{Ps} and E_{exc} and taking the difference $\sigma_t - \sigma_{el}$, shown as hatched areas. (From ref. 9)

far only <u>relative</u> differential cross sections, normalized to theory at some angle or energy, have been measured. The Detroit experiment is shown schematically in Fig. 3. At energies of 100 and 200 eV the electron cross sections show pronounced structures whereas the measured positron cross sections increase monotonically with decreasing scattering angle. Only the theoretical predictions for positron scattering show some structure at very small angles (Fig. 4, diagrams on the right side).

The situation is different at lower energies where different theoretical approximations led to the prediction of different structures. The Detroit group extended their measurements at 45° down to 6 eV and found, as for their data of $\sigma_{el}(\Theta)$ at 20 eV, a much better agreement with the predictions

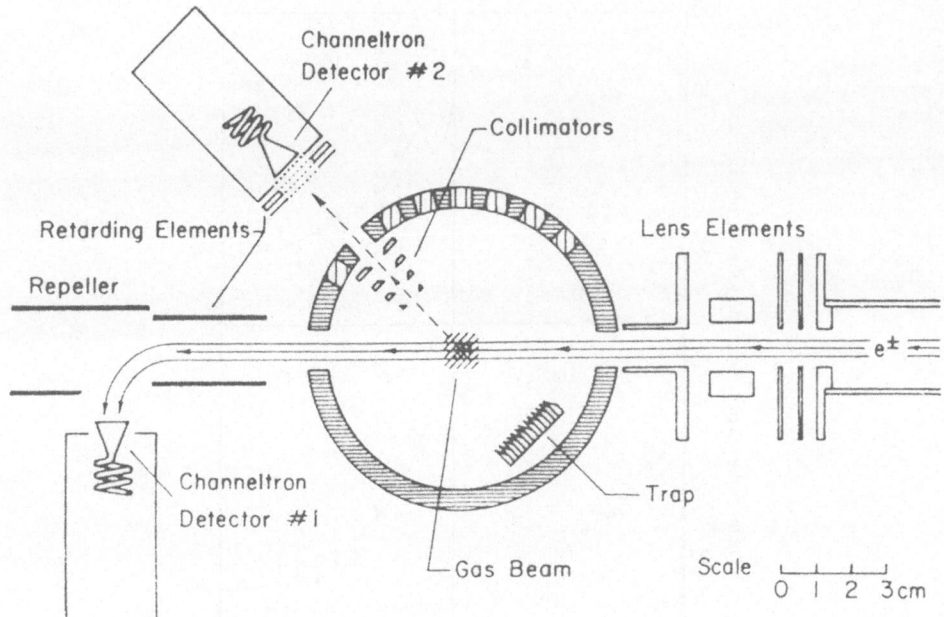

Fig. 3. Layout of the Detroit crossed-beam experiment for measurement of differential elastic scattering cross sections. (From ref. 9)

of Nahar and Wadehra[11] than with those of McEachran and Stauffer.[12] The very recent Bielefeld data at 30 eV, however, deviate from both theories for small scattering angles (Fig. 5). But both theories will be amended by their respective authors taking into account the loss of flux from the elastic channel, a procedure required by unitarity and apparently essential here, as was discussed by Khare et al. and also by Joachain and Potvliege.[14]

Of considerable interest is the convergence behavior of the elastic scattering. Experimental results for the angle-integrated helium cross sections exist for electrons over a large energy range, but for positrons only up to E_{Ps} (from transmission experiments). These experimental results are compared with various theoretical predictions in Fig. 6. The distorted-wave second-Born approximation of Dewangan and Walters,[16] which is considered to be reliable at higher energies, indicates ·convergence of elastic scattering above 3000 eV.

The existing spin polarization of the ß-decay positrons suggests polarization experiments. Since exchange is absent and the spin dependence of Ps formation is simply governed by the statistical weights of 1:3 for the singlet and triplet state, the only interesting process is low-energy Mott scattering. The polarization effects due to spin-orbit interaction are for positrons orders of magnitude smaller than for electrons, as shown in the calculations of Stauffer and McEachran.[17] This is so because it takes place at small radial distances which are not reached in positron scattering (Fig. 7).

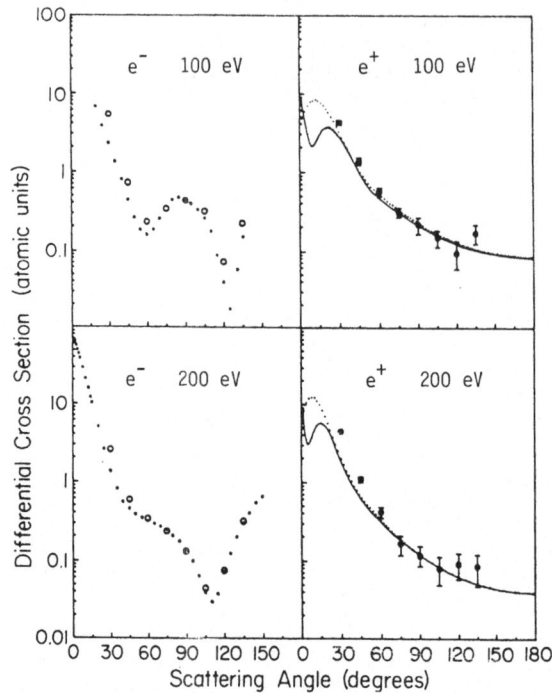

Fig. 4. Electron and positron elastic differential scattering
from argon atoms at 100 and 200 eV. Open circles (for
electrons) and full circles (for positrons) are
experimental results of the Detroit group, normalized
to other work at 90°. Dots (electrons) are the other
experimental results. The dotted and solid curves
(positrons) are the theoretical predictions of Nahar
and Wadehra[11] and of McEachran and Stauffer,[12] respectively. (From ref. 10)

INELASTIC SCATTERING

Crossed-beam scattering experiments have not yet progressed to inelastic measurements. Experimental information thus
far has come from gas-target transmission experiments with
magnetically guided positron beams and time-of-flight as well
as retarding potential methods which do not yield reliable
cross sections. The first observation of atomic excitation by
positrons is shown in Fig. 8.

An interesting theoretical result for the 2P, 3P and 4P
excitation of helium is the tendency of positron excitation
cross sections to exceed the corresponding electron cross
sections at high energies.[20]

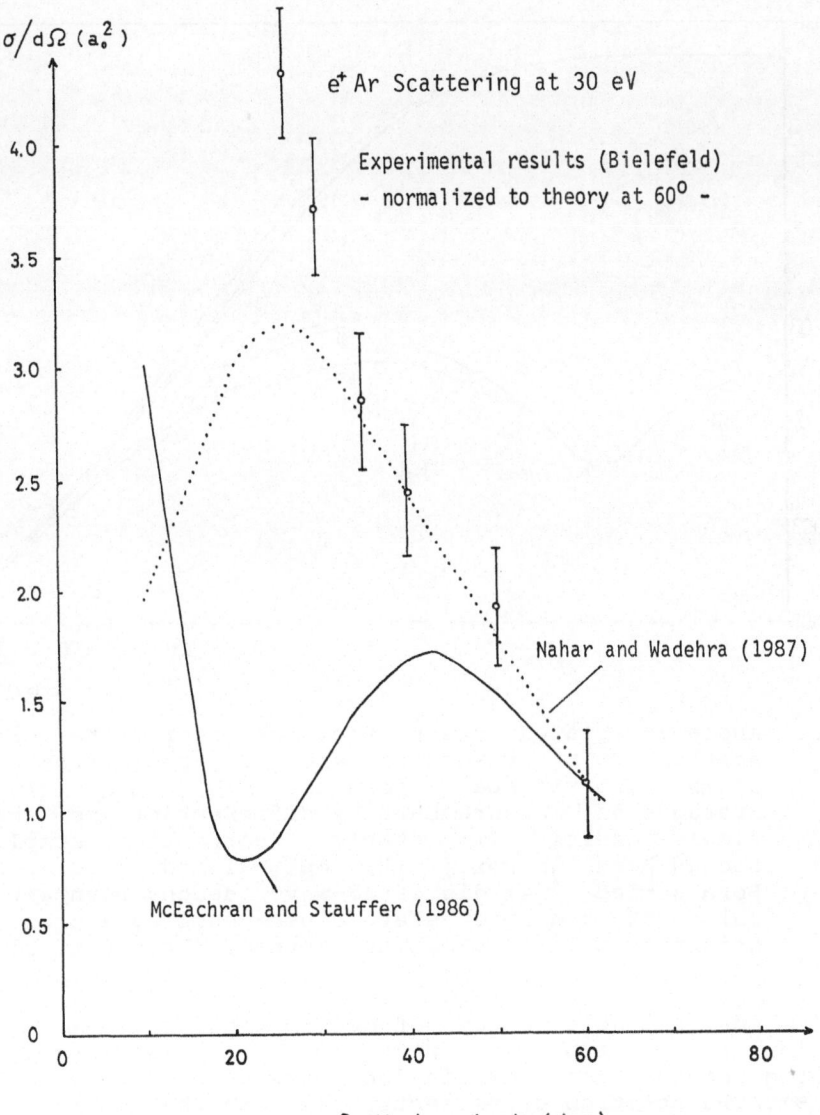

Fig. 5. Experimental results of the Bielefeld group for positron-argon differential elastic scattering at 30 eV, normalized to theory at 60°. (From ref. 13)

IONIZATION

The term "ionization" is meant to refer to the process

$$\sigma_{ion}^+: \qquad e^+ + A \longrightarrow A^+ + e^+ + e^-$$

and not to include Ps formation (which, of course, is also an ionizing reaction channel).

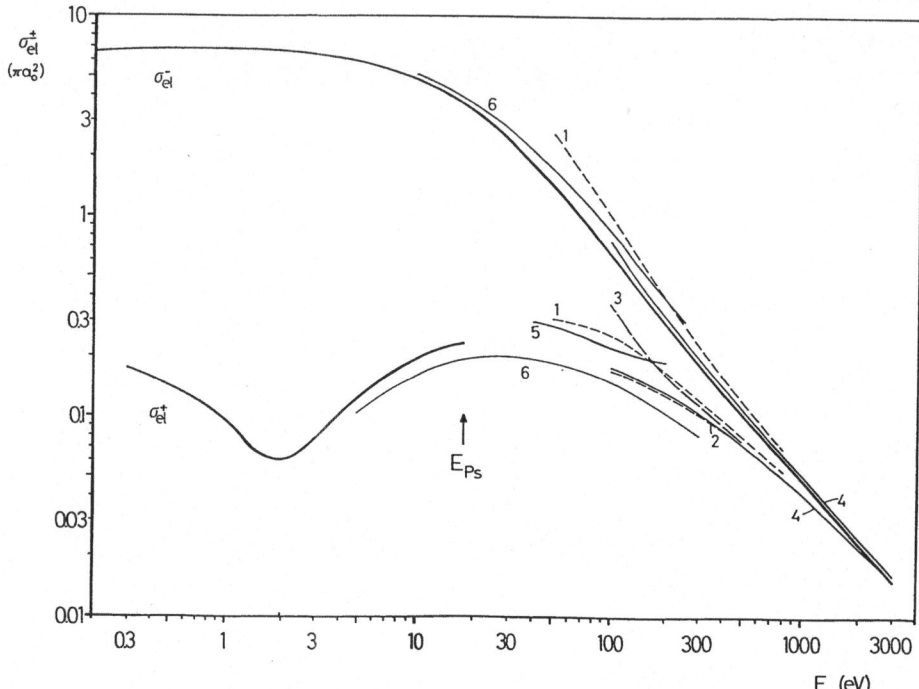

Fig. 6. Angle-integrated cross sections for the elastic scattering of electrons and positrons from helium atoms. Experimental results: σ_{el}^- and σ_{el}^+ are averages of measurements by different groups. Theoretical results: (1) static, corrected, simplified second-Born approx.; (2) optical model; (3) eikonal Born series; (4) distorted-wave second Born approx.; (5) three and five state close-coupling approx.; (6) adiabatic polarized-orbital method. (From ref. 15)

The most reliable method of determining σ_{ion}^+ is one in which ion and positron are extracted from a gas target, mass separated and the time correlation between them is measured (Fig. 9). The positron cross section exceeds the electron cross section at intermediate energies but both merge above 600 eV. Several different theoretical approximations describe the data quite well (Fig. 10). Some of those computations were made for electron ionization "without exchange". This indicates that the difference between σ_{ion}^+ and σ_{ion}^- is mainly due to exchange which lowers σ_{ion}^-.

The Wannier theory of threshold behavior, which ignores Ps formation, predicts $\sigma_{ion}^+ < \sigma_{ion}^-$ close to threshold[22] according to

$$\sigma_{ion}^{\pm} \; \alpha \; (E - E_{ion})^{n\pm}$$

with $n^+ = 2.651$ and $n^- = 1.127$. Such behavior is barely indicated for helium in Fig. 10 (lower diagram) but more clearly for H_2 in Fig. 11. This observation of threshold behavior is only a qualitative result; the energy resolution does not suffice for determining the power-law exponent.

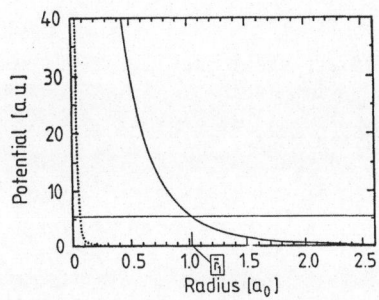

—————: Radial dependence of the p-wave potential
without spin-orbit term: $U + \ell(\ell+1)/(2r^2)$
with $\ell = 1$ for elastic e^+ - Hg scattering

......: Spin-orbit term $1/(4rc^2)\ dU/dr$

The horizontal line indicates the scattering energy
of 150 eV. $r \leqslant \tilde{r}_\ell$ ($\ell=1$) defines the classically
forbidden region.

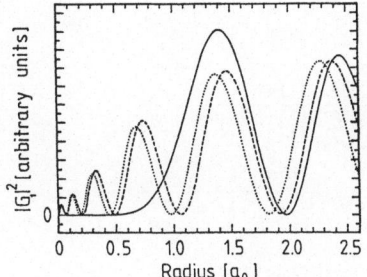

—————: $|G_\ell|^2$ ($\ell=1$, $j=1/2$) for e^+ - Hg scattering

......: $|G_\ell|^2$ ($\ell=1$, $j=1/2$) for e^- - Hg scattering

- - - - - : $|G_\ell|^2$ ($\ell=1$, $j=3/2$) for e^- - Hg scattering

Scattering energy: 150 eV

Fig. 7. Calculations for positron and electron elastic scattering. The results explain why the spin-polarization effects in low-energy Mott scattering of positrons are negligibly small. (From ref. 18)

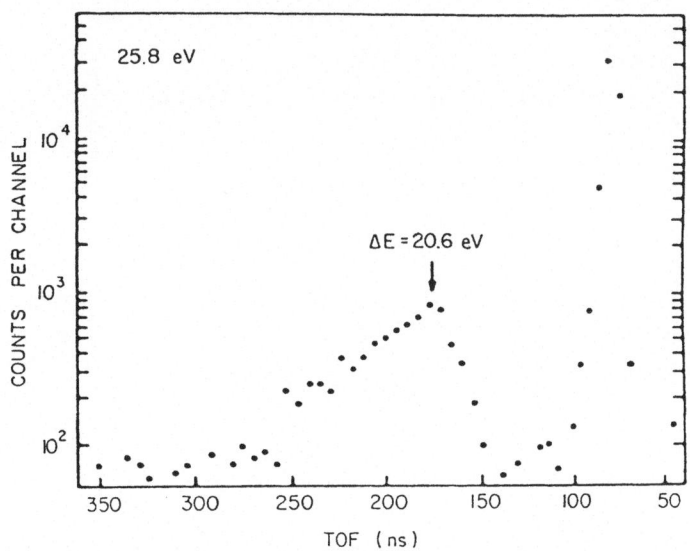

Fig. 8. Time-of-flight spectrum of a magnetically guided 25.8 eV positron beam which traversed a helium gas target. Positrons which excited helium atoms and suffered energy losses of $\Delta E \geq 20.6$ eV arrive later. The 2^3S_1 state, like all the other triplet states, cannot be excited by positrons because of the absense of exchange. (From ref. 19)

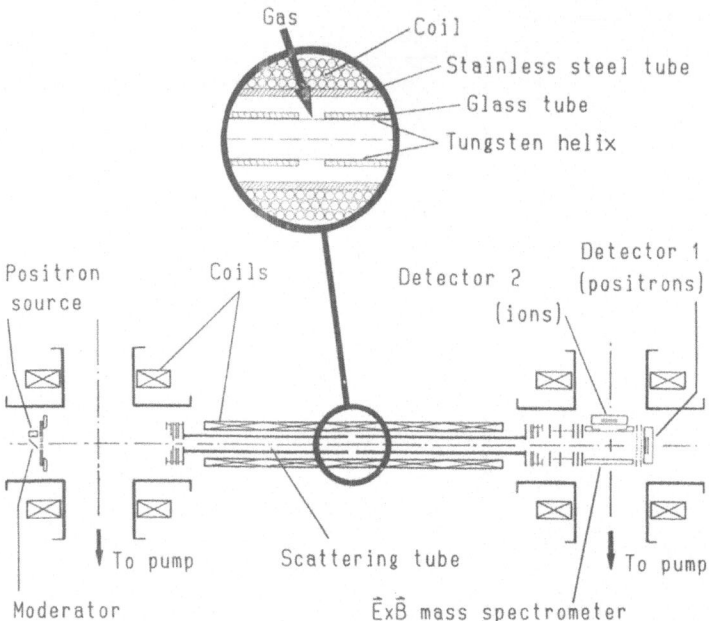

Fig. 9. Experimental arrangement used at Bielefeld for measuring ionization and Ps formation cross sections with gas targets. (From ref. 21)

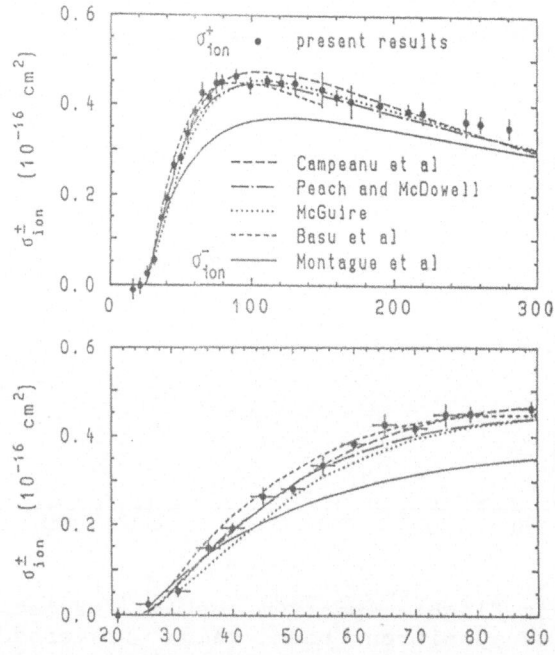

Fig. 10. Comparison of the positron ionization cross sections for helium measured at Bielefeld (dots) with theoretical results and with the experimental electron cross section. (From ref. 15)

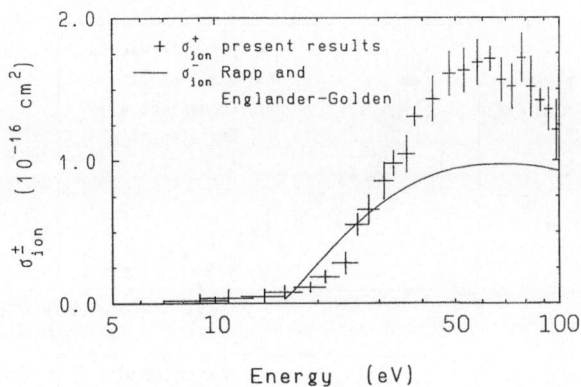

Fig. 11. The positron ionization cross section of H_2 near threshold measured at Bielefeld (crosses) compared with the experimental electron cross section (solid curve). (From ref. 23)

POSITRONIUM FORMATION

Three entirely different methods have been employed to measure σ_{Ps}. The results obtained are also quite different (Fig. 12). The Bielefeld results were obtained from ($\sigma_{ion}^+ + \sigma_{Ps}$) measured by detection of all ions and σ_{ion}^+ measured by detection of the time-correlated ions. The agreement with theory is not yet satisfactory. The measured σ_{Ps} include Ps formation into excited bound states but not into continuum states. The latter would contribute to the measurement of σ_{ion}^+. The electron capture into continuum states – for proton-atom as well as positron-atom collisions – were calculated by Brauner and Briggs.[24] They predicted a peak in the secondary-electron energy distribution at $E = (E_0 - E_{ion})/2$, where E_0 is the primary positron energy, due to electrons which have nearly zero relative velocity with respect to the outgoing positron (Fig. 13). Such a peak in the differential energy distribution should show up as a shoulder in the integral energy distribution given by a retarding-potential curve; a first indication of that has been observed by the London group.[25]

Ps(n=2) formation in positron collisions with noble-gas atoms has been detected by the London group observing the emission of Ps-Lyman-α which sets in at the appropriate threshold.[26]

Differential Ps-formation measurements showed a pronounced peak in forward direction. This has already been utilized for Ps-beam formation.[27] Ps-atoms emerging from moderators (cf. Fig. 1) are of low energy and do not travel very far within the lifetime of 142 ns (for the triplet state, only 125 ps for the

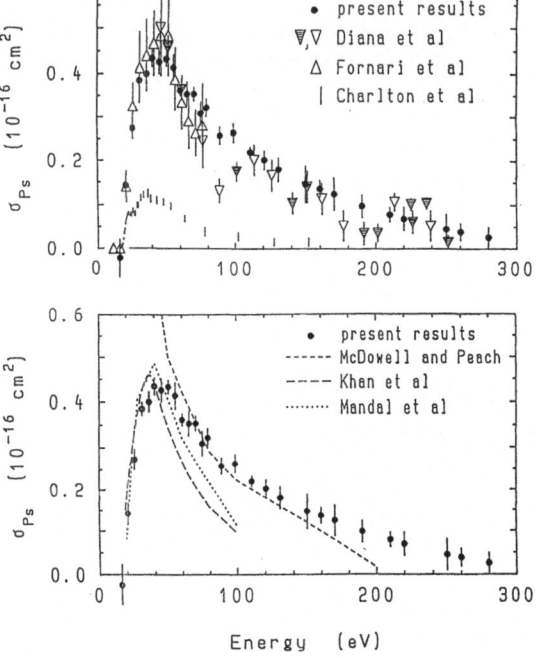

Fig. 12. Ps formation cross section for helium measured by the Bielefeld group (dots) compared with other results. Above – with experimental data of the groups at Arlington (triangles) and London (vertical bars), below – with theoretical predictions. (From ref. 21)

singlet state). As a rule of thumb the long-lived ortho-Ps travels "6 cm times square-root of E in eV" during one life-time. Therefore, it is very appealing to form a Ps beam of tens of eV energy by positron-gas collisions. The Brookhaven group anticipates an $e^+ \rightarrow$ Ps conversion efficiency of 1.3% for 50 eV positrons on helium with the Ps going into a 5° cone in forward direction.[28] Another important achievement is the production of a time-tagged Ps beam (Fig. 14).

Ps is a peculiar "atom"; it has zero net charge (as any neutral atom) but in the absence of an electric field it has also zero <u>charge density</u>. In its ground state both electron and positron are in an S-state and their charge densities compensate each other. Scattering experiments with Ps beams are of great interest. For example, the transfer of a positron from Ps to an antiproton is planned to be used to form antihydrogen.[30]

The proton collision leading to H formation

$$\sigma_H: \quad p^+ + A \longrightarrow A^+ + H$$

is analogous to

$$\sigma_{Ps}: \quad e^+ + A \longrightarrow A^+ + Ps.$$

472

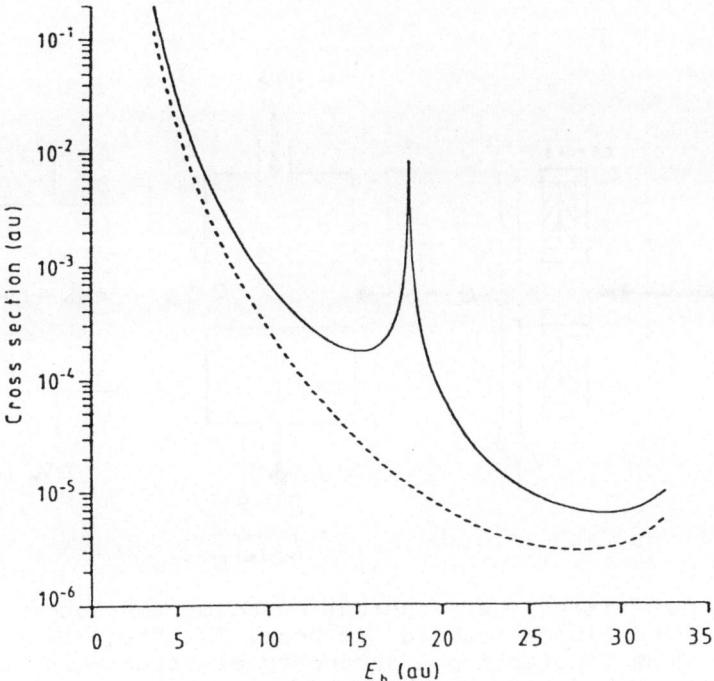

Fig. 13. Energy distribution of electrons ejected in 1 keV positron-impact ionization of neutral atoms, calculated by Brauner and Briggs. The peak consists of electrons which have near-zero relative velocities with respect to the outgoing positron. The broken curve is for an electron-positron plane wave as the final state; the full curve is for a Coulomb wave as the final state. (From ref. 24)

As is customary for comparing electron and proton scattering the cross sections are not plotted vs. energy but rather vs. velocity, because the cross sections are expected to be proportional to the collision time and, therefore, to $1/v$. For protons, initial and final velocities, v_i and v_f, are almost identical because the acceleration of the pick-up electron takes only a very small portion of the proton's energy. For positrons, on the other hand, v_f is about $v_i/\sqrt{2}$ and the question arises which one is more appropriate for the comparison (Figs. 15,16). The fact that $\sigma_{Ps} > \sigma_H$ at high initial velocities, v_i, as shown in Fig. 15, has been predicted by Deb et al.[31] But in the plot versus v_f (Fig. 16) this feature has disappeared and instead the cross sections exhibit the expected high-energy convergence behavior. This indicates dominance of post-collision interaction and can be explained by the large dielectric polarizability of Ps which is 26 times that of He. Consequently, the long-range dynamic Coulomb interaction between the outgoing particles is much larger than between the incoming particles.

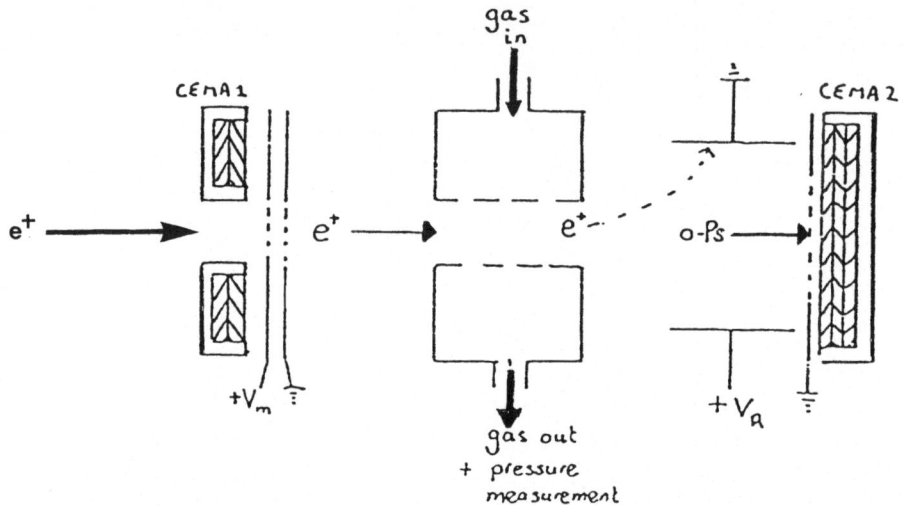

Fig. 14. Experimental arrangement used by the London group for
producing a "tagged" Ps beam. The "tag" is provided by
CEMA 1 detecting a secondary electron which is emitted
backwards from the moderator when the high-energy
positron, which will later form Ps in a gas collision,
is moderated. (From ref. 29)

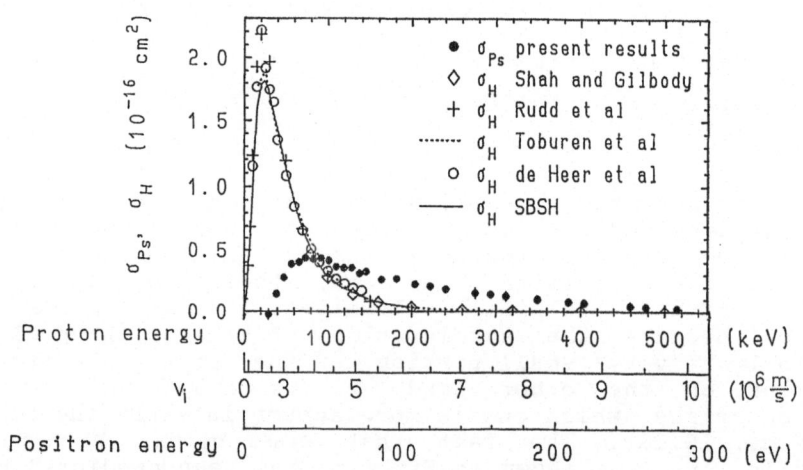

Fig. 15. Ps formation cross sections for helium measured at
Bielefeld (dots) compared with literature values of
the H-formation cross section in proton-helium
collisions versus <u>initial</u> velocity. (From ref. 23)

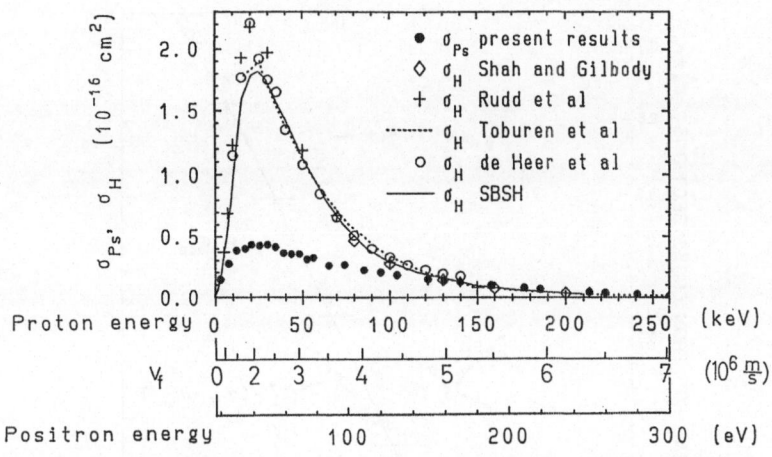

Fig. 16. Same as Fig. 15 but now plotted versus __final__ velocity.
(From ref. 23)

TOTAL SCATTERING

Total scattering cross sections can be measured with
relative ease in gas-target transmission experiments. They were
the first positron cross sections to be measured. Of particular
value are measurements of positron and electron cross sections
in the same apparatus. Many target gases were studied by the
Detroit group employing a positron source with a boron moderator which has a very small negative work function for positrons
and, therefore, yields a positron beam with less than 0.1 eV
energy width (but this advantage comes with a terribly low
moderation efficiency). The extensive searches for resonances
in positron scattering from atoms and molecules have not yet
been successful.

With data on σ_{Ps} and σ_{ion}^+ now available, it is possible to
deduct those partial cross sections from σ_t^+ and obtain the
angle-integrated elastic cross section, σ_{el}^+, above E_{Ps}. This
procedure led to the discovery of a cusp in $\sigma_{el}^+(E)$ at E_{Ps}
(Fig. 17).

Rather astonishing and not yet understood is the observed
equality of $\sigma_t^+(E)$ and $\sigma_t^-(E)$ for potassium[33]. Alkali atoms are
exceptional because for them $E_{ion} < 6.8$ eV, which implies that
Ps formation can occur even at zero energy.

The noble-gas atoms show different convergence behavior
for the total cross section (Fig. 18). For helium the convergence of σ_t at 200 eV is in contrast to that of σ_{el} above
3000 eV (Fig. 6) and of σ_{ion} at 600 eV (Fig. 10). Apparently
the early convergence of the total cross section is not an
indication of the validity of the FBA but rather is caused by a
surprisingly exact compensation of the electron/positron excess
for several partial cross sections.

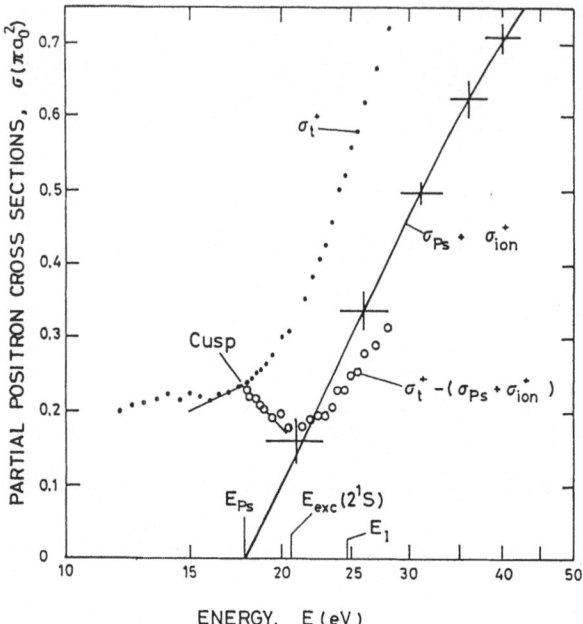

Fig. 17. Total positron-helium cross section, σ_t^+, measured at Detroit (dots) and ($\sigma_{Ps} + \sigma_{ion}^+$) measured at Bielefeld (crosses, fitted by a solid curve going to zero at E_{Ps}) used to obtain the angle-integrated elastic cross section, σ_{el}^+, at energies above E_{Ps}. This evaluation exhibits a cusp in $\sigma_{el}^+(E)$ at E_{Ps} and proves that the early estimates of σ_{Ps} (cf. Fig. 2) were too small. (From ref. 32)

For electron-atom scattering a sum rule follows from the forward-dispersion relation of Gerjuoy and Kroll and has for noble gases (which do not form bound states with the projectile) the form

$$-A - f_{FBA}(E=0,\theta=0) + g_{FBA}(E=0,\theta=0) = (2\pi)^{-1}\int_0^\infty \sigma_t(k)\,dk,$$

where A is the scattering length, f and g are the direct and exchange amplitudes in FBA and $\sigma_t(k)$ is the total cross section as a function of the particle momentum k; A, f and g are in units of the Bohr radius a_0 and σ_t is in units of πa_0^2. Due to singularities in the exchange amplitude the sum rule fails for electrons, but should hold for positrons (g=0). By using all the reliable experimental and theoretical information Kauppila et al.[34] showed that, at least for positron scattering from He and Ne, the sum rule is satisfied (Table 1).

Measured total cross sections have been used repeatedly for partitioning, with the aim of employing well known partial cross section for getting information about less well known

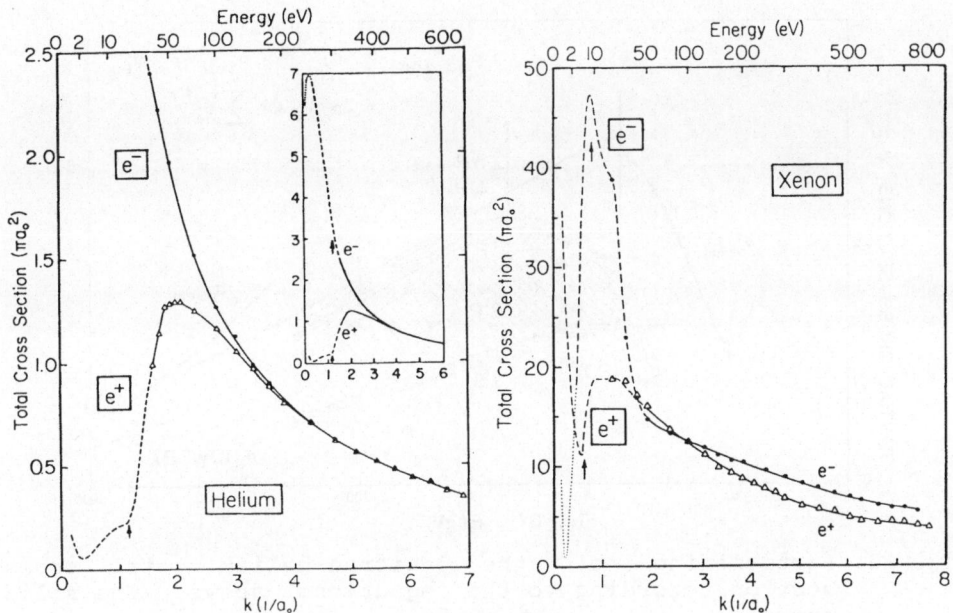

Fig. 18. Total electron and positron cross section for helium and xenon measured at Detroit. Of all noble gases, only helium exhibits the early convergence of the total cross sections at 200 eV, and only xenon exhibits the curve crossings at intermediate energies. (From ref. 34, 35)

Table 1. Test of Zero-Energy Forward-Scattering Dispersion Relation

Atom	Projectile	$-A-f+g$	$(2\pi)^{-1}\int_0^\infty \sigma_t(k)dk$
He	e^-	1.97	2.26
	e^+	1.28	1.25
Ne	e^-	1.89	5.10
	e^+	3.82	3.88

ones. For positron-helium scattering the partitioning which is most revealing at present is to compare

$$\sigma_{no\ ion}^+ = \sigma_t^+ - \sigma_{ion}^+ - \sigma_{Ps}$$

as obtained experimentally, with

$$\sigma_{no\ ion}^+ = \sigma_{el}^+ + \sum \sigma_{exc}$$

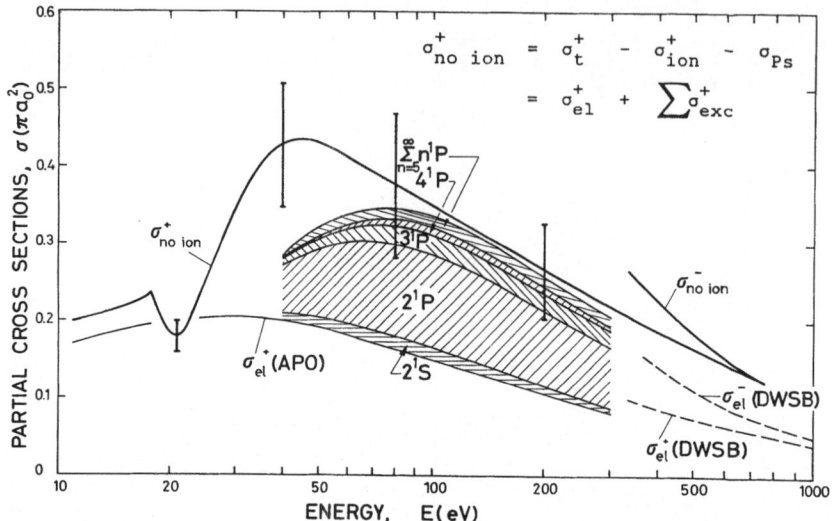

Fig. 19. Partitioning of the positron-helium total cross section according to the equations above. The solid curve $\sigma_{no\ ion}{}^+$, obtained from measurements of $\sigma_t{}^+$, $\sigma_{ion}{}^+$, and σ_{Ps}, is compared with theoretical results for elastic scattering and excitation. APO = adiabatic polarized-orbital method, DWSB = distorted-wave second-Born approximation. The hatched areas represent excitation cross sections for the states indicated, obtained with a distorted-wave approximation. To indicate the convergence behavior the electron cross sections $\sigma_{no\ ion}{}^-$ and $\sigma_{el}{}^-$(DWSB) are shown on the high-energy side. (Adapted From ref. 32)

as predicted theoretically (Fig. 19). The partitioning shows satisfactory agreement of experiment and theory at energies of 100 to 300 eV but also indicates that better calculations are needed for excitation cross sections near threshold. Calculations on excitation at high energies are now under way.[20]

CURRENT DEVELOPMENTS

In crossed-beam experiments (Detroit, Bielefeld) the elastic-scattering studies will be combined with searches for resonances. A crossed-beam e⁺H experiment (Bielefeld-Brookhaven) will start with measurements on ionization. Ionization measurements with a gas target (Bielefeld) will be performed with improved energy resolution in order to test the threshold behavior. Total scattering experiments will concentrate on the alkali atoms (Detroit). The rapidly improving Ps beams will soon be used in scattering experiments (London, Brookhaven).

In the early days of positron scattering the calculations (mainly for H) and the measurements (on easy-to-work-with gases) hardly ever overlapped. Now, theories and experiments have become much more sophisticated. The experimental developments mentioned above all overlap with theoretical studies.

ACKNOWLEDGEMENTS

The Bielefeld experiments have been supported by the Deutsche Forschungsgemeinschaft and the University of Bielefeld. Several colleagues carefully read this manuscript and made valuable suggestions: Professor M. Fink, Dr. G. Sinapius, Dr. D. Fromme and A. Schwab.

REFERENCES

1. "Proceedings of the International Conference on Positron Scattering and Annihilation in Gases", Can.J.Phys. 60:461 (1982).
2. "Positron Scattering in Gases", J.W. Humberston and M. R. C. McDowell, eds., Plenum Press, New York, 1984.
3. "Positron(Electron)-Gas Scattering", W. E. Kauppila, T. S. Stein, and J. M. Wadehra, eds., World Scientific, Singapore, 1986.
4. "Atomic Physics with Positrons", J. W. Humberston, ed., Plenum Press, New York, 1988 (in print).
5. G. Sinapius, G. Spicher and H. L. Ravn, Positron sources for atomic physics experiments, J.Phys.E 19:987 (1986).
6. A. P. Mills, Jr., Brightness Enhancement of Slow Positron Beams, Appl.Phys. 23:189 (1980).
7. A. P. Mills, Jr. and E.M. Gullikson, Solid Neon Moderator for Poducing Slow Positrons, Appl.Phys.Lett. 49:1121 (1986).
8. A. Rich, P. W. Zitzewitz, J. Van House, and D. W. Gidley, Spin Polarized Low Energy Positron Beams and Their Applications, in: Ref. 3, p. 190.
9. W. E. Kauppila and T. S. Stein, Positron-gas cross-section measurements, in: Ref. 1, p. 471.
10. G. M. A. Hyder, M. S. Dababneh, Y.-F. Hsieh, W. E. Kauppila, C. K. Kwan, M. Mahdavi-Hezaveh, and T. S. Stein, Positron Differential Elastic-Scattering Cross Section Measurements for Argon, Phys.Rev.Lett. 57:2252 (1986); S. J. Smith et al., Low Energy Positron-Argon Differential Elastic Scattering Cross Section Measurements, Abstracts XV ICPEAC (Brighton 1987), p. 404.
11. S. N. Nahar and J. M. Wadehra, Elastic scattering of positrons and electrons from argon, Phys.Rev.A 35:2051 (1987).
12. R. P. McEachran and A. D. Stauffer, Differential Cross-Sections for Positron Noble Gas Collisons, in: Ref. 3, p. 122.
13. A. Schwab, P. Höner, W. Raith and G. Sinapius, Elastic Differential Positron Scattering from Argon, in: Ref. 4 (in print).

14. S. P. Khare, A. Kumar, and K. Lata, Elastic Scattering of Positrons & Electrons by Argon Atoms, Indian J. Pure and App. Phys. 20:379 (1982); C. J. Joachain and R. M. Potvliege, Differential Cross Sections for Elastic Positron-Argon Scattering, Abstracts, XV ICPEAC (Brighton, 1987), p. 406.

15. D. Fromme, Ionisation von Helium durch Positronen, Dissertation, Bielefeld 1987.

16. D. P. Dewangan and H. R. Walters, The elastic scattering of electrons and positrons by helium and neon: the distorted-wave second Born approximation, J.Phys.B 10:637 (1977).

17. A. D. Stauffer and R. P. McEachran, Spin Dependent Effects in the Elastic Scattering of Electrons and Positrons from Heavy Atoms, in: Ref. 3, p. 166.

18. K. Hasenburg, Spin polarization in elastic positron-atom scattering, J.Phys.B 19:L499 (1986).

19. P. G. Coleman and J. T. Hutton, Excitation of Helium Atoms by Positron Impact, Phys.Rev.Lett. 45:2017 (1980).

20. R. P. McEachran, L. A. Parcell, A. D. Stauffer, and T. Zuo, Excitation of Helium by Positron Impact and a Comparison with Electron Data, in: Ref. 4 (in print).

21. D. Fromme, G. Kruse, W. Raith, and G. Sinapius, Partial-Cross-Section Measurements for Ionization of Helium by Positron Impact, Phys.Rev.Lett. 57:3031 (1986).

22. H. Klar, Threshold ionisation of atoms by positrons, J.Phys.B 14:4165 (1981).

23. D. Fromme, G. Kruse, W. Raith, and G. Sinapius, to be published in J.Phys.B.

24. M. Brauner and J. S. Briggs, Ionisation to the projectile continuum by positron and electron collisions with neutral atoms, J.Phys.B 19:L325 (1986).

25. M. Charlton, G. Laricchia, N. Zafar, and F. M. Jacobsen, Inelastic Positron Collisions in Gases, in: Ref. 4 (in print).

26. G. Laricchia, M. Charlton, T. C. Griffith, and G. Clark, Excited State Positronium Formation in Low Density Gases, Phys.Lett. 109A:97 (1985).

27. G. Laricchia, M. Charlton, S. A. Davies, C. D. Beling, and T. C. Griffith, The production of collimated beams of o-Ps atoms using charge exchange in positron-gas collisions, J.Phys.B 20:L99 (1987).

28. K. G. Lynn, A. P. Mills, Jr., L. O. Roellig, and M. Weber, Description of the Intense, Low Energy, Monoenergetic Positron Beam at Brookhaven, in: Electronic and Atomic Collisions, D. C. Lorents, W. E. Meyerhof, and J. R. Peterson, eds., North-Holland, Amsterdam, 1986, p. 227.

29. G. Laricchia, S. A. Davies, M. Charlton, T. C. Griffith, and C. D. Beling, On the production of a timed positronium beam by positron gas scattering, Abstracts, XV ICPEAC (Brighton, 1987), p. 414.

30. F. M. Jacobsen, L. H. Andersen, B. I. Deutch, P. Hvelplund, H. Knudsen, M. Charlton, G. Laricchia, and M. Holzscheiter, On Antihydrogen Production, in: Ref. 4 (in print).

31. N. C. Deb, J. McGuire, and N.C. Sil, Electron Capture by Positron in He, Abstract, XV ICPEAC (Brighton, 1987), p. 402.

32. R. I. Campeanu, D. Fromme, G. Kruse, R. P. McEachran, L. A. Parcell, W. Raith, G. Sinapius, and A. D. Stauffer, Partitioning of the positron-helium total scattering cross section, J.Phys.B 20:3557 (1987).

33. T. S. Stein, M. S. Dababneh, W. E. Kauppila, C. K. Kwan, and Y. J. Wan, Positron-(and Electron-) Alkali Atom Total Scattering Measurements, <u>in</u>: Ref. 4 (in print).

34. W. E. Kauppila, T. S. Stein, J. H. Smart, M. S. Dababneh, Y. K. Ho, J. P. Downing, and V. Pol, Measurements of total scattering cross sections for intermediate energy positrons and electrons colliding with helium, neon, and argon, Phys.Rev.A 24:725 (1981).

35. M. S. Dababneh, Y.-F. Hsieh, W. E. Kauppila, V. Pol, and T. S. Stein, Total-scattering cross-section measurements for intermediate-energy positrons and electrons colliding with Kr and Xe, Phys.Rev.A 26:1252 (1982).

36. P. G. Coleman and J. D. McNutt, Measurement of Differential Cross Sections for the Elastic Scattering of Positrons by Argon Atoms, Phys.Rev.Lett. 42:1130 (1979).

ATOMS IN INTENSE LASER FIELDS

Pierre Agostini

Service de Physique des Atomes et des Surfaces

CEN Saclay, 91191 Gif sur Yvette, France

INTRODUCTION

During the past few years, the interest in laser-atom interaction has been spurred by the fast progress of laser technology: it is now rather common to talk about intensities of 10^{15} W.cm^{-2} and the number of systems reaching 10^{17} W.cm^{-2} will be of the order of 10 within a year or so. The prospect for the next few years is to build systems with a peak power of one petawatt (10^{15}W) which could be focussed down to the diffraction limit (10^{-8} cm^2) yielding an incredible intensity of 10^{23} W.cm^{-2} (Mouroux 1987). Such intensities will be avalaible at 1064 nm (Nd^3 YAG), and 248 nm (KrF). An atom submitted to an intense electromagnetic field will be ionized whatever the light frequency by multiphoton absorption.

"Multiphoton Ionization of Atoms" is a topic which emphasizes the interaction between light and atoms. Multiphoton Ionization (MPI) of atoms has been investigated for about twenty years from the radio frequency range to vacuum UV. I would lack both the time and the expertise to review all the aspects that have been studied. As an introduction to the field, I have selected a few topics because of their tutorial character or current interest. For more complete reviews, see Lambropoulos, 1976 and Mainfray and Manus, 1984.

The lectures are organized as follows: Section I is devoted to the problem of one-electron atoms in strong fields: in the first part I summarize the results of Lowest Order Perturbation Theory (LOPT), which has been, for years, the reference framework. In the second part, I consider the corrections to LOPT and the problem of Above-Threshold Ionization (ATI) in intense fields. In Section II, I review the situation of two-electron atoms in moderate fields with special emphasis on the role of doubly excited states in MPI of Sr. In Section III, I discuss (rather briefly) the problem of multielectron excitation in strong fields.

I - SINGLE ELECTRON EXCITATION IN STRONG FIELDS

Consider an atom with one active electron irradiated by a quasi monochromatic ($d\omega \ll \omega$) electromagnetic field. The electric field of the em wave is related to the intensity (average of the Poynting vector) through the equation:

$$E_e = \sqrt{\frac{2I}{\varepsilon_0 C}} \tag{1.1}$$

The typical strength of the Coulomb field, E_c, (1 a.u = 5.7×10^9 V.cm^{-1}) is much larger than E_1 even for fairly high intensities. Several parameters can be defined to characterize the relative strengths of the Coulomb and the electromagnetic interactions. Comparing E_c and E_1 provides a naturel parameter for perturbation expansion:

$$\alpha_1 = \frac{E_e}{E_c} \tag{1.2}$$

Other useful parameters (found with different names in the literature) are related to the motion of the free electron in the em field: comparing the average kinetic quiver energy to the photon energy yields:

$$\alpha_2 = \frac{q^2 E_e^2}{4m\hbar\omega^3} \tag{1.3}$$

while comparing the amplitude of this motion to the atomic unit of length provides

$$\alpha_0 = \frac{q\,E_e^2}{a_0\,m\omega^2} \tag{1.4}$$

If both a_1 and a_2 are small compared to 1 ($I < 10^{12}$ W.cm^{-2} for optical frequencies) the problem can be treated within the framework of LOPT at least far from resonances, as explained below.

I.1 - Lowest-order perturbation theory

Using the Coulomb gauge, the vector potential for a plane wave linearly polarized in the z direction and propagating along the y direction can be written as:

$$\vec{A} = \vec{\varepsilon}_z \frac{\vec{E}_e}{\omega} [\sin(ky-\omega t)] \tag{1.6}$$

where $\vec{\varepsilon}_z$ is the unit vector along the z direction, $\vec{k}$ the wave vector. Since the electron is confined in a very small region of space during the interaction, the dipole approximation can be used and the vector potential will be taken as:

$$\vec{A} = -\vec{\varepsilon}_z \frac{\vec{E}_e}{\omega} \sin \omega t \tag{1.5}$$

The hamiltonian of the system is:

$$H = H_0 + V_A \tag{1.7}$$

where:

$$H_0 = \frac{p^2}{2m} + V(r) \tag{1.8}$$

and where V_A is the interaction hamiltonian:

$$V_A = -\frac{q}{m}\, \vec{p} \cdot \vec{A} - \frac{q^2 A^2}{2m} \tag{1.9}$$

By choosing a gauge in which $A = 0$ (in the dipole approximation), the interaction hamiltonian takes the form:

$$V_E = -qz \tag{1.10}$$

Generalized cross-sections

It is well known that, within the framework of time-dependent perturbation theory, the probability per unit time for a one-photon ionization can be obtained by the Fermi golden rule:

$$W = \frac{2\pi}{\hbar}\left(\frac{2q^2}{\varepsilon_0 c}\right)\, I\ |\langle \varepsilon |z| g \rangle|^2 \tag{1.11}$$

where the value of E_1 form Eq. (1.1) has been substituted in (1.10) and the continuum wavefunction $|\varepsilon\rangle$ has been energy normalized. Expression (1.11) can be generalized (Gontier and Trahin, 1984) to N-photon ionization as:

$$W_N = \frac{2\pi}{\hbar}\left(\frac{2q^2}{\varepsilon_0 c}\right)^N\, I^N \sum_\varepsilon \left(\sum_{i,j,k} \frac{\langle g|z|i\rangle\langle i|z|j\rangle \ldots \langle k|z| \rangle}{(E_g + \hbar\omega - E_i)(E_g + 2\hbar\omega - E_j) + \ldots} \right) \tag{1.12}$$

where $|i\rangle$, $|j\rangle$, $\ldots |k\rangle$ are the atomic states (discrete + continuum), I the intensity and $.\rangle$ the continuum states with energy $E_g + N\hbar\omega$, E_g being the energy of the ground state $|g\rangle$. Usually the generalized cross-section is defined by:

$$\tilde{\sigma}_N = \frac{W_N}{F^N} \tag{1.13}$$

where F is the photon flux $I/\hbar\omega$. Using $q_2/c = 4n\alpha h$ ($\alpha = 1/137$) one gets:

$$W_N = \frac{2\pi}{\hbar}\,(8\pi\alpha\hbar^2)M_N\ \omega^N \tag{1.14}$$

where M_N is the Nth-order matrix element (Crance, 1984a):

$$M_N = \sum_\varepsilon \left(\sum \frac{\langle g|z|i\rangle\langle i|z|j\rangle \ldots \langle k|z|\varepsilon\rangle}{\delta E_i\ \delta E_j\ \ldots\ \delta E_k} \right) \tag{1.15}$$

where δE_i, δE_j, $\ldots$ are defined by Eq. (1.12). The dimensions of $\tilde{\sigma}_N$ are (obviously) $L^{2N}T^{N-1}$.

The computation of M_N is quite a difficult problem which will not be considered here (Gontier and Trahin, 1971). However, an order of magnitude for σ_N can easily be obtained by using approximations (Crance, 1984b, Lambropoulos and Tang, 1987) based on the following idea (Bebb and Gold, 1966): in (1.12) all the energy denominators are replaced by a common value, chosen more or less arbitrarily: the denominators then factor out and the summations over i, j, k, vanish and one gets:

$$\tilde{\sigma}_N \simeq 2\pi \ (8\pi\alpha)^N \ |\langle\varepsilon|z^N|g\rangle|^2 \qquad (1.16)$$

Just to get order of magnitude, we will use here a rough approximation of the overlap integral (taken equal to a_0^{2N}), which yields suprisingly good numbers for low N:

$$\tilde{\sigma}_N \simeq 2\pi \ (8\pi\alpha)^N \ a_0^{2N} \ \omega^{-N+1} \qquad (1.17)$$

Using (1.17), one gets for the first N's:

$$\tilde{\sigma}_1 = 2.9 \ 10^{-15} \ cm^2$$

$$\tilde{\sigma}_2 = 1.3 \ 10^{-49} \ cm^4 s$$

$$\tilde{\sigma}_3 = 4.6 \ 10^{-82} \ cm^6 s^2$$

in reasonable agreemant with exact values (Morellec et al., 1982). It is well known (Gontier and Trahin, 1971) that $\tilde{\sigma}_N$ exhibits maxima (corresponding to intermediate resonances with atomic states such that one denominator in (1.16) vanishes) and minima. The resonant cases are considered farther on with the corrections to LOPT. The numbers provided by (1.17) are close to the value of $\tilde{\sigma}_N$ around the minima.

Antiresonant case: destructive interference

The minima of σ_N, particularly sharp for small N's, are due to destructive interference of different quantum paths leading to the same final state. One transparent case of such destructive interference is provided in two-photon ionization of Cs when the energy of one photon almost equal to the energies of two states coupled to the ground state. The infinite summations in (1.16) may then be reduced to the sum of two terms describing the two main quantum paths leading to ionization. Since these terms have comparable amplitudes and opposite signs, they interfere destructively and the generalized cross-section becomes very small (Morellec et al., 1982). Experimentally, the main difficulty in the measurement of such minima is to get rid of any background signal. In the case of Cs, considerable background comes from the dissociation and ionization of dimers (Morellec et al., 1982). Minima of generalized cross-sections provide excellent tests for calculations since both the amplitude and position of the minima critically depend on the values of the matrix elements.

"Slope" - Saturation intensity

According to Eq. (1.12), the transition rate w_N, for an N-photon process is proportional to the Nth power of the intensity. Of course, this is valid only for small probabilities (or short times). When valid, the quantity $d \log(w_N) / d \log(I)$, called "slope", is just equal to the number of photons absorbed in the transition. In practice, the slope is determined by plotting w_N in log-log coordinates. This simple consequence of LOPT has been experimentally verified for N up to 22 (Lompre et al., 1977) and $I = 10^{15}$ W.cm^{-2}. This has often been taken as a proof of the validity of LOPT. However, it is known now that at such intensities, other processes take place (as discussed hereafter). Therefore the power law I^N is not a sufficient proof of the validity of LOPT.

In a given experiment, when the intensity is increased beyond a certain value, the dependence of w_N on I changes because a significant fraction of the ground state population has been ionized. This "saturation intensity", I_S, is determined through the equation:

$$\tilde{\sigma}_N \ \tau_N \ I_S^N = 1 \tag{1.18}$$

where τ_N is the effective pulse duration (Morellec et al., 1982). Two points must be emphasized about saturation intensity; the first one is rather obvious: beyond I_S, the intensity dependence of the probability does no longer reflext the MPI law but rather the intensity dependence of the interaction volume. The second one has been often overlooked (Lambropoulos, 1985): the maximum intensity I_M, in a realistic pulse, is not reached instantaneously. If $I_M > I_S$ for a given MPI process, the atom may never be exposed to I_M since it may be ionized before the intensity reaches this value. This will happen if the time necessary to reach I_S is such that (1.18) is satisfied. Hence the necessity to use risetimes as short as possible in order to really utilize the peak intensity of the pulse. The condition under which the atom will be exposed to I_M (reached aften τ_N is that $I_M > I_S$ (calculated for τ). The saturation fluxes, for a given atom and increasing N's can be shown (Lambropoulos and Tang, 1987) to tend towards a limit roughly equal to the atomic unit of flux 3.3×10^{32} cm^{-2}s^{-1}. This property can easily be deduced from Eq. (1.17) and generalized to an other atom using the scaling law (Lambropoulos and Tang, 1987):

$$L_N^{(A)} = L_N^{(H)} \ \frac{R_A^2 \ E_i(H)}{R_H^2 \ E_i(A)} \tag{1.19}$$

where L_N is $(\tilde{\sigma}_N)^{1/N}$, R_A the atomic "radius" and $E_i(H)$, $E_i(A)$ are the ionization potentials of hydrogen and atom respectively. This will be used to interpret the multiple ionization observations in Sect. III. Finally, it should be pointed out that, experimentally one cannot choose freely the intensity at which a given order MPI process will be studied. The minimum intensity is the one at which the signal appears and the maximum one is I_S.

Angular distributions of photoelectrons

It can be shown (Lambropoulos, 1976, Leuchs, 1984) that, when the light is linearly polarized, the angular distribution of the photoelectrons produced by N-photon ionization depends only on the angle θ between the polarization vector and the electron wave vector. The differential cross-section can be writtent as:

$$\frac{d\tilde{\sigma}_N}{d\Omega} = \frac{\tilde{\sigma}_N}{4\pi} \sum_{j=0}^{N} b_{2j} \ P_{2j}(\cos \theta) \tag{1.19'}$$

where P_{2j} (cos θ) is the Legendre polynomial of oder 2j and b_{2j} are coefficients which depend on the transition matrix elements. Experimentally it is very simple to record such distributions just by rotating the laser polarization. In these lectures, only two aspects will be considered: (i) in some cases the angular distributions may become intensity dependent (due to ac-Stark shits); (ii) in ATI, the angular distributions are more and more peaked along the laser polarization when the number of absorbed photons increased.

Above-threshold Ionisation (moderate intensity)

Between the midsixties and the end of the seventies, most of the experimental investigations of MPI were focussed on the determination of the total cross-sections through the measurement of the total <u>ion</u> current. Since, as mentioned above, LOPT seemed to account for all that could be observed, it came as a surprise that recording of the <u>electron</u> spectra revealed electron energies much larger than allowed by the basic law of MPI:

$$E_N = N\hbar\omega - E_i \tag{1.20}$$

N being the minimum number of photons of energy $\hbar\omega$ required to ionize the atom with an ionization potential E_i ($E_i > 0$). After some improvements of the experimental techniques, it was soon realized (Agostini et al., 1979) that the energies in the electron spectra were not continuously distributed but were concentrated in lines separated by the photon energy, as shown on Fig. 1. The simplest way to understand such spectra without invoking collisions (ruled out under the usual experimental conditions) is to consider transitions involving (N+S) photon absorption which couple the atom ground state to higher continuum states. The energy conservation law for such a process writes:

$$E_{N+S} = (N+S)\hbar\omega - E_i \tag{1.21}$$

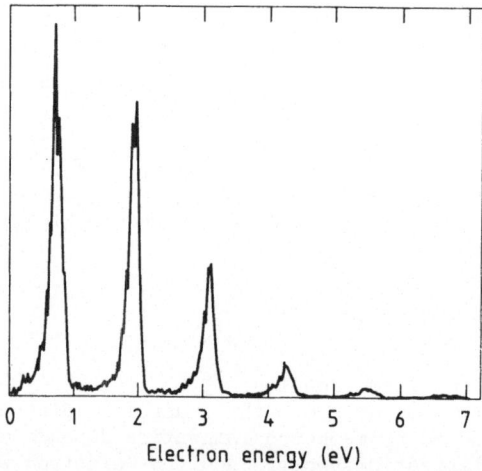

Electron energy (eV)

Figure 1. Typical electron energy spectrum from 11-photon ionization of Xe at 1064 nm displaying the characteristic $\hbar\omega$ structure

As a matter of fact, such a process has been calculated as soon as 1965 by Zernik and Klopfenstein for N = 1, S ≈ 1. It was then generalized by Gontier and Trahin in 1980 and named Above-Threshold Ionization (ATI). The reason to discuss ATI in this section rather than in the following one is that, in spite of the photons absorbed in excess to the minimum value N, each of the (N+S)-photon MPI can be described in terms of "LOPT" (Agostini et al., 1984): the intensity dependence of each ATI transitions being governed by a power law I^{N+S} with generalized cross-sections quite "normal", in the sense of Eq. (1.17), at least far from any resonances. Further proofs of a "perturbative" behavior of ATI under moderate intensity was provided by measurements of the ratios of amplitudes of consecutive electron "peaks". This ratio was found (i) to depend linearly on intensity (Kruit et al., 1981) and (ii) to saturate when $I = I_s$ (Petite et al., 1984). The latter effect was another proof of the direct character of the ATI transition. There is an apparent discrepancy between the total ionization power law I^N and the partial ionization laws I^{N+S} since the former must be the sum of the latter. However, this problem is easily solved when the amplitudes of the ATI transitions are rapidly decreasing with S and give small contributions to the total rate. ATI seemed therefore to fit within the framework of LOPT (up to 10^{11} W.cm^{-2}) until an experiment performed at higher intensity (Kruit et al., 1983), revealed clearly a non perturbative behavior and renewed the interest for this problem. This question will be addressed in the following sub-sections.

I.2 - Corrections to LOPT

So far we have considered only the lowest non-vanishing order of perturbation theory to described the transitions induced by the em field: we have neglected all the higher order terms of the perturbation expansion. For a given order N these terms are obtained by considering all possible combinations of arbitrary numbers of absorption and emission of photons in such a way that the net number of absorption remains N (Gontier et al., 1976). There are circumstances, however, where the higher order terms cannot be ignored. For instance, when there is a resonance (one of the denominators in (1.12) vanishes), LOPT fails to reproduce the experimental results even at low intensities. The physical effect that must be taken into account and added to the lowest-order transition is the perturbation of the atomic levels due to the interaction with the em field or ac-Stark shifts. In the next sub-section, this problem is addressed in the most simple way, by calculating the shifts to the second order of perturbation without trying to incorporate this effect in the ionization calculations. This is addressed properly by using more elaborate techniques (like the Resolvent Operator (Gontier and Trahin, 1984)) which will not be discussed here. For more details the reader is referred to the literature and to another paper in this book (Giusti-Suzor).

First and second order corrections to energies

In order to evaluate the lowest-order <u>corrections</u> to the level energies, we rewrite the interaction hamiltonian (1.9) as:

$$V_A = V_1 + V_2 \qquad (1.22)$$

with

$$V_1 = -\frac{q}{m}\vec{p}\cdot\vec{A} \quad \text{and} \quad V_2 = \frac{q^2 A^2}{2m}$$

The physical process giving rise to the correction we are seeking is the absorption and emission of one photon. As easily seen by defining the vector potential A in terms of creation and destruction operators a^+ and a: V_1, which is linear in a and a^+ must be taken to second order to describe one adbsorption and one emission while V_2 which contains aa^+ and a^+ operators describes such a process to first order. Using perturbation theory, the light shift for a state $|i\rangle$ (energy E_i) can be written as:

$$\Delta E_i = \Delta E_i^{(1)} + \Delta E_i^{(2)} \tag{1.23}$$

where:

$$\Delta E_i^{(1)} = \frac{q^2 E_e^2}{4m\omega^2} \tag{1.24}$$

and:

$$\Delta E_i^{(2)} = \frac{1}{2}\left(\frac{qE_e}{m\omega}\right)^2 \sum_j \left|\langle i|\vec{p}|j\rangle\right|^2 \left[\frac{1}{E_i - E_j + \hbar\omega} + \frac{1}{E_i - E_j - \hbar\omega}\right] \tag{1.25}$$

It is clear from (1.24) that $\Delta E_i^{(1)}$ does not depend on $|i\rangle$, as expected from the fact that A^2 does not act on atomic variables. V_2 shifts all the levels by the same amount which is equal to the average kinetic energy of a free electron oscillating in an em field. An order of magnitude of this shift is 1 eV at 1064 nm and $I = 10^{13}$ W.cm^{-2}. Note the dependence of V_2 on the photon energy (ω^{-2}). For the same intensity, in the beam of a CO_2 laser (10.6 μm), the shift would be 100 eV! Note also the A^2 term cannot account for shifts in energy differences and consequently of any change of the ionization potential. I will come back to this in the ATI sub-section. The discussion hereafter follows Pan et al., 1986.

Low frequency approximation

If $\hbar\omega \ll E_i - E_j$, (1.25) can be expanded in terms of $\hbar\omega/E_i - E_j$ and, after straightforward calculations, one gets:

$$\Delta E_i^{(2)} = -\frac{q^2 E_e^2}{4m\omega^2} - \frac{1}{2}\alpha_i E_e^2 \tag{1.26}$$

where:

$$\alpha_i = q^2 \sum_j \frac{X_{ij}^2}{E_i - E_j} \tag{1.27}$$

is the usual linear polarizability of state $|i\rangle$ in which we have used:

$$X_{ij} = \langle i|x|j\rangle = \frac{\hbar}{m}\frac{\langle i|\vec{p}|j\rangle}{E_i - E_j} \tag{1.28}$$

The first term in (1.26) is just the negative of (1.24) and the second term is equivalent ot a dc-stark shift. The total shift in the low frequency approximation is:

$$\Delta E_i^L = -\frac{1}{2} \alpha_i \, E_e^2 \tag{1.29}$$

For instance the ground state of Xe atoms irradiated by an em field at 1064 nm will have a negative shift given by (1.29).

High frequency approximation

If $\hbar\omega \gg E_i - E_j$, (1.23) can be expanded in terms of $\left(E_i - E^j\right)/\hbar\omega$. The result is that, to second order, the shift is essentially zero. Neglecting the second order shift, the total shift is in this approximation:

$$\Delta E_i^H = \frac{q^2 E_e^2}{4m\omega^2} \tag{1.30}$$

The physical interpretation of the Stark shift, in this case is very simple: the high-frequency approximation holds for regions of the spectrum where the eigenvalues of the unperturbed hamiltonian lie very close to each other, typically for Rydberg states. For such states, the electron motion is quasi classical and the orbit period is much longer than the optical period. The effect of the light is to add an oscillatory motion to be Kepler motion and therefore to add an (average) kinetic energy given by Eq. (1.24). This prediction has been experimentally checked in Rb (Liberman et al., 1983). However, several restrictions to applicability of the high-frequency approximation must be considered (Avan et al., 1976) and Eq. (1.14) must not be used without caution for all Rydberg states.

General case

If the photon energy is comparable with atomic energy differences, it is necessary to perform the full calculation in order to evaluate the ac-Stark shift. This is especially the case when the photon energy is comparable to atomic energy differences (alkali and alkaline earth atoms with UV or visible photons).

Note that the shift may be calculated using the length form of the interaction hamiltonian (1.10). The final result, which is gauge-independent for exact calculations, may depend on the choice of gauge if one resorts to approximate methods. The next two sub-sections deal with effects which are direct consequences of ac-Stark shift corrections. The question of the shift of the ionization potential will be treated separately.

Resonant Multiphoton Ionization

When the photon energy is such that one of the energy denominators in (1.12) vanishes (i.e. when the energy of an integer number of photons coincides with the energy of an allowed excited state), LOPT is no longer valid. Experiment shows that the ionization probability, as a function of the wavelength, shows a sharp maximum (resonance) but that this resonance is broadened and shifted when intensity is increased (Fig. 2). The problem must then be treated with more powerful formalisms (Gontier and Trahin, 1984; Grance, 1984) which will not be examined here. I will

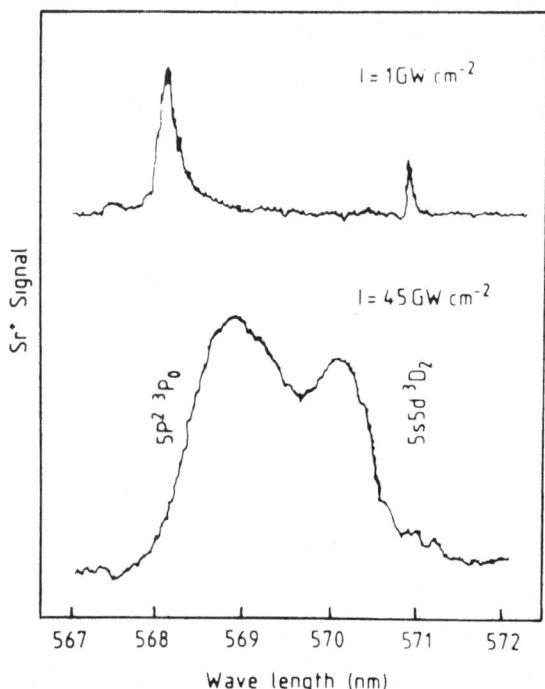

Figure 2. Resonance shifts and broadenings in MPI of Sr
Two-photon resonant, three-photon ionization for
two intensities

rather emphasize the physical effects which complicate the MPI processes. I will only consider the case where the coupling between the reso-
nant state $|r\rangle$ and the continuum is much larger than the coupling bet-
ween the ground state $|g\rangle$ and the resonant state. It is generally so
when the number of photons needed to reach the resonant state, say P is
larger than the number of photons absorbed to make the transition from
the excited state to the continuum N-P. In this case, the broadening is
mainly due to the ionization and it is still possible to define and
ionization rate which is given, around the resonance by:

$$W = \frac{2\ G\ R^2}{D^2 + G^2}$$

(1.31)

Where G is the ionization width of the resonant state (proportional to
i^{N-P}), R describes the coupling between $|g\rangle$ and $|r\rangle$ (R is proportional
to I^P) and D is the dynamic detuning which take into account the
ac-Stark shifts of $|g\rangle$ and $|r\rangle$. Eq. (1.31) describes a Lorentzian
centered on D = 0 and with a full width (half-maximum) G. The first
consequence of the level shifts is therefore that the resonance occurs
when the dynamic detuning is zero rather than for $P\hbar\omega = E_r - E_g$. The
second consequence is that the intensity dependence of the rate is
strongly modified around the resonance. This situation has been
throughly studied at the end of the seventies (Morellec et al., 1976). I
would like to stress that, although the correct theoretical treatment

of the problem is rather involved, the dominant physical effect is the ac-Stark shifts of the ground state and resonant level.

Quasi-resonant MPI/ATI

Even far away from resonances, the level shifts may play a role if the intensity is high enough. Such a situation was met in 4- and 5-photon ionization of Cs at 1064 nm, 10^{11} W.cm^{-2} (Petite et al., 1984). The closest resonances are a two-photon resonance with the 7s state or three-photon resonances with either the 6f states or np states with n form 9 to 13 (Fig. 3). The static detuning for any of these resonances is around 300 cm^{-1}. As shown on Fig. 4a, for I = 10^9 W.cm^{-2}, the generalized cross-sections do not depend on I, as expected from LOPT. When the intensity is about 10^{11} W.cm^{-2} the cross-sections first increase to a broad maximum before dropping by several orders of magnitude for I = 10^{12} W.cm^{-2}. This latter effect is easily traced back to the interference between the quantum paths leading to the 9p and the 10p states which have static detunings roughly equal and of opposite sign. The maximum is due to the Stark shifts of the 6s and 7s states which reduce the detuning and, as a consequence, increase the cross-section. Due to the level shifts, the angular distributions of the photoelectrons also become intensity dependent as shown on Fig. 4b in which the distributions are displayed, in polar coordinates, for different intensities. For instance the 4-photon angular distribution shows a minimum along the laser polarization and maxima at about 30 degres and 90 degres at I = 10^9 W.cm^{-2}. At I = 10^{11} W.cm^{-2}, the maximum at 30 degres and the minimum at 0 degre have disappeared.

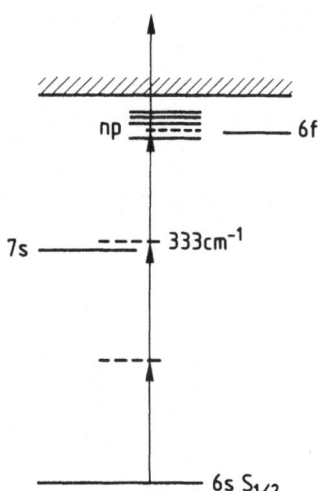

Figure 3. Cs partial energy level diagram pertaining to quasi-resonant four-photon ionization

I.3 - <u>Above-Threshold Ionization (high intensity)</u>

The first osbervation of a non-perturbative behavior of ATI (Kruit et al., 1983) came as a surprise since all non-resonant MPI experiments seemed, up to 10^{15} W.cm^{-2}, to fit into the LOPT framework. If theoretical models have been blooming for the last four years, it was only in 1985 that other experimental observations confirmed the first one (Humpert et al., 1985, Yergeau et al., 1986, Bucksbaum et al., 1986). All these observations have been made in the ATI spectra of the 11-photon ionization of Xe at 1064 nm and intensities between 2×10^{12} W.cm^{-2} and 3×10^{13} W.cm^{-2}. I will not describe here the different setups which have been used (lasers, electron spectrometers, etc...) but rather make a list of the main characteristics of the ATI spectra (Fig. 5).

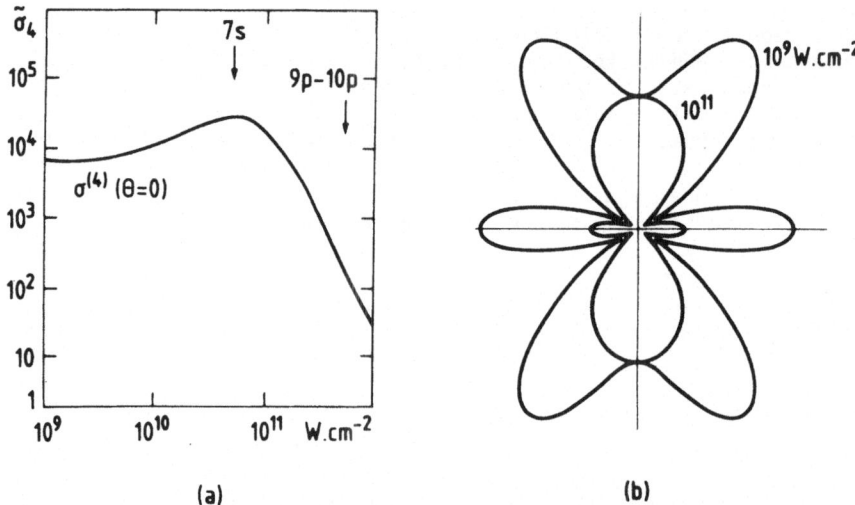

(a)   **(b)**

Figure 4. (a) Intensity dependent 4-photon ionization generalized
                      cross-section
         (b) Intensity dependent angular distribution

(i) The spectra consist, as in the low intensity case described above, in a series of lines (peaks evenly spaced by the photon energy, 1.165 eV. But: the number of peaks can be very large (S-values up to 40 have been reported (Feldmann, 1986) and the first peaks are relatively <u>suppressed</u>. In other words, the largest probability is not at the lowest possible order.

(ii) The peaks are found where expected (no shifts).

(iii) The intensity dependence of the peak amplitudes is at variance with the power laws predicted by LOPT: I^{N+S}. The "slope" of the first few peaks are smaller than N+S while for high order peaks the slopes seem to tend towards N. However the slope measurements must be taken with some caution since they carry rather large uncertainties.

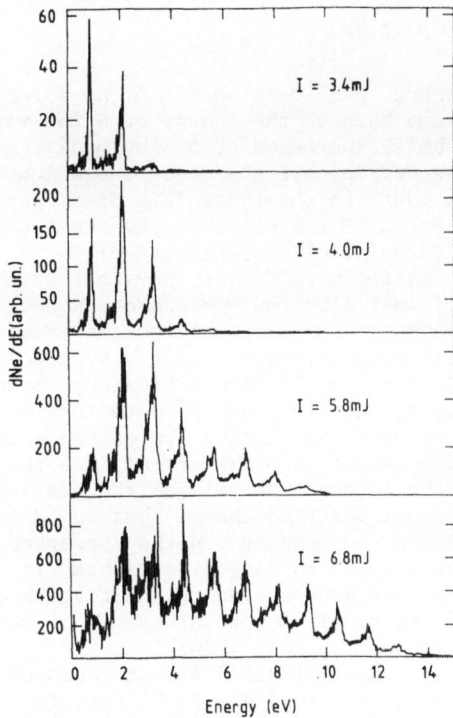

Figure 5. Above threshold Ionization in Xe at 1064 nm

(iv) The peak width increases rapidly with the intensity. The maximum observable width is about .4 eV.

(v) The peak suppression occurs at intensities well below I_s (Yergeau et al., 1986).

(vi) Circular polarization is more efficient for the peak suppression (Bucksbaum et al., 1986).

Theoretical models

First it should be pointed out that, at such intensities, LOPT should hold since $\alpha_1 \ll 1$ and resonances are not expected to play much of a role. It is out of the scope of these lectures to review all the theoretical models which have been produced to interpret this behavior, or even to quote them all. For instance I will leave aside the attempts to extend LOPT to the high intensity limit by using the dressed atom approach, by treating exactly the continuum-continuum transitions etc... (Guisti-Suzor, 1987, in this volume). The interested reader is also referred to the reference lists in the May issue of the J. Opt. Soc. Am. B, 1987. I will rather stress the physical ideas which underline some of them:

(i) The first explanation, proposed by Muller et al., 1983, is based on the idea that, because of the different Stark-shifts for the atom ground state and Rydberg states, the ionization potential (IP) is actually increased by a quantity close to the average quiver energy of the electron (1.24). This idea is supported by a calculation for a model-atom. Using this picture, the energy conservation for the (N+S)-photon ATI writes:

$$E_{N+S} = (N+S)\hbar\omega - (E_i + \Delta) \tag{1.32}$$

where the quiver energy has been noted Δ. As a result, the first few peaks may be suppressed because the energy of N (or more) photons is not enough to overcome the IP increased by Δ. Obviously, a shift of the same amount should be observed for all the electrons which have been actually ejected contrary to what was observed. This discrepancy was resolved by invoking the acceleration of the electrons by the ponderomotive force, a somewhat misleading formulation (see hereafter).

(ii) A second, entirely different, physical effect has been proposed by Crance (1986): most electron spectrometers have a small acceptance angle, which leads to very low collection efficiency whenever the electron angular distribution is not strongly peaked in the direction of the detector. Inefficient collection efficiency means that many more electrns have to be created than can be detected, so that space charge could also be the culprit for making low energy electron peaks disappear. After ionization, there is a mixture of ions and electrons of different energies in the interaction volume; the fast ones get away rapidly, leaving behind a net positive charge that could trap or deviate the slow electrons, leading to changes in the observed spectra. Although this effect must be present in many experiments, it has been ruled out as an explanation for the suppression of the first peaks since this has been observed with an extremely low total number of charges (< 10) created per laser shot (Yergeau et al., 1986).

(iii) The third treatment I would like to comment about is related to the Keldysh idea of MPI (Keldysh, 1965) that the final state of the transition should be described, a priori, as a Volkov state (the solution of the Schroedinger equation for a free electron in an em field) (Reiss, 1987). Leone et al., 1987, Kupersztych, 1987, have proposed models in which the Volkov states appears more naturally. I will now briefly summarize the Kupersztych model. It is based on the following simple argument: after absorption of N photons, the electron emerges in the continuum. During this part of the transition, the problem may be treated by LOPT since $\alpha_1 \ll 1$. Neglecting all levels shifts the electron has de drift kinetic energy given by (1.20) usually of the order of $\hbar\omega$. The electron remains coupled to the Coulomb field and the em field. However, in a very short time (about one optical period), the electron has travelled a distance large enough for the Coulomb field to be much smaller than the em field and therefore it is now the Coulomb field which appears as a perturbation. This will happen typically at 10^{12} W.cm^{-2} at 1064 nm (note that the ratio R of the Coulomb force to the em force is inversely proportional to the square root of I). The rest of the process should then be described as the scattering of an electron on the Coulomb potential in the presence of the em field. This process is known as "inverse Bremsstrahlung". Here, since the electron scatters in the field of its own parent ion, the process has been named half inverse Bremsstrahlung. As a matter of fact, there is a strong ressemblance between the appearance of additional electron peaks in ATI and the multiphoton inverse Bremsstrahlung (Weingartshofer and Jung, 1984) and it was necessary to investigate the pressure dependence of ATI to rule out the usual Bremsstrahlung (Agostini et al., 1984). The first outcome of the model is that the peak suppression is explained without need of a shift of the IP: since the photoelectron produced after absorption of N photons is still in the em field, it must be produced with a kinetic energy at least to Δ. The energy conservation must therefore be written as:

$$(N + S)\hbar\omega - E_i = E_N + \Delta \tag{1.33}$$

where E_N is given by Eq. (1.20). Eq. (1.33) predicts obviously the same energies as (1.32) with two important differences: first, the term Δ has e different physical meaning now since it is actually coming from the quiver motion of the electron which adds to the drift kinetic energy as endowed by the N-photon adsorption. In (1.32), the term Δ was a quantity which had to be substracted from the drift energy because of the shift of the IP. Second, in order to recover the expected ernergies, it is no longer necessary to invoke an acceleration of the photoelectron by the ponderomotive force but, as shown in the next subsection, a conversion of the quiver energy into drift energy. The peak of order S will be suppressed from the spectra if:

$$ S < \frac{E_i + \Delta}{\hbar\omega} - N \tag{1.34} $$

For Xe, E_i = 12.127 V. If Δ = 2 eV, $\hbar\omega$ = 1.17 eV and N = 11, S = 0,1 will be suppressed.
For He at 10^{14} W.cm^{-2} S up to 100 will be suppressed.
The condition (1.34) can be derived from (1.32) or (1.33) and therefore the two pictures appear as somehow equivalent: this is not suprising since both pictures express the fact that the electron undergood a forced oscillation in the field (more about this at the end of the section).
The probability per unit time corresponding to the $(S+1)^{th}$ electron peak has the form:

$$ W_S = \tilde{\sigma}_N I^N (2e^2/\alpha_0) E_N^{-3/2} E^{1/4} \beta_S^{-4} J_S^2 (\alpha_S) \tag{1.35} $$

where $\tilde{\sigma}_N I^N$ is the "weak-field" ionization probability per unit time, $E_S = E_N + S\hbar\omega - \Delta$, $\beta_S = (E_S/E_N)^{1/2} - 1$ and:

$$ \alpha_S = 2\beta_S (2\Delta E_N)^{1/2}/\hbar\omega $$

and J_S is the Bessel function of order S. The model predicts slopes for the different electron peaks which are in good agreement with experimental observations (Lompre et al., 1987).

Is there a shift of the ionization potential ?

As discussed above, the idea that the IP is increased by Δ is based on the Stark-shifts calculations to second order in perturbation. First it is not granted that this result holds at intensities 4 orders of magnitude larger than the intensity at which it was checked experimentally. Second, it is not clear that the high frequency approximation will hold for all the Rydberg states which could be involved in the MPI process. Finally, there is one direct experimental determination of the IP. In MPI of Kr with a tunable laser, Lompre et al., 1977, have measured the slope as a function of the laser wavelength in the range 1060 nm to 1065 nm. In this range, the threshold from 12 to 13-photon ionization is crossed. The slope measurement changes at a certain wavelength form 12 to 13. Comparing this wavelength to the position expected from the known value of the IP allows a direct measurement of the shift. The result was that the shift was 86 cm^{-1} $<<$ Δ = 16,140 cm^{-1} at the intensity used for the measurement: 2×10^{13} W.cm^{-2}. The conclusionis that the IP is not shifted as much as Δ. This is further supported by other slope measurements always close to N even wehn the shift would have been much larger than the photon energy. The conclusion is that there is no expe-

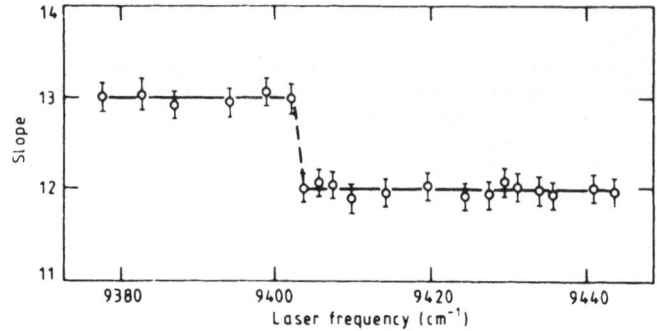

Figure 6. Is there a shift of the ionization potential of Kr at 10^{13} W.cm^{-2} ?

rimental evidence of an important shift of the IP that can be derived from ion detection. It should be stressed that this is consistent with the "two-step" half-bremsstrahlung picture in which the ionization process always involves N photons.

Extreme cases of peak suppression

Two measurements have extended to high intensity and long wavelength the observations of peak suppression. Both utilized a low resolution retarding potential electron spectrometer. Lompre et al., 1985, have reported that about 30 "peaks" are suppressed in the ATI spectrum

of He at 10^{15} W.cm^{-2} ($\hbar\omega$ = 1.165 eV) with a Nd^{3+} laser. Xiong et al., 1987, have reported that about 2500 (!) "peaks" are suppressed in ATI of Xe 8.8×10^{13} W.cm^{-2} ($\hbar\omega$ = .11 eV), with a CO_2 laser. The values of the quiver energies are respectively 100 eV and 190 eV, in qualitative agreement with the observations, owing to the limited precision of the intensity determinations. These results confirm spectacularly the idea that the peak suppression is primarily due to the fact that the electron must acquire at least the quiver energy in excess of the ionization energy to escape freely to the detector. In particular, the expression of the quiver energy (1.24) contains an explicit ω^{-2} dependence which accounts for the much larger effect observed with the CO_2 wavelength. Note also that the Stark-shift of the Rydberg states cannot be calculated using the high-frequency approximation at 10.6 μm. Examining what energy changes occur during this travel is the object of the subsection "ponderomotive effects" hereafter, but let us consider first the question of angular distributions.

Angular distributions in ATI

The angular distributions of ATI electrons have been measured by Humpert et al., 1985, Feldmann, 1986 and Freeman et al., 1986. The first group reported that, as S increases, the angular distributions be more sharply peaked in the direction of the laser polarization, as expected in the classical limit. This characteristic will be very useful in the next subsection, to calculate the work of the em force. The second group has discovered a very interesting feature when S is such that the final state lies close to doubly excited states of Xe: the angular distribution undergo qualitative changes with respect to the general trend defined above. No clear explanation of this observation has been proposed so far. The third group has studied the modifications of the angular distributions due to the ponderomotive effects.

Ponderomotive effects: direct evidence in ATI spectra

In this section we examine the so-called "ponderomotive" effects which are the effects of the space and time variations of the em field on the electron average velocity. We are only interested in the energy changes when the electron, which is "born" inside the em field, exits into free space. For earlier works on the ponderomotive forces or their influence on the angular distributions, the reader is referred to the literature. The energy of the electron outside the beam E_{out} is related to the energy inside the beam E_{in} through the equation:

$$E_{out} = E_{in} + \left\langle \int_{in}^{out} \vec{F} \cdot d\vec{x} \right\rangle_{av} \tag{1.36}$$

where $F = -e(x,t) \cos t$ is the em force. (x,t) is the em field amplitude, which is supposed to vary slowly both in time and space. We are interested in the situation of ATI electrons whose energies are given by (1.21). We identify this energy with the classical average electron kinetic energy inside the beam so that:

$$(N + S)\hbar\omega - E_i = \left\langle \frac{1}{2} m \left(\vec{V} + \vec{V}_{osc} \right)^2 \right\rangle_{av} \tag{1.37}$$

where $\vec{V}_{osc}$ is the oscillation velocity of the free electron in the laser field and $\vec{V}$ its drift velocity. The time dependence of the electron coordinate is supposed to be characterized by two time scales, a high-frequency one (h) and a low-frequency one (1), so that:

$$X - X_e + X_h \qquad \text{with } X_e = \langle X \rangle \tag{1.38}$$

in this approximation, which is one of the basic hypothesis of the ponderomotive force formalism, to derive the equation of motion X_1, the em force can also be separated in two components:

$$\vec{F} = \vec{F}_h + \vec{F}_e \tag{1.39}$$

where:

$$\vec{F}_h = - q \, \vec{E}_e (X_e, t) \, \cos \omega t \tag{1.40}$$

and where:

$$\vec{E}_e = - \left(\frac{q^2}{4m\omega^2}\right) \vec{\nabla}\left[\vec{E}_e^2 (X_e, t)\right] \tag{1.41}$$

is just the usual ponderomotive force. Eq. (138) and (1.39) can be substituted into Eq. (1.36) to yield:

$$E_{out} = E_{in} - \Delta + \Xi \tag{1.42}$$

with:

$$\Delta = - \left\langle \int^{.} \vec{F}_h \cdot \vec{V}_{osc} \, dt \right\rangle_{av} = \frac{1}{2} m \left\langle V_{osc}^2 \right\rangle_{av} \tag{1.43}$$

and:

$$\Xi = \int^{.} \vec{F}_e \cdot \vec{V}_e \, dt \tag{1.44}$$

Expression (1.44) can be approximated neglecting the slow temporal variations of V_1, which can be identified with the drift velocity of the electron, so that:

$$\Xi = - \int \frac{q^2}{4m\omega^2} \nabla\left[E_e^2 (X_e, t)\right] \vec{V} \, dt \tag{1.45}$$

We have seen in the previous section that most of the electrons are emitted along the laser polarization. It is then clear from Eq. (1.44) that the intensity gradient contributions to the electron energy outside the laser field essentially comes from the spatial inhomogeneity of the laser intensity along a direction (r) perpendicular to the propagation direction of the laser (z). An exact analytical solution of Eq. (1.44) can be derived in the simple case described hereafter. The spatial dependence of the laser field amplitude is supposed to be gaussian, and the temporal dependence is approximated by an exp (-yt) function with $y \ll w$. If Ω_0 is the laser beam waist, such a field can be described by:

$$E_0 \exp\left[- r^2/\Omega_0^2\right] \exp(-\gamma|t|)\cos \omega t \tag{1.45'}$$

then, for an electron released at t = 0 in the center of the focal volume, a lengthy but straightforward calculation yields for Eq. (1.44)

$$\Xi = \Delta[1 - \theta] \tag{1.46}$$

with:

$$\theta = \sqrt{\pi} \, \beta \, \exp(\beta^2) \, erf_c (\beta) \tag{1.47}$$

and:

$$\beta = \gamma \frac{\Omega_0}{\sqrt{2} \, V} = 1.2 \, 10^{-12} \, \gamma \, \Omega_0 \, E^{-\frac{1}{2}} \tag{1.48}$$

where:

$\Delta = \left(q^2 \, E_e^2/4m\omega^2\right) = 9.3 \, 10^{-14} \, I \, \lambda^2$ in eV with I in W.cm^{-2}, Ω_0 and λ in μm.

$E = (1/2)m\ V^2$ (in eV) is the initial drift kinetic energy of the electron, that is $E = E_{in} - \Delta$. Noting that the quantity $\gamma/\sqrt{2}$ represents the inverse of the pulse duration the physical meaning of β is obvious: it represents the ratio of the time $T_0 = \dfrac{\Omega_0}{2v}$ taken by the electron to exit the focal spot to the laser pulse duration. Finally, Eq. (1.36) reduces to:

$$E_{out} = E_{in} - \theta\Delta \qquad\qquad (1.49)$$

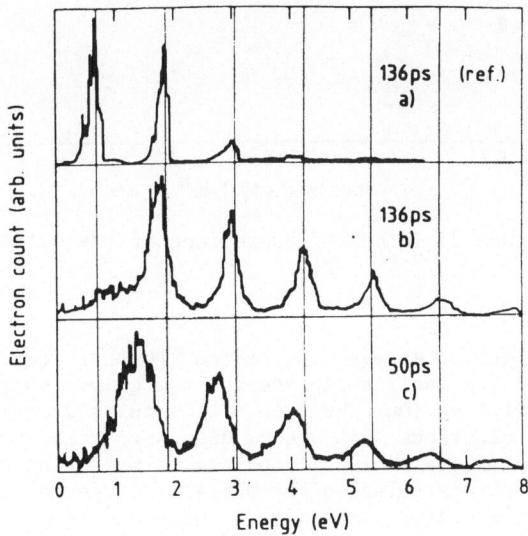

Figure 7. Shift of ATI peaks due to pondermotive effects in short pulses

Eq. (1.49) yields very simple results in the following two limit cases:

(i) $\beta \ll 1$ (case of very long pulses). Since $\theta(\beta = 0) = 0$, there is not change in the electron kinetic energy: the electron exits the laser beam before it is turned off and entirely feels the intensity gradient force F_1 along the laser polarization. In such a situation, the quiver energy is <u>converted</u> into drift energy while the <u>total energy is conserved</u>.

(ii) $\beta \gg 1$ (case of very short pulses): since $\theta(\beta \gg 1) = 1$, $\Xi = 0$ and:

$$E_{out} = E_{in} - \Delta \qquad\qquad (1.50)$$

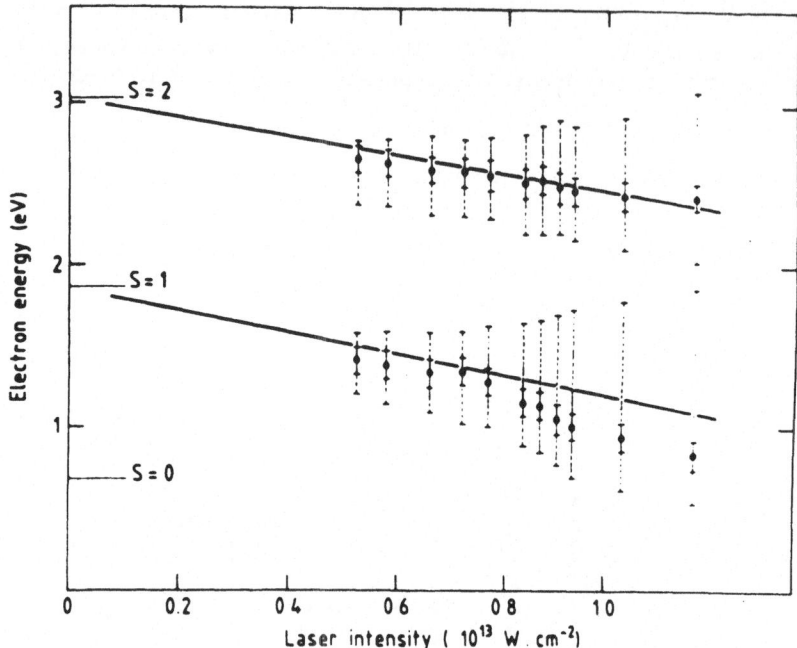

Figure 8. Intensity dependence of the shifts

that is, the electron energy is shifted towards lower energies by an amount equal to its oscillation energy: the quiver energy has not been converted into drift motion. The role of the pulse duration on the final energy of the electrons has been discussed also by Fiordilino and Mittleman (1985). The experiment consists in measuring the energy spectrum of the electrons released by MPI/ATI of Xenon and studying its dependence on the laser intensity, pulse duration and polarization. Details about the experiment can be found elsewhere (Agostini et al., 1987). Other experimental investigations of this effect have been carried out by Luk et al., 1987. The experimental results are summarized on Figs. 7a,b,c and 8. Figs. 7a,b,c show a series of spectra taken at different intensities and pulse durations. Spectrum a) is taken at an intensity of $2.8 \ 10^{12}$ W.cm^{-2}, for a laser pulse duration of 136 ps, and a linearly polarized laser. Spectrum b) is taken with the same pulse duration and laser polarization, but at an intensity of $7.5 \ 10^{12}$ W.cm^{-2}. Spectrum c) is taken at the latter laser intensity but for a pulse duration of 50 ps. Energy shifts and broadenings of the electron peaks are clearly seen on spectrum c). They depend on the laser intensity in a way represented on Fig. 8, which shows the position of the first two detectable ATI peaks (S = 1 and 2) maxima of a spectrum obtained with 30 ps laser pulse, at different intensities. The Full Width at Half Maximum (FWHM) of these two peaks is also shown on this figure as dashed bars around the experimental points. On Fig. 9, the comparison between the measured shifts (dots) and the calculation (solid line) is displayed, as θ versus β. For a pulse duration of 136 ps theory predicts a very small energy shift of the peaks, when the experiment finds all the peaks except may be the first one at their unshifted position (within the

502

apparatus resolution). For a 50 ps pulse duration, significant shifts are both predicted by theory and measured in the experiment. In both cases, they are found to decrease for peaks of increasing order. Theory predicts a linear dependence with the laser intensity which is in reasonable agreement with the result of the experiment (Fig. 8). To conclude this discussion of ATI in intense em fields, I would liket to stress the following:

(i) LOPT fails to describe what is observed at intensities where it should work since $\alpha_1 \ll 1$. Our present understanding is that LOPT fails because the relative amplitudes of the Coulomb and em interactions are <u>reversed</u> during the ATI process.

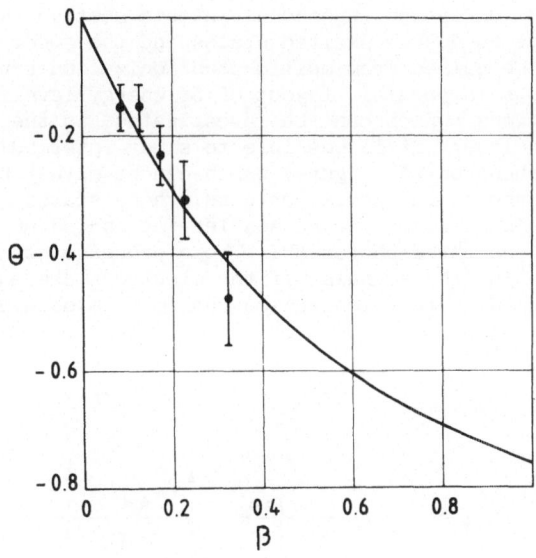

Figure 9. Comparison between theoretical (solid line) and experimental shifts

(ii) There is no evidence for a shift of the IP other than the peak suppression which is rather explained by the necessity to communicate to the electron, at least, the quiver energy. This is fully consistent with calculations and observations of the shifts detected with short pulses.

(iii) The picture of the ac-Stark shift increase of the IP, is based, as we have seen, on the high frequency approximation, which obviously fails at the CO_2 wavelength. Furthermore, the concept of ac-Stark shift is of limited interest when the sate under consideration is strongly mixed with other states.

(iv) The two-step model of the half-inverse Bremsstrahlung, by separating the production of the pair e^--ion from the acceleration of the electron, solves the puzzle of the ion signal slope (N) which is observed even when ATI is important.

(v) Calculations performing resummations of the perturbation series
are currently in progress (Gontier and Trahin, 1987) and should provide
soon quantitative predictions for ATI in strong fields. Other methods
are also being worked out: numerical integration of the time dependent
Schroedinger equation (Kulander, 1987; Javaneinen and Eberly, 1987);
scattering formulation of the ATI for Rydberg states along the line of
thought of MQDT (Giusti-Suzor and Zoller, 1987).

II - TWO-ELECTRON EXCITATION IN MODERATE FIELDS

Atoms two valence electron present interesting features when compared to Rare Gases or to alkali-like atoms. Compared to Rare Gases, they
are easier to ionize (i.e. moderate power tunable lasers can be used)
and their spectra are much simpler (except He). Compared to alkali-like
atoms, they possess doubly-excited states close to, and sometimes even
below, the first ionization threshold. In this section, we discuss some
of the effects due to doubly-excited states and the spectroscopic information which can be gained from multiphoton ionization experiments.

Fig. 10 shows a partial layout of Sr energy levels together with
some MPI transitions which can be investigated in the visible range.
Using a tunable laser, it is possible to study intermediate resonances
on states (not shown on the figure) which can be either below the ionization limit, or above. In the latter case, these states are intermediate states for higher order processes leaving behind an excited ion or
final states of the three-photon MPI. These processes can be studied on
the electron or the ion signals. If the electrons are energy-analysed,
information on the different ionization channels is obtained.

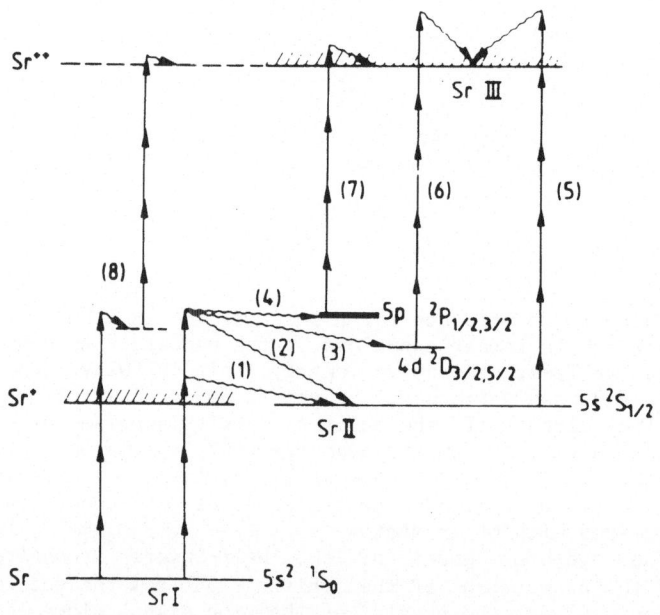

Figure 10. MPI of Sr: energy levels and some observed processes

Fig. 11 shows an electron energy spectrum obtained at 562 nm and an intensity of 3×10^{12} W.cm^{-2}. All the processes shown on Fig. 10 yield electron peaks which are labelled according to the channel numbering used on Fig. 10. Absorption of three photons allows to leave the ion in the ground state and an electron with an energy of .9 eV is released (channel 1). The corresponding peak is the largest one in the spectrum. After absorption of a fourth photon, the ion can be left in thress different states: (i) the ground states; this is an ATI process as discussed in Sect. I and it yields a peak (channel 2) which is small compared to the lowest order one due to the moderate intensity. (ii) the 4D state releasing an electron of 1.25 eV (channel 3). (iii) the 5p state; the electron is then released with an energy of 0.1 eV only and is not seen on this spectrum. Using an acceleration (1 V) this peak is easily detected and can be studied in the same conditions as peak (3). Another feature is clearly seen on Fig. 11 in the 1.8 - 2.0 eV region. This peak was shown (Agostini and Petite, 1985a) to be due to electrons released in the sequential double ionization processes using the 5s or 4d ion states as intermediate steps. Returning to the four-photon processes, several remarks can be made. First, ATI leaving the ion in the ground state is less probable than the other ones, in spite of the fact that they involve excitation of the core electron. This is probably due to the presence, in the energy region reached by absorption of three photons, of autoionizing states belonging to the 4dnl or pnl configurations. The influence of these states can be studied in a channel-selective way, by measuring separately the wavelength dependence of the different peaks amplitudes. The result is shown on Fig. 12.

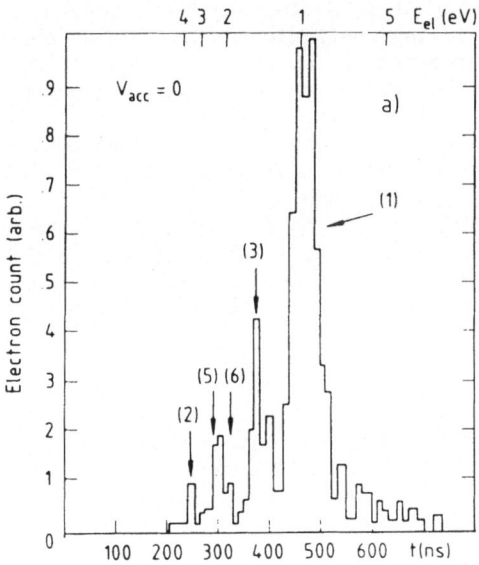

Figure 11. Electron spectrum from MPI of Sr

Resonances are clearly seen in both linear and circular polarizations. Some of them could be assigned to states known from absorption spectroscopy. In the independent electron picture, since the dipole transition operator only acts on one electron, the intermediate 3-photon resonant state must have a core corresponding to the ion state obtained in the 4-photon ionization. In this picture a resonance appearing in channel 3 (ion in state 4d) should have no effect on channel 4 (ion in state 5p). Of course, all resonances appear in channel 1, since autoionization is not subject to the dipole selection rules. A glance at Fig. 12 is enough to see that this picture is by far too simple, which was expected for a heavy atom like Sr. All resonances byt one (labelled V on Fig. 12) appear in channels 3 and 4, because of the strong configuration mixing. How much configuration mixing is there is precisely one of the possible outcomes of such experiments, provided one has good control of the MPI theory, including autoionization (Tang, 1987). Additional information can be gained by studying the ac-Stark shifts of the resonances.

Stark shift of two-photon resonances in Sr

Another group of resonance are easily observed at slightly longer wavelengths. These are two-photon resonances on state $5p^2\ ^1P_0$ on $5p^2\ ^3P_{0,2}$ and $5s\ 5d\ ^3D_{1,2}$. ac-Stark shifts of these resonance experimentally studied (see Fig. 2) and the result compared to ab-initio calculations using Multiconfigurations Hartree-Fock wavefunctions (Agostini et al., 1986). Such wavefunctions can reproduced the energies with an accuracy of a few percent which is not enough for the MPI calculations. Therefore the experimental energies were used with MCHF wavefunctions in the calculations. An additional aspect that is indispensable to the calculation is the singlet-triplet miwing of the atomic states. Its significance is obviously reflected upon the fact that 2-photon resonant transitions to triplet states exhibit prominent peaks. Given that the ground state $5s^2$ is singlet, only intercombination transitions could account for the presence of these resonances, which can, in turn, be used as information input in evaluating the strength of the singlet-triplet mixing. The mixing coefficients have been obtained from angular distribution measurements (Feldmann and Welge, 1982) and, when used in the MPI calculations yield a very good agreement with experiment (Agostini et al., 1986). The different data (resonances and Stark-shifts) allow to propose assignments to the various resonances, according to which the resonances labelled III and V on Fig. 12 are 3-photon resonances with $5p\ 5d\ ^3D_3$ and $5p\ 5d\ ^3D_1$ respectively. Note that these states had eluded observations so far. Of course, the pure configuration used above is only a convenient label. For example the $5p^2$ states have, in fact, almost 50% of $4d^2$ in them, as found from MVHF calculations and recently confirmed by R-matrix calculations (Aymar, 1987).

Double ionization

Multiphoton double ionization of alkaline earths was demonstrated for the first time more than ten years ago (Delone et al., 1984). Sr was studied by Feldmann et al., 1984, and, more recently by Agostini and Petite, 1985b. In these lectures, I will only briefly summarize the main conclusions of these investigations:

(i) Informations obtained from electron spectrometry and from resonances in the Sr^{2+} signal led to the conclusion that multiphoton double ionization is essentially a sequential process: after absorption of 3 or 4 photons, Sr is ionized, possibly leaving the ion in an excited state, as discussed above (Fig. 13). The ion is subsequently ionized by another

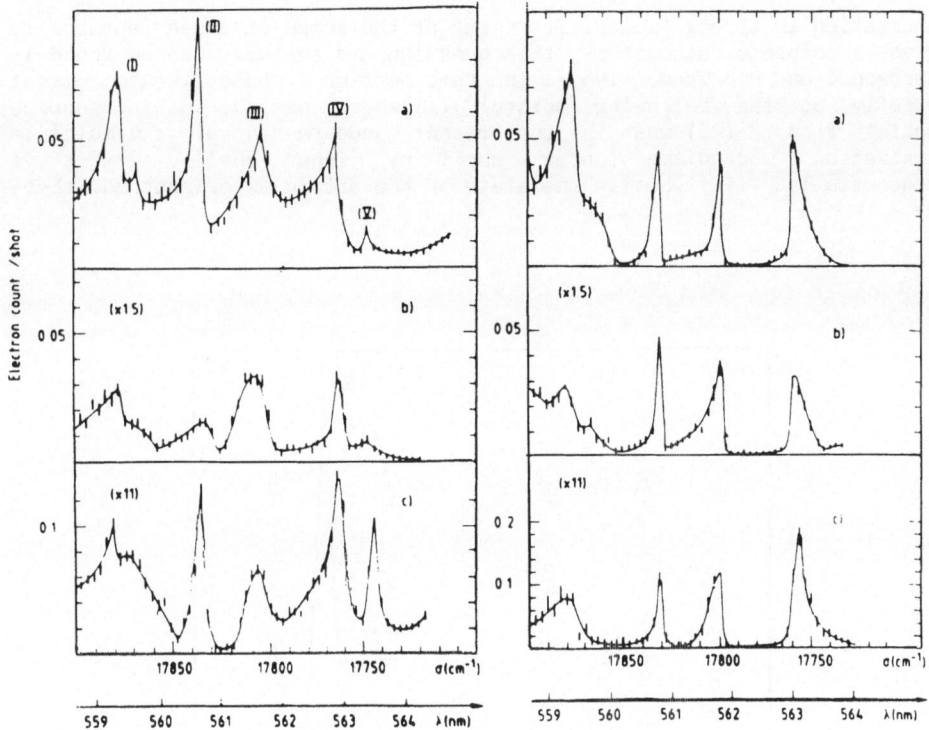

Figure 12. Wavelength dependence of different MPI channels

multiphoton step which can be identified by the ionic resonances (i.e. multiphoton resonances between ionic states). This is confirmed by electron spectra which reveal energy lines which must be assigned to the different ionization steps. In some circumstances, there is a very clear indication of the sequential character of the double ionization process: it is when double ionization only appears at intensities above the saturation intensity of the single ionization process. In this case there are no neutral atoms left in the interaction volume and only ionization of the ions can take place.

(ii) Direct double ionization is the process which would drive the two outer electron "together" up to the second ionization limit. The resulting electron spectrum is a continuum extending from 0 to E:

$$E = N\hbar\omega - (E_1 + E_2) \tag{2.1}$$

where E_1 and E_2 are the Sr and Sr$^+$ ionization energies respectively, N the minimum number of photons required to have E > 0. Such a channel, which is usually extremely weak, has not yet been observed.

III - MULTIELECTRON EXCITATION IN STRONG FIELDS

The discovery that multiphoton multiple ionization of Rare Gases was relatively "easy" (L'Huiller et al., 1982) came as a surprise (almost as a shock) in the multiphoton community. This finding triggered a lot of excitement and speculations about possible collective behavior of the whole electron outer shell. The reasons for this must be found in the difficulty to handle theoretically the problem of multielectron excitation in strong fields. It is out of the scope of these lectures to give a complete account of this question (a summary can be found in Lambropoulos and Tand, 1987). In this section I would like to comment briefly on the following points: (i) where was the initial suprise coming from ? (ii) what is the present undestanding of the multiple ionization mechanisms (outer shell vs inner shell; direct vs sequential) ? (iii) what is the state of the art in experiment and theory ?

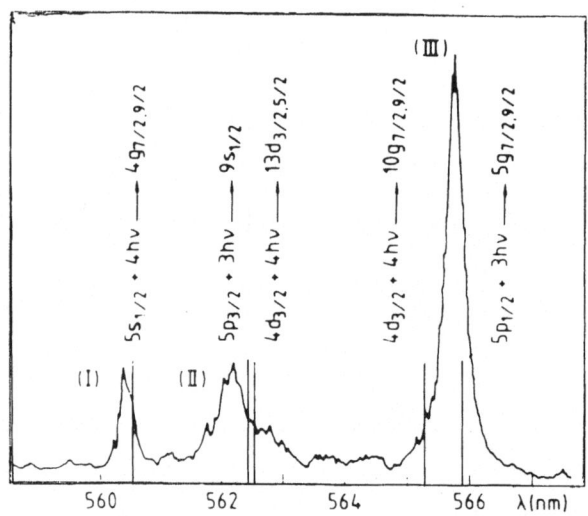

Figure 13. Doubly charged ion signal versus wavelength

Why is multiphoton multiple ionization so easy ?

The idea that multiphoton multiple ionization must have an extremely low probability (unless the intensity is extremely high) comes from extrapolation of MPI experiments involving a small number of photons. For small numbers of photons, the saturation intensity (see Sect. I), which is also roughly the intensity at which the process under study becomes observable, varies rapidly with N. For example, a two-photon process requires 10^7 W.cm^{-2}, a 4-photon process 10^9 W.cm^{-2}, a 6-photon process 10^{11} W.cm^{-2} etc... (typical values for ns or ps pulses). However, it is not difficult, using (1.17) and (1.18) to realize that I_s rapidly tends towards a limit when N > 10 (Fig. 14). This very simple calculation is confirmed by exact calculations (Lambropoulos and Tang, 1987) and can be extended to other atoms using (1.19) as already discussed. The consequence of this behavior is that, when I approaches this limit, no matter how many photons are involved, the transition will be saturated and, for instance, Xe^{5+} is obtained at almost the same intensity at which Xe^+ appears. Therefore, even within the framework of LOPT, it is impossible to account (qualitatively) for the production of multiply-charged ions at "moderate" intensities. We will see in the next subsection that LOPT also predicts that this process is mainly sequential.

Multiple ionization Mechanisms in Xe

Multiple ionization can occur through innner-shell ionization processes or through outer-shell ionization. In the latter case, the question of direct vs sequential ionization arises. It is known that inner-shell ionization is responsible for single-photon multiple ionization of Xe as soon as the photon energy is larger than the 4d-shell-ionization energy. Thus some competition between inner and outer shell ionizations is expected in multiphoton ionization. However, it can be shown (Wendin et al., 1987; L'Huiller, 1987) that outer-shell ionization largely dominates for the following reasons: (i) screening by the outer-shell significantly reduces the external field in the region of the 4d-shell. The screening results from the well known high polarizability of Xe. (ii) Due to its lower ionization energy, the 5p-shell ionizes much more rapidly during the risetime of the laser pulse. If this process saturates after a time t_{5p}, a Xe atom cannot survive after this time and cannot be exposed (with a complete outer shell) to intensities larger than $I(t_{5p})$: outer-shell ionization is completed before the intensity is high enough for inner-shell ionization to be significant.

As far as outer-shell ionization is concerned, everybody agrees now that this process is mainly sequential for essentially the same reason as discussed just above. A more quantitative theory of sequential ionization has been developed (Lambropoulos and Tang, 1987) based on the scaling law (1.19) to estimate the generalized cross-sections. The time evolution of the various ion population is then calculated using the rate approximation.

Recent progress

Experimentally, multiple ionization of Rare Gases has been observed from 193 nm to 10.6 µm. The latest results are Xe^{11+} at 248 nm and 3×10^{17} W.cm^{-2} with a pulse length of .5 ps (Kyrala et al., 1987) and Xe^{6+} at 10.6 µm, 10^{14} W.cm^{-2} with a pulse length of 1 ns (Yergeau et al., 1987). In the case of X^{11+}, all the outer 5p and 5s electrons plus three 4d electrons have been removed. In the case of the CO_2 laser more than 2400 photons are involved in the transition ! (this tremendous number probably indicates that the "photon" language is not adequate in this situation).

In the theoretical domain, a promising new approach, developed by Kulander (1987) is to solve numerically the time-dependent Hartree-Fock equation (applied so far to He). Faisal (1987) obtains surprisingly good agreement with the experiment of Yergeau et al. (1987) using a dynamical model in which the electron cloud is treated as an inhomogeneous gas.

SUMMARY

In these lectures, I have considered, in a rather elementary way, some aspects of laser-atom interaction with special emphasis on recent results in multiphoton ionization. Multiphoton ionization of atoms has evolved recently from studies of alkali-like atoms at moderate intensity, when it is well described by lowest order perturbation theory, to more complicated situations. In the case of single electron excitation, a non perturbative behavior is observed at relatively low intensity, i.e. intensities at which the em interaction is still much weaker than the Coulomb interaction (when the electron is in the ground state). It seems that the reason of this premature breakdown must be found in the reversal which occurs between these two interactions when the electron emerges in the continuum. In the case of two-electron excitation, multiphoton ionization provides spectroscopic informations even in the case of strong perturbation of the atomic structure. In the specific case of Sr, studies of angular distributions or ac-Stark shifts resulted in valuable spectroscopic informations especially about autoionizing states. Multiphoton double ionization of alkaline eaths is mainly a sequential process. This is due to the presence of doubly excited states close (and above) the first ionization limit which autoionize very rapidly. In the case of multielectron excitation in strong field, although the relative easiness with which atoms are multiply ionized is qualitatively explained, the theoretical understanding of the problem is still far from being complete. New methods are being developed which are expected to unravel this difficult problem in a not too far future.

REFERENCES

Agostini P., Fabre F., Mainfray G., Petite G. and Rahman N.K. 1979, Phys. Rev. Lett. 42, 1127.
Agostini P., Fabre F. and Petite G. 1984, Multiphoton Ionization of Atoms ed S.L. Chin and P. Lambropoulos (New York: Academic Press).
Agostini P. and Petite G. 1985a Phys. Rev. A 32, 3800.
Agostini P. and Petite G. 1985b J. Phys. B 18, L281.
Agostini P., L'Huiller A., Petite G., Tang X. and Lambropoulos P., 1986 Resonance Ionization Spectroscopy 1986 Ed G.S. Hurst and C. Grey Morgan, Institute of Physics Conference series 84 and 1988, Phys. Rev. A (to be published).
Agostini P., Kupersztych J., Lompre L.A. and Yergeau F. 1987, Phys. Rev. A
Avan P., Cohen-Tannougji C., Dupont-Roc J. and Fabre C., 1976, J. Phys. (Paris) 37, 993.
Aymar M., 1987 J. Phys. B (in Press).
Bebb H.B. and Gold A., 1966, Phys. Rev. 143, 1.
Boyer K. and Rhodes C.K., 1985, Phys. Rev. Lett. 54, 1490.
Bucksbaum P., Bashkansky M., Freeman R.R., Mc Illrath T.J. and Dimauro L.F., 1986, Phys. Rev. Lett. 56, 2590.
Bucksbaum P., Bachkansky M. and McIllrath T.J., 1987, Phys. Rev. Lett. 58, 349.
Crance M., 1984 Multiphoton Ionization of Atoms ed S.L. Chin and P. Lambropoulos (Neaw York: Academic Press).
Crance M., 1984b, J. Phys. B, 17, L635.
Crance M., 1986, J. Phys. B, 19, L267.
Faisal F.H.M., 1987, J. Phys. B, 20, L299.
Feldmann D. and Welge K.H., 1982, J. Phys. B, 15, 1651 and private communication.
Feldmann D., Krautwald H.J. and Welge K.H., 1984 Multiphoton ionization of Atoms ed S.L. Chin and P. Lambropoulos (New York: Academic Press).
Freeman R.R., McIllrath T.J., Buksbaum P.H. and Bachkansky M., 1986, Phys. Rev. Lett., 57, 3156.
Feldmann D. 1986 (unpublished).
Fiordilino E. and Mittleman M.H., 1985, J. Phys. B, 18, 4425.
Giusti-Suzor A., 1987, this book.
Giusti-Suzor A. and Zoller P., 1987, to be published.
Gontier Y. and Trahin M., 1971, Phys. Rev. A, 4, 1896.
Gontier Y., Rahman N.K. and Trahin M., 1986, Phys. Rev. A, 14, 2109.
Gontier Y. and Trahin M., 1980, J. Phys. B, 13, 4381.
Gontier Y. and Trahin M., 1984, in Multiphoton Ionization of Atoms Ed S.L. Chin and P. Lambropoulos (New York: Academic Press).
Gontier Y. and Trahin M., 1988 (to be published).
Humpert H.J., Hippler R., Schwier H. and Lutz H.O., 1985 Fundamental Processes in Atomic Collision Physics Ed H. Kleinpoppen, J.S. Briggs and H.O. Lutz (New York: Plenum).
Javanainen J. and Eberly J., 1987 (to be published).

Keldysh L.V., 1965 Sov. Phys. JETP 20, 1307.

Kruit P., Kimman J. and Van Der Wiel M.J., 1981, J. Phys. B 1981, 14, L597.

Kruit P., Kimman J. and Van Der Wiel M.J., 1983, Phys. Rev. A 28, 248.

Kulander K., 1987, Phys. Rev. A (to be published).

Kupersztych J., 1987 Eur. Phys. Lett., 4, 23.

Kyrala G.A., Casperson D.E., Lee P.H.Y., Jones L.A., Taylor A.J. and Schappert G.T., 1987, Multiphoton Processes, Ed S. Smith and P. Knight (Cambridge).

Lambropoulos P., 1976 Adv. At. Mol. Phys., 12, 87.

Lambropoulos P., 1985, Phys. Rev. Lett. 55, 2141.

Lambropoulos P. and Tang X., 1987 J. Opt. Soc. Am. B, 4, 821.

Leuchs G., 1984 in Multiphoton Ionization of Atoms Ed S.L. Chin and P. Lambropoulos (New York: Academic Press).

L'Huiller A., Lompre L.A., Mainfray G. and Manus C, 1982, Phys. Rev. Lett., 48, 1814.

L'Huiller A., Lompre L.A., Mainfray G. and Manus C., 1983, Phys. Rev. A 27, 2503.

Liberman S., Pinard J. and Taleb A., 1983, Phys. Rev. Lett., 50, 888.

Lompre L.A., Mainfray G., Manus C. and Thebault J., 1977, Phys. Rev. A 15, 1604.

Lompre L.A., L'Huiller A., Mainfray G. and Manus C., 1985, J. Opt. Soc. Am. B, 2, 1902.

Lompre L.A., Mainfray G., Manus C. and Kupersztych J., 1987 J. Phys. B. 20, 1009.

Leone C., Burlon R., Trometta F., Basile S. and Ferrante G., 1987, Nuovo Cimento (to be published).

Luk T.S., Graber T., Jara H., Johan U., Boyer K. and Rhodes C.K., 1987, J. Opt. Soc. Am. B 4, 847.

Mainfray G. and Manus C. in Multiphoton Ionization of Atoms Ed S.L. Chin and P. Lambropoulos (New York: Academic Press).

Morellec J., Normand D. and Petite G., 1976, Phys. Rev. A 14, 300.

Morellec J., Normand D. and Petite G., 1982, Adv. At. Mol. Phys. 18, 98.

Mouroux G., 1987, Multiphoton Processes, Ed S. Smith and P. Knight, Cambridge.

Muller H., Tip A. and Van Der Wiel M.J., 1983, J. Phys. B 16, L679.

Pan L., Armstrong L. and Eberly J.H., 1986, J. Opt. Soc. Am. B 3, 1319.

Petite G., Fabre F., Agostini P., Crance M. and Aymar M., 1984, Phys. Rev. A 29, 2677.

Petite G. and Agostini P., 1986, J. Phys. (Paris) 47, 795.

Rahman N.K. and Faisal F.H.M., 1976, J. Phys. B 9, L275.

Reiss H., 1987, J. Opt. Soc. Am. B 4, 726.

Suran V.V. and Zapesochnyi L.P., 1975 Sov. Tech. Phys. Lett. 1, 420.

Tang X., 1987 Z. Phys. D 1987 6, 255.

Weingartshofer A. and Jung C., Multiphoton Ionization of Atoms Ed S.L. Chin and P. Lambropoulos (New York: Academic Press).

Wendin C., Jonson L. and L'Huiller A., 1987, Jour. Opt. Soc. Am. B 4, 833.

Xiong W., Yergeau F. and Chin S.L., Proceedings of ICOMP IV, Boulder, Colorado, 1987 (unpublished).

Yergeay F., Petite G. and Agostini P., 1986 19, 1669.

Yergeau F., Chin S.L. and Lavigne P., 1987, J. Phys. B 20, 273.

Zernik W. and Klopfenstein R.W., 1965, J. Math. Phys. 6, 262.

PROBING EXCITED STATES WITH MULTIPHOTON IONIZATION[*]

J. L. Dehmer, P. M. Dehmer, S. T. Pratt, M. A. O'Halloran,
and F. S. Tomkins

Argonne National Laboratory
Argonne, Illinois 60439, U.S.A.

I. INTRODUCTION

Resonance enhanced multiphoton ionization (REMPI) utilizes tunable
dye lasers to ionize an atom or molecule by first preparing an excited
state by multiphoton absorption and then ionizing that state before it
can decay. This process is highly selective with respect to both the
initial and resonant intermediate states of the target, and it can be
extremely sensitive. In addition, the products of the REMPI process can
be detected as needed by analyzing the resulting electrons, ions,
fluorescence, or by additional REMPI. This points to a number of
opportunities for exploring excited state physics and chemistry at the
quantum-state-specific level. Here we will begin with a brief overview
of the large variety of experimental approaches to excited state
phenomena made possible by REMPI. Then we will examine in more detail
several examples which illustrate some of these approaches: First, we
will discuss three photon resonant, four photon (3+1) ionization of H_2
via the C $^1\Pi_u$ state. Strong non-Franck-Condon behavior in the
photoelectron spectra of this simple Rydberg state has led to the
examination of a variety of dynamical mechanisms. Of these, the role of
doubly excited autoionizing states now seems decisive. Second, we will
discuss recent progress on photoelectron studies of autoionizing states
in H_2, excited in a (2+1) REMPI process via the E,F $^1\Sigma_g^+$ state. Third, we
will describe recent use of three photon, non-resonant excitation to
probe autoionizing levels in Xe and Kr. Fourth, studies of
photoionization and predissociation of excited states of Xe_2, selectively
excited in a supersonic expansion of atomic Xe, will be reviewed.
Finally, we will describe studies of autoionization and Rydberg-Rydberg
interactions in atomic I, which is formed by photodissociation of CH_3I.
We wish to stress that these examples were selected from our own work for
convenience, and that they are but a few examples of a sizable literature
that has sprung up over the last ten years. Since we will focus

[*]Work supported by the U.S. Department of Energy, Office of Health & Environmental Research, under Contract W-31-109-Eng-38, and by the Office of Naval
Research. The U.S. Government retains a nonexclusive, royalty-free license
to publish or reproduce the published form of this contribution, or allow
others to do so, for U.S. Government purposes.

primarily on photoionization dynamics of excited molecular states using
REMPI together with photoelectron spectroscopy (PES), we have given a
fairly extensive bibliography for REMPI-PES studies on molecules in Refs.
1-97. Note that this entire subfield is a product of the 1980's. It is
beyond our scope to treat all MPI-related studies, but we partially fill
this gap by citing a few general reviews[98-106] on related topics.

To illustrate the broader potential of REMPI, we will outline
several different types of experiments that can be carried out using the
REMPI excitation processes shown schematically in Figure 1. In Figure
1a, two photons from a "pump" laser with frequency ν_1 are used to excite
an individual rotational (not shown) and vibrational level of an excited

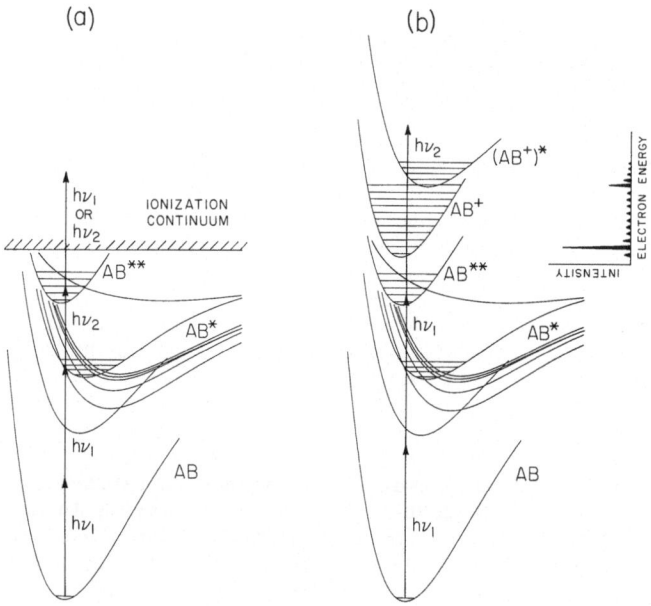

Fig. 1. Schematic potential energy diagram showing two different REMPI
processes in a diatomic molecule.

electronic state AB*. An independently tunable "probe" laser of
frequency ν_2 is used to further excite the AB* level to the manifold of
rotational-vibrational levels of a higher excited state AB**. A third
photon of frequency ν_1 or ν_2 is used to ionize AB**. In this case, one
is interested in the AB* → AB** transition rather than the ionization
step, so the continuum is represented simply by a structureless hatched
area. In Figure 1b, a similar process is indicated; however, in this
case, the state AB** is produced by two photons of the pump laser, with
the probe laser accessing the ionization continuum directly. Here, one
is interested in the ionization step itself, and so the accessible ionic
states AB+ and (AB+)* are shown explicitly.

The resonant multicolor excitation schemes represented schematically in Figure 1 permit us to address many problems in molecular physics and chemistry which were either very difficult or unimaginable with conventional excitation sources. These include the following. (a) By varying $h\nu_2$ (in Figure 1a) and detecting the total (or mass selected) ion current as a function of wavelength, one performs optical-optical double resonance (OODR) spectroscopy. This generates spectroscopic information on AB^{**} which typically lies in the vacuum ultraviolet (VUV) with single-photon sources, but in the visible or ultraviolet with multiphoton sources. This produces high-resolution spectroscopic information without the need for a large vacuum spectrograph. More importantly, the $h\nu_2$ transition originates from a single rotational level of AB^*, which greatly simplifies the spectrum. Use of OODR techniques also enables the direct study of states that are dipole forbidden in single-photon excitation. (b) If the excited state AB^{**} is predissociated, e.g., by the repulsive curve in Figure 1, it is possible to probe in detail the mechanisms of the dissociation process by analyzing both the internal energy distribution of the photofragments and the time dependence of their formation. In addition, photodissociation is often one of the simplest and most convenient methods of producing open-shell atoms, free radicals or transient species for further spectroscopic study. (c) Measurement of the photoelectron energy distribution (indicated by the inset in Figure 1b) will give the relative probabilities of producing alternative ionic states and thus will directly reflect the photoionization dynamics of individual excited quantum states. It is also possible to determine photoelectron branching ratios at various points within an autoionizing resonance, which will be an extremely sensitive probe of the interactions between the discrete state and the various ionization continua. (d) Since the ionization step in Figure 1b is performed with a visible or UV wavelength, simple rotation of a retardation plate will produce a photoelectron angular distribution, which accesses further dynamical information and also reflects the orientation of the excited state AB^{**} resulting from the multiphoton excitation process. (e) Preparation of an excited state AB^{**}, followed by a delayed laser probe can monitor the time evolution of intramolecular rearrangement and/or decay processes. In molecules more complicated than that indicated in Figure 1, this procedure can monitor the time evolution of vibrational energy redistribution. In this case, picosecond or subpicosecond lasers would be required to capture the normally very fast internal rearrangement. Use of a delayed probe beam can also be used to characterize collisional effects on a prepared state. (f) Using the high degree of selectivity and, hence, sensitivity of either excitation mechanism in Figure 1, it is possible to directly probe free radicals, clusters, ions and other transient species which are formed as minor components in complex mixtures. (g) Many possible chemical uses of the general scheme in Figure 1 can also be readily seen. For instance, by suitable selection of AB^{**} in Figure 1b, it is possible to produce AB^+ or AB^{+*} in particular vibrational and rotational quantum states in order to study the dependence of subsequent chemical transformations on varying degrees of internal energy in different electronic or nuclear modes. Also, by using the selectivity of the excitation process, it is possible to monitor the reactants and products of elementary chemical reactions at the quantum-state-specific level.

This list of possibilities is not exhaustive, but it is ample to show the great scientific potential of REMPI. In what follows, we will present specific examples of REMPI studies in order to illustrate the utility of multiphoton excitation in gaining new insight into both atomic and molecular photoionization dynamics.

II. PHOTOIONIZATION OF H_2 C $^1\Pi_u$, v', J'

Background

Several years ago, we reported the photoelectron spectra obtained following three photon resonant, four photon (3+1) ionization via the C $^1\Pi_u$, v' = 0 - 4 levels of molecular hydrogen[19]. The C $^1\Pi_u$ Rydberg state corresonds to the $1s\sigma_g 2p\pi_u$ configuration, and has a potential curve similar to that of the ground state of H_2^+.[107] Thus, the vibrational overlap integrals between a particular vibrational level, v', of the C $^1\Pi_u$ state and the vibrational levels of the X $^2\Sigma_g^+$ ion, v^+, will be nearly unity for v^+ = v', and nearly zero for $v^+ \neq v'$. On the basis of the Franck-Condon principle, the photoelectron spectra are therefore expected to show strong v^+ = v'peaks, with very little intensity in the $v^+ \neq v'$ peaks. Qualitatively, these expectations are fulfilled, as seen in Figure 2, which shows the photoelectron spectra obtained following (3+1) ionization via the C $^1\Pi_u$, v' = 0 - 4 Q(1) transitions. These spectra were obtained along the laser polarization axis ($\theta = 0°$) using an electrostatic energy analyzer. The Q(1) transitions were chosen to access the Π^- component of the C $^1\Pi_u$ state, which is unperturbed by the B $^1\Sigma_u^+$ state. In each spectrum the v^+ = v' peak is the most intense and the $v^+ \neq v'$ peaks are significantly weaker, in accord with the Franck-Condon arguments. However, a quantitative comparison of the relative intensities with theoretical Franck-Condon factors reveals significant discrepancies, particularly for v' = 3 and 4. For example, in the v' = 4 spectrum the observed v^+ = 3, 5, and 6 peaks are too large by factors of 3, 2, and 23, respectively.[19]

The discrepancy between the experimental results and the theoretical Franck-Condon factors could have a number of sources. These include: (1) a kinetic energy dependence of the electronic transition matrix element, which must be taken into account even within the Franck-Condon approximation; (2) an internuclear distance (R) dependence of the same electronic matrix element which, by definition, constitutes a breakdown of the Franck-Condon approximation; and (3) a v^+-dependence of the photoelectron angular distributions. Dixit et al.[26] have included all three effects in theoretical calculations of the $\theta = 0°$ spectra of Figure 2. However, the agreement with experiment, while improved, is still not good. Recently, we have measured angle-integrated branching ratios using two different techniques. In the first experiment,[62] the photoelectron angular distributions were determined for each spectrum and then integrated to give branching ratios. In the second experiment,[88] the integrated branching ratios were determined directly using a magnetic-bottle electron spectrometer with 2π steradian collection efficiency.[108] The two measurements are in good agreement. Figure 3 shows a comparison of the angle-integrated branching ratios calculated by Dixit et al.[26] with the results of O'Halloran et al.[88] The discrepancies are quite apparent, particularly for v' = 3 and 4, where the observed distributions are much broader than the theoretical predictions. This indicates that the photoionization of the C $^1\Pi_u$ state is more complicated than the direct excitation of the Rydberg electron into the X $^2\Sigma_g^+$ continuum, and suggests that another mechanism is important for higher v'. It is worth noting that the photoelectron angular distributions[62] for the $v^+ \neq v'$ photoelectron bands are generally more isotropic than those for the v^+ = v' bands, which also suggests another mechanism is responsible for the observed intensity of these bands.

Additional evidence for the complexity of the photoionization process is found in the rotational structure of the vibrational bands in

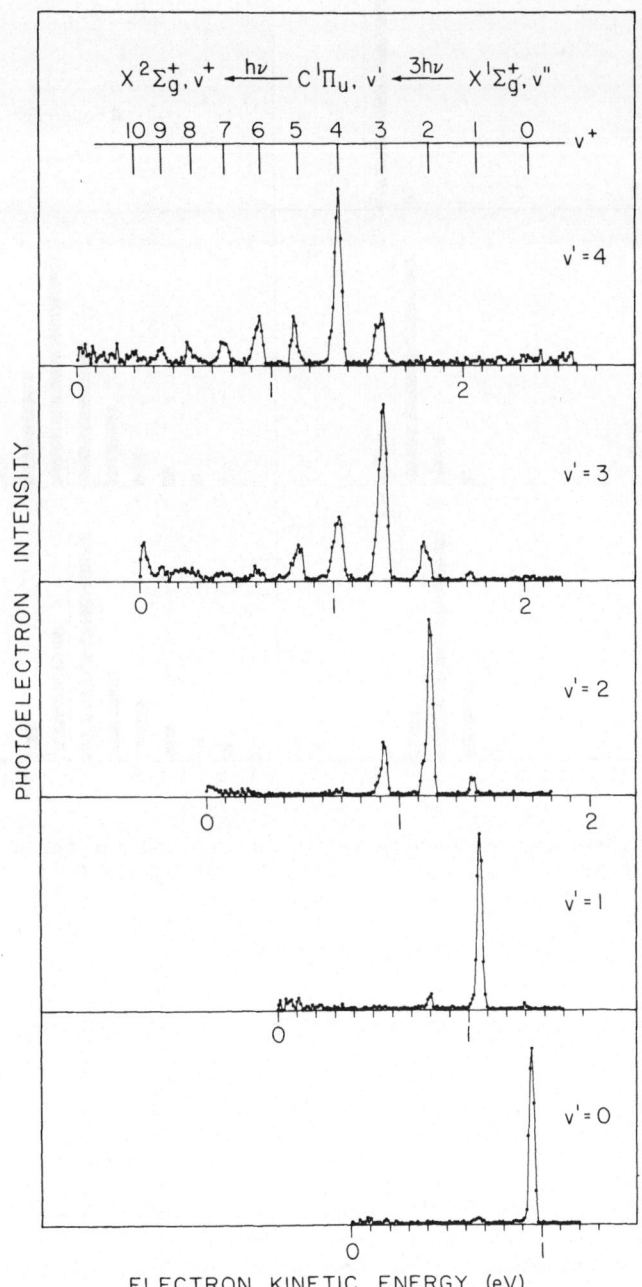

Fig. 2. Photoelectron spectra of H_2 determined along the laser polarization axis ($\theta = 0°$) at the wavelengths of the resonant three photon $C\ ^1\Pi_u$, $v' = 0 - 4 \leftarrow X\ ^1\Sigma_g^+$, $v'' = 0$, Q(1) transitions.

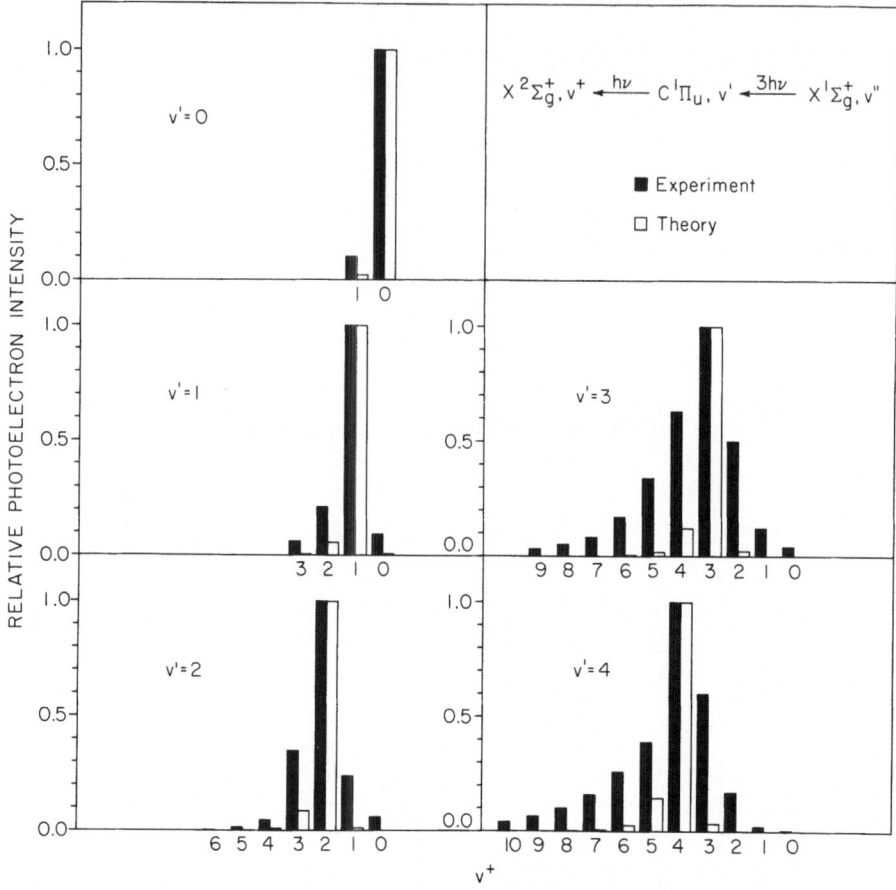

Fig. 3. Vibrational branching ratios determined for three photon resonant, four photon ionization of H_2 via $C\ ^1\Pi_u$, v', $Q(1)$ transitions. The vibrational level of the $C\ ^1\Pi_u$ state is denoted by v', and that of the ion by v^+. The calculation is that of Dixit, Lynch, and McKoy.[26]

the $C\ ^1\Pi_u$, $v' = 0 - 4$ spectra. Assuming that only s and d partial waves contribute to the outgoing electron, the selection rules for photoionization from the $C\ ^1\Pi_u$, $J' = 1$ state[55,62] indicate that the H_2^+ ion can only be formed in the rotational levels $N^+ = 1$ or 3. This is confirmed by the observation of only $N^+ = 1$ and 3 photoelectron peaks in the $C\ ^1\Pi_u$, $v' = 0 - 4$ spectra shown in Figure 4. These spectra were obtained at somewhat higher resolution than those of Figure 2 by using the magnetic-bottle electron spectrometer. (Note that the spectra shown in Figure 4 are angle-integrated spectra.) In these spectra, the relative intensity of the $N^+ = 3$ photoelectron peak increases dramatically with increasing v', both for $v^+ = v'$ and for $v^+ \neq v'$. In addition, for a given intermediate level, v', the $N^+ = 3$ rotational peaks tend to be larger relative to the $N^+ = 1$ peaks in the $v^+ \neq v'$ bands. This indicates that the photoionization mechanism is changing with increasing vibrational

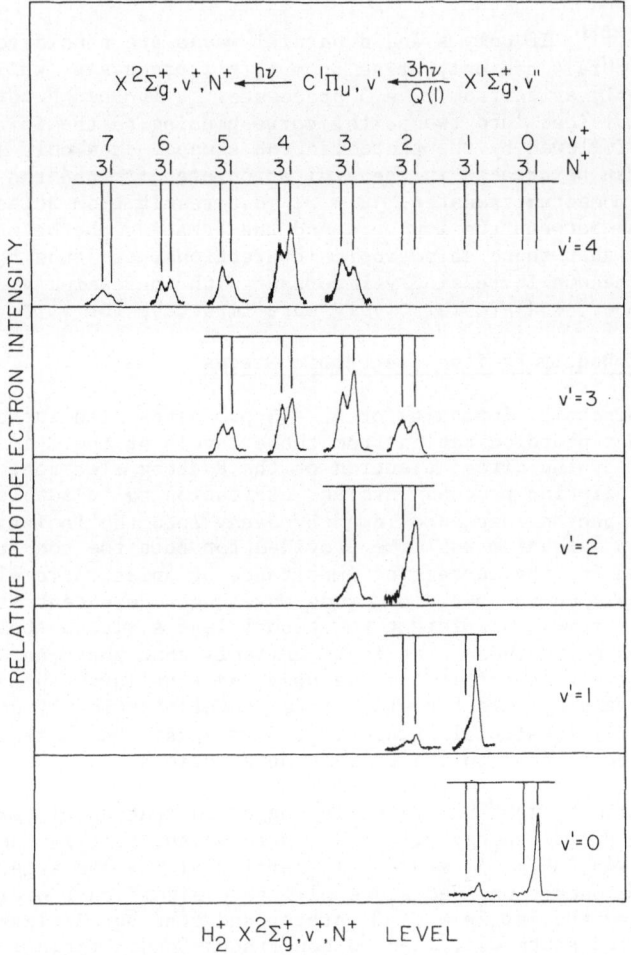

Fig. 4. Photoelectron spectra determined at the wavelengths of the three photon resonant H_2 C $^1\Pi_u$, v' ← X $^2\Sigma_g^+$, v" = 0 Q(1) transitions. The spectra of individual ionic vibrational bands were recorded separately, with retarding voltages chosen so as to achieve comparable energy resolution for each vibrational band. Note that the horizontal scale does not indicate energy, although the individual vibrational bands are plotted with the same energy scale. The integrated areas of the vibrational bands are set equal to the ionic vibrational branching ratios.

quantum number, v', and that this mechanism contributes to the intensity of the $v^+ \neq v'$ bands.

It is sometimes useful to describe photoionization dynamics in terms of the angular momentum transfer, j_t, which is defined as the angular momentum exchanged between the unobserved initial and final angular momenta.[109-111] If only s and d partial waves are considered, the $N^+ = 1$ peaks in Figure 4 can only arise from $j_t = 1$ processes, while the $N^+ = 3$ peaks can only arise from $j_t = 3$ processes.[88] If the photoionization process is divided into two parts, corresponding to the initial excitation followed by the photoelectron escape, then only $j_t = 1$ processes can be created in the excitation step,[110] and the higher value of angular momentum transfer, $j_t = 3$, must result from anisotropic interactions between the ion core and the escaping photoelectron. Figure 4 indicates that these anisotropic interactions, and thus the $j_t = 3$ processes, become increasingly important with increasing vibrational quantum number, and are relatively more important for $v^+ \neq v'$.

The Role of Doubly Excited Electronic States

The increasing intensity of $j_t = 3$ processes with increasing v' suggests that photoionization from these levels of the C $^1\Pi_u$ state does not proceed by the direct ejection of the Rydberg electron. If, instead, the photoionization process involves excitation to an autoionizing level at the four photon energy followed by decay into the ionization continuum, a mechanism would be provided for both the non-Franck-Condon behavior and for the increasing importance of anisotropic electron-ion interactions. On the basis of crude wavelength dependent studies, performed by pumping different rotational levels within the C $^1\Pi_u$, v' bands, we have concluded that it is unlikely that sharp autoionizing resonances are responsible for the observed behavior. In addition, the $\Delta v = -1$ propensity rule for vibrational autoionization[112] and the energetics for rotational autoionization suggest that neither of these processes contributes to the present observations.

However, Chupka[80] has recently suggested that doubly excited states at the four photon energy will play an important role in the (3+1) ionization via the C $^1\Pi_u$ state. In particular, he has argued that the $2p\sigma_u 2p\pi_u$ $^1\Pi_g$ doubly excited state will have significant oscillator strength from the $1s\sigma_g 2p\pi_u$ C $^1\Pi_u$ state, and that autoionization of this doubly excited state will lead to non-Franck-Condon vibrational branching ratios. Cornaggia et al.[89] have also suggested that doubly excited states will be more important for multiphoton ionization via the C $^1\Pi_u$ state into the gerade continua than for ionization via the E,F $^1\Sigma_g^+$ state into the ungerade continua, for which they performed calculations. Independently, Hickman[79,90] has performed model calculations of the vibrational branching ratios following autoionization of the $2p\sigma_u 2p\pi_u$ $^1\Pi_g$ doubly excited state accessed by (3+1) excitation via the C $^1\Pi_u$ state. Using the $2p\sigma_u 2p\pi_u$ $^1\Pi_g$ potential curve of Guberman,[113] Hickman[79,90] has obtained very encouraging agreement with the experimental results.

In a semi-classical time dependent framework,[80] such a process could be viewed as the production of a wave-packet on the repulsive curve of the doubly excited state. As it evolves in time, the wave-packet can be decomposed into outgoing and incoming components; the latter will be reflected and subsequently interfere with the originally outgoing component. Because the doubly excited state has a finite width for autoionization, it will have some probability for transitions into the $^2\Sigma_g^+$ continuum as the wave-packet evolves. In this model, the production of $v^+ > v'$ photoelectron bands arises from autoionization as the wave-packet evolves to dissociation products, while those with $v^+ < v'$ arise from autoionization of the incoming component propagating to smaller R.

This model also introduces the possibility that some of the molecules in the doubly excited state will not autoionize, but rather will dissociate into a ground state atom and an excited state atom. Excited states having n = 3 - 5 have been observed[63,64,88,91] and may result from curve crossings of the repulsive $2p\sigma_u 2p\pi_u$ $^1\Pi_g$ state with singly excited Rydberg states at large internuclear distance. In general, the dissociation processes for the C $^1\Pi$ levels are much weaker than the ionization processes. However, the (3+1) spectra via high-lying vibrational levels of the B $^1\Sigma_u^+$ state in the same energy region exhibit nearly complete dissociation, and the C $^1\Pi$ levels, which interact with the B $^1\Sigma^+$ state, generally display significantly more dissociation than the corresponding C $^1\Pi$ levels. The increased dissociation for the B $^1\Sigma_u^+$ levels may arise from two sources. First, the B $^1\Sigma_u^+$ state samples a much larger range of internuclear distance than the C $^1\Pi_u$ state, and may have a significant direct photodissociation cross section. Second, the $(2p\sigma_u)^2$ doubly excited state, which is "configurationally" allowed from the $1s\sigma_g 2p\sigma_u$ B $^1\Sigma_u^+$ state, could produce more dissociation than the $2p\sigma_u 2p\pi_u$ $^1\Pi_g$ state.

The relative positions of the C $^1\Pi_u$ and $2p\sigma_u 2p\pi_u$ potential curves indicate that the transition will occur from the outer turning point of the C $^1\Pi_u$ state to the inner turning point of the $2p\sigma_u 2p\pi_u$ $^1\Pi_g$ state. Increasing the vibrational quantum number has two effects: The four photon energy is increased, and the outer turning point of the lower state is moved to larger R. For v' = 0, both the total energy and the outer turning point are too small for transitions to the doubly excited state to be very important. As v' is increased, both the overlap with the doubly excited state and the energy requirement improve, and the effects attributable to the doubly excited state become more noticeable, as is the case for v' = 3 and 4. Eventually, as v' is increased further, the energy will be too high and the overlap again will be too poor for the $2p\sigma_u 2p\pi_u$ $^1\Pi_g$ state to play a role. In the region just above v' = 4 the qualitative model described above corresponds to excitation to somewhat larger R than the classical inner turning point, with considerable amplitude for autoionization at smaller internuclear distances. This would lead to the observed increase in the population of vibrational levels of H_2^+ with $v^+ < v'$.

As seen in Figure 5, these arguments are further supported by the photoelectron spectra obtained at $\theta = 0°$ following (3+1) ionization via the C $^1\Pi_u$, v' = 5, 6 Q(1) transitions. In both spectra, the v^+ = v' peak is the largest, with much smaller $v^+ \neq v'$ peaks. Comparison with Figure 2 reveals the distribution of $v^+ \neq v'$ peaks shifts to smaller values of v^+ with increasing v'. In addition, the sum of the intensities of the $v^+ \neq v'$ peaks relative to the v^+ = v' peak decreases monotonically as v' is increased from 4 to 6. The C $^1\Pi_u$, v' = 7, 8 ← X $^1\Sigma_g^+$, v' = 0 bands are overlapped by the much more intense B' $^1\Sigma_u^+$, v' = 1,2 ← X $^1\Sigma_g^+$, v' = 0 bands,[114,115] and were not studied. Although the C $^1\Pi_u$, v' ≧ 9 Q(1) transition is blended with the D $^1\Pi_u$, v' = 1 Q(1) transition,[114,115] some information can nevertheless be obtained regarding ionization of the C $^1\Pi_u$, v' = 9 level. The photoelectron spectrum following (3+1) ionization at this wavelength[78] is shown in the center frame of Figure 6. The D $^1\Pi_u$ state corresponds to the $1s\sigma_g 3p\pi_u$ Rydberg state, and the v' = 1 photoelectron spectrum is similar to that of the C $^1\Pi_u$, v' = 1 level. However, the small v' = 9 peak (at ~ 1.7 eV) almost certainly corresponds to ionization via the C $^1\Pi_u$, v' = 9 level. Although the v' = 9 peak is weak, it is interesting to note that no intensity is observed for v^+ = 5 - 12, which suggests that, at this energy and internuclear distance, the $2p\sigma_u 2p\pi_u$ $^1\Pi_g$ state no longer plays an important role in the ionization process.

Photoelectron Spectra of D_2

Doubly excited states at the four photon energy are also expected to play a role in the (3+1) ionization via the C $^1\Pi_u$ states of the heavier

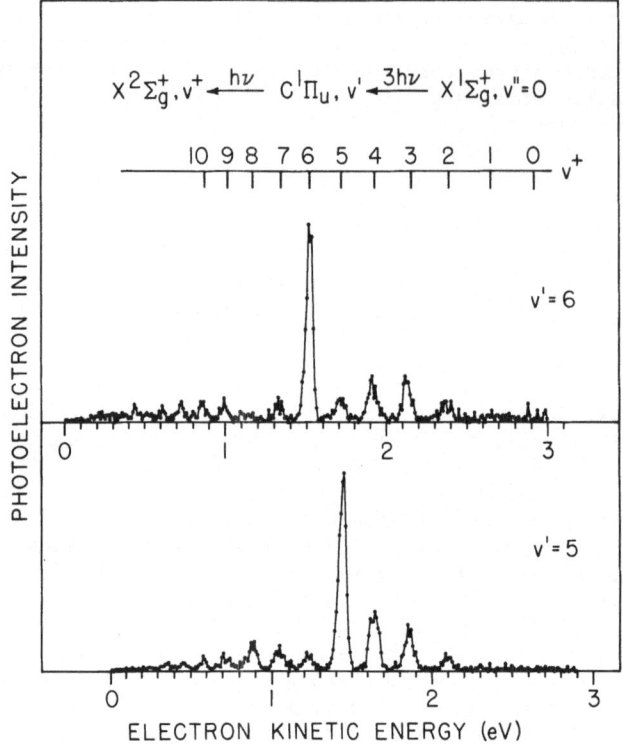

Fig. 5. Photoelectron spectra of H_2 determined along the laser polarization axis ($\theta = 0°$) at the wavelengths of the resonant three photon $C\ ^1\Pi_u$, $v' = 5, 6 \leftarrow X\ ^1\Sigma_g^+$, $v'' = 0$, Q(1) transitions.

isotopes of H_2. In particular, we have recently recorded the photoelectron spectra following (3+1) ionization via the $C\ ^1\Pi_u$, $v' = 0 - 4$ levels of D_2.[93] The spectra for the $C\ ^1\Pi_u$, $v' = 0 - 3 \leftarrow X\ ^1\Sigma_g^+$, $v'' = 0$, Q(3) transitions are shown in Figure 7. As in H_2, the $v^+ = v'$ peak dominates each spectrum. The most striking difference between the H_2 and $D_2\ C\ ^1\Pi_u$ photoelectron spectra is that for $v' = 3$ and 4, the H_2 spectra extend to $v^+ = v' + 6$, while the D_2 spectra show significant intensity only for $v^+ - v' \leqslant 2$. Although there are several possible explanations for this observation, they are all consistent with the model involving the $2p\sigma_u 2p\pi_u\ ^1\Pi_g$ doubly excited state. The doubly excited potential curves for H_2 and D_2 will be nearly identical, and the mass effect on the electronic autoionization width should be small. However, because of the difference in vibrational spacings in the $C\ ^1\Pi_u$ and $X\ ^1\Sigma_g^+$ states, the four photon energies for $v' = 3$ and 4 are smaller (by $\sim$0.26 and 0.34 eV, respectively) in D_2 than in H_2, which will make excitation to the doubly excited curve energetically less favorable. In addition, for lower values of v', the smaller range of R sampled by the D_2 vibrational wavefunctions in the $C\ ^1\Pi_u$ state will decrease the vibrational overlap with the $2p\sigma_u 2p\pi_u\ ^1\Pi_g$ state. Finally, if the D_2 is excited to the same position on the doubly excited potential curve as H_2, the D_2 wave-packet will propagate at only $1/\sqrt{2}$ of the speed of the H_2 wave-packet. Thus, with the same autoionization rate, a much narrower envelope is expected for D_2. Of course, detailed calculations of the $C\ ^1\Pi_u$ photoelectron

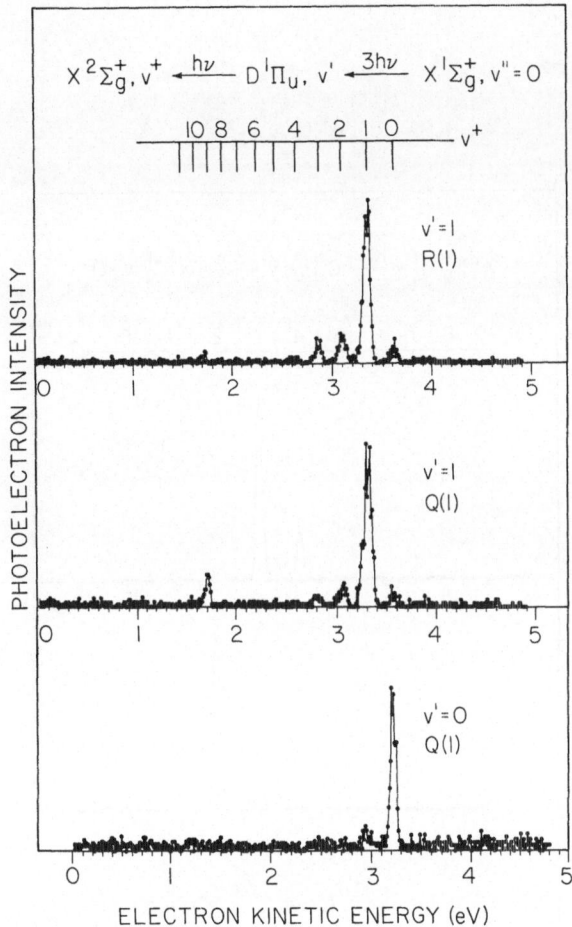

Fig. 6. REMPI-PES spectra of H_2 determined at the wavelengths of the resonant three photon D $^1\Pi_u$, v' = 0 ← X $^1\Sigma_g^+$, v" = 0, Q(1); D $^1\Pi_u$, v' = 1 ← X $^1\Sigma_g^+$, v" = 0, Q(1); and D $^1\Pi_u$, v' = 1 ← X $^1\Sigma_g^+$, v" = 0 R(1) transitions.

spectra are necessary to determine the validity of these arguments for D_2. However, at least qualitatively, it appears that the D_2 spectra can be explained in a manner consistent with the H_2 spectra.

Conclusions

The (3+1) ionization of H_2 via the C $^1\Pi_u$ state has been discussed in light of the existing experimental and theoretical data. It does not appear that the existing experimental data on the (3+1) ionization of H_2 via the C $^1\Pi_u$ state can be explained in terms of the simple direct photoionization of the Rydberg electron into the H_2^+ X $^2\Sigma_g^+$ continuum. As suggested by Chupka[80] and Hickman,[79,90] excitation and subsequent autoionization of the $2p\sigma_u 2p\pi_u$ $^1\Pi_g$ doubly excited state appear to strongly influence the vibrational branching ratios, particularly for C $^1\Pi_u$, v' = 3 - 6. Although a direct experimental study of the $2p\sigma_u 2p\pi_u$ state remains to be performed, the present data serve to bracket the position of this state. When coupled with more detailed calculations of the vibrational branching ratios, these data should improve our understanding of the doubly excited states of H_2.

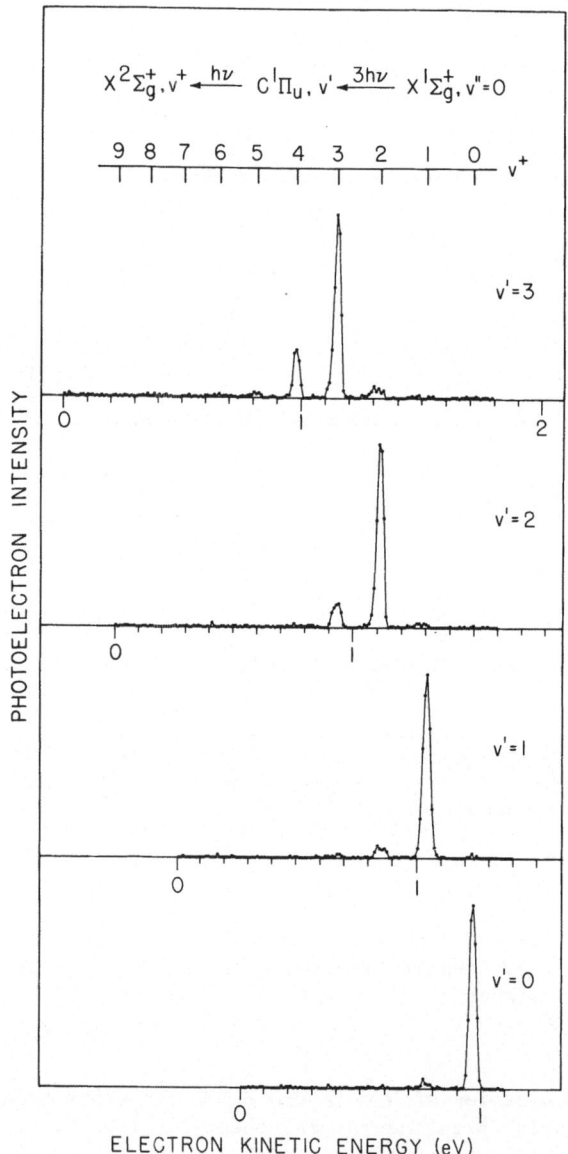

Fig. 7. Photoelectron spectra of D_2 determined along the laser polarization axis at the wavelengths of the resonant three photon C $^1\Pi_u$, $v' = 0 - 3 + X$ $^1\Sigma_g^+$, $v'' = 0$, Q(3) transitions.

III. TWO COLOR (2+1) REMPI-PES OF H_2 VIA E,F $^1\Sigma_g^+$

The next level of sophistication in REMPI-PES studies of molecular photoionization utilizes an independently tunable laser to photoionize the excited state. Such two-color REMPI processes permit one to examine the interaction and decay mechanisms of autoionizing states with unprecedented detail. Furthermore, compared to the established VUV techniques used to study such processes, two-color REMPI-PES has several useful advantages: ability to select the quantum state of the excited target in many cases, very high wavelength resolution, and access to nonoptical channels.

We have just completed initial measurements combining a two-step excitation process with photoelectron energy analysis to investigate rotational and vibrational autoioinization processes in molecular hydrogen. The 4th anti-Stokes component of a Raman-shifted, doubled dye laser provides 5-20 μJ of light at ~193 nm, which populates the state H_2 E,F $^1\Sigma_g^+$, $v'=2$, $J'=1$ through a two photon transition. A third photon at ~ 400 nm (the 1st anti-Stokes component of a Raman-shifted excimer pumped dye laser) then excites from this level to a region near the thresholds for production of H_2^+ $^2\Sigma_g^+$, $v^+=2$, $N^+=1,3$. The kinetic energies of the resulting photoelectrons are determined by time-of-flight analysis in a magnetic-bottle electron spectrometer.[108] The high collection efficiency (50%) of the magnetic-bottle permits us to follow individual ionic vibrational levels, through gated detection of individual photoelectron peaks, as the excitation wavelength of the H_2^+ $^2\Sigma_g^+$, v^+, N^+ ← H_2 E,F $^1\Sigma_g^+$, $v'=2$, $J'=1$ transition is scanned.

This experimental approach makes it possible to directly determine the final vibrational states produced by vibrational autoionization of individual rotational states as a function of position within the autoionization profile. (Analogous experiments on NO have also been performed by Achiba et al.[34] and Kimman et al.[65]) Moreover, the vibrational branching ratios between the thresholds for production of H_2^+ $^2\Sigma_g^+$, $v^+=2$, $N^+=1,3$ reflect the competition between the rotational and vibrational autoionization mechanisms. Thus it is now possible to measure quantities that characterize autoionization dynamics at the level at which they are calculated by the most sophisticated theories, for example, the MQDT calculations of Raoult and Jungen.[116]

IV. THREE PHOTON EXCITATION OF AUTOIONIZING STATES OF ATOMIC XENON BETWEEN THE $^2P_{3/2}^o$ AND $^2P_{1/2}^o$ FINE STRUCTURE THRESHOLDS

Background

Recently, Gangopadhyay et al.[117] reported calculations of two and three photon ionization spectra of Xe in the energy region between the Xe^+ $^2P_{3/2}^o$ and $^2P_{1/2}^o$ fine structure thresholds. Both the two and three photon calculations show strong resonances corresponding to autoionizing Rydberg series converging to the $^2P_{1/2}^o$ limit. In particular, the theoretical three photon ionization spectrum showed structure corresponding to $J = 1$ and $J = 3$ Rydberg series. However, a recent experimental study of the three photon ionization of Xe by Feldmann et al.[118] found no such resonance structure in the total ionization cross section in this region, although their photoelectron angular distributions did show a strong variation around a predicted autoionizing resonance at $3h\nu = 13.134$ eV.

The lack of structure in the experimental three photon ionization spectrum was somewhat surprising. The single photon absorption[119] and ionization[120] spectra of ground state Xe($^1S_0$) in the region between the

Xe$^+$ fine structure thresholds display the well known $(^2P^0_{1/2})$ns' and $(^2P^0_{1/2})$nd' J = 1 Rydberg series converging to the $^2P^0_{1/2}$ threshold. These series are also allowed in the electric dipole approximation for three photon absorption, although the relative intensities may be different than the one photon intensities. Additional Rydberg series with J = 3 are also expected in the three photon spectrum. For example, Wang and Knight[121] have recently observed the $(^2P^0_{1/2})$ns' and nd' Rydberg series with J ≠ 1 using two-step laser excitation of a Xe metastable beam; one of the reported series is the $(^2P^0_{1/2})$nd' series with J = 3. Also, the even parity Rydberg states in this region were recently observed by four photon excitation by Blazewicz et al.[122] Thus, one expects to see a number of Rydberg series in the three photon ionization spectrum of Xe between the Xe$^+$ $^2P^0_{3/2}$ and $^2P^0_{1/2}$ fine structure thresholds, and for this reason we have reexamined the spectrum in this region. Four autoionizing Rydberg series are observed using linearly polarized light, corresponding to the $(^2P^0_{1/2})$ns'$[1/2]^0_1$, $(^2P^0_{1/2})$nd'$[3/2]^0_1$, $(^2P^0_{1/2})$nd'$[5/2]^0_3$, and $(^2P_{1/2})$ng'$[7/2]^0_3$ Rydberg series. Here, the notation $(^2P^0_j)$ nℓ'$[K]^0_J$ corresponds to $j_c\ell$ coupling,[123] in which the angular momentum of the ion core, j_c, is coupled to the orbital angular momentum of the Rydberg electron, ℓ, to give K, which is then coupled to the Rydberg electron's spin to give J. In what follows the notation will be abbreviated nℓ'$[K]^0_J$, with the prime denoting the $^2P^0_{1/2}$ ion core. As expected, with circularly polarized light only the J = 3 levels are observed.

In addition to the three photon ionization spectrum, the photoelectron angular distributions have also been determined for the 9s'$[1/2]^0_1$ and the 7d'$[3/2]^0_3$ autoionizing resonances using linear polarized light. The angular distributions can be compared with the predictions of Gangopadhyay et al.[117]

The Three Photon Ionization Spectrum

Figure 8 shows the three photon ionization spectrum of Xe obtained using linear polarized light between 2791.00 Å and 2800.75 Å. Two members of three different Rydberg series are clearly discernible; these correspond to the sharp ns'$[1/2]^0_1$ and nd'$[5/2]^0_3$ series, and the broad nd'$[3/2]^0_1$ series. As will be demonstrated below, in this spectral range the intense ns'$[1/2]^0_1$ peaks obscure the members of the fourth allowed series, ng'$[7/2]^0_3$. The section of the spectrum shown in Figure 8 is representative of the three photon spectrum between 2920 Å (3hν = 12.74 eV) and the $^2P^0_{1/2}$ threshold at 2768.26 Å (3hν = 13.436 eV). In particular, the portion of the three photon ionization spectrum between 2850 Å and 2810 Å, in which Feldmann et al.[118] observed no resonant structure, displays structure completely analogous to that of Figure 8, but corresponding to lower principal quantum numbers (for example, the n = 10, 11, 12 members of the ns'$[1/2]^0_1$ series are observed at 2848.9 Å, 2827.0 Å, and 2812.9 Å, respectively). The explanation for the differences in the two observations is not clear to us at this time but may be due to the higher power employed in the earlier study. The structure in the present spectrum is quite regular; however, due to the modest resolution of the dye laser (~ 0.3 cm^{-1}) and the high order of the process (6 dye laser photons), the series are only resolved to n ~ 20.

The appearance of the three photon ionization spectrum is not determined solely by the structure at the three photon energy. The transition amplitude for three photon excitation is given by[117]

$$M_{fg} = \sum_{a_1,a_2} \frac{\langle f|\vec{r}\cdot\vec{\epsilon}|a_2\rangle \langle a_2|\vec{r}\cdot\vec{\epsilon}|a_1\rangle \langle a_1|\vec{r}\cdot\vec{\epsilon}|g\rangle}{(E_{a_2} - E_g - 2h\nu)(E_{a_1} - E_g - h\nu)},$$ (1)

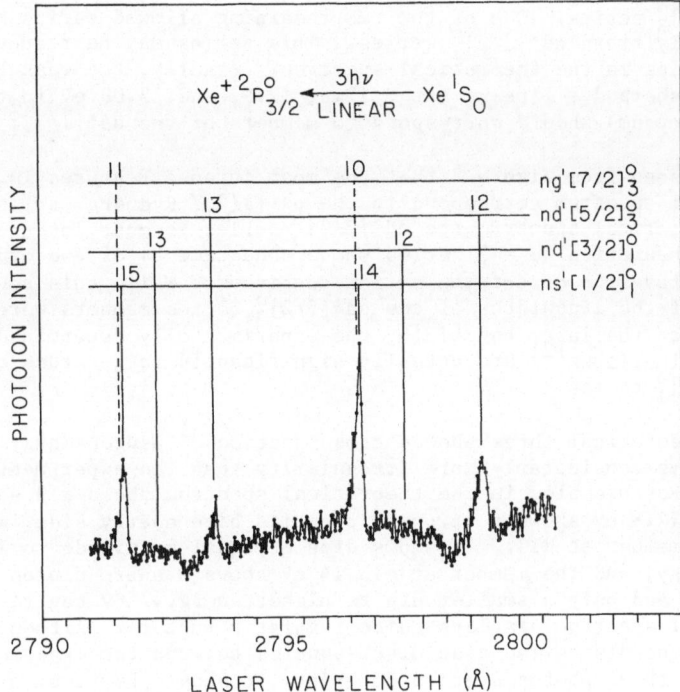

Fig. 8. The three photon ionization spectrum of atomic Xe between 2800.75 Å and 2791.00 Å obtained using linearly polarized light.

where $\vec{\epsilon}$ is the unit polarization vector of the light, the a_i's represent all possible intermediate levels allowed by selection rules, E_i is the energy of the ith level, and $h\nu$ is the photon energy. It is clear from Equation 1 that, as the laser wavelength is scanned, the three photon transition amplitude will change as different energy levels contribute more or less to the summation over states at the one and two photon energies. In addition, the interference between different terms in the summation must be considered when the amplitude M_{fg} is squared to obtain the transition probability. However, in the present case, the one and two photon energies are always over 4 eV and 1 eV away from the nearest allowed levels, respectively.[124] Thus, the important terms in Equation 1 are not expected to vary dramatically in the wavelength region of the present study. This allows one to understand the observed regularity of the Rydberg structure and indicates that interference effects do not dramatically distort the observed spectrum.

The comparison of the experimental spectrum with the theoretical three photon cross section of Gangopadhyay et al. (Figure 2 of Reference 117) requires some care. The calculation was not intended to provide precise energy levels, and it is not surprising that there is some discrepancy between the theoretical and experimental resonance positions. No labels other than the total angular momentum J are given to the resonances in the theoretical spectrum. Because the ng partial waves were not included in the theoretical calculations,[117] the ng'[7/2]$_3^o$ resonances will not be present in the theoretical cross section. Therefore, the most intense series, labelled J = 3, must correspond to the nd'[5/2]$_3^o$ series. One of the two remaining allowed series should be the extremely broad nd'[3/2]$_1^o$ series. This series may be responsible for the broad dips in the theoretical spectrum. Finally, the weak but relatively sharp J = 1 peaks (at ~ 12.6, 12.9, and 13.08 eV in the theoretical spectrum) should correspond to members of the ns'[1/2]$_1^o$ series.

It is seen from Figure 8 that the most intense features in the experimental spectrum correspond to the ns'[1/2]$_1^o$ Rydberg series, while in the theoretical spectrum the nd'[5/2]$_3^o$ series is most intense. In fact, the theoretical J = 3 series shows enhancements of two orders of magnitude above the direct ionization continuum. While this could be explained if the linewidths of the nd'[5/2]$_3^o$ series members were much narrower than the laser bandwidth, the experimentally determined nd'[5/2]$_3^o$ linewidths[121] are actually significantly larger than those of the ns'[1/2]$_1^o$ series.

The theoretical three photon cross section of Gangopadhyay et al.[117] also displays considerably more irregularity than the experimental spectrum. For example, in the theoretical spectrum the J = 3 series member at ~12.8 eV shows a strong dip on its high energy side, while the next J = 3 member at ~13.0 eV shows dips of equal magnitude to lower and higher energy, and the member at ~13.14 eV shows a sharp dip on its low energy side and only a smaller dip to higher energy. By contrast, the experimental spectrum displays quite regular structure. Although it is clear from the discussion that discrepancies between the experimental and theoretical three photon spectra do exist, it should be remembered that the calculations represent the first attempt at understanding multiphoton ionization via autoionizing levels in this complex system.

The linewidths for the ns'[1]$_{1/2}^o$ and nd'[5/2]$_{3,1}^o$ Rydberg series are in reasonable agreement with those of Wang and Knight.[121] Although the statistics make an analysis of the linewidths of the broad nd'[3/2]$_1^o$ series rather difficult, some qualitative remarks can be made on the lineshapes. In particular, the nd'[3/2]$_1^o$ Rydberg states show positive Fano q parameters[125] in three photon (and one photon[119,120]) excitation from the ground state, and negative q parameters in single photon excitation from the 6p'[3/2]$_1^o$ level.[121] That is, the peaks are asymmetric in opposite senses for the two different excitation pathways. Such a dependence of the line profile on the excitation pathway is not surprising. For example such behavior has been observed previously and is discussed in detail by Ganz et al.[126]

The lower half of Figure 9 shows the portion of the linearly polarized, three photon ionization spectrum that contains the 9s'[1/2]$_1^o$ resonance. The small peak to higher energy corresponds to the lowest member (n = 5) of the ng'[7/2]$_3^o$ Rydberg series. This is demonstrated in the upper half of Figure 9 by using circularly polarized light, which eliminates the 9s'[1/2]$_1^o$ peak through the ΔJ = 3 selection rule. The two spectra are plotted to show the approximate relative intensities for the two different laser polarizations. Higher members of the ng'[7/2]$_3^o$ series, which are overlapped by the more intense ns'[1/2]$_1^o$ series, can be uncovered by recording the entire spectrum using circularly polarized light. As expected for a high-ℓ, non-penetrating Rydberg series, the quantum defect of the ng' series is extremely small, having a value μ ~ 0.01.

Photoelectron Angular Distributions

The photoelectron angular distributions for three photon ionization of atomic Xe via the 9s'[1/2]$_1^o$ and 7d'[5/2]$_3^o$ autoionizing levels are shown in Figure 10. The photoelectron angular distributions for three

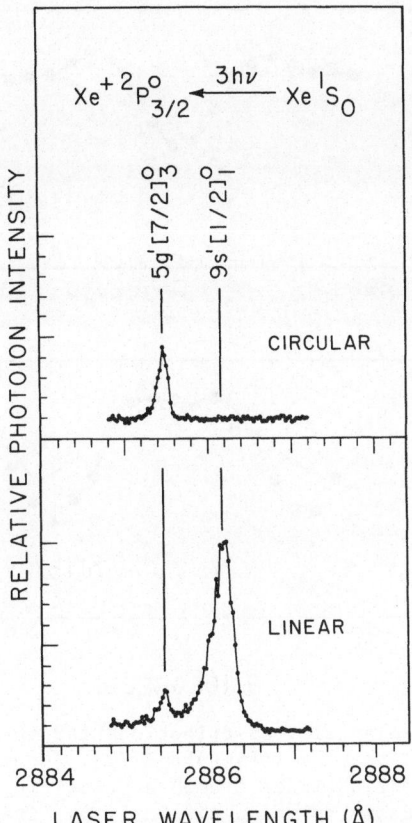

Fig. 9. The three photon ionization spectra of atomic Xe around the
$(^2P^o_{1/2})9s'[1/2]^o_1$ autoionizing resonance obtained using linearly
polarized light (lower half) and circularly polarized light
(upper half). The spectra are plotted so that the relative peak
heights in the two spectra reflect the approximate relative
intensities of the peaks.

photon ionization from the $^1S_0$ ground state using linearly polarized
light must have the functional form[127,128]

$$I(\theta) \propto \sum_{i=0}^{3} A_i \cos^{2i} \theta. \qquad (2)$$

Here, I is the photoelectron intensity, θ is the angle between the
polarization axis of the light and the detector, and the A_i are asymmetry
coefficients. Figure 10 shows both the best fits of the experimental
data to the functional form of Equation (2) as well as best fits to the
form

$$I(\theta) \propto A_o + A_2 \cos^2 \theta. \qquad (3)$$

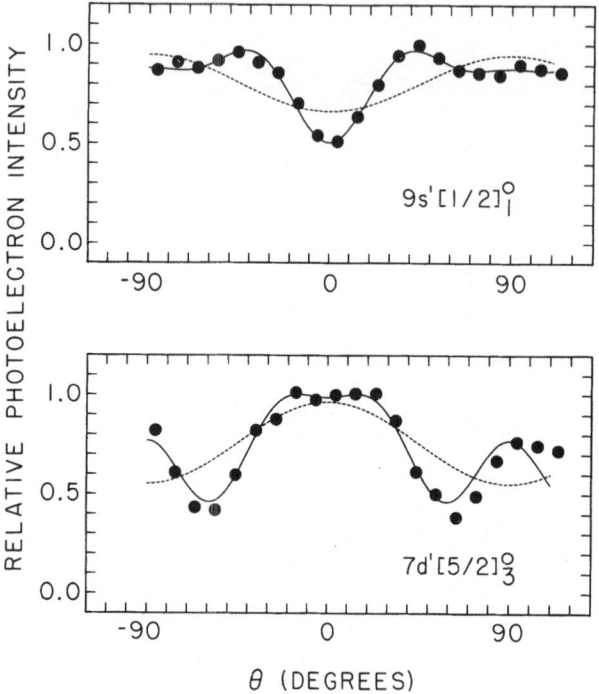

Fig. 10. Photoelectron angular distributions for three photon ionization
of atomic Xe via the $(^2P^o_{1/2})9s'[1/2]^o_1$ (upper frame) and the
$(^2P^o_{1/2})7d'[5/2]^o_3$ (lower frame) autoionizing resonances. Data
points are shown as solid circles (●). The dashed (---) and
solid (——) lines are best fits to the functional forms
$A_0 + A_2\cos^2(\theta)$ and $A_0 + A_2\cos^2(\theta) + A_4\cos^4(\theta) + A_6\cos^6(\theta)$,
respectively.

The latter is appropriate for single photon ionization of an unaligned
sample or for ionization when the final angular momentum is restricted to
$J = 1$. Although the $9s'[1/2]^o_1$ autoionizing level fulfills the latter
condition, it is clear that neither of the angular distributions in
Figure 10 can be satisfactorily fit with the functional form of Equation
3. The nonsesonant ionization background at the wavelength of the
$9s'[1/2]^o_1$ autoionizing resonance is substantial, and the higher order
terms necessary to fit the angular distribution at this wavelength must
arise from direct three photon ionization into the $J = 3$ continua.

Gangopadhyay et al.[117] have calculated photoelectron angular
distributions following three photon ionization for the resonances at
13.068 and 13.134 eV, which correspond to the $10s'[1/2]^o_1$ and $9d'[5/2]^o_3$
resonances, respectively. The $9s'[1/2]^o_1$ resonance was chosen in the
experimental study because it is better resolved from the corresponding
$ng'[7/2]^o_3$ level. In addition, the signal levels for the $9s'[1/2]^o_1$ and
$7d'[5/2]^o_3$ resonances are considerably higher than for the corresponding
$10s'[1/2]^o_1$ and $9d'[5/2]^o_3$ resonances. The differences in principal
quantum numbers notwithstanding, a comparison of the angular distributions from states in the same Rydberg series is certainly appropriate.

Both of the theoretical angular distributions[117] show minima at $\theta = 0°$ and $\pm 90°$ and maxima at $\sim \pm 45°$. The minima are most pronounced in the J = 3 distribution, for which the intensities at both 0 and $\pm 90°$ are nearly zero; the $\pm 90°$ minima in the J = 1 distribution are more pronounced than that at $\theta = 0°$, but none of the minima are as deep as those in the J = 3 distribution. In contrast, the experimental angular distributions are somewhat less dramatic in the degree of variations and show qualitative departures from theory. The $9s'[1/2]_1^0$ distribution shows a minimum at $\theta = 0°$, with maxima at $\sim \pm 45°$ but with very little fall off between $\pm 45°$ and $\pm 90°$. The $7d'[5/2]_3^0$ distribution displays a broad, flat maximum between $\pm 20°$ with minima at $\pm 60°$ and secondary maxima at $\pm 90°$.

Conclusions

The three photon ionization spectra of atomic Xe between the Xe^+ $^2P_{3/2}^0$ and $^2P_{1/2}^0$ ionization thresholds reveals four Rydberg series of autoionizing resonances, corresponding to the $ns'[1/2]_1^0$, $nd'[3/2]_1^0$, $nd'[5/2]_3^0$, and $ng'[7/2]_3^0$ series. The results are in qualitative agreement with the theoretical three photon cross section of Gangopadhyay et al.,[117] which predicts resonance structure due to Rydberg series converging to the $^2P_{1/2}^0$ limit. However, quantitative agreement between experiment and theory is rather poor, indicating that substantial improvements in the theoretical calculations are called for. In particular, the inclusion of g ($\ell = 4$) partial waves is clearly essential. Photoelectron angular distributions are reported for the $9s'[1/2]_1^0$ and $7d'[5/2]_3^0$ autoionizing resonances, and both exhibit very anisotropic behavior. More extensive angular distribution studies of three photon ionization of Xe in this wavelength region are desirable in order to map out the variation of angular distributions within resonant profiles, analogous to those observed in single photon ionization studies.[129-131]

V. PHOTOELECTRON SPECTRA OF Xe_2^* OBTAINED BY RESONANTLY ENHANCED MULTIPHOTON IONIZATION

Resonantly enhanced multiphoton ionization--photoelectron spectroscopy has proven to be an effective technique for the determination of photoelectron spectra of van der Waals molecules.[28,29,36,37,77] Since the REMPI technique provides selective ionization, interference from photoelectron peaks of other species is eliminated. This is particularly important for the study of van der Waals molecules, since they are usually a minor component of the molecular beam. We have recently reported the photoelectron spectra of $KrXe$[36] and $ArXe$[37] using this technique. Here we will focus on a recent paper[77] in which we described the REMPI-PES of Xe_2. In contrast to the HeI-PES of Xe_2,[132] which shows only a few of the six molecular dimer ion states that result from the combination Xe $^1S_0$ + Xe^+ $^2P_{3/2}^0$ or $^2P_{1/2}^0$, the REMPI-PES of Xe_2 allows the observation of all six states. As an example of the differences observed in the spectra obtained using these two techniques, Figure 11 shows both the HeI-PES and a REMPI-PES of Xe_2 obtained in the energy region near the Xe $^1S_0$ + Xe^+ $^2P_{3/2}^0$ dissociation limit. Four molecular dimer ion bands are predicted to occur in this energy region, and all four bands are seen in the REMPI-PES; however, as is seen from the data shown in the lower trace in Figure 11, three of these bands in the HeI-PES are either partially or completely obscured by the much more intense atomic Xe^+ $^2P_{3/2}^0$ photoelectron peak.

Two factors require that REMPI-PES of Xe_2 be obtained via a number of different resonant intermediate states in order to study all six dimer ion states. First, since each resonant intermediate state has a unique

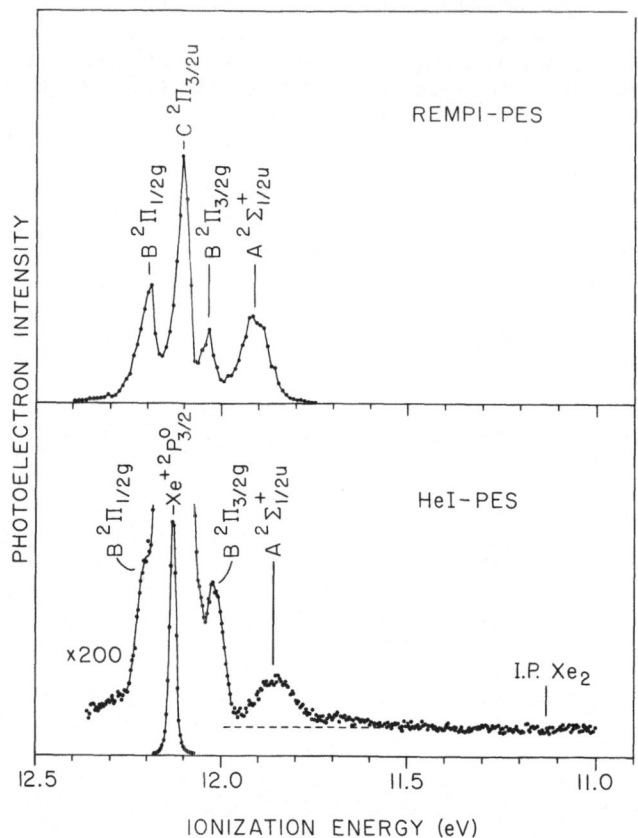

Fig. 11. Illustrative example of photoelectron spectra of Xe_2 obtained
using HeI ionization (lower trace) and resonantly enhanced
multiphoton ionization (upper trace) in the region of the
Xe^+ $^2P^o_{3/2}$ ionization limit. Note that in the HeI-PES, atomic
ionization is more than 100 times as intense as the molecular
ionization, but that no atomic ionization is observed in the
REMPI-PES.

core (which may actually be a mixture of two or more ion states) and
since core switching transitions are unfavored in the ionization step, it
is not usually possible to observe all of the dimer ion bands in a single
REMPI-PES. Indeed, the observation of all four molecular bands in the
energy region near the Xe $^1S_0$ + Xe^+ $^2P^o_{3/2}$ dissociation limit in the
REMPI-PES shown in Figure 11 is atypical. More often, a REMPI-PES of Xe_2
obtained in this energy region will show from one to three molecular
bands. In general, it is possible to study all the ionic states of a
given molecule by determining REMPI-PES via several resonant intermediate
states having different ion cores; in recent work,[77] REMPI-PES of Xe_2
were obtained via five different resonant intermediate states. Together,
these spectra provide information on all six dimer ion states, two of
which have not been observed previously. Second, as discussed
previously,[36,77] REMPI-PES may contain photoelectron peaks due to
predissociation of the resonant intermediate state followed by
photoionization of the excited atomic fragment, i.e., $Xe_2^* \rightarrow Xe + Xe^* \rightarrow$

Xe + Xe$^+$ + e. Some of the photoelectron spectra of Xe$_2$ in Ref. 77 show evidence for partial or even complete predissociation of the resonant intermediate state, while others show no evidence of predissociation of the resonant intermediate state (e.g., the REMPI-PES shown in Figure 11). In the limiting case of complete predissociation of the resonant intermediate state, the photoelectron spectrum will show no evidence of molecular dimer bands, and consequently, such spectra are not useful for the study of the dimer ion properties.

VI. REMPI STUDIES OF OPEN SHELL ATOMS

As mentioned in the Introduction, REMPI provides a powerful means for preparing and studying open-shell atoms, free radicals, and transient species via photodissociation. For illustration, we will briefly discuss some recent work on atomic iodine,[133,134] although several other atoms (e.g., Refs. 82, 135, 136) have already been studied and the technique is thought to be quite broadly applicable. During the past few years, considerable progress has been made in the experimental study of photoionization of halogen atoms;[137-141] however, the best resolution attained to date in such experiments ($\sim$ 20 cm^{-1}) is not sufficient to resolve all of the features of interest. For example, the spectrum of atomic iodine between 1120 Å and 1090 Å contains Rydberg series converging to two ionization limits within this region as well as members of a third Rydberg series converging to a limit at higher energy. Therefore, it is extremely difficult, if not impossible, to sort out the complex structure.[138] For this reason, we investigated[133] the possibility of using laser techniques for the study of this problem. We used two-color multiphoton ionization mass spectrometry to determine the spectra of the optically allowed autoionizing states of atomic iodine. In these experiments, the first laser is used to produce atomic iodine by the photodissociation of methyl iodide and to pump the iodine atoms to a low lying ...5p^{4}6p state via a two photon transition. A second laser is used to probe single photon transitions from these ...5p^{4}6p states to autoionizing ...5p^{4}ns and nd Rydberg states converging to the $^1D_2$ ionization limit. Because a total of three photons is used, this process accesses states of the same parity as those accessed by single photon excitation. However, the resolution in the present experiments is limited only by the linewidth of the laser ($\sim$ 0.05 – 0.3 cm^{-1}). Thus, it is possible to study the photoionization spectrum of atomic iodine and other open-shell atoms with unprecedented detail.

The choice of the resonant intermediate state will determine the manifold of autoionizing states that may be observed. Two primary considerations are useful in deciding which resonant intermediate state should be used. First, transitions that preserve the ion core of the resonant intermediate state are, in general, stronger than those that require a change in the ion core. For example, the transitions ($^1D_2$)6p $\rightarrow$ ($^1D_2$)ns, nd are expected to be stronger than the transitions ($^3P_1$)6p $\rightarrow$ ($^1D_2$)ns, nd. Second, by specifying the values of K and J for the resonant intermediate state, it is possible to limit the range of values of K and J of the autoionizing Rydberg states owing to the angular momentum selection rules.

Spectra were obtained by pumping the two photon transition from the $^2P_{1/2}$ state to the ($^1D_2$)6p[3]$_{5/2}$ and ($^1D_2$)6p[1]$_{1/2}$ states and probing transitions to the ($^1D_2$)ns and nd Rydberg series converging to the $^1D_2$ ionization limit. These are shown in Figs. 12 and 13, respectively. The former spectrum displays three sharp series that can be resolved to high principal quantum numbers (n $\sim$ 35), while the latter spectrum displays a single sharp series and a single broad series. The symmetry of several of these series was deduced from angular momentum coupling rules.[133]

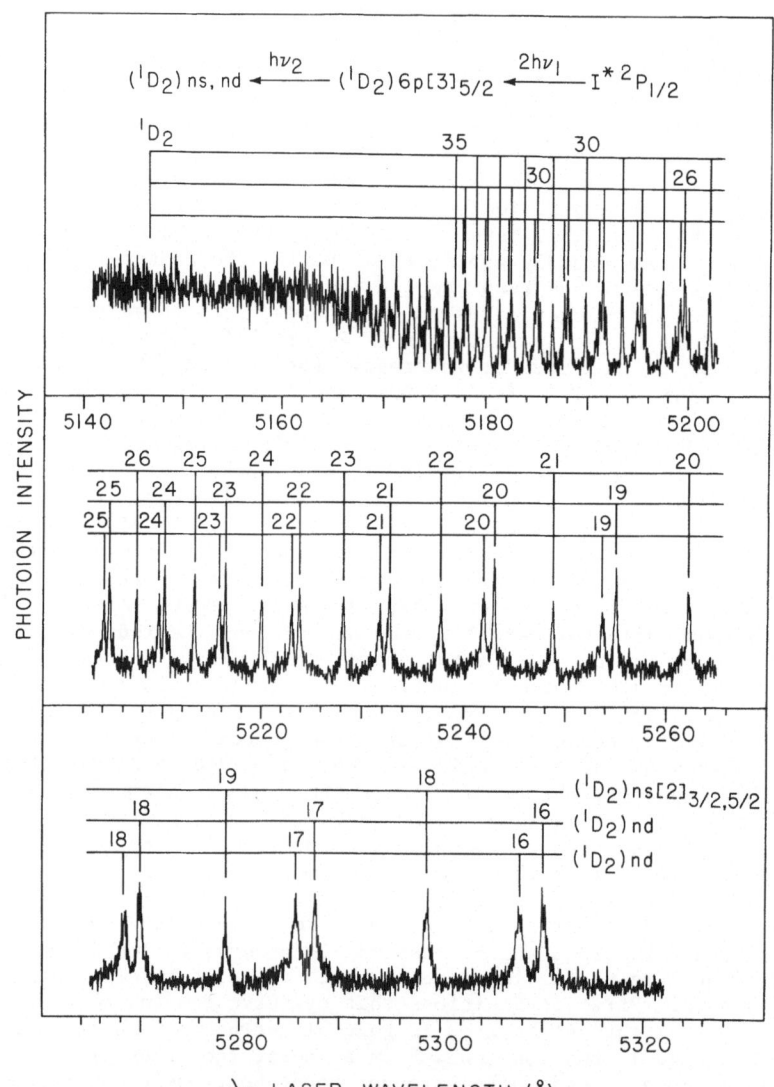

Fig. 12. The two-color photoionization spectrum of atomic iodine obtained by pumping the two photon $^2P_{1/2} \rightarrow (^1D_2)6p[3]_{5/2}$ transition with laser 1 and probing transitions from the $(^1D_2)6p[3]_{5/2}$ state to $(^1D_2)ns$ and nd Rydberg series with laser 2. The $I^{+\,1}D_2$ ionization limit occurs at $\sim$ 5146.75 Å (98022.3 cm^{-1}).

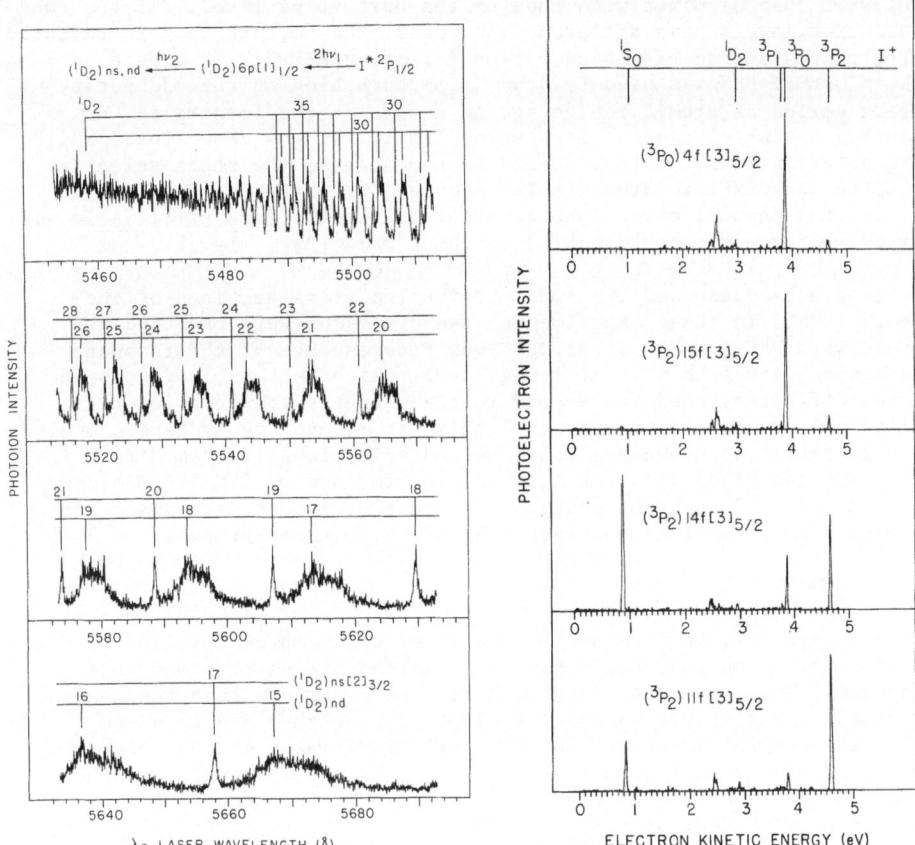

Fig. 13. The two-color photoionization spectrum of atomic iodine obtained
    by pumping the two photon $^2P_{1/2} \rightarrow (^1D_2)6p[1]_{1/2}$ transition with
    laser 1 and probing transitions from the $(^1D_2)6p[1]_{1/2}$ state to
    $(^1D_2)$ns and nd Rydberg series with laser 2. The $I^{+1}D_2$
    ionization limit occurs at ~ 5458.30 Å (98022.3 cm^{-1}).

Fig. 14. The photoelectron spectra of atomic iodine obtained following
    two photon resonant, three photon ionization via the $(^3P_0)4f[3]_{5/2}$,
    $(^3P_2)15f[3]_{5/2}$, $(^3P_2)14f[3]_{5/2}$, and $(^3P_2)11f[3]_{5/2}$ levels. The
    unlabeled photoelectron peaks are discussed in the text.

Several recent REMPI-PES studies have shown that ionization into a structureless continuum via an unperturbed Rydberg state usually proceeds by the ejection of the Rydberg electron without a change in the electronic or vibrational state of the ion core (see, e.g., Ref. 96). Thus, in many instances, REMPI can be used as a source of state-selected ions. However, the results can be quite different if the resonant intermediate level is perturbed, because the character of the unperturbed level will then be mixed with that of the perturbing level. If the two interacting levels have different ion cores, the perturbation is revealed in the photoelectron branching ratios following REMPI. In recent work,[134] REMPI-PES was used to study a perturbation in the odd parity Rydberg series of atomic iodine and to determine the feasibility of producing state-selected I^+ ions using REMPI. It was shown that, while state-selected ions can be produced in some cases, the state selecting capability of REMPI is dramatically reduced for perturbed levels. In Fig. 14, the photoelectron spectra obtained via the $(^3P_2)nf[3]_{5/2}$ levels show the effect of a perturbation by the $(^3P_0)4f[3]_{5/2}$ level.

As a consequence of the small ionization cross sections of the Rydberg levels in these experiments, two processes were observed that compete with REMPI. The first involves fluorescence of the resonant intermediate level to a lower lying level, which is then ionized with greater efficiency, and the second involves energy transfer (either collisional or radiative) between $I^{*2}P_{1/2}$ atoms and the Rydberg atoms. Although these two processes would be indistinguishable from direct REMPI using mass spectrometric techniques alone, the use of REMPI-PES allows the separation of the contributions of these different mechanisms and provides a more complete understanding of the overall process.

VIII. SUMMARY

To summarize, we have used a small set of examples to illustrate the use of multiphoton ionization to probe excited states of atoms and molecules. These examples were much narrower in scope than the general discussion of prospects in the Introduction, but they did serve to illustrate several important points: (1) REMPI makes it possible to prepare single excited quantum states of atoms and molecules and then to probe them in a variety of ways. (2) Using this technique one can exercise a great deal of control over the excitation pathway, i.e., by selecting particular excitation/ionization channels and probing different regions of R. (3) Experiments can be performed at high resolution, i.e. $0.3-0.01$ cm^{-1} with pulsed lasers. (4) The high sensitivity of REMPI permits performing experiments on minor components, open-shell atoms, dissociation fragments, etc. (5) Multiphoton excitation can be used to access non-optical channels. (6) Most importantly, REMPI has already shown its usefulness in providing new access and insight to numerous physical mechanisms, e.g., role of doubly excited states in photoionization, competition between autoionization mechanisms, dynamics of multiphoton excitation, effects of perturbations on excited state photoionization dynamics, and direct detection of predissociation products. Needless to say, these examples merely hint at the eventual scope of this field, whose main impact still lies in the future.

REFERENCES

1. J. C. Miller and R. N. Compton, J. Chem. Phys. 75:2020 (1981).
2. J. C. Miller and R. N. Compton, J. Chem. Phys. 75:22 (1981).
3. Y. Achiba, K. Sato, K. Shobatake, and K. Kimura, J. Photochem. 17:199 (1981).
4. J. H. Glownia, S. J. Riley, S. D. Colson, J. C. Miller, and R. N. Compton, J. Chem. Phys. 77:68 (1982).
5. M. G. White, M. Seaver, W. A. Chupka, and S. D. Colson, Phys. Rev. Lett. 49:28 (1982).
6. J. C. Miller, R. N. Compton, T. E. Carney, and T. Baer, J. Chem. Phys. 76:5648 (1982).
7. J. C. Miller and R. N. Compton, Chem. Phys. Lett. 93:453 (1982).
8. Y. Achiba, K. Sato, K. Shobatake, and K. Kimura, J. Chem. Phys. 77:2709 (1982).
9. J. T. Meek, S. R. Long, and J. P. Reilly, J. Phys. Chem. 86:2809 (1982).
10. J. Kimman, P. Kruit, and M. J. van der Wiel, Chem. Phys. Lett. 88:576 (1982).
11. S. L. Anderson, D. M. Rider, and R. N. Zare, Chem. Phys. Lett. 93:11 (1982).
12. S. T. Pratt, E. D. Poliakoff, P. M. Dehmer, and J. L. Dehmer, J. Chem. Phys. 78:65 (1983).
13. S. T. Pratt, P. M. Dehmer, and J. L. Dehmer, J. Chem. Phys. 78:4315 (1983).
14. S. T. Pratt, P. M. Dehmer, and J. L. Dehmer, J. Chem. Phys. 79:3234 (1983).
15. Y. Achiba, K. Sato, K. Shobatake, and K. Kimura, J. Chem. Phys. 78:5474 (1983).
16. Y. Achiba, K. Sato, K. Shobatake, and K. Kimura, J. Chem. Phys. 79:5213 (1983).
17. S. R. Long, J. T. Meek, and J. P. Reilly, J. Chem. Phys. 79:3206 (1983).
18. S. R. Long, J. T. Meek, P. J. Harrington, and J. P. Reilly, J. Chem. Phys. 78:3341 (1983).
19. S. T. Pratt, P. M. Dehmer, and J. L. Dehmer, Chem. Phys. Lett. 105:28 (1984).
20. S. T. Pratt, P. M. Dehmer, and J. L. Dehmer, J. Chem. Phys. 80:1706 (1984).
21. S. T. Pratt, P. M. Dehmer, and J. L. Dehmer, J. Chem. Phys. 81:3444 (1984).
22. A. M. Woodward, W. A. Chupka, and S. D. Colson, J. Phys. Chem. 88:4567 (1984).
23. M. G. White, W. A. Chupka, M. Seaver, A. Woodward, and S. D. Colson, J. Chem. Phys. 80:678 (1984).
24. Y. Achiba, A. Hiraya, and K. Kimura, J. Chem. Phys. 80:6047 (1984).
25. K. R. Kastidar and P. Lambropoulos, Phys. Rev. A 29:183 (1984).
26. S. N. Dixit, D. L. Lynch, and V. McKoy, Phys. Rev. A 30:3332 (1984).
27. J. C. Hansen and R. S. Berry, J. Chem. Phys. 80:4078 (1984).
28. K. Sato, Y. Achiba, and K. Kimura, J. Chem. Phys. 81:57 (1984).
29. K. Fuke, H. Yoshiuchi, K. Kaya, Y. Achiba, and K. Kimura, Chem. Phys. Lett. 108:179 (1984).
30. W. G. Wilson, K. S. Viswanathan, E. Sekreta, and J. P. Reilly, J. Phys. Chem. 88:672 (1984).
31. K. Müller-Dethlefs, M. Sander, and E. W. Schlag, Z. Naturforsch. 39a:1089 (1984).
32. K. Müller-Dethlefs, M. Sander, and E. W. Schlag, Chem. Phys. Lett. 112:291 (1984).
33. S. L. Anderson, G. D. Kubiak, and R. N. Zare, Chem. Phys. Lett. 105:22 (1984).
34. Y. Achiba, and K. Kimura, in: "Book of Abstracts, International Conference on Multiphoton Processes III," eds. P. Lambropoulos and S. J. Smith, University of Crete (1984), p. 13.

35. J. L. Durant, D. M. Rider, S. L. Anderson, F. D. Proch, and R. N. Zare, J. Chem. Phys. 80:1817 (1984).
36. S. T. Pratt, P. M. Dehmer, and J. L. Dehmer, Chem. Phys. Lett. 116:245 (1985).
37. S. T. Pratt, P. M. Dehmer, and J. L. Dehmer, J. Chem. Phys. 82:5758 (1985).
38. J. B. Pallix and S. D. Colson, Chem. Phys. Lett. 119:38 (1985).
39. A. Hiraya, Y. Achiba, N. Mikami, and K. Kimura, J. Chem. Phys. 82:1810 (1985).
40. M. Kawasaki, K. Kasatani, H. Sato, Y. Achiba, K. Sato, and K. Kimura, Chem. Phys. Lett. 114:473 (1985).
41. Y. Achiba, K. Sato, and K. Kimura, J. Chem. Phys. 82:3959 (1985).
42. J. T. Meek, E. Sekreta, W. Wilson, K. S. Viswanathan, and J. P. Reilly, J. Chem. Phys. 82:1741 (1985).
43. J. Kimman, M. Lavollee, and M. J. van der Wiel, Chem. Phys. 97:137 (1985).
44. W. E. Conaway, R. J. S. Morrison, and R. N. Zare, Chem. Phys. Lett. 113:429 (1985).
45. S. N. Dixit and V. McKoy, J. Chem. Phys. 82:3546 (1985).
46. S. N. Dixit, D. L. Lynch, V. McKoy, and W. M. Huo, Phys. Rev. A 32:1267 (1985).
47. S. L. Anderson, L. Goodman, K. Krogh-Jespersen, A. G. Ozkabak, R. N. Zare, and C. Zheng, J. Chem. Phys. 82:5329 (1985).
48. W. A. Chupka, A. M. Woodward, S. D. Colson, and M. G. White, J. Chem. Phys. 82:4880 (1985).
49. K. Müller-Dethlefs, M. Sander, and L. A. Chewter, in: "Laser Spectroscopy III," eds. T. W. Hänsch and Y. R. Shen, Springer-Verlag, Berlin (1985), p. 118.
50. J. Hager, M. A. Smith, and S. C. Wallace, J. Chem. Phys. 83:4820 (1985).
51. J. B. Pallix, P. Chen, W. A. Chupka, and S. D. Colson, J. Chem. Phys. 84:5208 (1986).
52. J. B. Pallix and S. D. Colson, J. Phys. Chem. 90:1499 (1986).
53. J. C. Miller and R. N. Compton, J. Chem. Phys. 84:675 (1986).
54. D. L. Lynch, S. N. Dixit, and V. McKoy, Chem. Phys. Lett. 123:315 (1986).
55. S. N. Dixit, and V. McKoy, Chem. Phys. Lett. 128:49 (1986).
56. R. L. Dubs, S. N. Dixit, and V. McKoy, J. Chem. Phys. 85:656 (1986).
57. H. Rudolph, D. L. Lynch, S. N. Dixit, and V. McKoy, J. Chem. Phys. 84:6657 (1986).
58. J. W. J. Verschuur, J. Kimman, H. B. van Linden van den Heuvell, and M. J. van der Wiel, Chem. Phys. 103:359 (1986).
59. C. Cornaggia, D. Normand, J. Morellec, G. Mainfray, and C. Manus, Phys. Rev. A 34:207 (1986).
60. A. M. Woodward, S. D. Colson, W. A. Chupka, and M. G. White, J. Phys. Chem. 90:274 (1986).
61. S. T. Pratt, P. M. Dehmer, and J. L. Dehmer, J. Chem. Phys. 85:5535 (1986).
62. S. T. Pratt, P. M. Dehmer, and J. L. Dehmer, J. Chem. Phys. 85:3379 (1986).
63. H.J.M. Bonnie, P. J. Eenschuistra, J. Los, and H. J. Hopman, Chem. Phys. Lett. 125:27 (1986).
64. H.J.M. Bonnie, J.W.J. Verschuur, H. J. Hopman, and H. B. van Linden van den Heuvell, Chem. Phys. Lett. 130:43 (1986).
65. J. Kimman, J.W.J. Verschuur, M. Lavollee, H. B. van Linden van den Heuvel, and M. J. van der Wiel, J. Phys. B 19:3909 (1986).
66. S. Ganguly, K. Rai Dastidar, and T. K. Rai Dastidar, Phys. Rev. A 33:337 (1986).
67. K. S. Viswanathan, E. Sekreta, E. R. Davidson, and J. P. Reilly, J. Phys. Chem. 90:5078 (1986).
68. K. S. Viswanathan, E. Sekreta, and J. P. Reilly, J. Phys. Chem. 90:5658 (1986).

69. K. Rai Dastidar, S. Ganguly, and T. K. Rai Dastidar, <u>Phys. Rev. A</u> 33:2106 (1986).

70. J. Hager, M. Ivanco, M. A. Smith, and S. C. Wallace, <u>Chem. Phys.</u> 105:397 (1986).

71. J. Hager, M. A. Smith, and S. C. Wallace, <u>J. Chem. Phys.</u> 84:6771 (1986).

72. A. Sur, C. V. Ramana, W. A. Chupka, S. D. Colson, <u>J. Chem. Phys.</u> 84:69 (1986).

73. D. Normand, C. Cornaggia, and J. Mosellec, <u>J. Phys. B</u> 19:2881 (1986).

74. S. Katsumata, K. Sato, Y. Achiba, and K. Kimura, <u>J. Electron Spectrosc.</u> 41:325 (1986).

75. J. R. Appling, M. G. White, T. M. Orlando, and S. L. Anderson, <u>J. Chem. Phys.</u> 85:6803 (1986).

76. K. Sato, Y. Achiba, and K. Kimura, <u>Chem. Phys. Lett.</u> 126:306 (1986).

77. P. M. Dehmer, S. T. Pratt, and J. L. Dehmer, <u>J. Phys. Chem.</u> 91:2593 (1987).

78. S. T. Pratt, P. M. Dehmer, and J. L. Dehmer, <u>J. Chem. Phys.</u> 86:1727 (1987).

79. A. P. Hickman, <u>Bull. Am. Phys. Soc.</u> 32:1252 (1987).

80. W. A. Chupka, <u>J. Chem. Phys.</u> 87:1488 (1987).

81. J. W. Hager and S. C. Wallace, <u>Comments At. Mol. Phys.</u> 20:63 (1987).

82. B. G. Koenders, K. E. Drabe, and C. A. DeLange, <u>Chem. Phys. Lett.</u> 138:1 (1987).

83. T. M. Orlando, S. L. Anderson, J. R. Appling, and M. G. White, <u>J. Chem. Phys.</u> 87:852 (1987).

84. P. Chen, J. B. Pallix, W. A. Chupka, and S. D. Colson, <u>J. Chem. Phys.</u> 86:516 (1987).

85. H. Rudolph, D. L. Lynch, S. N. Dixit, and V. McKoy, <u>J. Chem. Phys.</u> 86:1748 (1987).

86. L. A. Chewter, M. Sander, K. Müller-Dethlefs, and E. W. Schlag, <u>J. Chem. Phys.</u> 86:4737 (1987).

87. L. A. Chewter, K. Müller-Dethlefs, and E. W. Schlag, <u>Chem. Phys. Lett.</u> 135:219 (1987).

88. M. A. O'Halloran, S. T. Pratt, P. M. Dehmer, and J. L. Dehmer, <u>J. Chem. Phys.</u> 87:3288 (1987).

89. C. Cornaggia, A. Giusti-Suzor, and Ch. Jungen, <u>J. Chem. Phys.</u> 87:3934 (1987).

90. A. P. Hickman, <u>Phys. Rev. Lett.</u> 59:1553 (1987).

91. E. Y. Xu, T. Tsuboi, R. Kachru, and H. Helm, <u>Phys. Rev. A</u> 36:5645 (1987).

92. W. A. Chupka, P. J. Miller, and E. E. Eyler, <u>J. Chem. Phys.</u>, in press.

93. S. T. Pratt, P. M. Dehmer, and J. L. Dehmer, <u>J. Chem. Phys.</u> 87:4423 (1987).

94. J. P. Reilly, <u>Israel J. Chem.</u> 24:266 (1984).

95. K. Kimura, <u>Adv. Chem. Phys.</u> 60:161 (1985).

96. P. M. Dehmer, J. L. Dehmer, and S. T. Pratt, <u>Comments At. Mol. Phys.</u> 19:205 (1987).

97. K. Kimura, <u>Int. Rev. Phys. Chem.</u>, in press.

98. P. M. Johnson and C. E. Otis, <u>Ann. Rev. Phys. Chem.</u> 32:139 (1981).

99. R. W. Field, <u>Disc. Faraday Soc.</u> 71:111 (1981).

100. V. S. Antonov and V. S. Letokhov, <u>Appl. Phys.</u> 24:89 (1981).

101. S. R. Leone, <u>in</u>: "Dynamics of the Excited State," ed. K. P. Lawley, Wiley (1982), p. 255.

102. E. W. Schlag and H. J. Neusser, <u>Acc. Chem. Res.</u> 16:355 (1983).

103. R. M. Hochstrasser and H. P. Trommsdorff, <u>Acc. Chem. Res.</u> 16:376 (1983).

104. T. Baer, <u>Comments At. Mol. Phys.</u> 13:141 (1983).

105. W. Y. Cheung and S. D. Colson, <u>in</u>: "Advances in Laser Spectroscopy, Vol. 2," eds. B. A. Garetz and J. R. Lombardi, Wiley (1983), p. 73.

106. N. Bloembergen and A. H. Zewail, <u>J. Phys. Chem.</u> 88:5459 (1984).

107. K. P. Huber and G. Herzberg, <u>in</u>: "Constants of Diatomic Molecules," Van Nostrand Reinhold, New York (1979).

108. P. Kruit and F. H. Read, <u>J. Phys. E</u> 16:313 (1983).

109. D. Dill, Phys. Rev. A 6:160 (1972).
110. D. Dill, in: "Photoionization and Other Probes of Many Electron Interactions," ed. F. Wuilleumier, Plenum, New York (1976), p. 387.
111. U. Fano and D. Dill, Phys. Rev. A 6:185 (1972).
112. R. S. Berry, J. Chem. Phys. 45:1228 (1966).
113. S. L. Guberman, J. Chem. Phys. 78:1404 (1983).
114. T. Namioka, J. Chem. Phys. 40:3154 (1964).
115. T. Namioka, J. Chem. Phys. 41:2141 (1964).
116. M. Raoult and Ch. Jungen, J. Chem. Phys. 74:3388 (1981).
117. P. Gangopadhyay, X. Tang, P. Lambropoulos, and R. Shakeshaft, Phys. Rev A 34:2998 (1986).
118. D. Feldmann, G. Otto, D. Petring, and K. H. Welge, J. Phys. B 19:L141 (1986).
119. See, for example, K. Yoshino, and D. E. Freeman, J. Opt. Soc. Am. B 2:1268 (1985); K. D. Bonin, T. J. McIlrath, and K. Yoshino, J. Opt. Soc Am. B 2:1275 (1985).
120. See, for example, J. Berkowitz, "Photoabsorption, Photoionization, and Photoelectron Spectroscopy," Academic, New York (1979).
121. L. G. Wang and R. D. Knight, Phys. Rev. A 34:3902 (1986).
122. P. R. Blazewicz, J. A. D. Stockdale, J. C. Miller, T. Efthimiopoulos, and C. Fotakis, Phys. Rev. A 35:1092 (1987).
123. G. Racah, Phys. Rev. 61:537 (1942).
124. C. E. Moore, in: "Atomic Energy Levels. Vol. III," National Bureau of Standards, Washington, D.C. (1958).
125. U. Fano, Phys. Rev. 124:1866 (1961).
126. J. Ganz, M. Raab, H. Hotop, and J. Geiger, Phys. Rev. Lett. 53:1547 (1984).
127. P. Lambropoulos, Adv. At. Mol. Phys. 12:87 (1976); G. Leuchs and M. Walther, in: "Multiphoton Ionization of Atoms," eds. S. L. Chin and P. Lambropoulos, Academic, New York (1984), p. 109.
128. S. N. Dixit and P. Lambropoulos, Phys. Rev. Lett. 46:1278 (1981); S. N. Dixit and P. Lambropoulos, Phys. Rev. A 27:861 (1983).
129. D. Dill, Phys. Rev. A 7:1976 (1973).
130. J. A. R. Samson and J. L. Gardner, Phys. Rev. Lett. 31:1327 (1973).
131. Y. Morioka, M. Watanabe, T. Akahori, A. Yagishita, and M. Nakamura, J. Phys. B 18:71 (1985).
132. P. M. Dehmer and J. L. Dehmer, J. Chem. Phys. 68:3462 (1978).
133. S. T. Pratt, P. M. Dehmer, and J. L. Dehmer, Chem. Phys. Lett. 176:12 (1986).
134. S. T. Pratt, Phys. Rev. A 33:1718 (1986).
135. S. T. Pratt, Phys. Rev. A, in press.
136. S. Arepalli, N. Presser, D. Robie, and R. G. Gordon, Chem. Phys. Lett. 118:88 (1987).
137. S. Shahabi, A. F. Starace, and T. N. Chang, Phys. Rev. A 30:1819 (1984)
138. J. Berkowitz, C. H. Batson, and G. L. Goodman, Phys. Rev. A 24:149 (1981).
139. B. Ruscić and J. Berkowitz, Phys. Rev. Lett. 50:675 (1983).
140. B. Ruscić, J. P. Greene, and J. Berkowitz, J. Phys. B 17:L79 (1984).
141. B. Ruscić, J. P. Greene, and J. Berkowitz, J. Phys. B 17:1503 (1984).

SHAPE RESONANCES IN MOLECULAR FIELDS[*]

J. L. Dehmer,[†] D. Dill,[‡] and A. C. Parr[§]

[†]Argonne National Laboratory, Argonne, IL 60439
[‡]Department of Chemistry, Boston University, Boston, MA 02215
[§]National Bureau of Standards, Gaithersburg, MD 20899

I. INTRODUCTION

The last two decades have witnessed remarkable progress in characterizing dynamical aspects of molecular photoionization[1,2] and electron-molecule scattering[2,3] processes. The general challenge is to gain physical insight into the processes occurring during the excitation, evolution, and decay of the excited molecular complex. Of particular interest in this context are the uniquely molecular aspects resulting from the anisotropy of the molecular field and from the interplay among rovibronic modes. Throughout this work, special attention is invariably drawn to resonant processes, in which the excited system is temporarily trapped in a quasibound resonant state. Such processes tend to amplify the more subtle dynamics of excited molecular states and are often displayed prominently against non-resonant behavior in various physical observables.

One very vigorous stream of work has involved shape resonances in molecular systems. These resonances are quasibound states in which a particle is temporarily trapped by a potential barrier, through which it may eventually tunnel and escape. In molecular fields, such states can result from so-called "centrifugal barriers," which block the motion of otherwise free electrons in certain directions, trapping them in a region of space with molecular dimensions. Over the years, this basic resonance mechanism has been found to play a prominent role in a variety of processes in molecular physics, thus becoming a major theme in the study of molecular photoionization and electron-molecule scattering processes. As discussed more fully in later sections, the expanding interest in shape resonant phenomena has arisen from a few key factors:

First, shape resonance effects are being identified in the spectra of a growing and diverse collection of molecules and now appear to be active somewhere in the observable properties of most small (nonhydride) molecules. Examples of the processes which exhibit shape resonant

[*]Work supported in part by the U.S. Department of Energy, Office of Health and Environmental Research, under Contract W-31-109-Eng-38. Accordingly the U.S. Government retains a nonexclusive, royalty-free license to publish or reproduce the published form of this contribution, or allow others to do so, for U.S. Government purposes.

effects are x-ray and VUV absorption spectra, photoelectron branching ratios and photoelectron angular distributions (including vibrationally resolved), Auger electron angular distributions, elastic electron scattering, vibrational excitation by electron impact, and so on. Thus concepts and techniques developed in this connection can be used extensively in molecular physics.

Second, being quasibound inside a potential barrier on the perimeter of the molecule, such resonances are localized, have enhanced electron density in the molecular core, and are uncoupled from the external environment of the molecule. This localization often produces intense, easily studied spectral features, while suppressing non-resonant and/or Rydberg structure, and as discussed more fully below, has a marked influence on vibrational motion. In addition, localization causes much of the conceptual framework developed for shape resonances in free molecules to apply equally well to photoionization and electron scattering and to other states of matter such as adsorbed molecules, molecular crystals, and ionic solids.

Third, resonant trapping by a centrifugal barrier often imparts a well-defined orbital momentum character to the escaping electron. This can be directly observed, e.g. by angular distributions of scattered electrons or photoelectron angular distributions from oriented molecules, and shows that the centrifugal trapping mechanism has physical meaning and is not merely a theoretical construct. Recent case studies have revealed trapping of $\ell = 1$ to $\ell = 5$ components of continuum molecular wave functions. The purely molecular origin of the great majority of these cases is illustrated by the prototype system N_2 discussed in Section III.

Fourth, the predominantly one-electron nature of the phenomena lends itself to theoretical treatment by realistic, independent electron methods,[2,4-11] with the concomitant flexibility in terms of complexity of molecular systems, energy ranges, and alternative physical processes. This has been a major factor in the rapid exploration in this area. Continuing development of computational schemes also holds the promise of elevating the level of theoretical work on molecular photoionization and electron-molecule scattering and, in so doing, to test and quantify many of the independent-electron results and to proceed to other circumstances such as coupled channels, multiply excited states, etc. where the simpler schemes become invalid.

In the remainder of this article, we review the study of shape resonances in molecular fields with a fairly broad perspective, beginning with a discussion of the rudimentary concepts and ending with a comment on current challenges: Section II begins by presenting an example of dramatic shape resonance behavior, involving x-ray spectra of SF_6. Section III discusses the basic shape resonance mechanism in simple terms. Section IV treats shape-resonance-induced vibrational effects. Section V discusses connections between shape resonances in various settings, principally the connection between shape resonances in electron-molecule scattering and molecular photoionization. Finally, Section VI outlines the progress and prospects in this stream of work from a broader point-of-view, including some less transparent cases which represent current challenges to our understanding.

II. A DRAMATIC EXAMPLE OF SHAPE RESONANT BEHAVIOR

Among the earliest and still possibly the most dramatic examples of shape resonance effects in molecules are the photoabsorption spectra of

the sulfur K-[12,13] and L-shells[13-16] in SF_6. The sulfur L-shell
absorption spectra of SF_6 and H_2S are shown in Fig. 1 to illustrate the
type of phenomena that originally drew attention to this area. In Fig. 1
both spectra are plotted on a photon energy scale referenced to the
sulfur L-shell ionization potential (IP) which is chemically shifted by a
few eV in the two molecular environments, but lies near $h\nu \sim 175$ eV. The
ordinates represent relative photoabsorption cross sections and have been
adjusted so that the integrated oscillator strength for the two systems
is roughly equal in this spectral range, since absolute normalizations
are not known. The H_2S spectrum is used here as a "normal" reference
spectrum since hydrogen atoms normally do not contribute appreciably to
shape resonance effects and, in this particular context, can be regarded
as weak perturbations on the inner-shell spectra of the heavy atom.
Indeed, the photoabsorption spectrum exhibits what appears to be a

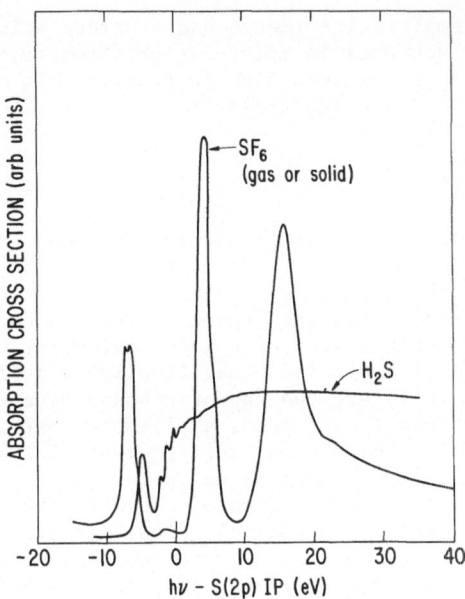

Fig. 1. Photoabsorption spectra of H_2S (taken from Ref. 15)
and SF_6 (taken from Ref. 16) near the sulfur $L_{2,3}$ edge.

valence transition, followed by partially resolved Rydberg structure,
which converges to a smooth continuum. The gradual rise at threshold is
attributable to the delayed onset of the "$2p \to \varepsilon d$" continuum which, for
second row atoms, will exhibit a delayed onset prior to the occupation of
the 3d subshell. This is the qualitative behavior one might well expect
for the absorption spectrum of a core level.

In sharp contrast to this, the photoabsorption spectrum of the same
sulfur 2p subshell in SF_6 shows no vestige of the "normal" behavior just
described. Instead three intense, broad peaks appear, one below the
ionization threshold and two above, and the continuum absorption cross
section is greatly reduced elsewhere. Moreover, no Rydberg structure is

apparent, although an infinite number of Rydberg states must necessarily be associated with any molecular ion. Actually, Rydberg states were detected[17] superimposed on the weak bump below the IP using photographic detection, but obviously these states are extremely weak in this spectrum. This radical reorganization of the oscillator strength distribution was interpreted[18] as potential barrier effects in SF_6, resulting in three shape-resonantly enhanced final state features of a_{1g}, t_{2g}, and e_g symmetry, in order of increasing energy. Another shape resonant feature of t_{1u} symmetry is prominent in the sulfur K-shell spectrum[12] and, in fact, is believed to be responsible for the weak feature just below the IP in Fig. 1. Hence, four prominent features occur in the photoexcitation spectrum of SF_6 as a consequence of potential barriers caused by the molecular environment of the sulfur atom. Another significant observation[16] is that the SF_6 curve in Fig. 1 represents <u>both</u> gaseous and solid SF_6, within experimental error bars. This is definitive evidence that the resonances are eigenfunctions of the potential well inside the barrier, and are effectively uncoupled from the molecule's external environment.

This beautiful empirical evidence had a strong stimulating effect in the study of shape resonances in molecular photoionization, just as early observations of the π_g shape resonance in elastic e-N_2 scattering did in the electron-molecule scattering field.[3,19]

III. BASIC PROPERTIES

The central concept in shape resonance phenomena is the single-channel, barrier-penetration model familiar from introductory quantum mechanics. In fact, the name "shape resonance" means simply that the resonance behavior arises from the "shape," i.e. the barrier and associated inner and outer wells, of a local potential. The basic shape resonance mechanism is illustrated schematically[20] in Fig. 2. There, an effective potential for an excited and/or unbound electron is shown to have an inner well at small distances, a potential barrier at intermediate distances, and an outer well (asymptotic form not shown) at large separations. In the context of molecular photoionization, this would be a one-dimensional abstraction of the effective potential for the photoelectron in the field of a molecular ion. Accordingly, the inner well would be formed by the partially screened nuclei in the molecular core and would therefore be highly anisotropic and would overlap much of the molecular charge distribution, i.e., the initial states of the photoionization process. The barrier, in all well-documented cases, is a so-called centrifugal barrier (other forces such as repulsive exchange forces, high concentrations of negative charge, etc., may also contribute, but have not yet been documented to be pivotal in the molecular systems studied to date). This centrifugal barrier derives from a competition between repulsive centrifugal forces and attractive electrostatic forces and usually (but not always) resides on the perimeter of molecular charge distributions where the centrifugal forces can compete effectively with electrostatic forces. Similar barriers are known for d- and f-waves in atomic fields,[21] however the ℓ (orbital angular momentum) character of resonances in molecular fields tends to be higher than those of constituent atoms owing to the larger spatial extent of molecular charge distributions, e.g., see discussion in connection with N_2 photoionization below. The outer well lies outside the molecule where the Coulomb potential ($\sim -r^{-1}$) of the molecular ion again dominates the centrifugal terms ($\sim r^{-2}$) in the potential. We stress that this description has been radically simplified to convey the essential aspects of the underlying physics. In reality effective barriers to electron motion in molecular fields occur for particular ℓ components of

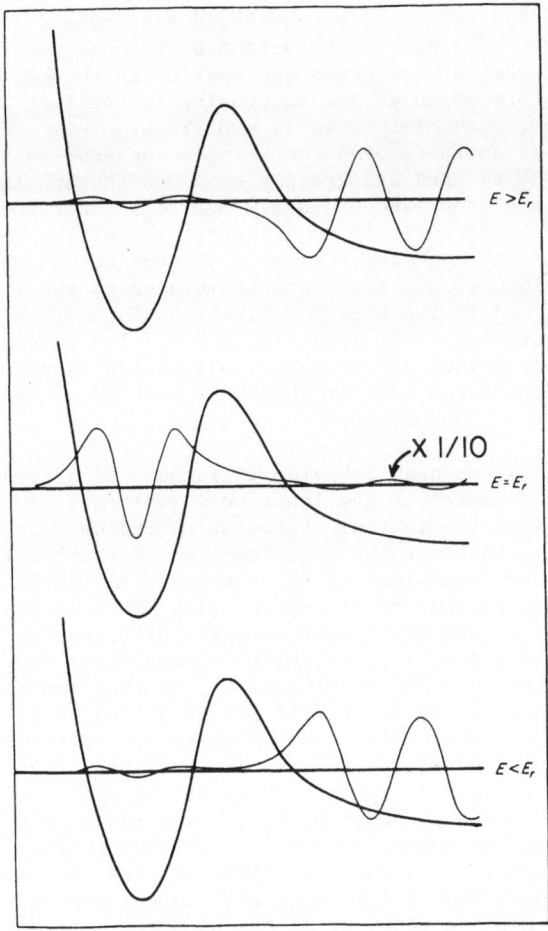

$E > E_r$

X 1/10

$E = E_r$

$E < E_r$

Fig. 2. Schematic of the effect of a potential barrier on an unbound wave function in the vicinity of a quasibound state at $E = E_r$ (adapted from Ref. 20). In the present context, the horizontal axis represents the distance of the excited electron from the center of the molecule.

particular ionization channels and restrict motion only in certain directions. Again, a specific example is described below.

Focusing now on the wave functions in Fig. 2, we see the effect of the potential barrier on the wave mechanics of the photoelectron. For energies below the resonance energy $E < E_r$ (lower part of Fig. 2), the inner well does not support a quasibound state, i.e., the wave function is not exponentially decaying as it enters the classically forbidden region of the barrier. Thus the wave function begins to diverge in the barrier region and emerges in the outer well with a much larger amplitude than that in the inner well. When properly normalized at large r, the amplitude in the molecular core is very small, so we say this wave function is essentially an eigenfunction of the outer well although small precursor loops extend inside the barrier into the molecular core.

At $E = E_r$, the inner well supports a quasibound state. The wavefunction exhibits exponential decay in the barrier region so that, if

the barrier extended to r → ∞, a true bound state would lie very near
this total energy. Therefore, the antinode that was not supported in the
inner well at $E < E_r$ has traversed the barrier to become part of a
quasibound wave form which decays monotonically until it reemerges in the
outer-well region, much diminished in amplitude. This "barrier
penetration" by an antinode produces a rapid increase in the asymptotic
phase shift by ∼ π radians and greatly enhances the amplitude in the
inner well over a narrow band of energy near E_r. Therefore, at $E = E_r$,
the wavefunction is essentially an eigenfunction of the inner well,
although it decays through the barrier and reemerges in the outer well.
The energy halfwidth of the resonance is related to the lifetime of the
quasibound state and to the energy derivative of the rise in the phase
shift in well-known ways. Finally, for $E > E_r$, the wavefunction reverts
to being an eigenfunction of the outer well as the behavior of the
wavefunction at the outer edge of the inner well is no longer
characteristic of a bound state.

Obviously this resonant behavior will cause significant physical
effects: The enhancement of the inner-well amplitude at $E ∼ E_r$ results
in good overlap with the initial states which reside mainly in the inner
well. Conversely, for energies below the top of the barrier, but not
within the resonance halfwidth of E_r, the inner amplitude is diminished
relative to a more typical barrier-free case. This accounts for the
strong modulation of the oscillator strength distribution in Fig. 1.
Also, the rapid rise in the phase shift induces shape resonance effects
in the photoelectron angular distribution. Another important aspect is
that eigenfunctions of the inner well are localized inside the barrier
and are substantially uncoupled from the external environment of the
molecule. As mentioned above, this means that shape resonant phenomena
often persist in going from the gas phase to the condensed phase, e.g.,
Fig. 1, and, with suitable modification, shape resonances in molecular
photoionization can be mapped[22] onto electron-scattering processes and
vice versa. Finally, note that this discussion was focussed on total
energies from the bottom of the outer well to the top of the barrier, and
that no explicit mention was made of the asymptotic potential that
determines the threshold for ionization. Thus, valence or Rydberg states
in this range can also exhibit shape resonant enhancement, even though
they have true bound state behavior at large r, beyond the outer well.

We will now turn, for the remainder of this section, to the specific
example of the well-known σ_u shape resonance in N_2 photoionization, which
was the first documented case[23] in a diatomic molecule and has since been
used as a prototype in studies of various shape resonance effects, as
discussed below. To identify the major final-state features in N_2
photoionization at the independent-electron level, we show in Fig. 3 the
original calculation[23-25] of the K-shell photoionization spectrum
performed with the multiple-scattering model. This calculation agrees
qualitatively with all major featues in the experimental spectrum,[26,27]
except a narrow band of double excitation features, and with subsequent,
more accurate calculations.[2,28] The four partial cross sections in Fig.
3 represent the four dipole-allowed channels for K-shell (IP = 409.9 eV)
photoionization. Here we have neglected the localization[29] of the K-
shell hole since it doesn't greatly affect the integrated cross section
and since the separation into u and g symmetry both helps the present
discussion and is rigorously applicable to the subsequent discussion of
valence-shell excitation. (Note that the identification of shape
resonant behavior is generally easier in inner-shell spectra since the
problems of overlapping spectra, channel interaction, and zeros in the
dipole matrix element are reduced relative to valence-shell spectra.)

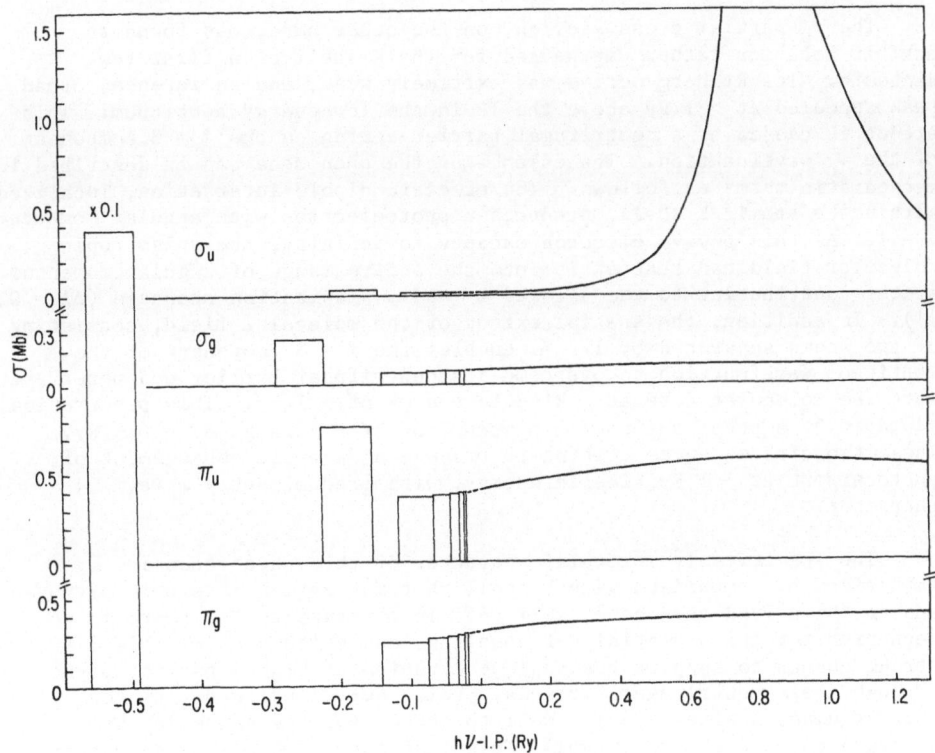

Fig. 3. Partial photoionization cross sections for the four dipole-
allowed channels in K-shell photoionization of N_2. Note
that the energy scale is referenced to the K-shell IP
(409.9 eV) and is expanded two-fold in the discrete part of
the spectrum.

The most striking spectral feature in Fig. 3 is the first member of
the π_g sequence, which dominates every other feature in the theoretical
spectrum by a factor of ~ 30. (Note the first π_g peak has been reduced
by a factor of 10 to fit in the frame.) The concentration of oscillator
strength in this peak is a centrifugal barrier effect in the d-wave
component of the π_g wave-function. The final state in this transition is
a highly localized state, about the size of the molecular core, and is
the counterpart of the well-known[3,19] π_g shape resonance in e-N_2
scattering at 2.4 eV. For the latter case, Krauss and Mies[30]
demonstrated that the effective potential for the π_g elastic channel in
e-N_2 scattering exhibits a potential barrier due to the centrifugal
repulsion acting on the dominant $\ell = 2$ lead term in the partial-wave
expansion of the π_g wavefunction. In the case of N_2 photoionization,
there is one less electron in the molecular field to screen the nuclear
charge so that this resonance feature is shifted[22] to lower energy and
appears in the discrete. It is in this sense that we refer to such
features as "discrete" shape resonances. The remainder of the π_g partial
cross section consists of a Rydberg series and a flat continuum. The π_u
and σ_g channels both exhibit Rydberg series, the initial members of which
correlate well with partially resolved transitions in the experimental
spectrum below the K-shell IP.

The σ_u partial cross section, on the other hand, was found to exhibit behavior rather unexpected for the K shell of a first-row diatomic. Its Rydberg series was extremely weak, and an intense, broad peak appeared at ~ 1 Ry above the IP in the low-energy continuum. This effect is caused by a centrifugal barrier acting on the ℓ = 3 component of the σ_u wavefunction. The essence of the phenomena can be described in mechanistic terms as follows. The electric dipole interaction, localized within the atomic K shell, produces a photoelectron with angular momentum ℓ = 1. As this p-wave electron escapes to infinity, the anisotropic molecular field can scatter it into the entire range of angular momentum states contributing to the allowed σ- and π- ionization channels ($\Delta\lambda$ = 0, ±1). In addition, the spatial extent of the molecular field, consisting of two atoms separated by 1.1 Å, enables the ℓ = 3 component of the σ_u continuum wavefunction to overcome its centrifugal barrier and penetrate into the molecular core at a kinetic energy of ~ 1 Ry. This penetration is rapid,[23] a phase shift of ~ π occurring over a range of ~ 0.3 Ry. These two circumstances combine to produce a dramatic enhancement of photocurrent at ~ 1 Ry kinetic energy, with predominantly f-wave character.

The specifically molecular character of this phenomenon is emphasized by comparison with K-shell photoionization in atomic nitrogen and in the united-atom case, silicon. In contrast to N_2, there is no mechanism for the essential p-f coupling, and neither atomic field is strong enough to support resonant penetration of high-ℓ partial waves through their centrifugal barriers. (With substitution of "d" for "f," this argument applies equally well to the d-type resonance in the discrete part of the spectrum.) Note that the π_u channel also has an ℓ = 3 component but does not resonate. This underscores the directionality and symmetry dependence of the trapping mechanism.

To place the σ_u resonance in a broader perspective and show its connection with high-energy behavior, we show, in Fig. 4, an extension of the calculation in Fig. 3 to much higher energy. Again the four dipole-allowed channels in $D_{\infty h}$ symmetry are shown. The dashed line is two times the atomic nitrogen K-shell cross section. Note that the modulation about the atomic cross section, caused by the potential barrier, extends to ~100 eV above threshold before the molecular and atomic curves seem to coalesce.

At higher energies, a weaker modulation appears in each partial cross section. This weak modulation is a diffraction pattern, resulting from scattering of the photoelectron by the neighboring atom in the molecule, or, more precisely, by the molecular field. Structure of this type was first studied over 50 years ago by de Kronig[31] in the context of metal lattices. It currently goes by the acronym EXAFS (extended x-ray absorption fine structure) and is used extensively[32,33] for local structure determination in molecules, solids, and surfaces. The net oscillation is very weak in N_2, since the light atom is a weak scatterer. More pronounced effects are seen, e.g., in K-shell spectra[34] of Br_2 and $GeCl_4$. Our reason for showing the weak EXAFS structure in N_2 is to show that the low-energy, resonant modulation (called "near-edge" structure in the context of EXAFS) and high-energy EXAFS evolve continuously into one another and emerge naturally from a single molecular framework, although the latter is usually treated from an atomic point-of-view.

Figure 5 shows a hypothetical experiment[24] which clearly demonstrates the ℓ character of the σ_u resonance. In this experiment, we first fix the nitrogen molecule in space and orient the polarization direction of a photon beam, tuned near the nitrogen K edge, along the molecular axis. This orientation will cause photoexcitation into σ final states, including the resonant σ_u ionization channel. (Again hole localization is neglected for purposes of illustration.) Figure 5 shows the angular distribution of photocurrent as a function of both excess energy above the K-shell IP and angle of ejection, θ, relative to the molecular axis. Very apparent in Fig. 5 is the enhanced photocurrent at the resonance position, KE ~ 1 Ry. Moreover, the angular distribution exhibits three nodes, with most of the photocurrent exiting the molecule along the molecular axis and none at right angles to it. This is an f-wave (ℓ = 3) pattern and indicates clearly that the resonant enhancement is caused by an ℓ = 3 centrifugal barrier in the σ_u continuum of N_2. Thus, the centrifugal barrier has observable physical meaning and is not merely a theoretical construct. Note that the correspondence between the dominant asymptotic partial wave and the trapping mechanism is not always

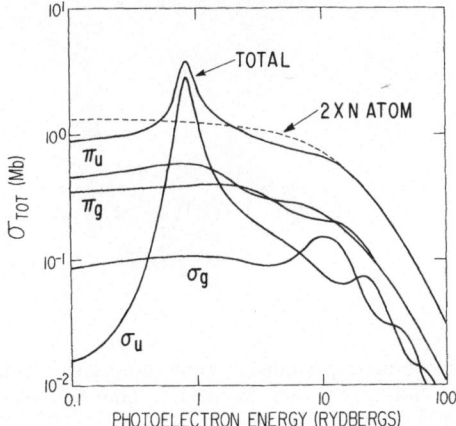

Fig. 4. Partial photoionization cross sections for the K shell of N_2 over a broad energy range. The dashed line represents twice the K-shell photoionization cross section for atomic nitrogen, as represented by a Hartree-Slater potential.

valid, especially when the trapping is on an internal or off-center atomic site where the trapped partial wave can be scattered by the anisotropic molecular field into alternative asymptotic partial waves, e.g., BF_3[35] and SF_6. Finally, note that the hypothetical experiment discussed above has been approximately realized by photoionizing molecules adsorbed on surfaces. The shape resonant features tend to survive adsorption and, owing to their observable ℓ-character, can even provide evidence[36,37] as to the orientation of the molecule on the surface.

The final topic in the discussion of basic properties of shape resonances involves eigenchannel contour maps,[38] or, "pictures" of unbound electrons. This is the continuum counterpart of contour maps of bound-state electronic wavefunctions which have proven so valuable as tools of quantum chemical visualization and analysis. Indeed, the present example helps achieve a physical picture of the σ_u shape resonance, and the general technique promises to be a useful tool for analyzing resonant trapping mechanisms and other observable properties in the future. The key to this visualization is the construction of those particular combinations of continuum orbital momenta that diagonalize the interaction of the unbound electron with the anisotropic molecular field. These combinations, known as eigenchannels, are the continuum analogues of the eigenstates in the discrete spectrum, i.e., the bound states.

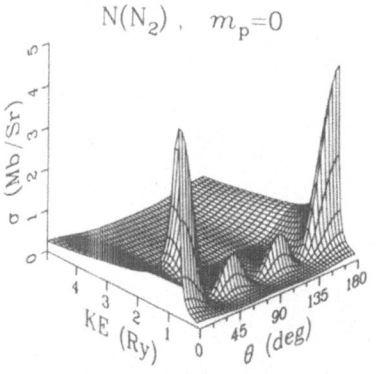

Fig. 5. Fixed-molecule photoelectron angular distribution for kinetic energies 0-5 Ry above the K-shell IP of N_2. The polarization of the ionizing radiation is oriented along the molecular axis in order to excite the σ continua, and the photoelectron ejection angle, θ, is measured relative to the molecular axis.

The f-dominated eigenchannel of σ_u symmetry in N_2 is an excellent case for which eigenchannel contour maps may be used to visualize shape resonant continuum states. In Fig. 6, this eigenchannel is plotted at two energies, one below the resonance energy and the other near the center of the resonance. In Fig. 6, the surfaces, whose contours have at large distance the three nodal planes characteristic of f orbitals, show clearly the resonant nature of the f-dominated σ_u eigenchannel. Below and above (not shown) the resonance energy, the probability amplitude and its distribution inside and outside the molecular core is typical of non-resonant behavior. But at the resonance energy there is an enormous enhancement of the wavefunction in the molecular interior; the wavefunction now resembles a molecular bound-state probability amplitude distribution. It is this enhancement, in the region occupied by the bound states, that leads to the very large increase in oscillator strength indicative of the resonance, and to the other manifestations

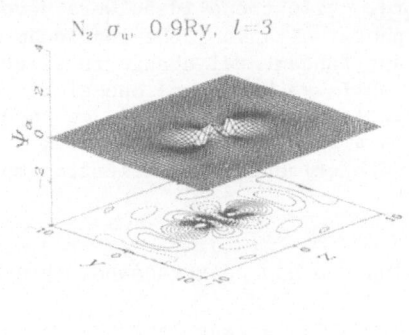

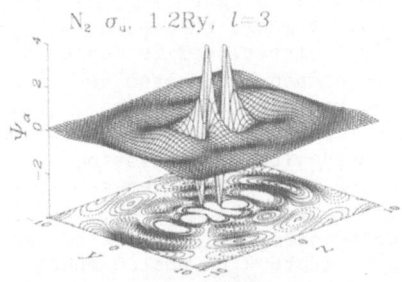

Fig. 6. F-wave-dominated eigenchannel wavefunctions for non-resonant (top) and resonant (bottom) electron kinetic energies in the σ_u continuum of N_2. The molecule is in the yz plane, along the z axis, centered at y = z = 0. Contours mark steps of 0.03 from 0.02 to 0.29; positive = solid, negative = dashed. The lack of contour lines for 1.2 Ry near the nuclei is because of the 0.29 cutoff.

discussed earlier and in the next section.

These eigenchannel plots are discussed more fully elsewhere;[38] however, before leaving the subject, several points should be noted. First, the N_2 example that we have chosen is somewhat special in that there is a near one-to-one correspondence between the eigenchannels and single values of orbital angular momentum. Orbital angular momentum is, however, not a good quantum number in molecules, and more generally we should not always expect such clear nodal patterns. More typically, earlier work[5,35,39-41] indicates that several angular momenta often contribute to the continuum eigenchannels (although a barrier in only one ℓ component will be primarily responsible for the temporary trapping that causes the enhancement in that and coupled components), and this means that the resulting eigenchannel plots will be correspondingly richer. Second, the dominant ℓ we have discussed pertains to the region outside the molecular charge distributions. The orbital momentum composition of these wavefunctions is more complicated in the molecular interior, as seen, e.g., in Fig. 6. Nonetheless, continuity and a dominant ℓ may, as in the case of N_2, cause the emergence of a distinct ℓ pattern, even into the core region. Third, while these ideas were developed[38] in the context of molecular photoionization, the continuum eigenchannel concept carries over without any fundamental change to electron–molecule scattering. Finally, while we have used one-electron wavefunctions here, obtained with the multiple-scattering model, we emphasize that the eigenchannel concept is a general one and we look foward to its use in the analysis of more sophisticated, many-electron molecular continuum wavefunctions.

IV. SHAPE-RESONANCE-INDUCED NON-FRANCK-CONDON EFFECTS

Molecular photoionization at wavelengths unaffected by autoionization, predissociation, or ionic thresholds was generally believed to produce Franck-Condon (FC) vibrational intensity distributions within the final ionic state and v-independent photoelectron angular distributions. We now discuss a prediction[42,43] from a few years ago, that shape resonances represent an important class of exceptions to this picture. These ideas are illustrated with a calculation of the $3\sigma_g \to \varepsilon\sigma_u$, $\varepsilon\pi_u$ photoionization channel of N_2, which accesses the same σ_u shape resonance discussed above at approximately $h\nu \sim 30$ eV, or ~ 14 eV above the $3\sigma_g$ IP. The process we are considering involves photoexcitation of N_2 $X^1\Sigma_g^+$ in its vibrational ground state with photon energies from the first IP to beyond the region of the shape resonance at $h\nu \sim 30$ eV. This process ejects photoelectrons leaving behind N_2^+ ions in energetically accessible states. As we are interested in the ionization of the $3\sigma_g$ electron, which produces the $X^2\Sigma_g^+$ ground state of N_2^+, we are concerned with the photoelectron band in the range 15.5 eV $\leqslant$ IP $\leqslant$ 16.5 eV. The physical effects we seek involve the relative intensities and angular distributions of the v = 0-2 vibrational peaks in the $X^2\Sigma_g^+$ electronic band, and, more specifically, the departures of these observables from behavior predicted by the FC separation.

The breakdown of the FC principle arises from the quasibound nature of the shape resonance, which, as we discussed in Section III, is localized in a spatial region of molecular dimensions by a centrifugal barrier. This barrier and, hence, the energy and lifetime (width) of the resonance are sensitive functions of internuclear separation R and vary significantly over a range of R corresponding to the ground-state vibrational motion. This is illustrated in the upper portion of Fig. 7 where the dashed curves represent separate, fixed-R calculations of the partial cross section for N_2 $3\sigma_g$ photoionization over the range $1.824a_0 \leqslant R \leqslant 2.324a_0$, which spans the N_2 ground-state vibrational wavefunction.

Of central importance in Fig. 7 is the clear demonstration that resonance positions, strengths, and widths are sensitive functions of R. In particular, for larger separations, the inner well of the effective potential acting on the $\ell = 3$ component of the σ_u wavefunction is more attractive and the shape resonance shifts to lower kinetic energy, becoming narrower and higher. Conversely, for lower values of R,

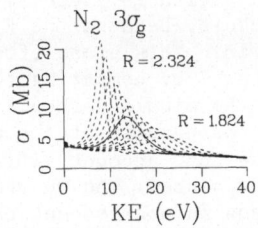

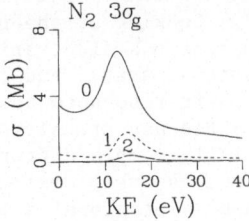

Fig. 7. Cross sections σ for photoionization of the $3\sigma_g$ ($v_i = 0$) level of N_2. Top: fixed-R (dashed curves) and R-averaged, vibrationally unresolved (solid curve) results. Bottom: results for resolved final-state vibrational levels, $v_f = 0 - 2$.

the resonance is pushed to higher kinetic energy and is weakened. This indicates that nuclear motion exercises great leverage on the spectral behavior of shape resonances, since small variations in R can significantly shift the delicate balance between attractive (mainly Coulomb) and repulsive (mainly centrifugal) forces which combine to form the barrier. In the present case, variations in R, corresponding to the ground-state vibration in N_2, produce significant shifts of the resonant behavior over a spectral range several times the fullwidth at half maximum of the resonance calculated at $R = R_e$. By contrast, nonresonant channels are relatively insensitive to such variation in R, as was shown by results[44] on the $1\pi_u$ and $2\sigma_u$ photoionization channels in N_2.

Thus, in the vicinity of a shape resonance, the electronic transition moment varies rapidly with R. This parametric coupling was estimated[42] in the adiabatic-nuclei approximation[45] by computing the net transition moment for a particular vibrational channel as an average of the R-dependent dipole amplitude, weighted by the product of the initial-

and final-state vibrational wavefunctions at each R,

$$D_{v_f v_i} = \int dR \chi_{v_f}^{\dagger}(R) D^-(R) \chi_{v_i}(R).$$

The vibrational wavefunctions were approximated by harmonic-oscillator functions and the superscript minus denotes that incoming-wave boundary conditions have been applied and that the transition moment is complex. Note that, even when the final vibrational levels v_f of the ion are unresolved (summed over), vibrational motion within the initial state $v_i = 0$ can cause the above equation to yield results significantly different from the $R = R_e$ result, because the R dependence of the shape resonance is highly asymmetric. This gross effect of R averaging can be seen in the upper half of Fig. 7 by comparing the solid line (R-averaged result, summed over v_f) and the middle dashed line ($R = R_e$). Hence, even for the calculation of gross properties of the whole, unresolved electron band, it is necessary to take into account vibrational-motion effects in channels exhibiting shape resonances. As we stated earlier, this is generally not a critical issue in nonresonant channels.

Effects of nuclear motion on individual vibrational levels are shown in the bottom half of Fig. 7. Looking at the partial cross sections in Fig. 7, we see that the resonance position varies over a few volts depending on the final vibrational state, and that higher levels are relatively more enhanced at their resonance position than is $v_f = 0$. This sensitivity to v_f arises because transitions to alternative final vibrational states preferentially sample different regions of R. In particular, $v_f = 1,2$ sample successively smaller R, governed by the maximum overlap with the ground vibrational state, causing the resonance in those vibrational channels to peak at higher energy than that for $v_f = 0$. The impact of these effects on branching ratios is clearly seen in Fig. 2 of Ref. 42, where the ratio of the higher v_f intensities to that of $v_f = 0$ is plotted in the resonance region. There we see that the ratios are slightly above the FC factors (9.3%, $v_f = 1$; 0.6%, $v_f = 2$) at zero kinetic energy, go through a minimum just below the resonance energy in $v_f = 0$, then increase to a maximum as individual $v_f > 0$ vibrational intensities peak, and finally approach the FC factors again at high kinetic energy. Note the maximum enhancement over the FC factors is progressively more pronounced for higher v_f, i.e., 340% and 1300% for $v_f = 1,2$, respectively.

Equally dramatic are the effects on $\beta(v_f)$ discussed in Ref. 42. Especially at and below the resonance position, the β's vary greatly for different final vibrational levels. Carlson first observed[46,47] that, at 584 Å, the $v_f = 1$ level in the $3\sigma_g$ channel of N_2 had a much larger β than the $v_f = 0$ level even though there was no apparent autoionizing state at that wavelength. This is in semiquantitative agreement with the theoretical calculation[42] which gives $\beta(v_f = 0) \sim 1.0$ and $\beta(v_f = 1) \sim 1.5$. Although the agreement is not exact, we feel this demonstrates that the "anomalous" v_f dependence of β in N_2 stems mainly from the σ_u shape resonance which acts over a range of the spectrum many times its own ~ 5 eV width. The underlying cause of this effect is the shape-resonance-enhanced R dependence of the dipole amplitude, just as for the vibrational partial cross sections. In the case of $\beta(v_f)$, however, both the R dependence of the phase and of the magnitude of the complex dipole amplitude play a crucial role, whereas the partial cross sections depend only on the magnitude.

The theoretical predictions discussed above were soon tested in two separate experiments. In Fig. 8 the branching ratio for production of the v = 0 and 1 vibrational levels of N_2^+ X $^2\Sigma_g^+$ is shown. The dash—dot line is the original prediction.[42] The solid dots are the recent

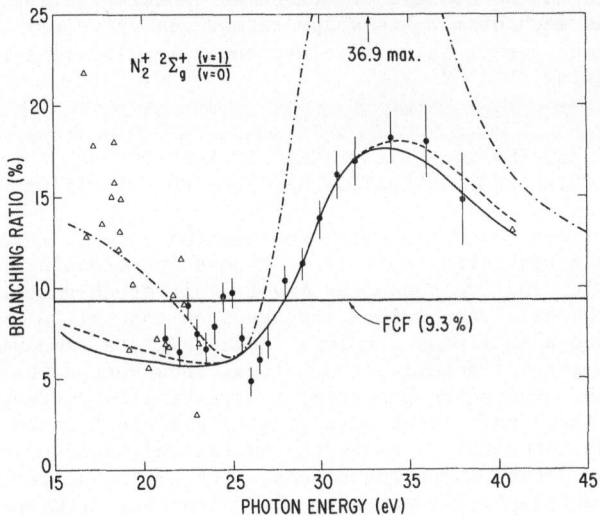

Fig. 8. Branching ratios for production of the v = 0,1 levels of N_2^+ X $^2\Sigma_g^+$ by photoionization of N_2: ●, Ref. 48; Δ, Ref. 49; —·—·—, multiple scattering model prediction from Ref. 42; ———, frozen—core Hartree—Fock dipole length approximation from Ref. 50; — — —, frozen—core Hartree—Fock dipole velocity approximation from Ref. 50.

measurements[48] in the vicinity of the shape resonance at hν ~ 30 eV. The conclusion drawn from this comparison is that the observed variation of the vibrational branching ratio relative to the FC factor over a broad spectral range qualitatively confirms the prediction; however, subsequent calculations[7,50] with fewer approximations have achieved better agreement based on the same mechanism for breakdown of the FC separation. The dashed and solid curves are results based on a Schwinger variational treatment[50] of the photoelectron wavefunction. The two curves represent a length and velocity representation of the transition matrix element, both of which are in excellent agreement with the data. This is an outstanding example of interaction between experiment and theory, proceeding as it did from a novel prediction, through experimental testing and final quantitative theoretical agreement in a short time. Also shown in Fig. 8 are data in the 15.5 eV ≤ hν ≤ 22 eV region which are earlier data[49] obtained using laboratory line sources. The apparently chaotic behavior arises from unresolved autoionization structure.

The angular distribution asymmetry parameters, β, for the v = 0,1 levels of N_2^+ X $^2\Sigma_g^+$ over roughly the same energy region are reported in Ref. 51. In the region above hν ~ 25 eV, this data also shows qualitative agreement with the predicted[42] v-dependence of β caused by the σ_u shape resonance. In this case, the agreement is somewhat improved in later calculations,[50] mainly for v = 1; however, the change is less dramatic than for the branching ratios.

Finally, note that we have illustrated the vibrational effects of shape resonances in the context of molecular photoionization; however, a rather analogous mechanism makes shape resonances extremely efficient in inducing vibrational excitation in electronically-elastic electron-molecule scattering.[3,19,43]

V. CONNECTIONS BETWEEN SHAPE RESONANCES IN ELECTRON-MOLECULE SCATTERING AND IN MOLECULAR PHOTOIONIZATION AND RELATED CONNECTIONS

At first glance, there is little connection between resonances in electron-molecule scattering (e + M) and those in molecular photoionization (hν + M). The two phenomena involve different numbers of electrons and the collision velocities are such that all electrons are incorporated into a collision complex. Hence, we are comparing a neutral molecule to a molecular negative-ion system. However, although the long-range part of the scattering potential is drastically different in the two cases, the inner part is not drastically different since it is dominated by the interactions among the nuclei and those electrons common to both systems. Thus, shape resonances which are localized in the molecular core substantially maintain their identity from one system to another, but are shifted in energy owing to the difference caused by the addition of an electron inside the molecule. This unifying property of shape resonances thus links together the two largest bodies of data for the molecular continuum -- hν + M and e + M -- and, although these resonances shift in energy in going from one class to another and manifest themselves in somewhat different ways, this link permits us to transfer information between the two. This can serve to help interpret new data and even to make predictions of new features to look for experimentally. Actually, this picture[22] was surmised empirically from evidence contained in survey calculations on e + M and hν + M and, in retrospect, from data. These observations can be summarized as follows: By and large, the systems hν + M and e + M display the same manifold of shape resonances, only those in the e + M system are shifted ~ 10 eV to higher electron energy. Usually, there is one shape resonance per symmetry for a subset of the symmetries available. The shift depends on the symmetry of the state, indicating, as one would expect, that the additional electron is not uniformly distributed. Finally, there is substantial proof that the ℓ character is preserved in this process, although interaction among alternative components in a scattering eigenchannel can alter the ℓ mixing present.

There are several good examples available to illustrate this point, e.g., N_2, CO, CO_2, BF_3, SF_6, etc. In general, one can start from either the neutral or the negative ion system, but, in either case, there is a preferred way to do so: In the hν + M case, it is better to examine the inner-shell photoabsorption and photoionization spectra. Shape resonances almost invariably emerge most clearly in this context. Additional effects, discussed briefly at the end of this section, frequently make the role of shape resonances in valence-shell spectra more complicated to interpret. In the e + M case, a very sensitive indicator of shape resonance behavior is the vibrational excitation channel. Vibrational excitation is enhanced by shape resonances,[3,19] and is typically very weak for non-resonant scattering. Hence, a shape resonance, particularly at intermediate energy (10-40 eV),[43,52] may be barely visible in the vibrationally and electronically elastic scattering cross section, and yet be displayed prominently in the vibrationally inelastic, electronically elastic cross section.

Two examples will help illustrate these points. In e − SF_6 scattering, the vibrationally elastic scattering cross section has been calculated theoretically[53] and shown to have four shape resonances of a_{1g}, t_{1u}, t_{2g}, and e_g symmetry at approximately 2, 7, 13, and 27 eV, respectively. The absolute total cross section measured by Kennerly et al.[54] shows qualitative agreement, except that no clear sign of the e_g is present. (This resonance could be more clearly seen in the vibrational excitation spectrum, which is not available.) Hence, using the guidelines given above, one would expect shape resonance features in the $h\nu$ + M case at −8, −3, 3, and 17 eV (on the kinetic energy scale) to a very crude, first approximation. Indeed, the K− and L−shell photo-absorption spectra of SF_6 show such intense features, as discussed in Section II.

Using N_2, we reverse the direction of the mapping, and start with $h\nu$ + N_2, which was discussed extensively in earlier sections. Here a "discrete" shape resonance of π_g symmetry and a shape resonance of σ_u symmetry are apparent in the K−shell spectrum,[26] e.g., Fig. 3. These occur at ∼ −9 and 10 eV on the kinetic energy scale (relative to the ionization potential). Hence, one would look for the same set of resonances in e−N_2 scattering at ∼ 1 and ∼ 20 eV electron scattering energies. The well-known π_g shape resonance[19] is very apparent in the vibrationally elastic cross section; however, there is only a very broad bump at ∼ 20 eV.[55] As noted above, the vibrationally inelastic cross section is much more sensitive to shape resonances, and, indeed, the σ_u shape resonance in e−N_2 scattering has been established theoretically and experimentally by looking in this channel (see e.g., Refs. 43, 56−60). Several other excellent examples exist, but we will conclude by pointing out that the connections between e−CO_2 and $h\nu$ − CO_2 have been recently discussed[41] in detail, including a study of the eigenphase sums in the vicinity of the σ_u shape resonance in the two systems.

Finally, we note similar connections and additional complications upon mapping from inner−shell to valence−shell $h\nu$ + M spectra. On going from deep inner−shell spectra to valence−shell spectra, shape resonances in $h\nu$ + M also shift approximately 1−4 eV toward higher kinetic energy, due to differences in screening between localized and delocalized holes as well as other factors. As mentioned above, several complications arise in valence−shell spectra which can tend to obscure the presence of a shape resonance compared to their more straightforward role in inner−shell spectra. These include greater energy dependence of the dipole matrix element, interactions with autoionizing levels, strong continuum−continuum coupling between more nearly degenerate ionization channels, strong particle−hole interactions, etc. So, for the most transparent view of the manifold of shape resonance features in $h\nu$ + M, one should always begin with inner−shell photoabsorption data.

VI. PROGRESS AND PROSPECTS

Summarizing, we have used prototype studies on N_2 to convey the progress made in the study of shape resonances in molecular fields, particularly in molecular photoionization. This included the identi-fication of shape resonant features in photoionization spectra of molecules and the accural of substantial physical insight into their properties, many of which were peculiar to molecular fields. One recent example has been the prediction and experimental confirmation of the role of shape resonances in producing non−Franck−Condon effects in vibrational branching ratios and photoelectron angular distribution.

What this discussion has failed so far to convey is the already extensive body of work that has developed around these basic themes. Even in an early interpretation[18] of shape resonance effects in x-ray spectra, it was obvious that the phenomena would be widespread as over ten molecules, or local molecular moieties, were already observed (see, e.g., Refs. 15 and 18) to have shape resonant behavior. At the present, it is not difficult to identify over two dozen examples of molecules exhibiting the effects discussed above in one or more final state symmetries (references for the following examples are cited in Ref. 1, and the inner-shell cases are listed according to molecule in the bibliography given by Hitchcock[61]). These include simple diatomics (N_2, O_2, CO, NO), polyatomics with subgroups related to the first-row diatomics (HCN, C_2N_2, CH_3CH), triatomics (CO_2, CS_2, OCS, N_2O, SO_2) and more highly coordinated molecules and local molecular environments (SF_6, $SO_4^=$, SF_5CF_3, SF_2O_2, SF_2O, BF_3, SiF_4, $SiCl_4$, $SiF_6^=$, SiO_2, NF_3, CF_4). There is no doubt that many cases have been overlooked in this list and that many examples will be identified in the future as the exploration of molecular photoionization dynamics continues, particularly with the increasing utilization of synchrotron radiation sources.

Several examples from this body of literature serve both to emphasize some of the interesting complications that can arise and to caution against assuming that manifestations of shape resonances will always conform to the independent-electron concepts used above to explain the fundamentals of the subject in connection with N_2 photoionization: (i) One recently documented case of many-body interactions causing deviations from the single-channel picture stressed above is continuum-continuum coupling in the $2\sigma_u^{-1}$ photoionization channel of N_2.[62-66] In the single-channel model, the σ_u shape resonance would not affect this channel; however, channel interaction[62] between the $2\sigma_u^{-1}$ and $3\sigma_g^{-1}$ ionization continua has been shown[62] to cause resonant structure in the β for the $2\sigma_u^{-1}$ channel at the energy of the σ_u shape resonance in the $3\sigma_g^{-1}$ channel. This mechanism had also been sited in the valence-shell photoionization of SF_6,[67] and should be a general phenomenon. (ii) In the case of O_2 photoionization, an analogous σ_u shape resonance is expected,[68-70] but its identification in the photoionization spectrum has been complicated by the existence of extensive autoionization structure in the region of interest. Recent work[71] using variable wavelength photoelectron measurements and a multichannel quantum defect theory (MQDT) analysis of the principal autoionizing Rydberg series has sorted out this puzzle, with the result that the σ_u shape resonance was established to be approximately where expected, but was not at all clearly identifiable without the extensive analysis used in this case. Several examples now exist which exhibit overlapping shape and autoionizing resonances, thus requiring great care in order to arrive at meaningful interpretation. This will be common, particularly in valence spectra where successive IP's are closely spaced. (iii) In the case of CO_2 a σ_u shape resonance of completely different origin was expected[39,72] for photoionization of the $4\sigma_g$ orbital, leading to the C $^2\Sigma_g^+$ state of CO_2^+. This resonance, however, was not apparent in partial cross section measurements on this channel.[73] Nevertheless predictions[40,72] of a shape-resonant feature in the corresponding β was confirmed[74] and work in several laboratories[39,40,72,74-80] have now converged to reasonable agreement in this observable. In addition, recent measurements have shown that this resonance is observable in the partial cross sections, but is shifted to lower energy and much diminished.[80] Future experimental work on vibrational branching ratios and v-dependent β's[40] would greatly aid in the further study of this case. Several other examples of shape resonances which are clearly displayed in some (usually inner-shell) spectra, but are "missing" in other (usually valence-shell) spectra are now known.[81-83] One of the remaining intriguing problems is to pinpoint the key interaction which erases the shape resonance in certain channels. (iv) One of the exciting new results involves studying resonances in inner-shell spectra with electron spectroscopy.[84,85] This type of measurement goes beyond the absorption type measurement and gives information on the <u>decay</u> of resonant features. In fact many resonances in SF_6, SiF_4, and other molecules decay by barrier penetration into the direct ionization continuum of the main single-hole configuration; however, Ferrett et al.[85] have noted exceptions, such as the e_g in SF_6 which decays with a finite branching ratio into satellite configurations. Hence, the enhanced excitation is caused by a simple shape resonant mechanism, but the decay involves multiparticle interactions, to a significant extent. Other observations include decay paths for discrete resonances excited from deep inner holes,[84] anisotropy,[85] or lack of it,[84] in Auger angular distributions following resonant excitation; and characterization of weak features in the K spectra of SF_6 as doubly excited states converging to satellite thresholds.[84,85] This is a rich and complicated stream of work, just getting underway. It is now feasible owing to developments in synchrotron radiation. (v) Returning to the case of N_2, we note that the photoionization of the $2\sigma_g$ should also access the σ_u shape resonance. However, for this inner-valence orbital, extensive vibronic coupling leads to a breakdown[86] of the single-particle model leading to the observation[87] of many "satellite" vibronic states in the photoelectron spectrum instead of a single peak due to ionization of the $2\sigma_g$ orbital. Nevertheless, if the sum of this complicated structure is plotted versus photon energy, the resonant enhancement reemerges.[87,88] (vi) Owing to the dependence of continuum resonances on internuclear separation, an examination of the utility of resonance positions in families of molecules to predict interatomic spacings was attempted.[89] Early results were very optimistic, claiming an accuracy of 0.05 Å; however, more conservative analyses[90] made it clear that, while the basic qualitative principal may be correct, the application to structure determination was not as promising as initially thought. (vii) Finally, in most molecules, a resonant channel will have only one resonance; however, in polyatomics, multiple resonances have been documented[91] in cases such as C_2N_2, where two resonances are found in the σ_u channel, resulting from the trapping on the C-C and C-N sites. In addition, N_2O exhibits[92] two resonances in the σ channel but without obvious correlation with different sites. In this case, two alternative ℓ's may resonate, or the higher energy feature could be the transition region between resonant trapping and diffractive scattering, where quantum mechanical ringing may be a more apt description. More detailed work is required to sort out this detail.

These eight cases are excellent examples of the additional types of challenges that can arise in the study of shape resonance phenomena. They should not diminish the simplicity and power of the fundamental shape resonance dynamics but, rather, should show how this fundamental framework showcases more complicated (and interesting) photoionization dynamics which, in turn, require a more sophisticated framework for full understanding.

Another form of progress is measured by the applicability of ideas to other observables, or, more broadly, to other subfields: (i) We have already touched upon the close connection[22] between shape resonance phenomena in molecular photoionization and electron-molecule scattering. (ii) Shape resonances in adsorbed molecules are now used rather extensively[36,37] as a probe of the geometry and electronic properties of adsorption sites. (iii) As discussed in connection with the inner-shell spectra of SF_6, free-molecule concepts concerning localized states carry over to the condensed phase. In such cases, a local "molecular" point-of-view can often provide more direct physical insight into photoexcitation dynamics of solids than can a band-structure approach. (iv) Also noted above, shape resonances are often low-energy precursors to EXAFS structure occurring from ~100 eV to thousands of eV above inner-shell edges. (v) An intimate connection also exists with antibonding valence states in quantum chemistry language.[93] This was dramatically demonstrated over ten years ago, when Gianturco et al.[94] interpreted the shape resonances in SF_6 using unoccupied virtual orbitals in an LCAO-MO calculation. This connection is a natural one since shape resonances are localized within the molecular charge distribution and therefore can be realistically described by a limited basis set suitable for describing the valence MO's. However, the scattering approach used in the shape-resonance picture is necessary for analysis of various dynamical aspects of the phenomena. (vi) Finally, shape resonances have been used as characteristic features in the analysis of such diverse subjects as electron optics[5] of molecular fields and hole localization[29] in inner-shell ionization, and as the cause for molecular alignment in photoionization, leading to anisotropy in the angular distribution[95,96] of Auger electrons from the decay of K-shell holes.

Clearly the subject of shape resonances in molecular fields has developed into a mature subfield. To a large extent the dynamics described above are now employed in a more or less routine fashion to interpret molecular photophysics and electron scattering processes. Nevertheless several challenges remain, mainly in the cases of significant departures from the simple concepts offered above. These will be due in large part to multiparticle mechanisms, such as those enumerated earlier in this section. Another interesting direction will be to study the effects of resonant excitations in more complicated vibrational modes and dissociative processes. In these and unforseen ways, the expansion, refinement, and unification of ongoing developments in the study of shape resonances in molecular fields should remain an active theme in molecular physics in the coming years.

References

1. J. L. Dehmer, A. C. Parr, and S. H. Southworth, in: "Handbook on Synchrotron Radiation, Vol. II," ed. G. V. Marr, North Holland, Amsterdam (1987), p. 241 and references therein.
2. "Resonances in Electron-Molecule Scattering, van der Waals Complexes, and Reactive Chemical Dynamics," ACS Symposium Series, No. 263, ed. D. G. Truhlar, American Chemical Society, Washington, DC (1984).
3. N. F. Lane, Rev. Mod. Phys. 52:29 (1980), and references therein.
4. D. Dill and J. L. Dehmer, J. Chem. Phys. 61:692 (1974).
5. J. L. Dehmer and D. Dill, in: "Electron-Molecule and Photon-Molecule Collisions," eds. T. Rescigno, V. McKoy, and B. Schneider, Plenum, New York (1979), p. 225.
6. P. W. Langhoff, in: "Electron-Molecule and Photon-Molecule Collisions," eds. T. Rescigno, V. McKoy, and B. Schneider, Plenum, New York (1979), p. 183.
7. G. Raseev, H. Le Rouzo, and H. Lefebvre-Brion, J. Chem. Phys. 72:5701 (1980).
8. R. R. Lucchese and V. McKoy, Phys. Rev. A 24:770 (1981).
9. R. R. Lucchese, G. Raseev, and B. V. McKoy, Phys. Rev. A 25:2572 (1982).

10. Z. H. Levine and P. Soven, Phys. Rev. A 29:625 (1984).
11. L. A. Collins and B. I. Schneider, Phys. Rev. A 29:1695 (1984).
12. R. E. LaVilla and R. D. Deslattes, J. Chem. Phys. 44:4399 (1966).
13. R. E. LaVilla, J. Chem. Phys. 57:899 (1972).
14. T. M. Zimkina and V. A. Fomichev, Dokl. Akad. Nauk SSSR 169:1304 (1966) [Sov. Phys. Dokl. 11:726 (1966)].
15. T. M. Zimkina and A. C. Vinogradov, J. Phys. (Paris) Colloq. 32:3 (1971).
16. D. Blechschmidt, R. Haensel, E. E. Koch, U. Nielsen, and T. Sagawa, Chem. Phys. Lett. 14:33 (1972).
17. M. Nakamura, Y. Morioka, T. Hayaishi, E. Ishiguro, and M. Sasanuma, in: "Third International Conference on Vacuum Ultraviolet Radiation Physics," Physical Society of Japan, Tokyo (1971), paper 1pA1-6.
18. J. L. Dehmer, J. Chem. Phys. 56:4496 (1972).
19. G. J. Shulz, Rev. Mod. Phys. 45:422 (1973).
20. M. S. Child, "Molecular Collision Theory," Academic, New York (1974), p. 51.
21. U. Fano and J. W. Cooper, Rev. Mod. Phys. 40:441 (1968).
22. J. L. Dehmer and D. Dill, in: "Symposium on Electron-Molecule Collisions," eds. I. Shimamura and M. Matsuzawa, University of Tokyo Press, Tokyo (1979), p. 95.
23. J. L. Dehmer and D. Dill, Phys. Rev. Lett. 35:213 (1975).
24. D. Dill, J. Siegel, and J. L. Dehmer, J. Chem. Phys. 65:3158 (1976).
25. J. L. Dehmer and D. Dill, J. Chem. Phys. 65:5327 (1976).
26. R. B. Kay, Ph. E. van der Leeuw, and M. J. van der Wiel, J. Phys. B 10:2513 (1977).
27. A. P. Hitchcock and C. E. Brion, J. Electron Spectrosc. 18:1 (1980).
28. T. N. Rescigno and P. W. Langhoff, Chem. Phys. Lett. 51:65 (1977).
29. D. Dill, S. Wallace, J. Siegel, and J. L. Dehmer, Phys. Rev. Lett. 41:1230 (1978); 42:411 (1979).
30. M. Krauss and F. H. Mies, Phys. Rev. A 1:1592 (1970).
31. R. L. de Kronig, Z. Physik 70:317 (1931); 75:191 (1932).
32. "Synchrotron Radiation: Techniques and Applications," ed. C. Kunz, Springer-Verlag, Berlin (1979).
33. "Synchrotron Radiation Research," eds. H. Winick and S. Doniach, Plenum, New York (1980).
34. B. Kincaid and P. Eisenberger, Phys. Rev. Lett. 34:1361 (1975).
35. J. R. Swanson, D. Dill, and J. L. Dehmer, J. Chem. Phys. 75:619 (1981).
36. T. Gustafsson, E. W. Plummer, and A. Liebsch, in: "Photoemission and the Electronic Properties of Surfaces," eds. B. Feuerbacher, B. Fitton, and R. F. Willis, J. Wiley, New York (1978).
37. T. Gustafsson, Surface Science 94:593 (1980).
38. D. Loomba, S. Wallace, D. Dill, and J. L. Dehmer, J. Chem. Phys. 75:4546 (1981).
39. J. R. Swanson, D. Dill, and J. L. Dehmer, J. Phys. B 13:L231 (1980).
40. J. R. Swanson, D. Dill, and J. L. Dehmer, J. Phys. B 14:L207 (1981).
41. P. M. Dittman, D. Dill, and J. L. Dehmer, Chem. Phys. 78:405 (1983).
42. J. L. Dehmer, D. Dill, and S. Wallace, Phys. Rev. Lett. 43:1005 (1979).
43. J. L. Dehmer and D. Dill, in: "Electronic and Atomic Collisions," eds. N. Oda and K. Takayanagi North-Holland, Amsterdam (1980), p. 195.
44. S. Wallace, Ph.D. thesis, Boston University (1980).
45. D. M. Chase, Phys. Rev. 104:838 (1956).
46. T. A. Carlson, Chem. Phys. Lett. 9:23 (1971).
47. T. A. Carlson and A. E. Jonas, J. Chem. Phys. 55:4913 (1971).
48. J. B. West, A. C. Parr, B. E. Cole, D. L. Ederer, R. Stockbauer, and J. L. Dehmer, J. Phys. B 13:L105 (1980).
49. J. L. Gardner and J.A.R. Samson, J. Electron Spectrosc. 13:7 (1978).

50. R. R. Lucchese and B. V. McKoy, J. Phys. B 14:L629 (1981).

51. T. A. Carlson, M. O. Krause, D. Mehaffy, J. W. Taylor, F. A. Grimm, and J. D. Allen, J. Chem. Phys. 73:6056 (1980).

52. D. Dill, J. Welch, J. L. Dehmer, and J. Siegel, Phys. Rev. Lett. 43:1236 (1979).

53. J. L. Dehmer, J. Siegel, and D. Dill, J. Chem. Phys. 69:5205 (1978).

54. R. E. Kennerly, R. A. Bonham, and J. McMillan, J. Chem. Phys. 70:2039 (1979).

55. R. E. Kennerly, Phys. Rev. A 21:1876 (1980).

56. J. L. Dehmer, J. Siegel, J. Welch, and D. Dill, Phys. Rev. A 21:101 (1979).

57. Z. Pavlovic, M.J.W. Boness, A. Herzenberg, and G. J. Shulz, Phys. Rev. A 6:676 (1972).

58. D. G. Truhlar, S. Trajmar, and W. Williams, J. Chem. Phys. 57:3250 (1972).

59. A. Chutjian, D. G. Truhlar, W. Williams, and S. Trajmar, Phys. Rev. Lett. 29:1580 (1972).

60. J. R. Rumble, D. G. Truhlar, and M. A. Morrison, J. Phys. B 14:L301 (1981).

61. A bibliography of inner-shell spectra is given by A. P. Hitchcock, J. Electron Spectrosc. 25:245 (1982).

62. J. A. Stephens and D. Dill, Phys. Rev. A 31:1968 (1985).

63. G. V. Marr, J. M. Morton, R. M. Holmes, and D. G. McCoy, J. Phys. B 12:43 (1979).

64. M. Y. Adam, P. Morin, P. Lablanquie, and I. Nenner, paper presented at International Workshop on Atomic and Molecular Photoionization, Fritz-Haber-Institut der Max-Planck-Gesellschaft, Berlin, W. Germany, 1983 (unpublished).

65. S. H. Southworth, A. C. Parr, J. E. Hardis, and J. L. Dehmer, Phys. Rev. A 33:1020 (1986).

66. B. Basden and R. R. Lucchese, Phys. Rev. A 34:5158 (1986).

67. J. L. Dehmer, A. C. Parr, S. Wallace, and D. Dill, Phys. Rev. A 26:3283 (1982).

68. A. Gerwer, C. Asaro, B. V. McKoy, and P. W. Langhoff, J. Chem. Phys. 72:713 (1980).

69. G. Raseev, H. Lefebvre-Brion, H. Le Rouzo, and A. L. Roche, J. Chem. Phys. 74:6686 (1981).

70. P. M. Dittman, D. Dill, and J. L. Dehmer, J. Chem. Phys. 76:5703 (1982).

71. P. Morin, I. Nenner, M. Y. Adam, M. J. Hubin-Franskin, J. Delwiche, H. Lefebvre-Brion, and A. Giusti-Suzor, Chem. Phys. Lett. 92:609 (1982).

72. F. Grimm, T. A. Carlson, W. B. Dress, P. Agron, J. O. Thomson, and J. W. Davenport, J. Chem. Phys. 72:3041 (1980).

73. T. Gustafsson, E. W. Plummer, D. E. Eastman, and W. Gudat, Phys. Rev. A 17:175 (1978).

74. T. A. Carlson, M. O. Krause, F. A. Grimm, J. D. Allen, D. Mehaffy, P. R. Keller, and J. W. Taylor, Phys. Rev. A 23:3316 (1981).

75. P. W. Langhoff, T. N. Rescigno, N. Padial, G. Csanak, and B. V. McKoy, J. Chem. Phys. 77:589 (1980).

76. F. A. Grimm, J. D. Allen, T. A. Carlson, M. O. Krause, D. Mehaffy, P. R. Keller, and J. W. Taylor, J. Chem. Phys. 75:92 (1981).

77. R. R. Lucchese and B. V. McKoy, J. Phys. Chem. 85:2166 (1981).

78. N. Padial, G. Csanak, B. V. McKoy, and P. W. Langhoff, Phys. Rev. A 23:218 (1981).

79. R. R. Lucchese and B. V. McKoy, Phys. Rev. A 26:1406 (1982).

80. P. Roy, I. Nenner, M. Y. Adam, J. Delwiche, M. J. Hubin Franskin, P. Lablanquie, and D. Roy, Chem. Phys. Lett. 109:607 (1984).

81. T. Gustafsson, Phys. Rev. A 18:1481 (1978).

82. H. J. Levinson, T. Gustafsson, and P. Soven, Phys. Rev. A 19:1089 (1979).

83. J. L. Dehmer, A. C. Parr, S. H. Southworth, and D. M. P. Holland, Phys. Rev. A 30:1783 (1984).

84. T. A. Ferrett, D. W. Lindle, P. A. Heimann, H. G. Kerkhoff, U. E. Becker, and D. A. Shirley, Phys. Rev. A 34:1916 (1986), and to be published.

85. T. A. Ferrett, Ph.D. Thesis, U. California, Berkeley (1986).

86. L. S. Cederbaum, W. Domcke, J. Schirmer, and W. von Niessen, Phys. Scr. 21:481 (1980).

87. S. Krummacher, V. Schmidt, and F. Wuilleumier, J. Phys. B 13:3993 (1980).

88. P. W. Langhoff, S. R. Langhoff, T. N. Rescigno, J. Schirmer, L. S. Cederbaum, W. Domcke, and W. von Niessen, Chem. Phys. 58:71 (1981).

89. F. Sette, J. Stöhr, and A. P. Hitchcock, Chem. Phys. Lett. 110:517 (1984); J. Chem. Phys. 81:4906 (1984).

90. M. N. Piancastelli, D. W. Lindle, T. A. Ferrett, and D. A. Shirley, J. Chem. Phys. 86:2765 (1987).

91. D. L. Lynch, S. N. Dixit, and V. McKoy, J. Chem. Phys. 84:5504 (1986).

92. M. Braunstein and V. McKoy, J. Chem. Phys. 87:224 (1987).

93. See, e.g., P. W. Langhoff in Chapter 7 of Ref. 2.

94. F. A. Gianturco, C. Guidotti, and U. Lamanna, J. Chem. Phys. 57:840 (1972).

95. D. Dill, J. R. Swanson, S. Wallace, and J. L. Dehmer, Phys. Rev. Lett. 45:1393 (1980).

96. U. Becker, R. Hölzel, H. G. Kerkhoff, B. Langer, D. Szostak, and R. Wehlitz, Phys. Rev. Lett. 56:1455 (1986).

RESONANCES IN ATOMIC SPECTROSCOPY

J.P. Connerdale

Blackett Laboratory Imperial College London SW7 2AZ
and Physikalisches Institut, Universitat Bonn, FRG

Work done in collaboration with A.M. Lane
Theoretical Physics Division AERE Harwell Laboratory
Oxfordshire OX11 ORA JK

1.INTRODUCTION

The subject of resonances in atomic spectroscopy is an old one, and the subject is too vast to be covered by a single review. Also, the purpose of the present courses is not to indulge in a long history, but rather to provide a fairly compact introduction to some recent developments. With due apologies for incompleteness, I therefore pick out only some essentials to lay the groundwork for the main theme, namely the recent studies of interacting resonances, or the influence which interactions between resonances have on their line shapes.

The present lectures describe progress in understanding the profiles of autoionising resonances, obtained in experimental studies, mostly by synchrotron radiation spectroscopy, but also by using lasers. We start from the celebrated Fano formula for an isolated resonance in a ´flat´ continuum, and then build up further levels of complexity. The first question one may ask is whether cases occur in which the continuum is not ´flat´ but presents some pronounced modulation. This will lead us to consider shape or giant resonances. One can also imagine a single resonance ´tuned´ through a broad modulation, a situation clearly equivalent to two resonances of different widths coupled by an interaction. There is of course no reason to stop at two resonances. Indeed, in atomic physics, one usually deals with a full Rydberg manifold with an infinite sequence of resonances which may or may not overlap in energy. It turns out that the shapes of resonances - their line profiles and symmetries - are markedly affected by such overlaps and interactions. A study of the line shapes is therefore a powerful method of understanding interchannel interactions - physical processes which couple resonances to each other.

Examples will be drawn from a range of experimental studies, involving both laser and synchrotron radiation spectroscopy of neutral atoms. The high resolution experiments are mainly conducted in photoabsorption, although thermionic diode detectors (the so-called ´hot wire´ systems) are increasingly used. Most of the experiments are conducted in the vacuum ultraviolet and soft x-ray ranges, above the ionisation-potentials of the atoms, but in a range in which instrumental resolution is sufficiently good to allow detailed studies of line profiles. When experiments are conducted with synchrotron radiation below about 1200 A in wavelength, there exist no suitable materials for use as windows and a direct vacuum connection between the absorption cell and the accelerator is required. For photoabsorption studies, the pressure differential to be maintained can be as high as seven orders of magnitude between the interior of the cell and the vacuum chamber of the accelerator. Special techniques have been developed for the dynamical containment of atomic vapours in windowless systems, and for the production of vapour comumns of even quite refractory elements in high temperature vacuum furnaces (see eg refs.1 for experimental details). Instead of measuring the attenuation of a beam, one may also count the ions produced with very high efficiency by the use of a hot wire detector, an approach which has mainly been applied in laser spectroscopy, where high signals can be obtained by space charge amplification. More recently, thermionic diodes have also been successfully applied to synchrotron radiation studies by using wiggler magnets to enhance the intensity of the beam. Our experimental examples will exploit all of these techniques.

2. The K-Matrix Formalism and the Fano Formula for an isolated Resonance

The K matrix formalism, described for nuclear scattering by Lane and Thomas and by Lane (2) in the present context provides a convenient and simple description of atomic resonances in photoabsorption. Expressions for autoionisation in terms of scattering phase shifts were given already by Fano in Appendix C of his earliest paper (3). Later, Shore (4) developed the scattering theory of absorption profiles and refractivity in terms of Wigner´s scattering or S. matrix, which has extremely wide application outside atomic physics.

The scattering or S matrix and the concept of photoionisation as a ´half scattering´ problem are both introduced in the overview presented by Dr J. Macek in the present volume.

In scattering theory, a projectile is imagined to move towards a target from the remote past (t → - oo) and from a large distance, where projectile and target are so far separated that the interactions between them tend to zero and the total wave function tends to a product of the projectile wave packet times the basis state of the target, in some definite state, say a. As time increases, an interaction takes place between the projectile and the target and, as $t \to +\infty$, the interaction again tends to zero, leaving the system in another definite state b of the target. Clearly, the states a and b must be states for which the system contains a projectile at an infinite separation from the target. These are referred to as open channel or continuum states. During a transient interaction, represented by an operator $U(t_1, t_2)$ other states referred to as closed channels or discrete bound states can become temporarily involved. The closed channels are responsible for the occurrence of resonances. The elements of the scattering or S matrix yield the proportion of b which emerges from a as a result of the interaction:

$$S_{ba} = \langle \psi_b \mid U(\infty, -\infty) \mid \psi_a \rangle \tag{2.1}$$

To introduce the K matrix, the S matrix is written as

$$\underline{\underline{S}} = \frac{1 + i \underline{\underline{K}}}{1 - i \underline{\underline{K}}} \tag{2.2}$$

the K-matrix is real and symmetric. It yields the essential part of the cross section as

$$\sigma_{ab} = \frac{4 K_{ab}^2}{\left| (1 - i K_{aa})(1 - i K_{bb}) + K_{ab}^2 \right|^2} \tag{2.3}$$

and K has especially convenient forms for all of the problems we are interested in here. It is of course the case that all the interactions involved in atomic physics are expressible in terms of matrix elements and that alternative descriptions exist (e.g. multichannel quantum defect theory) which are mathematically completely equivalent to the present approach. The connections between them have been explicitly stated (see 5, 6).

A specific advantage of using the K matrix approach in the present context is that the transition to a numerical treatment of the equations can be postponed until quite a late stage, thus allowing the full algebraic structure of the interacting resonances to be displayed.

It is useful to express photo absorption processes in terms of phase shifts, and to partition the phase shift into one part which is due to the resonances (Δo, say) and one part (δ, say) which is the phase shift of the background continuum. For just one open channel,

$$K = \tan(\delta + \Delta_0) \tag{2.4}$$

The form of Δ_0 near a single isolated resonance (see e.g. Landau and Lifshitz 7) is given by

$$\tan \Delta_0 = \frac{\Gamma}{2(E_o - E)} \qquad (2.5)$$

where Γ, E_o are the width and resonance energy.

From these expressions, we find:

$$K = \frac{\tan\delta + \tan\Delta_0}{1 - \tan\delta\,\tan\Delta_0} = \frac{(E_O - E)\tan\delta + \Gamma/2}{(E_o - E) - \frac{\Gamma}{2}\tan\delta} \qquad (2.6)$$

and, hence, that the cross section:

$$\sigma(E) = \frac{2K^2}{1 + K^2} = 2\sin^2\delta\;\frac{\{(E_o - E) + \frac{\Gamma}{2}\cot\delta\}^2}{(E_o - E)^2 + \left(\frac{\Gamma}{2}\right)^2} \qquad (2.7)$$

we write $q = \cot\delta$ and $|\tilde{D}|^2 = 2\sin^2\delta$ which yields the standard Fano formula for an autoionising line:

$$\sigma(\varepsilon) = |\tilde{D}|^2\,\frac{(q + \varepsilon)^2}{1 + \varepsilon^2} \quad \text{where } \varepsilon = \frac{2(E - E_o)}{\Gamma} \qquad (2.8)$$

This formula applies to an isolated line in a ´flat´ continuum, which means that δ is regarded as constant within the range of the resonance and only Δ_o varies.

The general shape of isolated autoionising lines as a function of the profile index is shown in Fig.1. The only excuse for reproducing this celebrated figure (apart from the presence of Professor Fano in Maratea) is that these curves will be our ´zeroth order´ lineshapes i.e. the ones which we expect to find for isolated autoionising lines in absence of any complications. A beautiful example of such a line, recorded in a recent laser experiment using a thermionic diode detector, is shown in Fig.2.

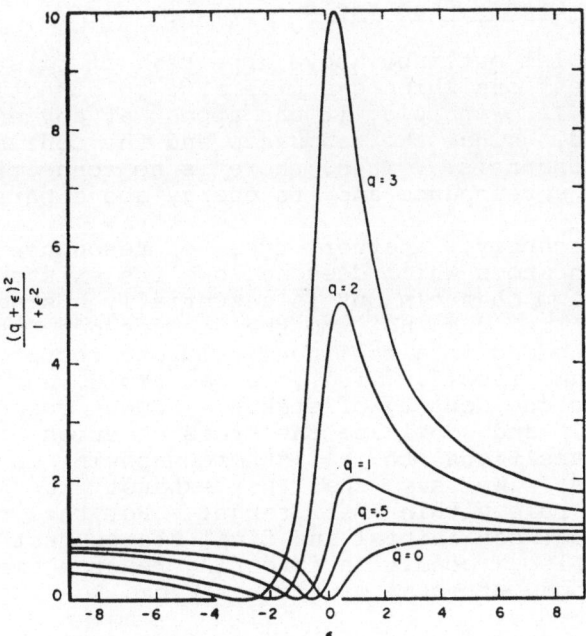

Fig.1. (After ref.3). The profiles of an isolated autoionising resonance in a flat continuum, for different values of the profile index or shape parameter q.

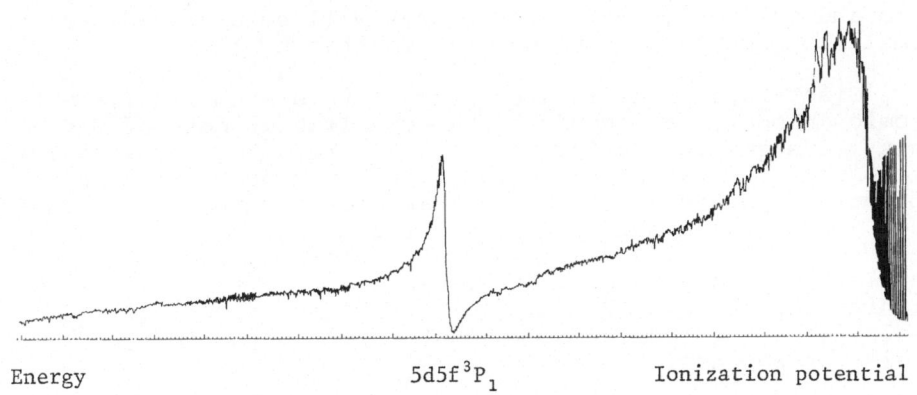

Energy $5d5f\,^3P_1$ Ionization potential

Fig.2. A portion of the barium spectrum including the first ionisation threshold (right of figure) and the $5d\,5f\ ^3P_1$, resonance, considered here as an isolated feature occurring sufficiently far above threshold for the continuum to be almost flat.

3. Shape or ´Giant´ Resonances

The problem outlined above appears, at first sight, to be the simplest one which can give rise to structure in the continuum. In principle, it can appear at any energy above the threshold, since the resonance and the continuum belong to distinct channels. Thus, there is no connection between the width of a resonance and its energy above threshold.

More recently, another type of resonance has been discovered in atoms which does not owe its existence to the presence of two channels but is essentially a single channel effect. It is due not to the occurrence of discrete structure embedded in a continuum, but to the properties of the continuum itself, i.e. to a rapid and intrinsic variation in the density of states. Such resonances can become huge, and dominate the cross section completely, dwarfing transitions to all other channels in the same energy range. We say that they exhaust the oscillator strength sum rule within their range. For this to be true, the overlap between initial and final states must be large, and this is also a condition favouring rearrangements of all kinds (many-body effects) within the atom.

By analogy with nuclear physics, where similar features occur in the collective oscillations of nucleons within the nucleus, the atomic resonances described above are referred to as ´giant resonances´. They have been the subject of a very recent NATO ASI (8) and therefore their properties will not be repeated in detail. Rather some differences between them and the Beutler-Fano resonances will be stressed, and a simple connection made to quantum scattering theory.

Giant resonances in atoms vary very systematically with atomic number, as shown in the experimental data of Fig.3. This systematic variation is due to the progressive deepening of a short range inner well in the double valley effective potential of the atom as atomic number increases (see 9, 10 for a more detailed discussion). It is simply a mean field effect, and although its detailed calculation may involve many-body theory, it is qualitatively present even within the Hartree-Fock model. This is demonstrated in Fig.4 for a sequence of atoms just prior to the emergence of a giant resonance.

The physical origin of the double well is readily understood. In hydrogen, if the centrifugal repulsive term is included within an effective potential then, since it grows as $\ell(\ell+1)\hbar^2/2mr^2$, there is a repulsive potential at small radius ($r < r_0$ say) which expels wavefunctions of high angular momentum into the outer, coulombic reaches. In the heavier atoms, a core develops within this radius r_0, so that the potential starts off as in hydrogen as a shallow coulombic well with a repulsive shoulder around r_0. However, within r_0, the potential becomes strongly attractive again because the electron penetrates the core and sees an incompletely screened nuclear charge, leading (for heavy atoms) to a deep inner well with a fast rising repulsive wall near the centre. The valley is a short range well and can therefore only support a finite number of bound states. If the well is 'squeezed' these states rise into the continuum and become virtual states or short lived resonances.

Now, it is well known in nuclear scattering theory (Schwinger's theorem) that the bound states and low energy scattering spectrum of a short range well are not dependent on its detailed shape, but merely on how binding it is. We can take advantage of this to 'mock up' a very simple potential which will give us an analytic formula for the phase shift Δ_0, namely a spherical square well with angular momentum, which possesses analytic solutions in the continuum (11). The resulting expression for the phase shift is

$$
\tan \Delta_0 = \frac{z\, j_\ell(z')j_{\ell-1}(z) - z'j_\ell(z)j_{\ell-1}(z')}{z\, j_\ell(z')j_{-\ell}(z) + z'j_{-\ell-1}(z)j_{\ell-1}(z')} \qquad (3.1)
$$

where ℓ is the angular momentum ($\ell > 1$) j_ℓ are the spherical Bessel functions, $z = ka$, $z' = k'a$, where k' and k are the wave vectors inside and outside the well and a is its radius.

The cross section curves in Fig.5a show how the giant resonance evolves as the binding strength of the short range well is progressively increased. Note the similarity between Fig.4, obtained from detailed Hartree-Fock calculations, and Fig.5a which is our very simple model based on the principles of effective range theory. In particular, the dashed curves in Fig.5a provides an immediate explanation for the corresponding feature of Fig.4: in the expression for the cross section

$$
\sigma(k) = \frac{4\pi}{k^2}\,(2\ell+1)\sin^2\Delta_0 \qquad (3.2)
$$

the presence of a resonance is signalled by Δ_0 passing through π as energy is increased. Thus, the cross section curves all lie below a $1/k^2$ envelope, known in nuclear scattering theory as the unitary limit.

An excellent example of this behaviour occurs in Dr Dehmer's lectures (present volume) where he discusses the influence of molecular bond length on a shape resonance (see page   ) see also Fig.5b.

Now, we can see an essential difference between the Beutler-Fano resonance and the giant resonance, which is due to the fact that the former is an interchannel errect

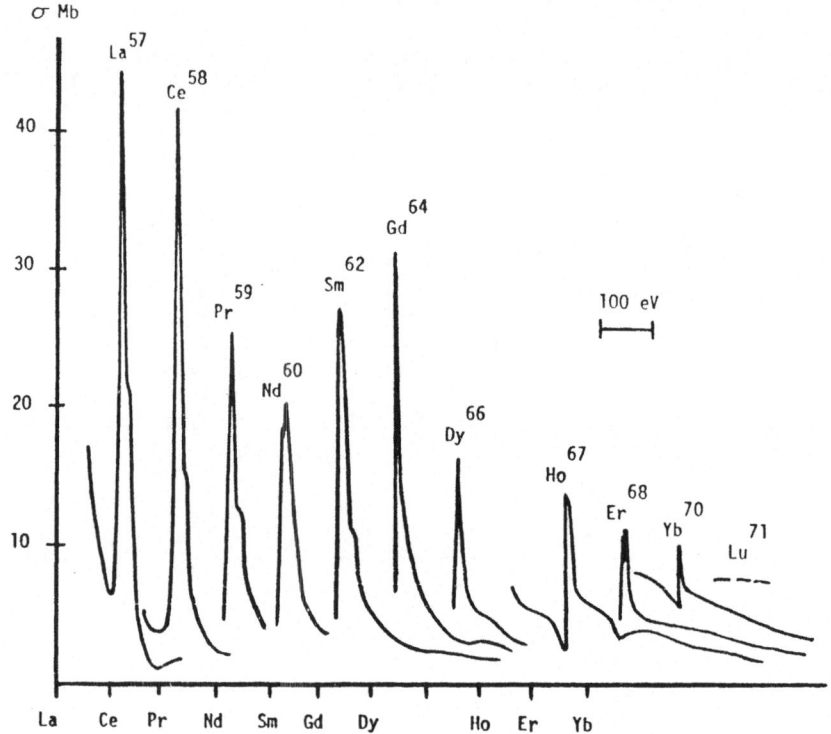

Fig.3 Showing the systematic variation of giant resonances as a function of atomic number as revealed by experimental data for solids in the lanthanide sequence.

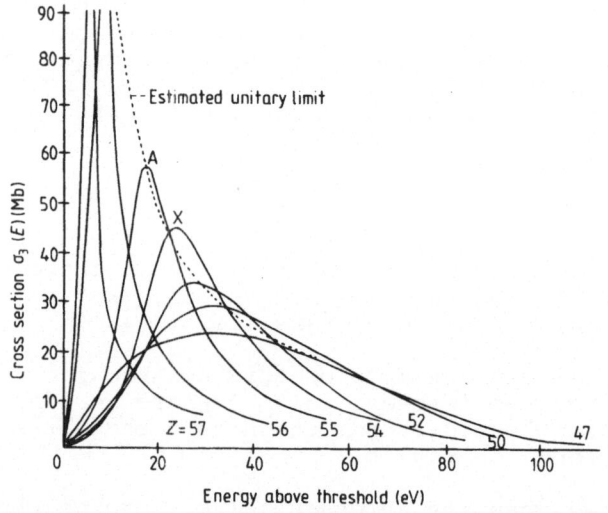

Fig.4. Ab initio calculations of 4d - f resonances calculated from an independent particle model. The curves have been translated in energy so that all thresholds coincide, in order to bring out the existence of an approximate unitary limit (cf Fig.5).

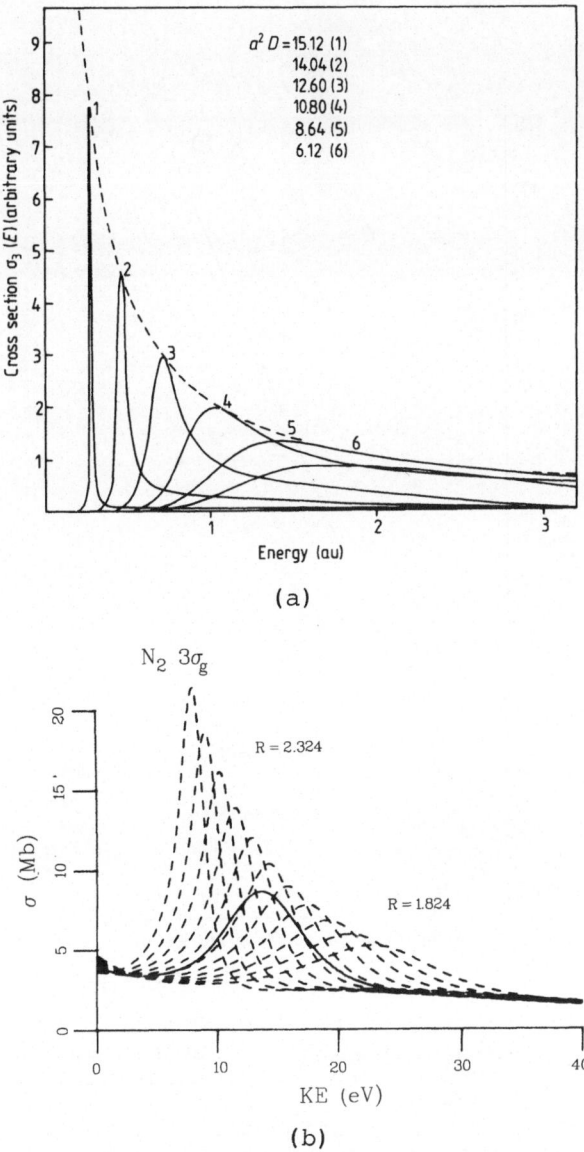

(a)

(b)

Fig.5(a) Showing the variation in the cross section for a
spherical square well with angular momentum 1=3 calculated from
(3.1) and for several values of the binding strength. The
dashed curve is the unitary limit. (b) (From J. Dehmer's
lectures, present volume). Showing the influence of a variation
in bond length on the shape resonance in N_2, as revealed by
calculations. In effect, modifying the bond length alters the
binding strength of the inner well, and the resulting curves
(dashed lines) exhibit a unitary limit. For further details,
see page

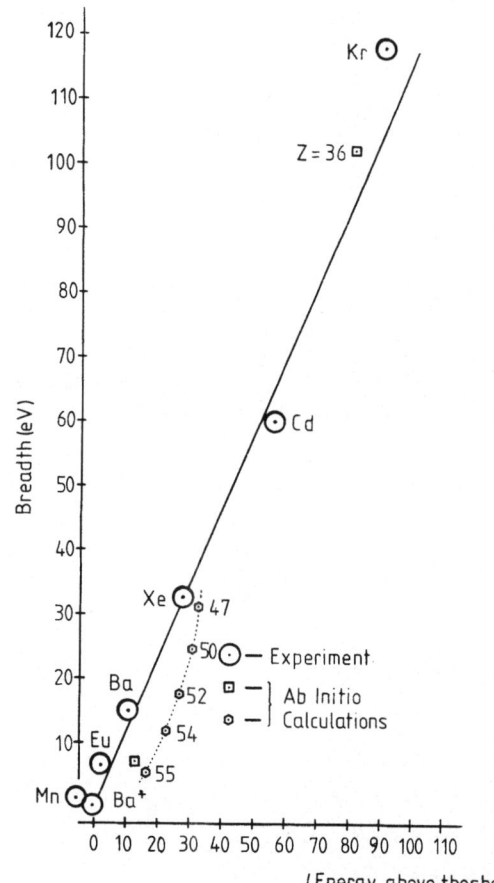

Fig.6. A plot of the breadths of atomic
        shape or giant resonances against
        their energy above threshold,
        showing the approximately
        linear relationship between the
        two, described in the text.

while the latter is <u>essential</u> a single channel phenomenon: in a giant resonance, the width of the resonance depends on how high the resonance lies above the associated threshold. If it is far above, the resonance is broad, and as the well becomes more binding and the resonance approaches the threshold, it becomes progressively sharper.

One can argue from the uncertainty principle (12) that there should be a linear relation between energies and widths for giant resonances, and this deduction is very well obeyed by experiment, as shown in Fig.6. In this respect, giant resonances and Beutler-Fano resonances behave very differently from each other.

Finally, in Fig.7, we show an experimental example of a giant resonance, together with a computed profile using the simple phase shift formula given above. The giant resonance has a characteristic asymmetric shape, with a low value of the cross section tied to the threshold and a long 'tail' extending towards high energies.

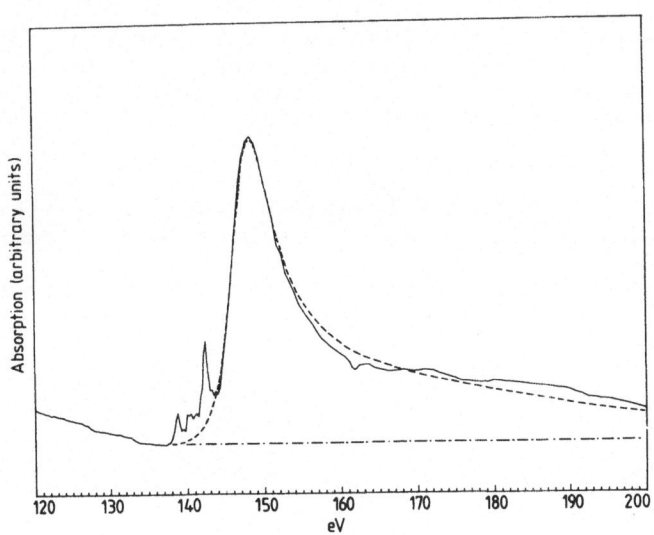

Fig.7. Showing the 4d - f resonance in gadolinium vapour (full curve-experimental data) and an approximate fit to the measured profile using the square well formula (3.1) described in the text.

4. Two Overlapping Resonances with a Flat Continuum

Returning now to the case of a flat continuum (constant δ), we can readily construct the case of two overlapping resonances, in which case the phase shift has a resonant part:

$$\tan \Delta_0 = \frac{\Gamma_1}{2(E_1-E)} + \frac{\Gamma_2}{2(E_2-E)} \tag{4.1}$$

and, by the same procedure as followed before:

$$\sigma(E) = \frac{2K^2}{1+K^2} = 2\sin^2\delta \; \frac{\{1 + \frac{\Gamma_1}{2(E_1-E)}\cot\delta + \frac{\Gamma_2}{2(E_2-E)}\cot\delta\}^2}{1 + \{\frac{\Gamma_1}{2(E_1-E)} + \frac{\Gamma_2}{2(E_2-E)}\}^2}$$

$$= 2\sin^2\delta \; \frac{\{1 + \frac{q\Gamma_1/2}{E_1-E} + \frac{q\Gamma_2/2}{E_2-E}\}^2}{1 + \{\frac{\Gamma_1/2}{E_1-E} \quad \frac{\Gamma_2/2}{E_2-E}\}^2}$$

such formulae have been used in the literature to analyse cases of two overlapping resonances. For resonances originating from two different channels, different values q_1 and q_2 should be introduced (13).

5. Generalisation to N Overlapping Resonances with One Flat Continuum and the Connection to Quantum Defect Theory

The result in the previous section can be instantly generalised to N resonances thus

$$\sigma(E) = \frac{2K^2}{1+K^2} = 2\sin^2\delta \; \frac{\{1 + q\sum_n \frac{\Gamma_n/2}{E_n-E}\}^2}{1 + \{\sum_n \frac{\Gamma_n/2}{E_n-E}\}^2} \tag{5.1}$$

One makes contact with quantum defect theory through a standard substitution:

$$\sum_n \frac{\Gamma_n/2}{E_n-E} \equiv x\cot\pi(\nu+\mu) \equiv x\cot\theta \tag{5.2}$$

where ν is the energy variable defined by $E_\infty - E = \dfrac{R}{\nu^2}$ and μ is the quantum defect. Thus

$$\sigma(\theta) = \frac{2\,K^2}{1+K^2} = 2\sin^2\delta\,\frac{(1 + qx\,\cot\theta)^2}{1 + (x\,\cot\theta)^2} \qquad (5.3)$$

By some simple algebra, one can show that

$$\sigma(\nu) = |\tilde{D}'|^2\,\frac{\tan^2\pi\nu + 2B\,\tan\pi\nu + B^2}{\tan^2\pi\nu + 2C\,\tan\pi\nu + D^2} \qquad (5.4)$$

where, as before, $|\tilde{D}|^2 = 2\sin^2\delta$ and there are now three shape parameters

$$B = \frac{qx + \tan\pi\mu}{1 - qx\tan\pi\mu}\ ;\ \ C = \frac{(1-x^2)\tan\pi\mu}{1 + x^2\tan^2\pi\mu}\ ;\ \ D^2 = \frac{x^2\tan^2\pi\mu}{1 + x^2\tan^2\pi\mu} \quad (5.5)$$

Of these, only B depends on q, and from the formula

$$q = \frac{1}{x}\tan\pi(\beta-\mu)\ \text{where}\ B = \tan\pi\beta \qquad (5.6)$$

For an isolated or Beutler-Fano resonance, q=0 implies a symmetric window. In the present case, when $B = \tan\pi\mu$, the transmission maxima also coincide with the resonance energies.

The formula for $\sigma(\nu)$ above (14, 15) is identical to the formulae given by Dubau & Seaton and (16) and, independently, by Giusti-Suzor and Fano (17). It provides us with a single simple expression for a complete Rydberg series of antoionising resonances in terms of just three shape parameters, which are constant for the whole series.

6. Study of the Lineshapes for a Rydberg Series of Autoionising Lines and a Flat Continuum

It is interesting to study the way in which the three shape parameters A,B and C affect the form of the line profiles. Since the numerator in the formula for $\sigma(\nu)$ is a perfect square, there must be a zero for $\tan\pi\nu = -B$. This defines the transmission window (see Fig.8). Likewise, one can readily show that the peak occurs for $\tan\pi\nu = (D^2-BC)/(B-C)$. There are also bounds on the value of C set by rearranging the formula thus:

$$\sigma(\nu) = |\tilde{D}'|^2\,\frac{(\tan\pi\nu + B)^2}{(\tan\pi\nu + C)^2 + (D^2 - C^2)} \qquad (6.1)$$

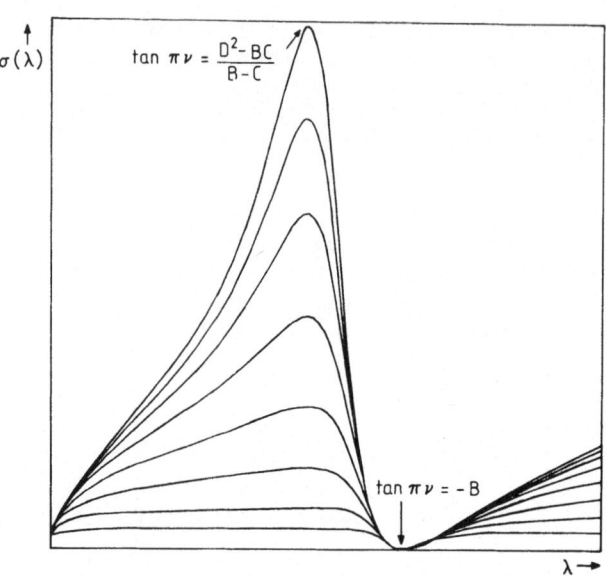

Fig. 8(a). Showing the family of curves generated by requiring the maxima and minima to remain fixed in energy in equation (5.4)

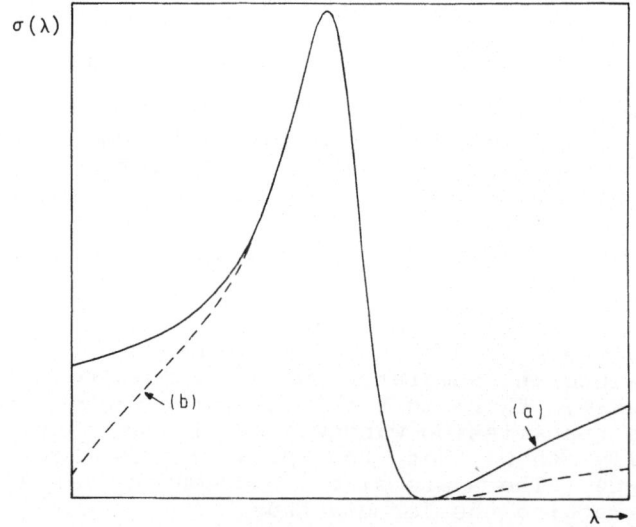

Fig. 8(b). Comparison between (5.4) and the Fano profile formula (2.8) near an isolated line. Curve (b) is obtained from (5.4) and departs most strongly from the Fano formula (a) in the wings of the profile, where the isolated line approximation breaks down.

Since a cross section is positive definite, and chosing D positive (no loss of generality) we have - D<C<D.

Since B is fully determined by a single point on the profile (the zero in the cross-section), it is convenient to vary the parameters C and D for fixed B.

We begin with C. Clearly, there exist just three points on the profile within any cycle which are independent of the values of C, namely (a) $\tan\pi\nu = -B$ (b) $\tan\pi\nu = 0$, where $\sigma = B^2/D^2$ and (C) $\tan\pi\nu \to \pm\infty$ where $\sigma = 1$. The points (a) and (c) are also independent of the value of D.

Points (b) correspond to $\nu = n$ where n is an integer and are simply the hydrogenic energies, while points (c) occur at $\nu = n + \frac{1}{2}$ and are conveniently described as ´half hydrogenic´ points. There is a special case B = C to which we return below.

Fig 9 shows the form of the profiles parametrised by C in a typical situation. There are two parts to the figure which correspond to changing the sign of B so that the zero in the cross section occurs either just above or just below a hydrogenic point. As expected, all the curves meet three times in a cycle, and once at the half hydrogenic point.

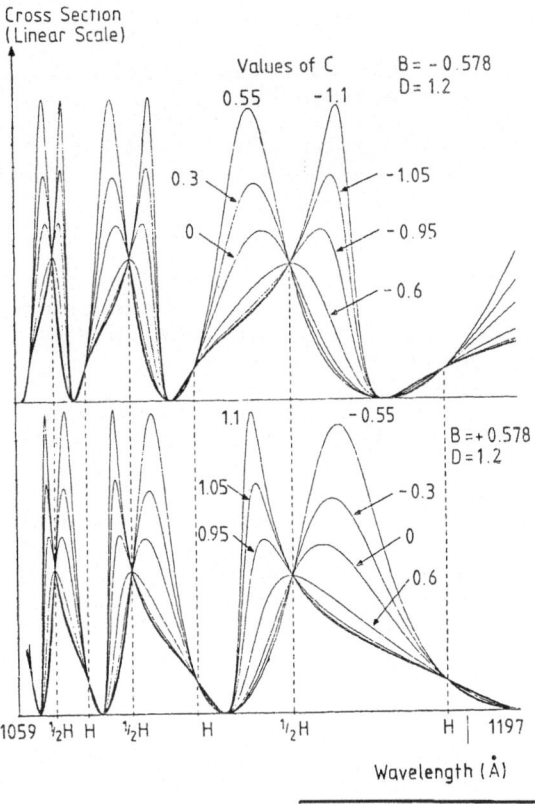

Fig. 9. Evolution of the profiles described by equation 5.4 as a function of the parameter C.

When the profile maximum occurs on the same side of the half hydrogenic point as the minimum, the appearance of the profile is similar to the Fano lineshape. However, if the maximum and minimum become sufficiently separated in energy to straddle the half hydrogenic point, then the shape of the profile changes to a more sinusoidal form.

Put in simple terms, if we are dealing with a Rydberg series of autoionising lines rather than just with an isolated line, then the profile will lie somewhere between two extreme situations: either the line width is much smaller than the separation between hydrogenic and half-hydrogenic points (Type 1 profile) or it is much larger (Type 2 profile). In the isolated line limit, type 1 profiles tend to the typical Fano shape. In Fig.10, we illustrate the geometrical distinction between the two types of profile by two experimental examples.

It is interesting to compare a type 1 profile with the corresponding Fano shape (Fig.8). The two are nearly coincident close to the resonance energy but become distinct from each other in the wings, where the more general formula for Rydberg series includes the influence of the adjacent members.

The type 2 profile is clearly one for which the profiles overlap in energy and become distorted from the Fano line shape. In this case, the original Fano parametrisation is of course inappropriate and individual q parameters become meaningless.

In the isolated profile limit of type 1, individual values are given by

$$q = \frac{B - C}{(D^2 - C^2)^{\frac{1}{2}}} \tag{6.2}$$

and q is therefore constant in a series for a given choice of the three shape parameters B, C and D. This expression is useful (see below) in discussing the interaction between a Rydberg series of autoionising lines and a shape or giant resonance.

The special case B = C yields q = 0 in the isolated profile limit and, more generally, gives a Lorentzian in $\tan \pi \nu$, i.e. an essentially symmetric profile. It is interesting in fact, to set B $\approx$ C and then vary D to study the evolution of profiles which are nearly symmetric. This is done in Fig.11 and shows another respect in which the generalised formula for a Rydberg series differs from the Fano-formula: profiles symmetric in $\tan \pi \nu$ can now occur either as transmission windows or as absorption peaks with no associated transmission maximum. In a Fano profile, pronounced absorption maxima are always associated with a

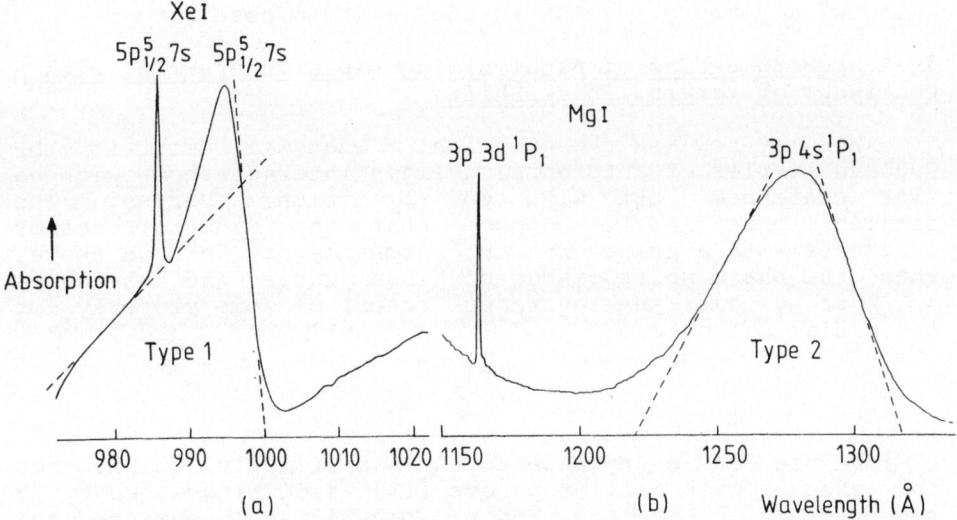

Fig. 10. Experimental examples, illustrating the difference between type 1 and type 2 profiles discussed in the text.

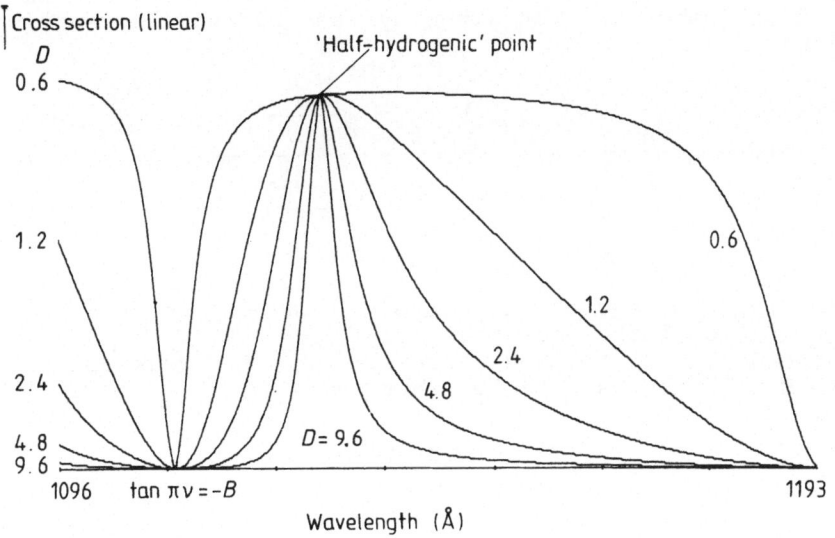

Fig.11. Showing the evolution of the line profile from a nearly symmetric transmission 'window' to a nearly symmetric absorption line as the parameter D is changed with B∿C (see text).

'window' asymmetry, except in the limiting case $q \to \pm \infty$

7. A Rydberg Series of Autoionising Lines Interacting with a Continuum of Varying Phase Shift

We may now ask the question: what will occur if the Rydberg series of autoionising lines interacts not with a flat continuum, but with one for which δ varies.   The simplest case is to suppose that the Rydberg series interacts with a shape or giant resonance, as defined above, when the phase shift of the continuum varies and changes by $\sim \pi$ in a small energy range because of some property of the radial potential.

We use Fano's approach to the K matrix (18) which, for consistency with earlier papers (19) is distinguished by a subscript F .   Let Γ_{BF} and E_{BF} be the particle width and resonance energy of the giant resonance.   We follow the treatment of Connerade and Lane (20).   The analysis considers only particle widths: strictly, it applies only to elastic scattering and may or may not extend to photoionisation, depending on the case considered.   It does, however, even in this simple form, reveal many crucial features of the problem.   For the Rydberg series of narrower antoionising lines, we use Γ_{nF}, E_{nF}.

The resonant part of the phase shift as defined in the previous cases is then given by (see 20 and also section 8 below):

$$\tan\Delta_0 = \sum_n \frac{\Gamma_{nF}/2}{E_{nF} - E} + \frac{1}{2} \frac{(\Gamma_{BF}^{\frac{1}{2}} - \sum_n \Gamma_{nF}^{\frac{1}{2}} \frac{H_{nF}}{E_{nF} - E})^2}{(E_{BF} - E - \sum_n \frac{H_{nF}}{E_{nF} - E})} \qquad (7.1)$$

where H_{nF} is the coupling between n and B.   It is a property of the coulomb potential that $\mu_F = (\Gamma_{nF}^{\frac{1}{2}} / H_{nF})$ is independent of n for a Rydberg series.   For weak coupling of n to B, μ_F^{-1} is also small.   Thus, writing

$$\Sigma \equiv \sum_n \frac{\Gamma_{nF}/2}{E_{nF} - E} \quad ,$$

$$\tan\Delta_0 = \Sigma + \frac{1}{2}(\Gamma_{BF}^{\frac{1}{2}} - 2\mu_F^{-1}\Sigma)(E_{BF} - E - 2\mu_F^{-2}\Sigma)^{-1} \qquad (7.2)$$

and to order μ_F^{-2}

$$\tan \Delta_0 \approx \frac{\Gamma_{BF}/2}{E_{BF}-E} + \sum_n \frac{\Gamma_{nF}/2}{E_{nF}-E} - \mu_F^{-1}\frac{\Gamma_{BF}^{\frac{1}{2}}}{E_{BF}-E} \sum_n \frac{\Gamma_{nF}}{E_{nF}-E} \qquad (7.3a)$$

$$= \frac{\Gamma_{BF}/2}{E_{BF}-E} + \sum_n \frac{\Gamma_{nF}/2}{E_{nF}-E} - \frac{\Gamma_{BF}^{\frac{1}{2}}}{E_{BF}-E} \sum_n H_{nF} \frac{\Gamma_{nF}^{\frac{1}{2}}}{E_{nF}-E} \qquad (7.3b)$$

i.e. the total resonant phase shift is the sum of the individual phase shifts of the resonances plus a mixed term representing the coupling of B to n.

This can be rearranged as:

$$\tan \Delta_0 \approx \frac{\Gamma_{BF}/2}{E_{BF}-E} + \sum_n (\Gamma_{nF}^{\frac{1}{2}} - \frac{H_{nF}\Gamma_{BF}^{\frac{1}{2}}}{E_{BF}-E})^2 (E_{nF}-E)^{-1} \qquad (7.4a)$$

$$= \frac{\Gamma_{BF}/2}{(E_{BF}-E)} + (1 - \frac{\mu_F^{-1}\Gamma_{BF}^{\frac{1}{2}}}{E_{BF}-E})^2 \sum_n \frac{\Gamma_{nF}/2}{(E_{nF}-E)} \qquad (7.4b)$$

We now follow the connection to quantum defect theory in the same manner as in section 5 above and find that Δ_0 can be written as

$$\tan \Delta_0 \approx \overline{\Delta} + x_1 x_2 \cot\theta \qquad (7.5)$$

where $\quad x_1 \equiv 1 - \dfrac{\mu_F^{-1}\Gamma_{BF}^{\frac{1}{2}}}{E_{BF}-E}$

$$x_2 \cot\theta \equiv \sum_n \frac{\Gamma_{nF}/2}{E_n - E} \qquad (7.6)$$

$$\overline{\Delta} \equiv \frac{\Gamma_{BF}/2}{E_B - E}$$

and the angle θ is as defined previously (see 5.2).

Again, we write the cross section as:

$$\sigma(\theta) = \frac{2 K^2}{1 + K^2} \qquad \text{(see 2.3)}$$

and, by the same route as previously (with a little more algebra) we find:

$$\sigma(\theta) = \frac{(\overline{\Delta} + \overline{\delta})^2}{(1+\overline{\Delta}^2)(1+\overline{\delta}^2)} \frac{\tan^2\theta + 2B\tan\theta + B^2}{\tan^2\theta + 2C\tan\theta + D^2} \qquad (7.7)$$

where:

$$B = \frac{x_1 x_2}{\overline{\Delta}+\overline{\delta}} \; ; \quad C = \frac{x_1 x_2 \overline{\Delta}}{1 + \overline{\Delta}^2} ; \quad D^2 = \frac{x_1^2 x_2^2}{1 + \overline{\Delta}^2} ; \; \overline{\delta} = \tan\delta \quad (7.8)$$

We instantly recognise the second factor in as identical to the general formula (5.4) for N overlapping resonances with one flat continuum. There is, however, an essential difference: B, C and D are no longer independent of energy. They become, mainly through the quantity $\overline{\Delta}$, functions of the detuning $E - E_{BF}$ from the centre of the giant resonance. Thus, the second factor in (7.7) represents a Rydberg series of autoionising lines, whose shapes (and in particular the profile index q) vary with energy in a manner which reveals the influence of the giant resonance on the individual Rydberg members.

The first factor in (7.7) is an amplitude term, which depends only on $\overline{\delta}$ and $\overline{\Delta}$, i.e. contains no information on the series structure, but only on the giant resonance and background phases. It is the profile of the giant resonance itself in absence of any series. For $\overline{\Delta}$ we can use the formula given in section 3 to characterise giant resonances.

For well separated Rydberg members, we find that

$$q = \frac{B - C}{(D^2 - C^2)^{\frac{1}{2}}} = \frac{1 - \overline{\delta}\,\overline{\Delta}}{\overline{\delta} + \overline{\Delta}} \qquad (7.9)$$

which means that, in the weak coupling limit (μ_F^{-1} was assumed to be small at the outset), q is determined entirely by the shape resonance and background phase shifts. The overall situation is illustrated in Fig.12. We note that $q = 0$ when $\Delta = 1/\overline{\delta}$, which is also the condition for a maximum in the first factor of (7.7) and occurs when

$$(E_{BF} - E) = \frac{\Gamma_{BF}}{2} \tan\delta \qquad (7.10)$$

i.e. is slightly detuned from the resonance energy E_{BF} for small values of the background phase shift. Thus, $q \to 0$ (window resonance) at the peak intensity of the giant resonance; about this value, Γ, and therefore q change sign.

We can also deduce the widths of Rydberg members by using $\Gamma_{nF} \alpha (D^2 - c^2)^{\frac{1}{2}}$ or more directly from equation (7.4a)

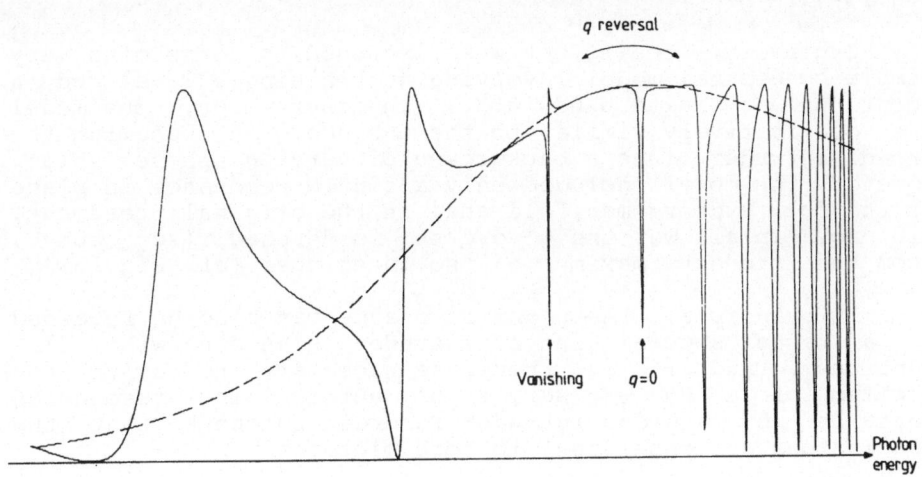

Fig.12 Calculation of the cross section of a Rydberg series interacting with a giant resonance in the limit of weak coupling (full curve). For vanishingly small μ_F^{-1}, the zero width coincides with E_B, but not with the maximum in the 'giant resonance' envelope, because a finite background phase shift was assumed. However, the $q \to 0$ window resonance and associated q-reversal do coincide with the maximum in the envelope of the giant resonance (dashed curve) as required by (7.10)

above. Either way, the width tends to zero (a vanishing width occurs) when the detuning

$$\varepsilon = (E - E_{BF}) = - H_{nF}\Gamma_{BF}^{\frac{1}{2}} / \Gamma_{nF}^{\frac{1}{2}} = -\mu_F^{-1}\Gamma_{BF}^{\frac{1}{2}} \qquad (7.11)$$

irrespective of the value of n. Both effects are clearly visible in Fig.12.

8. Perturbation by a Broad Intruder Level and the q-reversal Effect: General S-Matrix Formula

The change in sign of q discussed in the previous section is called a q-reversal (21) and is a much more general effect than the treatment just given implies. In fact, the ´q-reversal effect was discovered experimentally for atoms in a situation involving a Rydberg series of autoionising resonances traversing a very broad intruder, so broad indeed that it encompasses most of the series, including all of the higher members (see the data in Fig.13a).

Orginally, the effect was discussed in terms of a very simple conceptual model involving just a single level and a ´continuum of finite bandwidth.´ In other words, the model was qualitatively similar to the one above in subsuming the broad intruder into a background of varying phase shift. However, the model included only a single resonance in place of the full Rydberg manifold and, in the original treatment, did not explain why the q-reversal is detuned (see above) from the intensity maximum of the interloper (cf. Fig.13a).

In principle, the giant resonance can also be regarded as a rather special kind of intruder in an otherwise flat continuum, and it is therefore necessary to solve the problem of a Rydberg series of autoionising resonances perturbed by a broad intruder in order to check that the conclusions are consistent in both pictures.

The problem of a Rydberg series with a broad intruder was addressed by Lane (22) whose results have been related in detail to the q-reversal problem by Connerade Lane and Baig (23).

When an extra level B is added to a series of levels labelled by n, then the K matrix has the form

$$K_{ab} = \bar{K}_{ab} + \tfrac{1}{2} \sum_n \frac{\Gamma_{na}^{\frac{1}{2}} \Gamma_{nb}^{\frac{1}{2}}}{E_n - E} + \tfrac{1}{2} \frac{\hat{\Gamma}_{Ba}^{\frac{1}{2}} \hat{\Gamma}_{Bb}^{\frac{1}{2}}}{E_B - E - \sum_n \frac{H_n^2}{E_n - E}} \qquad (8.1)$$

where

$$\hat{\Gamma}_{Ba}^{\frac{1}{2}} = \Gamma_{Ba}^{\frac{1}{2}} - \sum_n \frac{H_n \Gamma_{na}^{\frac{1}{2}}}{E_n - E} \qquad (8.2)$$

the quantities Γ_{na} and Γ_{Ba} are the K matrix widths of levels n and B for the channel a, other quantities are as before and $\bar{K}_{ab}$ is a background constant.

For a general set of levels n, the inversion $(1 - K)^{-1}$ cannot be performed explicitly to calculate the cross sections. Atoms are interesting from this standpoint because the ratio $\mu_a \equiv (\Gamma_{na}^{\frac{1}{2}}/H_n)$ is independent of n: both H_n and $\Gamma_{na}^{\frac{1}{2}}$ are proportional to $(n*)^{-3/2}$ in a coulomb field, and it turns out that an exact result can be written down (23).

We can now treat the full problem involving two open channels, a strongly coupled particle channel a and a weakly coupled radiative channel γ, so that $K_{a\gamma}$ and $K_{\gamma\gamma}$ are of first and second order of smallness respectively. The exact result can then be specialised to yield the S matrix:

$$S_{a\gamma} = \frac{(2\,\overline{K}_{a\gamma}\varepsilon - \Gamma_{Ba}^{\frac{1}{2}}\Gamma_{B\gamma}^{\frac{1}{2}}) + (2\,\overline{K}_{a\gamma} + \mu_a\mu_\gamma\varepsilon + \Gamma_{Ba}^{\frac{1}{2}}\mu_\gamma + \Gamma_{B\gamma}^{\frac{1}{2}}\mu_a)\Sigma}{(i + \overline{K}_{aa})\varepsilon - \frac{1}{2}\Gamma_{Ba} + (i + \overline{K}_{aa} + \mu_a\Gamma_{Ba}^{\frac{1}{2}} + \frac{1}{2}\mu_a^2\varepsilon)\Sigma} \tag{8.3}$$

where $\varepsilon = (E - E_B)$ is the detuning from the broad level E_B . It is possible to redefine the quantities so that $\overline{K}_{aa}$ is absorbed, as in Fano's parametrisation, in which the expression for $S_{a\gamma}$ does not include $\overline{K}_{aa}$. One then has six real parameters, namely E_B , $\Gamma_{Ba}^{\frac{1}{2}}$, $\Gamma_{B\gamma}^{\frac{1}{2}}$, μ_a , μ_γ , $\overline{K}_{a\gamma}$ to characterise the resonances, the intruder and the interaction between them.

If the radiative widths $\Gamma_{B\gamma}^{\frac{1}{2}}$ and μ_γ are very small and the background constant $\overline{K}_{a\gamma}$ is neglected, then the formula simplifies to the following useful expression

$$\sigma(\varepsilon) = \frac{(\varepsilon + \frac{q\Gamma_B}{2})^2}{(\varepsilon + \Sigma)^2 + \{(\mu_a\Gamma_B^{\frac{1}{2}} + \frac{\mu_a^2}{2}\varepsilon)\Sigma - \frac{\Gamma_B}{2}\}^2} \tag{8.4}$$

which is closely related to the standard Fano formula for an isolated level. In fact, the numerator is the same as in the Fano formula for the resonance E_B, Γ_B, while an additional Rydberg series of resonances is introduced through the presence of Σ in the denominator.

From the simplified formula (8.4), one can calculate many practical cases for which radiative widths do not appear to be important. Thus, Fig.13b shows a ´skewed´ q-reversal effect similar to the one in Fig.13a, while Figs.14 shows enhancements of upper series members in observed and calculated spectra, and Figs.16 and 15 show a q reversal straddling one resonance of $q \sim 0$ in both calculated and experimental spectra. Fig.17 shows a strongly asymmetric perturber inducing a q reversal.

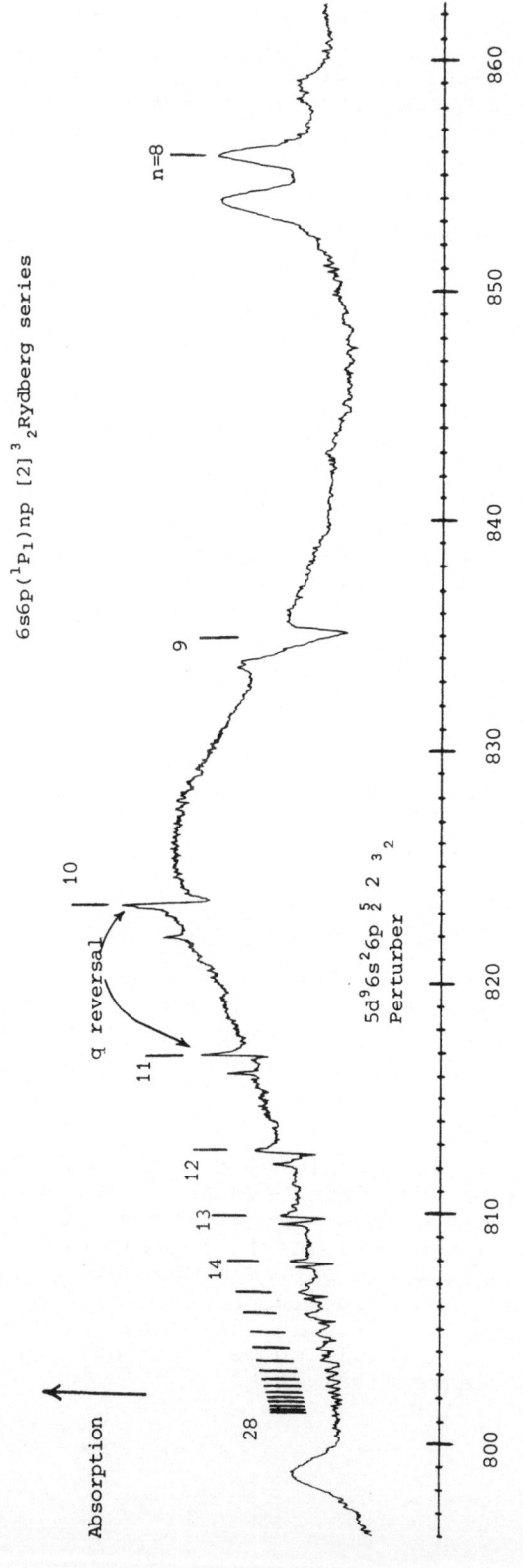

Fig. 13(a). Experimental example of the q-reversal effect observed in the photoabsorption spectrum of TII. Notice that the q reversal, in this case, does not coincide with the maximum of the broad perturber but is 'skewed'.

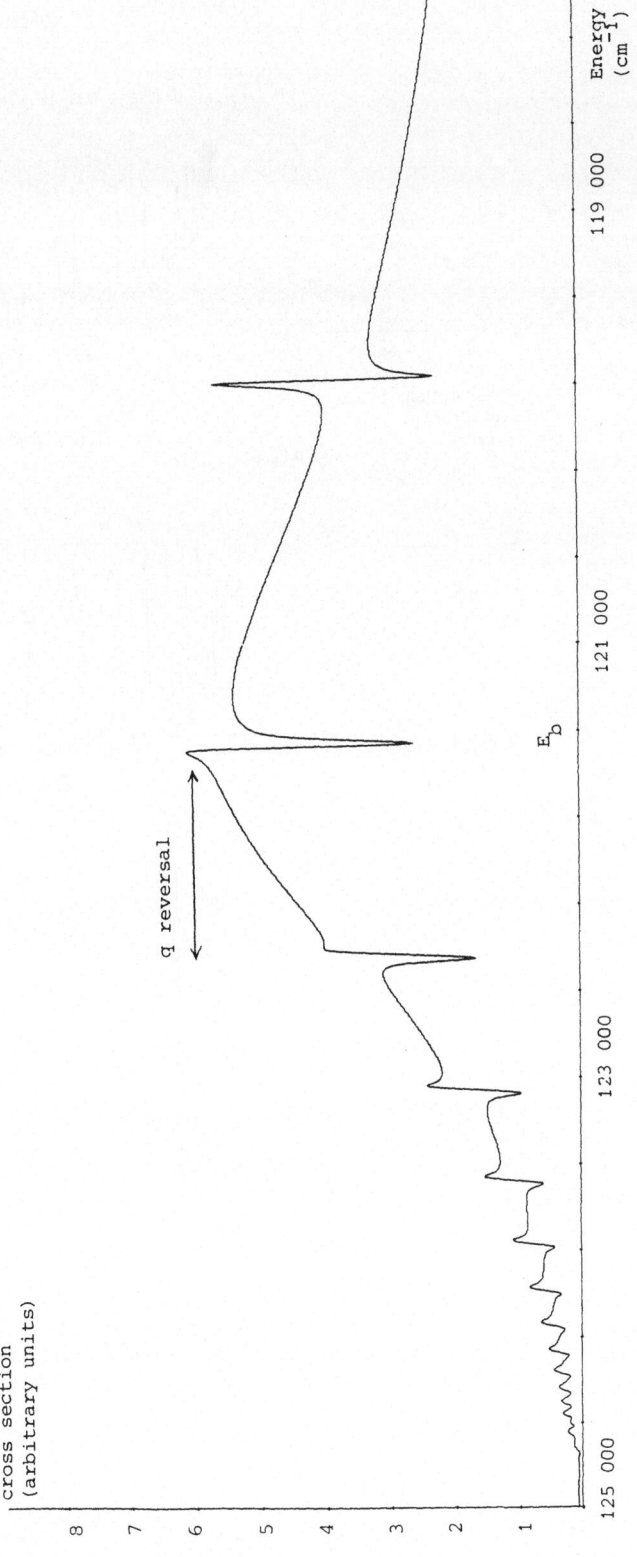

Fig. 13(b). A rough simulation of the 'skewed q-reversal' effect based on equation (8.4) - note that this is not a fit, merely an illustrative calculation.

589

Fig. 14. Example of the enhancement of upper series members which can occur as the result of a perturbation by a broad intruder. The upper curve shows an observed case, in the spectrum of PbI, while the lower curve is an illustrative calculation, based on equation (8.3).

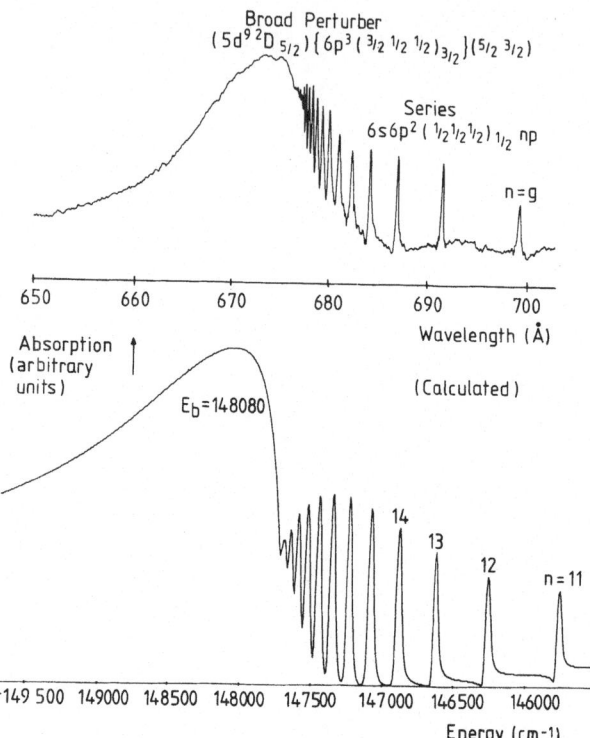

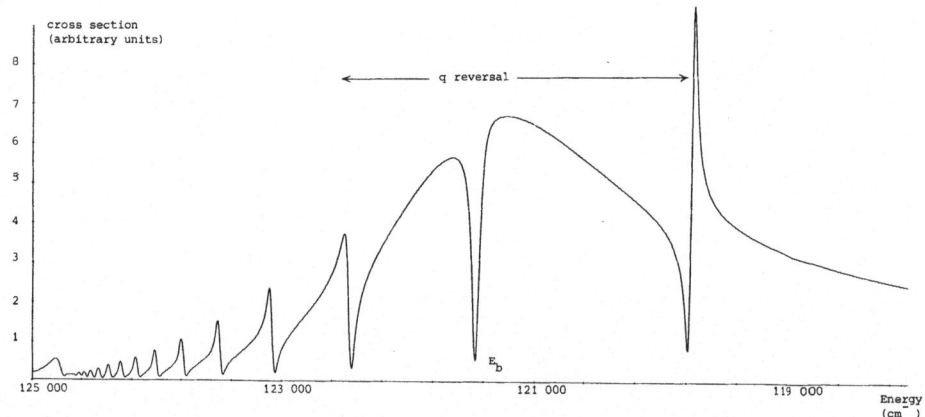

Fig. 15. Curve obtained by increasing the coupling strength above the value used to calculate Fig. 13(b) while keeping the other parameters unchanged. Notice how the q-reversal 'hops' over an intervening Rydberg series member with q=0. Such an effect is visible in the experimental spectrum of Fig. 16.

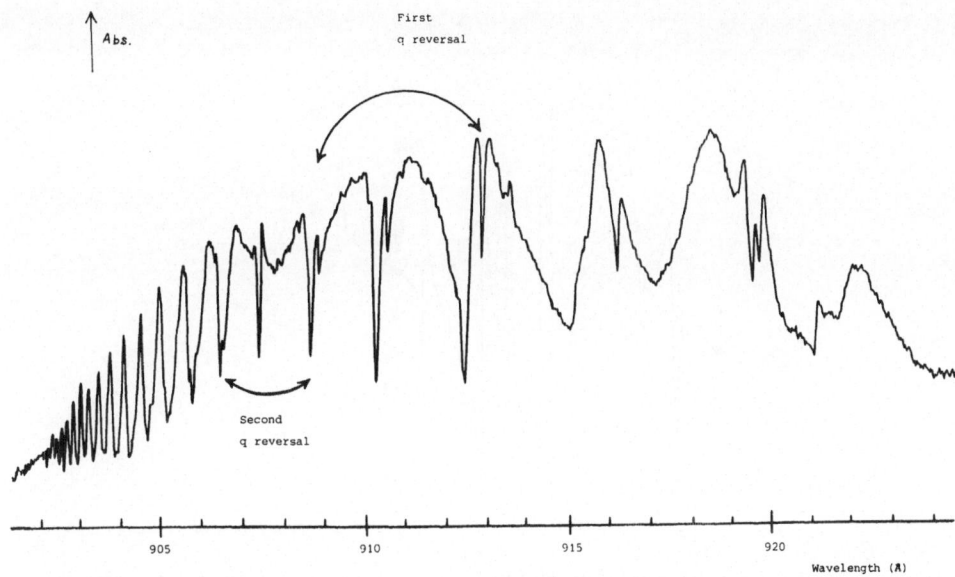

Fig.16. Example of a series of autoionising resonances in the spectrum of Tl I showing several additional effects which do not occur in the series of Fig.13 (a). Notice that this series exhibits two q reversals. The first of these is not between successive members, but hops over a nearly symmetric resonance, an effect which appears (see Fig.15) in the calculations using the parameters of Fig.13 (b) if the coupling strength is increased. The second q-reversal occurs in a similar fashion, but straddles a very narrow structure, above which the profiles change, apparently to a more sinusoidal shape.

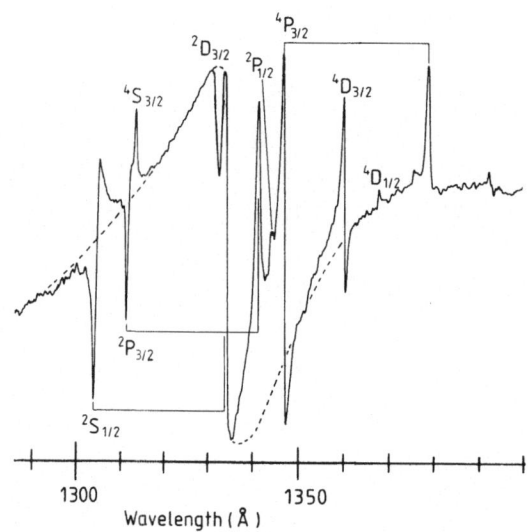

Fig.17. Experimental example of a q-reversal in a case in which the broad intruder state is itself strongly asymmetric. This example occurs in the spectrum of In I. The excited configuration is 5s 5p 6p, to which all the resonances belong. However, the $^2D_{3/2}$ resonance is much broader than all the others, and is apparently responsible for the change in symmetry between $^4D_{3/2}$ and $^4S_{3/2}$ visible in the figure.

A significant conclusion from the numerical studies in (23) was that q reversals are the hallmark of weak coupling, i.e. they tend to disappear as the interaction strength H_n is increased. A question which therefore arises is how many q-reversals one expects to find in a Rydberg series and the condition under which they may occur.

9. Variation of the Shape Parameter in a Rydberg Series of Autoionising Lines Perturbed by an Interloper Level

The spectrum of 16 indicates that, in experimental data, more than one q-reversal can occur in a given Rydberg series. General expressions have been derived within the K-matrix approach for the variation of q as a function of the detuning from the energy of the intruder, both in elastic scattering and in photoionisation (24). It turns out that as many as six q reversals can occur in the general case in photoionisation but that, in elastic scattering, there are at most two.

The K-matrix approach is a rather natural framework in which to study the behaviour of the q parameter, as evidenced by the fact that, in the case of a single channel $-q^{-1}$ is equal to the background K matrix.

For a single channel, the essential result is that

$$-q_a^{-1} = \overline{K}_{aa} + \mu_a(\Gamma_a^{\frac{1}{2}} + \frac{1}{2} \varepsilon\mu_a) - \frac{\frac{1}{2}(\varepsilon+\Sigma_\infty)(\Gamma_a^{\frac{1}{2}} +\varepsilon\mu_a)}{(\varepsilon+\Sigma_\infty)^2 + (\pi t)^2} \qquad (9.1)$$

for elastic scattering. In this expression, if

$$\Sigma(E) = \Sigma_n \frac{H_n^2}{E_n-E} , \qquad (9.2)$$

we write $\Sigma(E + i\delta) = \Sigma_\infty + i\pi t$ and $t = (H_n^2/ D_n)$ is independent of n, where D_n is the level spacing. Typical results for the variation of q as a function of energy and for coupling strengths ranging from weak to strong coupling are plotted in figure 18. These curves reveal the following characteristics:

(i)   The true form of the variation of q is not a Fano profile as some authors have inferred, but exhibits a pole at each q-reversal.

(ii)   All curves pass through the background point $q = q_B$ at $E = E_B$ (with $\Sigma_\infty = 0$); however, this is not necessarily observable unless there happens to be a fine level at $E = E_B$

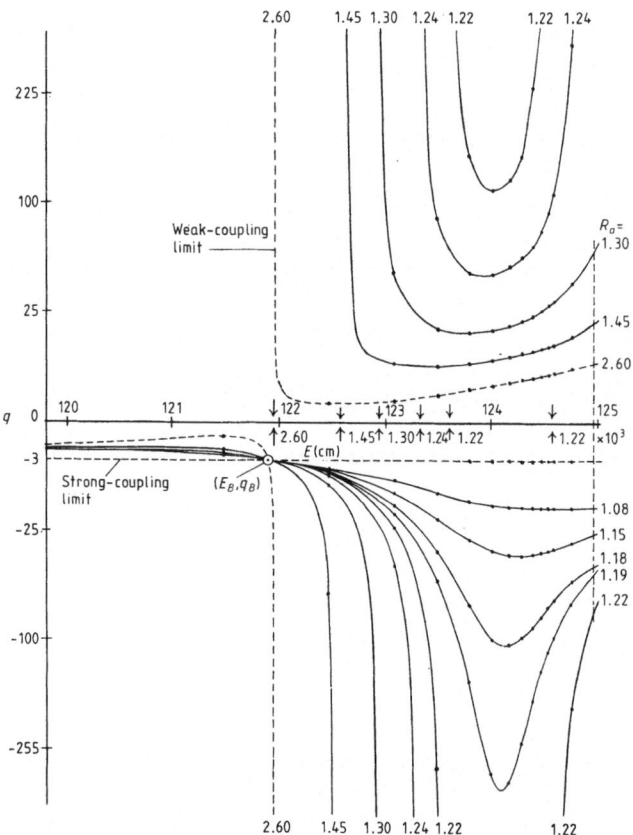

Fig.18. The variation in the shape index q_a for a parameter selection which corresponds to Figs.13 (b) and 15 (see also the caption to Fig.16) for different values of the coupling strength (parametrised by R_a - see ref.24 for details). In the weak coupling limit, only one q reversal is apparent and stays close to E_B. In the strong coupling limit, there is no q-reversal and the q of the series members become equal to the q of the background. For intermediate cases, the evolution is described in the text, and can involve two q-reversals.

(iii)   All q reversals disappear at large mixing
          strength, even when there is still appreciable
          variation in the value of q_a. The sign of q_a is
          the same as that of q_B. In the strong coupling
          limit, $q_a = q$ and so is constant.

(iv)   With the parameters selected, curves at weak
          mixing exhibit only one q-reversal within the
          energy range of the Rydberg series. However,
          within a narrow range of coupling strengths, two
          q-reversals occur. One, which can be regarded as
          the main one, is quite stable in energy and occurs
          close to E_B. It is always offset from the energy
          of the intruding level, moving closer to it as the
          mixing strength decreases, and therefore it
          remains in the range of the series over quite a
          wide range of parameters which we can define as
          the weak mixing range. The other pole occurs
          above the series limit in energy for much of the
          weak mixing range, but then travels very fast in
          energy as the mixing strength is varied and is
          actually visible only over a very narrow range of
          mixing strengths in the example shown.

(v)   The second q reversal appears at an interaction
          strength close to which there is a sudden switch
          between weak and strong mixing conditions.

For two channels γ and a, one of which (γ) is weakly
coupled and is the radiative channel (treated to leading
order), the situation is more complex. One can define a
quantity Q_γ which is a ratio of quantities linear in ε, and
in terms of which

$$\cot^{-1}q_\gamma = \cot^{-1}q_a - \cot^{-1}Q_\gamma \qquad (9.3)$$

which is a remarkably simple result connecting q_γ to q_a.
The form of the variation in q_γ is generally more
complicated, but it possesses a definite pole at $E = -\mu_a^{-1}\Gamma_{Ba}^{\frac{1}{2}}$
(cf 7.11)

10. The Vanishing Width Effect

As noted in section 7, a very narrow line occurs at a
detuning given by $\varepsilon = -\mu_a^{-1}\Gamma_{Ba}^{\frac{1}{2}}$ (cf 7.11) for a Rydberg
series of autoionising resonances interacting with a giant
resonance. The same result is readily obtained if the giant
resonance is treated explicitly as an intruder. For a
general intruder, two quantities are of interest in defining
the level sequence, namely

(i) the shape parameter

$$-q^{-1} = \text{Re } K (E + ie) \qquad (10.1)$$

(ii)   the strength function

$$\frac{\pi\Gamma}{2D} = \frac{\text{Im } K(E + ie)}{1 + \{\text{Re } K(E + ie)\}^2} \qquad (10.2)$$

where D is the average level spacing. The defining property of a giant resonance in this context is its large width
One can show (24) that

$$- q^{-1} = \tan\tilde{\Delta} \qquad (10.3)$$

where $\tilde{\Delta}$ is the total background phase shift including the giant resonance, so that this result is equivalent to (7.9) and also that the strength function:

$$\frac{\pi\Gamma}{2D} = \frac{\frac{1}{2}\pi t \ (\Gamma_{BF}^{\frac{1}{2}} + \varepsilon\mu)^2}{\Sigma_\infty^2 + (\overline{K}_{aa}\varepsilon - \frac{1}{2}\Gamma_{BF})^2} \qquad (10.4)$$

has a vanishing point at $\varepsilon = -\mu_F^{-1}\Gamma_{BF}^{\frac{1}{2}}$, which implies that the particle width vanishes.

The vanishing of particle widths of resonances as a result of an intruder state disturbing a sequence of levels is a well known effect in nuclear physics (25).   In atomic physics, under suitable conditions, either the radiative width (26) or the particle width (27) may go to zero. Several good examples of the latter instance have recently come to light. A particularly clear one is the perturbation of the 3dnp ´P, series by the interloper 4p5s ´P, recently analysed by Greene and Kim (28).

In Fig 19, we show some very recent date for this series which we have obtained using the wiggler beam line at DORIS II and the facilities at HASYLAB.   This spectrum is a remarkably good example of line narrowing within a Rydberg series of autoionising resonances.   Interestingly, the perturber is almost imperceptible otherwise than through the dramatic series perturbation for which it is responsible.

It is worth noting here that, whereas the data of Fig.2 were obtained by measuring on ion current in a ´hot wire´ detector, the data in Figs.3-17 were obtained in photo absorption using synchrotron radiation.   The data of Fig.19 appear to be the first ever obtained by the ´hot wire´ detection technique in combination with synchrotron radiation.   For further comments on the data, see the lectures by Dr Greene in the present volume.

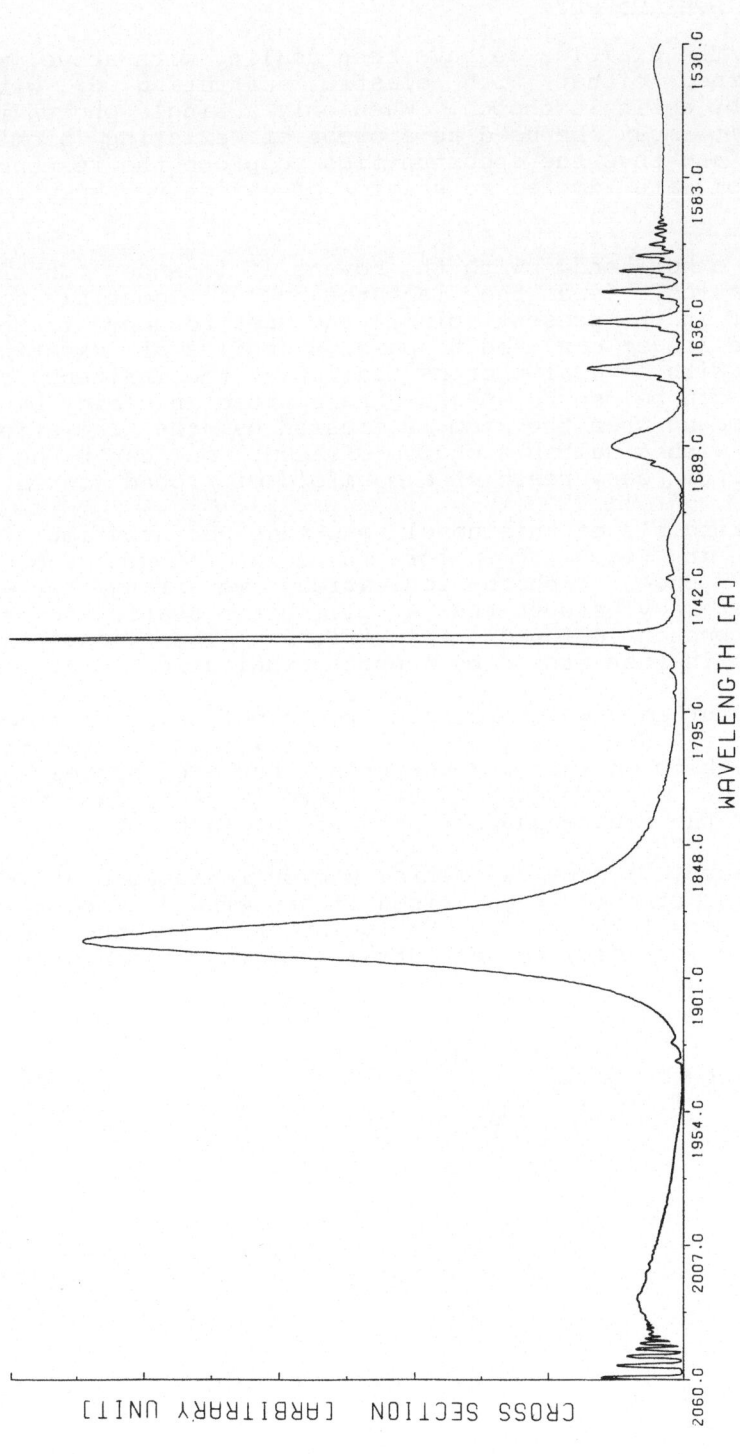

Fig. 19: Date on the 3dnp series of Ca I obtained at the wiggler port of the DORIS storage ring (HASYLAB facility) using the 'hot wire' detection technique. The 3dnp series is perturbed by a broad intruder, and the 6p member of the series exhibits the vanishing width effect discussed in the text. The peak of the first broad strong resonance above the ionization potential corresponds to a cross section of 50 Mb and the ordinate is linear in cross section above the ionization threshold. The perturber lies between 1689–1742 Å.

11. Perturbations and q-reversals in Resonant Multiphoton Ionisation Spectroscopy

In all the examples we have been dealing with above, we were concerned either with elastic scattering or with excitation by a single photon. When only a single photon is involved, it must be used as a probe of existing atomic structure, and thus the opportunities to probe the response of the atom are limited to studies of existing spectral structure.

On the other hand, with the advent of intense tunable laser radiation, (see the lectures by Drs Agostini and Giusti-Suzor in the present volume) the spectroscopy of the atom is no longer confined to passive studies of existing structure. With several photons available, the influence of a perturber can be, so to speak, transported in energy by a discrete amount when the atom is dressed by the radiation field and, with a suitable choice of atom, one can bring a perturber within easy reach of a manifold of probed states.

The principle of this novel man-made perturbation is illustrated in Fig. 20 for three and four photon ionisation. Under n-photon ionisation, we arrange for an energy degeneracy between the (n - 1) photon states and an (n - 2) photon allowed transition. The resulting interference is then probed by n photon ionisation.

The disturbance which results from superposing this new perturber on a Rydberg manifold is extremely similar to the manner in which an intruder perturbs a Rydberg series in single photon spectroscopy, but there is now far more control over the nature and strength of the interaction.

For example, consider three photon ionisation of KI vapour using circularly polarised light with $\hbar \omega$ between 12500 cm^{-1} and 16300 cm^{-1}. To lowest order in the atom laser interaction $e\mathcal{E}_z$, the essential term in the ionisation amplitude is (30):

$$T(\omega) = (e\mathcal{E})^3 \sum_{n_1,n_2} \frac{\langle \varepsilon|z|n_2 \rangle \langle n_2|z|n_1 \rangle \langle n_1|z|0 \rangle}{(E_{n_2} - E_0 - 2\hbar\omega)(E_{n_1} - E_0 - \hbar\omega)} \qquad (11.1)$$

where $|0\rangle$ is the s-ground state, $|\varepsilon\rangle$ is the ionisation state whose energy ($\varepsilon - E_0$) equals $3\hbar\omega$ and n_1, n_2 are the p, d states respectively excited by one and two photon absorption.

We follow Connerade and Lane (30), chosing $\hbar \omega$ to lie near a p state $n_1 = N$ and $2\hbar\omega$ near a d state $n_2 = n$ and then rewrite $T(\omega)$ to display these states explicitly:

$$T(\omega) = (e\mathcal{E})^3 <\varepsilon|z| \{ \frac{|n><n|}{2\hbar\omega-(E_n-\varepsilon_o)} + B_2 \} \ z \ \{ \frac{|N><N|}{\hbar\omega-(E_N-E_o)} + B_1 \} \ z|0>$$

where the background operators (11.2)

$$B_1 = \sum_{n_1 \neq N} \frac{|n_1><n_1|}{\hbar\omega - (E_{n_1}-E_o)}$$

(11.3)

$$B_2 = \sum_{n_2 \neq n} \frac{|n_2><n_2|}{2\hbar\omega - (E_{n_2}-E_o)}$$

If the detunings

$$\Delta_N \equiv \hbar\omega - (E_N-E_o) \tag{11.4}$$

$$\Delta_n \equiv 2\hbar\omega - (E_n - E_o)$$

are introduced, the ionisation yield has the form

$$W(\omega) = <\varepsilon|zB_2zB_1z|0>^2 \left(\frac{\Delta_n + Q_n}{\Delta_n} \right)^2 \left(\frac{\Delta_N + Q_N}{\Delta_N} \right)^2 \tag{11.5}$$

where Q_n, Q_N are generalised asymmetry parameters, which have dimensions of energy.

These expressions can be generalised explicitly to include a summation over many levels of the Rydberg manifold demoted by n. To display the energy variation more directly, we set $E_o = 0$, $E = \hbar\omega$ and write E_∞ for the series limit. This leads to

$$T(E) = \left(1 + \frac{Q_N}{\Delta_N} \right) \{ 1 + \frac{\left(1 + \frac{Z}{\Delta_N} \right)}{\left(1 + \frac{Q_N}{\Delta_N} \right)} \frac{Q_{n_o}^o}{\left(E_\infty - E_{n_o} \right)^{\frac{3}{2}}} \sum_{n=n_o}^{n_o} \frac{\left(E_\infty - E_n \right)^{\frac{3}{2}}}{2E-E_n} \tag{11.6}$$

where $Z = \frac{<n|z|N><N|z|0>}{<n|z \ B_1 \ z|0>}$ (11.7)

A typical calculation based on (11.6) is shown in Fig.21, with parameters appropriate for 3 photon ionisation of KI as studie by Rahman-Attia etal (31). Note the Q_n-reversal effect due to the influence of the perturber, transported by the laser field into the energy range of the Rydberg manifold.

This last example opens up a host of new possibilities for the studies of series perturbations, which are only just beginning to be realised.

12. Conclusion

The subject of interacting resonances, which we have addressed in the main part of the present lectures, is a lively field of contemporary atomic spectroscopy, whose importance is likely to increase over coming years through the developments in the adjacent areas of laser - implanted resonances and continuum - continuum transitions.

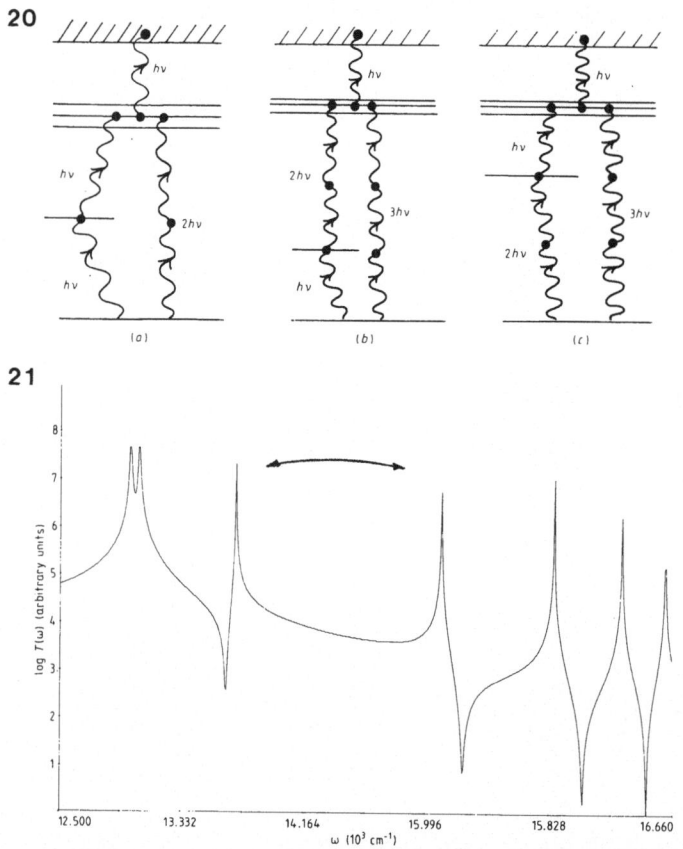

Fig. 20. Showing various schemes of (a) 3-and (b,c) 4-photon REMPI, any one of which can give rise to Q-reversals.

Fig. 21. The three photon ionisation of KI as calculated from (11.6) including nine Rydberg members. Note the Q reversal effect discussed in the text.(For further details see ref.30.)

References

(1)   CONNERADE J.P.   1978 Nuclear Instruments and Methods 152, 271

(2)   LANE A.M. and THOMAS R.G.
      1958 Rev. Mod. Phys. $\underline{30}$, 257 and

      LANE A.M.
      1984   J. Phys. B. At. Mol. Phys. $\underline{17}$, 2213
      1985   J. Phys. B. At. Mol. Phys. $\underline{18}$, 2339

(3)   FANO U.
      1961 Phys. Rev. $\underline{124}$, 1866

(4)   SHORE B.W.
      1967 Rev. Mod. Phys. $\underline{39}$, 440

(5)   J.P. CONNERADE and A.M. LANE
      1987 J. Phys. B. At. Mol. Phys $\underline{20}$, 1757 (Appendix)

(6)   LANE A.M.
      1986 J. Phys. B. At. Mol. Phys. $\underline{19}$, 253

(7)   LANDAU L.P. and LIFSHITZ E.M.
      1965   "Quantum Mechanics" (Reading USA; Addison-Wesley)p514

(8)   "Giant Resonances in Atoms Molecules and Solids"
      (Edited by J.P. CONNERADE J.M. ESTEVA and R.C. KARNATAK)
      1987 Plenum Press New York NATO ASI series.

(9)   Griffin D.C. Andrew KL and Cowan RD
      1969 Phys. Rev 177, 62
      1971 Phys Rev A3, 1233

(10) CONNERADE J.P.
      1978 Contemp Phys. 19, 415
      1984 Les Houches Session XXXVIII New Trends in Atomic Physics Vol 2 p643 Elsevier Science Amsterdam.

(11) MOTT N.F. and MASSEY H.S.W.
      1965 The Theory of Atomic Collisions
      3rd Edition (Oxford: Clarendon Press)

(12) CONNERADE J.P.
      1984 J. Phys. B. AT. Mol. Phys. $\underline{17}$, L165

(13) HEINZMANN V. HEUER H. AND KESSLER J.
      1976 Phys. Rev Lett $\underline{36}$, 1444

(14) CONNERADE J.P.
      1983 J. Phys. B. At. Mol. Phys. $\underline{16}$, L329

(15) CONNERADE J.P.
     1985   J. Phys. B. At. Mol. Phys. $\underline{18}$, L367

(16) DUBAU J. and SEATON M.J.
     1984   j. Phys. B. At. Mol. Phys. $\underline{17}$, 381

(17) GIUSTI-SUZOR A. and FANO U.
     1984   J. Phys. B At. Mol. Phys. $\underline{17}$, 215

(18) CONNERADE J.P. LANE A.M. and BAIG M.A.
     1985   J. Phys. B. At Mol. Phys. $\underline{18}$, 3507 (Appendix)

(19) e.g. CONNERADE J.P. and LANE A.M.
     1987   J. Phys. B. At. Mol. Phys. $\underline{20}$, L181
     or else ref.18 above

(20) CONNERADE J.P. and LANE A.M.
     1987   J. Phys. B At. Mol. Phys. $\underline{20}$, L181

(21) CONNERADE J.P.
     1978   Proc. Roy Soc. $\underline{A362}$, 361

(22) LANE A.M.
     1985 J. Phys. B At. Mol. Phys. $\underline{17}$, 2213

(23) CONNERADE J.P. LANE A.M. and BAIG M.A.
     1985   J. Phys. B. At. Mol. Phys $\underline{18}$, 3507

(24) CONNERADE J.P. and LANE A.M.
     1987 J. Phys. B. At. Mol. Phys. $\underline{20}$, 1757

(25) CONNERADE J.P. and LANE A.M.
     1985 J. Phys. B. At. Mol. Phys. $\underline{18}$, L605

(26) GARTON W.R.S.
     1965   Scientific Report Harvard College Observatory
     Cambridge MA

(27) RINNEBERG H.   JONSSON G. NEUKAMMER J. VIETZKE K.
     HIERONYMUS H. KONIG G. AND COOKE W.E.
     1985 Proc. 2nd. E. C.A.M.P. (Amsterdam) unpublished

(28) GREENE C.H. and KIM. L
     1987 Phys. Rev. (in the Press)

(29) GRIESMANN U., SHEN-NING, CONNERADE J.P.
     SOMMER K. and HORMES J.
     1987 J. Phys. B. At. Mol. Phys. $\underline{20}$, L363

(30) CONNERADE J.P. and LANE M.A.
     1987 J. Phys. B. At. Mol. Phys. $\underline{20}$, L363

(31) RAHMAN-ATTIA M. LAPLANCHE G. JAOUEN M.and RACHMAN A.
     1986 J. Phys. B. At. Mol. Phys. $\underline{19}$, 3669

CLUSTERS, OR THE TRANSITION FROM MOLECULAR TO CONDENSED MATTER PHYSICS

Hellmut Haberland

Fakultät für Physik
Universität Freiburg
Freiburg, Germany F.R.

INTRODUCTION

Scientists speak of clusters in very many different contexts indeed: structures in atomic nuclei, in liquid and solid materials, and in gases have been labeled with this title. Astronomers talk of clusters of stars and even clusters of galaxies, a computer company sells a VAX-cluster, there are musical clusters/*/, and so on....Obviously there exists something common among these widely different subjects. The Concise Oxford Dictionary defines a cluster as a "group of similar things".

In this lecture we will discuss clusters composed of atoms and molecules. By letting the clusters grow bigger and bigger, the transition from atomic or molecular physics and chemistry to the science of condensed matter will be traced. Cluster science bridges the gap between these traditionally separated fields, and it will be seen below that concepts and techniques from both fields are necessary to understand theories and experiments on clusters.

The outline of this lecture is as follows. After a brief introduction it will be discussed how some properties change with cluster size. In chapter 3 cluster production and detection is treated, and case studies of special types of clusters are given in chapter 4. This lecture is introductory by intent, more advanced and specialized reviews can be found under ref.1. Emphasis is laid on general concepts and results and on experimental techniques to obtain them. Readers with more interest in theoretical treatments will find them under ref.1, too. In keeping with the format of a school no attempt is made to cite the original papers. Rather readers are referred to review articles or the latest articles in the field, from which the older ones can be traced.

This lecture and the one given by J.Andrä generalizes the ideas presented by other authors in this volume to more complex situations. The lectures on photoionisation given by J.Dehmer, on electron scattering by W.Raith, and on Multi-Channel-Quantum-Defect Theory by R.Rau have a direct bearing on this subject.

1.1 Surface to volume ratio

One peculiar aspect of cluster physics is the large surface to volume ratio. If one takes r as the (van der Waals) radius of an atom and R the

---

*) A musical cluster is a generalisation of a triple accord.

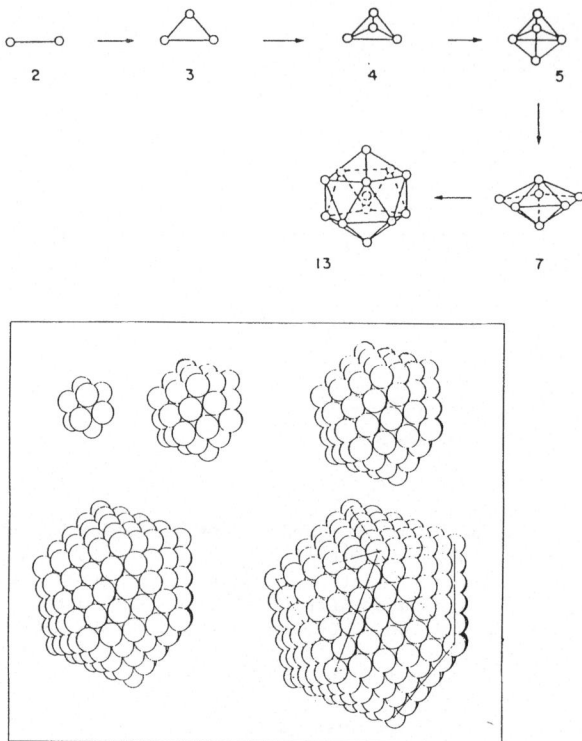

Fig.1. Growth of small(a) and large(b) rare gas clusters. The five fold symmetry, which is forbidden in the bulk appears for the first time for N=7, and is evident for the larger icosahedral clusters of (b). The nearest neighbour distances at the edges of the icosahedra are stretched compared to those inside the clusters. This leads to mechanical stresses, which increase with cluster size. If the stress becomes too large the fcc bulk structure becomes favoured (N>700).After Hoare/1i/ and Jortner/2/.

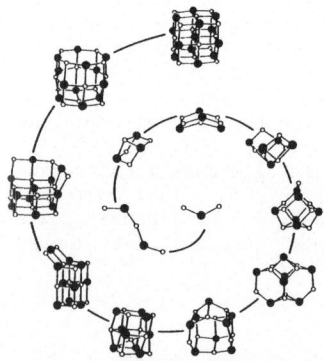

Fig.2. Calculated configurations for $Na^+(NaCl)$ clusters. Again crystalline structures are not preferred during the early stages of cluster growth. After Martin/4/.

radius of a spherical cluster, than the volume is given by $V = \frac{4}{3} \pi R^3$.
Neglecting all geometrical and packing effects, one can write

$$V = \frac{4}{3}\pi R^3 = N(\frac{4}{3}\pi r^3), \quad N = (\frac{R^3}{r})$$

where N is the number of atoms in the cluster. The next question one can ask
is: What is the number N_s of atoms sitting on the surface? Using the same
argument one has:

$$S = 4\pi R^2 = N_s \frac{1}{4}(4\pi r^2); \quad N_s = 4(\frac{R^2}{r})$$

The factor 1/4 accounts for the fact, that only about 1/4 of the surface of
an atom is on the surface of the cluster. This factor agrees reasonably with
computer simulations /1i/. The relative number of surface atoms is then:

$$N_s/N = 4\ (\frac{r}{R})$$

Table 1 lists some values obtained from these formulae assuming r=2Å, which
is valid for the case of krypton atoms. Other elements can have different
r-values, of course. Note the extremely large ratio of atoms sitting on the
surface of a cluster.

Table 1. A cluster of radius R contains
about N atoms, N_s of which are on the surface.
The values have been calculated for r=2Å.

R(Å)	N	N_s/N
10	125	0.8
20	10^3	0.4
10^2	10^5	0.08
10^3	10^8	0.008
10^8	10^{23}	10^{-7}

1.2 Classification

A large number of different words have been suggested and used for
clusters of various sizes. A classification which goes basically back to
Jortner /2/ is the following:
1. Microclusters, N=2-10 or 13. All atoms for N<13 are on the surface.
    Molecular concepts are useful to describe the cluster.
2. Small clusters, N = 10 or 13 to 100. Many isomeric structures
    exist. Molecular concepts cease to be useful.
3. Large clusters, 100 < N < 1000. A gradual transition to properties
    of the solid state occurs.
4. Small particles or microcrystals, N > 1000. Many but not all
    solid state properties are developed.

1.3 The growth of a solid

In the standard description of solid state physics a crystal is
described as an ordered, infinite array of atoms or molecules possessing
translational symmetry. This leads to the well known crystalline structures
as bcc (body centered cubic), fcc (face centered cubic), or hcp (hexagonal
closed packed), to name but the three most important. The problem how these
crystalline structures are obtained starting from the atom, dimer, trimer...
has been much studied. Experimental evidence is scarce. The electron

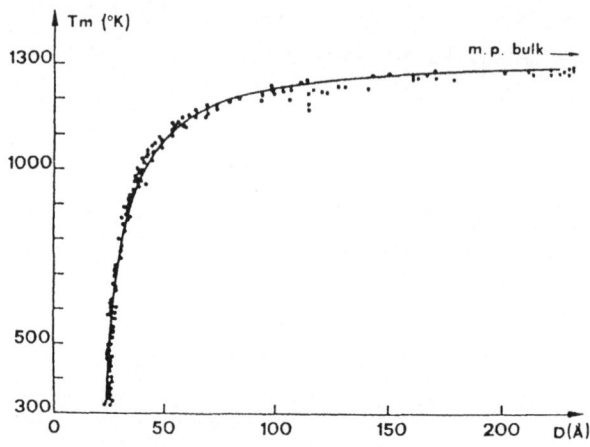

Fig.3. Experimental and theoretical values of the melting-point of large gold clusters. After Buffat and Borel/6/.

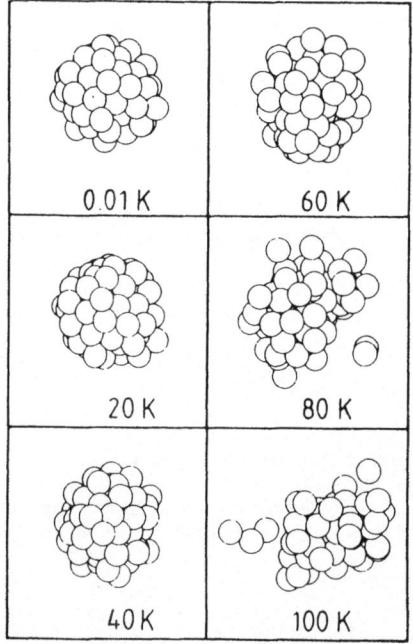

Fig. 4. Calculated melting and evaporating of argon clusters. After Abraham/12/.

diffraction experiments on argon by the Farges group /3a/ give some information. From their diffraction patterns they deduce that small clusters are amorphous, larger ones (N>100) have icosahedral structures. The transition to the fcc bulk structure occurs around N = 700 ± 150, where the error is due to the estimate of the size distribution. More often the transition to the bulk has been studied theoretically. An interaction potential is assumed (nearly always of the 2-body type) and the potential energy is minimized until the most favourable geometry is found. For the rare gases, shown in fig.1a, N = 3 corresponds to an equilateral triangle, N = 4 to a tetrahedron, N = 5 to triangular bipyramid. For N = 7 one has for the first time a pentagonal (five-fold) symmetry, a pentagonal pyramid which consists of a five membered ring and one atom above and below it. Pentagonal symmetry is forbidden for the standard infinite lattice of solid state physics, but is ubiquitous in cluster physics. For N = 13 one has the famous icosahedron, an interior atom capped by two pentagonal caps. For larger clusters, atoms are added to the faces of the icosahedron until the next complete shell is obtained for N = 55, 147, 309, 561. These growing sequences are shown in Fig.1a,b. Icosahedral structures are favoured for small clusters as more compact structures can be built with lower potential energy. For increasing N mechanical stress develops at the ridges and corners of the icosahedra. This stress leads to an increase of the potential energy, so that finally crystal structures become more favourable. In almost all cases the well known good interaction potentials of the rare gases (see below chapter 4.2.a) is not used, but the Lennard-Jones 12-6 potential is employed. Some of the results might therefore not apply to rare gas clusters, but rather to "Lennard-Jonesium" clusters.

The growth of alkali halide clusters has been studied theoretically by T.P. Martin /4/. One sees from fig. 2, that contrary to the case of rare gas clusters rectangular and chairlike structures are favoured. This picture is calculated for ionised clusters, but similar structures are obtained for the neutral ones. For larger clusters the number of potential minima grows so rapidly, that it is no longer possible to search the complete space for minima. Many of these minima are energetically very close. Therefore they play a large role at non zero temperatures.

Quite generally one can say, that for small clusters the geometric configuration of lowest energy is usually material and model dependent.

II) ALL PROPERTIES CHANGE WITH CLUSTER SIZE

In this chapter we will give some selected examples, how physical properties change with cluster size. We will start out with the melting temperature, a concept which is rigidly defined only in the solid state limit and obviously looses all meaning for a dimer. It is well known, that in finite systems the singularities of the thermodynamic phase transitions are rounded off. The singular behaviour at the critical points arises only in the limit that some length L of the cluster becomes large. A considerable literature exists on the finite size scaling theory, which calculates L for various geometries and dimensions, as reviewed by Barber/5/. The melting point is defined thermodynamically, as that temperature where the Gibbs free energy of the liquid and solid become equal. So far no theory has been derived giving an experimentally observable quantity using this definition. A more heuristic approach has to be taken.

2.1 Melting temperature

The melting temperature of gold clusters as measured by Buffat and Borel/6/ is shown in fig.3. They deposited their clusters on a thin support, and observed them in an electron microscope. The size of the cluster is thus

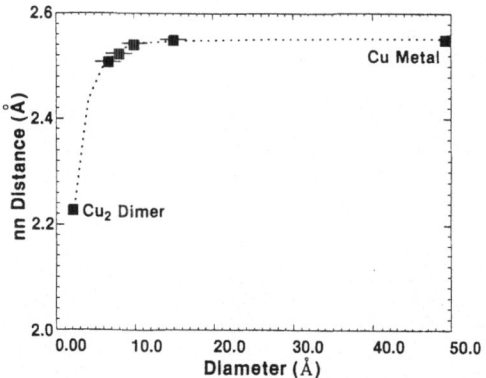

Fig. 5. Experimental nearest neighbour distances for copper clusters. Note the rapid convergence to the bulk limit. After Montano et al./13/.

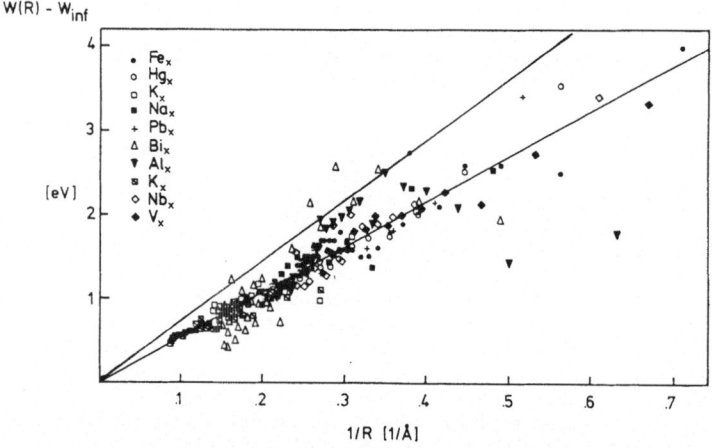

Fig.6. Difference between experimental ionisation potential and extrapolated bulk workfunction based on the spherical drop model. The two straight lines correspond to the 3/8 respective 4/8 value, as discussed in the text. The theoretically less sound value of 3/8 gives a better fit to the data. Adapted from ref./14/.

known. For modern electron microscopes a simple turning of a lever is sufficient to change from electron microscopy to electron diffraction. They used clusters with diameters of 20 Å or larger (N>350 atoms) for which the Debye-Scherrer lines of the diffraction patterns could be well resolved. For a macroscopic crystal these lines are sharp. They vanish into a broad background if the crystal melts. Buffat and Borel heated their clusters until the Debye-Scherrer rings vanished. This they took as a heuristic definition of a melting temperature. Newer data on other materials are discussed in reference /7/.

What is the physical reason for the precipitous drop at low temperatures? A theoretical treatment/6/ gives a very good fit to the data, as can be seen in Fig.3 We will use here a plausible argument based on the old Lindemann criterion for the melting of a solid/8,9/. Melting occurs if the root mean square (rms) of the thermal vibrational amplitude exceeds about 10% of the interparticle distance. This relation is reasonably well fulfilled for many solids, but it lacks theoretical backing. Experimentally the rms vibrational amplitude can be obtained from the temperature dependence of the intensity of the diffraction rings via the Debye-Waller factor, as explained in many solid state text books. Atoms in a solid are constrained by their neighbours in their free vibration, while an atom on a surface can move freely perpendicular to the surface. According to the Lindemann criterion one would therefore expect a lower surface than bulk melting temperature. A decrease by 40 K of the melting temperature of the first layer of a Pb-surface was indeed observed recently /8,9/. So it seems plausible that clusters containing fewer and fewer atoms, which are percentagewise more and more on the surface, and have a larger rms vibration will therefore melt earlier.

A pictorial example of cluster melting is shown in Fig.4. A 87-atom argon cluster is assumed to be bound by two-body Lennard-Jones potentials, and the classical equations of motion are assumed. Starting from a cluster in the lowest energy configuration at T = 0 K, this ensemble of atoms is carefully heated (on the computer) by slowly adding in kinetic energy. At low temperatures the structure is rigid, while for higher ones the atoms are free to move around the cluster, till they can finally evaporate. From a quantum-statistical model calculation, Berry and coworkers/10/ have predicted, that clusters exhibit unequal freezing and melting temperatures, and a coexistence range over which both solidlike and liquidlike forms can coexist. This has been observed in argon clusters surrounding an organic molecule/11/.

2.2 Nearest Neighbour Distance

A property which converges much faster to the bulk limit than the melting temperature is the interatomic distance. Fig.5 shows the result of an EXAFS (Extended X-ray Absorption Fine Structure) measurement on copper clusters embedded in solid argon/13/. For a measurement of a ground state property the argon matrix has probably a very small influence. One sees that the nearest neighbour distance changes from 2.23 Å for N = 2 to 2.51 Å for N = 13, while the bulk value of 2.55 Å is obtained nearly at N=200. Note that on the curve for the melting point there are even no data points for these small clusters. These measurements have been performed by condensing copper clusters from a gas aggregation source (see chapter 3.1) onto a cold target and measuring x-ray absorption spectra of the Cu K-shell for different cluster sizes. In first order the spectra show the familiar sawtooth-like behaviour of K-shell absorption. This is modified by interference of scattered electron waves originating from different atoms in the cluster. The interference of these waves produces an undulatory behaviour, whose wavelength can be used to calculate the interparticle distance/13/.

For metals, transition-metals, and semiconductors one observes an substantial increase of the nearest-neighbour distance in going from the

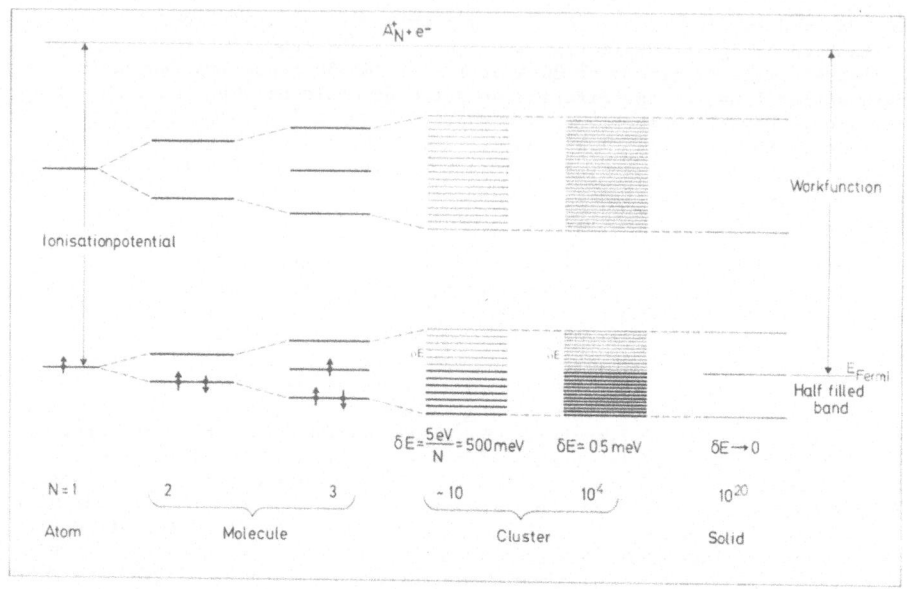

Fig.7. Evolution of electronic levels from the atom, molecule, cluster to a metallic bulk. The picture is highly schematic, assuming that the atom has only two bound states and one continuum. The variation of the ionisation potential with cluster size as shown in fig.6 has been neglected for clarity. The occupation of the electronic levels for N > 3 is indicated by the heavy shading.

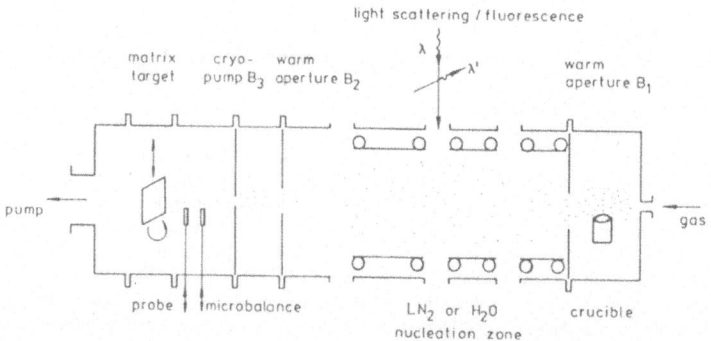

Fig.8. Gas aggregation source for cluster production. The material is evaporated from a crucible and entrained into an argon stream. The gas is collimated by an aperture, pumped by a cryopump, monitored by a microbalance and can be frozen out on a liquid He cooled matrix target. The data for fig.5 have been obtained with a source like this. (After Frank et al./31/).

dimer to the bulk. While this distance increases only very slightly for the rare gases (with the exeption of He due to its large zero point motion), it decreases for some atoms with a $d^{10}s^2$ or $p^6 s^2$ outer electronic shell (Mg, Ca, Sr, Ba, Hg).

The surface of macroscopic crystals often shows a different geometrical arrangements or a different nearest neighbour distance than the bulk. The two phenomena - surface relaxation and variation of interatomic distance in clusters - seem to have the same physical origin, as indicated in a recent theoretical paper/73/.

2.3 Ionisation Potentials

For metals (save for Al /22c/) a simple scaling holds for the variation of the ionisation potential (IP) with cluster size. Photoionisation of a cluster can be written:

$$A_N + h\nu \longrightarrow A_N^+ + e^-$$

Treating the remaining cluster in the spherical drop model as a classical conducting sphere of radius R, Wood has obtained:

$$IP = WF + \frac{3}{8}\frac{e^2}{R}$$

Using the same procedure one gets for the electron affinity(EA):

$$EA = WF - \frac{5}{8}\frac{e^2}{R}$$

where WF is the workfunction, i.e. the IP or the EA for $N \longrightarrow \infty$, and e the electronic charge. These relationships hold surprisingly well, as can be seen in fig.6. Two different kind of criticism have been levelled against these simple relations/16,17/. First, a more accurate treatment gives $\pm 4/8$ instead of $+3/8$ and $-5/8$. But this gives less good agreement with experimental results. Second, a cluster may break up in the ionisation process, so that the ionised cluster one observes in a mass spectrometer may have lost some of its constituents in the ionisation process. These break-up processes, generally called fragmentation for clusters, cannot always be ruled out completely. As will be discussed below, it is very difficult to make a beam of one cluster size only. One nearly always has a distribution. In this case a break-up process during ionisation can go undetected. The behaviour of the ionisation potentials for nonmetallic clusters is very different. It has been discussed earlier by Jortner /2/. A highly schematic picture of the evolution of electronic levels with cluster size for metal atoms is shown in fig. 7.

2.4 Other properties

Only very few properties can be discussed here. Very much more is known. The size dependence of the catalytic/18/, superconducting/19/, magnetic/20/, and optical/21,25/, chemical/22/, photographic/23/, and astrophysical/24/ properties are presently discovered or are partially known. The threshold for singly, doubly or higher charged clusters is studied in detail/1h,1k/. The transition to the physics and chemistry of solutions is traced/25/, a technological application towards thin film formation /26/ is already marketed. A way to produce antimatter clusters for the use of fuel for the next century has been proposed/27/.

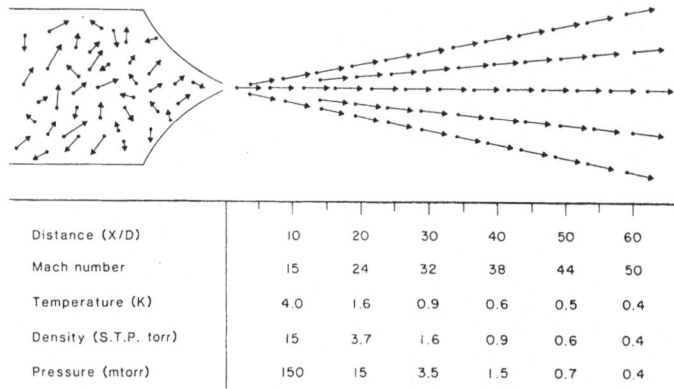

Distance (X/D)	10	20	30	40	50	60
Mach number	15	24	32	38	44	50
Temperature (K)	4.0	1.6	0.9	0.6	0.5	0.4
Density (S.T.P. torr)	15	3.7	1.6	0.9	0.6	0.4
Pressure (mtorr)	150	15	3.5	1.5	0.7	0.4

Fig.9. Schematic representation of a nozzle expansion. Several characteristics of the expansion are given as a function of the reduced distance X/D downstream from the nozzle, where D is the diameter of the nozzle hole, and X the distance downstream. After Miller/29/

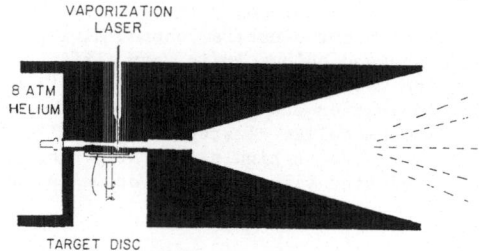

Fig.10. Cross section of a pulsed nozzle source for refractory materials. A vaporization laser irradiates a target disk. The ejected plasma is entrained in the pulsed helium flow, which leads to thermalization and cluster formation. The clusters formed are collimated by a 10 cm long 15° cone followed by a free expansion. After Liu et al./30/.

3) CLUSTER PRODUCTION and DETECTION

3.1 Gas-aggregation

The simplest and oldest method of cluster production is the gas-aggregation technique. A metal (or other solid) is evaporated into an inert gas atmosphere. The fast metal atoms are slowed down by collisions with the gas, and they start aggregating if their temperature is low enough. For gold clusters this occurs at about 35 to 40% of the bulk melting temperature. If the clusters are large enough they form visible smoke-like patterns during formation. Correspondingly these sources have also been called smoke-sources. Many experiments have been performed by looking at the resulting clusters with an electron microscope/28,29/. In another type of experiment "nanocrystals" are produced by collecting the clusters on a cold trap, and than compacting them in a mechanical press under UHV-conditions/30/. The properties of the resulting material are very different from those of an ordinary crystal, e.g. the elastic modulus can be a factor of 10 different. The gas aggregation-technique has been developed by W.Schulze to perfection/31/. His source is shown and explained in Fig.8.

3.2 Supersonic Expansion

If a gas is expanded from a high pressure through a small hole into vacuum, it undergoes an adiabatic expansion, which leads to a strong cooling of all degrees of freedom. Thermodynamically the expansion can be described by the Poisson equation:

$$T^\kappa \, P^{1-\kappa} = T_o^\kappa \, P_o^{1-\kappa} = \text{const}$$

Where T and P are temperature and pressure respectively, and κ the ratio of the specific heats ($\kappa = 5/2$ for the expansion of an atomic gas). The index o refers to the conditions before the expansion. The temperature T of the beam is that measured by a comoving observer/32/. The temperature corresponding to the relative velocity of the particles in the beam can become very low indeed, for He it can drop below 0.1 K. If this temperature is lower than the binding energy of a dimer, and if three-body collisions are still important, cluster formation starts and proceeds very rapidly. A typical example is the expansion of 10 bar Ar through a 0.05 mm nozzle to give a mean cluster size of N = 150 /33/. These sources have been discussed so often in the literature (ref./1, 32-35,43, 47/) that they will not be treated here any further. Valuable scaling laws for cluster formation have been given by Hagena/34/. A graphical representation of a nozzle expansion is given in Fig.9. As versatile as this source may be, it can only work with materials which have a vapour pressure of several hundred Torr at a temperature where one can still build a vessel to contain the hot gas.

3.3 Laser Evaporation Source

The laser evaporation source can be used for all solid materials. A pulsed (5 to 10 ns) laser beam of 10 to 20 mW is focused to a spot of about 1 mm. The high light intensity evaporates about 500 layers of atoms in a crater of 1mm diameter. As soon as the generated plasma expands a He gas pulse from a pulsed nozzle entrains it. The most recent version/36/ is shown in fig.10. This source is a hybrid of the sources mentioned above. A laser is used to evaporate material, and a supersonic expansion cools it, and induces clustering. This source has become quite popular recently, as it allows the production of a large variety of neutral and ionised clusters.

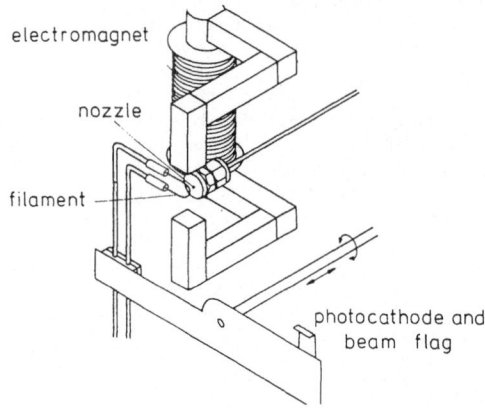

Fig.11. Source for cold positively or negatively charged cluster ions.
A discharge is ignited and collimated by a magnetic field directly in
front of the nozzle hole. This leads to high intensity beams of cold
cluster ions, some of which have not been synthesized before. The
discharge can alternatively be replaced by an electron beam, or a photo
electron source. The source can run CW or pulsed. After Haberland et
al./11 and 39/.

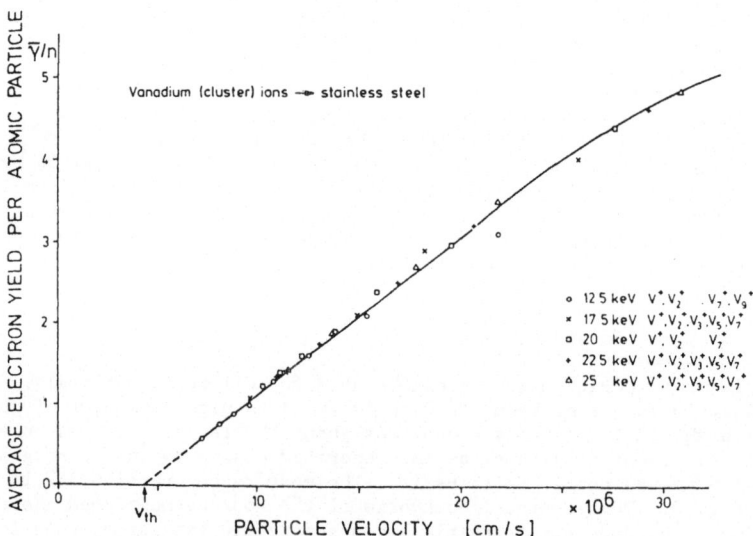

Fig.12. Secondary electron ejection coefficient as a function of the
velocity of impinging vanadium clusters. The yield is given per atom.
Only above a certain threshold velocity the probability for efficient
detection is large. After Thum and Hofer /40/.

3.4 Other Sources

If a keV ion beam impinges onto a solid target it can eject, among other things, atoms and clusters. The minor ionised fraction of the clusters can be formed into an intense beam, mass analyzed, and used for experiments. The clusters from this source are very "hot" probably "boiling" just after formation/37/. They continue to cool by evaporation in flight, but they are still much hotter, than those produced in the sources discussed above. Many other sources have been constructed, and the interested reader is referred to the ref./1,73/ and the current literature. They are generally various hybrids of the sources discussed above.

No source has been built yet which produces a neutral cluster beam of one size only, so that most workers in the field, who want to work with clusters of definite size, use mass selected ionised clusters. The only exception is the method employed by Buck to produce mass selected neutral clusters/38/. His method gives only a weak beam and works only with clusters with N < 8, but it is the only method available today.

3.5 Sources for cold cluster ions

The geometric and electronic structure of neutral and ionised clusters is different in general. The only exception might be not too small alkali clusters. Ionisation is a "vertical" process (i.e. position and velocities of the atoms do not change during the ionisation process). An isolated cluster after ionisation will therefor, in general, not be in an equilibrium configuration. While the cluster relaxes towards a potential minimum, potential energy is converted into vibrational excitation. The conclusion is, that ionised clusters will in general be "hot", i.e. they have a large internal energy content.

A possible way around this problem is to make the ions directly in the expanding gas, if possible in the presence of a large surplus of rare gas. Then an ionised atom or cluster can grow, and be subsequently cooled. The laser evaporation source can operate in this way. Fig.11 shows a source for which cluster temperatures below 20 K have been realised. Ref./1n/ describes other successful versions.

3.6 Problems of Cluster Detection

The most sensitive way of atomic ion detection is to let them impinge on a metal surface, and use the ejected electron for further amplification. A serious experimental problem arises if this scheme is applied to the detection of large clusters. Fig.12 shows the number of ejected electrons as a function of the velocity of impinging vanadium clusters/40/. Note that the yield is given per atom in the cluster. The uncomfortable message from this figure is, that one should use very high kinetic energies for efficient cluster detection. In fact Beuhler and Friedman have used energies of up to 250 keV /41/. In the majority of cluster experiments much lower energies are used, and very serious discrimination effect at high cluster sizes are then to be expected. Nevertheless large clusters can often been seen. For example low energy (3keV) argon clusters with N up to 2500, corresponding to 100.000 atomic mass units, have recently been observed in the author's laboratory with good intensity.

A way around the difficult detection problem is not to use the ejected electrons but the produced ions, which seem to be more abundant. These fragmented or sputtered particles seem to have a much smaller mass, than the original cluster and can therefore used efficiently for secondary electron ejection. An increase of the sensitivity by a factor of 100 has been obtained this way/42/. The method is not very suitable for time of flight

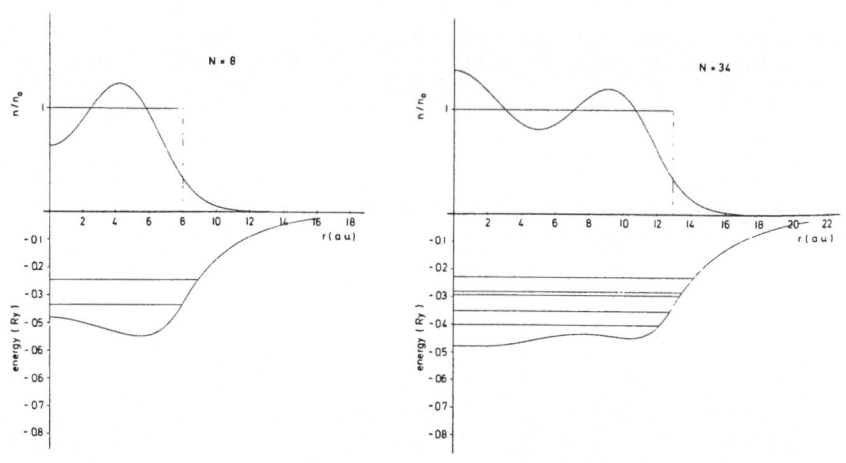

Fig.13. Evolution of the Jellium potential and charge density for clusters containing N=8 and N=34 atoms. The beginning of the level structure of the metal clusters can clearly be seen. After Eckart /16/.

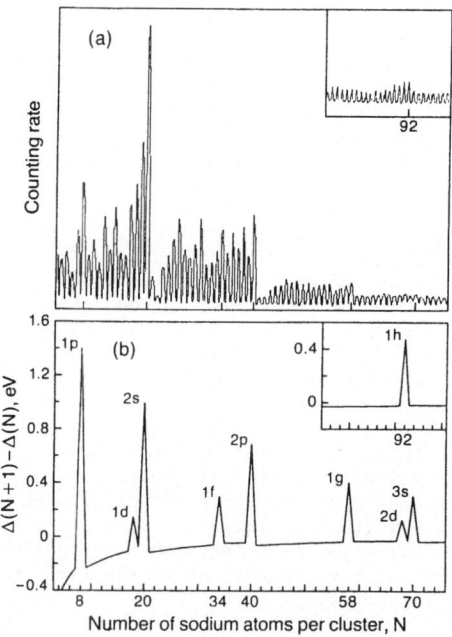

Fig.14. (a) Mass spectrum of sodium clusters, N=4 to 75. (b) The calculated change in electronic energy differences. Note the good correlation from which one can deduce the structure of the mass spectrum is due to the stability of the neutral clusters. After Knight et al./43/.

experiments, as the intermediate step of the detection process deteriorates the time resolution considerably. See ref./1m/ for more details.

4) CASE - STUDIES

To appreciate the amount of knowledge, which has already been accumulated in cluster physics, one has to look only at the references, and bear in mind, that only a tiny part of the literature could be cited. It is therefore impossible to treat all fields of cluster physics equally. Only three selected cases will be discussed below: In chapter 4.1 the salient features of alkali clusters are briefly summarized. The valence electrons are completely delocalised in this case, and many of the properties can be understood from a simple shell model for the electronic structure. Two selected examples of the completely opposite case of closed electronic shell systems are discussed in chapter 4.2 and 4.3. Specifically 4.2 deals with ionised rare gas clusters, while chapter 4.3 deals with negatively charged clusters, which do not have a bound negative ion state as a monomer. On each of these subjects one could write a long monograph by itself. An introductory overview will be attempted instead. Emphasis is laid not on experimental details. Instead the physics of the ionisation process is discussed, as it is necessary to understand and interpret the experimental mass spectra. It will become clear below, that in alkalis (and most metals) the charge is delocalised in the ionised clusters, while the localisation of the positive or negative charge in the two other cases gives rise to interesting additional phenomena, some of them well known in very different fields.

4.1 Alkali clusters

The completely delocalised nature of the conduction electrons of the alkalis is well treated in the so called "jellium"-model of solid-state physics. Also the noble metals (Cu, Ag, Au) can be treated to a certain extent with this model. Aluminum presents a special case. It is a good free electron metal in the bulk, but its variation of the ionisation potentials cannot be explained by the free electron model/22c/. In the jellium-model the valence electrons are assumed to be completely free to move within the metal. The charge of the positive ions is smeared out into a constant (jelly-like) background, which imposes a boundary at the surface. The physical reason for this - at first sight - egregious approximation is due to the Fermi statistics obeyed by the electrons and the very effective screening of the ionic Coulomb potentials by the free electrons as explained in many solid state text books. part of the r^{-1} Coulomb potential is screened by the electrons so that it is transformed into an $r^{-1} \exp(-r/\lambda)$ Yukawa potential. The screening length λ depends on the electron density and is for bulk sodium 0.67 Å compared to a nearest neighbour distance of 3.71 Å. This means that the electrons can move through the metal or cluster without "seeing" the ionic cores very much, or expressed in more scientific terms: the pseudo-potential is very weak . Note that the jellium model applies only to electrons at or below the Fermi surface, due to the vital role played by the Fermi-statistics in suppressing electron scattering. The interaction of higher energy electrons is very strong indeed, and their mean free path correspondingly very short. This can be important e.g. in above threshold ionisation of metal clusters, where the ejected electron can be expected to interact strongly with the remaining cluster. Assuming a spherical jellium cluster of radius R and solving self-consistently for the electronic wave function, Eckart/16/ and Cohen and coworkers/43/ have calculated many properties of alkali clusters. Clemenger has extended this treatment to include nonspherical shapes, along the lines of the Nilsson model of nuclear physics/44/. Koutecky and his group have performed very accurate ab initio calculations for Li clusters/45/, while Lindsay et al. have obtained interesting general results from an approximate treatment, which is called "Tight Binding Approximation" in solid-state physics and the "Hückel Molecular Orbital" method in molecular physics/46/.

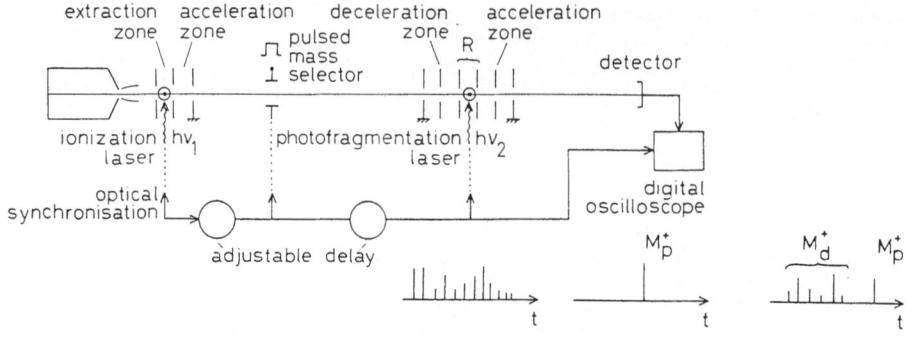

Fig.15. A tandem TOF (time of flight) mass spectrometer for the study of mass selected cluster ions. The alkali clusters are produced by supersonic expansion. A pulsed UV laser (hν 1) ionises the clusters. By applying a voltage pulse to the mass selector one cluster parent ion M_p^+ can be selected. In the second TOF the mass selected cluster can be photoexcited (by h ν2) within a decelerating-accelerating grid system. By proper adjustment of the delays one parent cluster ion M_p^+ can be selected, photoexcited and the daughter clusters (neutral or ionised) M_d can be recorded. After Bréchignac et al./48/.

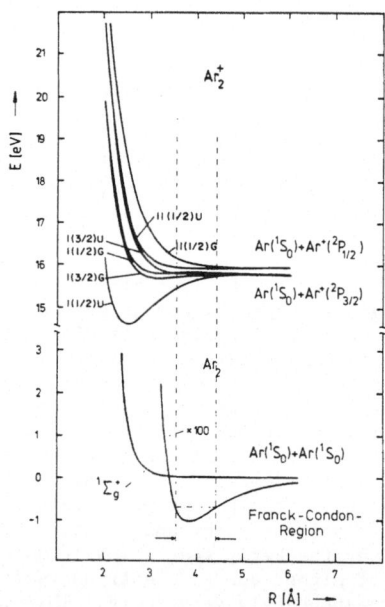

Fig.16. Potential for the neutral and ionised argon dimer. The ground state potential is also given multiplied by a factor 100, to show the shallow van der Waals well of 12.5 meV. The nearest neighbour distances of the neutral dimer, cluster or solid are very similar. The potential of the ion is split by the spin-orbit interaction. The lowest ionic potential has a well depth of 1.2 eV. This state is populated very rapidly in the gas, liquid or solid. After ionisation the argon dimer ion is highly vibrationally excited, as it is formed with an internuclear distance in the Franck-Condon region. Such an excited dimer (or similar ionic structure) is formed about 1 picosecond after ionisation in all rare gas clusters. The large excitation energy heats the cluster and leads to evaporation. A second contribution to the internal energy of the ionised cluster results from the nuclear part of the polarisation energy as discussed in the text.

Fig.13 shows the result of a calculation for a spherical jellium cluster, where the grouping of levels can clearly be observed. If the cluster would contain more atoms and therefore also more valence electrons, the level density would be higher and one can easily visualize the transition to a continuum as depicted schematically in fig.7. The total binding energies are smooth functions of cluster size/16,43,44,46/, but the second differences of the binding energies show large peaks at shell closures as can be seen in fig. 14. The first atom outside a closed shell is particularly weakly bound, very similar to the weak binding of an electron or nucleon outside a filled shell in atomic or nuclear physics. At the same mass numbers, sodium cluster mass spectra recorded by Knight et al./43/ show peaks of higher intensity, as can be seen in fig 14 a. The assumption is often made, that a higher intensity of some peak in a mass spectrum corresponds to a particular stable neutral structure, thus "explaining" these so-called "magic numbers". But even for alkalis, the correlation

$$\begin{bmatrix} \text{higher intensity in} \\ \text{a mass spectrum} \end{bmatrix} \longleftrightarrow \begin{bmatrix} \text{higher binding energy} \\ \text{of the neutral cluster} \end{bmatrix}$$

cannot be completely correct (and is wrong in general). This is shown by the data of the Schumacher group, which has published sodium mass spectra which are much less pronounced/47/. The reason for the differences can be different source conditions, but more probably they are due dissociation processes in the ionisation region, which go undetected in a standard mass spectrometer. The group of C.Bréchignac/48/ has clarified this situation, by decoupling the ionisation and fragmentation using the apparatus shown in fig.15. If the photon energy of the ionising laser is just above the ionisation potential of the cluster and if the laser fluence is small, one records a mass spectrum corresponding to the shell structure of the neutral clusters. Under these stringent conditions fragmentation during the ionisation process seems to be neglegible. (Note this is not true for the rare gas clusters treated below, as the Franck-Condon factors at threshold are vanishingly small, and the nuclear part of the polarisation force gives a large contribution). If the laser fluence is too high two or more photons can be absorbed from the same laser pulse. If the photon energy is too high, one or more atoms can evaporate after the ionisation process. In these cases one records a mass spectrum corresponding to the shell structure of the ionised clusters. By changing the photon energy and photon flux the appearance of the mass spectra can be changed nearly at will.

This behaviour may sound complicated. But it is still simple compared to nearly all nonmetallic cases. The simplification results from the metallic binding, which lets the electronic structure of the neutral and ionised clusters be rather similar. This is not the case for the clusters treated below.

4.2 Rare gas clusters

Contrary to the case of the alkalis, the electronic state of the neutral and ionised clusters of the rare gases (and of other atoms and molecules having closed electronic shells) are very different indeed and cannot be treated on the same footing. Properties of nonmixed rare gas clusters are discussed below.

4.2.a Information on neutral rare gas clusters: The only experimental results come from the electron diffraction experiments already discussed in chapter 1.3. The bonding in the neutral rare gases is very simple. The bond is characterized by a two-body potential, which is well known/49/, the three-body contribution amounts to only a few percent and is often neglected. Because of this simplicity rare gas clusters are often used for model calculations/1,2,10-12/.

4.2.b Information on ionised rare gas clusters: The simplicity of the bonding between rare gas atoms is lost, as soon as one of them is ionised. As will be discussed below in more detail, the positive charge (the hole in solid state language) does not remain delocalised over the entire cluster as in the case of metallic bonding. Rather the electronic energy can be lowered if the charge becomes localised on some smaller substructure. An Ar_n^+ cluster would then have an ionic part, Ar_i^+ with i=2, 3 or 4 surrounded by a neutral core of (n-i) argon atoms. During this localisation process electronic energy is converted into vibrational excitation, which heats the cluster and leads to evaporation. In molecular physics there exists no name for such a structure, in solid state physics it is named an "ionic defect". This simple picture is consistent with a wealth of cluster experiments and also data on condensed rare gases/50,53/.

The processes leading to charge localisation and cluster fragmentation will now be treated in more detail below. They are a fine example of the interplay between solid state and molecular physics necessary to explain cluster properties. One can set up a time-table of the events happening in a rare gas cluster after ionisation. The moment of ionisation is taken as time "zero".

Step 1. The charge is delocalised. One can assume, to begin with, that only one atom in the cluster is ionised, as the electrons are tightly attached to the atoms. For a linear chain cluster this can be pictorially represented as below, where each circle symbolizes an atom and the "+" sign the positive charge.

$$|n>$$
$$O \; O \; O \; O \; O \; O \; \oplus \; O \; O \; O \; O \; O \; O$$

This is obviously not an eigenstate of the corresponding Hamiltonian, because one would have a state of the same energy, if the charge would reside on a neighbouring atom. Denoting the charged state of atom n with $|n>$, one has an amplitude A for charge transfer to atom n + 1 given by

$$A = <n+1|H|n>$$

This problem is treated e.g. in the Feynman lectures Vol.III, chapter 13. It is shown there that the nonvanishing of the transfer integral A leads to a finite bandwidth W for the electronic levels.

$$W = 2 \, M \, A$$

where M = 2 in the linear case assumed by Feynman. As shown in ref. 51, M is

Table II. Number of nearest neighbours entering the equation directly above. Note that the same equation holds, from the diatomic molecule to the infinite lattice .

M	Physical situation
1	dimer
2	linear chain
6	simple cubic crystal
8	body centered cubic crystal
12	face centered cubic crystal

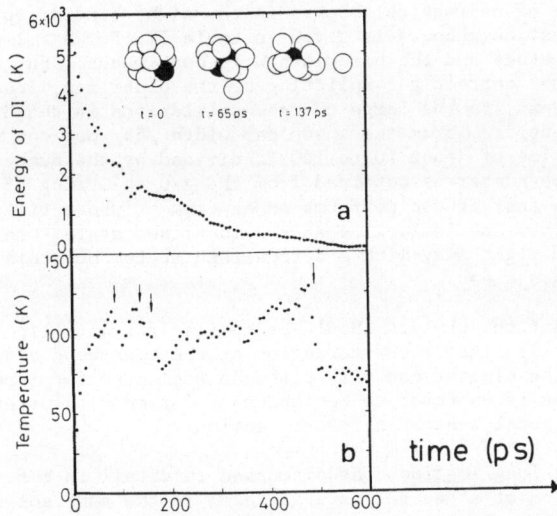

Fig.17. (a) Decrease of the vibrational energy of the dimer ion of
fig.16 inside a Xe$_{13}^+$ cluster initially at 1 K. Also shown are
characteristic snapshots at different times; the black balls are the
atoms forming the dimer ion. b) Evolution of the cluster temperature
for the same run. The arrows indicate the successive evaporations of the
atoms. After Soler et al./59/.

Timescale for the processes occurring in a rare gas cluster after ionisation		
R = He, Ne, Ar, Kr or Xe		
time (s)	process	
0	R_n	ionisation
	R_n^+	positive ion (hole) delocalised hopping frequency~10^{14} Hz
10^{-12}	$R_2^+R_{n-2}$	hole self trapping $D(Ar_2) = 0.012$ eV $D(Ar_2^+) = 1.2$ eV dissipation of about 1 eV heating and evaporation of the cluster
10^{-8}	$R_2^+R_{n-m}$	$+ (m-2)$ R
$>10^{-6}$	$R_2^+R_x$	x = ? detection in a mass spectrometer

Fig.18. Processes occurring in a closed shell cluster can be time ordered
as discussed in the text. Some evaporation is also occuring at a time scale of
10 ns or longer/59-61/, but it constitutes a minor fraction.

equal to the number of nearest neighbours in the tight binding approximation. The number of nearest neighbours is given in table II. For M > 1 one has an infinite number of atoms and the bandwidth W is continuous. But even for M = 1 the formulas give the correct g/u splitting of the dimer ion, leading, in a very different context, to the large charge oscillations in the Everhard scattering experiment/52/. From the known bandwidth /53/ one can now easily deduce a transfer time of about 10 to 100 fs divided by the number of nearest neighbours. The same number is obtained from the g/u splitting of Ar potentials. This means, if one performs an experiment whose time scale is long compared to this time, one is looking at a delocalised state. One could of course have started right away with a delocalised state, but this would have made the physics less clear.

Up to this point the physics should be similar for metallic bonding. The main difference is, that the interaction of the scattered electrons before they leave the cluster can very often be neglected for rare gas clusters/53/. But it is expected to be important for metal clusters, due to the large electron metal scattering cross section.

Step 2. Charge localisation. As discussed in detail in ref. /50,53/, an electronic excitation of a neutral rare gas atom in the same solid, liquid or gas becomes quickly localised on an excited neutral dimer. For the condensed phase the time for localisation is about one picosecond/53/. These excited states are the Rydberg states to the ionic states discussed here. They form excited neutral dimer molecules having potential curves which have also one deep well like the one shown in Fig.16. Without further knowledge it is assumed, that roughly the same localisation time applies also to the ionic case treated here.

For the condensed rare gases it seems to be common knowledge and consent, that localisation occurs on a dimer. For small ionic clusters this has recently been contested. Three calculations report that a trimer or larger ion is formed/54-56/, while a fourth one is consistent with a dimer ion/57/. In any case all authors agree on the idea of charge localisation on N = 2 ,3 or 4. For the sake of conciseness only the term dimer will be used below. For other cluster composed of atoms with closed electronic shell localisation on even larger structures is suggested by experiment/58/.

Step 3. Heating and fragmenting the cluster. There are two sources of energy which heat the cluster and lead to fragmentation. First, according to fig.16, the freshly localised charge resides in a highly vibrationally excited dimer ion with a vibrational energy of about 1 eV. This will be dissipated in the cluster leading to a rapid increase in temperature as dramatically shown in fig.17 /59/. Second, even photo-ionisation of a rare gas clusters at the adiabatic threshold of the dimer ion will lead to fragmentation. The potential between the newly created dimer ion and the neutral rest of the cluster changes from a van der Waals type r^{-6} potential to an ion induced polarisation r^{-4} behaviour. The resultant shift of the potential minima induces vibrations, and correspondingly heating of the cluster and evaporation of atoms. This nuclear part of the polarisation energy is as important as the vibrational energy of the dimer ion for rare gas clusters, and is expected to play a large role for all nonmetallic clusters. Once the dimer ion is vibrationally relaxed, the charge is effectively localised, as there is nearly vanishing overlap with geometrically nearby states. These states differ in energy by over 1 eV,and have a factor 1.3 too large an internuclear spacing, making the transfer amplitude A (see above step 1) vanishingly small.

Step 4. Detection. The detection of the cluster proceeds on a time scale of 1 to 100 microseconds, so the majority of the above processes have subsided. Metastable decays are still present, as studied by Märk et al./60/ and Stace and coworkers/61/. The different time steps are collected in Fig.18.

Most of the experiments reported so far, are in agreement with this picture, and it is commonly accepted today. The strong fragmentation of small argon clusters found by Buck et al. /32/, observed for N = 10 to 100 by Haberland et al./33,52/, for larger ones by Hagena et al./62/, and the electron stimulated desorption from argon films/63,64/ are all different aspects of the same process. The irregular structures in the mass spectra have been rationalised as being due to compact packings around the dimer ion/65/, that is: independent of the way of ionisation, rare gas mass spectra indicate ionic and never neutral stability of the cluster. A good overview of the relevant experimental data can be found in the review by Castleman and Märk/1h/. In several laboratories photofragmentation experiments on mass selected cluster ions are in progress, to test this model to its limits. The experimental set-ups are similar to that of fig.15 in principle, but differ considerably in detail/1n/. A photo-dissociation experiment on Ar_3^+ has already been published/66/. Very recent photo-fragmentation experiments by the Lineberger group also confirm the picture of charge localisation in ionised argon clusters.

4.3 Negatively charged clusters of atoms and molecules, which do not have a bound negative ion state as a monomer

a) General Introduction. All atoms and molecules with closed electronic shell do not have negative ion states as monomers. The closed shell is only weakly polarisable, so that there is just not enough polarisation potential for an extra bound electronic state. But a cluster of a certain magnitude can often support a bound state for an extra electron. Table III gives the minimal M necessary to form a stable negatively charged cluster.

The value for Mg in table 3 has been obtained by the group of Meiwes-Broer and Lutz/68/ with a source similar to that in Fig.10. The last three entries are from Kondow et al. /1r/. All other values have been obtained in the authors laboratory with different variants of the source shown in fig.11. Experimental details are given in ref./67/. Only for the case of water clusters there is some understanding of the details of the binding of the electron, as discussed below. For a theoretical introduction into electron localisation in clusters, see ref./1p/.

Table III. At least M atoms or molecules are necessary to form a stable negatively charged cluster of these closed shell atoms and molecules. For the three atomic cases the electron is delocalised in the bulk, while it is localised for the molecular ones. The last three entries are not expected to be the minimal M /1r/.

Species	Minimal M
Hg	3
Mg	3
Xe	not known
H_2O	2
D_2O	2
HCl	2
NH_3	35
ND_3	41
CD_3CN	11
C_5H_5N	4
N_2O	6

b) Introduction to the solvated electron problem. More than 125 years ago Weil discovered that one can dissolve sodium in liquid ammonia/69/. At low Na concentrations the formerly colourless liquid turns blue. Today we know, that Na is ionised in the solution, as the energy necessary to form $Na^+ + e^-$ is smaller, than the energy gained by forming in the liquid "clusters" of the form:

$$Na^+(NH_3)_N \text{ and } e^-(NH_3)_N$$

where the latter is conventionally written as

$$(NH_3)_N^-$$

The corresponding "hydrated electron", $(H_2O)_N^-$ was observed in the bulk only in 1962. A vast amount of literature has been published on the condensed matter properties of these solvated electrons, because of their special chemical nature. They occur in the interaction of ionising radiation with water and therefore with all biological matter. Once it has become possible to synthesize these clusters in a beam experiment, some of their physical properties could be elucidated in a fruitful cooperation of experimental and theoretical efforts. The binding of the excess electron results from a combination of short range and long range attraction and short range repulsion. It is accompanied by large rearrangements within the cluster, i.e. the neutral and negatively charged clusters do not have the same geometric arrangement. The blue colour cited above result from bound-bound and bound-free electronic transitions.

The theoretical calculations have become increasingly sophisticated, describing the excess electron by Feynman path integrals and using methods imported from polymer physics and the theory of liquids/69/. For $(H_2O)_2^-$ the electron orbit is extremely diffuse with a diameter of about 20 Å much larger than the two water molecules together. The electron affinity is very low, 3 to 20 meV/69,70/. The electron is so weakly bound, that a strong electric field can detach it from the dimer/70/. For clusters with N between 10 and 20 the electron occupies a surface state. In the lowest energy configuration it is wedged in some crevice of the cluster's surface. Beautiful colour picture of this strange entities are given in ref./69/. The transition to the bulk structure observed in the liquid occurs only around n=35. The theoretical results are in good agreement with photodetachment experiments on these solvated electrons/71/. A continuum model has been proposed/72/ for the ammonia clusters of table 3, giving good agreement with experiment. It has been proposed that the water surface states might not exist for ammonia, thereby explaining the large differences between the isoelectronic species water and ammonia/69/.

CONCLUSION and OUTLOOK

Many of the fascinating features of cluster science have only lightly been touched upon, others have not been mentioned at all. Some difficult problems have been simplified considerably, to make them tractable in the space available. But it is hoped, that the reader has tasted some of the flavour of the fascinating and interdisciplinary field of cluster science. The novice reader should bear in mind that many important fields of cluster science have not been touched at all. Carbon, semiconductor, transition metal, organic, metalorganic, supported clusters have not been treated. Also dynamic effects, chemistry with clusters, the problems of multiply charged clusters, and the wide area of technological application are missing. An up to date reference to the newest developments will probably be the Proceedings of the next International Meeting on Small Particles and Inorganic Clusters/74/ and an "Enrico Fermi" School/75/.

Nature itself does not know the distinction between atomic and molecular or solid state physics/chemistry, only we who have been trained in these traditionally separated fields. Cluster science encompasses all these fields and has also strong links into technology. So cluster science can help to transgress the traditional borders between this fields. This feature accounts for much of the liveliness and fascination of the field.

REFERENCES

1. General references on clusters:
    a) Surf. Sci. Vol. 106 (1980) and 156 (1985), Proceedings of the
       2. and 3. International Meeting on Small Particles and Inorganic
       Clusters.
    b) Z.Phys.D 3, Nr. 2 and 3 (1986)
    c) Ber. Bunsenges. Phys.Chem. 88 (1984)
    d) J.Phys.Chem. 91 Nr.10 (1987)
    e) The Physics and Chemistry of small clusters, NATO ASI series B158
       P.Jena, B.K.Rao and S.N.Khanna, eds. ,Plenum, New York 1987
    f) Microclusters, S.Sugano, Y.Nishina and S.Onishi, eds.,
       Springer 1987
    g) Metal Clusters, M.Moskovits,ed. J.Wiley, New York 1986
    h) T.D.Märk, A.W.Castleman,Jr. Experimental studies on cluster ions,
       Adv. in Atomic and Molecular Physics, Vol.20:65 (1985)
    i) M.R.Hoare, Structure and dynamics of simple microclusters,
       Adv.Chem.Phys. 40:49 (1979)
    k) Elemental and Molecular Clusters, Proceedings of the International
       School of Material Science and Technology, 13th Course,
       Erice 1987 (eds. G.Benedek and T.P.Martin), to be published by
       Springer Co. (Heidelberg), in Material Science Ser.
    l) "Large Finite Systems", J.Jortner, A. and B.Pullman,eds.
       D.Reidel, Dordrecht 1987 , Proc. of the 20. Jerusalem Symposium
       on Quantum Chemistry and Biochemistry
    m) J.E.Campana, Cluster ions. I. Methods,
       in Mass Spectrometry Review 6:395 (1987)
    n) M.A.Johnson and W.C.Lineberger, Pulsed methods for cluster ion
       spectroscopy
    o) Y.Imry, Physics of Mesoscopic Systems
    p) J.Jortner, D.Scharf and U.Landman, Energetics and dynamics of
       clusters, Int.Scool of Physics, "Enrico Fermi", Course XCVI
       Excited state spectroscopy in solids, North Holland 1987
    q) W.P.Halperin, Quantum size effects in metal particles,
       Rev.Mod.Phys.58:533 (1986)
    r) T.Kondow, "Ionisation of Clusters in collisions with high-Rydberg
       rare gas atoms, J.Phys.Chem. 91:1307 (1987)
    s) C.Hayashi, Ultrafine particles, Physics Today, December 1987
    t) J.Friedel, Small aggregates, Helv.Phys.Acta 56:507 (1983)
2. J.Jortner, Level structure and dynamics of clusters,
    Ber.Bunsenges.Phys.Chem.88:1 (1984)
3. a) J.Farges, M.F.de Feraudy, B.Raoult and G.Torchet, Structure and
       temperature of rare gas clusters in a supersonic expansion
       Surf.Sci. 106:95 (1981), and ref. 1 e+l
    b) Y.Z.Barshad and L.S.Bartlett, Electron diffraction of
       supersonically generated clusters, in ref. 1 e
4. T.P.Martin, Alkali-halide clusters, Physics Reports 95:167 (1983)
    and: Clusters what are they, in 1k
5. M.N.Barber, in "Phase Transitions and Critical Phenomena", Vol.8
    Eds. C.Domb and J.L.Lebowitz, Academic Press, New York 1983
6. Ph.Buffat, J.-P.Borel, Size effect on the melting temperature of
    gold clusters, Phys.Rev. A13:2287 (1976)

7. G.L.Allen et al. Small particle melting of pure metals
    Thin Solid Films 152:297 (1986)

8. J.W.M.Frenken and J.F.van der Veen, Observation of surface melting,
    Phys.Rev.Lett. 54:134 (1985)

9. J.W.M.Frenken and J.F.van der Veen, Dynamics and melting of
    surfaces, Surf.Sci. 178:382 (1986)

10. J.Jellinek, T.L.Beck and R.S.Berry, Solid-liquid
    like phase changes in simulated isoenergetic Ar_{13}
    J.Chem.Phys.84:2783 (1986), and ref. 1e and 1f.

11. J.Bösiger and S.Leutwyler, Surface melting transitions and phase
    coexistence in argon solvent clusters, Phys.Rev.Lett. 59:1895 (1987)

12. F.F.Abraham, Statistical surface physics: a perspective via computer
    simulations of microclusters, interfaces and simple fluids,
    Rep.Prog.Phys.45:1113 (82)

13. P.A.Montano, G.K.Shenoy, E.E.Alp, W.Schulze and J.Urban
    Structure of copper microclusters isolated in solid argon
    Phys.Rev.Lett. 56:2076 (1986)

14. E.Schumacher and M.Kappes, Isolated, bare metal clusters: abundances
    and ionisation, in ref.11.

16. W.Eckart, Workfunction of small metal particles:self-consistent
    spherical jellium background model, Phys.Rev. B29:1558 (1984),
    and priv.communication.

17. G.Makov, A.Nitzan and L.E.Brus. in 11, and J.Chem.Phys, 'submitted

18. W.F.Hoffman III, E.K.Parks, G.C.Riley, G.C.Nieman, L.G.Pobo, and
    S.J.Riley, The kinetics of reaction of nickel clusters with hydrogen
    and deuterium, Z.Phys.D7:83 (1987); see also ref.1d and 1e.

19. B.Mühlschlegel, Theoretical and experimental aspects of super-
    conducting small particles, Surf.Sci.106:350 (1981)

20. a) D.M.Cox, D.J.Trevor, R.L.Whetten, E.A.Rohlfing, and A.Kaldor,
    Aluminum clusters: Magnetic properties, J.Chem.Phys. 84:4651 (1986)
    b) I.S.Jacobs and C.P.Bean "Fine particles, thin films and exchange
    anisotropy" in Magnetism III, eds. G.T.Rado and H.Suhl, Academic
    Press, New York and London, 1966
    c) D.A.Garland and D.M.Lindsay, J.Chem.Phys. 80:4761 (1984)
    Matrix isolation of Li clusters
    d) K.W.Blazey, K.A.Müller, F.Blatter and E.Schumacher, Conduction
    electron spin resonance of caesium metallic clusters in zeolite X,
    Europhys.Lett. 4:857 (1987)

21. a) W.A. de Heer, K.Selby, V.Kresin, J.Masui, M.Vollmer, A.Châtelain,
    and W.D.Knight, Collective dipole oscillations in small sodium
    clusters, Phys.Rev.Lett.59,1805 (1987)
    b) G.Delacrétaz, E.R.Grant, R.L.Whetten, L.Wöste, and J.W.Zwanziger,
    Fractional quantisation of molecular pseudorotation in Na_3
    Phys.Rev.Lett.56:2598 (1986)
    c) D.G.Leopold, J.Ho, and W.C.Lineberger, Photoelectron
    spectroscopy of mass-selected metal cluster anions.I.Cu_n^- ,n=1-10,
    J.Chem.Phys.86:1715 (1987);
    d) C.L.Petiette, S.H.Yang, M.J.Craycraft, J.Conceicao, R.T.Laaksonen,
    O.Chesnovsky, and R.E.Smalley, Ultraviolet photoelectron spectroscopy
    of copper clusters,J.Chem.Phys.
    submitted Sept. 1987;
    e) G.Ganteför, K.H.Meiwes-Broer and H.O.Lutz, Photodetachment
    spectroscopy of cold aluminum cluster anions, Phys. Rev.A 1988 submitted
    and ref.1k;
    f) W.Schulze, K-P.Charlé and U.Kloss, in ref. 1a (1985)

22. a) M.L.Mandich, V.E.Bondybey and W.D.Reents, Reactive etching of
    positive and negative silicon clusters ions by nitrogen dioxide,
    J.Chem.Phys. 86:4245 (1987)
    b) Complex-forming reactions in neutral noble gas clusters
    D.J.Levandier, J.McCombie, R.Pursel and G.Scoles, Complex forming
    reactions in neutral noble gas clusters, J.Chem.Phys. 86:7239 (1987)

c) M.A.Jarrold and J.E.Bauer, A detailed study of the reactions of size selected aluminum clusters ions and oxygen.

23. a) M.Kawasaki, Y.Tsujimura and H.Hada, Oscillations of Photo-ionisation thresholds of small photolytic silver clusters on silver bromide grain surface, Phys.Rev.Lett. 57:2796 (1986)
b) P.Fayet, F.Granzer, G.Hegenbart, E.Moisar, B.Pischel and L.Wöste, Latent-image generation by deposition of monodisperse silver clusters, Phys.Rev.Lett.55:3002 (1985)

24. B.Donn, J.Hecht, R.Khanna, J.Nuth, D.Stranz and A.B.Anderson, The formation of cosmic grains: an experimental and theoretical study, Surf.Sci.106:576 (1981)

25. T.E.Gough, M.Mengel, P.A.Rowntree and G.Scoles, Infrared spectroscopy at the surface of clusters: SF on Ar, J.Chem.Phys. 83:4958 (1985), see also R.J.Le Roy et al. in ref.1 1.

26. T.Takagi, Ionised cluster beam technique for thin film deposition, Z.Phys.D3:170 (1986) and ref.1 a

27. W.C.Stwalley, The synthesis of large cluster ions from elementary constituents: A possible route to bulk antimatter. Proc. of the Rand corporation workshop on Antiproton science and technology, ed. Augenstein, Santa Monica, Cal. Oct.1987

28. K.Takayanagi, Growth of clusters studied by high resolution electron microscopy, in ref. 73.

29. a) A.K.Petford-Long, N.J.Long, D.J.Smith, L.R.Wallenberg, and J.O.Bovin, Atomic resolution study of structural rearrangements in metal clusters, page 127 in ref. 1e.
b) S.Iijima, Some experiments on structural instability of small particles of metals, ref. 1f.

30. R.Birringer, U.Herr, and H.Gleiter, Nanocrystalline materials - a first report,in "Grain Boundary and Related Phenomena", Proc. of JIMIS-4 (1986), Supplement to Transactions of the Japan Institute of Metals

31. F.Frank, W.Schulze, B.Tesche, J.Urban and B.Winter, Formation of metal clusters and molecules by means of the gas aggregation technique and characterisation of the size distribution, Surf.Sci. 156:90 (1985)

32. H.Haberland, U.Buck, and M.Tolle, Velocity distribution of supersonic nozzle beams, Rev.Sci.Instum.59:1712 (1985)

33. H.P.Birkhofer, H.Haberland, M.Winterer,and D.R.Worsnop, Penning, photo- and electron impact ionisation of argon clusters. Ber.Bunsenges.Phys.Chem.88:207 (1984)

34. O.F.Hagena, Condensation in free jets: Comparison of rare gases and metals, Z.Phys.D4:291 (1987) and ref. 1 a

35. T.A.Miller, Chemistry and chemical intermediates in supersonic free jet expansion, Science 223:545 (1984)

36. Y.Liu, Q.-L.Zhang, F.K.Tittel, R.F.Curl, and R.E.Smalley, Photodetachment and photofragmentation studies of semiconductor cluster anions, J.Chem.Phys.85:7434 (1986)

37. a) L.Wöste or K.H.Meiwes-Broer in ref.1e and 1k
b) I.Katakuse, T.Ichihara, Y.Fujita, T.Matsuo, T.Sakurai and H.Matsuda, Int.J.Mass Spectrom.Ion Phys. 67:229 (1985)

38. a) U.Buck, Properties of neutral clusters from scattering experiments J.Phys.Chem., feature article, accepted 1988
b) U.Buck and H.Meyer, Scattering analysis of cluster beams, formation and fragmentation of small Ar clusters, Phys.Rev.Lett.52:109 (1984)

39. H.Haberland, C.Ludewigt, H.G.Schindler and D.R.Worsnop, Clusters of water and ammonia with excess electrons, Surf. Sci. 156:157 (1985) more details to the source can be found in ref. 11.

40. F.Thum and W.O.Hofer, No enhanced electron emission from high density atomic collision cascades in metals, Surf.Sci. 90:331 (1979)

41. L.Friedman and G.H.Vineyard, Cluster ion impacts on solid surfaces, Comments on At.Mol.Phys. 15:251 (1984)

42. H.Haberland and M.Winterer, Improved detection of large rare gas cluster ions, Rev.Sci.Inst. 54:764 (1983)

43. W.D.Knight, K.Clemenger, W.A.de Heer, W.A.Saunders, M.Y.Chou, and M.L.Cohen, Electronic shell structure and abundances of sodium clusters, Phys.Rev.Lett. 52:2141 (1984)

44. K.Clemenger, Ellipsoidal shell structure in free electron metal clusters, Phys.Rev.B33:1359 (1985)

45. I.Boustani, W.Pewesdorf, P.Fantucci, V.Bonacic-Koutecky and J.Koutecky, Systematic ab initio CI study of alkali metal clusters Phys.Rev. B35:9437 (1987)

46. D.M.Lindsay, Y.Wang, and T.F.George, The Hückel model for small metal clusters. II. Orbital energies, shell structures, ionisation potentials, and extrapolation to the bulk limit, J.Chem.Phys. 85:3500 (1987)

47. M.M.Kappes, M.Schär, E.Schumacher and A.Vayloyan, On ionisation induced unimolecular dissociation of sodium clusters, Z.Phys. D5:359 (1987)

48. C.Bréchignac, Ph.Cahuzac, and J.Ph.Roux, Photoionisation of potassium clusters, J.Chem.Phys, 87:229 (1987)

49. J.A.Barker,"Interatomic potentials for inert gases from experimental data" in "Rare Gas Solids", eds. M.L.Klein and J.A.Venables, Academic Press 1977

50. H.Haberland, A model for the processes happening in a rare-gas cluster after ionisation, Surface Science 156,305 (1985)

51. C.Kittel, Introduction to Solid State Physics, Advanced Topic F 4th edition, Wiley, New York 1976

52. M.Alonso and E.J.Finn, Fundamental University Physics, Vol.3, chapter 5.2, example 5.1, Addison-Wesley publishing Co. 1974

53. a) N.Schwentner, E.E.Koch, and J.Jortner, "Electronic excitations in condensed rare gases", Springer Tracts in Modern Physics, Vol.107 Springer, Berlin 1986
b) G.Zimmerer, Creation, Motion and Decay of excitons in rare gas solids, Int.Scool of Physics, "Enrico Fermi", Course XCVI, p.37, Excited state spectroscopy in solids, North Holland 1987

54. H.U.Böhmer and S.D.Peyerimhoff, MRD-CI calculations for the ground state potential energy surface of Ar_3^+ , Z.Phys.D4:195 (1986)

55. M.Amarouche, G.Durand and J.P.Malrieu, Structure and stability of Xe_{19}^+ cluster, J.Chem.Phys. submitted

56. J.Hesslich and P.J.Kuntz, A diatomics-in-molecules model for singly ionised argon clusters, Z.Phys.D2:251 (1986)

57. H.H.Michels, priv.comm. 1987

58. H.Haberland, On the electronic structure of a singly ionised cluster composed of closed shell atoms or molecules, in ref. 1 e.

59. J.M.Soler, J.J.Sáenz, N.Garcia and O.Echt, The effect of ionisation on magic numbers of rare gas clusters, Chem.Phys.Lett. 109:71 (1984)

60. P.Scheier and T.Märk, Observation of sequential decay in metastable Ar cluster, Phys.Rev.Lett.59:1813 (1987)

61. A.J.Stace, A measurement of the average kinetic energy releases during the unimolecular decomposition of argon ion clusters, J.Chem.Phys.85:5774 (1986)

62. H.Falter, O.F.Hagena,W.Henkes, und H.V.Wedel, Einfluss der Elektronenenergie auf das Massenspektrum von Clustern in kondensierten Molekularstrahlen, Int.J.Mass Spec.Ion Phys. 4:145 (1970), see also ref. 27.

63. C.T.Reiman, R.E.Johnson and W.L.Brown, Sputtering and luminescence in electronically excited solid argon, Phys.Rev.Lett.53:600 (1984)

64. J.Schouch, P.Borgensen, O.Ellegard, H.Sorensen, C.Clausen, Erosion of solid neon by means of keV electrons, Phys.Rev.B34:93 (1986)

65. D.Kreisle, O.Echt, M.Knapp and E.Recknagel, Time dependent size distribution of xenon cluster ions, Phys.Rev. A33:786 (1986)

66. C.R.Albertoni, R.Kuhn, H.W.Sarkas and A.W. Castleman
    Photodissociation of rare gas cluster ions: Ar_3^+
    J.Chem.Phys.87:5043 (1987)

67. H.Haberland, C.Ludewigt, H.G.-Schindler and D.R.Worsnop, Clusters of
    water and ammonia with excess electrons, Surf.Sci. 156:157 (1985)

68. K.-H.Meiwes-Broer, priv. comm., and ref./21e/

69. R.N.Barnett, U.Landman, C.L.Cleveland, J.Jortner, surface states of
    excess electrons on water clusters, Phys.Rev.Lett.59,811 (1987)
    and ref. 11 and 2

70. H.Haberland,C.Ludewigt,H.G.-Schindler and D.R.Worsnop, Field
    detachment of $(H_2O)_2^-$ clustered with rare gases,
    Phys.Rev.A36:967 (1987)

71. G.H.Lee, J.Eaton, Ch.Ludewigt, H.Haberland, and K.H.Bowen,
    unpublished data

72. P.Stampfli and K.H.Bennemann, Theory for the electron affinity of
    small polar clusters $(NH_3)_N$, Phys.Rev.Lett.58:2635 (1987)

73. P.Jiang, F.Jona, P.M.Marcus, Surface effects in metal microclusters
    Phys.Rev.B36:6336 (1987)

74. Proceedings of the 4. International Meeting on Small Particles and
    Inorganic Clusters, Aix-en-Provence, France, July 1988.

75. 106th Course of the International School of Physics "Enrico Fermi"
    The Chemical Physics of Atomic and Molecular Clusters
    G.Scoles, editor

ELECTRONIC INTERACTION OF IONS(ATOMS) WITH METAL SURFACES

H.J. Andrä

LAGRIPPA – CEA-CNRS – CEN-Grenoble, France
and
Institut für Kernphysik, University of Münster, FRG

1) Introduction

The goal of this lecture is not to cover the whole histo-
rical development of the vast field of Ion Scattering at Surfaces; instead it will concentrate on a few more recent improvements of the understanding of the Electronic Interaction
of Ions(Atoms) with Metal Surfaces. References are given only
for the examples discussed and are therefore far from complete.

To be even more specific this lecture will be devoted only
to processes taking place in front of the metal surface at a
rather low vertical velocity component parallel to the normal
of the surface ($0.01 < v_v < 0.2$ a.u. corresponding to $2.5 < E < 1000$ eV/u). This allows an adiabatic (or quasi-adiabatic)
treatment of the electronic interation between the ion and the
metal surface ("ion" corresponds to ion or atom throughout the
lecture, if not otherwise stated). At these conditions the probability of the ion penetrating the last surface layer of atoms
of an ideal metal surface becomes very small (< 0.05). The ion
is thus practically fully reflected from the surface. As indicated in Fig. 1a the angular distribution of this reflection
depends, however, critically on the ion-surface interaction
potential (Coulomb repulsion due to imperfect screening by the
electrons at decreasing distances) and the vertical velocity
component. This interaction potential is obtained as the superposition of the contributions from all surface atoms. It is
planar at larger distances from the last atomic layer and
becomes more or less sinusoidal at smaller distances. These
distances depend slightly on the vertical velocity component of
the ion.

In the planar region one expects nearly perfect specular
reflection whereas in the sinusoidal region a wide angular
scattering distribution is expected. It is obvious from Fig. 1a
that this angular distribution (including interference effects)
carries information on the surface structure and indeed this
technique is being widely used to determine the structure of
surfaces with hyperthermal He beams (< 100 meV) and in particular the distribution of adsorbed atoms on such surfaces

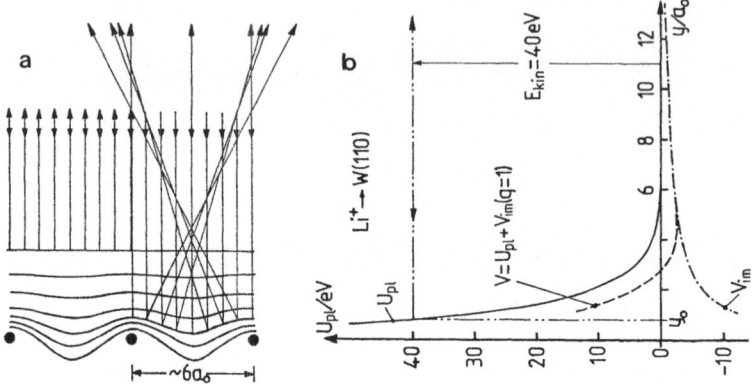

Fig. 1 a) Reflection of ions with two different
energies from a single crystal surface. b) Planar
potential for Li$^+$-W(110) interaction including the
image potential.

[1,2,3]. Without going into a lengthy discussion on the particular shape of the ion-surface interaction potentials at different relative velocities [4] it may be sufficient for this
lecture to accept the qualitative behaviour of Fig. 1a. This
behaviour may be specified for higher energies in the planar
region on the basis of a Molière-type ion-atom inter-action
potential with an important Z1 and Z2 dependence [5] where

$$U_{pl}(y) = 2\Pi \cdot Z_1 \cdot Z_2 \cdot \mu \cdot a \cdot f_{pl}(x=y/a); \quad a=0.8853 \cdot a_0/(\sqrt{Z_1}+\sqrt{Z_2})^{2/3}$$
$$(1.1)$$
$$f_{pl}(x) = [1.17 \cdot \exp(-0.3 \cdot x) + 0.46 \cdot \exp(-1.2 \cdot x) + 0.02 \cdot \exp(-6 \cdot x)],$$

μ is the density of surface atoms per unit area in a.u.. The
example for Li$^+$ on a W(110)-surface is shown in Fig. 1b. The
Li-ion approaches the surface with 40 eV up to the turning
point at about 1.4 a.u. and is nearly specularly reflected
because the turning point lies in a region where the planar
potential is a good approximation. It should be noted that the
momentum transfer to the solid in such a reflection is twice
the momentum of the projectile and corresponds to the order of
1000 phonons if this momentum is completely transferred to
phonons - an assumption which has to be corrected by a factor
equivalent to the Debye-Waller factor for neutron scattering
[6] due to the collective repulsion of all surface atoms.
With a typical phonon energy of 0.01 eV this would yield a
"vertical energy loss" of the order of 10 eV.

For the following discussion of the electronic interaction
of projectiles in front of the surface it is now assumed that
the projectile is either reflected from the surface or that
quantities are observed which are not very much affected when
the projectiles penetrate into the surface.

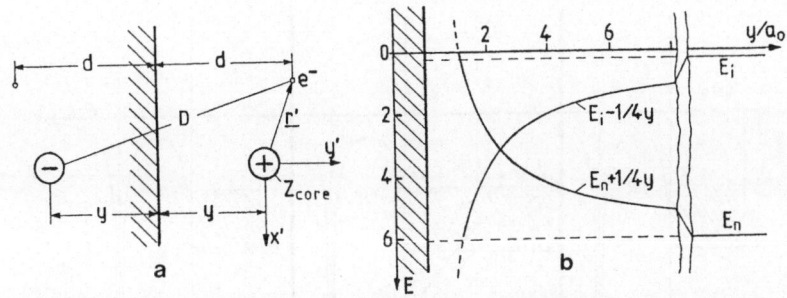

Fig. 2 a) Construction of the image potential of
a valence electron. b) Modification of terms of a
negative ion and of a neutral atom due to the image
potential.

2) Image potential seen by the atomic electron at intermediate and large distances

When an ion is approaching a metal surface the long range
Coulomb interaction causes a rearrangement of the metal electrons to form a surface charge distribution which acts as if an
image charge existed. This image charge not only creates an
attractive potential for the ionic core with an effective
charge Z_{eff} but also perturbs the ionic electrons.

The attractive potential for the ion core $V_{im} = -1/2y$ has
to be added to the repulsive potential in Fig. 1b to form a
shallow potential well of the order of eV's which may influence
the ion trajectory or may even lead to trapping into this
potential well in front of the surface under very specific conditions when the energy loss to phonons is high enough. For the
present discussion it can be neglected, however, but it should
be noted that highly charged ions may suffer a considerable
long range attraction which has to be taken into account in
some cases.

The concept of the image charge is an ingenious approximation for the description of the interaction of all electrons
of the conduction band with the ionic electrons [7]. In this
approximation a valence electron in the primed coordinate
system of the ionic core with Z_{core} sees besides the core-
potential an additional image potential $V_{im}(\underline{r}',y) = Z_{core}/D-
1/4d$ with $D=\sqrt{x'^2+z'^2+(y'+2y)^2}$ and $d=y+y'$ as defined in Fig.
2a. The term $1/4d$ is obtained when approaching the electron
from infinity with fixed ion in front of the surface. When
expanding $V_{im}(\underline{r}',y)$ in terms of y' [8] the image potential

$$V_{im}(y',y)= (2Z_{core}-1)/4y - (Z_{core}-1)\cdot y'/4y^2 + \ldots \qquad (2.1)$$

in the primed coordinate system of the atom is obtained which
has the following consequences on the atomic terms involved,
depending on Z_{core} which forms with one valence electron an ion
with charge q:

$$Z_{core}=0, \; q=-1 : \; V_{im}(y',y) = -1/4y + y'/4y^2 \qquad (2.2)$$

With $Z_{core}=0$ the valence electron forms a negative ion

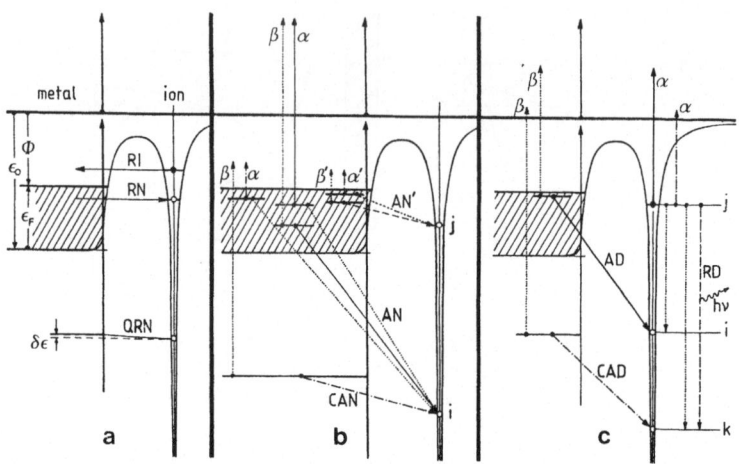

Fig. 3a,b,c   Ion(atom)-surface charge exchange processes

which is exposed to an electric field $1/4y^2$ and the eigenstates of which are shifted downwards by $-1/4y$ when the ion approaches the surface.

$$Z_{core}=1, \quad q=0 \ : \ V_{im}(y',y) = 1/4y \tag{2.3}$$

With $Z_{core}=1$ the valence electron forms a neutral atom, the eigenstates of which are shifted upwards by $1/4y$ when the atom is approaching the surface.

$$Z_{core}=2, \quad q=+1 \ : \ V_{im}(y',y) = 3/4y - y'/y^2 \tag{2.4}$$

With $Z_{core}=2$ the valence electron forms a positive ion which is exposed to an electric field $-1/y^2$ and the eigenstates of which are shifted upwards by $3/4y$ when the ion approaches the surface.

The unprimed and primed coordinate systems in Fig. 2a for the surface-(lab-)- and for the ion-frame, respectively, with the z- and z'-axis pointing towards the observer will be used throughout this lecture.

3) Ion-surface charge exchange processes

Charge exchange between ions and metal surfaces is a long known phenomenon. The different types of processes possible have been summarized by Hagstrum in an excellent review [9] and are depicted in Fig. 3. The electronic properties of the metal surface are described by a conduction band with ε_0 being the depth of the potential well for the free electron jellium model (independent free electrons imbedded into a continuous and constant positive charge distribution) and ϕ the work-function to yield the Fermi energy $\varepsilon_F=-(\varepsilon_0-\phi)$ as measured from the bottom of the well whereas ε_0 and ϕ are measured from the vacuum level.

If an ion at a given distance in front of the surface has an occupied term above the Fermi energy or Fermi-edge of the metal this term is facing an empty continuum and can consequently lose one electron into this continuum via an adiabatic single electron tunneling process which is called <u>Resonance Ionization (RI)</u>. If on the other hand the ion has an empty term below the Fermi edge this term is facing a completely filled continuum so that it may gain an electron via <u>Resonance Neutralization (RN)</u>. From this simple picture in Fig. 3a one can deduce at once that the neutralization of a scattered ion beam will strongly depend on the energetic position of the Fermi edge relative to the corresponding atomic ground state. The well known Langmuir-Taylor detector [10] for Alkali atoms at thermal energies may serve as the best example for a long existing and very successful application.

Also indicated in Fig. 3a is a charge exchange process between a filled core level of the metal and an empty deep lying ion term which are separated accidentally only by a very small energy gap. If the kinetic energy of the ion is sufficient to bridge this energy gap a <u>Quasi Resonance Neutralization (QRN)</u> can take place.

Besides the single electron resonance charge exchange phenomenon the electronic interactions involving two electrons, which are called in general Auger transitions, play an important role in the exploration of the ion surface interaction. One may distinguish two classes of Auger processes:

The <u>Auger-Neutralization (AN)</u> in Fig. 3b and the <u>Auger-Deexcitation (AD)</u> in Fig. 3c. In an AN process an electron from the conduction band of the metal jumps into some lower lying vacant ionic level (i or j). For reasons of energy conservation this jump is possible only when another electron of the conduction band takes up the released energy and is excited either into the continuum (AN) or into empty states of the metal above the Fermi energy (AN'). It is to be noted that the AN' process to the ionic term (j) occurs in competition to a RN process, the relative importance of which has to be discussed later on.

In an AD process an ion excited in a (metastable) term (j) is de-excited in front of the surface via an electron which jumps from the conduction band to some lower lying vacant ion-level (i) forcing the excited ionic electron (j) to leave into the continuum again for reasons of energy conservation. Also indicated in Fig. 3b is a <u>Core Auger Neutralization (CAN)</u>-process from a low lying core level of the metal to an empty ion term under excitation of a conduction electron into the continuum of the metal. This process occurs in competition to the QRN process in Fig. 3a. Furthermore a <u>Core Auger Deexcitation (CAD)</u>-process may occur in Fig. 3c which is, however, in strong competition to AD transitions, except for particular situations.

In all five cases of Auger transitions these so called direct processes (α) <u>cannot</u> be distinguished from the corresponding exchange processes (β) (dotted lines) where the initial electrons are interchanged.

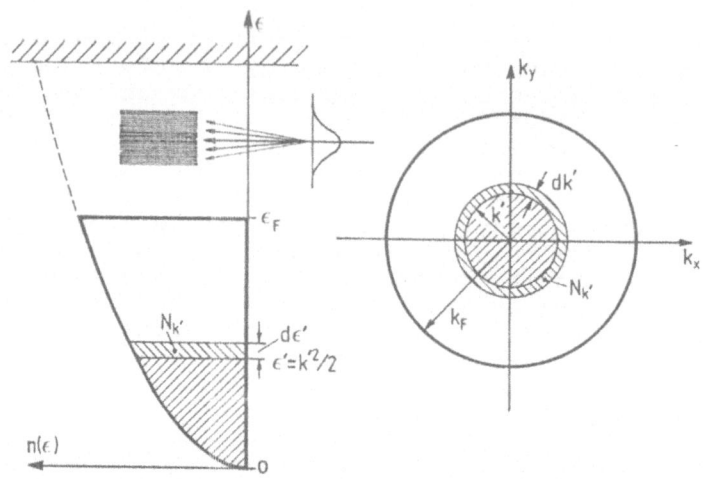

Fig. 4   Construction of the density of states
as function of energy ε from the Fermi-sphere
with constant density of states in momentum space.

In competition to the AD process one has to consider
Radiative Deexcitation (RD) which becomes possible after an RN-
or AN'-process if a very low lying empty ionic term (k) exists
so that the fluorescence yield may become important.

4) Transition rates of RN and RI processes

Due to the energy shift (induced by the image charge) of
the ionic energy terms such a term may be below the Fermi edge
and is therefore neutralized when the ion (atom) is at large
distance but may move above the Fermi edge and be ionized when
the ion is approaching the surface. This behaviour implies a
dynamic treatment of the interaction along the whole ion
trajectory in front of the surface.

The coupling matrix element between the electrons of the
ion and the metal is governed by the operator Z/r [11], where
Z is the effective charge of the ion "seen" by the electron to
be exchanged. For the case of RI the discrete term of the ion
interacts via this matrix element with the empty, near resonant
continuum states of the metal. The density of these states in
the free electron jellium model is obtained from the Fermi-
sphere with constant density in k-space as shown in Fig. 4b.

The transformation of the density of states from k-space
to energy-space is described in text books [12] but is visualized in Fig. 4 because it is used later on for a less usual
transformation. For a given k'-value the number of states
$dN'(k',dk')$ is simply counted in the spherical shell between k'
and k'+dk' and then transformed into $dN(\varepsilon,d\varepsilon)$ using the relation $\varepsilon=k^2/2$. The resulting density of states $n(\varepsilon)$ is proportional to $\sqrt{\varepsilon}$ and continues up to the ionization limit. It is only
the multiplication of $n(\varepsilon)$ by the Fermi function which dif-
ferentiates between occupied and unoccupied states.

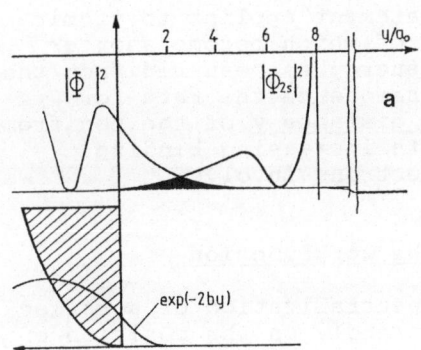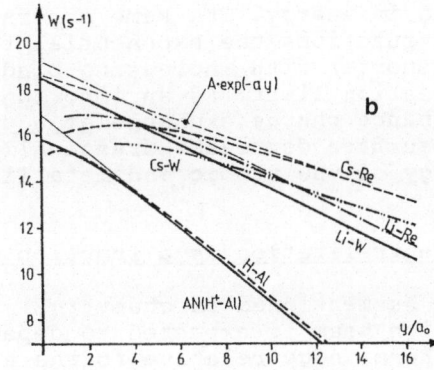

Fig. 5 a) Visualization of the tails of the squares of the metal wavefunctions and of their overlap with an atomic wavefunction. b) Transition rates for the charge exchange between some alkalis and metals.

$$n(\varepsilon) = (L^3/2\Pi) \cdot \sqrt{8\varepsilon} / (1+\exp[(\varepsilon-\varepsilon_F)/kT]) \qquad (4.1)$$

With this density of the continuum states the Golden rule may be applied to obtain the ionization rate $w(s^{-1})$ of the atomic state a

$$w(s^{-1}) = 2\Pi \ |\langle\varepsilon|Z/r|a\rangle|^2 n(\varepsilon). \qquad (4.2)$$

Since this ionization rate depends on the overlap of the metal and the atomic wave functions as indicated in Fig. 5a, which both decay exponentially with distance, it can immediately be deduced that the ionization rate has to depend exponentially on the distance y of the ion from the surface in Fig. 5b. In fact for many practical applications the rate can be approximated by $w(y)=A\cdot\exp(-a\cdot y)$ [9]. Taking a 2s-atomic wave function, however, as an example in Fig. 5a one can deduce in addition that this exponential behaviour is only valid for distances where the non-exponential internal structure of the atomic wave function is not overlapping yet with the tail of the metal wave function.

Furthermore the concept of unperturbed atomic wave functions is only valid for a small overlap with the potential well of the metal, i.e. for distances $y > \langle r_n,_l\rangle/2$. For smaller distances modified atomic wave functions have to be used so that deviations from the exponential behaviour of $w(y)$ may occur.

Without going into any detailed arguments it is said that the rate of electron exchange is independent of the population of the metal continuum. **The rate for resonance ionization can thus be set equal to the rate of resonace neutralization** when the ionic term is facing an occupied continuum with the same $n(\varepsilon)$.

One last but important feature can furthermore be deduced from the exponential tails of the metal wave functions into the vacuum. These tails decay proportional to $\exp(-2\cdot b\cdot y)$ where $b=1/\sqrt{2\cdot(\varepsilon_0-\varepsilon)}$ so that the exponential tails of the energetically lowest electrons are considerably shorter than those for electrons near the Fermi energy or those which are even higher

lying in energy. The same general statement applies to atomic
wave functions the exponential tails of which become shorter
and shorter with increasing binding energy as measured from the
ionization limit. As an important consequence the rate for
resonance charge exchange <u>at a given distance y</u> of the ion from
the surface decreases drastically with increasing binding
energy of the atomic and metallic electrons involved.

5) Neutralization as a function of the workfunction

As mentioned in chapter 2) the neutralization of a reflected ion beam is expected to depend strongly on the position of
the Fermi edge relative to the atomic ground state. This
concept has been verified in a series of impressive experiments
in Fig. 6a,b by Geerlings et al. [13,14].

In the apparatus sketched in Fig. 6a a 3 pA Li⁺ beam of
400 eV energy impinged at $\phi_{in}=15°$ on a W(110) surface with a
workfunction of 5.25 eV. The latter could be modified by evaporating Cs from a Cs-dispenser onto this W(110)-surface. With
increasing Cs-coverage the workfunction of a cesiated W-surface
is known [15] to decrease to a minimum of $\phi_{min}=1.4$ eV at
about 0.64 monolayers and then to increase back up again to the
bulk value of Cs of $\phi(Cs)=2.4$ eV as indicated in the left
corner of Fig. 7. During the experiment this change of the
workfunction could be monitored by a Kelvin probe. In a vacuum
of 10^{-7} Pa the scattered beam was observed at $\phi_{out}=10°$, charge
state analyzed and the positive charge state fraction was
plotted as a function of the workfunction of the surface. The
results in Fig. 6b show the expected strong dependence of the
neutralization (or of the survival of the positive charge
fraction as shown) on the workfunction. It follows, however,
not exactly the theoretical prediction (broken line).

This theoretical prediction is based on the concepts outlined in the chapters 1) to 3) which are summarized for this
specific experiment in Fig. 7. On the left is indicated the
Fermi edge of W with its modification due to increasing Cs-
coverage (line through crosses). When a Li⁺ ion is approaching

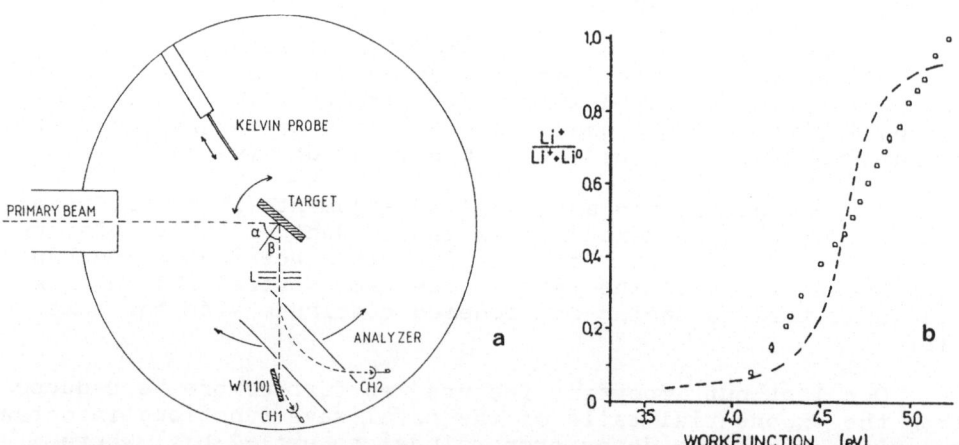

Fig. 6 a) Apparatus used in ref.[13]. b) Fraction of the
positive charge after interaction of Li⁺ with W(110). See
text.

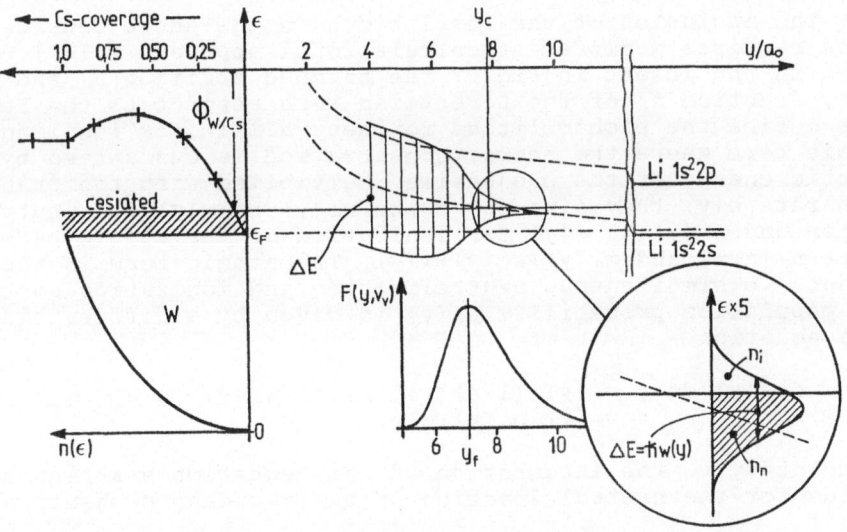

Fig. 7 Interaction of a Li⁺-ion with a cesiated
W(110)-surface. Details see text.

such a surface the ground and first excited terms of the Li-
atom have to be considered for the resonance neutralization of
which the ground term will contribute dominantly and will
therefore be discussed exclusively at first. At infinite distance from the surface this ground term has a binding energy of
-5.39 eV and lies therefore below the Fermi edge of W and even
further below a cesiated W-surface of course. When approaching
the surface the image potential is shifting the ground term
upwards as indicated by the broken line. At the distance y_c it
may cross the Fermi edge of a slightly cesiated surface so that
ionization dominates for $y<y_c$ and neutralization dominates for
$y>y_c$. During its approach to the surface the ion is thus
neutralized with a certain probability up to the point y_c and
then completely ionized again during the path from y_c to y_0,
the point of closest approach. In this way the ion loses
completely its memory of what happened during its approach and
turns back in a totally ionized state. It stays ionized up to
y_c where the neutralization starts with a rate [7,14] which
is approximated by the one for clean tungsten $w(y)=3.69 \cdot E18 \cdot$
$\exp(-1.07 \cdot y)$ s⁻¹ where y is to be measured in a.u..

Assuming that the width of the atomic level is negligible
the final fraction of ions which stays in the positive charge
state is given by the probability of survival in this charge
state

$$p_+ (\infty)=\exp\left[-\int_{y_c}^{\infty} w(y) \cdot dy/v_v \right]=\exp[-(A/a \cdot v_v) \cdot \exp(-a \cdot y_c)] \quad (5.1)$$

[14] from which the neutral fraction $p_n (\infty)=1-p_+ (\infty)$ can be
derived immediately. This calculated positive charge state
fraction $p_+ (\infty)$ is compared with the data in Fig. 6b and
exhibits an obvious discrepancy.

639

The inclusion of the level width in the above considerations requires a different calculational approach [16]. As shown in the insert in Fig. 7 the hatched fraction n_n and the empty fraction n_i of the Lorentzian with respect to the Fermi edge define the probabilities for neutralizing or ionizing the atomic term where the energy conservation is guaranteed by the kinetic energy of the projectile. Multiplied with the transition rate $w(y)$ they give the rates $w_n(y)=n_n \cdot w(y)$ for neutralization and $w_i(y)=n_i \cdot w(y)$ for ionization of the atomic term where $n_n + n_i = 1$ and $w_n(y) + w_i(y) = w(y)$. The atomic term is thus subject to simultaneous neutralization and ionization so that its population probability $p_n(y)$ is given by a differential rate equation

$$
\begin{aligned}
dp_n(y)/dt &= w_n(y) \cdot [1 - p_n(y)] - p_n(y) \cdot w_i(y) \\
&= w_n(y) - p_n(y) \cdot w(y).
\end{aligned} \tag{5.2}
$$

Using $dt = dy/v_v$ the integration of this equation starting at y_0 yields for the neutral fraction of the receding projectiles [16]

$$
p_n(\infty) = \int_{y_0}^{\infty} w_n(y) \cdot \exp\left[-\int_{y}^{\infty} w(y') \cdot dy'/v_v\right] \cdot dy/v_v. \tag{5.3}
$$

The integrand is composed of the neutralization rate at the position y multiplied by the survival probability of the electron in this atomic term from y to infinity. When approximating $w(y) = A \cdot \exp(-a \cdot y)$ [9] one obtains

$$
p_n(\infty) = \int_{y_0}^{\infty} n_n(y) \cdot (A/v_v) \cdot \exp\left[-a \cdot y - (A/a \cdot v_v) \cdot \exp(-a \cdot y)\right] \cdot dy \tag{5.4}
$$

$$
= \int_{y_0}^{\infty} n_n(y) \cdot F(y, v_v) \cdot dy. \tag{5.5}
$$

The function $F(y, v_v)$ is displayed in Fig. 7 and gives the spatial probability of where the neutralization takes place. If $n_n(y)$ is close to constant in the region where $F(y, v_v)$ is noticeably different from zero then $n_n(y_f)$ yields directly the neutral fraction of the receding projectiles because the integration of $F(y, v_v)$ yields 1. For such situations Overbosch et al. [16] have introduced the so called freezing distance y_f at the position of the maximum of $F(y, v_v)$ in order to define the mean capture distance which shifts towards the surface at higher v_v and away from the surface for lower v_v. If, however, $n_n(y)$ changes considerably in the region where $F(y, v_v)$ is noticeably different from zero then $p_n(\infty) \approx n_n(y_f)$ is still a good approximation but a numerical integration of $p_n(\infty)$ yields obviously better accuracy.

The application of this calculational procedure to the neutralization of Li+ does not, however, remove the discrepancies in Fig. 6b. It only shifts the total theoretical curve in Fig. 6b and modifies its shape slightly. Other approximations entering into the interpretation (e.g. T=0) don't explain the discrepancy either. Therefore the authors conclude from an extended experimental study that the Cs-coverage below one monolayer produces a strong local variation of the workfunction which is well seen by the Li ions but not by the

Kelvin probe or any other standard method for measuring workfunctions. In essence the Li atoms see at low coverages adsorbed Cs atoms on otherwise clean W so that the concept of a steadily varying workfunction is a very crude approximation. Qualitative estimates explain the observed deviations from the model calculation. [17].

These results are well corroborated by similar experiments by V. Kempter at al. [18] who detected not only the charge states but also the photon emission (2p -> 2s) of Li$^+$-projectiles reflected at 1000 eV and $\phi_{in}=3°$ from a cesiated W-surface. According to Fig. 7 the energy of the excited Li-1s^{2}2p term is by 1.63 eV higher than the ground term in the transition region. It can thus be populated in a similar manner as the ground state when the workfunction is reduced by an additional 1.63 eV. Indeed the experiment shows that with increasing Cs-coverage the (2p -> 2s) photon yield is increasing up to a maximum at about 0.45 monolayer and then drops down again while the positive charge fraction drops continuously from 1 to close to zero in the region from 0 to 0.5 monolayer. One should note, however, that the population of these two terms are competing processes so that a more elaborate treatment is required for the proper description of the 2p-term population which stays always much smaller than the ground term population.

6) Influence of the parallel velocity on the neutralization

When the energy of the incident Li$^+$ beam is varied a systematic trend in the data is observed in Fig. 8 which can only partially be explained by the modification of the small vertical velocity component (v_v [400 eV]= 2.7·E4 m/s). According to the formula for $p_n(\infty)$ the three different vertical velocities simply shift the function $F(\gamma, v_v)$ in Fig. 6b horizontally. Consequently the higher the vertical velocity the more the workfunction has to be reduced in order to obtain the same degree of neutralization. This conclusion implies a shift of

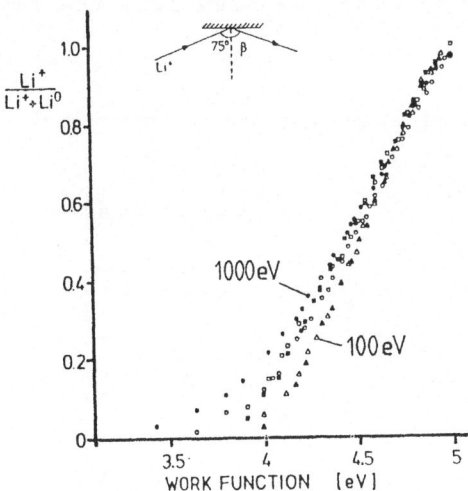

Fig. 8   Li+-fraction of Li-projectiles reflected from a cesiated W-surface. See text and ref.[20].

the data points to the left in Fig. 8 in qualitative agreement
with the data at intermediate degrees of neutralization. The
large shifts observed for the neutralization above 80% can,
however, not be accounted for by this simple argument. Instead
they can be attributed to a large extent to the modification of
the large velocity component parallel to the surface (v_P [400
eV]= 1·E5 m/s = 0.045 a.u.) which has been neglected so far.

From a theoretical point of view this parallel velocity
can be taken into account by the introduction of a Galilei-
translational factor in the atomic or metal wavefunctions.
Since the metal wavefunctions are plane waves this translational factor can be very well visualized in k(momentum)- and
energy-space when regarding the "moving" metal from the rest
frame of the ion. As first introduced by van Wunnik [19]
the Fermi-sphere is simply shifted in k(momentum)-space by
the momentum $m_e \cdot v_P$ (=v_P in a.u.) with respect to the ionic
k(momentum)-coordinates as indicated on the left in Fig. 9.

Using this picture and the transformation from k- to
energy-space as described in connection with Fig. 4 one obtains
the following construction for the density of states $n(\varepsilon)$ as
"seen" from the rest frame of the ion. The Fermi-sphere is cut
into a large number of spherical shells 1,2,3... of thickness
dk in which one finds N(k,dk) occupied states. As in Fig. 4
N(k,dk) is transformed into N(ε,dε) and is plotted as a function of ε in order to obtain $n(\varepsilon)$ on the right of Fig. 9. The
procedure has exactly the same result as in Fig. 4 up to the
spherical shell no.5 because it is still inside the borders of
the Fermi-sphere with constant density of states. For the
spherical shell no.6, however, due to the shift $m_e \cdot v_P$ (=v_P in
a.u.) the diagonally shaded area lies outside the Fermi-sphere
and is therefore also missing as shaded area in the $n(\varepsilon)$ distribution on the right of Fig. 9. This missing fraction occurs
of course as the horizontally shaded area of a partially filled
spherical shell no.7 which contributes on the right the horizontally shaded area of occupied states above the Fermi edge.
The density of states $n(\varepsilon)$ as seen from the rest frame of the
ion has thus the surprising shape on the right of Fig. 9 which
rises in the normal way from point A to point B and is then

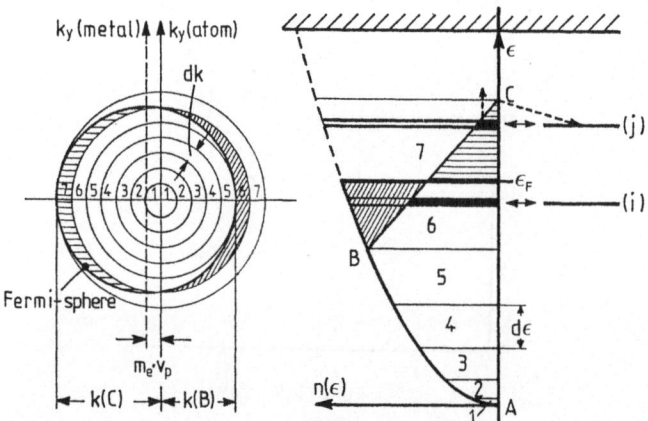

Fig. 9   The Fermi-sphere with constant
density of states in momentum- and energy-
representation as seen from the moving ion.

cut off to drop down close to linearly from point B to point C.
For $v_p = 0$ the normal $n(\varepsilon)$ distribution is of course restored
whereas for higher v_p-values the point B moves down and the
point C moves up and the linear drop from B to C is slightly
modified.

The consequences of this significantly altered density of
states as seen from the rest frame of the ion moving parallel
to the surface are also indicated in Fig. 9. An atomic term (i)
for instance which was for $v_p = 0$ well below ε_F is now in resonance not only with occupied (full bar) but also with empty
(empty double bar) states of the metal. Neutralization from the
occupied metal states to the atom and ionization from the atom
to the empty metal states will thus occur simultaneously and
one can only expect an equilibrium population at a given distance of the ion (see equations 5.3 to 5.5).

For the application of these features to the results in
Fig. 8 it is sufficient to note that full neutralization can
now only occur when the energy of the point B lies above the
atomic term (i) in the region where the function $F(y, v_v)$ is
noticeably different from zero. For Fig. 9 this means that full
neutralization for increasing parallel velocity can only occur
when the workfunction is more and more reduced, i.e. the points
B and C are both shifted upwards. This is not only in good
qualitative but also in good quantitative agreement with the
results in Fig. 8 .

A second aspect introduced by a parallel velocity is
indicated in Fig. 9 for an atomic term (j) which may be above
ε_F for $v_p = 0$ at all distances y from the surface. For $v_p = 0$ the
neutralization or population probability of this term is thus
zero. With increasing v_p, however, the point C moves above this
term (j) so that it may become populated by a small fraction
due to the neutralization rate $w_n(y)$ from a small number of
occupied metal states (full bar). The ionization rate $w_1(y)$ to
a large number of empty (empty bar) metal states is of course
counteracting so that in the simplest approximation $p_n(\infty) \approx$
$n_n(y_f)$ where $n_n(y_f)$ is the fraction of occupied resonant metal
states weighted by the transition probability which will turn
out in chapter 13) to be a function of the k-vector of the
metal state. Excited terms of ions well above the Fermi edge
can thus be populated with small fractions whenever the parallel velocity is high enough.

This latter argument also applies to the formation of
negative Li⁻ ions which according to V.Kempter [18] sets in
at a Cs-coverage where the 2p-term population starts dropping.

Extensive studies of the Li⁻ formation on cesiated W-
surfaces have been performed by J.Geerlings at al. [20,21]
who conclude, however, that the vertical velocity is still of
greater importance for the Li⁻ formation than the parallel
velocity.

It is interesting to note that the situation of $v_p > 0$
allows not only single electron resonance transitions to high
lying ionic terms (j) but also two-electron Auger transitions
of the type AN' of Fig. 3b to the term (j). This possibility is
shown in Fig. 9 with the dashed arrows, a phenomenon which has
not been taken into account as yet in any model calculation.

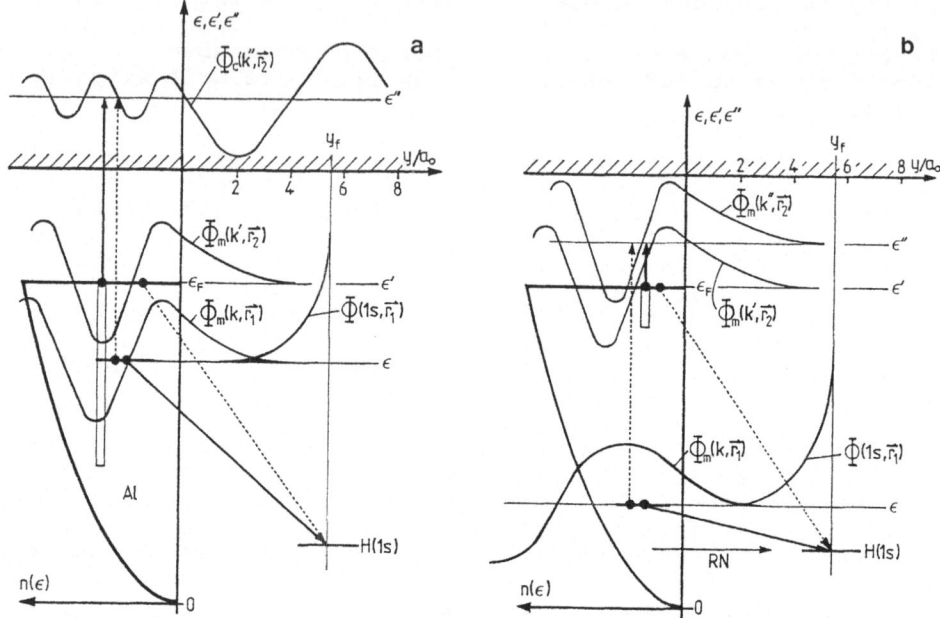

Fig. 10 Derivation of the AN-rates with the help of the matrix element in equ.(7.1) from the shapes of the wavefunctions; a) for a large energy gap between ε and H(1s), b) for a small energy gap between ε and H(1s).

7) Transition rates of AN- and AD- processes

A qualitative understanding of the transition rates of the AN/AD processes can again be derived from the structure of the transition matrix element and from the spatial extent of the metal- and ionic wavefunctions as shown in Fig. 10a,b for the specific case of H$^+$-neutralization on aluminium [22]. As for RN one can use the "Golden Rule" to express the transition rate W_{AN} as a function of the density of the final continuum states $n(\varepsilon")$ and of a matrix element

$$W_{AN}\,(s^{-1})=2\Pi\left|\langle\Phi_c\,(\underline{k}",\underline{r_2})\Phi(1s,\underline{r_1})\,|\,1/r_{12}\,|\,\Phi_m\,(\underline{k}',\underline{r_2})\Phi_m\,(\underline{k},\underline{r_1})\rangle\right|^2 n(\varepsilon").$$

$$(7.1)$$

Into this matrix element enters the overlap I of the wavefunctions $\Phi(1s,\underline{r_1})$ and $\Phi_m(\underline{k},\varepsilon,\underline{r_1})$ and the overlap II of the wavefunctions $\Phi_m(\underline{k}',\varepsilon',\underline{r_2})$ and $\Phi_c(\underline{k}",\varepsilon",\underline{r_2})$ "weighted" by the operator $1/r_{12}=1/(\underline{r_1}-\underline{r_2})$. The overlap II is non-zero because of this operator. Due to antisymmetrization direct (full lines-α) and exchange (dotted lines-β) matrix elements have to be taken into account. For the ion being at a given distance y_f from the surface the overlap I in Fig. 10a becomes largest when ε is approaching ε_F due to the increasing exponential tail of the metal wavefunction with increasing ε. At the same time the overlap II stays roughly constant because the energy difference $\varepsilon"-\varepsilon'=\varepsilon-E(1s)$ stays constant independent of ε' which may vary along the double bar which is proportional to the number of possible initial states for this given ε. In Fig. 10b ε is lowered and $\varepsilon-E_{1s}$ is reduced. Consequently the overlap I is

644

considerably reduced due to the shorter exponential tail of $\Phi_m(\underline{k}, \varepsilon, \underline{r_1})$ but the overlap II is strongly increased due to the smaller energy difference $\varepsilon''-\varepsilon'$ with respect to the situation in Fig. 10a. It is to be noted, however, that the double bar is much shorter. As for the RN-process the overlap I is responsible for the behaviour of the AN/AD-transition rates as a function of distance from the surface which thus drop off exponentially as for RN. Without discussing the modifications by the exchange contributions the AN-probability can now be set proportional to the square of [(overlap I)·(overlap II)]² times the number of initial states times the density of final states. With the arguments given above on the overlaps, one can thus roughly conclude that all states of the conduction band above the energy of the H(1s) ground term do contribute to the AN-transition rate with slightly increasing probability with ε up to a maximum at $\varepsilon=\varepsilon_F$. Although each particular $\varepsilon, \varepsilon', \varepsilon''$-AN-transition probability is clearly smaller than any RN-transition probability it is the large number of states involved in AN which may lead to comparable total transition rates for AN as for RN. This may be particularly the case when comparing with unfavourable RN-cases as indicated in Fig. 10b where the overlap I (responsible for RN) at the given distance is very small.

The comparison of RN- and AN-transition rates in Fig. 5b clearly supports these considerations since the resonance transition rates are observed to be systematically higher than the AN-rates except for the unfavourable case of H^+ on Al just mentioned which is of the same order as the corresponding AN-rate. Not included in this figure are favourable AN-cases when the energy difference $\varepsilon-E(b)$ becomes small as shown in Fig. 3b. Here again the AN- and RN-transition rates are expected to be of similar size but this has still to be shown in a proper calculation.

In contrast to the discussion of Fig. 7 and of equations (5.1) to (5.5) one can assume and deduce from numerous experimental results that the AN/AD transitions dominantly take place during the approach of the projectile towards the surface. In addition the re-ionization of a once AN/AD-neutralized atom is energetically blocked at velocities $v < 1$ a.u. so that the differential rate equation (5.2) reduces to

$$dp_n(y)/dt = w_{AN}(y) \cdot [1-p_n(y)] = w_{AN}(y) \cdot p_+(y), \qquad (7.2)$$

where $p_+(y)$ is the probability of survival of the low lying empty ionic term, to be filled by the AN/AD-process, from infinity up to the point y. Using $dt=dy/v_v$ and assuming $p_+(\infty)=1$ one obtains

$$p_+(y) = \exp\left[-\int_y^\infty w_{AN}(y') \cdot (dy'/v_v)\right]. \qquad (7.3)$$

The probability to make a transition at the position y in the element dy is then

$$p_t(y, v_v) \cdot dy = (w_{AN}(y)/v_v) \cdot p_+(y) \cdot dy,$$

which reduces with the approximation $w_{AN}(y)=A \cdot \exp(-a \cdot y)$ to

$$p_t(y, v_v) \cdot dy = F(y, v_v) \cdot dy \qquad\qquad (7.4)$$

so that the same function $F(y, v_v)$ as in equations (5.4) and (5.5) and in Fig. 7 is obtained. It gives as in chapter 5) the range of distance in which the AN/AD-transition will most probably take place with a maximum which shifts towards the surface with increasing vertical velocity v_v and vice versa.

8) Auger-Neutralization spectroscopy

Besides the famous calculation [24] and application of the AD-process as a detection scheme in the early Lamb-shift experiments [25] it was in particular the systematic study of the AN-process by H.D.Hagstrum [9,23,26,27] which can now be considered as the foundation of the field of <u>Ion Neutralization Spectroscopy (INS)</u>. Hagstrum's goal was to develop an alternative technique for the measurement of the density of states of metals by observing the electron spectrum following AN. To this end he used an experimental scheme [26] of which a modern version by M.Delaunay et al. [28] is shown in Fig. 11 which will be relevant for the next chapter too.

A polycrystalline W-target is placed at a pressure of 2-4·E-8 Pa in the center of three hemispherical grids G1-G3 and a hemispherical collector C (180°-LEED system) (a simple spherical collector in Hagstrum's original work). In 1953 the achievement of this vacuum was an extremely difficult task so that Hagstrum devotes several pages of reference [27] to vacuum problems ! Singly charged He+ ions from an electron impact ion source were mass and momentum analyzed and then directed via a lens system onto this target after it was sputter-cleaned with Ar+ ions. For the measurement of energy distributions the electric circuit on the right of Fig. 11 was used: The grids 1 and 3 were connected to the target (biased if necessary), a retardation potential was applied to grid 2 (to the collector in Hagstrum's original work), and the electron current emitted from the target surface was measured via the collector as a function of the retardation potential either biased or simply to ground. For the measurement of the total yield of electrons

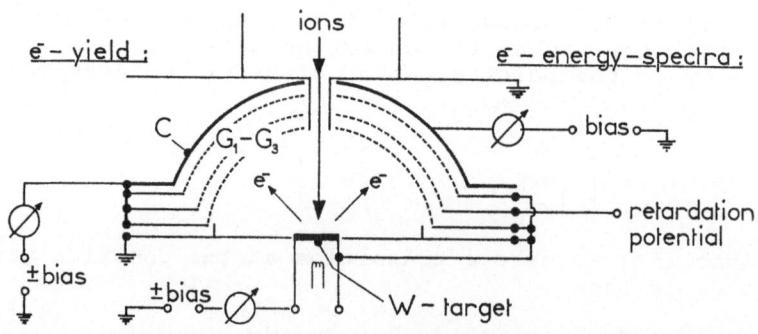

Fig. 11 Experimental apparatus for the measurement of yield and spectra of secondary electrons emitted when ions approach a metal surface.

emitted per ion the electric circuit on the left of Fig. 11 was used: The target was biased or grounded and all three grids were connected to the collector to form a simple collector as in Hagstrum's original work.

The electron spectra and their reconstruction obtained from He$^+$ interacting with a clean, polycrystalline W-surface are shown on the left side of Fig. 12. Assuming the jellium model for the conduction band of W with $\varepsilon_0 = 9.7$ eV, constant AN-transition rate for all values of ε (see Fig. 10) and no effects from level shifts or level broadenings one can construct the electron spectrum to be expected. An Auger-transition from ε_F to the He ground term yields the replica of the density of states, hatched vertically, with $\varepsilon''_{max} = 15.48$ eV while an Auger-transition from $\varepsilon = 0$ to the He ground term yields the replica of the density of states, hatched diagonally, with $\varepsilon''_{min} = -3.92$ eV. When allowing now Auger-transitions from all metal states $0 < \varepsilon < \varepsilon_F$ one expects the complete energy spectrum of the emitted electrons in full line which is a self-convolution of the density of states of the metal. Since ε''_{min} is negative the low energy part of the spectrum cannot escape from the metal.

For 40 eV of energy of incidence (for which the vertical velocity is neglected at first) this expected spectrum is astonishingly well reproduced by the experimental data on the right of Fig. 12 although the jellium model represents only a very crude approximation for the density of states of W. ε''_{max} is approximately obtained as upper limit of the corresponding spectrum and the cut-off at $\varepsilon'' = 0$ is well visible.

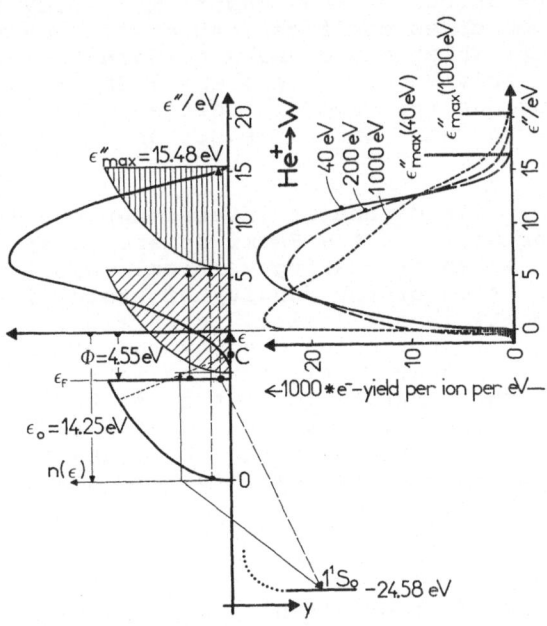

Fig. 12 Electron spectra observed when He$^+$ is approaching a W-surface at three different energies. Reconstruction of the electron spectrum for $v_v = 0$.

It was these observations which led Hagstrum to hope that the $n(\varepsilon)$-distribution could possibly be recovered by a deconvolution of the observed spectrum. One has to note, however, that for such a recovery of $n(\varepsilon)$ the AN-transition probability for all values of ε has to be known - an information which can be obtained neither experimentally nor theoretically without knowing $n(\varepsilon)$ already - even if one assumes that all other problems can be solved unambiguously. These principle problems are intimately related to the fact that the "Golden Rule" can only be applied to very smooth $n(\varepsilon)$-distributions when high transition rates (level widths) occur for projectiles coming very close to the surface. As soon as the level width becomes large compared to structures in the $n(\varepsilon)$-distribution the transition rate can only be calculated with the help of the known $n(\varepsilon)$-distribution [29]. In conclusion one has to state therefore that the initial goal of Hagstrum can only be achieved with rather limited energy resolution - if the deconvolution is possible at all.

One should not forget, however, that similar problems arise for the calculation of transition probabilities in photo electron spectroscopy, the well accepted technique for probing the electronic structure of solids and surfaces. In his excellent review on the comparison of both techniques [9] Hagstrum comes to the conclusion that both techniques should corroborate in order to obtain the best possible results.

Another striking feature in the experimental spectra, which spoils unfortunately the initial goal of Hagstrum, is the strong dependence of the spectral shape on the incident energy. The corresponding He velocities for energies between 40 and 1000 eV lie between 0.02 and 0.1 a.u. of which the upper value is somewhat above the limit of the validity of the adiabatic or Born-Oppenheimer approximation. Hagstrum [23,27] therefore attributes the observed modifications of the spectra to time dependent effects which may cause a certain energy broadening or to the level shifts due to very small freezing distances. It is, however, obvious from Fig. 12 that an upward shift of the He ground term can cause only a reduction in energy of the spectra.

Besides the time dependent interaction one may discuss the Galilei-transformation of the Fermi-sphere in order to visualize the effect of an increasing vertical velocity. One notes that the modification of the density of states in Fig. 9 due to a relative velocity is not limited to parallel velocities since the Fermi-sphere is isotropic. One may therefore try to use Fig. 9, with v_p replaced by v_v, for the interpretation of the energy shift of the upper tails of the spectra. With the point C in Fig. 12 as the new energy $\varepsilon_C = \varepsilon_F + 2.43$ eV for 1000 eV He$^+$ from which Auger transitions to the He ground term can occur one can derive immediately a new value for $\varepsilon''_{max}(1000$ eV$)$ which is shifted upwards by $d\varepsilon''_{max}(1000$ eV$)=4.86$eV with respect to $\varepsilon''_{max}(0$ eV$)$. The same procedure yields $d\varepsilon''_{max}(40$ eV$)=0.93$eV for 40 eV He$^+$. These shifts produce the new values of $\varepsilon''_{max}(40$ eV$)=16.41$ eV and $\varepsilon''_{max}(1000$ eV$)=20.34$ eV indicated in Fig. 12 which are to within 0.5 eV in excellent agreement with the observations. One may conclude therefore that the Galilei-translational factor, i.e. the shifted Fermi-sphere allows to visualize quite well the shift of the upper tail of the observed spectra with increasing velocity. This picture will, however, yield

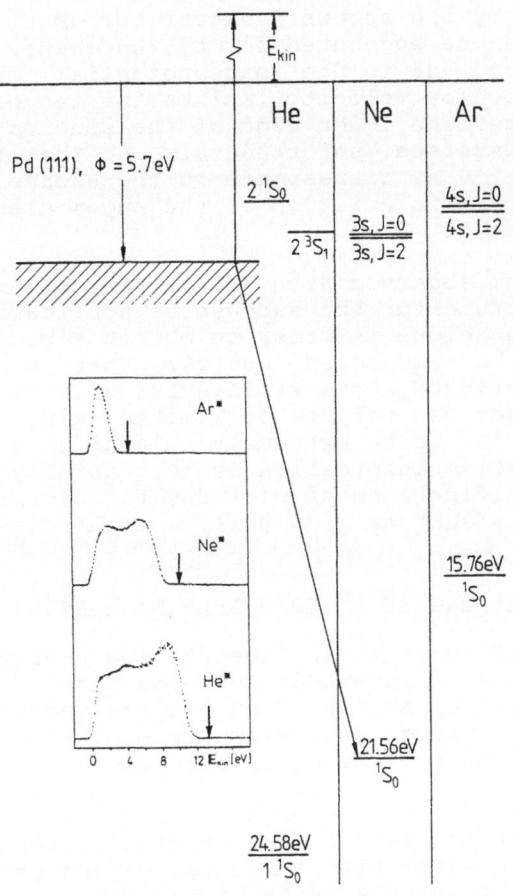

Fig. 13   AN-spectra observed when metastable rare
gas atoms are interacting with a Pd(111)-surface.

only a sharp upper limit of the energy to be observed while a
complete time dependent treatment of the interaction including
the Galilei-transformation is expected to yield a tail to
higher energies without an exact limit.

In order to avoid these problems connected to the incident
velocity of charged particles Hagstrum has used the lowest
possible ion energies of about 5 eV in his systematic study of
the INS. [9]. More recently G.Ertl et al. [30-33] have
further improved the INS by using neutral rare gas atoms in
metastable excited states at thermal velocities. In Fig. 13 the
ground terms and the metastable states of the light rare gas
atoms are shown with respect to the workfunction of a Pd(111)
surface. The metastable states lie above the Fermi edge so that
the atoms approaching this surface are resonance ionized at
large distance and a singly charged rare gas ion is continuing
to approach the surface so that an Auger neutralization can
take place at thermal energy plus the energy due to the attractive image force. Under such conditions one can expect the
velocity dependent effects to be as small as possible. Indeed
the upper edge of the electron spectra obtained in the insert
corresponds very well to the potential energy available in the

ions (indicated by the arrows), except for about 1 eV. This energy defect can be accounted for by the energy shift of the ionic ground terms due to the image potential, which is a minimum since the low velocity implies the largest possible freezing distance. The lower edge of the spectra cannot fully escape from the surface, unfortunately, so that the steep rise of the spectra at 0 eV corresponds to the escape function from a metal when isotropic emission of the Auger electrons is assumed in the metal.

The spectrum observed with He* is very clean and shows sufficient structure for the successful application of the deconvolution technique in order to obtain some information on $n(\varepsilon)$. It should be remembered, however, that the principle difficulties mentioned above still exist so that any $n(\varepsilon)$ obtained in this way can only be of limited value on an absolute scale but turns out to be extremely helpful on a relative scale. Such relative information is successfully employed for the study of modifications of $n(\varepsilon)$ due to adsorption of atoms or molecules on a surface [30-33].

9) Interaction of highly charged ions with metal surfaces

To the great surprise of Hagstrum low energy electrons were dominating the AN-spectrum when the charge state of Kr-projectiles impinging at 200 eV on a clean tungsten surface was increased from 1 through 4 as shown in Fig. 14 - a behaviour which he also found for the other rare gases [23,27] . He further observed:

i) that the electron yield Γ (number of electrons emitted per incident ion) increases with ionic charge and hence with total potential energy available in the ion;

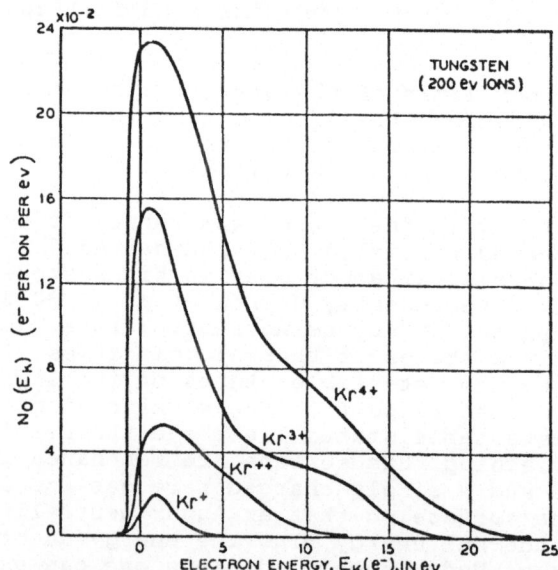

Fig. 14 AN-spectra observed when Kr-ions with various charges interact with a W-surface.

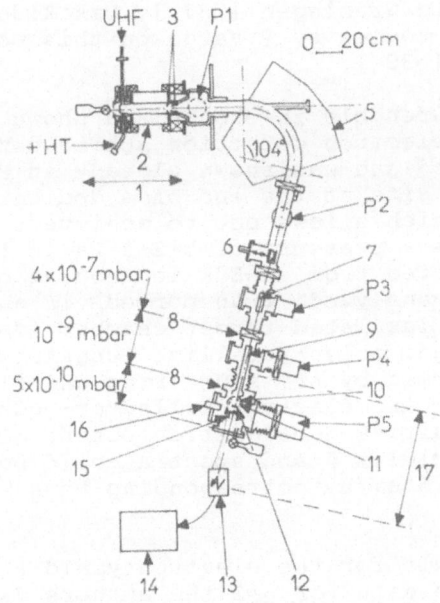

Fig. 15 Apparatus for studies on the
interaction of highly charged ions with
surfaces: (1) ECR-source, (9) retardation
lens, (10) electron-collector, (12) target.
See ref.[36] for more details.

ii) that faster and faster electrons are produced with more
highly charged ions but that the mean energy of the ejected
electrons is relatively independent of ionic charge;

iii) that Γ drops slowly with increasing ion energy.

From the observations i) and ii) he concluded that the
main features can be well interpreted if one assumes that
multiply charged ions are neutralized in a series of approximately isoenergetic steps of mean energy $d\varepsilon''$. This conclusion
was confirmed by a theoretical model calculation by Arifov et
al. [34] in 1973. These authors consider RN and subsequent
Auto Ionization (AI) as well as AN processes between a surface
and Hydrogen-like states of highly charged ions and obtain the
average energy of an ejected electron to be $(15-2\cdot\phi)eV < d\varepsilon'' <
(30-2\cdot\phi)eV$. On this basis they assume $d\varepsilon''$ to be a constant for
all ions so that the electron yield Γ has to become a linear
function of W_q – the total potential energy of the ion which is
set free when neutralizing it in free space or of W_q' when
neutralizing it in front of a metal with workfunction ϕ:

$$\Gamma = k\cdot W_q = k\cdot\Sigma_q U_B(q^+) \approx k'\cdot\Sigma_q [U_B(q^+)-2\cdot\phi] = k'\cdot W_q'. \qquad (9.1)$$

This relation for the so called Potential Emission (PE) could
be experimentally confirmed for highly charged ions of the rare
gases interacting with a molybdenum surface up to q=7 [34].

With the invention of powerful sources for the efficient
production of highly charged ions [35,36] a renaissance of
these studies became possible with further increasing charge

states up to q=12 in Groningen [37], Oak Ridge [38], and Grenoble [36]. A review by P.Varga on this more recent work has just appeared [39].

The setup in Grenoble is chosen and shown in Fig. 15 as an example since its electron detection scheme corresponds closely to Hagstrum's method and was shown already in Fig. 11. This detection assembly sits at the end of a dedicated ultra high vacuum beam line which allows one to achieve via two differential pumping stages a pressure of $5 \cdot E{-}8$ Pa in the target region. Ions extracted from an ECR ion source operating at 10 or 14 GHz were q/m-analyzed, transported via an electrostatic lens system to the insulated target chamber, decelerated, and focused on a cleaned, polycrystalline tungsten target. The cleaning was performed by Ar^+-sputtering and flash heating of the target. Target- and electron collector- currents were measured at the positive deceleration potential so that total electron emission yields Γ and spectra could be measured down to ion energies of 5 eV/u, corresponding to a vertical velocity $v_v \approx 3 \cdot E4$ m/s.

A typical result for the electron yield Γ is shown in Fig. 16 [40]. From the data for q=2 the authors deduce that the <u>Kinetic Emission (KE)</u> of electrons is not negligible so that the Γ observed is generally composed of $\Gamma = \Gamma(PE) + \Gamma(KE)$. Under the assumption that PE is negligible for q=2 the KE can be obtained from shaded area and can be accounted for in measurements with q>2 by subtracting this slowly increasing KE-contribution via a linear extrapolation. The data in Fig. 16 thus represent dominant PE which after subtraction of KE clearly increases with q and decreases slowly with vertical velocity above $1 \cdot E5$ m/s.

When plotting the data at $v_v = 4 \cdot E4$ m/s and at $v_v = 2 \cdot E5$ m/s as a function of W_q in Fig. 17a,b, respectively, one observes indeed in Fig. 17a the linear dependence of Γ on W_q as predicted by the formula. For Fig. 17b, however, a systematic deviation from the linear dependence is observed which the authors ascribe to the higher vertical velocity. They argue

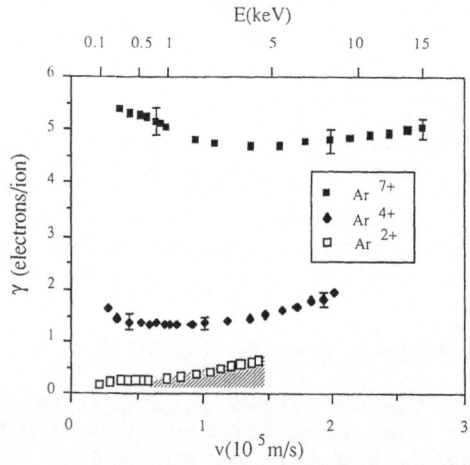

Fig. 16 Electron yield for Ar^{q+}-ions versus v_v. Shaded area indicates contribution from KE.

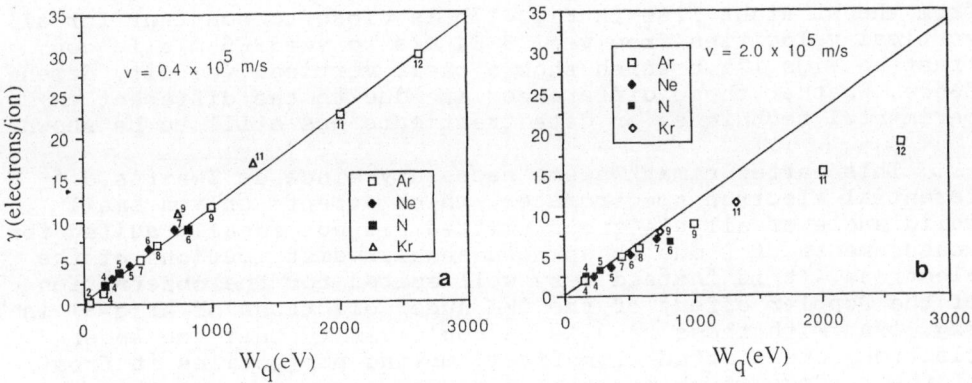

Fig. 17 Electron yield vs W_q, a) at $v_v = 0.4 \cdot E5$ m/s, b) at $v_v = 0.4 \cdot E5$ m/s.

that the time $T = s_0/v_v$ available for the ion within a given surface distance s_0 becomes shorter and shorter so that the multistep neutralization becomes less and less complete before the ion hits the surface. This implies that an increasing fraction of W_q cannot be converted into free electrons so that Γ has to decrease with increasing vertical velocity. This effect has to be more pronounced for high q-values because the number of isoenergetic steps is increasing with q.

A more detailed analysis of the electron spectra [41-45] sheds, however, more and different light on the data in Figs.17 through 20 and on the interpretation above. Hagstrum's data in Fig. 14 already imply a very slow but steady increase of the mean energy $d\varepsilon"$ of the emitted electrons with increasing q. This increase of $d\varepsilon"$ becomes more pronounced when 2p-holes are existing in the incoming Ar-ions for q>=9 which give rise to significant Ar-LMM Auger emission at around 200 eV as shown in Fig. 18. With a LMM-Auger-contribution of as low as 1% to Γ the mean energy $d\varepsilon"$ is nevertheless increased by 13 % from $d\varepsilon" = 14.9$ eV to $d\varepsilon" = 16.8$ eV. It is thus obvious that the reduction of the $\Gamma(Ar^{9+})$-value with respect to the straight line in Fig. 17b is at least partially due to this difference in $d\varepsilon"$ of 13 %. S.T.de Zwart [42,45] could indeed show that the deviation

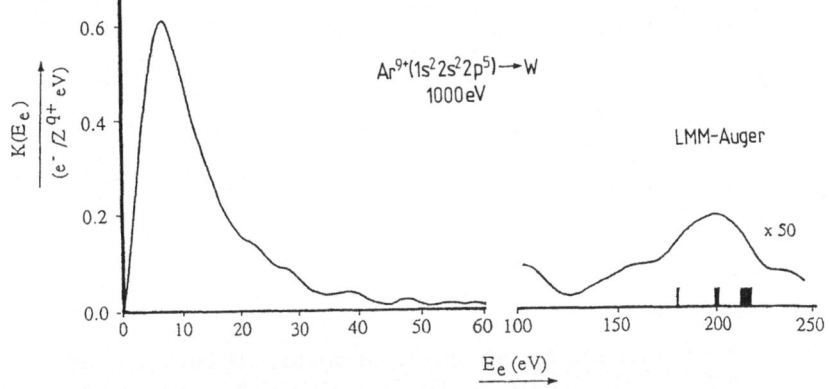

Fig. 18 Electron yield and spectrum per eV for Ar^{9+} approching a W-surface.

from the straight line in Fig. 17b is close to constant for all vertical velocities from $v_v = 1.3 \cdot E4$ m/s to $v_v = 3 \cdot E5$ m/s in contrast to Figs.17a,b which show a clear vertical velocity dependence. Whether these differences are due to the different experimental techniques or data treatments has still to be shown.

This latter remark seems necessary since de Zwart's differential electron spectrometer, which accepts only a small solid angle of all electrons emitted, is not ideally suited for measurements of Γ due to unknown angular distributions of the electrons. It is instead very well suited for the observation of the Doppler effect of the LMM Auger electrons of Ar(q=9) in Fig. 19a. With these results he could verify that the Auger electrons are emitted from freely moving projectiles in front of the surface which start to disappear and wash out, however, in Fig. 19b for vertical velocities higher than $3 \cdot E4$ m/s while saturating for vertical velocities below $3 \cdot E4$ m/s. This is related to the former noted interpretation that the time T of interaction with the surface starts becoming too short for a complete filling of the 2p-hole of Ar(q=9). One has to recall in addition that an LMM-Auger transition leaves two M-holes which have to be refilled for a complete neutralization after the LMM-Auger transition has occurred. The vertical velocity v_v required for a complete neutralization of Ar(q=9) has therefore

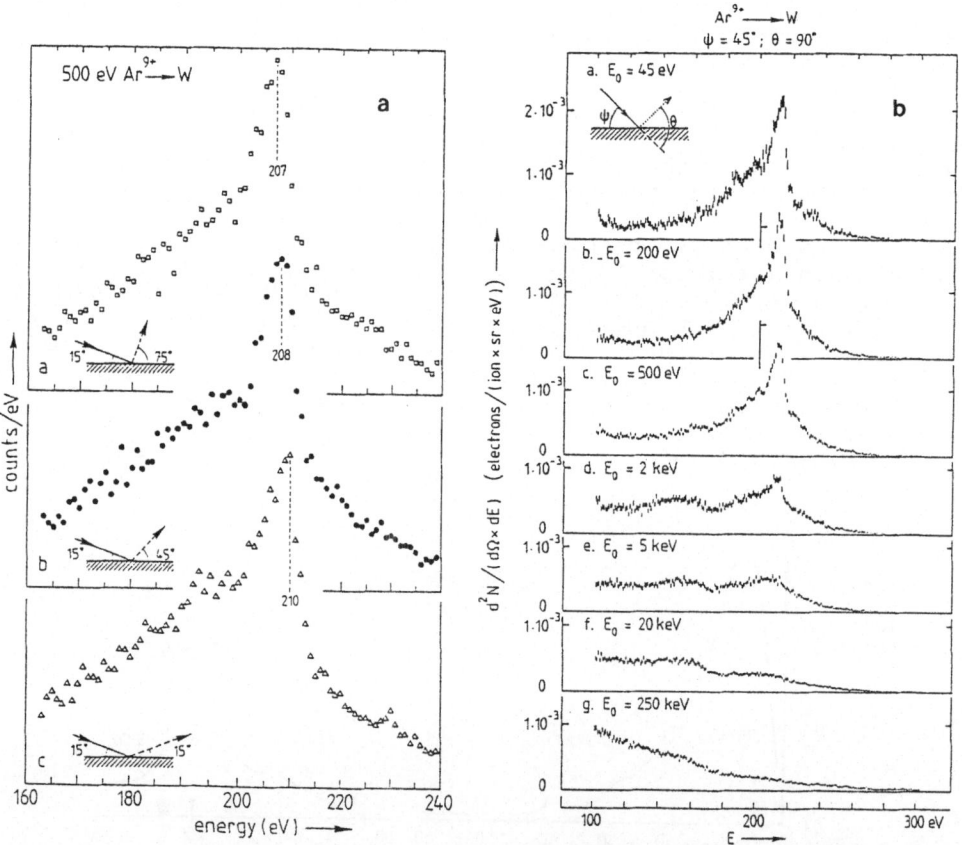

Fig. 19 a) Doppler-shift of the LMM-Auger emission into three different angles with respect to the incoming Ar[9+]-ions.
b) Attenuation of the Ar-LMM-Auger emission for increasing energy of the ioncoming Ar[9+]-ions.

to be smaller than 3·E4 m/s. These results and considerations corroborate very well the findings in Grenoble where a strong increase of Γ is found consistently in this low vertical velocity domain for N(q+), Ne(q+) [41], and Ar(q+) [46] without reaching the saturation of Γ at $v_v \approx 3 \cdot E4$ m/s.

In spite of some discrepancies in the data from the different groups a slightly renewed interpretation of the PE-phenomenon emerges due to an improved comprehension of the RN- and AN- processes and stimulated by the systematic measurements of the electron spectra.

For an ion approaching the surface with vertical velocity v < 0.05 a.u.= 1.1·E5 m/s the energetic situation of an ion with q=10 at a distance of about $65a_0$ in front of a surface with ⌀=5 eV is shown in Fig. 20. Note that this corresponds to an ion with q'=11 approaching the surface which is to be neutralized to an ion with q'-1=10. The ionic terms with 12 < n < 16 are in resonance with the conduction band but the n_s –wavefunctions, with their mean radius indicated, clearly favour the term energetically closest to the Fermi edge for RN.

This argument equally well applies to AN' which is most likely to occur with transitions of type (j) in Fig. 3b into terms closest to the Fermi edge. It is important to note that such AN'-processes do only produce electrons which heat up the metal but which are not ejected from the metal! AN-processes producing electrons with ε">0 are negligible in comparison to RN as is visible from the negligible overlap of the (n=11)-wavefunction with the metal wavefunction at the Fermi edge in Fig. 20. For the ion with q=10 the population of the terms with n=16,15,14 take therefore place dominantly via RN and AN' and

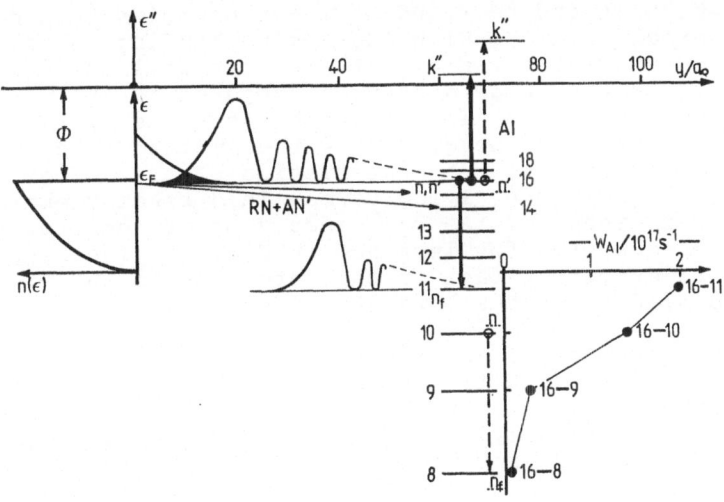

Fig. 20 Rydberg-levels of an ion with q=10 in front of a metal surface. The overlap of the squares of the n=16,s and of the k_F –metal-wavefunctions clearly indicate the dominant near resonant RN- and AN'-processes. The AI-rates for two electrons initially in the n=16 shell of which one is decaying to 11>n_f >8 are shown on the right.

sets in at distances as large as $s_0 = 70$ a_0. For this estimate
only geometric arguments have been used and the influence of
the image potential has been neglected. In a refined discussion
the latter has to be taken into account, however, because
Fig. 2 suggests important level shifts and polarizations which
tend to increase the distance where RN sets in.

Actually an ion with q'+1=11 which has captured already an
electron into a n'=16 term is shown in Fig. 20 which is now
ready to capture a second electron into the n=16 term. As soon
as this electron is captured strong correlation will cause
immediately an autoionization for which one can estimate the
rate depending on the final term n_f of the ion with q=10. Only
terms with $n_f < 12$ are considered which allow a free electron
to be ejected with energy ε">0. A model calculation is based on
the use of uncorrelated hydrogenic s-wavefunctions for the
bound states and Coulomb s-wavefunctions for q=10 for the free
electron. (Correlated terms would be embedded in a continuum
with equivalent AI-rates.) The result of this simple estimate
suggests two significant features as indicated on the right of
Fig. 20:
i) The rate of AI is extremely high for such Rydberg levels in
general.
ii) The rate of AI has a maximum when the energy separation
$dE(n=16,n_f)$ of the terms n,n_f involved is a minimum, yielding
$ε"=dE(n=16,n_f)-\phi$, and decreases rapidly with the increase of
this energy separation.

This latter observation holds in this hydrogenic approximation for all n,n_f terms. In correspondence to the arguments
used for the AN-rates in Fig. 10b this allows one to conclude
that the AI-rates will always have a maximum for $dE(k",n') =
dE(n,n_f) \approx ε"+ \phi$ being a minimum. This readily explains the
dominant maximum observed in the electron spectra at very low
energies which is indeed found at lower energies than predicted
by Arifov et al. [34]. With increasing binding energy of the
term with quantum number n direct AN-rates will stay negligible
at intermediate and large distances compared to their population via a series of RN-AI transitions, but the AI-rates will
decrease and the mean energy $dε"$ of the emitted electron will
slowly increase following the statistics of the energy separations $dE(n,n_f)$ for lower lying terms. This readily explains the
rather slow increase of the mean energy of the emitted electrons $dε"$ as observed for instance in Fig. 14. The AI-process
will thus be extremely fast in the outer shells and will become
slower and slower for the lower-lying shells.

The consequence of the first neutralization step in Fig.20
would be a discrete electron spectrum with a maximum close to
ε"=0. Discrete structures are, however, not to be expected
because of the extremely high transition rates correspond to
energy uncertainties of up to 130 eV !! In the outer shells one
has to discuss therefore a continuous flow of electrons from
the Fermi edge of the metal to an extremely broadened ensemble
of resonant terms of the ion. A large fraction of these electrons is emitted as free electrons with very low energy while
the rest drop down to lower-lying shells with an energy difference $dE(n,n_f)$ just above ϕ. This feeding of electrons from
the metal to the ion continues while electrons now in lower
shells continue to drop further down under emission of free

electrons with slightly increasing mean energy dε". Since the AI-rates for the lower-lying shells decrease the ion is thus completely filled from the outside such that intermediate multi-excited ion configurations exist (an exciting new feature in atomic physics) well before the lowest lying shells are filled with the lowest rate of the whole neutralization process. Due to energy uncertainty the quantum structure of the ion will therefore only start playing a role for AI(Auger)-transitions between the lowest lying shells so that discrete structures in the electron spectrum may occur. This phenomenon indeed becomes visible with the well resolved high energy Auger electrons which are particularly observed when highly charged ions with a hole in a low lying shell are used [28] for which Ar(q=9) with a 2p-hole may serve here as an example. The discrete LMM-Auger transitions, which fill this 2p-hole, are observed by Delaunay and de Zwart and shown in Figs.18 and 19, respectively. Their energies clearly indicate that the outer shells of these emitting ions are nearly completely filled before this Auger transition occurs with the lowest rate of all transitions involved in the complete neutralization process.

When the vertical velocity of the ion approaching the surface is increased it is therefore the innermost transitions which will disappear first when the time $T=s_0/v_v$ available for the whole neutralization process of the ion is reduced. This is in very good agreement with the observations in Fig. 19b by de Zwart or by Delaunay who clearly see the Auger electrons emitted from Ar(q=9) disappear much faster than the electrons at lower energy when v_v is increased. The higher the vertical velocity becomes the more inner shell holes survive up to very small distances from the surface. In such a situation the rates for direct AN- and AD-transitions and in addition for the QNR-, CAN-, and CAD-transitions of Fig. 3a,b,c, respectively, become non-negligible. All these transitions may therefore contribute at higher vertical velocities to the neutralization process in competition to the RN-AI-transitions which dominate at vertical velocities v_v < 5·E4 m/s. As a consequence more and more moderately resolved structures will occur in the electron spectra with increasing vertical velocity. Some of these structures may, however, also stem from pure intra-metal Auger transitions which occur to refill the core holes of the metal after QNR-, CAN-, or CAD-transitions. Most of the spectral features obtained in Oak Ridge [43,44] can so be explained when relatively high vertical velocities $v_v \approx 2 \cdot E5$ m/s are assumed. This may have been the case due to surface roughness in spite of the fact that a geometry with grazing incidence of 5° was used which was supposed to yield $v_v \approx 2 \cdot E4$ m/s. Kinetic emission (KE) may, however, spoil a lot of the overall structures contained in these spectra. KE certainly increases with respect to the simple relation derived from Fig. 16 when ions with inner shell holes (or residual charge q_r) penetrate the surface at these velocities of $v_v \approx 2 \cdot E5$ m/s so that the differentiation between PE and KE becomes more and more meaningless.

The appearance of discrete high energy structures in the electron spectra destroys the physical basis of the simple relation for the electron yield Γ to be proportional to W_q. Instead the total energy released by free electrons per ion should now be related to W_q. Defining $n(\varepsilon")=d\Gamma/d\varepsilon"$ one obtains for the total energy released by free electrons

$$E_q = \int_0^\infty n(\varepsilon'') \cdot \varepsilon'' \cdot d\varepsilon'' + 2 \cdot \phi \cdot \Gamma \qquad\qquad (9.2)$$

where the second term accounts for the fact that RN feeds the ion with electrons approximately from the Fermi edge and that AI has to overcome ϕ in order to produce a free electron. Not included in these definitions are the AN-processes which excite electrons into the continuum of the metal (E_{AN}), the RD-deexcitations (significant for high q-values) (E_{RAD}), and a geometric factor which has to be introduced due to the emission of 50 % of the free electrons with velocity vectors pointing into the metal of which an as yet unknown fraction is reflected. One can therefore relate E_q to W_q by

$$E_q = c_{geom} \cdot (W_q - E_{AN} - E_{RAD}), \qquad\qquad (9.3)$$

where $0 < E_{AN} < 0.4 \cdot \phi \cdot \Gamma$ (roughly estimated from the relative rates), $E_{RAD} << E_q$ for ions with q < 13, and $0.5 < c_{geom} < 1$. The expression (9.2) for E_q requires a rather precise absolute knowledge of the whole electron spectrum particularly at high energies. This region has so far been measured only in a few cases, unfortunately, with moderate absolute accuracy and with an intermediate part missing. An attempt with a linearly interpolated intermediate part is nevertheless being made here to integrate the spectra of N^{6+} impinging at $v_v = 1.1 \cdot E5$ m/s on W [28] and of Ar^{9+} impinging at $v_v = 7 \cdot E4$ m/s on W [47,48] to yield $E_q(N^{6+}) = 443$ eV $= 0.54 \cdot W_q$ and $E_q(Ar^{9+}) = 324$ eV $= 0.32 \cdot W_q$, respectively, for which uncertainties are difficult to estimate.

As striking result one notes that E_q is a larger fraction of W_q for N^{6+} than for Ar^{9+} although the interaction time T is shorter for N^{6+} because of the higher vertical velocity and because of the smaller s_0. More important radiative energy losses for Ar^{9+} are not likely to be responsible for this different behaviour because the radiative transition rates are on the contrary higher to fill the K-hole of N than to fill the L-hole of Ar. E_{RAD} and E_{AN} can therefore be neglected for the further discussion of these results. The key to a consistent interpretation may be the observation that for N^{6+} the largest contribution to E_q is stemming from values of ε'' above 60 eV while for Ar^{9+} more than 50 % of E_q stems from values of ε'' below 60 eV. This difference suggests that the K-hole of N is filled with a probability close to one during the interaction time T while the L-hole of Ar is filled only with a probability of the order of 0.5 in agreement with Fig. 19b. As a consequence the refilling of the L-holes in N after the KLL-Auger transition will set in at higher vertical velocities than the refilling of the M-holes in Ar after the corresponding LMM-Auger transition. Therefore the steep rise of Γ due to the filling of the L- or M- holes, respectively, under emission of low energy electrons will set in at higher velocities for N than for Ar in good agreement with the observations of Γ at the lowest vertical velocities [41,46]. From these observations and with this interpretation it is, however, also obvious that the neutralization is not completed in either case before the penetration of the surface occurs since otherwise a saturation of Γ would show up at the lowest vertical velocities similar to the saturation of the LMM-Auger yield in Fig. 19b. In order to better detect such a saturation effect, Γ and E_q should better be plotted as functions of $1/v_v$ for v_v being still smaller than all present data. It should, however, be noted that still other mechanisms might be responsible for the steep increase of Γ at the very low velocities.

One can therefore conclude that the interpretations given above allow a consistent description of nearly all published data on PE within some uncertainty set by the limited accuracy of some data. However, measurements at extremely low vertical velocities (well below $5 \cdot E4$ m/s) of Γ, the determination of E_q subsequent to an accurate absolute measurement of $n(\varepsilon")$ up to high energies $\varepsilon"$, and the extension of these measurements to ions in still higher charge states with significantly deeper inner shell holes (including observations of E_{RAD} if possible) are required as next experimental steps for the further verification of these interpretations. As next theoretical steps a much more profound study of the AI-processes under the RN-electron feeding conditions in front of a metal surface, i.e. of the thrilling physics of multiply excited atoms, are required for a conclusive understanding of the dynamics of the PE when a highly charged ion is neutralized in front of a metal surface at extremely low vertical velocities. Measurements at increasing vertical velocities above $5 \cdot E4$ m/s will be facing increasing difficulties in disentangling the PE and KE contributions so that much more experimental sophistication and theoretical work has to be invested for consistent interpretations of the great variety of possible phenomena to be observed.

As mentioned already first attempts have been made to further reduce the vertical velocities of multiply charged ions by using the geometry with incident beams at grazing angles of $\phi_{in} = 5°$ with respect to the surface [44]. This type of geometry had been developped for extreme grazing angles well below $\phi_{in} < 2°$ for the study of the electronic interaction of weakly charged ions at very low vertical but simultaneously high parallel velocity with extremely well polished single crystal surfaces [49,50] as outlined in the next chapters.

10) Motivation for the use of a geometry with grazing incidence

The initial motivation for the use of ion beams interacting with surfaces at extremely grazing angles of $\phi_{in} < 2°$ stems from the need to produce highly polarized excited terms in fast projectiles in order to apply the quantum beat technique in high resolution atomic spectroscopy [51]. This quantum beat technique was first employed in the field of beam foil spectroscopy [52] where an ion beam penetrates at a velocity of typically 1 a.u. through a thin carbon foil at normal incidence. In this geometry the axially symmetric ion-solid interaction allows the occurrence of actually observed linearly polarized light emission. It turned out, however, that the degree of this polarization was considerable only for the light elements so that this method could not generally be applied for all ions. A new era started in 1973 with the suggestion to tilt the carbon foil with respect to the ion beam axis [53,54]. This idea is based on the reflection symmetry with respect to a plane defined by the velocity vector $\underline{v}$ of the ion beam and the normal $\underline{n}$ to the surface in Fig. 21 if and only if the interaction takes place in the surface region at the

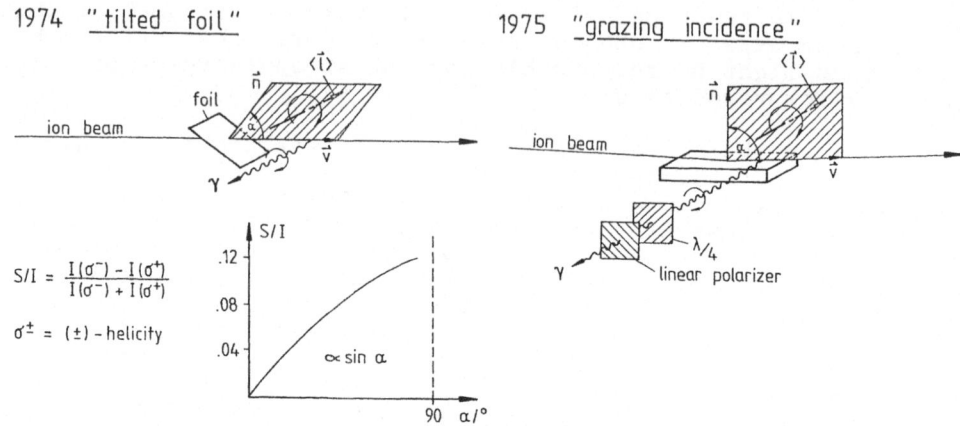

Fig. 21 Tilted-foil and grazing-incidence geometries. For details see text.

exit of the foil which defines $\underline{n}$. The symmetry operation of reflection on this plane keeps an axial vector, like the orbital angular momentum vector, invariant so that such a net orbital angular momentum may build up in excited terms during the ion-surface interaction. It shows up as circular polarization in the subsequent light emission perpendicular to this plane. Despite many doubts about the surface structure of ordinary carbon foils first experiments [55] immediately yielded considerable fractions of circular polarization S/I which increased approximately in proportion to the sine of the tilt angle α between $\underline{n}$ and $\underline{v}$. The normalized circular polarization (Stokes parameter) S/I is defined as $S/I=[I(\sigma^-)-I(\sigma^+)]/[I(\sigma^-)+I(\sigma^+)]$ where $I(\sigma^\pm)$ corresponds to the light intensity with $(\pm)$-helicity. Further experiments confirmed this general trend and encouraged in 1975 [56] an approach to $\alpha \approx 89°$ in Fig. 21 to maximize the circular polarization with a geometry of grazing incidence where the projectiles are essentially specularly reflected at angles $\phi_{in} \approx \phi_{out} < 1°$ with respect to the surface plane. Surprisingly this geometry yields values of S/I which are on the average, for more than ten different surface materials and for a great variety of projectiles, quite high and a factor of 2-10 larger than the tilted foil results extrapolated to $\alpha = 90°$. For its anticipated purpose this excitation mechanism is now being successfully exploited for high resolution quantum beat spectroscopy [57]. Furthermore it could be shown that the interaction is a pure orbital angular momentum effect [58] as long as the surface material is not magnetized.

From a theoretical point of view the main result of "tilted foil" and "grazing incidence" experiments was the fact that the interaction which leads to the circular polarization observed has to be well localized in the surface region for reasons of symmetry. This finding initiated considerable theoretical activity which led - after various approaches - to the physically very instructive "density gradient" model [59]. If one assumes in Fig. 22 an ion moving in an idealized surface plane with velocity $\underline{v}$ then it experiences in its restframe a beam of electrons with $-\underline{v}$ and with a density gradient $-\nabla\underline{n_e}$ across this beam, which represents the decrease of the electron

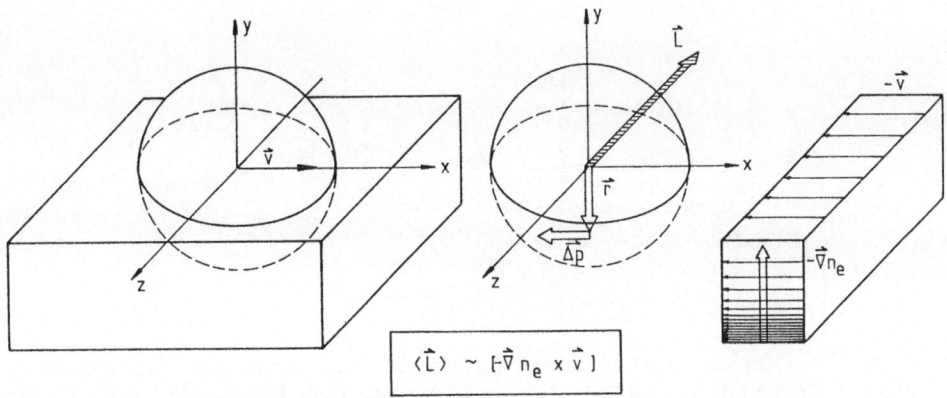

$$\langle \vec{L} \rangle \sim [-\vec{\nabla} n_e \times \vec{v}]$$

Fig. 22   The "density-gradient" model. See text.

density into the vacuum of a realistic metal surface. Consequently the ion experiences a stronger interaction in its "lower hemisphere" than in its "upper" one. If this interaction leads to a momentum transfer $\langle \underline{\Delta p} \rangle$ (e.g. in an electron capture process) to an electron in an excited term then an expectation value of orbital angular momentum $\langle \underline{L} \rangle$ is obtained in this excited term which points into the image plane due to the preferential occurrence of radius vectors $\langle \underline{r} \rangle$ pointing into the "lower hemisphere" of the ion, since there are simply more electrons available for the interaction.

As a result one always expects to observe the same direction of the orbital angular momentum polarization $\langle \underline{L} \rangle$ as has indeed been found in all experiments so far, except for one case to be discussed in chapter 12). With this "density gradient" model the basic physical comprehension of the polarization phenomenon is presented which will be worked out in detail in later chapters in connection with the charge exchange between a metal surface and an incident ion.

11) Special features of the grazing incidence geometry

In an ideal experiment with the condition of an atomically flat and absolutely clean single crystal surface, the projectile approaches the last atomic surface layer at small angles of incidence $0.2° < \phi_{in} < 2.0°$ against the screened ion-surface Coulomb-potential $U_{pl}(y)$ as shown in Fig. 23 and outlined in the introduction. Decomposing the velocity $\underline{v}$ into a large parallel component $v_p = v \cdot \cos(\phi_{in}) \approx v$ and a very small vertical component $v_v = v \cdot \sin(\phi_{in})$ one can derive a vertical energy of the projectile $E_v = (M/2) \cdot v_v^2 = E \cdot \sin^2(\phi_{in})$ which becomes extremely small $E \cdot E-5 < E_v < E \cdot E-3$ and corresponds to 0.3 eV $< E_v/u < 30$ eV for $E/u = 25$ keV. At these energies the use of a Molière type ion-atom potential [5] is well justified to derive the planar potential of equ.(1.1) and to calculate distances of closest approach $3a_0 > y_0 > a_0$ for moderate Z_1 and Z_2 of the projectile and the target, respectively. According to Fig. 1a the projectile sees at such distances y_0 a very smooth averaged potential so that it becomes specularly reflected ($\phi_{in} \approx \phi_{out}$). Conse-

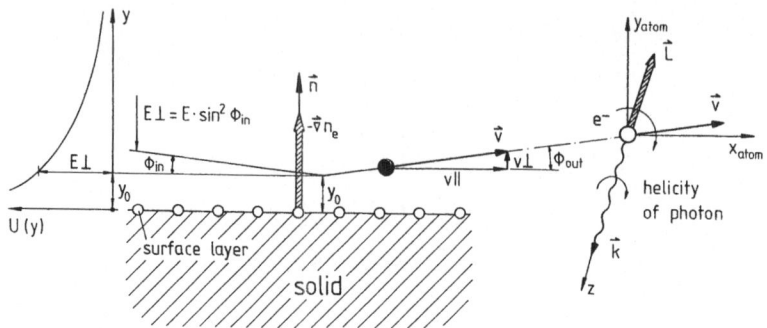

excitation in y-x-plane   emission $\vec{k} \perp$ y-x-plane

Fig. 23 Experimental conditions of the grazing incidence geometry with a specularly reflected ion beam off an atomically flat surface.

quently the energy and angular spread are observed to be rather narrow, of the order of dE/E < 0.01 and $d\phi_{out}$ < 1°, where the angular spread corresponds reasonably well to the estimated "vertical energy loss" in the introduction.

Under such ideal conditions for the interaction one may describe, to a very good approximation, all projectiles by one single averaged trajectory $\phi_{in}=\phi_{out}$, y_0, $|v_{in}| \approx |v_{out}|$ which can be controlled by the parameters of the incoming ion beam so that time consuming coincidence measurements are not required. By approximating this generalized trajectory by a straight line up to the point y_0 of closest approach one easily obtains at $\phi_{out} =0.4°$ a time T = 4·E-13 s which the projectile spends in a surface region of 2 nm. Due to the large distance of closest approach $y_0 > a_0$ this ideal geometry is completely insensitive to any close collisions with localized atoms in the last layer of the surface. Together with the long time T this geometry is thus very well suited for the study of long range interactions between the projectile and the surface electrons at very small vertical and simultaneously very large parallel velocity of the projectile with respect to the surface.

Unfortunately or fortunately – depending on one's interest – this geometry is, however, extremely sensitive to the adsorption of atoms and molecules even at very good vacuum conditions, p<=1·E-9 Pa, as well as to the slightest surface imperfections. The reason is of course the small angle ϕ_{in} which allows each ion to sample quite an elongated area of the surface. The probability for a projectile to suffer a collision with an adsorbed atom is given by

$$p_{coll} = 1 - \exp\left[-(\sigma\cdot\Theta)/(b^2 \cdot \sin[\phi_{out}])\right] \tag{11.1}$$

where σ is the cross section for a collision between a projectile and a single adsorbed atom(molecule), b^2 is the surface area covered by one adsorbed atom(molecule) in one monolayer, and Θ is the fraction of a monolayer covering the surface. According to this equation there exists a high probability for collisions with adatoms at angles $\phi_{out} < 1°$ even for $\Theta < 0.01$ depending on σ. One has to accept therefore a certain fraction of such collisions in any realistic experiment before the projectile recedes into a region of long range interactions $4a_0 < y < 20a_0$. It should be noted, however, that violent collisions with adatoms, i.e. with impact parameters $<< a_0$, will not be observed in the narrow reflection cone around $\phi_{out} \pm 0.5°$ since the scattering angle ϕ_{viol} will then be large. Consequently one can select the projectiles which have experienced to a good approximation the ideal interaction conditions outlined at the beginning of this chapter as soon as one can uniquely relate the quantities to be observed to projectiles receding from the surface in this narrow reflection cone. One has to add, however, that even these projectiles in the narrow reflection cone may nevertheless have experienced soft collisions with adatoms with high probability at intermediate or large impact parameters. Such collisions influence the trajectory only very little but may represent an important perturbation for the electronic structure of the projectile. The interaction history of a projectile appearing in the narrow reflection cone may thus be divided into three phases:

i)   Ideal interaction condition with the surface during the incoming trajectory.

ii) A certain probability to experience an ideal reflection but also a large probability to make a soft collision with an ad-atom in the region of closest approach. This alters the electronic properties of the projectile in a similar manner as a gas phase collision at equivalent impact parameters and scattering angles $< 0.5°$, so that the projectile loses its memory of the phase i).

iii) Ideal interaction condition with the surface during the receding trajectory with the starting conditions created during phase ii).

The selection of the projectiles in the reflection cone is most easily accomplished by a diaphragm at long distance from the surface as long as long lived quantities like the charge state distribution or some ground term properties (electronic polarization or nuclear spin polarization [57]) of the projectiles are to be investigated. Short lived excited term properties, like their photon emission (see chapters 12) and (14)), require already extremely narrow diaphragms at very short distances behind the surface or coincidence techniques where e.g. the photons are measured in coincidence with the projectiles in the reflection cone which are again selected and detected behind a diaphragm at long distance. The only way to relate prompt interaction products like the secondary electron emission of chapter 9) to the projectiles in the reflection cone is via a measurement of these secondary electrons in coincidence with the projectiles in the reflection cone. If such a coincidence technique is not applied in investigations at grazing incidence one takes the risk of spoiling the results by a strong contribution from violent collisions for which the projectiles are scattered into directions outside of the reflection cone and have thus experienced not the close to ideal interaction conditions desired.

This description of the consequences due to the existence of adsorbed atoms(molecules) applies equally well to the scattering centers created by imperfections of the surface structure. On an atomic scale these imperfections can easily have a much stronger negative influence on the ideal interaction conditions than the adsorbed atoms(molecules) so that the selection of the reflection cone becomes still more important. As soon as the slightest problems with the surface flatness on an atomic scale exist the application of coincidence techniques becomes inevitable for the investigation of prompt properties like the secondary electron emission. It may have been this problem which has contributed a lot of electrons from violent collisions to some of the electron spectra obtained in Oak Ridge at $\phi_{in} = 5°$ [43,44].

12) Distance of formation of excited terms in front of a surface

As outlined in chapter 4) the distance of formation of an excited term will be somewhat related to the spatial extension of the atomic and metal wavefunctions. It is thus reasonable to assume the distance of formation of an (n,l)-term to be proportional to $\langle r_{n,l} \rangle$ [60]. In chapter 5) the distance of formation of an excited term in a receding projectile was set equal to the freezing distance y_f. These are results derived from models and more sophisticated theoretical considerations which require experimental confirmation. The first and only experimental approach to this question is intimately related to the Post Collision Stark-Interaction (Stark-PCI) of the receding projectile in electric fields near the surface.

First indications for the existence of such a Stark-PCI stem from polarization measurements on the He II - n=4 -> n=3 - Paschen α - 468.6 nm line after He$^+$-interaction at grazing incidence with polycrystalline Cu [61] as well as Ni-single crystal surfaces under surface channeling conditions [62]. In contrast to all other emission lines observed with S/I > 0 in accordance with the "density gradient" model this line exhibited negative S/I-values for certain experimantal conditions. Due to the near degeneracies of the fine structure of hydrogen-like terms the He II n=4 term is very sensitive to electric fields, and it was readily assumed that electric fields near the surface could be responsible for the change of the sign of the polarization.

For a discussion of the Stark-PCI after the excitation by ion-surface interaction the model system of the n=2 manifold in hydrogen is used as an introductory example with the 2s- and the three 2p-sublevels with m = 1,0,-1 (spin neglected) shown in Fig. 24. The trajectory of the incoming proton and of the reflected hydrogen atom is basically directed along the x-axis, whereas the angular momentum (orientation) is along the -z-axis (axis of quantization), and the electric field $\underline{F}_y(y)$ is assumed to be along the y-axis. Since $\underline{F}_y(y)$ is perpendicular with respect to the axis of quantization, opposite parity levels with $\delta m = \pm 1$ are mixed by the electric field as indicated in the level scheme. This interaction modifies the initial level populations and thus changes the initial polarization.

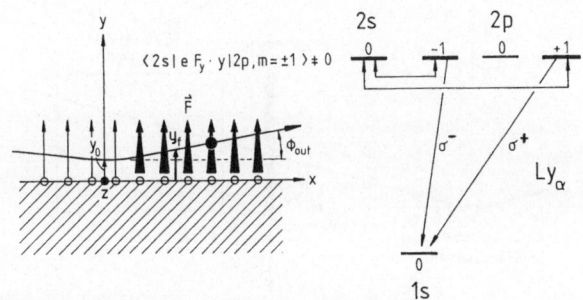

Fig. 24 Left: Ion-surface interaction in the presence of an electric image field. Right: Effect of the Stark-mixing on the populations of the n=2-manifold and on the polarization of the Ly-α line.

The effect of the electric field on the fraction of circular polarization S/I of the subsequent Ly-α emission has been worked out in detail [61,63-65]. For an introductory description one assumes that excitation coherences between the different n=2 - sublevels can be neglected, that the level splittings due to the electric field exceed fine-structures and Lamb-shifts, that n is still a good quantum number, and that only the 2p, m = 1 sublevel is initially populated by the surface interaction process. One obtains

$$S/I = 4 \cdot \cos(B)/[3+\cos(2 \cdot B)] \tag{12.1}$$

with

$$B=-(3 \cdot n)/(2 \cdot Z) \cdot (1/v_v) \cdot \int_{y_f}^{\infty} F_y(y) \cdot dy = C \cdot (1/v \cdot \sin \phi_{out}) \cdot A. \tag{12.2}$$

This result corresponds to an oscillatory quantum beat phenomenon for S/I which can readily be understood with Fig. 24 as a time dependent transfer of population from the initially populated 2p,m=1 level through the 2s-level to the 2p,m=-1 level and back again so that the circular polarization of the 2p→1s Ly-α emission has to change its sign periodically.

For a systematic study of the Stark-PCI one has thus to experimentally influence the phase B in order to obtain some information on the integral A which then is a measure of y_f if the field F_y is known. With the geometry of Fig. 25 an elegant modification of B is achieved by variation of the position of the aperture which varies for the transmitted and observed projectiles via ϕ_{out} the vertical velocity v_v and therefore the time the receding projectile spends in the electric field $F_y(y)$.

After these explanations the experimental results in Fig. 26a,b are very surprising since they yield at high fractions of S/I no effect at all for the Balmer-α line of hydrogen in Fig.

665

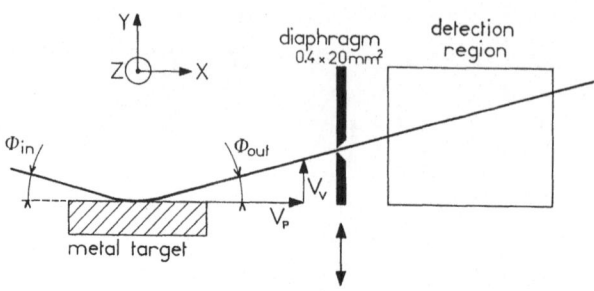

Fig. 25 Geometrical selection of the vertical velocity
component for the measurements in Fig. 26a,b.

26a but a convincing modulation of S/I of the Paschen-α line of
He+. It is in astonishingly good agreement with equ.(12.1)
adapted to the maximum of S/I and to the point of crossover at
$\phi_{out} \approx 1.36°$, represented by the full line in Fig. 26b. The
inclusion of the finite angular resolution in the experiment
would reduce the deep minimum of equ(12.1) and would thus
reproduce the data still better [64]. The maximum of S/I
indicates an initial polarization <1 (see chapter 14) which
requires a more elaborate theoretical treatment [65] yielding
a further reduction of the minimum.

The experiments were performed on a highly polished, polycrystalline Cu-target at a pressure of 10^{-7} Pa in a UHV chamber
and the data were recorded only after sputtering of about 15
hours by the incident ion beam at $\phi_{in}=1°$ and at an ion current
density of about 200 µA/cm². The He+ result at large angles
ϕ_{out} represents the initially produced polarization by the
surface interaction since the time, the He+-ion spends in the
field region after its formation, is too short to allow any
Stark-PCI. When ϕ_{out} becomes smaller the time the He+-ion
spends in the electric field region becomes longer and longer
so that stronger and stronger influence of the Stark-PCI can
develop which causes S/I to oscillate.

The striking difference between the neutral hydrogen and
the singly charged helium ion can only be understood when the
electric image field of equ.(2.4) is acting on the excited
terms involved, since it is zero for hydrogen and $F_y(y) =
1/(y+\Omega)^2$ for He+, where the screening length $\Omega \approx 0.055$ nm of the
Cu-target [60] has been added. With this knowledge of the
electric field equ.(12.2) can be integrated to yield

$$y_f = (3 \cdot n)/[2 \cdot \Pi \cdot v \cdot \sin(\phi_0)] - \Omega, \qquad (12.3)$$

where ϕ_0 is the largest angle observed where S/I=0. From the
data for n=4 in Fig. 26b and from similar data for n=5,6 the
distance of formation $y_f(n)$ of the excited terms of He+ could
be determined: $y_f(n=4)=0.7\pm0.1$ nm, $y_f(n=5)=1.2\pm0.3$ nm, and
$y_f(n=6)=1.7\pm0.6$ nm [65]. These y_f-values are in good agreement with earlier estimates [60] giving $y_f = \langle r_{n,1} \rangle + d/2$, where
d is the lattice constant of the target.

666

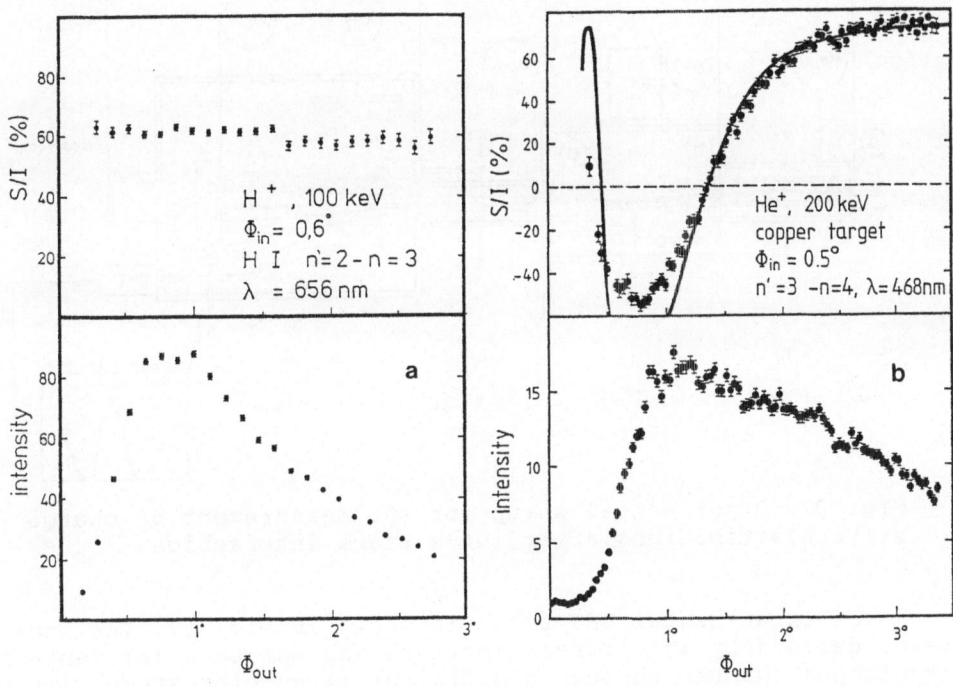

Fig. 26 a) S/I vs ϕ_{out} of the HI n'=2-n=3, 656 nm emission
at 100 keV energy. b) I and S/I of the HeII n'=3-n=4 468 nm
emission vs ϕ_{out} at 200 keV energy. The solid line is equ.
(12.1) adapted to the maximum of S/I.

It has to be stressed of course that the n=4,5,6 manifolds
of He⁺ require a much more complex theoretical treatment [65]
than the simple model presented here because of the large
number of n² sublevels involved. It turns out, however, that
the result of such a complete treatment is identical to the
simple model when the initial angular momentum expectation
value <$\underline{l}$>, created by the surface interaction, has a maximum
for each n-manifold, i.e. only the m=-(n-1) sublevels are populated. This is an assumption which is the main source of the
large error bars in the results because the unknown initial
density matrix of the n-manifold causes some ambiguities in the
interpretation of the data. In spite of these fundamental
difficulties these results represent the first and only successful experimental attempt to measure y_f. Unfortunately the
error bars don't allow to systematically measure y_f e.g. as
function of v_v.

13) Resonant charge exchange of ions at grazing incidence

In this chapter the extension of chapter 5) is considered
to considerably higher parallel velocities v_p and simultaneously smaller vertical velocities v_v due to the extreme grazing
incidence geometry. Instead of modifying the workfunction, as
in chapter 5), projectiles with significantly different electronic structures are used to demonstrate the ideas of Fig. 3,
initially developed for low energies, up to surprisingly high
particle energies when the grazing incidence keeps v_v small.

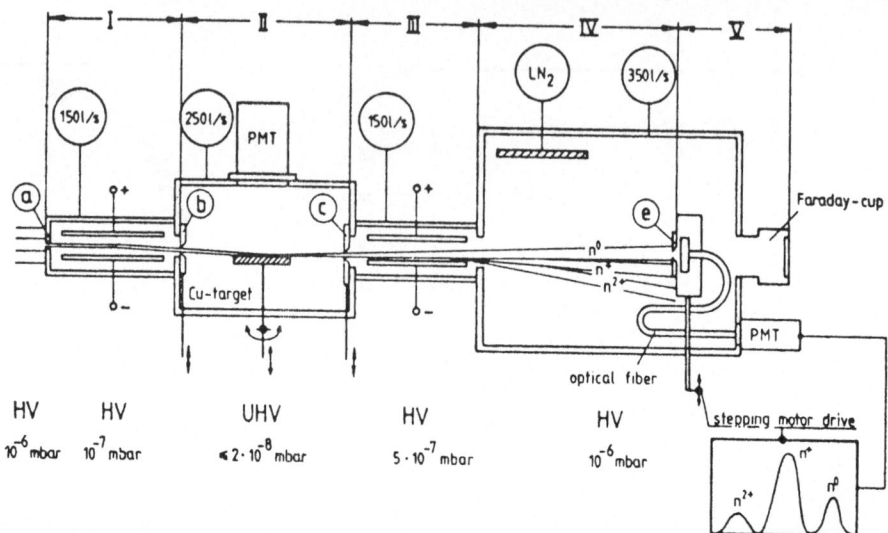

Fig. 27 Experimental setup for the measurement of charge state distributions after ion-surface interaction.

The experimental UHV-setup is shown in Fig. 27. The ion beam, defined by an upstream aperture and aperture (a), enters the target chamber through a differential pumping stage with a pair of field plates to direct it through an aperture (b) with $\phi_{in} = 0.4°$ on a highly polished, polycrystalline Cu-target or on a well cut and highly polished Ni(111)-surface. The reflection cone is selected with aperture (c) and so guarantees that only projectiles are observed which have experienced close to ideal surface interaction conditions (see chapter 11). The pair of field plates in the exit differential pumping stage serves to separate the charge states of the projectiles in the reflection cone so that they can be detected through an aperture (e) by a quartz plate the fluorescence of which is counted via a fiber optics and a photomultiplier (PMT-) tube. For the registration the assembly of quartz and aperture (e) is simply moved across the charge state distribution at fixed electric field. The data are taken only after several hours of Ar+ sputter cleaning of the surfaces at grazing incidence at current densities of about 200 µA/cm². Some of the results to be shown have been measured with the further improved apparatus in Fig. 34.

As examples of nearly opposite results the charge state distributions of nitrogen and rubidium are shown in Fig. 28a after the interaction of N+ and Rb+ at $v_p = 0.4$ a.u., $\phi_{in} = 0.4°$, and consequently $v_v = 0.0028$ a.u. with a Ni(111)-surface. While the N+ beam leaves the surface practically fully neutralized the Rb+ beam stays nearly unaffected and leaves the surface practically fully ionized, except for a small fraction of neutrals of about 4%. The comparison with the energy diagrams of the neutral atoms in Fig. 28b readily explains this strikingly different behaviour of the two projectiles. Neglecting the large parallel velocity v_p at first, all energy terms of Rb will stay at all distances y from the surface above the Fermi edge so that neutralization is impossible in excellent accord with the observation. On the contrary the ground terms of N are subject to RN- and AN-processes in a similar way as H+ interacting with a Al-surface in Fig. 10 so that full

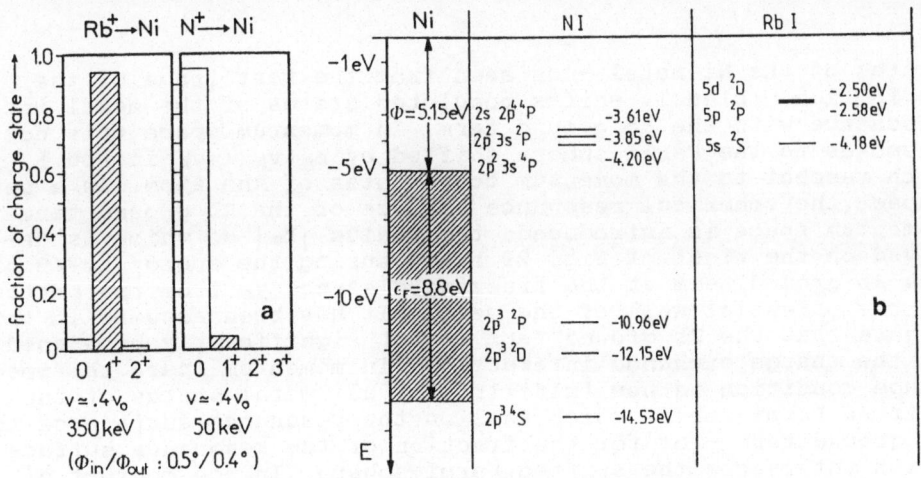

Fig. 28 a) Comparison of charge state distributions of Rb- and N-projectiles after interaction with a Ni-surface at the same velocity. b) Energy level diagrams for neutral N and Rb atoms in comparison to Ni, approximated by a free electron gas with sharp Fermi-edge at T=0.

neutralization is to be expected, again in excellent accord with the observation.

From Fig. 9 it is, however, evident that the parallel velocity v_p cannot be neglected. The corresponding situation is shown in Fig. 29 for the interaction of Rb$^+$ with a Ni(111)-surface under the above conditions. The modified density of

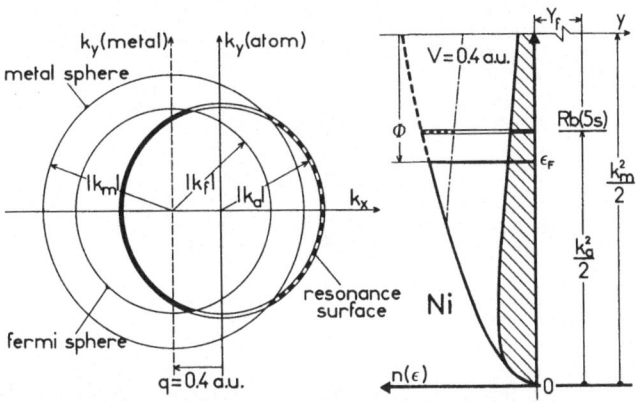

Fig. 29 Interaction of a Rb-atom moving at $v_p = 0.4$ a.u. parallel to a Ni-surface at a distance y_f.

states of the Ni-metal – as seen from the rest frame of the moving ion – clearly shifts populated states of the metal into resonance with the Rb ground term. In momentum space this corresponds to the Fermi-sphere shifted by $m_e \cdot v_p$ ($=v_p$ in a.u.) with respect to the momentum coordinates of the atom. As a new aspect the spherical resonance surface of the Rb ground term in momentum space is introduced, the radius $|k_a|$ of which is derived on the right of Fig. 29 by measuring the energy $k_a{}^2/2$ of the Rb ground term at the freezing distance y_f from the bottom of the potential well of the metal. It has been assumed in this figure that the Rb ground term is not significantly broadened by the charge exchange interaction. In momentum space the resonance condition is now fulfilled for all metal states of the shifted Fermi-sphere which fall on the resonance surface of the Rb ground term – or for the fraction of the resonance surface which intersects the shifted Fermi-sphere. In the context of chapter 5) the neutralization rate w_n could thus be simply related to the fraction of the resonance surface of the Rb ground term intersecting the shifted Fermi-sphere. The ionization rate w_i could be related to the remainder of the resonance surface which intersects the continuum above the Fermi-sphere. As another new feature the metal sphere is introduced in Fig. 29, the radius $|k_m|$ of which is derived on the right of Fig. 29 by measuring the energy $k_m{}^2/2$ of the ionization limit from the bottom of the potential well of the metal. This metal sphere is also shifted with respect to the atom so that above a certain parallel velocity v_p a fraction of the atomic resonance sphere lies outside the metal sphere. This fraction of the resonance sphere (broken fraction) allows thus ionization into states of free electrons, some of which may have a chance to

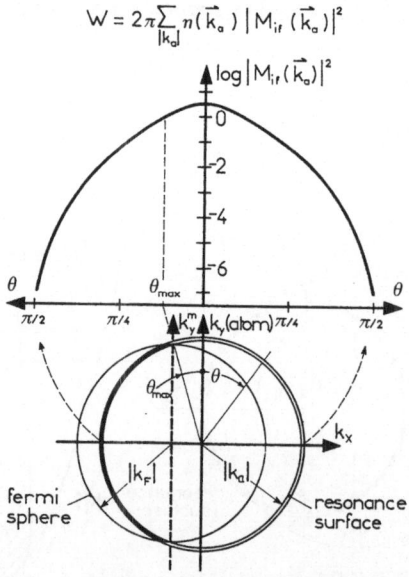

Fig. 30   View of the resonant charge exchange in momentum space and the angular dependence of the matrixelement.

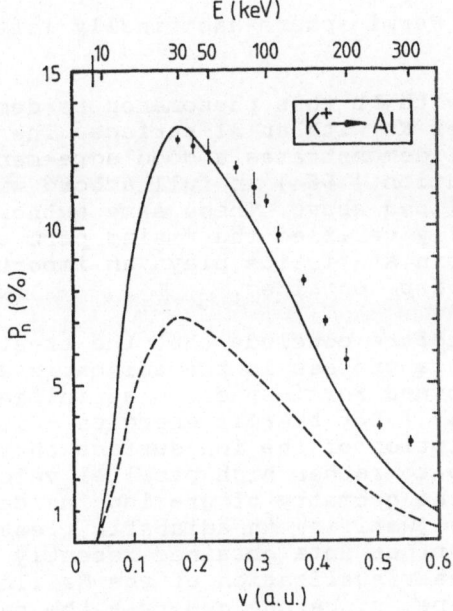

Fig.31: Neutral charge state fraction
vs parallel velocity after K⁺-Al
interaction at grazing incidence.

leave the metal surface after elastic collisions and do then
appear as secondary free electrons. The influence of this
phenomenon on the neutralization probability will, however,
stay negligible so that w_1 can be simply related to the
remainder of the resonance surface outside the Fermi-sphere.

In order to calculate the neutralization probability the
various fractions of the resonance surface have to be weighted
by the square of the interaction matrix element, displayed in
Fig. 30. It is strongly dependent on the Θ-direction of the
atomic $\underline{k}_a$-vector and obviously favours charge exchange normal
to the surface in a narrow cone with a halfwidth of about $\Theta <$
$2 \cdot \Pi/8$. With the neutralization rate w_n, so obtained, one can
calculate the neutral fraction of Rb at y_f which well explains
the small neutral fraction observed in the experiment.

Without entering into any detailed calculation one can
qualitatively derive from Fig. 30 the neutral fraction of Rb as
a function of the parallel velocity v_p. ϕ_{in} has to be adjusted
such that $v_v < 0.05$ a.u. in order to justify the adiabatic
treatment of the interaction used here. Since the Fermi-sphere
is smaller in radius than the resonance sphere of the Rb ground
term one has no overlap between the two at $v_p = 0$ and hence no
neutralization at all. At a given threshold of v_p the Fermi-
sphere touches the resonance sphere and neutralization sets in
at a very low level, however, because the squares of the matrix
elements are extremely small for $\Theta \approx \Pi/2$. With increasing v_p
the neutralization becomes stronger and stronger up to a
maximum when the broken k_y ᵐ-axis hits the intersection point of
the two spheres and thus defines Θ_{max}. When v_p is further
increased the neutral fraction p_n will steadily decrease again

to zero when the Fermi-sphere has finally left the resonance sphere.

Instead of with Rb this phenomenon is demonstrated with the interaction of K⁺ with an Al-surface. The experimental result in Fig. 31 demonstrates a good agreement with a complete ab initio calculation [66] in full accord with the qualitative picture outlined above. These same authors have recently also experimentally verified the rising part of p_n and have shown that the spin statistics plays an important role for the absolute values of p_n obtained.

One can therefore conclude that the treatment of the resonant charge exhange process in the adiabatic approximation with a Galilei-transformed Fermi-sphere - as initially introduced by van Wunnik [19,67] for thermal energies - is very well adapted to the description of the ion-surface charge exchange. It may be applied up to rather high parallel velocities $v_p < 1$ a.u. as long as the geometry of grazing incidence guarantees $v_v < 0.05$ a.u. which justifies an adiabatic treatment. The results in Fig. 31 and further data obtained recently [50,66] are indeed an impressive visualization of the Galilei-transformation of the Fermi-sphere, since one observes the passage of the Fermi-sphere through the resonance surface of the Rb-ground term.

Whether higher vertical velocities v_v can be admitted in such an adiabatic treatment by simply shifting the Fermi-sphere in Fig. 29 and 31 slightly downwards for the interaction with the receding projectiles (slightly upwards for the incoming projectile) has to be shown in the future. The successful interpretation of the high energy tail in Fig. 12 seems to justify such an approach.

14) Polarization of the projectile following grazing surface interaction

The concepts outlined in the chapter 13) for the charge exchange from the surface to an atomic ns-term under grazing interaction conditions can readily be extended to the capture into np, m=0,±1 - states. From the populations σ_+ , σ_0 , σ_- of these m=+1,0,-1 - states, respectively, one can directly deduce the fraction of circular polarization S/I, which becomes a very simple expression for a ¹P -> ¹S - transition observed against the direction of the positive z-axis in the optical convention:

$$S/I = (\sigma_- - \sigma_+)/(\sigma_- + \sigma_+) \tag{14.1}$$

When a fine structure or hyperfine structure exists, however, corrections to this expression have to be added due to the transfer of polarization from the orbital angular momentum to the electronic spin and the nuclear spin, respectively [51,53]. For a ³P -> ³S transiiton with nuclear spin I=0 one obtains

$$S/I = 27 \cdot (\sigma_- - \sigma_+)/(41 \cdot \sigma_+ + 26 \cdot \sigma_0 + 41 \cdot \sigma_-), \tag{14.2}$$

for a ²P -> ²S transition with a nuclear spin I=1/2

$$S/I = 43 \cdot (\sigma_- - \sigma_+)/(53 \cdot \sigma_+ + 38 \cdot \sigma_0 + 53 \cdot \sigma_-), \tag{14.3}$$

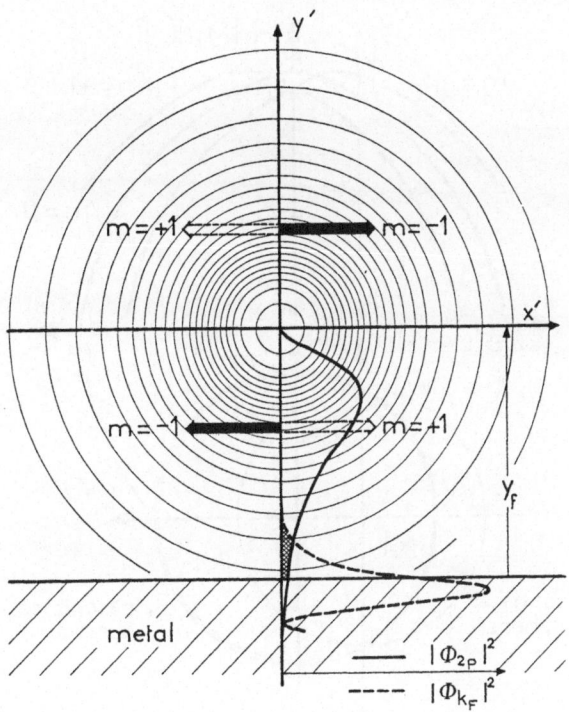

Fig. 32 Electron density of 2p, m=±1
states with their ⟨tangential momenta⟩
interacting with k_F metal electrons at
a distance y_f.

and for I=3/2

$$S/I = 325 \cdot (\sigma_- - \sigma_+)/(627 \cdot \sigma_+ + 546 \cdot \sigma_0 + 627 \cdot \sigma_-). \qquad (14.4)$$

Instead of proceding with a straight forward calculation
of the σ_i an attempt will be made to visualize the physics
involved in the interaction mechanism which is responsible for
the S/I observed as a function of v_p. At the typical capture
distances $y_f \approx \langle r_{n,1} \rangle + d/2$ the np-wavefunctions have such a
small overlap with the solid already, so that they can be
considered to a good approximation as unperturbed free orbitals
existing, however, only in the region y>0. As an example the
absolute squares of a 2p, m=±1 wavefunctions and of a metal
wavefunction with $k=k_F$ are shown in Fig. 32. Such free atomic
2p, m=±1 orbitals are characterized by an effective circular
electron current or by an expectation value for the tangential
momentum of the electron as indicated by the arrows. In the
cross hatched region of the overlap of the atom- and metal-
wavefunctions, which - when "weighted" with the operator -Z/r -
is responsible for the charge exchange, one can therefore
assume that only the electron densities corresponding to the
lower arrows will contribute to the exchange probability. The
calculation shows that this fact leads to a shift of the square
of the matrix element of Fig. 30 by an amount which corresponds
to the expectation value of the tangential momentum $\langle p_t \rangle$ evaluated in the region of overlap to yield the Fig. 33. The probability for charge exchange in momentum space is thus shifted

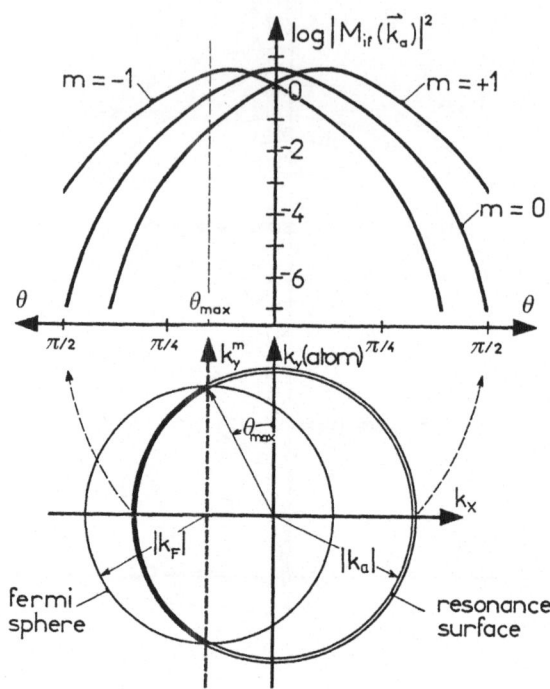

Fig. 33 Resonant charge exchange in momentum space an angular dependence of the matrix elements for the 2p, m=±1 states.

to the left for the m=-1 state, to the right for the m=+1 state, while the m=0 state stays unaffected. When now repeating the discussion of the excitation probability following Fig. 30 one can for the three substates at once deduce from Fig. 33 that the population of the m=-1 state will dominate through all velocities v_p when the Fermi-sphere is travelling through the resonance surface. This fact that $\sigma_- > \sigma_0 > \sigma_+$ is essentially the basis for an independent discussion of all three substates, since otherwise at similar σ for all three states a coupled rate equation would have to be solved which takes e.g. into account that y_f (m=0) < y_f (m=±1) due to the different spatial extensions of the wavefunctions in the y-direction.

From Fig. 33 one can now qualitatively construct $S/I(v_p)$. At the threshold velocity - when the Fermi-sphere touches the resonance surface - S/I will jump to its maximum value at a minimum of intensity emitted because σ_- will be very small but still orders of magnitude larger than σ_0 and σ_+. When increasing the parallel velocity the overall excitation probability will increase and S/I will slowly decrease a little bit because σ_0 and σ_+ will slowly grow at the expense of σ_-. At the maximum of the intensity (see Fig. 31) a shallow minimum of S/I will appear which can be estimated with equs. 14.1-14.4 after reading the σ_i-values from Fig. 33 at Θ_{max}. With further increasing parallel velocity the intensity will decrease again as in Fig. 31 while S/I remounts back to its maximum value right when the Fermi-sphere is leaving the resonance sphere.

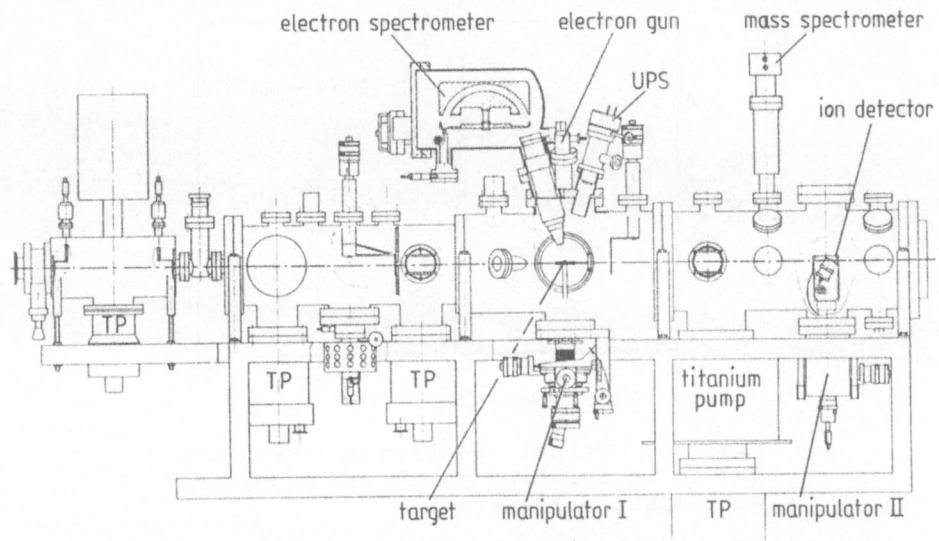

electron spectrometer   electron gun   mass spectrometer

UPS

ion detector

TP

TP   TP

titanium
pump

target   manipulator I   TP   manipulator II

Fig. 34 Experimental UHV-setup with surface diagnostics dedicated for the study of the ion-surface interaction at grazing incidence.

For the experimental demonstration of these ideas extreme care has to be devoted to the preparation of an atomically flat single crystal surface since the lifetimes of the excited terms are usually rather short so that a selection of the reflection cone (see chapter 11)) via a diaphragm behind the target surface becomes difficult for intensity reasons, in particular at the low parallel velocities near threshold. For Na^+ interacting with a Ni(111)-surface, chosen here as example, the distance travelled during one lifetime (16.4 ns) of the excited $3\,^2P$-term at the threshold velocity of about $2.5 \cdot E5$ m/s is only 4 mm so that a diaphragm could not be installed. Coincidence techniques with the detection of optical photons are likewise extremely time consuming so that, instead, the integrated light emission of all reflected(scattered) Na-projectiles was measured in an experimental UHV-setup in Fig. 34 which allowed the standard surface preparations and diagnostics in order to obtain the best Ni(111)-surface possible.

The Ni(111)-single crystal was well cut and polished to highest perfection by U.Linke of the KFA-Jülich before it was implanted into the apparatus and further prepared by cycles of Ar^+-sputtering at 5 keV at grazing incidence and careful annealing. The cleanness and the flatness of the surface were verified by Auger spectroscopy with a spherical analyzer and by measurement of the angular distribution of the reflection cone, respectively. The apparatus uses the same basic geometry for the incoming beam as explained with Fig. 27, except that the beam has to pass through two differential pumping stages in order to allow for a final pressure of about $3 \cdot E-9$ Pa in the target area and in the subsequent area of reflected beam diagnostics. Essentially some of the charge state measurements of chapter 13) have been obtained with this setup by making use of this diagnostic equipment. The volume above the target and slightly displaced to the rear of the target is viewed against

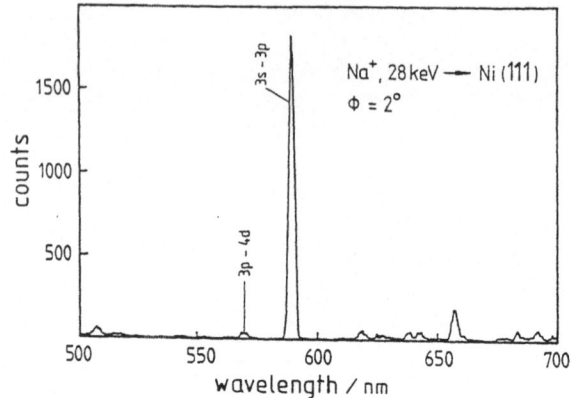

Fig. 35 Optical spectrum observed when
Na$^+$ is interacting at 28 keV and $\phi_{in}=2°$
with a Ni(111)-surface.

the +z-axis via a polarization analyzer, a spectrometer or
filter, and a photomultiplier.

The optical spectrum observed when Na$^+$ is interacting at
28 keV and $\phi_{in}=2°$ with a Ni(111)-surface is shown in Fig. 35.
The $3^2P \rightarrow 3^2S$ resonance line at 589 nm completely dominates
the spectrum in excellent agreement with the energy diagram in
Fig. 36 which shows the 3s, 3p, and 4d terms of Na at infinity
and at the freezing distances in front of a Ni-surface [68].

Without parallel velocity neither of these terms can be
populated. With increasing v_P, however, first the 3s ground
term, then at still higher v_P the 3p term, and finally at $E_P >$
20 keV the 4d term can be populated with probabilities which
decrease drasticaly from 3s to 3p to 4d. This fully explains
the spectrum including the very small intensity of the 4d -> 3p

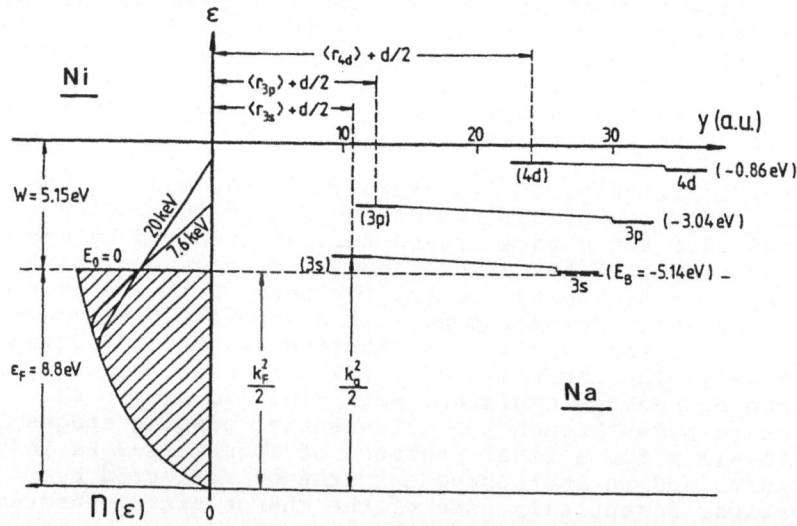

Fig. 36 Energy levels of neutral Na at three distances
in front of a Ni-surface approximated by a free electron
gas.

transition which is of great significance for a systematic
study of the resonance line since perturbations due to cascades
from 4d to 3p can practically be neglected.

For the measurements of the polarization the interaction
of Na⁺ with a Ni(111)-surface at 0.4° was chosen. In contrast
to the predictions the fraction of circular polarization S/I of
the Na-D-line in Fig. 37 does not follow the expected behaviour
as outlined above. Instead of jumping at the threshold velocity, corresponding to about 7.6 keV, to the maximum value one
observes a steady increase from zero to a saturation value at
about 150 keV of $S/I_{sat} \approx 0.3$. This value of S/I_{sat} has to be
compared with equ.(14.4) which yields $S/I_{max} = 0.52$ for $\sigma_- = 1.0$,
i.e for complete orbital angular momentum polarization initially created by the ion-surface interaction along the $-z$-
axis. Allowing for $\sigma_- < 1.0$ and for $\sigma_0, \sigma_+ > 0$ one can readily
explain S/I_{sat} observed in approximate agreement with the value
deduced from Fig. 33 but no interpretation of the steady rise
of S/I is possible within the model when the ideal interaction
conditions are assumed. After the impressive success in the
description of the neutral charge state fraction in Fig. 31
there is actually no reason to discard the model, in particular
since the calculation of the absolute matrix elements for the
charge exchange is more difficult than the evaluation of their
relative values for the polarization. The use of a coupled rate
equation for the calculation of the population of all states
involved, as mentioned above, is not likely to bridge the discrepancy to the experiment since its impact is expected to be a
minimum at the threshold velocity where the discrepancy to the
experiment is largest.

For the time being one is thus left with the critical
evaluation of possible experimental difficulties which might
explain the observed steady increase of S/I as a function of

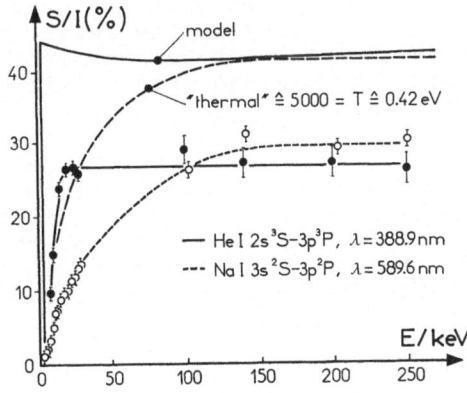

Fig. 37 Energy dependence of S/I of the NaI
3s-3p, 589.6 nm and of the HeI 2s-3p, 388.9 nm
emissions in comparison to the model calculations and to a calcualtion assuming surface
roughness.

v_P. The way this experiment was performed does not relate the photons observed to the projectiles in the reflection cone. Although great care has been devoted to the preparation and to the cleanness of the Ni-surface it is thus possible that residual surface imperfections and adsorbed species may cause sufficient violent collisions (see chapter 11)) with scattering angles well outside the reflection cone. Unfortunately, for the experimental reasons given above, the light emission of such projectiles was accepted by the detection optics and can thus fully obscure the results expected from the model. In particular near the threshold velocity this contribution may fully dominate the intensity observed because the light emission due to the resonance capture starts out at zero at threshold. Besides the violent collisions the surface imperfections may cause still another effect. If the receding projectile sees a micro-structured surface, it sees essentially a potential varying rapidly in time, the Fourier transform of which may be approximated by the introduction of a broadening of the energy levels involved or by an effective temperature. The broken line in Fig. 37 shows such an attempt of interpretation in much better agreement already with the observations, but the drop to zero of the broken line will always stay too sharp at all temperatures. The behaviour observed may thus be due to both the violent collisions and an energy broadening.

From the theoretical point of view former simulations have shown [49] that a tail of occupied states above the Fermi edge could readily explain the steady rise of S/I observed. As origin for such occupied states one could think of the production of electron-hole pairs by the ion itself, so that a fraction of the charge exchange could proceed via a second order process through these occupied states. Another theoretical aspect is related to the use of unperturbed ionic- and metal-wavefunctions. On the one hand side it is an open question whether the local action of the ion on the metal, creating the image charge, is or is not locally affecting those metal wavefunctions which enter into the charge exchange process. On the other side the distortion of the ionic wavefunction due to the cut-off at the metal boundary in Fig. 32 has not yet been taken into account. Besides the introduction of more sophisticated experimental techniques in order to uniquely relate the observed S/I to the projectiles in the reflection cone it is thus obvious that further theoretical sophistication is also called for in order to understand this sensitive test of the ion-surface interaction.

The measured rapid rise of S/I of the HeI $3p^3 P -> 2s^3 S$ - 388.9 nm - line as a function of energy also shown in Fig. 37 is indeed steeper than the one of the Na-D line when plotted as a function of velocity, but it still is not sharp enough to be explained only by a temperature. The value of $S/I_{sat} = 0.27$ of this line is way below the value of 0.66 calculated with equ.(14.2) under the assumption of $\sigma_- = 1$. This large difference indicates significantly higher σ_0 and σ_+ values than in the case of the Na-3^2P polulations. A calculation of these values within the framework of the model taking into account the correct spatial wavefunctions of the $3p^3 P$-term has not yet been performed, however. Such a comparison of a complete calculation [69] with experimantal data is possible for the HI $2^2P -> 1^2S$ - 121.6 nm Lyman-α emission after interaction of H^+ with a Ni(111)-surface in Fig. 38. At maximum of initial polarization

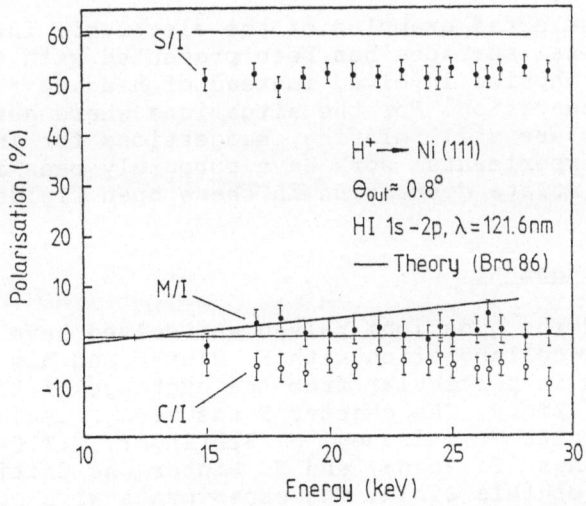

Fig. 38 Energy dependence of the Stokes
parameters of the HI-Ly-α emission after
H⁺-Ni(111)-surface interaction at grazing
incidence.

($\sigma_- = 1$) one calculates with equ.(14.3) $S/I_{sat} \approx 0.81$ while the
complete treatment yields $S/I_{sat} \approx 0.6$ and the data represent a
value of $S/I_{sat} \approx 0.5$. The other two normalized Stokes parameters
[51] M/I and C/I stay close to zero and will therefore not be
discussed here. The rather good agreement between the theoretical and experimental values for S/I_{sat} is quite satisfactory.
It has to be noted, however, that the population of the HI-2²P
term represents only a very small fraction of the total neutralization of H⁺ due to the existence of the low lying ground
term. When interacting with the surface this ground term is
dominantly populated via AN near the surface ($y_f \approx \langle r_{1s} \rangle + d/2$) so
that only a very small fraction of H⁺ survives in the receding
projectiles which can then be populated via RN at larger dis-
tance $y_f \approx \langle r_{2p} \rangle + d/2$. The neutralization via AN is so efficient
that sufficient Lyman-α emission for a polarization measurement
is observed only above 15 keV. This is a strong indication for
violent collisions which reduce the AN-efficiency and allow a
small fraction of H⁺ to recede from the near surface region
into the region of RN processes. In this situation one takes
thus advantage of the violent collisions due to adsorbates or
surface imperfections in order to observe the RN-process into a
highly excited term. One has to be very careful, however, with
the interpretation of the data obtained since a large angular
scattering distribution will result from the violent collisions
so that a unique correlation between the data and an ideal
trajectory of chapter 11) is impossible. One has to conclude
therefore that all polarization measurements performed so far
suffer from the imperfection of not uniquely defined trajectories so that measurements with a clearcut relation between
trajectory and S/I are urgently required.

15) Summary

A few selected examples of the electronic interaction of ions with metal surfaces has been presented with the aim of showing the physics involved instead of heading for a perfectly elegant presentation. For the situations where definite physical pictures are still missing, suggestions for interpretations and future experimental work have purposely been included in order to stimulate discussion in these open fields.

16) Acknowledgement

Many of the questions raised and solved have profited from the fruitful collaboration with H. Winter and his crew in Muenster and in particular from the exchange of theoretical aspects with R.Zimny. The chapter 9 has been stimulated by lively discussions with M. Delauney, M. Fehringer, S.T.DeZwart, J.J.C.Geerlings, P. Varga, and H. Winter who initiated the renaissance of this field. The experiments with single crystals have only been possible with the generous support by Prof.G. Comsa and the great skills of U.Linke, both at KFA Jülich.

17) References

1) G.Comsa and B.Poelsema, Appl.Phys. A38, 153 (1985).

2) J.P.Toennies, Phys.Scr. T19A, 39 (1987).

3) J.Lapujoulade, Surf.Sci. 178, 406 (1986).

4) J.P.Biersack and J.F.Ziegler, Nucl.Instr.Meth. 194, 93 (1982).

5) G.Molière, Z.Naturf. A2, 133 (1947). For a general discussion of planar potentials see D.S.Gemmel, Rev.Mod.Phys. 46, 129 (1974).

6) R.L.Mößbauer and W.H.Wiedemannn, Z.Phys. 159, 33 (1960) and references therein.

7) R.Remy, J.Chem.Phys. 53, 2487 (1970); Compt.Rend.Acad.Sc. Paris 287, C235 (1978) and references therein.

8) R.Zimny, H.Winter, B.Becker, A.Schirmacher, and H.J. Andrä, Nucl.Instr.Meth. B2, 252 (1984).

9) H.D.Hagstrum, in "Electron and Ion Spectroscopy of Solids", edited by L.Fiermans, J.Vennik, and W.Dekeyser (Plenum, New York, 1978), and references therein.

10) W.Schroen, Z.Phys. 176, 237 (1963).

11) J.W.Gadzuk, Surf.Sci. 6, 133 (1967); Phys. Rev. B1, 2110 (1970).

12) N.W.Ashcroft nad N.D.Mermin,"Solid State Physics", Holt-Saunders Intern. Ed., 1981.

13) J.J.C.Geerlings, L.F.Tz.Kwakman, and J.Los, Surf.Sci. 184, 305 (1987).

14) J.J.C.Geerlings, Thesis, F.O.M.Amsterdam, (1987).

15) J.L.Desplat and C.A.Papageorgopoulos, Surf.Sci. 92, 97 (1980).

16) E.G. Overbosch, B. Rasser, A.D. Tenner and J.Los, Surf. Science 92, 310 (1980).

17) J.J.C.Geerlings, L.F.Tz.Kwakman, and J.Los, Surf.Sci. 184, 305 (1987).

18) J.Hermann, J.Gehring, and V.Kempter, Surf. Sc. 171, 377 (1986), and private communication.

19) J.N.M Van Wunnik, R.Brako, K.Makoshi, and D.M.Newns, Surf.Sci. 129, 618 (1983).

20) J.J.C.Geerlings and J.Los, Vacuum 53, 866 (1983).

21) J.J.C.Geerlings, R.Rodink, J.Los, and J.P.Gauyacq, Surf.Sci.Lett. (1986).

22) R.Hentschke, K.J.Snowdon, P.Hertel, and W.Heiland, Surf.Sci. 173, 565 (1986).

23) H.D. Hagstrum, Phys.Rev.96, 336 (1954).

24) A.Cobas and W.E.Lamb,Jr., Phys.Rev. 65, 327 (1944) and references therein.

25) W.E.Lamb,Jr. and R.C.Retherford, Phys.Rev. 72, 241 (1947); ibid. 79, 549 (1950); ibid. 81, 222 (1951); ibid. 85, 259 (1952); ibid. 86, 1014 (1952).

26) H.D.Hagstrum, Rev.Sci.Instr. 24, 1122 (1953).

27) H.D.Hagstrum, Phys.Rev. 96, 325 (1954); ibid. B8, 107 (1973).

28) M.Delaunay, M.Fehringer, R.Geller, P.Varga, and H.Winter, Europhys. Lett. 4, 377 (1987). See also J.A. Simpson, Rev.Sci.Instr. 32, 1283 (1961).

29) J.Burgdörfer, Phys.Rev. A35, 4963 (1987).

30) H.Conrad, G.Ertl, J.Küppers, W.Sesselmann, and H.Haberland, Surf.Sci. 121, 161 (1982).

31) W.Sesselmann, H.Conrad, G.Ertl, J.Küppers, B.Wondraschek, and H.Haberland, Phs.Rev.Lett. 50, 446 (1983).

32) B.Wondraschek, W.Sesselmann, J.Küpers, G.Ertl, and H.Haberland, Phys.Rev.Lett. 55, 611 (1985); ibid. 55, 1231 (1985).

33) B.Wondraschek, W.Sesselmann, J.Küppers, G.Ertl, and H.Haberland, Surf.Sci. 180, 187 (1987).

34) U.A.Arifov, L.M.Kishinevskii, E.S.Mukkamadiev, and
    E.S.Parilis, Zh.Tekh.Fiz. 43, 181 (1973); Sov.Phys.
    -Tech.Phys. 18, 118 (1973).

35) R.Geller, IEEE Trans.Nuc.Sci., NS-23 (2) (1972).

36) M.Delaunay, S.Dousson, R.Geller, B.Jacquot, D.Hitz,
    P.Ludwig, P.Sortais, ans S.Bliman, Nucl.Instr.Meth. B23,
    177 (1987) and refernces therein.

37) A.G.Drentje, Nucl.Instr.Meth. B9, 526 (1985).

38) F.W.Meyer, Nucl.Insr.Meth. B9, 532 (1985).

39) P.Varga, Appl.Phys. A44, 31 (1987).

40) M.Delaunay, M.Fehringer, R.Geller, D.Hitz, P.Varga,
    H.Winter, Phys.Rev. B35, 4232 (1987).

41) M.Fehringer, M.Delaunay, R.Geller, P.Varga, and H.Winter,
    Nucl.Instr.Meth. B23, 245 (1987).

42) S.T.de Zwart, Nucl.Instr.Meth. B23, 239 (1987).

43) F.W.Meyer, C.C.Havener, S.H.Overbury, K.J.Snowdon,
    D.M.Zehner, W.Heiland, and H.Hemme, Nucl.Instr.Meth. B23,
    234 (1987).

44) D.M.Zehner, S.H.Overbury, C.C.Havener, F.W.Meyer, and
    W.Heiland, Surf.Sci. 178, 359 (1986).

45) S.T.de Zwart, Thesis, Groningen (1987).

46) M.Delaunay, C.Benazeth, N.Benazeth, R.Geller, and
    C.Mayoral, Surf. Sci. 195, 455 (1988).

47) M.Fehringer, Thesis, T.U. Wien (1987).

48) P.Varga, M.Delauney, M.Fehringer, R.Geller, and H.Winter,
    15. Intern.Conf.on the Physics of Electronic and Atomic
    Collisions, Brighton,U.K., Book of Contributed Papers,
    J.Geddes, H.B.Gilbody, A.E.Kingston, C.J.Latimer, and
    H.J.R.Walters eds., p.824 (1987).

49) H.J.Andrä, R.Zimny, H.Winter, and H.Hagedorn,
    Nucl.Instr.Meth. B9, 572 (1985).

50) H.Winter and R.Zimny, in: Coherence in Atomic Collision
    Physics, J.Beyer, K.Blum, and R.Hippler eds. (Plenum Press,
    London, 1988) p.283; see also H.Winter, Habilitations-
    schrift, Univ. Muenster (1987).

51) H.J.Andrä, in: Physics of Atoms and Molecules, Vol.B,
    W.Hanle and H.Kleinpoppen eds. (Plenum Press, New York,
    1979) p.829, and references therein.

52) H.J.Andrä, Phys.Rev.Lett. 25, 325 (1970).

53) U.Fano and J.Macek, Rev.Mod.Phys. 45,553 (1973).

54) D.G.Ellis, J.Opt.Soc.Am. 63, 1232 (1973).

55) H.G.Berry, L.J.Curtis, D.G.Ellis, and R.M.Schectman,
Phys.Rev.Lett 32, 751 (1974).

56) H.J.Andrä, Phys.Lett. 54A, 315 (1975).

57) H.Winter and H.J. Andrä, Hyperf.Int. 24, 277 (1985).

58) H.J.Andrä, R.Fröhling, H.J.Plöhn, and J.D.Silver,
Phys.Rev.Lett 37, 1212 (1976).

59) H.Schröder and E.Kupfer, Z.Phys. A279, 13 (1976).

60) T.P.Grozdanov and R.K.Janev, Phys.Lett A65, 396 (1978).

61) R.Fröhling and H.J.Andrä, Z.Phys.A320, 207 (1985).

62) H.J.Andrä, H.Winter, R.Fröhling, N.Kirchner, H.J.Plöhn,
W.Wittmann, W.Graser, and C.Varelas, Nucl.Instr.Meth. 170,
527 (1980).

63) J.Burgdörfer, Thesis, F.U. Berlin 1981.

64) H.Hagedorn, H.Winter, R.Zimny, and H.J.Andrä, Nucl.Instr.
Meth. B9, 637 (1985).

65) R.Zimny, H.Hagedorn, H.Winter, and H.J.Andrä,
Nucl.Instr.Meth. B13, 601 (1986).

66) R.Zimny, Thesis, Univ. Muenster (1988) and
H.Nienhaus, Diploma-Thesis, Univ. Muenster (1988).

67) J.N.M. van Wunnik, Thesis, F.O.M. Amsterdam (1983).

68) R.Brako and D.M.Newns, 15. Intern.Conf.on the Physics of
Electronic and Atomic Collisions, Brighton,U.K., Book of
Contributed Papers, J.Geddes, H.B.Gilbody, A.E.Kingston,
C.J.Latimer, and H.J.R.Walters eds., p.783 (1987).

EPILOGUE

Since the Grand Old Man had left early and we had no
elder statesman to produce a summary of our deliberations,
someone suggested (perhaps unwisely) that I should perform
this function after dinner, because I had been espied taking
notes.

Of course, nobody could guess what I was really writing
down during the lectures. This is a standard ploy, well known
to all students. One should always appear to be writing
something, whatever it may really be. The problem was that
some of these 'notes' had been leaked, or at least an
inaccurate version was circulating on the beach. It was
therefore imperative to put the record straight.

My notes really concerned a philosophical problem, to
which the lecturers provided incidental contributions. The
question was the following: what is the perfect NATO school?
Two weeks in Maratea came close to providing an answer.

The solution seems to be that one should scatter a
coherent beam of complete (but unprepared) students off an
idealised cluster of lecturers, each one in his own,
well-defined quantum state. There are, as you will recognise,
several inherent problems in this approach. After ten days
(even if Professor Kleinpoppen is present) the beam of
students should not become completely polarized, nor indeed
(which might be a good deal worse) completely neutralised.

Fortunately, nature is on our side. Several effects
appear, which conspire to make the interactions interesting.
Down the lifts are the Langevin baths, which have always
introduced a certain randomness into our proceedings. Then,
there is spontaneous party production at all hours of the
night. Observations on the two big Joes, and on the lightest
of the ladies present revealed no Z-dependence at all, and so
we are forced to conclude that it is not a relativistic
effect. This was a shame, because the whole Institute nearly
had to give up diving.

In order to achieve a better understanding, perhaps the easiest way to think about it all is to introduce a radius r_0, without which nobody ever seems to get anywhere. Within this radius, anything can, and usually does, happen, sometimes, if you are lucky, by shapely resonances and wavefunction collapse. In fact, as Ugo Fano often points out, if you grab the hand of a lady (an experiment to be conducted in the dark) and it turns out to be the left hand, then you know instantly that the other one must be the right hand, so the velocity of light does not come into the problem at all.

If somehow, instead of the lady, you had managed to grasp a Jost function and step outside the sphere r_0, then the next cause for concern is Agostini's ponderomotive force, about which you must be rather careful, because it could (and often does) change sign as you walk away from the bar.

The best advice I can give students here is to approach the doorway states with an adjustable impact parameter, in order to avoid a sudden q-reversal as you go through them. An interesting alternative would be to get back to your room inside Chris Greene's box, but the only person who could let you out into a single channel is Mireille Aymar, who probably would not want to travel with you outside r_0 (although one should always try to push the theory to its limits).

As I have indicated, a complete NATO school must include a degree of randomness. After all, complete experiments only test a complete theory in the asymptotic limit in which everything becomes boring. Rather, one should apply Raith's Principle, as enunciated in his first lecture "Of course (this is the only principle I know which actually begins with the words: of course), if there is something in the theory which you did not put in, then you should not be surprised to find something different."

Indeed, one should always leave plenty of room for such surprises. The best way is to ensure from the beginning that things are not completely coherent. How you choose to achieve this, whether by the density matrix, or by the mindless approach, or by hyperspherical coordinates is entirely a matter of taste - what Professor Kleinpoppen likes to call a question of aesthetics. A helpful contribution towards aesthetics was made on the way by Annick Giusti-Suzor, who introduced the students to various states, both dressed and undressed, with which they have been wrestling ever since near the pool.

Experimenters, in their own way, can also be relied upon, especially if they insist on impressing their colleagues by showing diagrams of the central heating system annotated in German, with scales marked in nanometers and costs in Italian lire.

But it comes as a shock, at the end of a perfect NATO interaction, to realise that the incoherent component will necessarily grow and that, in spite of useful rail strikes, we must all eventually scatter away into our asymptotic forms. I am sure this was also a problem for Professor Briggs, and for his co-organisers Professors Kleinpoppen and Lutz, who had given so much time and thought to our well-being, and to whom we all expressed our heartfelt thanks at the end of the NATO Institute.

J.-P. Connerade

(Post-Banquet speech given in Maratea on October 3rd 1987)

Participants in the NATO ASI 'Fundamental Processes of Atomic Dynamics' held at Maratea, Italy, Sept. 21st - Oct. 2nd 1987.

PARTICIPANTS

Agostini, P., France
Agren, H., Sweden
Andersen, N., Denmark
Andersson, H., Sweden
Andrä, J., France
Anthony, J.M., USA
Arcuni, P.W., USA
Aymar, M., France
Aynacioglu, A., Germany
Bopp, P., Germany
Bordas, Ch., France
Botero, J. Germany
van den Brink, J., Netherlands
Broad, J.T., Germany
Bukhari, M. A.-H., United Kingdom
Burgdörfer, J., USA
Campbell, E., Germany
Carravetta, V., Italy
Cavagnero, M., USA
Chrysos, M., Greece
Claeys, W., Belgium
Colle, R., Italy
Connerade, J.P., United Kingdom
Cunha, M. A.C.M.I., Portugal
Daniele, R., Italy
Dehmer, J.L., USA
Dietz, K., Germany
Eichmann, U., Germany
Elsener, K., Switzerland
Falcone, G., Italy
Fano, U., USA
Feagin, J.M., USA
Ferrett, T. A., USA
Finck, K., Germany
Fink, M., Germany
Ford, M.J., United Kingdom
Gargaud, M., France
Gervat, M., France
Giusti-Suzor, A., France

Gozzini, S., Italy
Graulich, K., Germany
Greene, Ch. USA
Greenland, P.T., United Kingdom
Guardala, N., USA
Haberland, H., Germany
Hagmann, S., USA
Harvey, J.F., USA
Hendriks, B., Netherlands
Hinze, J., Germany
Huetz, A., France
Iga, I., Brasil
Inokuti, M., USA
de Jong, R., Netherlands
Jungen, Ch., France
Kempter, V., Germany
Kleinpoppen, H.A., United Kingdom
Kökten, H., Turkey
Kolsuz, N., Turkey
Komninos, Y., Greece
Kulander, K.C., USA
Laranjeira, M., Portugal
LeBrun, T., France
Lin, C.D., USA
Lindle, D. W., USA
Lourenco, J.M.C., Portugal
Lutz, H.O., Germany
Macek, J., USA
Martin, P., France
Marxer, H., Germany
Masche, C., Germany
Merz, H., Germany
Molmer, K., Denmark
Mota Furtado, F., United Kingdom
Müller, U., Germany
Nicolaides, C.A., Greece
Niemeier, R., Germany
O'Mahony, P., United Kingdom
Prunelè, E., France

689

Raith, W., Germany
Raoult, M., France
Rau, A.R.P., USA
Rechtien, H., Germany
Robicheaux, F. J., USA
Roller-Lutz, Z., Germany
Rost, J.-M., Germany
Royer, T., France
Sanna, G., Italy
Segal, D.M., United Kingdom
Seip, R., Germany
Sepp, W.-D., Germany
Sonnek, D.B., Sweden
Starace, A.F., USA
van der Straten, P., Netherlands

Theodosiou, C., USA
Thies, B., Germany
Thumm, U., Germany
Tomassetti, G., Italy
Tosi, P., Italy
Tsatis, D., Greece
Veniard, V., France
Ward, S., Canada
Weber, W., Germany
Wintgen, D., Germany
Wörmann, T., Germany
Zarcone, M., Italy
Zheng Zhen, USA
Zucatti, S., Germany

INDEX